OCULAR SURFACE

Anatomy and Physiology, Disorders and Therapeutic Care

OCULAR SURFACE

Anatomy and Physiology, Disorders and Therapeutic Care

Editors

Raul Martin Herranz
IOBA–Eye Institute, Optometry Research Group
School of Optometry
University of Valladolid (Spain)

Rosa M. Corrales Herran
Ocular Surface Center
Department of Ophthalmology
Baylor College of Medicine
Houston, TX (USA)

CRC Press is an imprint of the
Taylor & Francis Group, an **informa** business
A SCIENCE PUBLISHERS BOOK

CRC Press
Taylor & Francis Group
6000 Broken Sound Parkway NW, Suite 300
Boca Raton, FL 33487-2742

First issued in paperback 2019

Cover illustrations provided by the Editors, Dr. Raul Martin Herranz and Dr. Rosa M. Corrales Herran and reproduced by their permission.

ISBN-13: 978-1-57808-740-2 (hbk)
ISBN-13: 978-0-367-38077-9 (pbk)

Visit the Taylor & Francis Web site at
http://www.taylorandfrancis.com

and the CRC Press Web site at
http://www.crcpress.com

Preface

This book has been conceived to aid eye care practitioners in their daily-clinical practice. However, this book also includes brief and selective background information for appropriate patient management. Therefore, this book will not only be useful to general practitioners, ophthalmologists, and optometrists but also to other eye care or research professionals such as biologists, biochemists and medical doctors interested in the ocular surface.

For this reason, this text is structured in four main sections that involve 20 chapters. The first section is dedicated towards the anatomy and physiology of the ocular surface; defining the concept, describing the tear film structure and its effect on the optical function of the eye, the corneal and conjunctival anatomy. This section provides the background to the reader to understand ocular surface disorders and their clinical management described in later sections.

The second section of this book covers the most common ocular surface disorders, namely: dry eye, anterior and posterior blepharitis, keratitis, ocular allergy, cicatricial conjunctivitis, and the ocular surface alterations related with Glaucoma treatment.

Following the description of the common ocular surface disorders, the third section of this book delineates the therapeutic care of these disorders, including topical, systemic and surgical treatment options.

Finally, the fourth section addresses the relationship of the ocular surface and contact lens-wear, specifically to describe indications for contact lenses, complications that are related to contact lens-wear and managing the adverse effects of contact lens-wear.

The contributors to this book are internationally recognized investigators in various aspects of ocular surface physiology, pathology and inflammation research. They were asked to use the ocular surface as the central theme when writing their chapters and to use simple language to explain their work in order to make this text appropriate for a wide variety of readers as well as update the state-of-the-art in the ocular surface.

We feel that the strength of this book lies in the wide variety of specialities of the contributors, thus providing this range to our readers. It is our hope that you will use this text as a guide to understanding the pathophysiology, diagnosis and treatment of ocular surface diseases.

Finally, we also wish to express our profound gratitude to our esteemed authors for find the immense amount of time and effortdedicated to the evolution of this book, without them this book would not have been possible.

Raul Martin Herranz &
Rosa M. Corrales Herran

Contents

Section III: Care of Ocular Surface Disorders

Section IV: Ocular Surface and Contact Lens

Contributors

Alvim Bonfioli, Adriana
Oculoplastics Department, Visio Clinica, Address Av. do Contorno 4747 suite 1705, Belo Horizonte, Minas Gerais, Brazil, Zip Code: CEP 30110-090.
E-mail: dribonfioli@gmail.com

Argüeso, Pablo
Schepens Eye Research Institute, Massachusetts Eye and Ear, Harvard Medical School, 20 Staniford Street, Boston, MA 02114.
E-mail: pablo.argueso@schepens.harvard.edu

Rafael I. Barraquer
Intitut Universitari Barraquer, C/Muntaner 314 08021, Barcelona, Spain.
E-mail: rib@barraquer.com

Campbell, Stephanie
School of Optometry and Vision Sciences, Cardiff University, Maindy Rd. CF24 4LU Cardiff Wales, UK.
E-mail: campbells3@cardiff.ac.uk

Carrasco, Belen
Hospital Clínico Universitario Valladolid—University of Valladolid, Paseo Ramón y Cajal 3, 47011 Valladolid, Spain.
E-mail: belencarras@gmail.com

Carreño, Ester
Instituto Universitario de Oftalmobiología Aplicada (IOBA-Eye Institute), University of Valladolid. Paseo de Belén, 17-Campus Miguel Delibes, 47011 Valladolid, Spain.
E-mail: estherc@ioba.med.uva.es

de Juan Herraez, Victoria
Instituto Universitario de Oftalmobiología Aplicada (IOBA-Eye Institute) Optometry Research Group, School of Optometry, University of Valladolid, Paseo de Belén, 17-Campus Miguel Delibes, 47011 Valladolid, Spain.
E-mail: victoria@ioba.med.uva.es

Cintia S. de Paiva
Ocular Surface Center, Department of Ophthalmology, Cullen Eye Institute, Baylor College of Medicine, Houston, Texas, 6565 Fannin Street, NC 205-Houston, TX 77030.
E-mail: cintiadp@bcm.edu

Dieckow, Julia
Martin Luther University Halle-Wittenberg, Department of Anatomy and Cell Biology, Grosse Steinstraße 52, 06097 Halle (Saale), Germany.
E-mail: julia@dieckow.de

Doughty, Michael J.
Department of Vision Sciences, Glasgow-Caledonian University, Glasgow G4 OBA, United Kingdom.
E-mail: M.Doughty@gcu.ac.uk

de la Paz, María F.
Intitut Universitari Barraquer, C/Muntaner 314 08021, Barcelona, Spain.
E-mail: mpaz@barraquer.com

Freitas Bernardes, Taliana
Oculoplastics Department, Visio Clinica, Address Av. do Contorno 4747 suite 1705, Belo Horizonte, Minas Gerais, Brazil, Zip Code: CEP 30110-090.
E-mail: taliana.freitas@gmail.com

Galarreta, David J.
Hospital Clínico Universitario Valladolid—University of Valladolid, Paseo Ramón y Cajal 3, 47011 Valladolid, Spain; and Instituto Universitario de Oftalmobiología Aplicada (IOBA-Eye Institute)—University of Valladolid, Paseo de Belén, 17—Campus Miguel Delibes, 47011 Valladolid, Spain.
E-mail: davidgalarreta@hotmail.com

Giancarlo, Montani
Centro di Ricerche in Contattologia, Università del Salento, Via per Arnesano, Lecce, 73100 Italy.
E-mail: montani.gc@libero.it

Giuseppe, Maruccio
Dipartimento di Matematica e Fisica, Università del Salento, Via per Arnesano, Lecce, 73100 Italy.
E-mail: giuseppe.maruccio@unisalento.it

González-Méijome, José Manuel
Clinical & Experimental Optometry Research Lab (CEORLab), University of Minho, Campus de Gualtar 4710-057 Braga–Portugal.
E-mail: jgmeijome@fisica.uminho.pt

Henriksson, Johanna T.
Ocular Surface Center, Department of Ophthalmology, Cullen Eye Institute, Baylor College of Medicine, Houston, Texas, 6565 Fannin Street, NC 205-Houston, TX 77030.
E-mail: henrikss@bcm.edu

Herreras, Jose M.
Hospital Clínico Universitario Valladolid—University of Valladolid, Paseo Ramón y Cajal 3, 47011 Valladolid, Spain; and Instituto Universitario de Oftalmobiología Aplicada (IOBA-Eye Institute)—University of Valladolid, Paseo de Belén, 17—Campus Miguel Delibes, 47011 Valladolid, Spain.
E-mail: herreras@ioba.med.uva.es

Lan, Wanwen
Singapore Eye Research Institute. 11 Third Hospital Avenue, 168751, Singapore.
E-mail: lan.wanwen@gmail.com

Lopes-Ferreira, Daniela
Clinical & Experimental Optometry Research Lab (CEORLab), University of Minho, Campus de Gualtar 4710-057, Braga–Portugal.
E-mail: dlopesferreira@fisica.uminho.pt

Martin Herranz, Raul
Instituto Universitario de Oftalmobiología Aplicada (IOBA-Eye Institute) Optometry Research Group, School of Optometry, University of Valladolid, Paseo de Belén, 17-Campus Miguel Delibes, 47011 Valladolid, Spain.
E-mail: raul@ioba.med.uva.es

Martín Torres, Rodrigo
Centro de Ojos Dr. Lódolo, Asociación Entrerriana de Oftalmología, Las Calandrias 4789 Barrio Las Acaias, Colonia Avellaneda, 3107, Entre Ríos–Argentine.
E-mail: romator7@hotmail.com

Martínez-Osorio, Hernan
Intitut Universitari Barraquer, C/Muntaner 314 08021, Barcelona, Spain.
E-mail: hernanophth@hotmail.com

Motterle, Laura
Ophthalmology Unit, Department of Neuroscience, University of Padova, Via Giustiniani 2, 35128, Padova, Italy.
E-mail: laura.motterle@yahoo.it

Pelegrino, Flavia S.A.
Ophthalmologic Clinic, 292 Querubim Uriel St., Zip code 13024-470
Campinas, São Paulo, Brazil.
E-mail: flaviapelegrino@gmail.com

Perez-Gomez, Inma
c/o Alcon SA, Avenue Louis-CasaÏ 58, CH 1216 Cointrin-Geneva, Switzerland.
E-mail: inma.perez@cibavision.com

Petznick, Andrea
Singapore Eye Research Institute. 11 Third Hospital Avenue, 168751, Singapore.
E-mail: andre_petznick@yahoo.com.au

Portero, Alejandro
Instituto Universitario de Oftalmobiología Aplicada (IOBA-Eye Institute), University of Valladolid. Paseo de Belén, 17-Campus Miguel Delibes, 47011 Valladolid, Spain.
E-mail: aporterob@ioba.med.uva.es

Ribeiro Miguel F.
Clinical & Experimental Optometry Research Lab (CEORLab), University of Minho, Campus de Gualtar 4710-057, Braga–Portugal.
E-mail: mig.afr@gmail.com

Rodriguez Zarzuelo, Guadalupe
Instituto Universitario de Oftalmobiología Aplicada (IOBA-Eye Institute) Optometry Research Group, School of Optometry, University of Valladolid, Paseo de Belén, 17-Campus Miguel Delibes, 47011 Valladolid, Spain.

Schweizer, Helmer
c/o Alcon SA, Avenue Louis-CasaÏ 58, CH 1216 Cointrin-Geneva, Switzerland.
E-mail: helmer.schweizer@cibavision.com

Tong, Louis MG
Singapore National Eye Center, Singapore Eye Research Institute, Duke-NUS Graduate Medical School, and Department of Ophthalmology, Yong Loo Lin School of Medicine, National University of Singapore, 11, Third Hospital Avenue Singapore 168751 Singapore.
E-mail: louis.tong.h.t@snec.com.sg

Valentina, Arima
NNL Institute Nanoscience-CNR, Via per Arnesano, Lecce, 73100 Italy.
E-mail: valentina.arima@unile.it

Yáñez Álvarez, Bety
Department of Ophthalmology, Hospital Nacional Dos de Mayo
Cdra. 13 Av. Grau–Cercado de Lima, Lima, Peru.
E-mail: byanez@hotmail.com

Section I: Anatomy and Physiology of the Ocular Surface

1

Definition of the Ocular Surface

Louis Tong,[1,*] *Wanwen Lan*[2] and *Andrea Petznick*[2]

SUMMARY

The ocular surface is a very important part of the eye. It consists of the conjunctiva and the cornea, together with elements such as the lacrimal gland, lacrimal drainage apparatus and associated eyelid structures. This part of the eye has unique properties and is associated with special physiological mechanisms, for example tear production and drainage, as well as predisposition to specific diseases. Certain diseases affect only this part of the eye, due to its functional requirement for vision, exposed anatomical location and its proximity to the nasal mucosa and sinuses. For these reasons, diseases such as allergic keratoconjunctivitis affecting the ocular surface primarily are common.

This chapter summarizes important terms and root words used in conjunction with the ocular surface in the scientific literature. Understanding of the nomenclature is essential for any research discussion or clinical practice related to the ocular surface.

INTRODUCTION TO THE OCULAR SURFACE

The ocular surface, an integrated unit comprising the cornea, conjunctiva, lacrimal glands and eyelids, was first described by Thoft in 1987 (Thoft 1978). Gipson extended the description of the ocular surface system in her Friedenwald lecture in 2007 (Gipson 2007): *'the ocular surface...*

[1]Singapore National Eye Center, Singapore Eye Research Institute, Duke-NUS Graduate Medical School, and Department of Ophthalmology, Yong Loo Lin School of Medicine, National University of Singapore, 11 Third Hospital Avenue, 168751, Singapore.
E-mail: louis.tong.h.t@snec.com.sg

[2]Singapore Eye Research Institute. 11 Third Hospital Avenue, 168751, Singapore.
*Corresponding author

includes the surface and glandular epithelia of the cornea, conjunctiva, lacrimal gland, accessory lacrimal glands, meibomian glands and their apical (tears) and basal (connective tissue) matrices; the eyelashes with their associated glands of Moll and Zeis; those components of the eyelids responsible for the blink and the nasolacrimal duct'. Together, these components are interconnected through a continuous epithelium, as well as the nervous, vascular, immune and endocrine systems. Figure 1 illustrates a cross-sectional view showing some components of the ocular surface system. The lacrimal functional unit is defined by the 2007 International Dry Eye Work Shop (2007) as *'an integrated system comprising the lacrimal glands, ocular surface (cornea, conjunctiva and meibomian glands), lids, and the sensory and motor nerves that connect them'*. The anatomical components of the lacrimal functional unit will be described under 'Lacrimal glands' below.

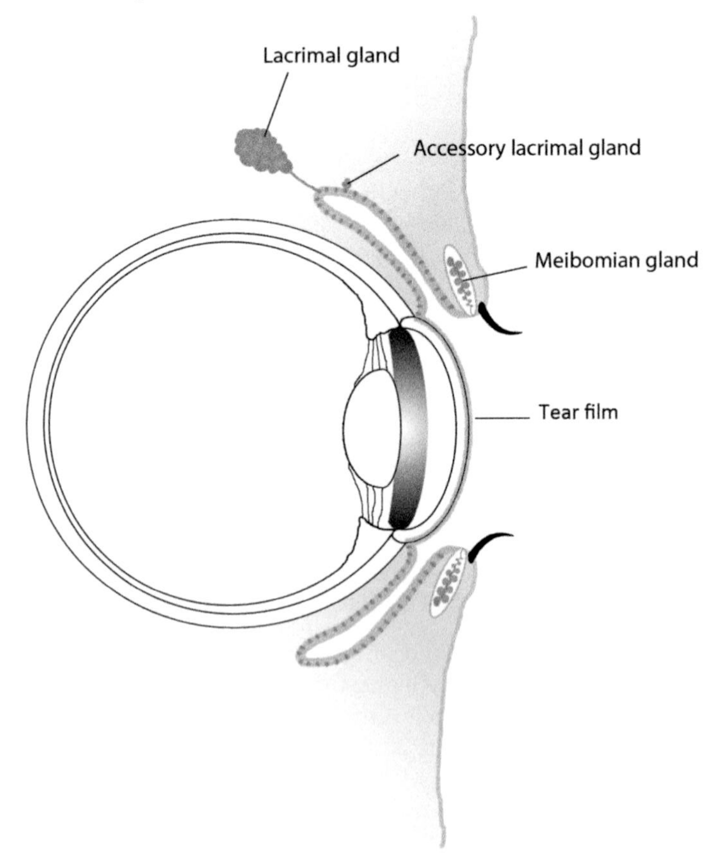

Figure 1 This figure shows a cross-sectional view of the ocular surface system with the continuous epithelium highlighted in pink and the tear film in blue.

Color image of this figure appears in the color plate section at the end of the book.

General Appearance of the Ocular Surface

In the normal ocular surface, the cornea occupies the approximate centre of the exposed surface. The outer limit of the cornea adjacent to the bulbar conjunctiva is the limbus. At both sides of the limbus are two triangular sclero-conjunctival areas, visualised as the white part of the sclera. When a person is looking straight ahead, the lower eyelid height is normally 1–2 mm higher or lower than the lower corneal limbus (see section on Cornea), whereas the superior eyelid is normally 1–2 mm higher than the visual axis but just lower than the superior corneal limbus (Lens et al. 2007).

The upper and lower eyelids meet at the medial (inner) and lateral (outer) canthi which are the angles of the palpebral fissures. The medial canthus is near to the nose, whereas the lateral canthus is located temporally. The medial canthus is usually positioned slightly lower than the lateral canthus. The positions of the canthi have significance in oculoplastic surgery. They facilitate the flow of the tear film from the lateral to medial direction, and are youthful-looking aesthetically.

The Eyelids

The eyelid or palpebrae refers to a movable fold of skin, muscle and cartilage that can be closed or opened over the eyeball. The upper and lower eyelids form a covering over the globe protecting against excessive light or injury. When the eyelids are open, the margins or the palpebral fissures form an almond-shaped structure (Lens et al. 2007).

The eyelid structure consists of four layers. The first or outermost layer includes the skin, eyelashes and associated glands. The second layer comprises the muscular layer, namely the orbicularis oculi, the circular sphincter-like muscles responsible for closing the eyelids. The third fibrous layer, important for mechanical stability of the eyelid, consists mainly of the tarsal plate. The innermost layer of the eyelid is the palpebral conjunctiva. The first two layers are sometimes termed anterior lamella of the eyelid, whereas the last two layers are termed posterior lamella of the eyelid. In oculoplastic surgery, a substantial full- thickness eyelid defect may need to have the anterior and posterior lamellae reconstructed separately, as these lamellae have different mechanical requirements.

The orbicularis oculi are innervated by cranial nerve VII. When there is pathology of cranial nerve VII, such as in Bell's palsy, the eyelids may not be able to close. On the other hand, some conditions irritate the ocular surface, resulting in secondary blepharospasm and tonic contraction of orbicularis muscle. The levator palpebrae muscles are innervated by a branch of the cranial nerve III. In the event that the cranial nerve III function is compromised, there may be drooping of the upper eyelid or, in more extreme cases, inability to open the lid.

Palpebral Fissures

The fusiform lid fissure represents the shape of a spindle. However, the curvature of the upper eyelid fissure is often greater than that of the lower eyelid. Certain dynamic features of the palpebral fissures are unique: when the eyes are blinking or closing, the inferior eyelid moves only minimally and almost all the movement is done by the descending sweep of the upper lid. To facilitate this downward motion of the upper eyelid, the medial and lateral insertions of the palpebral ligaments are lower than the level of the pupil, and as a result, the upper eyelid is quite bowed.

The piriform lid fissure (like a pear lying flat on the side) is similar to the fusiform one, except that the maximal height of the interpalpebral fissures is not aligned with the vertical diameter of the cornea but slightly lateral to that.

Eyelid Margin

The mucocutaneous junction, or gray line, is located just posterior to the eyelashes. The gray line demarcates the eyelid into anterior and posterior lamellae, and contains the muscle of Riolan (Wulc et al. 1987, Lipham et al. 2002). This is also called the Marx's line and can be visualised with fluorescein dye. Along the posterior lid margin, the single or double row of meibomian gland orifices is found. This forms the white line, which corresponds to the free border of the tarsal plate, where the bulk of the meibomian gland elements islocated.

Tarsal Plate

The tarsal plate or tarsus is a thin flat plate of dense connective tissue (one in each eyelid), which supports the eyelid structure. The tarsus extends from the orbital septum to the lid margin. The upper tarsal plate has a D shape structure lying on its side and is much larger than the lower tarsal plate. The height is 11 mm centrally, whereas the corresponding height in the lower oblong tarsus is only 5 mm. Each tarsus is about 3 cm long and 1 mm thick. The meibomian glands (refer to 'Tear producing glands: meibomian glands') are located within the tarsus (Jester et al. 1981, Obata 2002).

Eyelashes

The eyelashes are hairs growing at the edge of the eyelids. They are positioned in the anterior portion of the lid margin, lined by skin epithelium. The eyelashes are surrounded by the glands of Zeis (modified sebaceous glands) and the glands of Moll (modified sweat glands) (Stephens et al. 1989). The eyelashes help to keep foreign particles from entering the ocular surface system, as well as increase the sensitivity of the eye to touch. Refer to Table 1 for the terms related to the eyelids or eyeball.

Table 1 This table summarizes words related to the eyelids or eyeball.

	Eyelid-related definitions	**Examples of use of term**
Ankyloblepharon/ blepharosynechiae	Fusion of the upper to the lower eyelids	
Blepharo-	Related to the eyelid	Blepharospasm
Blepharitis	Eyelid inflammation	Anterior, posteriorblepharitis
Blepharo-conjunctiva	Related to the eyelid and conjunctiva	Blepharo-conjunctivitis
Enophthalmos	Decreased palpebral fissures associated with posterior movement of the eyeball into the orbit	Traumatic enopthalmos
Epiblepharon	Congenital horizontal fold of skin over the upper eyelid which can result in inward-turning of the eyelashes of the medial upper eyelid, common in people of Asian origin	
Exophthalmos	Increased palpebral fissures associated with bulging of the eyeball from the orbit	Intermittent, specific exophthalmos
Entropion	Abnormal inward-turning of the eyelid margin	Cicatricial, involutional, acuspastic, congenitalentropion
Ectropion	Abnormal outward-turning of the eyelid margin	Involutional, cicatricial, paralytic, punctual, mechanical, congenitalectropion
Lagophthalmos	Condition of inadequate eyelid closure	Nocturnal, paralytic lagopthalmos
Marx's line	Refers to the muco-cutaneous junction at the eyelid margin	
Ptosis	Drooping of the eyelids	Involutional or paralytic ptosis
Symblepharon	Condition where the palpebral conjunctiva of the upper or lower eyelids fusing with the bulbar conjunctiva of the eyeball	Anterior, posterior, totalsymblepharon
Trichiasis	Condition where backwardly directed lashes cause irritation, usually due to scarring of the eyelids	Trachomatoustrichiasis

THE CORNEA

Gross Anatomy and Corneal Limbus

The cornea is an avascular, transparent tissue that provides the majority of optical power to refract the light entering the eye. It is generally ovoid in nature, with a steeper or shorter radius of curvature in the vertical than the horizontal plane. It occupies a central position in the ocular surface and consists of several layers including the epithelium, stroma and endothelium.

A general overview on the corneal structure is shown in Fig. 2. The stroma is separated from the epithelium and endothelium by the Bowman's membrane and Descemet's membrane respectively.

The surgical limbus is an important landmark for surgical incisions as well as the anatomical position of ocular surface progenitor or stem cells important for regeneration of the corneal epithelium (Lens et al. 2007). The surgical limbus can be differentiated into a bluish zone anteriorly, and a more whitish zone posteriorly. The line dividing the two zones corresponds to the Schwalbe's line, which is the termination of the Descemet's membrane (see below) (Preziosi 1968).

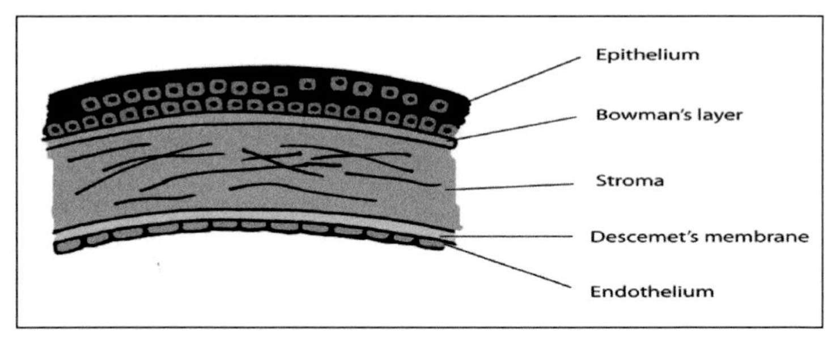

Figure 2 This figure shows a schematic illustration of the different layers of the cornea (not drawn to scale).

Color image of this figure appears in the color plate section at the end of the book.

Corneal Epithelium

The epithelium is the outermost layer of the cornea. The corneal epithelium is approximately 50 μm thick (Li et al. 1997, Haque et al. 2004) and is composed of five to seven cell layers. The stratified epithelium of the cornea is non-keratinized and composed of several morphologically distinct layers of cells including two to three layers of superficial flattened squamous cells, several layers of wing cells and a single layer of columnar basal cells (Beuerman and Pedroza 1996).

Bowman's Layer

The Bowman's layer, produced by the corneal epithelium, is a thin layer separating the corneal epithelium from the stroma. It is usually 10–17 μm thick (Li et al. 1997, Hayashi et al. 2002)and comprises collagen fibrils found in random distribution. Damage to the Bowman's layer may confound adherence of corneal epithelium to the stroma and disrupt structural integrity of the ocular surface (Wilson and Hong 2000).

Corneal Stroma

The stroma accounts for approximately 90% of the total corneal volume. It is predominantly composed of hydrated fibrils of collagen, glycoproteins (such as fibronectin and laminin) and proteoglycans (Linsenmayer et al. 1998). It has the tendency to imbibe fluid and controlled dehydration is essential to maintain transparency of the cornea. Damage to the cellular limiting layers, with subsequent fluid influx, interferes with the orderly fibrillar arrangement of the stromal lamellae increasing light scatter, resulting in loss of transparency of the cornea (Maurice 1957). Infiltration of the cornea with immune cells such as macrophages can also reduce corneal transparency.

Descemet's Membrane

The Descemet's membrane, which is a basement membrane secreted by the endothelium, separates the stroma from the endothelium. The Descemet's membrane is approximately 10 μm thick in an adult human cornea and its thickness increases throughout life (Joyce 2003).

Corneal Endothelium

The corneal endothelium is a cellular monolayer of approximately 5 μm thickness and 15 to 20 μm width (Binder et al. 1991). Endothelial cells forming this layer are joined by interdigitations and leaky tight junctions. In order to keep the cornea dehydrated, the endothelial cells serve as a pump which removes fluid by active transport from the corneal stroma to the aqueous (Maurice 1972). The number of endothelial cells in humans decreases with age (Joyce 2003). Following injury, the non-replicative endothelium in humans repairs the endothelial wound by enlargement and migration of existing cells rather than mitosis (Chan-Ling et al. 1988b, Joyce 2003). Refer to Table 2 for terms related to the cornea.

Table 2 This table summarizes words related to the cornea.

	Cornea-related definitions	**Examples of use of term**
Corneal degeneration	Acquired age-related condition of the cornea	Fuchs' endothelial dystrophy, keratoconus
Corneal dystrophy	Inherited developmental disease of the cornea	Fuchs' dystrophy, keratoconus, lattice dystrophy
Corneal ectasia/melt/ descemetoele	Abnormal thinning of the cornea	Post-LASIK corneal ectasia
Corneal epitheliopathy	Pathology of the corneal epithelium	Punctate corneal epitheliopathy
Corneal guttae	Abnormal spots or indentations in the corneal endothelium	
Corneal infiltrates	Abnormal deposits in the cornea	
Corneal striae/folds	Abnormal lines in the cornea	
Corneal ulceration	Condition with a defect in the corneal epithelium. Often associated with infection	Bacterial keratitis, fungal keratitis, Chlamydia trachomatis
Keratitis	Inflammation of the cornea	Infectious, marginal keratitis
Keratoconjunctivo-	Related to the cornea and the conjunctiva	Keratoconjunctivitis
Keratolysis	Thinning or melting of cornea	Paracentral, rheumatoid keratolysis
Keratopathy	Any disease of the cornea	Band, exposure, lipid, filamentous keratopathy
Keratomalacia	Melting of the cornea usually related to xerosis	

THE CONJUNCTIVA

The conjunctiva is a thin, transparent tissue that lines the inner surface of the eyelids, fusing with the eyelid epithelium at the eyelid margin and the corneal epithelium at the limbus. It covers the sclera up to the limbus and is continuous with the corneal epithelium. The conjunctiva consists of six or more non-keratinized epithelial cell layers near the limbus and can be up to 12 layers near the fornix (Wanko et al. 1964). The conjunctival fold (cul-de-sac) is open at the palpebral fissure and is closed when the eyelids are apposed.

The conjunctiva is divided into 3 parts: The forniceal conjunctiva, bulbar conjunctiva and palpebral conjunctiva. These will be described in the sections below.

Bulbar Conjunctiva

This is the part of the conjunctiva that lines the globe of the eye. It consists of 2 parts: the limbal conjunctiva which is fused with the episclera at the limbus, and the scleral conjunctiva which extends from the limbus to the forniceal conjunctiva.

Forniceal Conjunctiva

This is the intermediate portion of the conjunctiva which is not attached to the eyelids or the eyeball. This lines the bottom of the conjunctival sac (fornix) and joins the bulbar and palpebral portions.

Palpebral Conjunctiva

The palpebral conjunctiva lines the posterior surface of the eyelids. It is divided into 3 portions: the marginal conjunctiva which extends from the margin of the eyelids to the tarsus, the subtarsal conjunctiva which extends over the tarsal plate and the orbital conjunctiva which extends from the tarsus to the fornix (Lens et al. 2007).

Caruncle

The lacrimal caruncle is the reddish structure or eminence that is found in the medial angle of the eye. It contains sebaceous and sweat glands.

Plica semilunaris

The Plica semilunaris is asmall crease of the bulbar conjunctiva found at the medial canthus (Arends and Schramm 2004). It produces a watery/mucoid secretion with fatty substances that traps dirt and foreign particles from the ocular surface to form rheum. The rheum forms the crust from the canthus during sleep. Refer to Table 3 for terms related to the conjunctiva.

THE SCLERA

The sclera is a tough fibrous coat of the eye composed of mainly collagen and elastic fibrous tissue. It has 3 poorly defined layers from superficial to deep, called the episclera, the sclera proper and the lamina fusca (Lens et al. 2007). Refer to Table 4 for definitions of some conditions involving the sclera.

Table 3 This table summarizes words related to the conjunctiva.

	Conjunctiva-related definitions	Examples of use of term
Blepharoconjunctiv	Related to eyelid and conjunctiva	Blepharoconjunctivitis
Conjunctivalisation	Abnormality of the cornea where the epithelium resembles that of the conjunctiva. Usually related to limbal stem cell deficiency	
Conjunctivitis	Inflammation in the conjunctiva, usually infectious in nature, but sometimes due to rare immune-mediated conditions	Ligneous, bacterial, viral, allergy-based, chlamydial, chemical/irritant conjunctivitis
Keratoconjunctivitis	Inflammation involving the cornea and conjunctiva	Allergic keratoconjunctivitis —atopic and vernal keratoconjunctivitis, keratoconjunctivitis sicca (dry eye disease)
Membranous conjunctivitis	A special type of conjunctivitis where a 'membrane' is visualised over the ocular surface	Diphtheritic, pseudo-membranous conjunctivitis

Table 4 This table summarizes words related to the sclera.

	Sclera-related definitions
Episcleritis	Inflammation of the episclera
Scleritis	Inflammation of the sclera
Sclerokeratitis	Inflammation of the sclera and cornea
Scleromalacia	Degenerative thinning of the sclera, commonly associated with rheumatoid arthritis and collagen diseases

THE TEAR FILM

Preocular Tear Film

The tear film is a mixture of ocular surface secretions from the main lacrimal and accessory lacrimal glands, meibomian glands and the corneal and conjunctival epithelium. The tear film, as part of the ocular surface system, has several roles that include the provision of a smooth refractive surface, protection of the ocular surface with its antibacterial and immune functions, supply of oxygen and removal of metabolites such as carbon dioxide, lubrication, and clearance of cells, debris and foreign bodies in conjunction with the eyelids (Van Haeringen 1981, Tiffany 2008).

The preocular tear film is the first structure that incident light encounters on reaching the eye; so the air-tear interface is the first refractive surface for focusing of light rays. Irregularities of the tear film therefore affect quality of vision. The tear film in human eyes is structured in three layers: the outer lipid layer, the intermediate aqueous layer and the inner mucin layer. However, the aqueous and mucin layers have been described as a continuum (Chen et al. 1997, Spurr-Michaud et al. 2007). The lipid layer of the tear film is secreted by the meibomian glands. This layer can be evaluated by interferometry, and may be deranged in meibomian gland disease. There have been studies which suggest that this layer consists of an outer layer of neutral lipids and an inner layer of amphipathic polar lipids (Linsen and Missotten 1990). The functions of this layer include the prevention of aqueous tear evaporation and aiding the spreading of the tear film.

The aqueous layer of the tear film is basally secreted by the Kraus and Wolfring accessory lacrimal glands, and reflex secretion is by the main lacrimal gland. This layer forms the main bulk of the volume of the tear film.

The mucin layer is secreted by the goblet cells and also the corneal and conjunctival epithelial cells (Asbell and Lemp 2006). This layer merges with the glycocalyx of the conjunctival epithelial cells. The glycocalyx, being anterior to the corneal epithelium, holds the tear film on to the ocular surface via membrane-spanning mucins attached to the microvilli and small filaments of the outermost superficial epithelial cells (Beuerman and Pedroza 1996, Gipson 2007).

Tear Meniscus

The upper and lower tear menisci are strips of tear found on the upper and lower lid margins respectively. They account for 75–90% of the total tear volume on the ocular surface (Mishima et al. 1966, Holly 1985). During blinking, tears are distributed from the menisci onto the ocular surface (Lens et al. 2007).

The radius of curvature of the tear menisci has been found to be directly correlated to the total tear volume (Yokoi et al. 2004), and the absence of tear menisci is associated with dry eye disease. In ocular surface practice, the menisci may be evaluated using the anterior segment optical coherence tomography (Qiu et al. 2010).

Tear Turnover

Tear turnover is defined as the percentage decrease in fluorescein concentration of tears per minute and is clinically used to assess the lacrimal functional unit as well as tear quality (Mishima et al. 1966, Nelson 1995). Reduced tear turnover is usually associated with ocular irritation and inflammation (de Paiva and Pflugfelder 2004).

TEAR PRODUCING GLANDS

Tear Production

There are multiple factors governing tear production. Tear production may be classified as basal, reflex and emotional.

The lacrimal reflex is the process where tear is produced after irritation of the cornea and conjunctiva. This is in contrast to the basal tear production which is constant in the resting state and added to the reflex production in the stimulated state. Psycho-emotional tears are related to emotional states such as sadness, anger or happiness (Murube 2009).

Lacrimal Glands

The main lacrimal gland is located in the upper, outer quadrant of the orbit and is made up of two lobes, the palpebral and orbital lobes. It is composed of acinar, ductal and myoepithelial cells. The main lacrimal gland is approximately 15 to 20 mm long, 10 to 12 mm wide and 5 mm thick (Lorber 2007). The lacrimal gland is innervated by the smallest branch of the ophthalmic nerve, the lacrimal nerve (Burton 1992). Any stimulation to the ocular surface activates afferent sensory nerves in the cornea and conjunctiva, which then activate afferent sympathetic and parasympathetic nerves that subsequently stimulate the lacrimal gland to secrete proteins, electrolytes and water (Zoukhri 2006).

With age, the morphology of the lacrimal gland changes and infiltration of inflammatory cells into the lacrimal gland tissue occurs. This, in turn, leads to inflammation of the tissue and reduces protein and tear production (Draper et al. 1998, Nagelhout et al. 2005).

Accessory Lacrimal Glands

The smaller accessory glands of Krause and Wolfring are part of the lacrimal system and are located close to the superior fornix of the conjunctiva. Their microanatomy is identical to that in the main lacrimal gland (Lemp and Wolfley 1992).

Meibomian Glands

The meibomian glands are large, tubuloacinar structures embedded within the tarsal plate of the eyelids (Jester et al. 1981, Obata 2002). There are about 32 glands in the upper eyelid and 25 in the lower eyelid (Greiner et al. 1998).

The sebaceous meibomian glands consist of branched, round-shaped acini that secrete lipids into a long single duct.

The meibomian gland lipids, also referred to as meibum, comprise waxy esters, sterols, cholesterol, polar lipids and fatty acids which are transported towards the ductal orifice by blinking (Lemp and Wolfley 1992). The muscle of Riolan at the gray line of the eyelid margin may also regulate meibomian gland secretion. Once the lipids are secreted into the ocular surface, they form the superficial lipid layer of the tear film.

Goblet Cells

Conjunctival goblet cells are specialized glandular epithelial cells. They are integrated in the conjunctiva epithelium and secrete gel-forming mucin (Gipson 2004). The highest density of conjunctival goblet cells in humans is found in the nasal and nasal inferior area of the conjunctiva (Kessing 1968, Rivas et al. 1991). The number of goblet cells decreases in humans in some conditions such asa loss of vascularisation after chemical injury, or inflammation in the ocular surface, e.g. conjunctivitis (Tseng et al. 1984, Grahn et al. 2005). The innermost mucin layer of the tear film is hydrophilic, accounting for its adherence to the ocular surface.

LACRIMAL APPARATUS

Tear Drainage

The nasolacrimal drainage system consists of the lacrimal puncta, canaliculus, lacrimal sac and nasolacrimal duct. This is the tear drainage pathway from the lacrimal lake to the nasal cavity. The lacrimal lake is the space or recess between the eyelids at the nasal commissure of the eye. The tears collect from the upper and lower tear menisci and the preocular surface into the lacrimal lake before draining into the lacrimal punctum.

The concept of the lacrimal pump is that the orbicularis oculi muscle creates a pressure on the lacrimal sac. On relaxation, there is a negative pressure which draws tears from the lacrimal lake into the sac via the canaliculi.

Lacrimal Puncta

The lacrimal puncta are small openings located on the medial aspect of the eyelid margins and sit on an elevated structure called the papilla lacrimalis. The openings are usually directed posteriorly against the globe.

Lacrimal Canaliculus

The lacrimal canaliculi are the tubular structures linking the lacrimal puncta to the lacrimal sac. There is a vertical 2mm segment and a horizontal 8mm segment in each canaliculus. The upper canaliculus and the lower canaliculus join to form the common canaliculus.

The common canaliculus enters the lacrimal sac obliquely forming the valve of Rosenmüller, which prevents backflow of tears from the sac to the canaliculi. Sometimes, the common canaliculus dilates slightly before entering the lacrimal sac, forming the sinus of Maier.

Lacrimal Sac

This structure sits on the part of the bony orbit called the lacrimal fossa. The superior part of the sac is the fundus and the inferior part of the sac, the body, is where it extends to the osseous opening of the nasolacrimal canal which contains the nasolacrimal duct.

Nasolacrimal Duct

The nasolacrimal duct consists of a superior (intraosseous) portion travelling within the maxillary bone and a shorter inferior (membranous) portion along the nasal mucosa. The duct ultimately opens into the inferior meatus of the nasal cavity under the inferior nasal turbinate. There may be a valve of Hasner at the opening of the nasolacrimal duct just before the tear drains into the nasal cavity (Snell and Lemp 1998). Refer to Table 5 for terms related to the lacrimal apparatus or unit.

The Blood Supply of the Ocular Surface

The cornea is avascular except for the limbal area, which has a similar blood supply as the conjunctiva. The blood supply to the conjunctiva is mainly derived from the anterior ciliary and the palpebral artery branches. The palpebral arteries also supply the eyelids and associated structures. The corresponding veins drain these structures. The main lacrimal gland is supplied by the lacrimal artery derived from the ophthalmic artery, with veins draining to the superior ophthalmic vein (Snell and Lemp 1998).

The Lymphatics of the Ocular Surface

The conjunctival and eyelid lymphatic vessels drain to the submandibular lymph glands medially and to the superficial preauricular nodes laterally.

Table 5 This table summarizes words related to lacrimal production and drainage.

	Lacrimal-related definitions	Examples of use of term
Dacryo-adenitis	Inflammation of the lacrimal gland	Granulomatous, tuberculoid, necrotizingdacryo-adenitis
Dacryocystitis/ lacrimal sac abscess	Pus-filled infected swelling in the lacrimal sac	
Dacryocystogram	Imaging of the nasolacrimal duct	
Dacryocysto-rhinostomy	Surgical procedure where anastomosis is established between nasal cavity and lacrimal sac	Endoscopic dacryocystorhinostomy
Epiphora	Excessive tearing onto the face, i.e. caused by obstructed drainage of tears	
Lacrimal cysts	Congenital displacement of the lacrimal tissue resulting in subconjunctival cysts	
Lacrimal sacmucocele/ amniotocele/dacryocele	Swelling of the lacrimal sac due to congenital malformation or, if infected, called lacrimal sac abscess	

After certain types of infection or inflammatory processes, corneal lymphatics can be observed (Tang et al. 2010, Nakao et al. 2011, Zhang et al. 2011).

THE INNERVATION OF THE OCULAR SURFACE

Cornea

The cornea is the most densely innervated structure in the human body. The nerve branches supplying the cornea are derived from the long posterior ciliary nerves, which are from the ophthalmic and maxillary division of the trigeminal nerve. At the peripheral cornea, the nerves from the conjunctiva, episclera and sclera penetrate the cornea at various depths.

The nerve fibers become unmyelinated when they penetrate the cornea and run parallel to its surface. The corneal nerves turn abruptly 90 degrees and proceed towards the corneal surface where they branch into the dense subepithelial plexus (Marfurt et al. 2010). Unsheathed nerve endings arise from the subepithelial plexus and continue superficially into the wing and squamous cell layers of the epithelium (Beuerman and Pedroza 1996, Müller et al. 2003).

Conjunctiva, Eyelids and Lacrimal Gland

The conjunctiva and eyelids are innervated by the ophthalmic and maxillary branches of the trigeminal nerve (cranial nerve V). Refer to "Lacrimal gland section" for lacrimal gland innervation.

Immunology

Immune Privilege of Cornea

Since the cornea is avascular, the corneal stroma is normally not accessible to the immune system. However, components of the innate immunity, such as dendritic cells, have access to the limbal region. The corneal epithelial cells possess pathogen recognition receptors, form a barrier to pathogens and constitute an important part of innate defence of the ocular surface (Kiel 2010).

In addition, components of the immune system, such as the complement system and immunoglobulins, are present in the tear film, thereby exposing the corneal epithelium to these elements of the immune system. During inflammation of the cornea or when the cornea epithelium is breached by trauma or microbes, cytokines and immune cells such as neutrophils and macrophages will have access to the deeper layers of the cornea.

Mucosal Associated Lymphoid Tissues/Conjunctiva Associated Lymphoid Tissues

The mucosal associated lymphoid tissue (MALT) is an important component of the immunity in many mucosa of the body, for example, the Peyer's patches in the intestinal mucosa. Embedded into the palpebral conjunctiva are lymphoid aggregates, the conjunctival associated lymphoid tissues (CALT), which are an ocular surface form of MALT. The CALT are composed of T and B lymphocytes, macrophages, plasma cells and dendritic cells (Knop and Knop 2005) and serve as an induction site for immune defence responses of the ocular surface (Astley et al. 2003, Liang et al. 2010).

CONCLUSION

The ocular surface is a well-integrated unit comprising the different parts summarized above. Being a relatively exposed anatomical unit, the ocular surface is prone to infections that can seriously affect the quality of lifestyle of patients. In this chapter, we have summarized the main functions of these different units and their associated commonly encountered clinical diseases. We hope that this will help readers better understand the ocular surface and its related pathology.

ACKNOWLEDGMENT

We would like to thank Lee Man Xin for aid in diagrams and research relevant for the review.

REFERENCES

Arends, G. and U. Schramm. 2004. The structure of the human semilunar plica at different stages of its development—a morphological and morphometric study. Ann Anat 186: 195–207.

Asbell, P.A. and M.A. Lemp. 2006. Dry eye disease: the clinician's guide to diagnosis and treatment. New York, Thieme Medical Publishers.

Astley, R.A., R.C. Kennedy and J. Chodosh. 2003. Structural and cellular architecture of conjunctival lymphoid follicles in the baboon (Papio anubis). Exp. Eye Res. 76: 685–694.

Beuerman, R.W. and L. Pedroza. 1996. Ultrastructure of the human cornea. Microsc. Res. Tech. 33: 320–335.

Binder, P.S., M.E. Rock, K.C. Schmidt and J.A. Anderson. 1991. High-voltage electron microscopy of normal human cornea. Invest Ophthalmol. Vis. Sci. 32: 2234–2243.

Burton, H. 1992. Somatic Sensations from the eye. Adler's Physiology of the Eye. W. M. Hart. St. Louis, Mosby-Year Book 71–100.

Chan-Ling, T.L., A. Vannas and B.A. Holden. 1988b. Long-term changes in corneal endothelial morphology following wounding in the cat. Invest Ophthalmol Vis. Sci. 29: 1407–1412.

Chen, H.B., S. Yamabayashi, B. Ou, Y. Tanaka, S. Ohno and S. Tsukahara. 1997. Structure and composition of rat precorneal tear film. A study by an *in vivo* cryofixation. Invest Ophthalmol. Vis. Sci. 38: 381–387.

de Paiva, C.S. and S.C. Pflugfelder. 2004. Tear clearance implications for ocular surface health. Exp. Eye Res. 78: 395–397.

Draper, C.E., E. Adeghate, P.A. Lawrence, D.J. Pallot, A. Garner and J. Singh. 1998. Age-related changes in morphology and secretory responses of male rat lacrimal gland. J. Auton. Nerv. Syst. 69: 173–183.

Gipson, I.K. 2004. Distribution of mucins at the ocular surface. Exp. Eye Res. 78: 379–388.

Gipson, I.K. 2007. The ocular surface: the challenge to enable and protect vision: the Friedenwald lecture. Invest Ophthalmol. Vis. Sci. 48: 4390; 4391–4398.

Grahn, B.H., S. Sisler and E. Storey. 2005. Qualitative tear film and conjunctival goblet cell assessment of cats with corneal sequestra. Vet. Ophthalmol. 8: 167–170.

Greiner, J.V., T. Glonek, D.R. Korb, A.C. Whalen,E. Hebert, S.L. Hearn, J.E. Esway and C.D. Leahy. 1998. Volume of the human and rabbit meibomian gland system. Adv. Exp. Med. Biol. 438: 339–343.

Haque, S., D. Fonn, T. Simpson and L. Jones. 2004. Corneal and epithelial thickness changes after 4 weeks of overnight corneal refractive therapy lens wear, measured with optical coherence tomography. Eye Contact Lens 30: 189–193.

Hayashi, S., T. Osawa and K. Tohyama. 2002. Comparative observations on corneas, with special reference to Bowman's layer and Descemet's membrane in mammals and amphibians. J. Morphol. 254: 247–258.

Holly, F.J. 1985. Physical chemistry of the normal and disordered tear film. Trans. Ophthalmol. Soc. UK 104: 374–380.

Jester, J.V., N. Nicolaides and R.E. Smith. 1981. Meibomian gland studies: histologic and ultrastructural investigations. Invest Ophthalmol. Vis. Sci. 20: 537–547.

Joyce, N.C. 2003. Proliferative capacity of the corneal endothelium. Prog. Retin Eye Res. 22: 359–389.

Kessing, S.V. 1968. Mucous gland system of the conjunctiva. A quantitative normal anatomical study. Acta. Ophthalmol. 95: 91.

Kiel, J.W. 2010. The Ocular Circulation. San Rafael, Morgan and Claypool Publishers.
Knop, E. and N. Knop. 2005. The role of eye-associated lymphoid tissue in corneal immune protection. J. Anat. 206: 271–285.
Lemp, M.A. and D.E. Wolfley. 1992. The Lacrimal Apparatus. Adler's Physiology of the Eye, Mosby Year Book 18–28.
Lens, A., S.C. Nemeth and J.K. Ledford. 2007. Ocular Anatomy and Physiology. Basic Bookshelf for Eyecare Professionals Series, Slack Incorporated.
Li, H.F., W.M. Petroll, T. Moller-Pedersen, J.K. Maurer, H.D. Cavanagh and J.V. Jester. 1997. Epithelial and corneal thickness measurements by *in vivo* confocal microscopy through focusing (CMTF). Curr. Eye Res. 16: 214–221.
Liang, H., C. Baudouin, B. Dupas and F. Brignole-Baudouin. 2010. Live conjunctiva-associated lymphoid tissue analysis in rabbit under inflammatory stimuli using *in vivo* confocal microscopy. Invest Ophthalmol. Vis. Sci. 51: 1008–1015.
Linsen, C. and L. Missotten. 1990. Physiology of the lacrimal system. Bull. Soc. Belge Ophtalmol. 238: 35–44.
Linsenmayer, T.F., J.M. Fitch, M.K. Gordon, C.X. Cai, F. Igoe, J.K. Marchant and D.E. Birk. 1998. Development and roles of collagenous matrices in the embryonic avian cornea. Prog. Retin Eye Res. 17: 231–265.
Lipham, W.J., H.A. Tawfik and J.J. Dutton. 2002. A histologic analysis and three-dimensional reconstruction of the muscle of Riolan. Ophthal. Plast. Reconstr. Surg. 18: 93–98.
Lorber, M. 2007. Gross characteristics of normal human lacrimal glands. Ocul. Surf. 5: 13–22.
Marfurt, C.F., J. Cox, S. Deek and L. Dvorscak. 2010. Anatomy of the human corneal innervation. Exp. Eye Res. 90: 478–492.
Maurice, D.M. 1957. The structure and transparency of the cornea. J. Physiol. 136: 263–286.
Maurice, D.M. 1972. The location of the fluid pump in the cornea. J. Physiol. 221: 43–54.
Mishima, S., A. Gasset, S.D. Klyce, Jr. and J.L. Baum. 1966. Determination of tear volume and tear flow. Invest Ophthalmol. 5: 264–276.
Müller, L.J., C.F. Marfurt, F. Kruse and T.M. Tervo. 2003. Corneal nerves: structure, contents and function. Exp. Eye Res. 76: 521–542.
Murube, J. 2009. Hypotheses on the development of psychoemotional tearing. Ocul. Surf. 7: 171–175.
Nagelhout, T.J., D.A. Gamache, L. Roberts, M.T. Brady and J.M. Yanni. 2005. Preservation of tear film integrity and inhibition of corneal injury by dexamethasone in a rabbit model of lacrimal gland inflammation-induced dry eye. J. Ocul. Pharmacol. Ther. 21: 139–148.
Nakao, S., S. Zandi, Y. Hata, S. Kawahara, R. Arita, A. Schering, D. Sun, M.I. Melhorn, Y. Ito, N. Lara-Castillo, T. Ishibashi and A. Hafezi-Moghadam. 2011. Blood vessel endothelial VEGFR-2 delays lymphangiogenesis: an endogenous trapping mechanism links lymph- and angiogenesis. Blood 117: 1081–1090.
Nelson, J.D. 1995. Simultaneous evaluation of tear turnover and corneal epithelial permeability by fluorophotometry in normal subjects and patients with keratoconjunctivitis sicca (KCS). Trans. Am. Ophthalmol. Soc. 93: 709–753.
Obata, H. 2002. Anatomy and histopathology of human meibomian gland. Cornea 21: S70–74.
Preziosi, V.A. 1968. The Periphery of Descemet's Membrane: A Study by Light Microscopy. Arch. Ophthalmol. 80: 197–201.
Qiu, X., L. Gong, X. Sun and H. Jin. 2011. Age-related Variations of Human Tear Meniscus and Diagnosis of Dry Eye With Fourier-domain Anterior Segment Optical Coherence Tomography. Cornea 30: 543–549.
Rivas, L., M.A. Oroza, A. Perez-Esteban and J. Murube-del-Castillo. 1991. Topographical distribution of ocular surface cells by the use of impression cytology. Acta. Ophthalmol. (Copenh) 69: 371–376.
Snell, R.S. and M.A. Lemp. 1998. Clinical Anatomy of the Eye. Malden, Blackwell Science. USA.

Spurr-Michaud S., P. Argueso and I. Gipson 2007. Assay of mucins in human tear fluid. Exp. Eye Res. 84: 939–950.

Stephens, L.C., T.E. Schultheiss, K.J. Vargas, D.M. Cromeens, K.N. Gray and K.K. Ang. 1989. Glands of the eyelids of rhesus monkeys (Macaca mulatta). J. Med. Primatol. 18: 383–396.

Tang, X.L., J.F. Sun, X.Y. Wang, L.L. Du and P. Liu. 2010. Blocking neuropilin-2 enhances corneal allograft survival by selectively inhibiting lymphangiogenesis on vascularized beds. Mol. Vis. 16: 2354–2361.

The International Dry Eye Disease Workshop. 2007. The definition and classification of dry eye disease: report of the Definition and Classification Subcommittee of the International Dry Eye WorkShop. Ocul. Surf. 5: 75–92.

Thoft, R.A. 1978. Role of the ocular surface in destructive corneal disease. Trans. Ophthalmol. Soc. UK 98: 339–342.

Tiffany, J.M. 2008. The normal tear film. Dev. Ophthalmol. 41: 1–20.

Tseng, S.C., L.W. Hirst, A.E. Maumenee, K.R. Kenyon, T.T. Sun and W.R. Green. 1984. Possible mechanisms for the loss of goblet cells in mucin-deficient disorders. Ophthalmology 91: 545–552.

Van Haeringen, N.J. 1981. Clinical biochemistry of tears. Surv. Ophthalmol. 26: 84–96.

Wanko, T., B.J. Lloyd, Jr. and J. Matthews 1964. The Fine Structure of Human Conjunctiva in the Perilimbal Zone. Invest. Ophthalmol. 3: 285–301.

Wilson, S.E. and J.W. Hong. 2000. Bowman's layer structure and function: critical or dispensable to corneal function? A hypothesis. Cornea 19: 417–420.

Wulc, A.E., R.M. Dryden and T. Khatchaturian. 1987. Where is the gray line? Arch. Ophthalmol. 105: 1092–1098.

Yokoi, N., A.J. Bron, J.M. Tiffany, K. Maruyama, A. Komuro and S. Kinoshita. 2004. Relationship between tear volume and tear meniscus curvature. Arch. Ophthalmol. 122: 1265–1269.

Zhang, H., X. Hu, J. Tse, F. Tilahun, M. Qiu and L. Chen. 2011. Spontaneous lymphatic vessel formation and regression in the murine cornea. Invest Ophthalmol. Vis. Sci. 52: 334–338.

Zoukhri, D. 2006. Effect of inflammation on lacrimal gland function. Exp. Eye Res. 82: 885–898.

2

The Human Tear Film

Julia Dieckow[1,*] and *Pablo Argüeso*[2,*]

SUMMARY

The human tear film is a multifunctional fluid secreted to protect the ocular surface from injury, such as chemical, physical (e.g. drying, mechanical damage) and microbial stress. The fluid itself is produced mainly by the lacrimal gland in the upper temporal orbita and in minor amounts by the accessory glands of Wolfring and Krause. Other functionally important components are produced by conjunctival goblet cells, which produce and secrete mucins, and by the acinous, lipid producing meibomian glands found at the rim of the tarsal plate.

For a long time, the tear film was thought to consist of three separate layers: lipid, aqueous and mucinous, each of which plays its unique role in tear film stability and function. However, more recent evidence suggests a two-layer tear film model with a lipid and an aqueous layer overlying the epithelial glycocalyx.

This chapter presents an updated overview of the human tear film characteristics, structure, components and function.

INTRODUCTION

The ability to produce tear fluid to protect the eye from physical, chemical or pathogenic harm is found in almost every species, including mammals, birds, and reptiles. A human being produces between 0.5 and 2.0 μl of tear

[1]Martin Luther University Halle-Wittenberg, Department of Anatomy and Cell Biology, Grosse Steinstraße 52, 06097 Halle (Saale), Germany; E-mail: julia@dieckow.de

[2]Schepens Eye Research Institute, Massachusetts Eye and Ear, Harvard Medical School, 20 Staniford Street, Boston, MA 02114; E-mail: pablo.argueso@schepens.harvard.edu
*Corresponding author

fluid per minute (Clinch et al. 1983). It is spread over the ocular surface during blinking, which happens every three to six seconds while awake (Sforza et al. 2008). The permanently renewed tear film has to fulfill a variety of tasks. In the following chapter, the structure, composition and functions of this unique fluid will be further described.

TEAR FILM THICKNESS AND VOLUME

The total surface area of the eye that has to be covered by tear fluid is not only the surface directly exposed to the environment, meaning the cornea, limbus and perilimbal conjunctiva. Approximately half of the wet-surfaced area is hidden under the lids and in the conjunctival fornix, serving as a lubricating layer for the moving lids (Holly and Holly 1994). The total area of the conjunctival sac, including the cornea, has been estimated as 16 cm^2 (Ehlers 1965).

The thickness of the precorneal film is a variable that can be directly assessed. Depending on the author and assessment method, measurements vary between 40 nm (two wavelength technique) and 46 µm (confocal microscopy) (King-Smith et al. 2004). Given the 16 cm^2 of estimated spreading area and varying fluid height in the exposed and hidden areas (e.g. no lipid layer in unexposed areas, see below), the common estimate of tear film volume has been calculated to be about 4–10 µl, comprising the under-lid volume (~5–6 µl), the upper and lower menisci (~3 µl) and the classical preocular tear film (~1 µl) (Yokoi et al. 2004, Bron et al. 2002).

STRUCTURE OF THE TEAR FILM

The human tear film is not a homogenous fluid, but is composed of several definite layers, each one with its own distinct function. Our understanding of the tear film composition has changed over the last decade. In this chapter, both the traditional three-layer model, and the revised two-layer model, will be further described.

The Traditional Model

Traditionally, the human tear film has been described as a three-layered liquid film with each layer deriving from a distinct origin: (1) a superficial lipid film that inhibits evaporation of the aqueous components and is secreted by meibomian glands at the rim of the eyelids, (2) an intermediate aqueous layer secreted largely by the main lacrimal gland, and (3) a mucous layer resting directly upon the corneal and conjunctival epithelia, being secreted by goblet cells of the conjunctiva (Fig. 1).

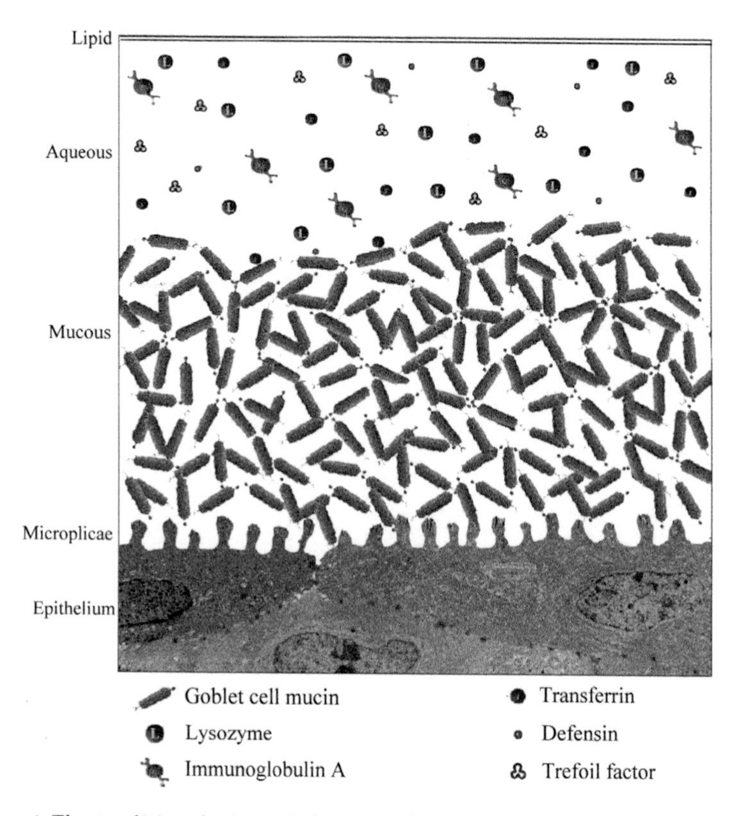

Figure 1 The traditional view of the tear film in which there are three distinct layers: the superficial lipid layer, the aqueous layer and the mucous layer. Modified from "Gipson I.K. and P. Argüeso. 2003. Int. Rev. Cytol. 231: 1–49."

The superficial lipid layer coats the tear fluid and faces the outer environment. It prevents evaporation of aqueous components and plays a special role in inhibiting early tear film breakup. It has been assumed to cover only the directly exposed ocular surface, the interpalpebral area. In the closed eye, the lipid layer is compressed between the lid edges, and spreads again directly with the opening of the eye (Holly and Holly 1994).

The intermediate aqueous phase, the tear fluid itself, is a major fraction of the tear film. It contains several soluble components (see below) and is responsible for most general functions of the tear film, such as hydration of the ocular surface.

In the traditional tear film model, the mucous layer is understood to serve as a coating for the microplicae of the underlying epithelium, creating a hydrophilic lubricating cover that allows an even distribution of the tear film over the eye. It is formed by mucins, highly glycosylated glycoproteins, secreted by conjunctival goblet cells.

The Two-layer Model

According to recent findings in tear film research, the traditional model of the tear film structure has been revised, and the existence of boundaries between the aqueous and mucin layers disputed. With today's knowledge, it appears more likely that the aqueous and mucin layers comprise a single phase (Fig. 2).

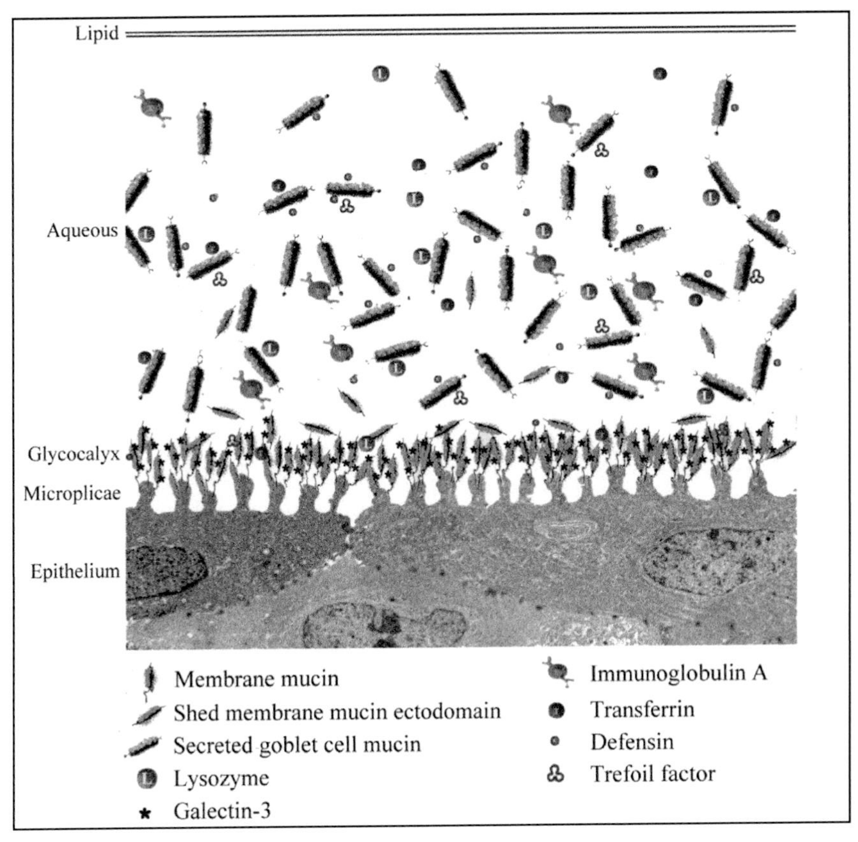

Figure 2 The two-layer model of the tear film includes a superficial lipid layer and an aqueous layer containing mucins and other tear proteins such as antimicrobials. Modified from "Gipson I.K and P. Argüeso. 2003. Int. Rev. Cytol. 231: 1–49."

The presumed presence of insoluble mucins, which would be necessary to create a separate mucus layer, has not been detected (Argüeso et al. 2002, Spurr-Michaud et al. 2007). In fact, the only place in the body where insoluble mucins are found as a distinct layer is in the gastro-intestinal tract, including the colon, where it is formed by MUC2 complexes (Herrmann et al. 1999). Mucins secreted into the tear film by conjunctival goblet cells are soluble and disperse into the aqueous layer (Argüeso et al. 2002, Spurr-Michaud et al. 2007, Jumblatt et al. 1999).

Moreover, cell surface-associated mucins are present in two forms; first, attached to the microplicae of corneal and conjunctival epithelial cells, where they extend up to 500 nanometers into the tear film, second, shed into the tear film (Spurr-Michaud et al. 2007).

The understanding of an overlying lipid layer and its composition resembles the traditional model.

COMPONENTS OF THE TEAR FLUID

In the following section, a model containing two separate layers (lipid and aqueous) and an underlying glycocalyx will be assumed and the different compositions of the layers described.

The Lipid Layer

Lipids of the tear film belong into two separate categories: Lipids of the polar fraction affect the distribution of the lipid film at the aerial interface; lipids of the non-polar fraction make up the barrier to water and contaminants as well as control the layer's melting temperature (Tiffany and Marsden 1986) (Table 1).

Table 1 Major lipid components of the tear film and their main functions (Tiffany 2008).

Fraction	Function	Examples
Polar **(Surface active)**	Ensures optimal spreading over aqueous layer Facilitates spreading of non-polar fraction	Phosphatidylcholine Phosphatidylethanolamine Glycolipids Free fatty acids* Cholesterol*
Non-polar	Barrier to water and contaminants Determines melting temperature of lipid layer	Wax esters (fatty acid + long-chain fatty alcohol) Cholesterol esters

*only present in small amounts.

Diseases in which this composition is disturbed and/or the amount of lipids produced is insufficient (e.g. in meibomian gland dysfunction) result in altered anti-evaporative function and tear film instability.

The Aqueous Layer

The aqueous layer constitutes the major fraction of the tear film. It contains not only oxygen and electrolytes, but also antimicrobial peptides and soluble immunoglobulins to protect the ocular surface from parasitic infection (Table 2). More than 400 proteins have been detected in this layer using mass spectrometry (de Souza et al. 2006).

Table 2 Major components of the aqueous layer and their concentrations in the tear fluid.

Component	Examples	Concentration
Proteins		
Mucins	MUC1; MUC4; MUC16; MUC5AC	
Antimicrobials	sIgA	1.93 mg/ml
	IgG	4 μg/ml
	Lysozyme	2.07 mg/ml
	Lipocalin	1.55 mg/ml
	Lactoferrin	1.65 mg/ml
	Transferrin	1 μg/ml
	Defensins	
Cytokines	IL-1α	
	IL-1β	
	IL-6	
	IL-8	
Albumine	-	0.04 mg/ml
Surfactant proteins	SP-A, SP-D	
	SP-B, SP-C	
Trefoil Factors	TFF2, TFF3	
Electrolytes	Na^+	128.70 mmol/l
	K^+	17.00mmol/l
	Ca^{2+}	0.32 mmol/l
	Mg^{2+}	0.35 mmol/l
	HCO^{3-}	12.40 mmol/l
	Cl^-	141.30 mmol/l

Mucins found in the aqueous layer are either secreted (MUC5AC) or shed from the cell surface (MUC1, MUC4, MUC16) (Spurr-Michaud et al. 2007). Antimicrobials present in the aqueous compartment include immunoglobulins A and G (IgA, IgG), as well as cytokines such as IL-1α, IL-1β, IL-6 and IL-8, which promote immunological reactions to infection (Nakamura et al. 1998). The iron-binding proteins lipocalin, lactoferrin and transferrin act bacteriostaticly by removing free iron that is necessary for bacterial replication (Fluckinger et al. 2004,Fullard and Snyder 1990).

Other proteins include surfactant proteins SP-B and -C, which affect spreading properties and the film's stability (Bräuer et al. 2007a), and SP-A and D, which are immune modulating peptides (Bräuer et al. 2007b). As the tear film contains oxygen to supply the cornea, reactive oxygen species (ROS) have to be rendered harmless by enzymes like catalase, superoxide dismutase and glutathione peroxidase (Crouch et al. 1991). Lacritin is a tear protein that promotes basal tear secretion and stimulates corneal epithelial cell growth (Samudre et al. 2011).

The Glycocalyx

The glycocalyx is comprised of cell surface-associated mucins that extend into the tear fluid with large extracellular domains. It is assumed that the glycocalyx is highly organized, most likely through interactions of surface mucins with lectin-like proteins such as galectin-3 (Argüeso et al. 2009). Other components of the glycocalyx include proteoglycans with their associated glycosaminoglycans.

FUNCTIONS OF TEAR FLUID

Protection

The ocular surface epithelium is constantly confronted with disturbances from the environment, many of which can be repelled by instinctive blinking or averting the face from the source of danger. Drying, infection and minor injuries, however, are some of the ever-present threats to the eye that may be prevented by a healthy tear film.

Physical Protection

Superficial lipids repel dust particles and some types of bacteria (Brauninger et al. 1972). Hydrophilic particles cannot penetrate through this lipid layer, while many of those hydrophobic particles able to penetrate are absorbed and immobilized by mucins lying underneath. During the blink, the trapped particles are removed toward the lower fornix and finally drained through the lacrimal puncta (Adams 1979).

As an excess of lipids at the ocular surface may lead to the development of hydrophobic, non-wettable spots and tear film breakup, lipocalin, the major lipid-binding protein in the tear film, scavenges lipids in the tear fluid and delivers them to the aqueous phase. Thereby, it increases the tear film's surface pressure and preserves its integrity (Glasgow et al. 1999). Surfactant proteins found in tears perform similar stabilizing functions (Bräuer et al. 2007).

Antimicrobial Protection

The tear film also contains major antimicrobial peptides and proteins. While IgA and IgG are produced by plasma cells and secreted through the epithelium, the cell-wall-degrading enzyme lysozyme is provided by the lacrimal gland and fends off gram-positive bacteria through its muraminidase activity (Fullard and Snyder 1990). Siderophilic bacteria, which require free iron, are attenuated by lactoferrin and lipocalin, two peptides with iron-sequestering functions (Fluckinger et al. 2004). The highly conserved molecules α- and β-defensin have also been found in the tear film and have a broad range of antimicrobial activities (Haynes et al. 1999).

Trefoil factors like TFF2 and -3 have antiapoptotic properties and promote wound healing in case of minor injuries (Paulsen et al. 2008). The glycocalyx itself is assumed to be of high importance in preventing pathogen penetrance. Recent data has shown that cell surface O-glycans contribute to the prevention of bacterial adherence to the apical surface of corneal epithelial cells (Ricciuto et al. 2008).

Lubrication

To guarantee a smooth and undisturbed blinking process, all moving parts are covered by the thin tear fluid that serves as a lubricating layer. Drying of the eye can lead to discomfort, epithelial erosions, even ulcerations. Moreover, the glycocalyx and its associated O-glycans guarantee boundary lubrication of the epithelium, which persists during blinking (Sumiyoshi et al. 2008).

Refraction

The tear film coats small corneal irregularities, thus smoothing the refractive surface and preventing light scattering and blurred vision (Ridder et al. 2009).

Nutrition

Due to the necessity for transparency, the cornea is an avascular tissue. Conjunctival vessels may only provide nourishment to corneal epithelial cells close to the limbus. For many years it has been known that the apical epithelium is a barrier for most nourishing components. Accordingly, glucose supply has been shown to happen mostly through the endothelial side (Thoft and Friend 1975). However, transportation of oxygen into the cells is one of the primary functions of the tear film. Excretion of metabolites can be evidenced by the amount of carbon dioxide, lactate and pyruvate found in tear samples (Tiffany 2008).

CONCLUSION

The overall function of tear fluid is to protect, lubricate and nourish the ocular surface. It may not only trap physical hazards like dust particles or fine hairs, but also prevent microbial penetrance and ensuing infection of the underlying cornea and conjunctiva.

Disturbed tear film composition or otherwise impaired function may lead to ocular surface diseases like dry eye syndrome. These diseases may be severe, in that they may cause serious irritation, and even be extremely painful. Subsequent ocular infection may be long lasting and difficult to cure.

ACKNOWLEDGMENT

Supported by NIH/NEI Grant R01 EY014847 (PA).

REFERENCES

Adams, A.D. 1979. The morphology of human conjunctival mucus. Arch. Ophthalmol. 97: 730–734.

Argüeso, P., A. Guzman-Aranguez, F. Mantelli, Z. Cao, J. Ricciuto and N. Panjwani. 2009. Association of cell surface mucins with galectin-3 contributes to the ocular surface epithelial barrier. J. Bio. Chem. 284: 23037–23045.

Argüeso, P., M. Balaram, S. Spurr-Michaud, H.T. Keutmann, M.R. Dana and I.K. Gipson. 2002. Decreased levels of the goblet cell mucin MUC5AC in tears of patients with Sjögren syndrome. Invest Ophthalmol. Vis. Sci. 43: 1004–1011.

Bräuer, L., M. Johl, J. Börgermann, U. Pleyer, M. Tsokos and F.P. Paulsen. 2007a. Detection and localization of the hydrophobic surfactant proteins B and C in human tear fluid and the human lacrimal system. Curr. Eye Res. 32: 931–938.

Bräuer, L., C. Kindler, K. Jäger, S. Sel, B. Nölle, U. Pleyer, M. Ochs and F.P. Paulsen. 2007b. Detection of surfactant proteins A and D in human tear fluid and the human lacrimal system. Invest. Ophthalmol. Vis. Sci. 48: 3945–3953.

Brauninger, G.E., D.O. Shah and H.E. Kaufman. 1972. Direct physical demonstration of oily layer on tear film surface. Am. J. Ophthalmol. 73: 132–134.

Bron, A.J., J.M. Tiffany, N. Yokoi and S.M. Gouveia. 2002. Using osmolarity to diagnose dry eye: a compartmental hypothesis and review of our assumptions. Adv. Exp. Med. Biol. 506: 1087–1095.

Clinch, T.E., D.A. Benedetto, N.T. Felberg and P.R. Laibson. 1983. Schirmer's test. A closer look. Arch. Ophthalmol. 101:1383–1386.

Crouch, R.K., P. Goletz, A. Snyder and W.H. Coles. 1991. Antioxidant enzymes in human tears. J. Ocul. Pharmacol. 7: 253–258.

de Souza, G.A., L.M. Godoy and M. Mann. 2006. Identification of 491 proteins in the tear proteome reveals a large number of proteases and protease inhibitors. Genome Biol. 7: R72.

Ehlers, N. 1965. On the size of the conjunctival sac. Acta Ophthalmol. 43: 205–210.

Fluckinger, M., H. Haas, P. Merschak, B.J. Glasgow and B. Redl. 2004. Human tear lipocalin exhibits antimicrobial activity by scavenging microbial siderophores. Antimicrob. Agents Chemother. 48: 3367–3372.

Fullard, R.J. and C. Snyder. 1990. Protein levels in nonstimulated and stimulated tears of normal human subjects. Invest. Ophthalmol. Vis. Sci. 31: 1119–26.

Glasgow, B.J., G. Marshall, O.K. Gasymov, A.R. Abduragimov, T.N. Yusifov and C.M. Knobler. 1999. Tear lipocalins: potential lipid scavengers for the corneal surface. Invest Ophthalmol. Vis. Sci. 40: 3100–3107.

Haynes, R.J., P.J. Tighe and H.S. Dua. 1999. Antimicrobial defensin peptides of the human ocular surface. Br. J. Ophthalmol. 83: 737–741.

Herrmann, A., J.R. Davies, G. Lindell, S. Martensson, N.H. Packer, D.M. Swallow and I. Carlstedt. 1999. Studies on the "insoluble" glycoprotein complex from human colon. Identification of reduction-insensitive MUC2 oligomers and C-terminal cleavage. J. Bio. Chem. 274: 15828–15836.

Holly, F.J. and T.F. Holly. 1994. Advances in ocular tribology. Adv. Exp. Med. Biol. 350: 275–283.

Jumblatt, M.M., R.W. McKenzie and J.E. Jumblatt. 1999. MUC5AC mucin is a component of the human precorneal tear film. Invest Ophthalmol. Vis. Sci. 40: 43–49.

King-Smith, P.E., B.A. Fink, R.M. Hill, K.W. Koelling and J.M. Tiffany. 2004. The thickness of the tear film. Curr. Eye Res. 29: 357–368.

Nakamura, Y., C. Sotozono and S. Kinoshita. 1998. Inflammatory cytokines in normal human tears. Curr. Eye Res. 17: 673–676.

Paulsen, F.P., C.W. Woon, D. Varoga, A. Jansen, F. Garreis, K. Jäger, M. Amm, D.K. Podolsky, P. Steven, N.P. Barker and S. Sel. 2008. Intestinal trefoil factor/TFF3 promotes re-epithelialization of corneal wounds. J. Bio. Chem. 283: 13418–13427.

Ricciuto, J., S.R. Heimer, M.S. Gilmore and P. Argüeso. 2008. Cell surface O-glycans limit Staphylococcus aureus adherence to corneal epithelial cells. Infect. Immun. 76: 5215–5220.

Ridder, W.H., J. LaMotte, J.Q. Hall Jr, R. Sinn, A.L. Nguyen and L. Abufarie. 2009. Contrast sensitivity and tear layer aberrometry in dry eye patients. Optom. Vis. Sci. 86: 1059–1068.

Samudre, S., F.A. Lattanzio Jr, V. Lossen, A. Hosseini, J.D. Sheppard Jr., R.L. McKown, G.W. Laurie and P.B. Williams. 2011. Lacritin, a novel human tear glycoprotein, promotes sustained basal tearing and is well tolerated. Invest. Ophthalmol. Vis. Sci. 52: 6265–6270.

Sforza, C., M. Rango, D. Galante, N. Bresolin and V.F. Ferrario. 2008. Spontaneous blinking in healthy persons: an optoelectronic study of eyelid motion. Ophthalmic. Physiol. Opt. 28: 345–353.

Spurr-Michaud, S., P. Argüeso and I.K. Gipson. 2007. Assay of mucins in human tear fluid. Exp. Eye Res. 84: 939–950.

Sumiyoshi, M., J. Ricciuto, A. Tisdale, I.K. Gipson, F. Mantelli and P. Argüeso. 2008. Antiadhesive character of mucin O-glycans at the apical surface of corneal epithelial cells. Invest. Ophthalmol. Vis. Sci. 49: 197–203.

Thoft, R.A. and J. Friend. 1975. Permeability of regenerated corneal epithelium. Exp. Eye Res. 21: 409–416.

Tiffany, J.M. 2008. The normal tear film. Dev. Ophthalmol. 41: 1–20.

Tiffany, J.M. and R.G. Marsden. 1986. The influence of composition on physical properties of meibomian secretion. pp. 597–608. In: Holly FJ [ed]: The Preocular Tear Film in Health, Disease, and Contact Lens Wear. Lubbock/Tex, USA.

Yokoi, N., A.J. Bron, J.M. Tiffany, K. Maruyama, A. Komuro and S. Kinoshita. 2004. Relationship between tear volume and tear meniscus curvature. Arch. Ophthalmol. 122: 1265–1269.

3

Comparative Anatomy and Physiology of the Cornea and Conjunctiva

Michael J. Doughty

SUMMARY

The clinical appearance and common methods of assessment of the eyelid margin, conjunctiva and cornea are reviewed within the context of objective anatomical, fine structure and ultrastructure studies of the same. Data from *in vivo* imaging methods such as confocal or specular microscopy are also considered. The physiological properties of the conjunctival and corneal epithelia will be considered, especially from their inter-relationship with the pre-ocular tear film. The anatomy, ultrastructure and physiology are considered in a comparative way to highlight the similarities as well as the differences.

INTRODUCTION TO ANATOMY AND PHYSIOLOGY OF THE OCULAR SURFACE

The ocular surface is a mucous membrane comprising the visible portions of the corneal surface, the exposed bulbar conjunctiva as well as hidden portions composed of the bulbar conjunctiva covered by the eyelids (within the conjunctival sac) and the palpebral conjunctiva (on the underside of the eyelids) (Fig. 1A). The depth (extent) of the conjunctival sac is not symmetrical, with the superior and inferior aspects being dominant (Fig. 1A). Typical quoted values, for adults and probably of Caucasian origin,

Department of Vision Sciences, Glasgow-Caledonian University, Glasgow G4 OBA, United Kingdom; E-mail: M.Doughty@gcu.ac.uk are 14 to 15 mm for the superior sac, 10 to 12 mm for the inferior sac, and 5 to 8 mm for the lateral (temporal) aspect. The medial (nasal) aspect of the conjunctival sac has no depth since it terminates at the plica semilunaris (Hogan et al. 1971).

In the open eye in primary gaze, the vertical distance between the edges of the upper and lower eyelids is best referred to as the palpebral aperture height (PAH) and this distance (in mm) can be used to estimate the exposed ocular surface area (EOSA) (Zaman et al. 1998); in Caucasians at least, as PAH increases there is a very predictable increase in EOSA (Fig. 1B).

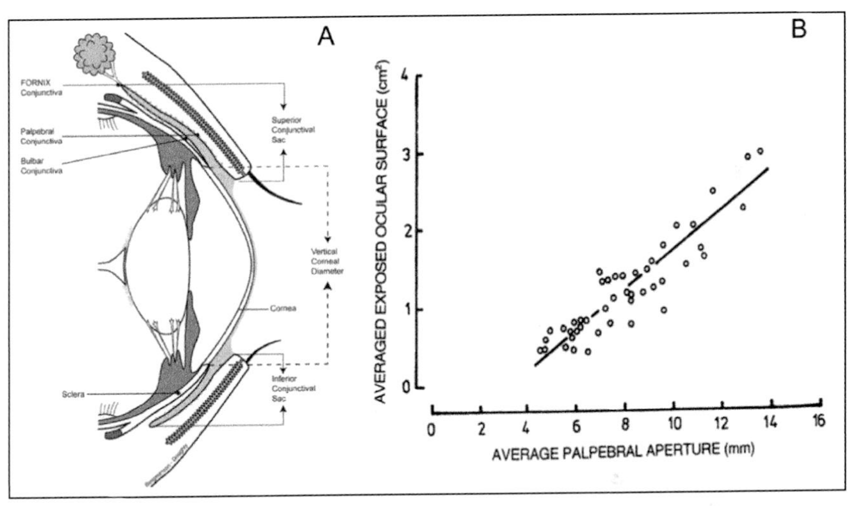

Figure 1 Schematic diagram of the anterior aspects of the human eye as viewed from the side (A) and the relationship between the palpebral aperture height (PAH) and (B) the exposed ocular surface area (EOSA). (A) Modified from Bergmanson and Doughty 2005, copyright Lippincott-Williams & Wilkins (B) modified from Zaman et al. 1998; copyright Elsevier Scientific.

Traditionally, it has been the exposed ocular surface that receives most attention in the clinical assessment of conditions such as keratoconjunctivitis sicca (KCS), but it is not just the cornea and bulbar conjunctiva that are affected in ocular surface disease. While the pre-corneal tear film can only be sustained across a healthy corneal surface, the same applies to the exposed bulbar conjunctiva and the health of the eyelid margins plays an important role in determining the tear meniscus characteristics (Cui et al. 2011). In addition, the apparent tear meniscus height (TMH) decreases as the PAH increases (Doughty et al. 2001a).

This review will cover the essential anatomical and ultrastuctural features of the eyelid margins, the conjunctiva and cornea, but with emphasis wherever possible on those characteristics thought to be most relevant to the development and clinical assessment of dry eye.

THE MARGINAL ZONE OF THE CONJUNCTIVA

Gross Organization and Clinical Features of the Marginal Zone

The marginal zone extends from the eyelashes inward to include the orifices of the main tarsal (Meibomian) glands and then a specialized zone of stratified epithelium (Wirtschafter et al. 1999, Knop et al. 2011). Anterior to this zone, and usually extending across part of the Meibomian orifices, is a different type of stratified epithelia that is covered with a superficial keratinized layer as it extends from the skin tissue that covers the outer surface of the eyelid (Jester et al. 1981, Knop et al. 2011). The specialized zone of stratified epithelium of the marginal zone proper is not usually keratinized and is clearly composed of basal cells (with vertically aligned nuclei), layers of intermediate cells and then layers of flattened superficial cells (Fig. 2A). These superficial cells appear similar to those seen at the corneal epithelial surface. The squamous nature of these cells can also be seen if a sample of cells is taken off the surface of this marginal zone. The cells appear quite large (note the scale marker of 100 μm), some have normal sized nuclei but some have pyknotic (shrunken) nuclei and are sometimes even anucleate (Doughty 2010a) (Fig. 2B).

This region of the marginal zone stains clinically with both lissamine green (Fig. 3) and rose bengal (Norn 1985), and the recognition of this

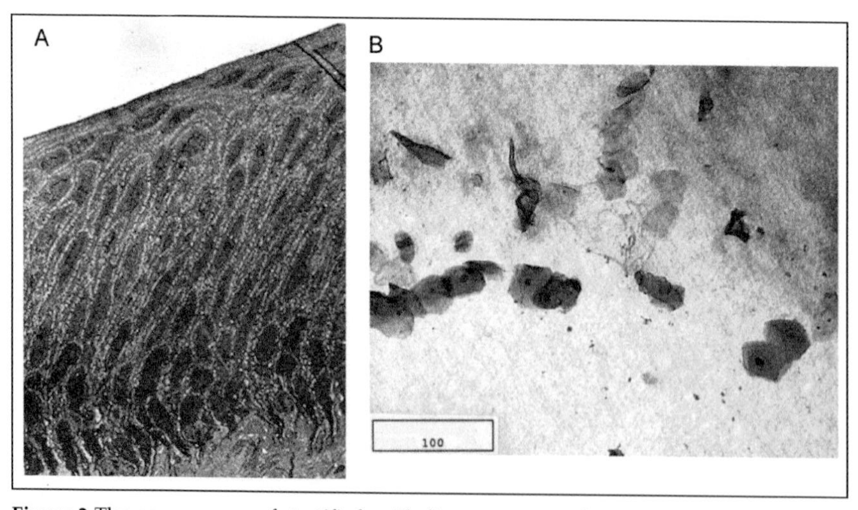

Figure 2 The narrow zone of stratified epithelium present at the eyelid marginal zone. (A) Lower magnification electron micrograph from rabbit, from Doughty and Bergmanson 1999, (B) Impression cytology sample (Biopore membrane) taken from across marginal zone of a healthy human subject after staining with rose bengal and Giemsa. Cells are clearly squamous in nature sometimes with pyknotic nuclei or even anucleate. From Doughty 2010a, copyright Tiva publications.

feature is usually attributed to Marx (Marx 1924) after whom the staining line is often referred (i.e. as Marx's line) (Donald et al. 2003, Doughty et al. 2004). The darker staining of the cells in Fig. 2B is because these cells were double stained with rose bengal and Giemsa. In normal human subjects, this line of specialized epithelium lies just posterior to the orifices of the main tarsal (Meibomian) glands, and is normally very narrow. Measurements taken from the upper lid of Caucasian subjects indicate an average width of just 0.10 ± 0.09 mm (Donald et al. 2003), with the anterior edge (margin) located an average of 0.13 mm behind (i.e. posterior to) the Meibomian gland orifices; similar staining can be seen across the marginal zone of the lower eyelid. The staining of the marginal zone on either eyelid can become very substantially wider when there is some degree of ocular irritation and frequent eye-blink activity (Fig. 3).

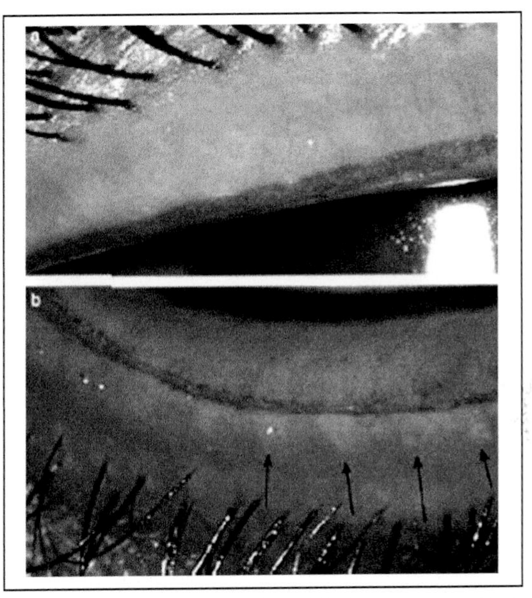

Figure 3 Highlighting of the marginal zone of stratified squamous epithelium in the living eye with lissamine green. For upper (a) and lower (b) eyelids. Arrows (b) show location of Meibomian gland orifices just anterior to Marx's line. Images show atypical appearance often seen in those with some degree of ocular irritation and frequent eye-blink activity. From Doughty et al. 2004; copyright College of Optometrists/Wiley.

Physiological Functions of the Marginal Zone

Some aspect of the marginal zone is not only in contact with the rest of the ocular surface (cornea and bulbar conjunctiva) but the contact is considered substantial enough to exert physical pressure (Shaw et al. 2009). The function of this marginal zone of specialized stratified squamous epithelium has been a subject of considerable debate perpetuated by what appear to be marked differences in opinion as to the type of correlations that might be drawn between clinically-evident features (in most instances) and the much greater detail visible in histopathology or electron microscopy.

Some consider the marginal zone to be the natural contact point between the edge of the lids and the bulbar and corneal surfaces (Kessing 1967, Fatt 1992, Donald et al. 2003, Doughty et al. 2004, Shaw et al. 2009), and attribute its structural characteristics (Figs. 2 and 3) to it being in frequent pressure-related frictional motion against these surfaces. Stated another way, this very narrow zone could be considered as having these features because it is adapted to playing this specialized role. Others disagree (Bron et al. 2011, Knop et al. 2011), not only considering that Marx's line cannot be the contact point but that the special structural features (especially the staining with lissamine green and rose bengal) are the result of desiccation of the cells. The latter aspect is not without its merits since, in the normal eye, this part of the marginal zone would be at the distal edge of the lower tear meniscus (or lacrimal river or lacrimal lake as it is also sometimes referred to).

Beyond this specialized marginal zone of squamous epithelium, there is a progressive and gradual transition to the palpebral conjunctiva proper (Wirtschafter et al. 1999, Knop et al. 2011). The transition includes a slight depression or partial fold referred to as the marginal sulcus (sometimes referred to as the ciliary sulcus because of its relative proximity to the eyelashes).

The eyelid marginal zone is sensitive to touch (a tactile sensitivity) (McGowan et al. 1994), but is not expected to be as sensitive as the cornea.

THE PALPEBRAL CONJUNCTIVA INCLUDING THE FORNIX

The Palbebral Conjunctival Epithelium

The palpebral conjunctiva and that covering the fornix are considered to be 'columnar' epithelia that contain goblet cells. This cell layer forms the inner lining of the eyelids and is normally hidden from view (Fig. 1A).

The palpebral conjunctiva is normally coated with mucus. While not normally visible (i.e. it is transparent), it can be visualized post-mortem with the use of special chemical fixatives such as cetylpyridinium chloride (CPC), which essentially preserve the mucus in its hydrated gel state. Across the palpebral conjunctiva, at least in rabbits, this fixed gel layer can be seen to be very thick in that it is interrupted by large crypt-like openings (Doughty 1997) (Fig. 4). These clearly extend deep down to the accessory lacrimal glands (of Wolfring) and are numerous in the rabbit. In the human eyelid, these glands are more organized into large clusters and while large openings of the type shown Fig. 4 have not been reported for humans, a possible location is just behind the tarsal plate (Knop and Knop 2010). This thick mucus layer is probably important in providing a robust lubricating system to facilitate regular eye-blink activity (Cher 2008). The mucus-based material likely fills the space between the palpebral surface and the bulbar conjunctiva (Fig. 1A), a space that can be imaged using X-ray (and sometimes referred to as Kessing's space) (Kessing 1967).

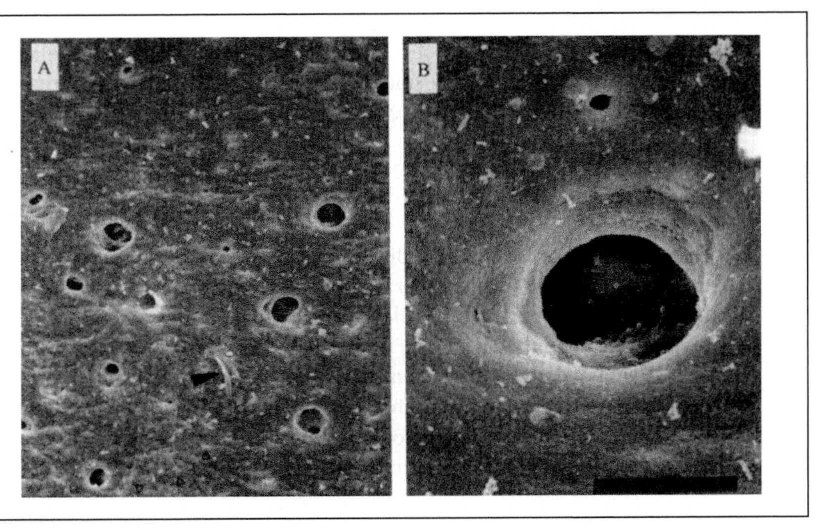

Figure 4 Scanning electron microscopy to show specially-fixed mucus layer across the palpebral conjunctiva, rabbit (A), including large crypts leading to the accessory lacrimal glands (B). From Doughty 1997, copyright Karger.

Without the layer of mucus, the surface of the palpebral conjunctiva can be seen by scanning electron microscopy to be composed of small polygonal cells (i.e. having a range of shapes) but of relatively consistent size (Doughty 1997) (Fig. 5A and B). At least in rabbits, the surface of these cells in the tarsal region is predominantly decorated with microplicae (fine micro-ridges) (Fig. 5A), so resemble corneal epithelial cells. However, across the orbital portion of the palpebral conjunctiva there are more distinctive microvilli on the surface (Doughty 1997) (Fig. 5B). The cell layer also includes numerous very small openings (orifices) of the goblet cells especially visible in the upper left of Fig. 5B and in the middle of Fig. 5C (see later).

Underneath the mucus layer and away from either the marginal zone or the large crypt-like openings, the palpebral conjunctiva is composed of 3 to 5 layers of cells, with the fewest layers likely being in the fornix regions. There is little obvious difference between cells at the surface (in contact with the mucus/tear film) and those basal cells lower down (and in contact

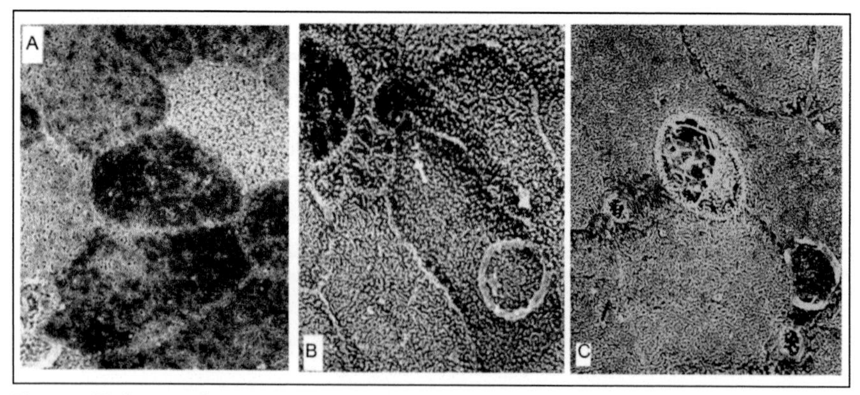

Figure 5 High magnification scanning electron microscopy to illustrate features of the surface cells of the tarsal palpebral conjunctiva (A), orbital region (B) and bulbar (C). A and C from Doughty 1997, copyright Karger, while C is Doughty (unpublished).

with the connective tissue underlying the conjunctival epithelia) (Weyrauch 1983, Wirtschafter et al. 1999, Doughty and Bergmanson 2004a). The basal conjunctival cells (across the palpebral and bulbar regions) undergo discrete cycles of division, originating from a special population of stem cells (Peter et al. 1996, Nagasaki and Zhao 2005); these stem cells and the division cycle may also be linked to that in the vicinity of the marginal zone of conjunctiva (Peter et al. 1996, Wirtschafter et al. 1999). The new daughter cells then can migrate across the conjunctiva, probably in the direction of the fornix (Peter et al. 1996). The overall process is presumably linked to cell replacement, but the characteristics of the cells that are routinely lost from the palpebral conjunctival surface remains to be clearly established although some, at least, undergo apotosis (programmed cell death) (Dogan et al. 2004).

The conjunctival epithelial cells do not form a continuous layer, being frequently interrupted by the mucus-producing (goblet) cells (Doughty and Bergmanson 2004a) (Fig. 6). As illustrated, some of these are located deep within the epithelial cell layers, while others are in intimate contact with the surface and can even be found to be discharging their mucin content. Individual goblet cells contain an actin cytoskeleton (Gipson and Tisdale 1997) surrounding the content of mucin granules, which can be clearly seen in relatively high magnification transmission electron microscopy (Doughty and Bergmanson 2004a) (Fig. 6B). Once having 'contracted' and discharged their contents, the goblet cells leave a small void within the epithelium (see upper right of Fig. 6A), the actin cytoskeleton of which presumably then collapses in on itself. The orifices of the goblet cells can also be seen in scanning electron microscopy (Doughty 1997) (Fig. 5B).

The presence and distribution of these goblet cells across the palpebral conjunctiva can also be revealed in the living human eye by a technique called conjunctival impression cytology (CIC) (Saini et al. 1990). A small

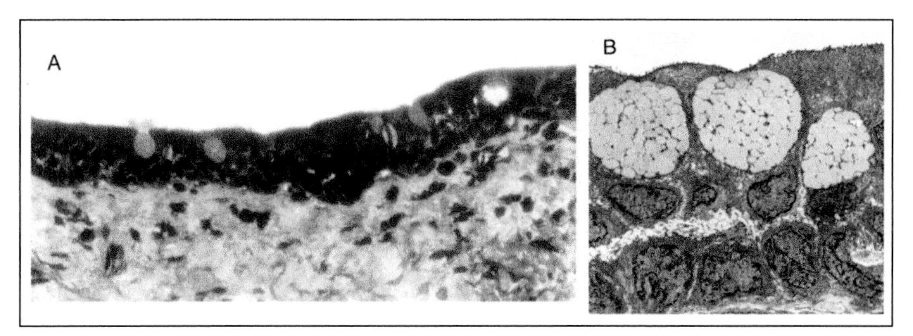

Figure 6 Palpebral conjunctiva and its goblet cells. Light microscope image (A) and that from higher magnification transmission electron microscopy (B) to show tarsal portion of palpebral conjunctiva with goblet cells. Underlying the tarsal conjunctiva is connective tissue (the parenchyma). Modified from Doughty and Bergmanson 2004a, copyright Australian Association of Optometrists/Wiley.

piece of absorbent filter paper is pressed onto the tarsal surface of the everted upper eyelid, removed and then stained to detect the mucus content of the goblet cells. The same principle can also be applied to post-mortem specimens using what is referred to as a flat mount technique (Kessing 1968, Gipson and Tisdale 1997). In this case, the entire conjunctiva is dissected from post-mortem (cadaver) eyes, stretched and pinned out onto a suitable support and then stained for the mucus-containing cells with periodic acid–Schiff (PAS) stain. With any of these techniques (light microscopy, electron microscopy, CIC or flat mounts), it can be readily demonstrated that the normal palpebral conjunctiva contains numerous goblet cells, with densities likely exceeding some 500 cells/mm^2.

While the palpebral conjunctival surface might be described in a clinical grading as satin smooth (Saini et al. 1990), examination at higher magnification and with the aid of fluorescein dye reveals the surface to be organized into numerous arrays of polygonal features (Fig. 7) (Potvin et al. 1994, Doughty et al. 1995, Doughty and Bergmanson 2003). The area of these fluorescein-highlighted features is normally on the order of 0.025 mm^2, but can attain sizes in excess of 1 mm^2 in chronic inflammatory conditions (Doughty et al. 1995).

While the development of these arrays can be associated with an allergic reaction (and a follicular-papillary reaction is possible), rubbing of the palpebral surface against the anterior surface of a contact lens can also produce such changes, a condition referred to as contact lens wear associated papillary conjunctivitis (CLPC) or giant papillary conjunctivitis (GPC). In some contact lens-wearing individuals, the most distal parts of the palpebral conjunctival surface, adjacent to and abutting onto the marginal zone, can be transformed by the contact with the contact lens surface. A change occurs from a normal mucous membrane to one which stains with

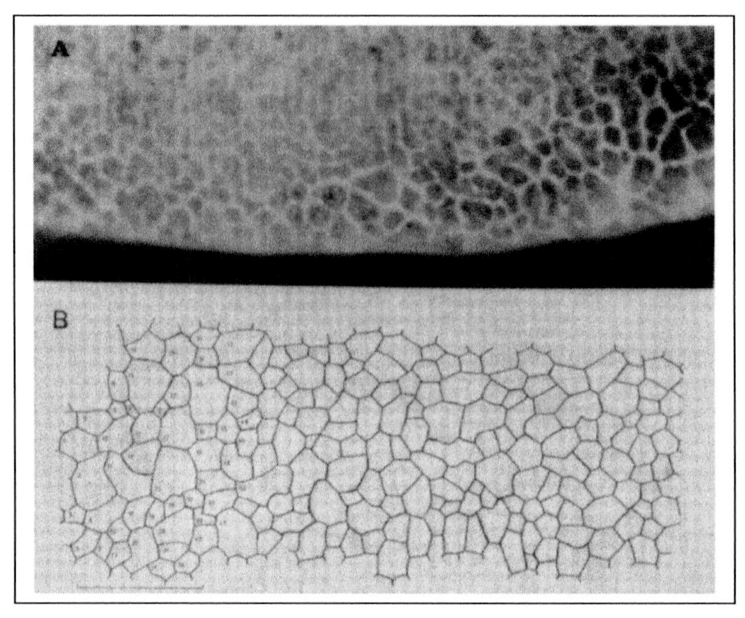

Figure 7 Reticular patterns across the tarsal portion of the palpebral conjunctiva. (A) example of original fluorescein-highlighted patterns and (B) overlay generated from such an image for measurements. From Potvin et al. 1994. Copyright Butterman-Heinemann.

rose bengal or lissamine green; this condition has been referred as 'lid wiper epitheliopathy' (Knop et al. 2011).

The Sub-epithelial Parenchyma for the Palpebral Conjunctiva

Directly underneath the palpebral conjunctiva is the supporting connective tissue or parenchyma (Fig. 6A). It consists of collagen-based connective tissue with a rich blood supply. Associated with the vasculature is a lymphatic vessel system which provides the natural mechanism by which the fluid balance within these tissues is maintained, i.e. the eyelid tissue can become grossly swollen as part of an inflammatory response, but will be restored to its normal volume by the action of the lymphatics. The rich blood supply also means that some white blood cells can be found within the conjunctival parenchyma and even migrate to be located within the conjunctival epithelial cell layer (Doughty and Bergmanson 2004a). When present, the inflammation-mediating exogeneous cells can also change the essential features of the structure of the conjunctival cells and the resident population of goblet cells (Doughty and Bergmanson 2004a).

Some of the collagen bundles under the tarsal portion of the palpebral conjunctiva are further specialized to form a fibrous sheath which surrounds the main tarsal gland (the Meibomian gland) (Jester et al. 1981); these are very densely packed and this feature is more developed in the upper than lower eyelid and allows for the eversion of the upper eyelid during clinical examination. The rest of the eyelid tissue, including the other accessory glands and eyelid muscles, are also embedded in connective tissue and also covered by the eyelid skin on the most anterior aspect.

Measures of the light transmission through the eyelid *in vivo* indicate that less than 1% of shorter wavelengths of visible light should be transmitted, but up to 5% of longer wavelength red light (Ando and Kripke 1996).

THE BULBAR CONJUNCTIVA

The Exposed Conjunctival Epithelium

Within the normal palpebral aperture in the open eye (primary gaze) position, part of the bulbar conjunctiva is habitually exposed. As with palpebral conjunctiva, the mucus membrane of the bulbar conjunctiva is also covered with mucus. As across the palpebral conjunctiva, this mucus (mucin) coat is largely produced by the goblet cells (Rohen and Lutjen-Drecoll 1992, Gipson and Inatomi 1997), and can be considered to be functionally attached to the surface cells of the bulbar conjunctiva. Transmission electron microscopy studies of the bulbar conjunctiva after special chemical fixation with cetylpyridinium chloride (CPC) indicate an a relatively thin amorphous surface layer that is some 0.8 µm thick (Nichols et al. 1985). The principal component of this mucin, originating from the goblet cells, is the MUC5AC glycoprotein (Inatomi et al. 1996). The conjunctival epithelial cells may also produce 'secretory' mucin independently of the goblet cells, a contribution that is considered to function to anchor the goblet cell mucins to the epithelial cell surfaces. Numerous small vesicles can be seen in the superficial bulbar conjunctival cells by transmission electron microscopy (Takakusaki 1969, Dilly and Mackie 1983), and these cells can also be shown to contain mucins (Inatomi et al. 1996). The combined mucin components, especially perhaps those produced and secreted by the conjunctival cells, are thought to provide a protective covering that prevents these cells, in the living eye, from staining with 'vital' dyes (stains) such as rose bengal (Feenstra and Tseng 1992).

As viewed by scanning electron microscopy, the bulbar conjunctival surface (like the palpebral conjunctival surface) can be seen to be made up of a layer of small-sized polygonal cells with similar sizes and shapes (Fig. 5C) interrupted with numerous orifices of the goblet cells. The exposed bulbar conjunctival cells have been reported to show a range of surface appearances (Greiner et al. 1981, Gilbard et al. 1988, Kern et al. 1988), indicating that these surfaces are very susceptible to even small changes in preparative procedures for electron microscopy (Doughty 1997). Shown in Fig. 5C are cells largely decorated with microplicae, but others have reported observing cells decorated with peculiarly 'tufted' microvilli. The surface cells of the bulbar conjunctiva can also be viewed in the living eye by confocal microscopy or samples of the surface cells taken by impression cytology (see later). *In vivo* confocal microscopy studies have provided a wide range of average cell sizes from c. 212 to 592 μm^2; these average cell area values are equivalent to cell densities of between 4708 to 1689 cells/mm^2 (Messmer et al. 2006, Kojima et al. 2010, Efron et al. 2010, Zhu et al. 2010).

Both light microscopy and electron microscopy can be used to show that the bulbar conjunctiva, like the palpebral conjunctiva, includes large numbers of goblet cells (see below). As compared to the corneal epithelium (see later), only sparse tight junctions (*zonnula occludens*) are evident between the superficial cells (Ruskell 1991, Rohen and Lutjen-Drecoll 1992), i.e. they are not commonly observed. The deeper-lying cells of the bulbar conjunctiva make contact with each other, with occasional special cell-cell linking/adhesion junctions called desmosomes, with some preparative techniques indicating only loose contact between either superficial or deeper-lying bulbar conjunctival cells. There is a progressive change in the bulbar conjunctiva where it is in close proximity to the cornea. This transition is the limbus, the most anterior manifestation of the sclero-cornea junction. As viewed in section with light or electron microscopy, the bulbar conjunctiva right at the sclero-corneal (limbal) junction may be 6 or more cell layers thick (Wanko et al. 1964, Rohen and Lutjen-Drecoll 1992). At all locations, an undulating and very thin basement membrane underlies the bulbar (limbal and palpebral) conjunctiva.

By light microscopy or electron microscopy, the goblet cells can be seen to be located between the epithelial cells of the bulbar conjunctiva (Wanko et al. 1964, Kessing 1968, Ganley and Payne 1981, Nichols et al. 1985, Nichols 1996, Steul 1989, Rohen and Lutjen-Drecoll 1992), and so some of their orifices should be visible at the surface (Ganley and Payne 1981). If staining of a flat mount tissue is used to visualize the goblet cells (Kessing 1968), then goblet cells can also be seen at all depths of the conjunctival epithelial layer. As with the palpebral conjunctiva (see earlier, and Fig. 6A), it is generally accepted that conjunctival goblet cells (like those in other mucous membranes in the body), are formed in the deeper epithelium and then migrate to the surface, where they discharge their entire mucin granule contents. Any stressful stimulus will likely result in such goblet cell discharge out onto the ocular surface (Verges Roger et al. 1986). This discharge, if substantial enough, can result in mucus strands being evident on the ocular surface and in the cul-de-sac (fornix) regions.

The goblet cells of the bulbar conjunctiva have been extensively assessed by a technique called conjunctival impression cytology (CIC), a technique that also allows for visualization of the non-goblet cells of the conjunctival epithelium. Using appropriately stained specimens (e.g. with Giemsa; Fig. 8) (Doughty 2011a), the bulbar conjunctival cells are readily seen with their relatively large and densely stained nuclei. By contrast, the goblet cells appear of similar size, lack notable staining (so appear as empty balloon-like cells) and have a very small but still densely staining nucleus. The cells shown in Fig. 8A are all relatively small in size, as can be judged from the 100 μm scale marker rectangular box on the image. Current estimates of this cell size from CIC specimens indicate that these epithelial (non-goblet) cells normally have a surface area of approximately 250 μm^2 (Doughty and Naase 2008, Doughty 2010b, 2011b). Such an average cell area value translates to a cell density of approximately 4000 cells/mm^2. The surface cells of the bulbar conjunctiva can however change quite considerably under the influence of any form of stress, e.g. contact lens wear (Doughty and Naase 2008, Doughty 2011b). The changes include cell enlargement (a condition referred to as squamous metaplasia) to values in excess of 2000 μm^2, to give cell densities that would be just 500 cells/mm^2.

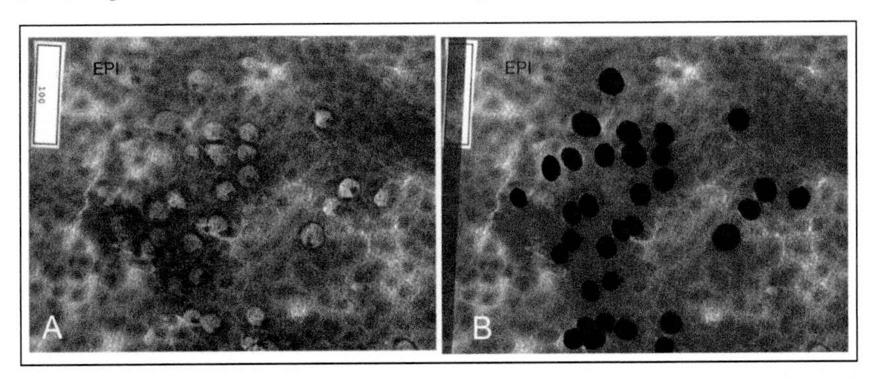

Figure 8 Example of a conjunctival impression cytology (CIC) specimen taken from the nasal bulbar conjunctiva of a young adult (from Doughty 2011a, copyright Elsevier Scientific) to show small conjunctival epithelial cells (epi) (A) and the balloon-like features of the goblet cells which are marked in (B). The goblet cell density in this image is 230 cells/mm^2.

Using impression cytology specimens, Nelson and colleagues (Nelson and Wright 1984, 1986) reported that the goblet cell density (GCD) across the bulbar surface was > 500 cells/mm^2, and a specific value of 443 ± 266 cells/mm^2 was reported for inter-palpebral bulbar conjunctiva. Since that time, much lower values (e.g. 65 cells/mm^2) (Satici et al. 2003) or much higher values (e.g. 1318 cells/mm^2) (Matsumoto et al. 2008) have been reported for GCD estimates at this location. The average estimate of about 500 cells/mm^2 is perhaps reasonable, but nominally healthy individuals can apparently show very substantial differences in GCD. Part of the reason for this is likely related to the development of conjunctival cell abnormalities. When slight (grade 1) squamous metaplasia is present, it has been proposed that the GCD (presumably across the inter-palpebral conjunctiva) would be between 350 and 500 cells/mm^2, would be between 100 and 350 cells/mm^2 for grade 2 cell changes, and less than 100 goblet cells/mm^2 when extreme squamous metaplasia (grade 3) was present (Nelson and Wright 1984).

In the living eye, the bulbar conjunctival surface can appear to be remarkably smooth yet, just as easily show with numerous fine wrinkles or microfolds across it (Doughty and Bergmanson 2003). Under certain conditions (e.g. different types of ocular surface desiccation?), these microfolds can show preferential organization or patterns, now sufficiently visible even without fluorescein staining. One form of these has been referred to as lid-parallel conjunctival folds (LIPCOF) (Hoh et al. 1995), but all likely represent as some form of excess proliferation of the conjunctival epithelium and/or the underlying parenchyma, a condition more generally known as conjunctivochalasis (Hughes 1942). This propensity for the conjunctival tissue to form infoldings can lead to another feature of the conjunctival epithelium, the so-called 'crypts of Henle' or the 'mucus glands' of Henle (Ruskell 1991, Rohen and Lutjen-Drecoll 1992).

The structural organization of the bulbar (as well as the palpebral) conjunctiva is indicative of a high para-cellular permeability for the conjunctival cell layers, at least as based on measurements of drug permeation (Huang et al. 1989, Hamalainen et al. 1997). As with any cuboidal epithelium, bulbar conjunctiva preparations (supported on the underlying connective tissue) have been widely reported to exhibit the typical properties of ion-transporting cell layers (Turner et al. 2000). By covering the apical (normally exposed) surface with various physiologically relevant solutions (with chemical composition having similarities to that of the tear film) with and without the presence of various drugs known to inhibit different ion transport processes, one can conclude that the mammalian conjunctiva has the ability to absorb Na^+ (Maurice 1973), a transport system that is coupled to the uptake of glucose and amino acids and a range of cationic drugs (Ueda et al. 2000). There is also a transport system for Cl^- that, indirectly, supports fluid transfer from the underlying parenchyma through to the conjunctival surface (Li et al. 2001, Candia 2004). While any specific and clinically-observable measure of these ion and fluid transport systems has yet to be made, it can be expected that any general and chronic perturbation of the ionic balances of these cells can result in clinically-observable changes in the cells. So, for example, if the tears become chronically hypertonic (even by just 30 mOsm/L) (Sullivan et al. 2010) in severe dry eye, then there is a greater chance that the bulbar conjunctival cells will stain noticeably using rose bengal (Doughty et al. 2007a). This reflects a change to a squamous (and stratified) phenotype for the conjunctival cells. Part of the reason for the cell transformation is considered to be the result of reduction in the quantity, quality or integrity of a mucus coating on the cell surfaces (Pflugfelder et al. 1997, Rolando et al. 1990, Paschides et al. 1991, Feenstra and Tseng 1992).

Electron microscopy studies indicate that bulbar (and limbal) conjunctiva contains complex sensory nerve endings called corpuscles (Ruskell 1991). These are served mostly by single myelinated nerve fibers, derived from nerve fiber bundles deeper in the conjunctiva or even within the episclera. Such nerve endings, at least in part, determine the sensitivity of the conjunctiva to various physiological stimuli (see below). Linked to these and other sensory endings/terminals, widely distributed within both the epithelial layers and also within the underlying connective tissue, are therefore numerous conjunctival sensory nerve branches. The epithelial and sub-epithelial distributions have a sensory origin from both the ophthalmic and maxillary divisions of the trigeminal nerve.

The nerve ending corpuscles in the conjunctiva are thought to have a sensory role (Acosta et al. 2001). The sensitivity to physiological stimuli is fairly predictable and can range from a very subtle (psychophysiological) sensation of dryness (Varikooty and Simpson 2009) to burning and pain when there is excessive desiccation of the conjunctiva (as in dry eye) (Liu et al. 2010). The sensory perception is a complex mixture of responses from different types of nerve endings, the activities of which can be probed with a range of stimulus modalities.

Recent research indicates that the relative sensitivity and the sensations experienced for the human conjunctiva to mechanical/touch stimulation can be expected to be somewhere between that for the eyelid margin and the cornea (McGowan et al. 1994, Acosta et al. 2001, Doughty et al. 2009). The conjunctival sensitivity, for a mechanical (tactile) stimulus, is slightly less across the exposed portion of the bulbar conjunctiva as compared to the cornea (Goleblowski et al. 2008, Doughty et al. 2009), and substantially less close to the plica and under the upper eyelid (Draeger 1984).

Conjunctival tactile sensitivity, in common with corneal tactile sensitivity, has been reported to decline with age (Norn 1973), but the same has not been found for measures using a pulse of air (to detect a 'pneumatic' mechanical sensitivity) (Situ et al. 2008). Sensations of cooling and warmth can also be detected using pulses of air, sensitivities that might be also linked to changes in blood flow in the parenchyma (Gallar et al. 2003).

The Subepithelial Structures under the Bulbar Conjunctiva

Similar to the palpebral conjunctiva, there are layers of connective tissue containing a rich blood supply under the bulbar conjunctiva. The connective tissue appears to be much more organized than that under the palpebral conjunctiva with a more anterior layer of small bundles of collagen fibrils (the episclera) and the anterior aspect of a capsular sheath for the eyeball (i.e. Tenon's capsule) (Koornneef 1977).

The collagen fibrils of the episclera, at least in rabbits, are arranged in flattened bundles. There is a discrete spacing between the fibrils (Fig. 9A), and they appear to have a consistent size (diameter) of c. 65 nm (Fig. 9B) (Doughty and Bergmanson 2004b). These fibrils are thus very much smaller than those of scleral tissue underneath (Doughty and Bergmanson 2004b), but somewhat larger than those in the cornea (see later). There are also a few strands (fibers) of elastin interspersed within the collagen fibril matrix.

Mast cells are normally found in the parenchyma, just below the interface with the epithelium (Iwamoto and Smelser 1965), identifiable because their cytoplasm contains numerous metachromatic round-to-oval shaped granules, which can stain reddish purplish with the appropriate dyes, and appear electron-dense in transmission electron microscopy. These granules can have slightly different ultrastructural features (Wanko et al. 1964, Iwamoto and Smelser 1965, Hogan et al. 1971, Nichols 1996), but distinctly different from the ultrastructure of electron-dense granules that can be found in goblet cells, melancytes or various types of white blood cells. The mast cell granules contain both vasodilation-inducing substances (e.g. histamine) and other mediators of the inflammatory responses (prostaglandins, interleukins and enzymes such as chymase and tryptase, etc.) (Welle 1997). The mast cells can migrate towards the conjunctival surface and release their granules into the surrounding tissue and into the tear film.

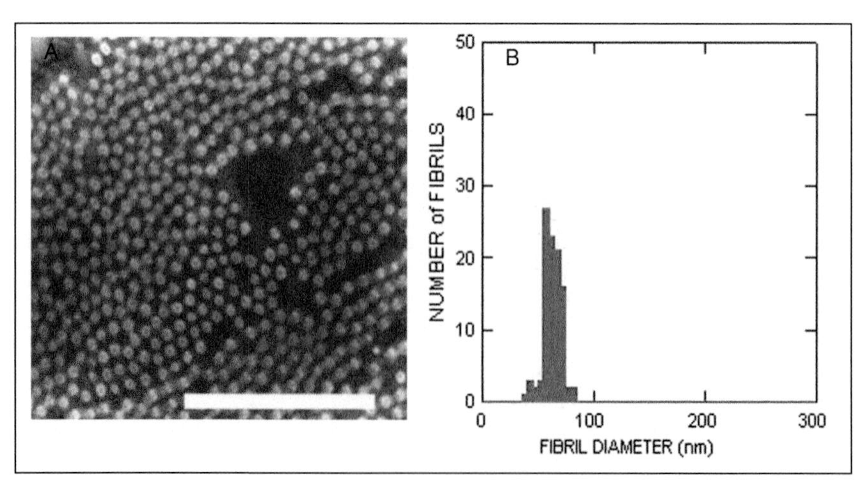

Figure 9 Transmission electron microscopy of the sub-conjunctival episcleral tissue to show part of a collagen bundle with uniform size of fibrils and spacing between the fibrils. The histogram (B) shows the distribution of fibril diameter values. Modified from Doughty and Bergmanson 2004b, copyright Australian Association of Optometrists/Wiley.

Vasculature and Lymphatic Vessels under Bulbar Conjunctiva

The subepithelial parenchyma (with anterior and deep fibrous layers) underlying the bulbar conjunctiva is richly supplied with blood vessels. The vessels closer to the surface can be visualized in the living eye by fluorescein angiography (Meyer and Watson 1987), allowing for the arterial and venous aspects of the blood flow to be measured. The arterial supply is mainly from the anterior ciliary arteries (Iwamoto and Smelser 1965, Hogan et al. 1971). The conjunctival veins, which appear more numerous than arteries, ultimately drain into superior and inferior ophthalmic veins. The substantial networks of capillaries found in these layers have slightly different ultrastructural features with some of them being non-fenestrated and others being fenestrated (Hogan et al. 1971, Ruskell 1991). As viewed in transmission electron microscopy, the fenestrations are locations in the endothelial lining of the blood vessels that are very substantially attenuated and only the two membranes are evident (Hogan et al. 1971). The blood vessels are supplied with nerves and display distinct pharmacological characteristics, principally alpha1-adrenoceptors for vasoconstriction and muscarinic cholinergic receptors for vasodilation.

The oxygen levels in the conjunctival (and episcleral) vessels can be measured with special surface electrodes and can be shown to decrease with altitude (Mader et al. 1987), and can change dramatically with systemic oxygen levels (Nisam et al. 1986). The conjunctival vasculature allows for efficient diffusion of systemically-administered drugs (medications) into the tear film (Lifshitz et al. 1999), a process that can (at least indirectly) result in changes to the conjunctival surface cells (Doughty et al. 2007b). As noted earlier for palpebral conjunctiva, the vasculature under the epithelial cell layers will contain a balancing lymphatic vessel system. This has been better characterized for the parenchyma under the bulbar conjunctiva.

Lymphatic vessels can be noted in tissue sections viewed by light microscopy, often partly collapsed, because they lack obvious contents (i.e. red blood cells) (Ruskell 1991, Nichols 1996). They form a complex network of lymph-filled vessels or 'channels' that are located both immediately under the epithelium and sometimes deeper (Sugar et al. 1957, Iwamoto and Smelser 1965, Collin 1966, Hogan et al. 1971, Doughty and Bergmanson 2003). The lymphatic vessels often follow the route of the blood vessels, principally the venous capillaries, and tend to be more organized under the perilimbal and limbal zones around the cornea.

As viewed by higher magnification transmission electron microscopy, there is no obvious and continuous basement membrane to support the (endothelial) cells lining the lymphatic vessels and so these tend to have a more irregular outline as compared to most blood vessels (Doughty and Bergmanson 2003) (Fig. 10); the endothelial cell-cell contacts are more in the

Figure 10 Transmission electron micrograph to show part of a large conjunctival lymphatic vessel with irregular endothelial lining (from Doughty and Bergmanson 2003), copyright American Optometric Association.

form of an overlap, without obvious cell-cell junctional complexes (Iwamoto and Smelser 1965, Nichols 1996, Cursiefen et al. 2002). Examination of the outer edger edge of lymphatic vessels reveals no obvious supportive cells (pericytes), but there are characteristic discrete bundles of collagen fibrils that appear to serve the role of anchoring the lymphatic vessels to surrounding connective tissue.

The lymphatic system under the bulbar conjunctiva contains valve-like structures (Grüntzig et al. 1990), which are presumed to control the net flow. This is generally in a temporal direction towards the lateral extremes of the eyelids (to the parotid node) with some via a medial route (leading to the submandibular lymph nodes). Any abrupt change (reduction) in the flow (perhaps as a result of a special regulation of these valves) can result in dramatic changes in the fluid content of the sub-epithelial parenchyma (e.g. manifest as chemosis). Detailed information of the location and mechanism of control of these valves has yet to be described.

THE CORNEAL EXTERNAL APPEARANCE AND SURFACE FEATURES

External Dimensions, Curvature and Surface Appearance

As based on measures on post-mortem human eyes, the corneal area is approximately 1/10th of the surface area of the globe (Olsen et al. 1998) and only about 1/15th of the surface area of the conjunctiva (Watsky et al. 1988). When viewed from the outside, the cornea has a certain diameter, most commonly now assessed by measures of the horizontal corneal diameter (HCD) (Jonuscheit and Doughty 2009). This is also known as the visible iris diameter or white-to-white diameter since it is essentially the transition from the visible edge of the iris (through the transparent cornea) to the white sclera surrounding the cornea. The cornea undergoes rapid growth—as assessed by an increase in its diameter—during the very early post-natal years, to achieve adult values of close to 11 mm usually by the age of 2 or 3 years (Müller and Doughty 2002) (Fig. 11A). More recent cross-sectional age-related studies indicate no significant further change in HCD in relation to adult age (Fig. 11B). Anatomically, while the human cornea might be considered as round when viewed from the outside, it is slightly oval so that the vertical diameter measures are slightly smaller.

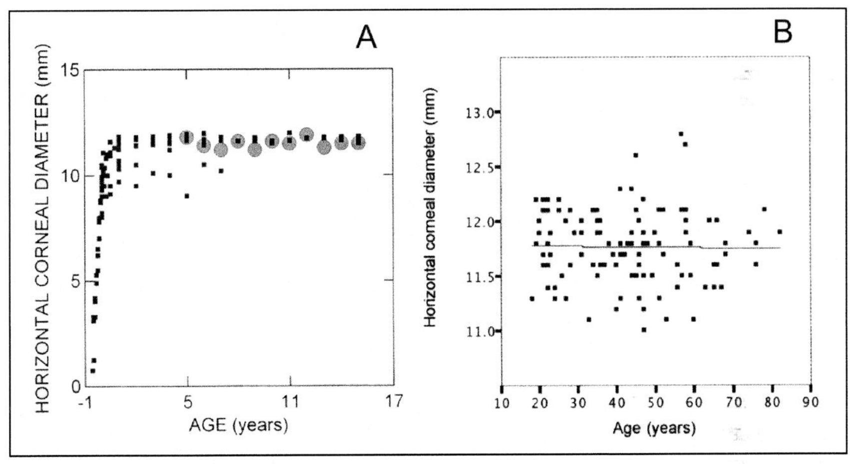

Figure 11 Age-related changes in the human horizontal corneal diameter (visible iris diameter) in (A) the early post-natal years and into childhood (modified form Müller and Doughty 2002, copyright College of Optometrists/Wiley) and (B) in the adult years (Jonuscheit and Doughty, unpublished).

The anterior surface curvature of the cornea is relatively steep at birth, with average values at 1 year of age reported as 7.58 mm (von Reuss 1881); various contemporary studies have yielded similar values. The cornea then starts to flatten rapidly so that, for adults, values at the central location are generally considered to be closer to an average of 7.8 mm. For adults, the posterior curvature is steeper with values of c. 6.5 mm (Wichterle et al. 1991). Measures of anterior and posterior curvature are readily obtained by modern instrumentation, e.g. averages of 7.75 mm ± 0.28 and 6.34 ± 0.28 mm using a special Scheimpflug photography (Pentacam) system (Ho et al. 2008). Based on central keratometry measures, the curved anterior corneal surface is capable of providing substantial refractive power in the order of 43 D. The characteristics of the curvature of the anterior surface of the central corneal region (optical zone) has long been of interest because of its importance as a refractive surface (von Ruess 1881), but it is the anterior surface curvature in the more peripheral parts of the cornea that are more relevant to the tear film and ocular surface (Wichterle et al. 1991). The peripheral anterior corneal curvature, in relation to the slope (curvature) of the eyelid margin tissue, presumably plays an important role in determining the meniscus profile of the tears along the eyelid margin (Fatt 1992, Cui et al. 2011).

In the healthy eye, there is a notable reflection of light off the corneal surface (Goto et al. 2011), a reflection (the corneal 'lustre') that presumably owes its existence to a stable tear film (covering the corneal surface) that is underpinned by a uniform (homogeneous) mucin layer that is in intimate contact with surface cells of the corneal epithelium. The mucin-corneal epithelial cell interface is formed by specialized glycosylated protein molecules, known collectively as the 'glycocalyx' (Dilly and Mackie 1981). The mucin layer is normally invisible in the healthy eye, whether this be being viewed by slit lamp or specular microscopy. However, in some types of interference microscopy this highly reflective layer at the air/tear film/corneal surface interface can resolve into layers. This lead some investigators to propose that the pre-corneal tear film or, more specifically, the mucin component of the pre-corneal tear film has thickness values approaching that of the corneal epithelium (see later), i.e. of c. 40 μm (Prydal et al. 1992). Using special post-mortem preparation techniques for corneal tissue, a presumed mucin layer can be seen in scanning (Doughty 2003a) (Fig. 12A), or transmission electron microscopy (Nichols et al. 1985). As viewed at very

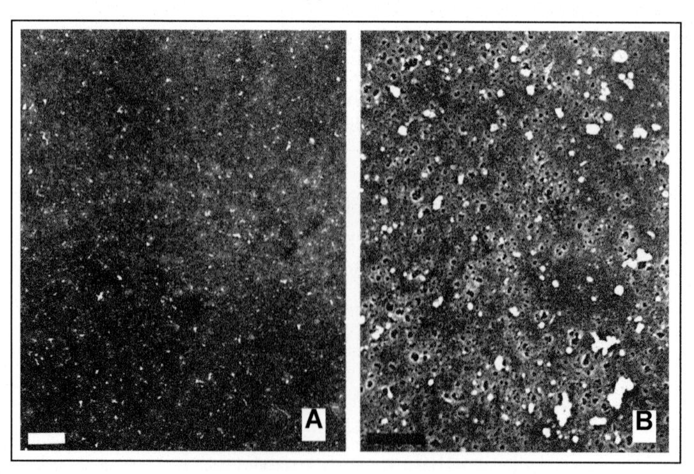

Figure 12 Scanning electron microscopy to show specially-fixed mucus layer across the corneal surface, rabbit, at low magnification (A) to show amorphous nature of the overall surface and (B) at very high magnification to show pores in the matrix. From Doughty 2003a, copyright Informa.

high magnification in scanning electron microscopy, a distinct texture can be observed in the precipitated and desiccated mucous layer, and there appear to be numerous small pores in the matrix (Fig. 12B).

The apical surfaces of the corneal epithelial cells that are in contact with the tears are specially formed into very high numbers of special micro-projections. For the normal corneal epithelium, these projections from the cell surfaces appear to principally be microscopic ridges (called microplicae) as opposed to individual features (called microvilli). As viewed in conventional tissue sections, these projections are hard to distinguish but when viewed with scanning electron microscopy, a complex pattern of these microridges is clearly evident (Doughty 1990a) (Fig. 13A).

At extremely high magnification (50,000X at the microscope stage), the ridges can be seen to be decorated with small globular protrusions (Fig. 13B), which may be the glycocalyx (Doughty 1999). The cell surfaces also have other features, namely the presence of numerous round or 'ring-shaped' structures, also known as 'epithelial craters'. While their function is, at this time, unknown, they are visible by scanning electron microscopy (Doughty 2006, 2008), as well on the corneal epithelial cell surfaces in the living eye as viewed with specular microscopy or scanning laser confocal microscopy (Ganem 1999).

From a morphological perspective, it is generally observed that the most superficial cells of the corneal epithelium have difference sizes, shapes and appearances. This mosaic of cells can be seen in the living eye by

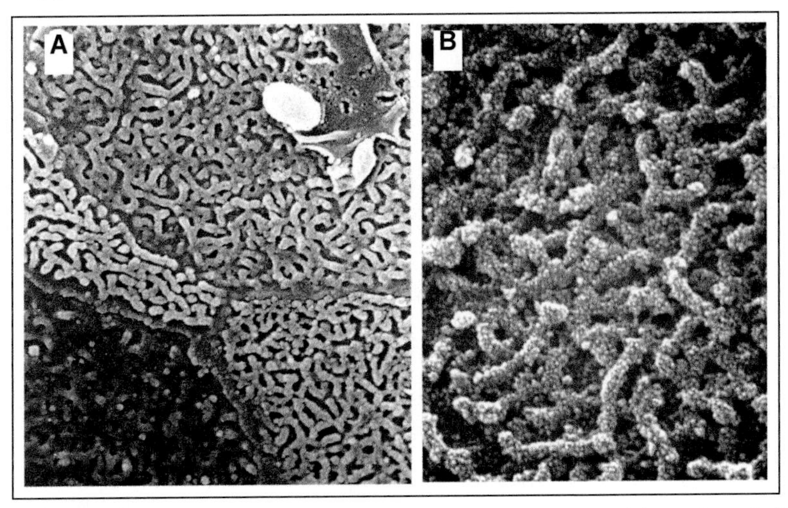

Figure 13 Very high magnification scanning electron microscope views of the surface of rabbit corneal epithelial cells to show (A) high density of microplicae (microridges) on the lighter appearing cells (from Doughty 1990a, copyright American Academy of Optometry / Williams & Wilkins), and (B) the presumed glycocalyx on the surfaces of the microplicae (from Doughty 1999, copyright and British Contact Lens Association/Elsevier Scientific).

specular microscopy (Mathers and Lemp 1992) and some types of confocal microscopy (Erdelyi et al. 2007, Zhang et al. 2011), and in post-mortem eyes by scanning electron microscopy (Doughty 1990a, 1997, 2008, Doughty and Fong 1992). As a result of differences in density of the microplicae (Fig. 13A), the cells viewed at low magnification can have very distinct light, medium and dark appearances in scanning electron microscopy. This distinguishes them from the limbal epithelial cells at the periphery of the cornea (Goes et al. 2008), as well as most cells of the bulbar and palpebral conjunctiva (Fig. 5). There are likely to be topographical differences in the size, shape and appearance of these surface cells of the cornea with studies on rabbits indicating an average cell area of 468 ± 54 μm^2 in the central optical region compared to the mid-periphery (at 583 ± 85 μm^2) and peripheral zone of the cornea (at 669 ± 87 μm^2) (Doughty and Fong 1992); these average cell area values would translate into cell density values of 2136, 1715 and 1495 cells/mm^2 respectively. Overall, it appears that the size of the cells can be predicted according to the cell shape, with there being an obvious relationship between the number of cell apices (sides) and the surface area (Doughty 1990b, 2008).

Measures of the sizes of the most superficial cells on living human corneas have proved more difficult to achieve and, as a result, there are differences in opinion as to how large or small these surface cells are. In specular (reflection) microscopy, average values of between 463 and 596 μm^2 for cell area have been reported (Lemp et al. 1989, Mathers and Lemp 1992, Tsubota et al. 1992), area values which would translate to cell densities of between 2160 and 1678 cells/mm^2. However, for confocal microscopy, if the first cells images are those just behind the corneal 'lustre' reflex, then these are likely to be surface cells with relatively large areas compared to deeper lying cells. Average values of between c. 814 and 1006 μm^2 have been reported, values that translate to cell density values between 1228 and 994 cells/mm^2 (Erdelyi et al. 2007, Zhang et al. 2011).

THE CORNEA—GROSS ANATOMY AND STRUCTURE

The Corneal Epithelium

The cells that are visible on the surface of the cornea represent only the most superficial squamous cells. In the healthy eye, the corneal epithelium is formed by four to six, or even seven, layers of tightly packed cells in humans and primates (Calmettes et al. 1956, Ehlers 1970, Ehlers et al. 2010). Histological studies report the total thickness of the human corneal epithelium to be between 30 and 50 μm, i.e. 0.03 to 0.05 mm (Calmettes et al. 1956, Ehlers 1970, Hogan et al. 1971). Similar, but perhaps slightly larger, values have been reported from *in vivo* confocal microscopy imaging of the human cornea (Li et al. 1997, Tao et al. 2001, Chan et al. 2011).

The corneal epithelium, unlike most of the conjunctival epithelium, is distinctly stratified in that it is organized into three layers based on the obvious differences in the shapes of the cells as viewed from the side in tissue sections, i.e. superficial squamous cells, intermediate wing cells and the deepest layer of basal cells (Hogan et al. 1971, Walsh et al. 2008) (Fig. 14A). The basal cells are notable for their large and usually vertically oriented nuclei. There is a distinct resemblance to the marginal zone epithelium (Fig. 3A), at least as studied for rabbit tissue. Cells removed from the corneal surface by impression cytology are clearly very much enlarged (Fig. 14B, note 100 µm scale marker) and squamous in nature.

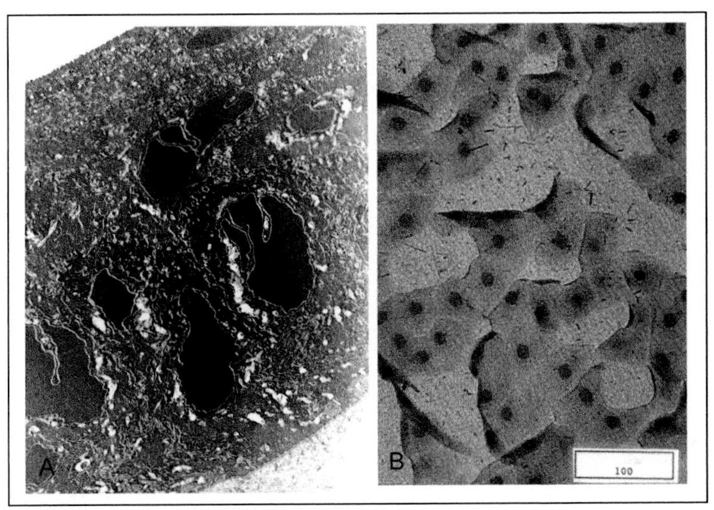

Figure 14 Stratification of corneal epithelium with the presence of apical squamous cells. (A) Lower magnification transmission electron micrograph, rabbit, to show distinct stratification of cells in squamous, intermediate and basal layers (from Walsh et al. 2008) Copyright IOP Publishing, and (B) Impression cytology sample (Biopore membrane) from the surface of the peripheral cornea of a healthy human subject stained with Giema. Cells are clearly squamous in nature (Doughty, unpublished).

The cells that make up these layers are not generally considered to represent three different classes of cells, but rather the same cell lineage captured at different stages in a cycle between genesis and their eventual loss from the corneal surface (see later). In the normal healthy cornea, all cells in these layers have numerous contacts and connections between them. Tight junctions (zonula occludenes) are only found in the superficial squamous cell layer, with adherens (or gap) junctions, which are between cells in all layers for inter-cellular attachment (Sugrue and Zieske 1997). There are also numerous desmosomes formed between neighboring cells throughout the intermediate and basal cell layer, effecting substantial cell-to-cell adhesion and thus a minimal para-cellular space. Further enhancement of this adhesion is achieved through slight interdigitations between neighboring cells.

The basal cells, in contact with the thin (underlying) basement membrane are usually firmly attached via a large number of hemidesmosomes to the underlying basement membrane. Following division of deeper-lying cells, these contacts and connections will need to undergo a progressive alteration to allow repositioning and migration of cells from the basal layer to suprabasal layers and ultimately to the ocular surface (see later).

The relative proportion of each cell type (squamous, intermediate and basal) seems relatively consistent, but there are likely subtle differences that could be very relevant to *in vivo* imaging of the corneal epithelium. Part of the cell stratification relates to a progressive change in the shape of the cells as they migrate from the basal layer through to the corneal surface. The term squamous literally means flat, but is generally used to also imply that these special cells also routinely undergo a desquamation from the ocular surface. Only a certain proportion of the basal cells actually migrate through the other cell layers to eventually reach the surface and, in the process get larger and larger. This progressive enlargement, and thus a distinct layer-by-layer difference in cell 'size' has been observed by confocal microscopy, e.g. 1228 ± 248 cells/mm^2 for superficial cells, 6974 ± 662 cells/mm^2 for intermediate layers and 11307 ± 1876 cells/mm^2 for basal cells (Zhang et al. 2011); these density values would translate to average cell areas of 814, 143 and 88 μm^2 respectively.

The most superficial cells of the corneal epithelium can be seen to have different appearances (i.e. the light, medium and dark shades), which could reflect the period of time for which these cells have been exposed at the surface after having migrated from deeper levels of the corneal epithelium. The largest and darkest appearing cells with fewer surface microplicae have been considered to be those that have been at the surface for the longest period of time and would be ready to desquamate from the surface (Hazlett et al. 1980). However, it has also been noted that desquamating cells can have a very different appearance in that they can be 'bright' cells with abundant surface projections (Doughty 1996). The different character of the desquamating cells could indicate several phenotypes of surface epithelial cells, with a subtly different lineage and differentiation (Doughty 1996). It is also recognized that some of the most superficial epithelial cells do not simply desquamate but undergo a special type of degeneration or programmed cell death, called apoptosis (Ren and Wilson 1996, Ladage et al. 2002). The differences in appearance of the superficial cells that are undergoing desquamation may therefore be because some are apoptotic and others are not.

The cornea has a rich supply of sensory nerve fibers, with their main origin being the ophthalmic division of the trigeminal nerve (CNV). The corneal sensory nerve endings form 'intraepithelial terminals' connected to slender nerve fibres extending from the basal to more anterior layers (Müller et al. 1996). Recent studies on post-mortem human corneas indicate that central nerve fibers form a network that extends uniformly in all directions across the epithelium with estimated densities of some 600/mm^2 in more anterior layers (Marfurt et al. 2010).

As a result of the differences in nerve terminals and the organization of the nerve fibers, the touch sensitivity of the central corneal zone is the greatest, with slightly less sensitivity being found for the peripheral cornea. This mechanical (tactile) sensitivity of the cornea (or conjunctiva or eyelid margin) is most often determined with a Cochet-Bonnet aesthesiometer (Cochet and Bonnet 1960), where the maximum length of a nylon filament that can be detected has been used as a measure of sensitivity. For central corneal zone, the maximum sensitivity is typically observed, as being 60 mm, while that of the peripheral cornea is likely closer to 30 mm. This tactile sensitivity appears to develop to its maximum very early in life (Snir et al. 2002, Dogru et al. 2004), but then declines with age (Millodot 1977, Roszkowska et al. 2004, Acosta et al. 2006). Presenting a pulse of air to the cornea also exerts a mechanical (Belmonte et al. 1999, Acosta et al. 2006) as well as a slight cooling effect (Murphy et al. 2004), a sensation that (unlike the tactile sensation elicited by a nylon filament) can be shown to be dependent on the duration of the stimulus (Jalavisto et al. 1951). Such time-related sensations to an air pulse have been reported to be notably different in elderly individuals (Jalavisto et al. 1951). However, when using a constant duration air pulse, this pneumatic 'mechanical' sensation appears to show a less obvious decline with age (Murphy et al. 2004, Situ et al. 2008). In addition to such a thermo-mechanical stimulus, the cornea can also detect a chemical-mechanical stimulus such as that elicited by using CO_2 instead of air in the test pulse, with both sensory systems likely merging to produce sensations of stinging and burning. The corneal (and conjunctival) sensory system is also able to detect ambient changes in air temperature (Kolstad 1970), changes in tear film osmolality (Liu et al. 2009) as well as air-borne irritants (Doughty et al. 2002, Ernstgard et al. 2006, Cometto-Muñiz et al. 2007).

The Anterior Limiting Lamina (Bowman's Membrane)

Below the basement membrane of the corneal epithelium is an acellular layer usually known as Bowman's membrane (Bowman 1847), but better referred to as the Anterior Limiting Lamina (ALL) since this describes its essential features and highlights the fact that it has no resemblance to other true (cell) membranes in the cornea and conjunctiva. The ALL is substantial in human and non-human primate corneas as well as in some birds, but notably inconspicuous in many mammals such as rabbits. In transverse section, the ALL has a certain thickness at the central region of the cornea and then progressively thins close to the corneal periphery as the limbus is approached (Iwamoto and Smelser 1965, Hogan et al. 1971, Ruskell 1991). In human corneas, the ALL has been measured to be 8 to 14 μm in thickness using histology and transmission electron microscopy (Hogan et al. 1971), and average values of between 16.6 and 17.7 μm have been reported from *in vivo* confocal microscopy (Li et al. 1997, Tao et al. 2011). It is possible to measure this thickness in the living eye because the ALL scatters light rather differently to the epithelium lying over it and corneal stroma underlying it, so giving a unique band across the cornea.

As viewed at the electron microscope level, the ALL can be seen to be formed by a dense network of very small diameter, randomly-oriented collagen fibrils (Kayes and Holmberg 1960, Mathew et al. 2008). Within this network of fibrils are also some more organized arrays that appear to provide what appear to be anchoring sites for epithelial basement membrane as well as at least two types of terminal arrays that provide an interface with the corneal stroma proper (Mathew et al. 2008).

The normal ALL is acellular, although it is punctuated by numerous fine neural branches that traverse it linking the sub-epithelial neural network to the nerve and receptors within the corneal epithelium (Marfurt et al. 2010). Recent studies indicate some 200 penetrations through the ALL (Bowman's) over the central 10 mm zone of the cornea (Marfurt et al. 2010). These 'holes' through the ALL can also be visualized by scanning electron microscopy (Kaji et al. 1998). The nerve fibres merge to form bundles in the anterior stroma (the sub-epithelial plexus) that exit across the corneo-scleral boundary to make a network of some 70 fibre bundles distributed uniformly around the limbus (Marfurt et al. 2010).

The Corneal Stroma

The corneal stroma, or *substantia propria*, comprises around ninety percent of the corneal thickness in humans and it is considered to be approximately 0.5 mm thick based on older traditional views on the cornea. However, if it were a relatively constant 10% of total thickness, then even for normal human corneas the corneal stroma could be between 0.42 and 0.54 mm as based on contemporary central corneal thickness measures (Doughty and Zaman 2000) (see later). The corneal stroma is a 3-D matrix of fibrils (principally type I collagen), an interstitial proteoglycan ground substance (that both coats the collagen fibrils and fills the space between them) (Doughty 2001) and the overall matrix is populated by numerous cells (Doughty et al. 2001b, Mathew et al. 2008).

The type I collagen forms the basis of small diameter fibrils (see later) which in turn, are arranged in a highly organized pattern in flattened bundles called lamellae; flattened cells, called keratocytes, are usually found between the lamellae (see later). In more recent studies on human corneas, the more anterior stromal lamellae have been reported to have average thickness values of 1.29 ± 0.23 µm, while the posterior lamellae were thicker with an average of 2.92 ± 1.54 µm; the overall average lamella thickness value was 1.66 ± 0.11 µm (Takahashi et al. 1990). In another study, it was noted simply that the more anterior lamellae varied in thickness from about 0.2 to 1.2 µm while a range from 0.2 to 2.5 µm was found in the posterior stroma (Komai and Ushiki 1991). Similar ranges of values have been reported for a variety of animal corneas as well (Doughty et al. 2001b).

When viewed by transmission electron microscopy at very high magnification, the collagen fibrils within the stromal lamellae of the normal mammalian cornea are all of very similar diameter and are relatively uniformly spaced. The clarity of the collagen fibrils and the finer details of the surface of the individual fibrils probably depend on the quality of the post-mortem tissue. While it is likely that images of the type shown in Fig. 15A can be fairly easily obtained from very recent post-mortem corneal samples (Doughty and Bergmanson 2004b), images of the type shown in Fig. 15B will more likely be obtained from an eyebank sample that had been stored for a few days (Bergmanson and Doughty 2005).

As shown in Fig. 15, the stromal fibrils are specially arranged so that there is a certain space occupied by the fibrils. For posterior corneal stroma

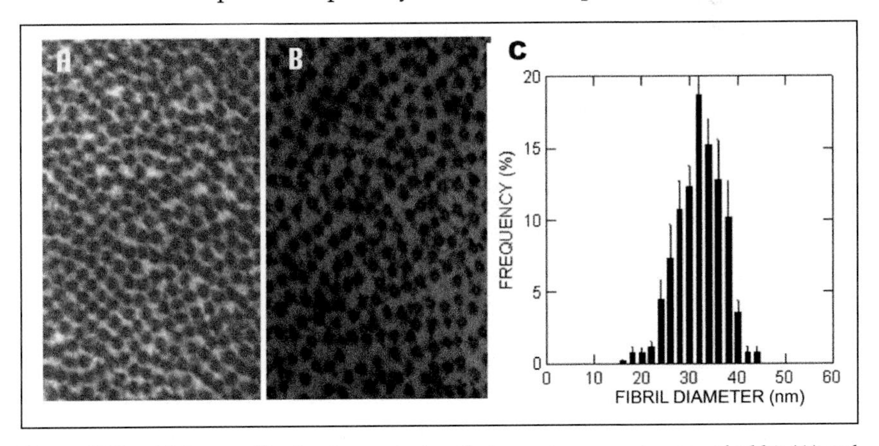

Figure 15 Very high magnification transmission electron microscopy images of rabbit (A) and human (B) stroma to show fibrils, and (C) the distribution of fibril diameters in the rabbit. (A) and (B) from Bergmanson and Doughty 2005, copyright Lippincott-Williams & Wilkins, (C) from Doughty and Bergmanson 2005, copyright Elsevier.

of the rabbit, the relative space occupied by the fibrils (the fibril volume fraction) appears to have a predictable value of close to 0.4 (Doughty 2011c). This value would be expected to decrease if there was any post-mortem oedema in the tissue (as illustrated in Fig. 15B). The remainder of the stroma between the fibrils is filled by an amorphous proteoglcyan 'ground substance' or extracellular matrix (ECM), with fibrils not being in contact with each other and separated by discrete inter-fibril distances, which are also predictable (Doughty 2011c, d).

Studies on post-mortem human eyes by transmission electron microscopy indicate that there are some 200 lamellae/cornea in the central zone (Bergmanson et al. 2005). These are stacked on top of each other with, in successive lamellae, the fibrils being oriented in different directions, so one set will be seen in cross section and the next set in transverse section or at an angle between these. Scanning electron microscopy studies clearly show that some lamellae can cross each other at various angles (Komai and Ushiki 1991, Radner et al. 1998), arrangements that can also be observed by interference light microscopy (Krauss 1937, Kokott 1938), through silver staining of the stroma (Polack 1961) and is implied from both polarised light transmission (Christens-Barry et al. 1996) and X-ray diffraction studies (Daxter and Fratzl 1997).

A range of average fibril diameters for corneal stroma from 17.8 to 40.6 nm has been reported (Doughty and Bergmanson 2005); these are 'small' collagen fibrils. The variability in values is also probably related to the post-mortem condition of the tissue and fixation methods used. For an individual specimen, however, a discrete range of fibril diameters can be expected (Doughty and Bergmanson 2004b, 2005) (see Fig. 15C). The average size in the set illustrated (Fig. 15C) was close to 32 nm, i.e. half the size of the episcleral fibrils (Doughty and Bergmanson 2004b). The size and spacing of the fibrils is considered important for the corneal stroma to be transparent (Maurice 1957). The size (diameter) of the individual collagen fibrils is such that they are considered to be too small to scatter significant light (Benedek 1971, Vaezey and Clark 1991). It has been proposed that as long as the fibers remain of fairly uniform small size and are not spaced further apart than λl/2n (λl is the wavelength and n is the refractive index of the media), then this would generally satisfy the requirements for models of corneal stroma transparency (Farrell and Hart 1969).

The corneal stroma, unlike the anterior and posterior limiting laminae, is normally populated with cells, the keratocytes (Doughty et al. 2001b). These cells, which synthesize and are continuously attached to the fibrils of the corneal stroma, are visible in conventional histology or in lower magnification transmission electron microscopy, but their extended flattened (spread-out) characteristics are best viewed in coronal sections through the corneal stroma (Fig. 16A). The extent of the flattening can be assessed by

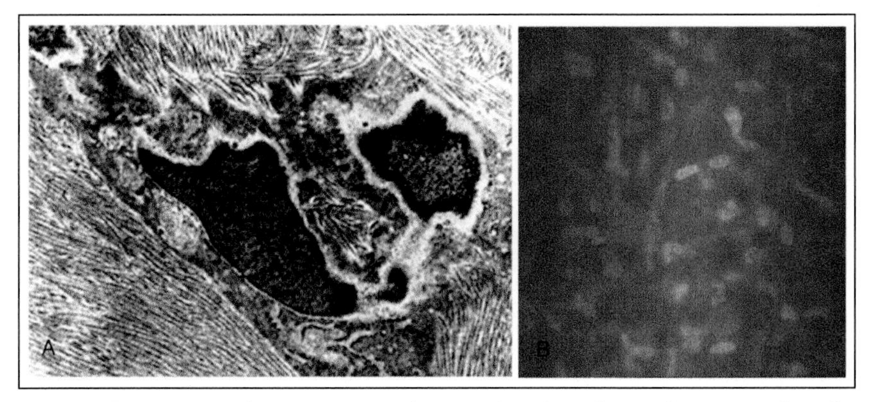

Figure 16 Transmission electron micrograph, coronal section, of corneal stroma to show the cell body with prominent nuclei of a flattened keratocyte between the collagen lamellae, rabbit, with extensive attachments of the cell to the surrounding collagen fibrils (from Doughty et al. 2001b, Copyright Churchill-Livingstone, Edinburgh), (B) confocal microscopy image, coronal perspective, of human corneal stroma to show light-reflective keratocite nuclei (from Doughty 2003b, copyright British Contact Lens Association/Elsevier Scientific).

measures of the thickness in relation to the adjacent lamellae (Doughty et al. 2001b). These cells also densely stain with silver (Jester et al. 1994), as well as scatter light in a different way to the surrounding lamellae (Poole et al. 1993, Hahnel et al. 2000, Doughty 2003b). The silver-based former method uses what are termed whole mount preparations within which the flattened cells can be seen to extend in all directions with the cell processes appearing to form a closely linked network. The special light scattering properties of the keratocytes can be seen with *in vivo* confocal microscopy, including some visualization of the cell processes in horizontal or vertical directions (Poole et al. 1993, Doughty 2003b) (Fig. 16B). The light reflected off the nuclei of the keratocytes has been used to count the number of these cells in living human corneas, and confirm that the highest density of keratocytes is found in the more anterior stroma and from there the density declines in a posterior direction by as much as thirty percent (Hollingsworth et al. 2001, Patel et al. 2001, Berlau et al. 2002).

The Posterior Limiting Lamina (Descemet's Membrane)

The corneal stroma merges into a slightly different collagenous layer, the Posterior Limiting Lamina (PLL, also widely referred to as Descemet's membrane) immediately adjacent to the posterior (inner) aspect of the stroma. It is however composed of different collagens to the corneal stroma (mainly type IV) forming a flat and largely amorphous layer. In humans, the PLL is only around 3 μm thick at birth and then progressively thickens throughout life (Johnson et al. 1982), to reach values of around 20 μm in the adult cornea (Leuenberger 1978). The PLL is synthesized and secreted by the corneal endothelium (see below) so that the most anterior portion is the oldest and the least uniform in texture. This is the foetal part of the PLL and it is known as the banded layer, whereas the post- natal component of the PLL is uniform in texture and is also the portion that thickens with age.

The corneal endothelium

A single layer of cells forms the corneal endothelium across the most posterior (innermost) aspect the cornea (Leuenberger 1978, Doughty 1989a, Edelhauser 2000). The distinct organization of this cell layer (cell mosaic) is now a well-known clinical feature of the cornea, with images being readily obtainable with a specular microscope (Doughty 1998, Edelhauser 2000). The corneal endothelium can also be viewed by confocal microscopy (Jonuscheit et al. 2011). With either method, the endothelial cells can form a rather uniform or regular mosaic (i.e. all cells seem to have similar sizes) or be non-uniform in which the cells exhibit a range of sizes; the latter appearance is widely referred to as polymegethism. It can be shown that a relatively large proportion of the endothelial cells in a regular mosaic are in fact 6-sided, but that there are notable numbers of other polygons, especially 5- and 7-sided cells (Doughty 1998). While it might be considered that the normal adult cornea has a slight deficiency of 6-sided cells (i.e. all cells should be 'hexagonal'), detailed studies of the corneal endothelia of children also clearly indicate that there are quite a large number of non-6-sided cells (Müller et al. 2000). In younger adults (c. 20 years of age), the average area of the endothelial cells, regardless of shape (polygonality), is around 330 μm^2; this translates to a calculated endothelial cell density (ECD) of close to 3000 cells/mm^2.

After birth, it is generally considered that these cells do not normally replicate in the human or primate cornea (Edelhauser 2000). Since human corneal endothelial cells have apparently lost their ability to replicate *in situ*, any cells lost by stress, trauma, or damage will not be replaced (Edelhauser 2000). Therefore, there is a natural age-related decline in the number of corneal endothelial cells. At birth, the ECD is likely to be around 5500 cells/mm^2 (Doughty et al. 2000, Müller and Doughty 2002). In the early post-natal years, there is a dramatic decline in cell density by around 1000 to 2000 cells to reach values of close to 4000 cells/mm^2 by childhood (Doughty et al. 2000, Müller and Doughty 2002) (Fig. 17). This initial apparent reduction is associated with the dramatic growth (enlargement) of the cornea that occurs during this period (Müller and Doughty 2002) (Fig. 11A). Thereafter, although somewhat debatable, there is a predictable further decline in cell density as a result of cell loss so that ECD values that are rather less than 3000 cells/mm^2 and even closer to 2000 cells/mm^2 may well be observed

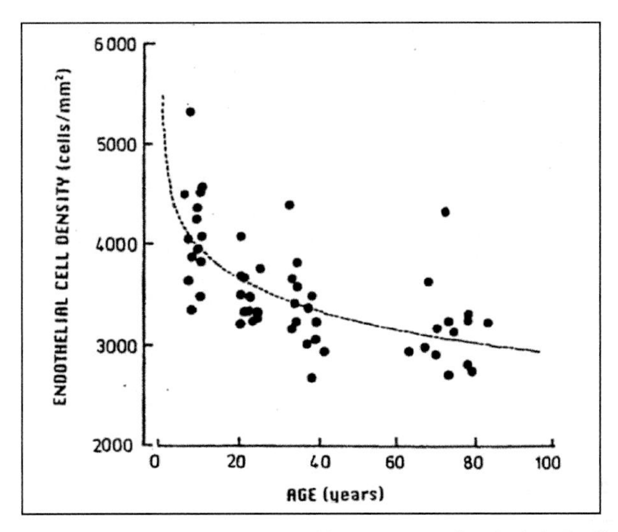

Figure 17 Age-related changes in the density of human corneal endothelial cells as observed with the specular microscope. From Doughty et al. 2000, copyright Raven Press, New York.

in the elderly (Edelhauser 2000, Doughty et al. 2000, Hollingsworth et al. 2001, Abib and Barreto 2001) (Fig. 17). As the cell density decreases, the area of some of the individual cells will increase and so, with age, the resultant endothelium can contain a mixture of some smaller and some larger cells. The range of cell areas is thus greater than in a younger endothelium, i.e. some degree of polymegethism has developed (Doughty 1989a, Doughty et al. 2000). Associated with the decline in cell density it is likely that the percentage of 6-sided cells also declines with age, with there being increased proportions of non-6-sided cells. The range of cell shapes thus increases, and this type of change is generally referred to as pleomorphism.

Corneal Thickness, Transparency and Optical Properties

In addition to diameter, the human cornea is now routinely characterized by its thickness, especially the anterior-to-posterior thickness measured within the central optical zone of the cornea, i.e. the central corneal thickness or CCT. Different instruments and techniques will produce slightly different values for CCT (Doughty and Zaman 2000) (Fig. 18). In general, a historical review of published data indicates that slit-lamp based techniques tend to yield slightly lower CCT values compared to ultrasound-based techniques, with specular microscopy techniques yielding values similar to those with ultrasound. A recent analysis of the data from scanning light measures (the Orbscan™) indicates that the CCT measures were slightly greater than those from ultrasound pachymetry. An option, for this or similar discrepancies,

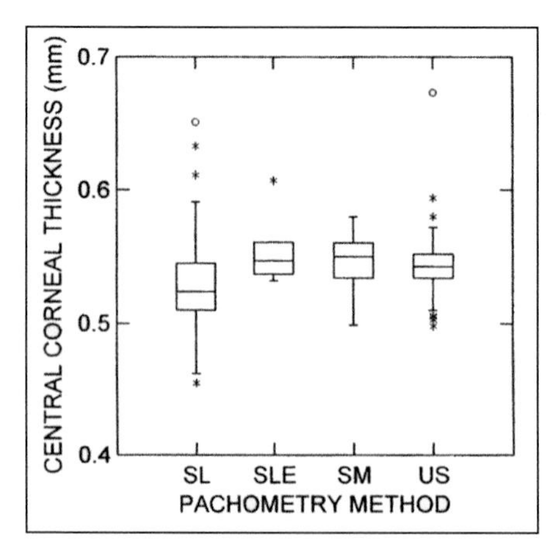

Figure 18 Box plot to show central corneal thickness values in adult humans as assessed by optical slit-lamp pachometry (pachymetry, SL and SLE), specular microscopy (SM) and ultrasound pachometry (US). From Doughty and Zaman 2000, copyright Elsevier Scientific.

would be to apply some to sort of correction factor to adjust the measures to those obtained by ultrasound, e.g. the acoustic correction factor applied to Orbscan measures (Doughty and Jonuscheit 2010).

In the early post-natal period, central corneal thickness values may well exceed 0.600 mm, but a progressive decrease in CCT values is expected over the first year of life (Fig. 19A). Adult CCT values appear to be reached during childhood (Müller and Doughty 2002). As measured for white Northern European subjects with ultrasound pachymetry or Orbscan™ (Jonuscheit and Doughty 2009), CCT does not appear to normally decline with age (Fig. 19B). In those of non-Caucasian (e.g. Asian) origin, a slight decline in CCT with age has been reported (Doughty and Zaman 2000). Based on an analysis of literature reports for adult human corneas over a 30 year period, a population mean value for CCT was calculated to be 0.536 mm (i.e. 536 μm). Perhaps of greater significance is that calculation of a 95% confidence interval (95% CI, based on ±1.96 x SD, reported from 230 separate clinical studies) was broad, being of between 0.473 and 0.595 mm (Doughty and Zaman 2000). A normal cornea can apparently have a wide range of thickness values.

For the human cornea, the central region is thinner than more peripheral regions. Numerous studies have been reported of these regional differences and, as with CCT, the range of peripheral thickness values appears to be highly variable, with a range of at least 109 to 152% of central thickness values (average 121%) (Doughty and Zaman 2000). The difference between peripheral and central locations can be calculated as a ratio, also referred to as the cornea thickness profile index (Jonuscheit and Doughty 2009). At

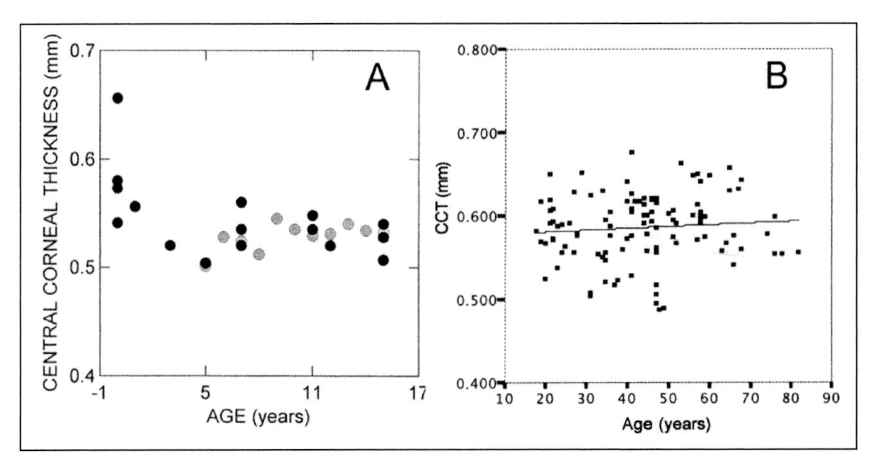

Figure 19 Changes in human central corneal thickness (A) in the post-natal period (modified from Doughty et al. 2002, copyright Williams and Wilkins), and (B) over adult years (Jonuscheit and Doughty 2009, copyright Association for Vision and Research in Ophthalmology). The data points in (A) represent averaged values from groups of subjects, while those in (B) are from individual subjects. The values in (B) are generally higher than in (A) since a scanning slit (Orbscan) method was used.

least for those of white Northern European origin, this ratio decreases very slightly with age, i.e. there appears to be a slight relative thinning of the peripheral cornea with age.

The normal human cornea appears to be transparent. For the normal cornea, this can be assessed from measurements of the light transmittance at different wavelengths. Such spectra, usually obtained on post-mortem samples, indicate at least 90% throughput in the visible range above 400 nm and between 70 and 80% above 320 nm (Beems and Van Best 1990, Ventura et al. 2000, Walsh et al. 2008) (Fig. 20). A combination of epithelial and endothelial compromise post-mortem can however lead to dramatic decreases in the spectral transmission profile, especially at shorter wavelengths (Walsh et al. 2008) (Fig. 20).

The refractive characteristics of the cornea result from a combination of the tissue composition and the character of the curved surfaces in contact with the tear film and aqueous humour. Recent estimates of the refractive indices of the corneal epithelium, anterior stroma and posterior corneal surface yield values of 1.401, 1.380 and 1.373 respectively (Patel et al. 1995). The refractive indices for the collagen fibrils that make up the corneal stroma appear to have values higher that of the corneal epithelium (i.e. over 1.400), while the extracellular matrix (ground substance) within the stroma have lower values than the tissue as a whole (i.e. close to 1.350) (Leonard and Meek 1997).

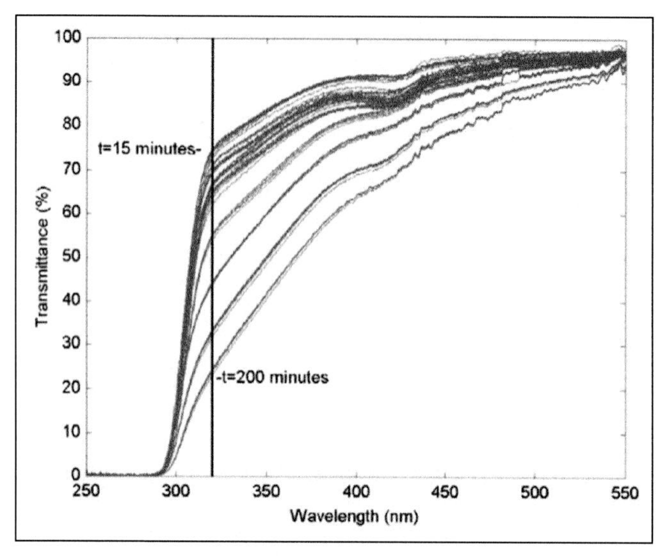

Figure 20 Spectra to show UV-visible light transmission through corneas, rabbit, before and over time after damage to the epithelium and endothelium between 15 min and 200 min. From Walsh et al. 2008, copyright IOP Publishing.

THE CORNEA AND ITS RELATIONSHIP TO THE TEAR FILM AND AQUEOUS HUMOUR

Overall Corneal Properties and their Maintenance

The normal cornea has a certain thickness, light transmission and optical properties. That these properties exist is dependent on there being normal structure and ultrastructure of the corneal tissue. Any perturbation of these characteristics can be expected to change corneal thickness, usually resulting in a net increase (Doughty 1995, Doughty and Zaman 2000), and a deterioration in the light transmission especially at shorter wavelengths (Walsh et al. 2008).

The cornea-tear film interface and/or the tear film-air interface can be shown to contribute to optical distortion; this can be readily demonstrated by measuring wavelength-dependent aberrations before and after instillation of artificial tears (Montes-Mico et al. 2010). A healthy corneal epithelium appears to contribute little to the transparency profile (Walsh et al. 2008), so it is the organization of the corneal stromal fibrils that is considered important in determining normal corneal transparency. Since corneal epithelial cells, per se, are generally considered to be transparent, they are not a source of optical distortion. However, if oedema develops in the epithelial cells, then optical distortion will occur (Caldicott and Charman

2002). Mild epithelial oedema is most likely to develop concurrently with compromise of the corneal epithelial 'barrier' (e.g. with application of very hypotonic solutions to the corneal surface) while substantial epithelial oedema (bullous keratopathy) can develop secondarily to endothelial 'barrier' compromise.

The Corneal Epithelial Barrier

The corneal epithelium, as a unit, is considered to be a 'tight' epithelial membrane and forms an effective biological barrier to fluid as a result of the specialized cell-cell junctional complexes. Various functions have been attributed to the corneal epithelial cell layer, namely barrier, optical, and protective properties. The corneal epithelial surface can be considered as a tear film stabilizer with the microplicae (and glycocalyx) along the epithelial surface promoting tear film interaction with the corneal surface and so contributing to the stability of the overlying tear film. Similarly, the corneal epithelium can be considered as a protective shield against microorganisms normally present in the tear fluid so minimize the risk of microbial invasion (and infection) of the corneal tissue.

With the corneal surface being continuously exposed to the environment (via a potentially unstable tear film) and at risk from stress from airbourne irritants (including volatile organic compounds) or allergens, some sort of defense and repair process is needed. After exposure to dust and/or repeated rubbing of the ocular eyes, some mechanical damage to the ocular surface is inevitable and so a process of epithelial cell renewal is present. This is essentially accomplished through a simple process which involves mitotic cell division at the basal cell level followed by cell migration to the surface to replace lost or damaged cells (Thoft and Friend 1983).

The overall process of cell migration and enlargement is considered important for the maintenance of a healthy epithelium. Early histological studies identified dividing and migrating cells with [3H]-thymidine and concluded that the life cycle of an epithelial cell was around seven days (Hanna and O'Brien 1960), i.e. a basal epithelial cell would take this period of time to slowly migrate to the ocular surface. There are regional differences across the cornea and the overall process has been described in a model proposed by Thoft and Friend (Thoft and Friend 1983). This is the X,Y,Z hypothesis by which proliferation of basal epithelial cells (X) together with a centripetal movement of peripherally located corneal epithelial cells (Y) equals the desquamation rate (Z) at the corneal epithelial surface. The hypothesis suggests that to remain healthy the central cornea can not solely depend on mitosis of cells within the cornea but requires a net flux of cells from the periphery. More recent studies, using brdU labelling to detect cell mitoses, indicate that following epithelial wounding it is the peripheral basal cells that divide at a faster rate (Ladage et al. 2002). The overall rate and/or extent of renewal of the corneal epithelial cells is also dependent on a special population of cells, the stem cells, that are usually located deep in the limbal region (Sun and Lavker 2004). These stem cells are considered to have an unrestricted capacity for cell division and a special asymmetric division occurs so that one of the cells (referred to as a daughter cell) stays to provide replenishment to the stem cell pool. The other cell from the division (also known as transient amplifying cells or transitional cells) then migrates into the cornea and upwards through the corneal epithelial cell layers, and may further divide before reaching the surface to desquamate.

The corneal epithelium also serves as a fluid barrier with the surface layer of epithelial cells containing special cell-cell linking structures, the *zonula occludens*. These serve as a barrier to reduce permeability (Hamalainen et al. 1997), and therefore limit any influx of fluid (water) into the corneal epithelium from the tear film. If the epithelial barrier is removed *in vivo*, then quite substantial corneal stroma swelling occurs; this will slowly reverse once epithelial repair is completed (Maurice and Giardini 1951). Over the shorter term, damage to the superficial cells and some corneal swelling can readily occur, even with just the presentation of non-physiological solutions to the epithelial surface (Doughty 1995); this includes hypotonic solutions and even some clinically-relevant solutions such as used in artificial tears.

In contrast, with any chronic tear film instability and associated damage to the superficial cells, slight corneal thinning (as opposed to thickening) can be expected (Liu and Pflugfelder 1999). This is because the tear film is now likely to be chronically slightly hypertonic (e.g. as in advanced keratoconjunctivitis sicca, KCS) (Sullivan et al. 2010), so contributing to an osmotically-driven efflux of fluid out of the corneal epithelium. With chronic tear film instability, there is likely to be a net increase in the rate of evaporation of the pre-corneal tear film, so contributing to a progressive hypertonic shift in tear film osmolality.

As a tight epithelium, numerous ion transport mechanisms have been demonstrated in the normal corneal epithelium, the most notable of which are linked to an inward flux of Na^+ coupled to an outwardly directed flux of Cl^- which, indirectly, supports an osmotic fluid flow (Candia 2004). Such osmotically driven flows can be expected to adapt to tear film osmolality changes (Bildin et al. 2000). The corneal epithelium also contains a special set of water channel proteins called aquaporins, that both facilitate water balance between the intact epithelium and the overlying tear film and osmotic balance in new cells arriving at the corneal surface (Levin and Verkman 2006). In addition, the corneal epithelial cells have coupled transport systems for lactate/H^+, which serve to help counteract any osmotic effects of metabolism-related excess, lactate production (Leung et al. 2011).

The Corneal Endothelial Barrier and the Intraocular Pressure

A lesser fluid barrier is present at the level of the corneal endothelium where the cell-cell junctions result in a 'leaky' membrane; it is likely that there is routinely some leakage of aqueous humor from the anterior chamber into the relatively dehydrated corneal stroma. If the endothelial barrier is severely damaged, then rapid fluid flow into the stroma will occur, with a very substantial increase in corneal thickness (Maurice and Giardini 1951).

The normal leakage (posterior-to-anterior) is likely dependent upon intraocular pressure (IOP), with increased posterior-to-anterior flow occurring under some conditions of clinically-relevant ocular hypertension (e.g. acute-onset elevated IOP). Overall, the function of the endothelium can be considered as a means to regulate the fluid balance between the corneal stroma and the aqueous humour. It is generally accepted that the endothelium accomplishes this by acting both as a 'leaky' fluid barrier and as a fluid pump (Edelhauser 2000). In the laboratory and using isolated corneal stroma-endothelial preparations, is possible to measure a net fluid pump activity, in the stroma-to-aqueous (i.e. anterior-to-posterior) direction (Doughty and Maurice 1988); the 'leak' is in the opposite direction and dominates when the endothelial function is severely compromised.

This 'pump' is often considered to be a means by which fluid could be removed from the corneal stroma in the living eye (Leung et al. 2011), a feature that has been attributed to an active metabolism and ability to actively transport ions such as bicarbonate.

However, fluid can apparently be 'pumped' (transported) in the nominal absence of ions such as bicarbonate (Doughty and Maurice 1988, Doughty et al. 1993, Diecke et al. 2007), so is not obviously exclusively dependent on a 'bicarbonate pump'. A set of water transport proteins, the aquaporins may be responsible for such fluid movement (Levin and Verkman 2006). A case can also be made that some of the bicarbonate-pH effects could be on the corneal stroma per se, and that those cellular metabolic activities that appear to regulate corneal endothelial function can be equally applied to the keratocytes. Since these cells are attached to the collagen fibrils, they could serve to reorganize the collagen matrix and lamellae of the corneal stroma and so indirectly change the fluid distribution (hydration) within the stroma (Doughty 1989b, 2001, 2003b).

Biomechanical Properties of the Cornea, Including in Relation to Surrounding Sclera

Overall, the nature of the interface between the various levels of the corneal stroma and the surrounding episclera and sclera, along with the differences in inter-lacing of the lamellae, are considered to be important determinants of the shape and mechanical strength of the anterior segment of the eye and its resistance to change (Müller et al. 2001, Asejczyk-Widlicka et al. 2011). The corneo-scleral tissue is also largely non-swelling and so exerts a 'clamping' effect on the peripheral parts of the corneal stroma to limit stromal swelling (Doughty and Bergmanson 2004b). As usually, when corneal swelling does occur, the increases in thickness are greater in the central region of the cornea.

Anatomical differences between anterior and posterior lamellae of the stroma may be part of the explanation to a substantial tendency for the swelling of the central regions of the corneal stroma to occur in the posterior direction (Cogan 1951). In contrast, the anterior stroma changes relatively little even when there is substantial posterior swelling (Müller et al. 2001). Therefore, when corneal (stromal) oedema develops, the posterior surface of the cornea (i.e. the endothelium) moves inwards (i.e. towards the surface of the crystalline lens). This process can result in compression of the posterior stroma, the PLL and the endothelial cell layer with the development of small wrinkles or folds, usually referred to as striae (Schoessler and Loewther 1971, Skaff et al. 1995). For contact lens wear associated corneal swelling (i.e. increase in corneal thickness), the compression of the corneal endothelium can be manifest as a small (but predictable) increase in the apparent cell density (Doughty et al. 2004).

REFERENCES

Abib, F.C. and J. Barreto. 2001. Behavior of corneal endothelial density over a lifetime. J. Cataract Refract. Surg. 27: 1574–1578.

Acosta, M.C., M.E. Tan, C. Belmonte and J. Gallar. 2001. Sensations evoked by selective mechanical, chemical, and thermal stimulation of the conjunctiva and cornea. Invest. Ophthalmol. Vis. Sci. 42: 2063–2067.

Acosta, M.C., M.L. Alfaro, F. Borras, C. Belmonte and J. Gallar. 2006. Influence of age, gender and iris color on mechanical and chemical sensitivity of the cornea and conjunctiva. Exp. Eye Res. 83: 932–938.

Ando, K. and D.F. Kripke. 1996. Light attenuation by the human eyelid. Biol. Psychiat. 39: 22–25.

Asejczyk-Widlicka, M., W. Srodka, R.A. Schachar and B.K. Pierscionek. 2011. Material properties of the cornea and sclera: A modeling approach to test experimental analysis. J. Biomechanics 44: 543–546.

Beems, E.M. and J.A. Van Best. 2000. Light transmission of the cornea in whole human eyes. Exp. Eye Res. 50: 393–395.

Belmonte, C., M.C. Acosta, M. Schmelz and J. Gallar. 1999. Measurement of corneal sensitivity to mechanical and chemical stimulation with a CO_2 esthesiometer. Invest. Ophthalmol. Vis. Sci. 40: 513–519.

Benedek, G.B. 1971. Theory of transparency of the eye. Appl. Optics 10: 459–473.

Bergmanson, J.P.G., J. Horne, M.J. Doughty, M. Garcia and M. Gondo. 2005. Assessment of the number of lamellae in the central region of the normal human corneal stroma, at the resolution of the transmission electron microscope. Eye Contact Lens 31: 282–287.

Bergmanson, J.P.G. and M.J. Doughty. 2005. Anatomy, morphology and electron microscopy of the cornea and conjunctiva. Chapter 2. In: Bennettt, E. and B.A. Weissman. [eds.]. Clinical Contact Lens Practice. Philadelphia, Lippincott-Williams & Wilkins, pp. 11–39.

Berlau, J., H-H. Becker, J. Stave, C. Oriwol and R.F. Guthoff. 2002. Depth and age-dependent distribution of keratocytes in healthy human corneas. J. Cataract Refract. Surg. 28: 611–616.

Bildin, V.N., H. Yang, R.B. Crook, J. Fischbarg and P.S. Reinach. 2000. Adaptation by corneal epithelial cells to chronic hypertonic stress depends on upregulation of Na:K:2Cl cotransporter gene and protein expression and ion transport activity. J. Membrane Biol. 177: 41–50.

Bowman W. 1847. Lectures delivered at the Royal London Ophthalmic Hospital, Moorfields. London Medical Gazette 5: 743–753.

Bron, A.J., N. Yokoi, E.A. Gaffney and J.M. Tiffany. 2011. A solute gradient in the tear meniscus. I. A hypothesis to explain Marx's line. Ocular Surf. 9: 70–91.

Caldicott, A and W.N. Charman. 2002. Diffraction haloes resulting from corneal oedema and epithelial cell size. Ophthal. Physiol. Opt. 22: 209–213.

Calmettes, L., F. Deodati, H. Planel and P. Bec. 1956. Étude histologique et histochimique de l'épithélium antérior de la cornée et de ses basales. Arch. d'Ophtalmol. 16: 481–506.

Candia, O.A. 2004. Electrolyte and fluid transport across corneal, conjunctival, and lens epithelia. Exp. Eye Res. 78: 527–535.

Chan, K.Y., S.W. Cheung, A.K. Lam and P. Cho. 2011. Corneal sublayer thickness measurements with the Nidek ConfoScan 4 (Z Ring). Optom. Vis. Sci. 88: E1240-4.

Cher, I. 2008. A new look at the lubrication of the ocular surface: Fluid mechanics behind the blinking eyelids. Ocular Surf. 6: 79–86.

Christens-Barry, W.A., W. Green, P.J. Conolly, R.A. Farrell and R.L. McCally. 1996. Spatial mapping of polarized light transmission in the central rabbit cornea. Exp. Eye Res. 62: 651–662.

Cochet, P. and R. Bonnet. 1960. L'Esthesie corneenne. Sa measure clinique. La Clin. Ophtalmol. 4: 3–27.

Cogan, D.G. 1951. Applied anatomy and physiology of the cornea. Trans. Am. Acad. Ophthalmol. Otolargyngol. 55: 329–359.

Collin, H.B. 1966. Ocular lymphatics. Am. J. Optom. Amer. Acad. Optom. 43: 96–106.

Cometto-Muñiz, J.E., W.S. Cain, M.H. Abraham and R. Sanchez-Morento. 2007. Concentration-detection functions for eye irritation evoked by homologous n-alcohols and acetates approaching a cut-off point. Exp. Brain Res. 182: 71–79.

Cui, L., M. Shen, J.Wang, J. Jiang, M. Li, D. Chen, Z. Chen, A. Taoand and F. Lu. 2011. Age-related changes in tear menisci imaged by optical coherence tomography. Optom. Vis Sci. 88: 1214–9.

Cursiefen, C., U. Schlötzer-Schrehardt, M. Küchle, L. Sorokin, S. Breiteneder-Geleff, K. Alitalo and D. Jackson. 2002. Lymphatic vessels in vascularized human corneas: immunohistochemical investigation using LYVE-1 and podoplanin. Invest. Ophthalmol. Vis. Sci. 43: 2127–2135.

Daxter, A. and P. Fratzl. 1997. Collagen fibril orientation in the human corneal stroma and its implications in keratoconus. Invest. Ophthalmol. Vis. Sci. 38: 121–129.

Diecke, F.P.J., L. Ma, P. Iserovich and J. Fischbarg. 2007. Corneal endothelium transports fluid in the absence of net solute transport. Biochim. Biophys. Acta 1768: 2043–2048.

Dilly, P.N. and I.A. Mackie. 1981. Surface changes in the anaesthetic conjunctiva in man, with special reference to the production of mucus from non-goblet-cell source. Br. J. Ophthalmol. 65: 833–842.
Doğan, A.S., M. Orhan, F. Söylemezoğlu, M. İrkeç and B. Bozkurt B. 2004. Effects of topical antiglaucoma drugs on apoptosis rates of conjunctival epithelial cells in glaucoma patients. Clin. Exp. Ophthalmol. 32: 62–66.
Dogru. M., H. Karakaya, M. Baykara, A. Ozmen, N. Koksal, E. Goto, Y. Matsumoto, T. Kojima, J. Shimazaki and K. Tsubota. 2004. Tear function and ocular surface findings in premature and term babies. Ophthalmology 111: 901–905.
Donald, C., L. Hamilton and M.J. Doughty. 2003. A quantitative assessment of the location and width of 'Marx's line' along the marginal zone of the human eyelid. Optom. Vis. Sci. 80: 564–572.
Doughty M.J., K.A. Blades and N. Ibrahim. 2002. Assessment of the number of eye symptoms and the impact of some confounding variable for office staff in non-air-conditioned buildings. Ophthal. Physiol. Opt. 22: 143–155.
Doughty, M.J. 1989a. Toward a quantitative analysis of corneal endothelial cell morphology: A review of techniques and their application. Optom. Vis. Sci. 66: 626–642.
Doughty, M.J. 1989b. Evidence for a direct effect of bicarbonate on rabbit corneal stroma. Optom. Vis. Sci. 68: 687–698.
Doughty, M.J. 1990a. On the evaluation of the corneal epithelial surface by scanning electron microscopy. Optom. Vis. Sci. 67: 735–756.
Doughty, M.J. 1990b. Morphometric analysis of the surface cells of rabbit corneal epithelium by scanning electron microscopy. Am. J. Anat.189: 316–328.
Doughty, M.J. 1995. Evaluation of the effects of saline versus bicarbonate-containing mixed salts solutions on rabbit corneal epithelium *in vitro*. Ophthal. Physiol. Opt. 15: 585–599.
Doughty, M.J. 1996. Evidence for heterogeneity in a small squamous cell type (light cells) in the rabbit corneal epithelium—a scanning electron micocroscope study. Doc. Ophthalmol. 92: 117–136.
Doughty, M.J. 1997. Scanning electron microscopy study of the tarsal and orbital conjunctival surfaces compared to peripheral corneal epithelium in pigmented rabbits. Doc. Ophthalmol. 93: 345–371.
Doughty, M.J. 1998. Prevalence of 'non-hexagonal' cells in the corneal endothelium of young Caucasian adults and their inter-relationships. Ophthal. Physiol. Opt. 18: 415–422.
Doughty, M.J. 1999. Re-wetting, comfort, lubricant and moisturising solutions for the contact lens wearer. Contact Lens Anterior Eye 22: 116–126.
Doughty, M.J. 2001. Changes in hydration, protein and proteoglycan composition of the collagen-keratocyte matrix of the bovine corneal stroma *ex vivo* in a bicarbonate-mixed salts solution, compared to other solutions. Biochim. Biophys. Acta 1525: 97–107.
Doughty, M.J. 2003a. Impact of brief exposure to balanced salts solution or cetylpyridium chloride on the surface appearance of the rabbit corneal epithelium—a scanning electron microscopy study. Curr. Eye Res. 26: 335–346.
Doughty, M.J. 2003b. A physiological perspective on the swelling properties of the mammalian corneal stroma. Contact Lens Anterior Eye 26: 117–129.
Doughty, M.J. 2006. Quantitative assessment of ring-shaped (crater-like) features at the tear film-epithelial interface of the rabbit cornea as assessed by scanning electron microscopy. Curr. Eye Res. 31: 999–1010.
Doughty, M.J. 2008. Normal features of superficial non-desquamating cells of the rabbit corneal epithelium assessed by scanning electron microscopy. Vet. Ophthalmol. 11: 81–90.
Doughty, M.J. 2010a. Ocular therapeutics 3. Management options for UK optometrists for eyelid and eyelid margin disorders. Optom. Today 50: 36–44.
Doughty, M.J. 2010b. Assessment of agreement for assignment of a normal grade to human conjunctival impression cytology samples. Cytopathology 21: 329–335.
Doughty, M.J. 2011a. Contact lens wear and the goblet cells of the human conjunctiva—a review. Contact Lens Anterior Eye 34: 157–163.

Doughty, M.J. 2011b. Contact lens wear and the development of squamous metaplasia of the surface cells of the conjunctiva. Eye Contact Lens 37: 274–281.

Doughty, M.J. 2011c. Assessment of collagen fibril spacing in relation to selected region of interest (ROI) on electron micrographs—application to mammalian corneal stroma. Microsc. Res. Techn. 14: 1–10.

Doughty, M.J. 2011d. Options for determination of the 2-D organization of collagen fibrils in transmission electron micrographs—application to corneal stroma. Microsc. Res. Techn. 74: 184–195.

Doughty, M.J., A. Müller. and M.L. Zaman. 2000. Assessment of the reliability of human corneal endothelial cell-density estimates using a noncontact specular microscope. Cornea 19: 148–158.

Doughty, M.J., B.M. Aakre, A.E.Ystanaes and E. Svarerud. 2004. Short-term adaptation of the human corneal endothelium to continuous wear of silicone hydrogel (Lotrafilcon A) contact lenses after dialy hydrogel lens wear. Optom. Vis. Sci. 82: 473–480.

Doughty, M.J., C-A. Lee, S. Ritchie and T. Naase. 2007. An assessment of the discomfort associated with the use of rose bengal 1% eyedrops on the normal human eye: a comparison with saline 0.9% and a topical ocular anaesthetic. Ophthal. Physiol. Opt. 27: 159–167.

Doughty, M.J. and D.M. Maurice. 1988. Bicarbonate sensitivity of rabbit corneal endothelium fluid pump *in vitro*. Invest. Ophthalmol. Vis. Sci. 29: 216–223.

Doughty, M.J. and J.P.G. Bergmanson. 1999. Reassessment of the conjunctival ocular surface at the muco-cutaneous junction and the ultrastructural search for Marx's line. Optom. Vis. Sci. 76: 167–175.

Doughty, M.J. and J.P.G. Bergmanson. 2003. New insights into the surface cells and glands of conjunctiva and their relevance to the tear film. Optometry 74: 485–500.

Doughty, M.J. and J.P.G. Bergmanson. 2004a. Heterogeneity in the ultrastructure of the mucous (goblet) cells of the rabbit palpebral conjunctiva. Clin. Exp. Optom. 87: 377–385.

Doughty, M.J. and J.P.G. Bergmanson. 2004b. Collagen fibril characteristics at the corneo-scleral boundary and rabbit corneal swelling. Clin. Exp. Optom. 87: 81–92.

Doughty, M.J. and J.P.G. Bergmanson. 2005. Resolution and reproducibility of measures of the diameter of small collagen fibrils by transmission electron microscopy-application to the rabbit corneal stroma. Micron 36: 331–343.

Doughty, M.J., K. Newlander and O. Olejnik. 1993. Effect of bicarbonate-free balanced salt solutions on fluid pump and endothelial morphology of rabbit corneas *in vitro*. J. Pharm. Pharmacol. 45: 102–109.

Doughty, M.J., M. Laiquzzaman and N.F. Button. 2001a. Video-assessment of tear meniscus height in elderly Caucasians and its relationship to the exposed ocular surface. Curr. Eye Res. 22: 420–426.

Doughty, M.J., M. McIntosh, S. McFadden and N.F. Button. 2007b. Impression cytology of a case of conjunctival metaplasia associated with oral carbamazepine use?. Contact Lens Anterior Eye 30: 254–257.

Doughty, M.J. and M.L. Zaman. 2000. Human corneal thickness and its impact on intraocular pressure measures: A review and meta-analysis approach. Surv. Ophthalmol. 44: 367–408.

Doughty, M.J., R. Potvin, N. Pritchard and D. Fonn. 1995. Evaluation of the range of areas of the fluorescein staining patterns of the tarsal conjunctiva in man. Doc. Ophthalmol. 89: 355–371.

Doughty, M.J. and S. Jonuscheit. 2010. The Orbscan acoustic (correction) factor for central corneal thickness measures of normal human corneas. Eye Contact Lens 36: 106–115.

Doughty, M.J., T. Naase and N.F. Button. 2009. Frequent spontaneous eyeblink activity associated with reduced conjunctival (trigeminal nerve) tactile sensitivity. Von Graefes Arch. Clin. Exp. Ophthalmol. 247: 939–946.

Doughty, M.J., T. Naase, C. Donald, L. Hamilton and N.F. Button. 2004. Visualisation of 'Marx's line' along the marginal eyelid conjunctiva of human subjects with lissamine green dye. Ophthal. Physiol. Opt. 24: 1–7.
Doughty, M.J. and T. Naase. 2008. Nucleus and cell size changes in human bulbar conjunctival cells after soft contact lens wear, as assessed by impression cytology. Contact Lens Anterior Eye 31: 131–140.
Doughty, M.J., W. Seabert, J.P.G. Bergmanson and Y. Blocker. 2001b. A descriptive and quantitative transmission electron microscopy study of the keratocytes of the corneal stroma of albino rabbits. Tissue Cell 33: 408–422.
Doughty, M.J. and W.K. Fong. 1992. Topographical differences in cell area at the surface of the corneal epithelium of the pigmented rabbit. Curr. Eye Res. 11: 1129–1136.
Draeger, J. 1984. Corneal Sensitivity. Measurement and Clinical Importance. Springer-Verlag, New York.
Edelhauser, H.F. 2000. The resiliency of the corneal endothelium to refractive and intraocular surgery. Cornea 19: 263–273.
Efron, N., M. Al-Dossari and N. Pritchard. 2010. Confocal microscopy of the bulbar conjunctiva in contact lens wear. Cornea 29: 43–52.
Ehlers, N. 1970. Morphology and histochemistry of the corneal epithelium of mammals. Acta Anat. 75: 161–198.
Ehlers, N., S. Heegaard, J. Hjortda, A. Ivarsen, K. Nielsen K and J.U. Prause. 2010. Morphological evaluation of normal human corneal epithelium. Acta Ophthalmol. 88: 858–861.
Erdelyi, B., R. Kraak, A. Zhivov, R. Guthoff and J. Nemeth. 2007. *In vivo* confocal laser scanning microscopy of the cornea in dry eye. Graefe's Arch. Clin. Exp. Ophthalmol. 245: 39–44.
Ernstgard, L., A. Iregren, B. Sjogren and G. Johanson. 2006. Acute effects of exposure to vapours of acetic acid in humans. Toxicol. Lett. 165: 22–30.
Farrell, R.A. and R.W. Hart. 1969. On the theory of the spatial organization of macromolecules in connective tissue. Bull. Math. Biophys. 31: 727–760.
Fatt, I. 1992. Architecture of the lid-cornea juncture. CLAO J. 18: 187–192.
Feenstra, R.P.G. and S.C.G. Tseng. 1992. What is actually stained by rose bengal? Arch Ophthalmol. 110: 984–993.
Gallar, J., M.C. Acosta and C. Belmonte. 2003. Activation of scleral cold thermoreceptors by temperature and blood flow changes. Invest. Ophthalmol. Vis. Sci. 44: 697–705.
Ganem, S. 1999. La microscopie confocale et la cornee. J. Fr. Ophtalmol. 22: 262–265.
Ganley, J.P. and C.M. Payne. 1981. Clinical and electron microscopic observations of the conjunctiva of adult patients with Bitot's spots. Invest. Ophthalmol. Vis. Sci. 20: 632–643.
Gilbard, J.P., S.R. Rossi, K.L. Gray, L.A. Hanninen and K.R. Kenyon. 1988. Tear film osmolarity and ocular surface disease in two rabbit models for keratoconjunctivitis sicca. Invest. Ophthalmol. Vis. Sci. 29: 374–378.
Gipson, I.K. and A.S. Tisdale. 1997. Visualization of conjunctival goblet cell actin cytoskeleton and mucin content in tissue whole mounts. Exp. Eye Res. 65: 407–415.
Gipson, I.K. and T. Inatomi. 1997. Mucin genes expressed by the ocular surface epithelium. Progr. Retinal Eye Res. 16: 81–98.
Goes, R.M., F.L. Barbosa, J. De Faria-e-Sosa and A. Haddad. 2008. Morphological and autoradiographic studies on the corneal and limbal epithelium of rabbits. Anat. Rec. 291: 191–203.
Goleblowski, B., E.B. Papas and F. Stapleton. 2008. Factors affecting corneal and conjunctival sensitivity measurement. Optom. Vis. Sci. 85: E241–E246.
Goto, E., M. Dogru, E.A. Sato, Y. Matsumoto, Y. Takano and K. Tsubota. 2011. The sparkle of the eye: The impact of ocular surface wetness on corneal light reflection. Am. J. Ophthalmol. 151: 691–696.
Greiner, J.V., A.S. Henriques, H.I. Covington, T.A. Weidman and M.R. Allansmith. 1981. Goblet cells of the human conjunctiva. Arch. Ophthalmol. 99: 2190–2197.

Greiner, J.V., H.I. Covington and M.R. Allansmith. 1977. Surface morphology of the human upper tarsal conjunctiva. Am. J. Ophthalmol. 83: 892–905.

Grüntzig, J., D. Uthoff, P. Schad, G. Arnold and G. Schwinger. 1990. Zur Frage der kolbenformigen Initialsegmente des Lymphgefäβ systems. Klin. Monatsbl. Augenheilkd. 197: 404–409.

Hahnel, C., S. Somodi, D.G. Weiss and R.F. Guthoff. 2000. The keratocyte network of human cornea: A three-dimensional study using confocal laser scanning fluorescence microscopy. Cornea 19: 185–193.

Hämäläinen, K.M., K. Kananen, S. Auriola, K. Kontturi and A. Urtti. 1997. Characterization of paracellular and aqueous penetration routes in cornea, conjunctiva, and sclera. Invest. Ophthalmol. Vis. Sci. 38: 627–634.

Hanna, C. and J.E. O'Brien. 1960. Cell production and migration in the epithelial layer of the cornea. Arch. Ophthalmol. 64: 536–539.

Hazlett, L.D., P. Wells, B. Spann and R.S. Berk. 1980. Epithelial desquamation in the adult mouse cornea. A correlative TEM-SEM study. Ophthal. Res. 12: 315–323.

Ho, J.D., C.Y. Tsai, R.J. Tsai, L.L. Kuo, I.L. Tsai and S.W. Liou. 2008. Validity of the keratometric index: evaluation by the Pentacam rotating Scheimpflug camera. J. Cataract Refract. Surg. 34: 137–145.

Hogan, M.J., J.A. Alvarado and J.E. Weddell. 1971. Histology of the human eye. W.B. Saunders, Philadelphia.

Höh, H., F. Schirra, C. Kienecker and W. Ruprecht. 1995. Lid-parallel conjunctival fold (LIPCOF) and dry eye: A diagnostic tool for the contactologist. Contactologia 17: 104–117.

Hollingsworth, J., I. Perez-Gomez, H.A. Mutalib and N. Efron. 2001. A population study of the normal cornea using an *in vivo*, slit-scanning confocal microscope. Optom. Vis. Sci. 78: 706–711.

Huang, A.J.W., S.C.G. Tseng and K.R. Kenyon. 1989. Paracellular permeability of corneal and conjunctival epithelia. Invest. Ophthalmol. Vis. Sci. 30: 684–689.

Hughes, W.L. 1942. Conjunctivochalsis. Am. J. Ophthalmol. 25: 48–51.

Inatomi, T., S. Spurr-Michaud, A.S. Tisdale, Q. Zhan, S.T. Feldman and I.K. Gipson. 1996. Expression of secretory mucin genes by human conjunctival epithelial cells. Invest. Ophthalmol. Vis. Sci. 37: 1684–1692.

Iwamoto, T. and G.K. Smelser. 1965. Electron microscopic studies on the mast cells and blood and lymphatic capillaries of the human corneal limbus. Invest. Ophthalmol. 4: 815–834.

Jalavisto E., E. Orma and M. Tawast. 1951. Ageing and relation between stimulus intensity and duration of corneal sensibility. Acta Physiol. Scand. 23: 224–233.

Jester, J.V., N. Nicolaides and R.E. Smith. 1981. Meibomian gland studies: histologic and ultrastructural investigations. Invest. Ophthalmol. 20: 537–547.

Jester, J.V., P.A. Barry, G.J. Lind, W.M. Petroll, R. Garana and H.D. Cavanagh. 1994. Corneal keratocytes: *In situ* and *in vitro* organization of cytoskeletal contractile proteins. Invest. Ophthalmol.Vis. Sci. 35: 730–743.

Johnson, D.H., W.M. Bourne and R.J. Campbell. 1982. The ultrastructure of Descemet's membrane: I. Changes with age in normal corneas. Arch. Ophthalmol. 100: 1942–1947.

Jonuscheit, S., M.J. Doughty and K. Ramaesh. 2011. *In vivo* confocal microscopy of the corneal endothelium: comparison of three morphometry methods after corneal transplantation. Eye 25: 1130–1137.

Jonuscheit, S. and M.J. Doughty. 2009. Evidence for a relative thinning of the peripheral cornea with age in white European subjects. Invest. Ophthalmol. Vis. Sci. 50: 4121–4128.

Kaji, Y., H. Obata H and T. Usui. 1998. Three-dimensional organization of collagen fibrils during corneal stromal wound healing after excimer laser keratectomy. J. Cataract Refract. Surg. 24: 1441–1446.

Kayes, J. and A. Holmberg. 1960. The fine structure of Bowman's layer and the basement membrane of the corneal epithelium. Am. J. Ophthalmol. 50: 1013–1021.

Kern, T.J., H.N. Erb, J.M. Schaedler and E.P. Dougherty. 1988. Scanning electron microscopy of experimental keratoconjunctivitis sicca in dogs: Cornea and bulbar conjunctiva. Vet. Pathol. 25: 568–474.

Kessing, S.V. 1967. A new division of the conjunctiva on the basis of X-ray examination. Acta Ophthalmol. 45: 680–683.

Kessing, S.V. 1968. The mucus gland system of the conjunctiva. Acta Ophthlamol. Suppl. 95: 1–19.

Knop, E., N. Knop, A. Zhivov, R. Kraak, D.R. Korb, C. Blackie, J.V. Greiner and R. Guthoff. 2011. The lid wiper and muco-cutaneous junction anatomy of the human eyelid margins: an *in vivo* confocal and histological study. J. Anat. 218: 449–461.

Knop, N. and E. Knop. 2010. Regulation of the inflammatory component in chronic dry eye disease by eye-associated lymphoid tissue (EALT). Dev. Ophthalmol. 45: 23–29.

Kojima, T., Y. Matsumoto, M. Dogru and K. Tsubota. 2010. The application of *in vivo* laser scanning confocal microscopy as a tool of conjunctival *in vivo* cytology in the diagnosis of dry eye ocular surface disease. Mol. Vis. 16: 2457–2464.

Kokott, W. 1938. Uber mechanisch-functionelle Strukturen des Auges. Albrecht v. Graefes Arch. Ophthalmol. 138: 424–485.

Kolstad, A. 1970. Corneal sensitivity by low temperatures. Acta Ophthalmol. 48: 789–793.

Komai, Y. and T. Ushiki. 1991. The three-dimensional organization of collagen fibrils in the human cornea and sclera. Invest. Ophthalmol. Vis. Sci. 32: 2244–2258.

Koornneef, L. 1977. Details of the orbital connective tissue system in the adult. Acta Morphol. Neerl-Scand. 15: 1–34.

Krauss, R. 1937. Der konstruktive Bau der Cornea. Z. Mikrosk. 53: 420–434.

Ladage, P.M., K. Yamamoto, L. Li, D.H. Ren, W.M. Petrol, J.V. Jester and H.D. Cavanagh 2002. Corneal epithelial homeostasis following daily and overnight contact lens wear. Contact Lens Anterior Eye 25: 11–21.

Lemp, M.A., W.D. Mathers and J.B. Gold. 1989. Surface cell morphology of the anesthetic human cornea. A color specular microscopic study. Acta Ophthalmol. 67: 102–107.

Leonard, D.W. and K.M. Meek. 1997. Refractive indices of the collagen fibrils and the extracellular material of the corneal stroma. Biophys. J. 72: 1382–1387.

Leuenberger P.M. 1978. Morphologie functionnelle de la cornée. Adv. Ophthalmol. 35: 94–166.

Leung, B.K., J.A. Bonanno and C.J. Radke. 2011. Oxygen-deficient metabolism and corneal edema. Progr. Retinal Eye Res. 30: 471–92.

Levin, M.H. and A.S. Verkman. 2006. Aquaporins and CFTR in ocular epithelial fluid transport. J. Membrane Biol. 210: 105–115.

Li, H.F., W.P. Petroll, T. Møller-Pedersen, J.K. Maurer, H.D. Cavanagh and J.V. Jetser. 1997. Epithelial and corneal thickness measurements by *in vivo* confocal microscopy through focusing (CMTF). Curr. Eye Res. 16: 214–221.

Li, Y., K. Kuang, B. Yerxa, Q. Wen, H. Rosskothen and J. Fischbarg. 2001. Rabbit conjunctival epithelium transports fluid, and P2Y22 receptor agonists stimulate Cl^- and fluid secretion. Am. J. Physiol. Cell Physiol. 281: C595–C602.

Lifshitz, M., O. Weinstein, V. Gavrilov, G. Rosenthal and T. Lifshitz. 1999. Acetaminophen (paracentamol) levels in human tears. Therap. Drug Monitor. 21: 544.

Liu, H., C. Begley, M. Chen, A. Bradley, J. Bonanno, N.A. McNamara, J.D. Nelson and T. Simpson. 2009. A link between tear instability and hyperosmolarity in dry eye. Invest. Ophthalmol. Vis. Sci. 50: 3671–36719.

Liu, Z. and S.C. Pflugfelder. 1999. Corneal thickness is reduced in dry eye. Cornea 18: 403–407.

Mader, T.H., K.E. Friedl, L.C. Mohr and W.N. Bernhard. 1987. Conjunctival oxygen tension at high altitude. Aviat. Space Environm. Med. 58: 76–79.

Marfurt, C.F., J. Cox J, S. Deek and L. Dvorscak. 2010. Anatomy of the human corneal innervation. Exp. Eye Res. 90: 478–492.

Marx, E. 1924. Über vitale Farbung des Auges und der Augenlider. I. Über Anatomie, Physiologie und Pathologie des Augenlidrandes und der Tranenpunkte. Graefes Archiv. F. Ophthalmol. 114: 465–482.
Mathers, W.D. and M.A. Lemp. 1992. Morphology and movement of corneal surface cells in humans. Curr. Eye Res. 11: 517–523.
Mathew, J.H., J.P.G. Bergmanson and M.J. Dougthy. 2008. Fine structure of the interface between the anterior limiting lamina and the anterior stromal fibrils of the human cornea. Invest. Ophthalmol. Vis. Sci. 49: 3914–3918.
Matsumoto, Y., M. Dogru, E. Goto, Y. Sasaki , H. Inoue, I. Saito, J. Shimazaki and K. Tsubota. 2008. Alterations of the tear film and ocular surface health in chronic smokers. Eye 22: 961–968.
Maurice, D.M. 1957. The structure and transparency of the cornea. J. Physiol. 136: 263–286.
Maurice, D.M. 1973. Electrical potential and ion transport across the conjunctiva. Exp. Eye Res. 15: 527–532.
Maurice, D.M. and A.A. Giardini. 1951. Swelling of the cornea *in vivo* after the destruction of its limiting layers. Br. J. Ophthalmol. 35: 791–797.
McGowan, D.P., J.G. Lawrenson and G.L. Ruskell. 1994. Touch sensitivity of the eyelid margin and palpebral conjunctiva. Acta Ophthalmol. 72: 57–60.
Messmer, E.M., M.J. Mackert, D.M. Zapp and A. Kampik. 2006. *In vivo* confocal microscopy of normal conjunctiva and conjunctivitis. Cornea 25: 781–788.
Meyer, P.A.R and P.G. Watson. 1987. Low dose fluorescein angiography of the conjunctiva and episcelera. Br. J. Ophthalmol. 71: 2–10.
Millodot, M. 1977. The influence of age on the sensitivity of the cornea. Invest. Ophthalmol. Vis. Sci. 16: 240–242.
Montés-Micó, R., A. Cerviño, T. Ferrer-Blasco, S. García-Lázaro and D. Madrid-Costa. 2010. The tear film and the optical quality of the eye. Ocular Surf. 8: 185–192.
Müller, A., M.J. Doughty and L. Wright. 2000. Reassessment of the corneal endothelial cell organisation in children. Br. J. Ophthalmol. 84: 692–696.
Müller, A. and M.J. Doughty. 2002. Assessments of corneal endothelial cell density in growing children, and its relationship to horizontal corneal diameter. Optom. Vis. Sci. 79: 762–770.
Müller, L.J., E. Pels and G.F.J.M. Vrensen. 1996. Ultrastructural organization of human corneal nerves. Invest. Ophthalmol. Vis. Sci. 37: 476–488.
Muller, L.J., E. Pels and G.F.J.M. Vrensen. 2001. The specific architecture of the anterior stroma accounts for maintenance of corneal curvature. Br. J. Ophthalmol. 85: 437–443.
Murphy, P.J., S. Patel, N. Kong, R.E. Ryder and J. Marshall. 2004. Noninvasive assessment of corneal sensitivity in young and elderly diabetic and nondiabetic subjects. Invest. Ophthalmol. Vis. Sci. 45: 1737–1742.
Nagasaki, T. and J. Zhao. 2005. Uniform distribution of epithelial stem cells in the bulbar conjunctiva. Invest. Ophthalmol. Vis. Sci. 46: 126–132.
Nelson, J.D. and J.C. Wright. 1984. Conjunctival goblet cell densities in ocular surface disease. Arch. Ophthalmol. 102: 1049–1051.
Nichols, B.A. 1996. Conjunctiva. Microsc. Res. Techn. 33: 296–319.
Nichols, B.A., M.L. Chiappino and C.R. Dawson. 1985. Demonstration of the mucus layer of the tear film by electron microscopy. Invest. Ophthalmol. Vis. Sci. 26: 464–473.
Nisam, M., T.E. Albertson, E. Panacek, W. Rutherford and C.J. Fischer. 1986. Effects of hyperventilation on conjunctival oxygen tension in humans. Crit. Care Med. 14: 12–15.
Norn M. 1985. Meibomian orifices and Marx's line studied by triple vital staining. Acta Ophthalmol. 63: 698–700.
Norn, M.S. 1973. Conjunctival sensitivity in normal eyes. Acta Ophthalmol. 51: 58–66.
Olsen, T.W., S.Y. Aaberg, D.H. Geroski and H.F. Edelhauser. 1998. Human sclera: Thickness and surface area. Am. J. Ophthalmol. 125: 237–241.
Paschides, C.A., G. Petroutsos and K. Psilas. 1991. Correlation of conjunctival impression cytology results with lacrimal function and age. Acta Ophthalmol. 69: 422–425.

Patel, S., J. Marshall and F.W. Fitzke. 1995. Refractive index of the human corneal epithelium and stroma. J Refract. Surg. 11: 100–105.

Patel, S.V., J.W. McLaren, D.O. Hodge and W.M. Bourne. 2001. Normal human keratocyte density and corneal thickness measurement using confocal microscopy *in vivo*. Invest. Ophthalmol. Vis. Sci. 42: 333–339.

Peter, J., G. Zajicek, H. Greifner and M. Kogan. 1996. Streaming conjunctiva. Anat. Rec. 245: 36–40.

Pflugfelder, S.C., S.C. Tseng, K. Yoshino, D. Monroy, C. Felix and B.L. Reis. 1997. Correlation of goblet cell density and mucosal epithelial membrane mucin expression with rose bengal staining in patients with ocular irritation. Ophthalmology 104: 223–235.

Polack, F.M. 1961. Morphology of the cornea. I. Study with silver stains. Am. J. Ophthalmol. 51: 1051–1056.

Poole, C.A., N. Brookes and G.M. Clover. 1993. Keratoctye networks visualized in the living cornea using vital dyes. J. Cell Sci. 106: 685–692.

Potvin, R.J., M.J. Doughty and D. Fonn. 1994. Tarsal conjunctival morphomteyr of asymptomatic soft contact lens wearers and non-lens wearers. Int. Contact Lens Clin. 21: 225–231.

Prydal, J.I., P. Artal, H. Woon and F.W. Campbell. 1992. Study of human precorneal tear film thickness and structure using laser interferometry. Invest. Ophthalmol. Vis. Sci. 33: 2006–2011.

Radner, W., M. Zehetmayer, R. Aufreiter and R. Mallinger R. 1998. Interlacing and cross-angle distribution of collagen lamellae in the human cornea. Cornea 17: 537–543.

Ren, D.H., W.M. Petroll, J.V. Jester and H.D. Cavanagh. 1999. The effect of rigid gas permeable contact lens wear on proliferation of rabbit corneal and conjunctival epithelial cells. CLAO J. 25: 136–141.

Ren, H. and G. Wilson. 1996. Apoptosis in the corneal epithelium. Invest. Ophthalmol. Vis. Sci. 37: 1017–1025.

Rohen, J.W., E. Lütjen-Drecoll. The Dry Eye. pp. 38–63. Functional morphology of the conjunctiva. In: M.A. Lemp and R. Marquardt [eds.]. 1992. Springer-Verlag, Berlin. Germany.

Rolando, M., F. Terragna, G. Giordano and G. Calabria. 1990. Conjunctival surface damage distribution in keratoconjunctivitis sicca. An impression cytology study. Ophthalmologica 200: 170–176.

Roszkowska, A.M., P. Colosi, F.M.B. Ferreri and S. Galasso. 2004. Age-related modifications of corneal sensitivity. Ophthalmologica 218: 350–355.

Ruskell, G.L. 1991. The conjunctiva, Chapter 3. In: E.S. Bennett and B.A. Weissman [eds.]. Clinical Contact Lens Practice. J.D. Lippincott, Philadelphia. USA, pp. 1–18.

Saini, J.S., A. Rajwanshi and S. Dhar. 1990. Clinicopathological correlation of hard contact lens related changes in tarsal conjucntiva by impression cytology. Acta Ophthalmol. 68: 65–70.

Satici, A., M. Bitiren, I. Ozardali, H. Vural, A. Kilie and M. Guzey. 2003. The effects of chronic smoking on the ocular surface and tear characteristics: a clinical, histological and biochemical study. Acta Ophthalmol. Scand. 81: 583–587.

Schoessler, J.P. and G.E. Loewther. 1971. Slit lamp observations of corneal edema. Am. J. Optom. Arch. Am. Acad. Optom. 48: 666–671.

Shaw, A.J., M.J. Collins, B.A. Davis and L.G. Carney. 2009. Eyelid pressure: Inferences from corneal topographic changes. Cornea 28: 181–188.

Situ, P., T.L. Simpson, D. Fonn and L.W. Jones. 2008. Conjunctival and corneal pneumatic sensitivity is associated with signs and symptoms of ocular dryness. Invest. Ophthalmol. Vis. Sci. 49: 2971–2976.

Skaff, A., A.P. Cullen, M.J. Doughty and D. Fonn. 1995. Corneal swelling and recovery following wear of thick hydrogel contact lenses in insulin-dependent diabetics. Ophthal. Physiol. Opt. 15: 287–297.

Snir, M., R. Axer-Siegel, D. Bourla, I. Kremer, Y. Benjamini and D. Weinberger. 2002. Tactile corneal reflex development in full-term babies. Ophthalmology 109: 526–529.

Steul, K-P. 1989. Ultrastructure of the conjunctival epithelium. Dev. Ophthalmol. 19: 1–104.
Sugar, H.S., A. Riazi and R. Schaffner. 1957. The bulbar conjunctival lymphatics and their clinical significance. Trans. Am. Acad. Ophthalmol. Ototolaryngol. 61: 212–223.
Sugrue, S.P. and J.D. Zieske. 1997. ZO1 in corneal epithelium: Association to the zonula occludens and adherens junctions. Exp. Eye Res. 64: 11–20.
Sullivan, B.D., D. Whitmer, K.K. Nichols, A. Tomlinson, G.N. Foulks, G. Geerling, J.S. Pepose, V. Kosheleff, A. Porreco and M.A. Lemp. 2010. An objective approach to dry eye disease severity. Invest. Ophthalmol. Vis. Sci. 51: 6125–6130.
Sun, T.T. and R.M. Lavker. 2004. Corneal epithelial stem cells: past, present, and future. J. Invest. Dermatol. Symp. Proc. 9: 202–207.
Takahashi, A., K. Nakayasu, S. Okisaka and A. Kanai. 1990. Quantitative analysis of collagen fiber in keratoconus. Acta Soc. Ophthalmol. Jpn. 94: 1068–1073.
Takakusaki, I. 1969. Fine structure of the human palpebral conjunctiva with special reference to the pathological changes in vernal conjunctivitis. Arch. Histol. Jap. 30: 247–282.
Tao, A., J. Wang, Q. Chen, M. Shen, F. Lu, S.R. Dubovy and M.A. Shousha. 2011. Topographic thickness of Bowman's layer determined by ultra-high resolution spectral domain-optical coherence tomography. Invest. Ophthalmol. Vis. Sci. 52: 3901–3907.
Thoft, R.A. and J. Friend. 1983. The X,Y,Z hypothesis of corneal epithelial maintenance. Invest. Ophthalmol. Vis. Sci. 24: 1442–1443.
Tsubota, K., M. Yamada and S. Naoi. 1992. Specular microscopic observation of normal human corneal epithelium. Ophthalmology 99: 89–94.
Turner, H.C., L.J. Alvarez and O.A. Candia. 2000. Cyclic AMP-dependent stimulation of basolateral K^+ conductance in the rabbit conjunctival epithelium. Exp. Eye Res. 70: 295–305.
Ueda, H. and Y. Horibe, K-J. Kim and W.H.L Lee. 2000. Functional characterization of organic cation drug transport in the pigmented rabbit cornea. Invest. Ophthalmol. Vis. Sci. 41: 870–876.
Vaezy, S. and J.I. Clark. 1991. A quantitative analysis of transparency in the human sclera and cornea using Fourier methods. J. Microsc. 163: 85–94.
Varikooty, J. and T.L. Simpson. 2009. The interblink interval I: the relationship between sensation intensity and tear film disruption. Invest. Ophthalmol. Vis. Sci. 50: 1087–1092.
Ventura, L. and S.J.F. Sousa, A.M.V. Messias and J.M. Bispo. 2000. System for measuring the transmission spectrum of '*in vitro*' corneas. Physiol. Meas. 21: 197–207.
Verges Roger, C., D. Pita Salorio, M.F. Refojo and M.T. Sainz de la Maza Serra. 1986. Cambios en la poblacion de celulas calciformes conjunctivales tras la aplicacion de una solucion hiperosmolar. Arch. Soc. Esp. Oftalmol. 51: 403–406.
Versura, P., M.C. Maltarello, R. Caramazza and R. Laschi. 1989. Mucus alteration and eye dryness. A possible relationship. Acta Ophthalmol. 67: 455–464.
Von Reuss, A. 1881. Untersuchungen ueber den Einfluss der Lebenssalers auf die Kruemmung der Hornhaut nebst einigen Bemerkungen ueber die Dimensionen der Lidspalte. Albrecht v. Graefes Arch. Ophthalmol. 27: 27–53.
Walsh, J.E., J.P.G. Bergmanson, L.V. Koehler, M.J. Doughty, D. Fleming and J.H. Harmey JH. 2008. Optimising the methodology for measuring corneal spectral transmittance. Physiological Meas. 29: 375–388.
Wanko, T., B.J. Lloyd and J. Mathews. 1964. The fine structure of the human conjunctiva in the perilimbal zone. Invest. Ophthalmol. 3: 285–301.
Watsky, M.A., M.M. Jablonski and H.F. Edelhauser. 1988. Comparison of conjunctival and corneal surface areas in rabbit and human. Curr. Eye Res. 7: 483–486.
Welle, M. 1997. Development, significance, and heterogeneity of mast cells with particular regard to the mast cell-specific proteases chymase and typtase. J. Leuk. Biol. 61: 233–245.
Weyrauch, K.D. 1983. The conjunctival epithelium in domestic ruminants. Lightmicroscopic investigations. Z. Mikrosk.-anat. Forsch. 97: 565–572.

Wichterle, K., J. Vodañský and O. Wichterle. 1991. Shape of the cornea and conjunctiva. Optom. Vis. Sci. 68: 232–235.

Wirtschafter, J.D., J.M. Ketcham, R.J. Weinstock, T. Tabesh and L.K. McLoon. 1999. Mucocutaneous junction as the major source of replacement palpebral conjunctival cells. Invest. Ophthalmol. Vis Sci. 40: 3138–3146.

Zaman, M.L., M.J. Doughty and N.F. Button. 1998. The exposed ocular surface and its relationship to spontaneous eyeblink rate in the elderly Caucasians. Exp. Eye Res. 67: 681–686.

Zhang, X., Q. Chen, W. Chen, L. Cui, H. Ma and F. Lu. 2011. Tear dynamics and corneal confocal microscopy of subjects with mild self-reported office dry eye. Ophthalmology 118: 902–907.

Zhu, W., J. Hong, T. Zheng, Q. Le, J. Xu and X. Sun. 2010. Age-related changes of human conjunctiva on *in vivo* confocal microscopy. Br. J. Ophthalmol. 94: 1448–1453.

Section II: Ocular Surface Disorders

Dry Eye

Johanna T. Henriksson and *Cintia S. de Paiva*[a,*]

SUMMARY

Dry eye is a disease that affects millions of people worldwide. Important aspects of the disease such as demographics, definition, clinical features, categories and tests utilized for diagnosis are discussed in this chapter.

DRY EYE PREVALENCE

Dry eye is a disease that affects between 4 and 34% of the population (Lekhanont et al. 2006, Schaumberg et al. 2009). The prevalence increases with age in both females and males (Moss et al. 2008, Schaumberg et al. 2003, Schaumberg et al. 2009) however, women are approximately 2.4 times more likely to develop the disease (Galor et al. 2011). It is estimated that roughly 5 million Americans, 3.23 million women and 1.68 million men over the age of 50 have dry eye (Schaumberg et al. 2003, Schaumberg et al. 2009). However, it has been suggested that an additional tens of millions of people have less severe symptoms of the disease, which only manifest during stressful conditions for the ocular surface, e.g. contact lens wear, airline travel or drafty environmental conditions (Dry Eye Workshop 2007b). The variety of definitions and criteria utilized to describe and diagnose the disease, as well as the different experimental designs used to obtain the data are factors likely to explain the large discrepancies in prevalence found

Ocular Surface Center, Department of Ophthalmology, Cullen Eye Institute, Baylor College of Medicine, Houston, Texas; 6565 Fannin Street, NC 205-Houston, TX 77030.
[a]E-mail: cintiadp@bcm.edu
*Corresponding author

between different studies. For example, some studies used a subjective method (questionnaires) whereas others used objective methods (clinical tests) or a combination of both.

Systemic diseases, e.g. Thyroid disease, Rheumatoid arthritis, Post traumatic stress disorder or different medications such as antihistamines, antidepressants, antianxiety medications, oral steroids, β-blockers, diuretics have been reported as risk factors for developing dry eye. Contact lens wear, refractive surgeries or a non balanced diet lacking in intakes of vitamins and fatty acids are other risk factors for developing this disease (Galor et al. 2011, Dry Eye Workshop 2007b, Uchino et al. 2011). A few studies have indicated an increased prevalence of dry eye in Asians (Lee et al. 2002, Lekhanont et al. 2006, Lin et al. 2003) and Japanese (Uchino et al. 2011) compared to Caucasian Americans (Schaumberg et al. 2003), however further studies are needed to confirm if ethnicity is a risk factor for developing dry eye.

Contact lens related dry eye is a significant problem for the contact lens wearer as well as for the contact lens industry. The discontinuation of contact lens use has been reported as one of the major contributing factors for the lack of increase in contact wearers in Europe. In fact, 40% of contact lens wearers' claim discomfort due to dryness as the main reason for the drop out (Young et al. 2002). According to a recent study, the waist majority of contact lens wearers report to experience dry or marginally dry eye symptoms at a reasonably high frequency and intensity at some point during the day (Young et al. 2011). Recent studies have shown that environmental factors such as visual display terminal (VDT) work, exposure to heaters or air conditioners in combination with contact lens wear increases the risks of developing dry eye (Kojima et al. 2011, Uchino et al. 2008).

Dry eye has been reported as one of the chief complaints after laser *in situ* keratomileusis (LASIK), affecting 50% of the patients one week after surgery (de Paiva et al. 2006). However, approximately 20% of the LASIK population still suffers from dry eye 6 months or more after the procedure (Levinson et al. 2008, Shoja et al. 2007). The risk of developing post LASIK dry eye has been reported to increase with the amount of myopic refractive error prior to surgery (Albietz et al. 2004, Shoja et al. 2007). A decreased corneal sensation as a consequence of the cut corneal nerves associated with decreased tear production, an alteration of corneal curvature (Battat et al. 2001) or damage to the goblet cells by the suction ring (Rodriguez-Prats et al. 2007) are some of the mechanisms suggested to be involved in dry eye after LASIK. LASIK induced Neurotrophic Epitheliopathy (LINE) —characterized by punctuate epithelial staining likely caused by damaged corneal nerves is another factor proposed to cause dry eye after LASIK (Wilson et al. 2001).

DEFINITION, SYMPTOMS AND CLINICAL FEATURES

Dry eye is defined as *"a multifactorial disease of the tears and ocular surface that results in symptoms of discomfort, visual disturbance and tear film instability, with potential damage to the ocular surface. It is accompanied with increased osmolarity of the tear film and inflammation of the ocular surface"* (Dry Eye Workshop 2007a).

There are many symptoms associated with dry eye disease. The chief complaints are dryness and ocular irritation however, a person suffering from dry eye might complain about eyes that are burning, stinging, tired, sore, scratchy, red or achy. Pain sensation, photophobia, blurry vision, watery eyes or swollen eyelids are other symptoms frequently reported from dry eye sufferers (Schaumberg et al. 2003).

Some clinical features observed in dry eye include increase in ocular staining, hyperemia and tear instability. Decreased tear meniscus height, tear break up time and corneal sensitivity are also commonly noted. A decrease in number of bulbar conjunctival goblet cells, loss of nasal-lacrimal reflex and tarsal (meibomian) gland orifice metaplasia are other signs reported in correlation with a dry eye (Afonso et al. 1999, Lee et al. 1997, Pflugfelder et al. 1997, Pflugfelder et al. 1998, Prabhasawat et al. 1998, Pult et al. 2011, Uchino et al. 2006). Elevated levels of inflammatory components such as pro-inflammatory cytokines interleukin (IL-1, IL-6), tumor necrosis factor (TNF)-α, immune activation molecules (ICAM-1), and human leucocyte antigen (HLA-DR) are other ocular surface alterations observed in the dry eye (Brignole et al. 2000, Jones et al. 1994, Pflugfelder et al. 1999, Solomon et al. 2001).

CATEGORIES

Dry eye disease can be divided into several main and subcategories (illustrated in the flow chart below). Tear or (aqueous) deficiency is caused by a decreased amount of tears available to lubricate the ocular surface. In evaporative dry eye, the tear production and secretion rate is normal however the tears evaporate (disappear) too quickly from the ocular surface (Fig. 1).

Aqueous or (tear) Deficiency

Tear deficient dry eye is divided into two subcategories, Sjögrens syndrome related and Non-Sjögrens syndrome related dry eye.

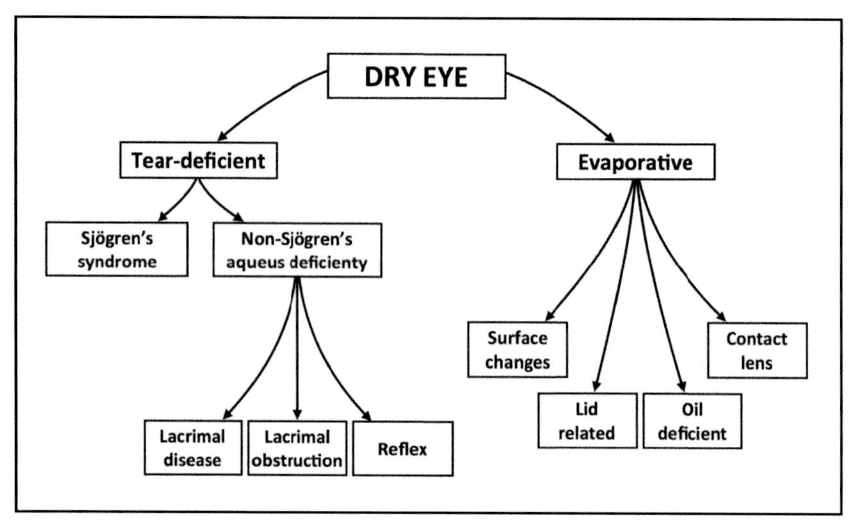

Figure 1 Flow chart illustrates the different categories of dry eye disease. Adapted from International Dry Eye Workshop.

Aqueous or (tear) Deficient Sjögrens Syndrome Related Dry Eye

Sjögrens syndrome is an autoimmune disease affecting the salivary and lacrimal glands, leading to clinically dry eye (keratoconjunctivitis sicca) and dry mouth (xerostomia) so-called primary Sjögrens syndrome. However, dry eye can also be present secondarily to other autoimmune diseases such as rheumatoid arthritis, so secondary Sjögrens syndrome (Vitali et al. 2002). Very limited epidemiological information is available in regards to Sjögrens syndrome. The prevalence have been reported to be between 0.1–3.0% depending on the study (Alamanos et al. 2006, Bowman et al. 2004). The disease has been reported in all age groups; however, the first symptoms normally occur between 40–60 years of age. The disease has a higher prevalence in women and increases with age (Jonsson et al. 2011). Female gender and first grade relatives diagnosed with an autoimmune disease have been identified as risk factors for developing the disease (Priori et al. 2007).

Aqueous or (tear) Deficient Non-Sjögrens Syndrome Related Dry Eye

Diseases of the lacrimal glands may lead to aqueous deficient Non-Sjögrens syndrome related dry eye. These diseases result in decreased tear production and can be due to congenital conditions or age related. Age has been reported to play an important role as an increase in pathologies of the lacrimal ducts has been noted in an aging population. In some systemic diseases such as sarcoidosis, lymphoma or AIDS the lacrimal gland may get infiltrated with inflammatory cells and thus, the inflammation eventually leading to failed lacrimal secretion (Dry Eye Workshop 2007a).

Obstructions of the main, palpebral and accessory lacrimal gland ducts are other factors that can result in Non-Sjögrens syndrome related aqueous deficient dry eye. These obstructions are mainly due to diseases, e.g. trachoma (a bacterial infection causing inflammation of the conjunctiva leading to tarsal gland obstruction and severe conjunctival scaring) or cicatricial and mucous membrane pemphigoid (diseases causing severe blistering of the conjunctiva that can potentially lead to blindness) (Dry Eye Workshop 2007a).

In the open eye there is an increased reflex sensory input from the ocular surface. A decrease in sensory input decreases both the sensory input to the lacrimal glands and the blink rate, leading to increased evaporation from the ocular surface and evaporative dry eye as a consequence (Battat et al. 2001). This can be caused by for example contact lens wear or diabetes mellitus both which have proven to decrease corneal sensitivity (Kaiserman et al. 2005, Situ et al. 2010).

The lacrimal gland is innervated by parasympathetic fibers from CN VII (facial nerve). A motor block in this pathway can cause hyposecretion from the lacrimal gland and lagophthalmos (incomplete lid closure) which also results in aqueous or (tear) deficient non-Sjögrens syndrome related dry eye. Systemic drugs, e.g. antihistamines, beta-blockers; antidepressants are other factors that can cause aqueous deficiency and dry eye (Galor et al. 2011, Uchino et al. 2011).

Evaporative Dry Eye

An abundance of factors can cause evaporative dry eye; oil (lipid) deficiency, lid related syndromes, contact lens or ocular surface changes.

The tarsal (meibomian glands) are sebaceous glands, producing an oily (lipid secretion), contributing to the most superficial layer of the tear film. This layer's main function is to, decrease evaporation from the surface. 30–40 glands are located in the superior lid and 20–30 in the inferior lid (Bergmanson 2010). A dysfunction in their secretion, either due to obstruction of the orifices or lack of production causes a deficient lipid layer of the tear film, leading to evaporative dry eye. Not properly functioning tarsal (meibomian) glands is one cause for evaporative dry eye (Tomlinson et al. 2011). A reduced number of glands due to congenital disorders, blepharitis, acne rosacea, psoriasis are some other causes of tarsal (meibomian) gland dysfunctions (Tomlinson et al. 2011).

Another factor that has the potential to cause evaporative dry eye are lid related disorders. These can be due to endocrine disorders, e.g. hyperthyroidism causing proptosis (bulging of the eye) or any kind of lid deformity. These diseases are in one way or another increasing the area of the ocular surface exposed to the environment and thus causing a higher rate of evaporation.

Contact lens wear is also involved in causing evaporative dry eye. A few of the mechanisms potentially create evaporative dry eye are, pre-lens tear film thinning (faster thinning of the tear film with dry eye as a result), lens wettability (a lower wettability is correlated with a dryer eye) or reduced corneal sensitivity (reduced sensitivity leading to lower blink frequency and a dry eye) (Cheng et al. 2004, Nichols et al. 2005, Situ et al. 2010).

Changes to the normal structure of the ocular surface are yet another factor that plays an important role in evaporative dry eye. Chronic ocular surface diseases and allergies, e.g. conjunctivitis are two examples that alters the normal homogenizes of the ocular surface with a destabilized tear film and a dry eye as a result. Mucous deficiency is another example. The mucin produced by the epithelial and goblet cells, forms the inner most layer of the tear film and functions as a wetting agent, aiding in spreading the tears over the ocular surface (Bergmanson 2011). In for example vitamin A deficiency and Sjögrens syndrome a loss of goblet cells has been reported (Pflugfelder et al. 1990, Wittpenn et al. 1986). Thus, a lack of mucin leads to a non-smooth ocular surface where the tears don't spread as easily, with a faster break up time and a dry eye as a result. Tear film instability is yet another example. The lacrimal fluid is in its normal state hypotonic (a liquid that contains less solute and more water). In cases of excess evaporation (evaporative dry eye) the fluid becomes hypertonic. Hypertonicity is considered the major contributor to ocular surface inflammation with tear film instability as a result. However, tear film stability can also be the initiating factor to a dry eye. If the tear break up time is less than the blink frequency local drying and hyperosmolarity occurs thus creating tear film instability with dry eye as a consequence. The latter example can be caused by a disturbance of the ocular surface mucins (Dry Eye Workshop 2007a).

DIAGNOSIS

The biggest challenge when diagnosing a dry eye is that the disease has a very low correlation between objective signs (clinical tests, appearance) and the subjective symptoms reported by patients (Hay et al. 1998, McCarty et al. 1998, Schein et al. 1997) especially for a mild or moderately dry eye (Narayanan et al. 2005). To aid to the challenge in the diagnosis, not all symptoms are present at all times, a whole battery of different tests are needed and to date there is no consensus of which tests to utilize (Korb 2000, Turner et al. 2005).

The clinical tests that are commonly utilized to determine and diagnose a dry eye are described below:

Questionnaires

Clinical history and questionnaires the most reliable and repeatable methods available today for dry eye diagnosis (Dry Eye Workshop 2007b). Today, there are a number of questionnaires available for diagnosis; which all varies in length, intended use and validity. Therefore, it is important for every practitioner to find a questionnaire that works in their practice and once they have done so to be consistent, since different questionnaires potentially could yield different diagnosis.

Quantitative Tests

The tests included in this category measures ***quantity*** of tears.

***Schirmer test*:** A paper strip is placed below the lower lid and the amount of wetting that occurs over a 5 minutes period is evaluated. Result of less than 15 mm/5 minutes is considered abnormal. This test can be performed without (Schirmer 1) or with (Schirmer II) anesthetic. One drawback with the Schirmer test is that the paper strip inserted below the lid is quite uncomfortable and thus can stimulate to reflex tearing (Bergmanson 2011, Hom et al. 2000).

***Phenol red thread (cotton thread test)*:** This test is similar to the Schirmer test in regards to measuring quantity of tear production, however instead of a paper strip a cotton thread is inserted under the lower lid. The thread is impregnated with phenol red and will change color from yellow to red due to the pH of the tears. Wetting of less than 20 mm/15 seconds is considered abnormal. The thread is much less irritating than the paper strip used in the Schirmer test and thus the phenol red thread test is considered to measure basal tear secretion and be much more reliable (Bergmanson 2011).

***Tear meniscus (tear prims) height*:** Is measured at lower lid margin in a biomicroscope (slit lamp). A height of less than 0.75 mm is considered abnormal (Bergmanson 2011).

***Anterior segment optical coherence tomography (OCT)*:** OCT is an instrument utilized for high resolution *in vivo* ocular measurements. A recent study has shown that the OCT can be a valuable tool in tear meniscus height measurements because dry eye patients had a small tear height (Gumus et al. 2010, Koh et al. 2010).

Evaporimetry: An evaporimeter is used to measure tear evaporation rate (how quickly the tears disappear) from the ocular surface (Guillon et al. 2010).

Qualitative Tests

The tests included in this category measures ***quality*** of tears.

Tear break up time (BUT): This test measures how fast the tears break up. Sodium flourescein is installed to the eye and the break up time is measured in a slit lamp. A yellow wratten 12 filter together with blue light is used to enhance the visualization of the stain.

Noninvasive tear break up time (NIBUT): Tear break up time can also be measured without installing fluorescein. This test is often evaluated looking at the keratometer mires. Longer than 10 seconds is considered normal for both TBUT and NIBUT (Bergmanson 2011).

Tear clearance: This test is utilized to investigate how long it takes for sodium fluorescein to disappear from the ocular surface and can be performed by various methods. One option is to investigate how quickly the dye reaches a Q tip placed in the nasal cavity (Jones 1 test) (Jones 1966). Another option is to evaluate the time the sodium fluorescein stays on the ocular surface as described in TBUT above using Schirmer strips (Macri et al. 2000).

Tear osmolarity: Tear osmolarity is measured with an osmometer. Values between 295–310 mOsm/kg are considered normal and values over 320 mOsm/kg are considered severely dry eye (Bergmanson 2011) and it has correlated with signs and symptoms of dry eye (Moore et al. 2011).

Tear thinning time: Tear thinning time is measured with an interferometer *in vivo*. This instrument measures wavelength dependent interference between surfaces, e.g. between the cornea and the tear film and from these values a tear thinning time is calculated. This methodology is not readily available in a clinical setting and is therefore currently only utilized in research settings (Nichols et al. 2003, Nichols et al. 2005).

High performance liquid chromatography: This technique utilizes a chromatographer, an instrument capable to identify, quantify and purify individual components in a mixture. The chromatographer has shown to be beneficial when investing changes in tear protein patterns. A recent study have shown that this instrument can be utilized in dry eye diagnosis, however at this point this technique is mainly utilized in research settings (Grus et al. 2001).

Ocular Surface Health

Test included in this category assesses the ***health and integrity*** of the ocular surface.

Biomicroscopy (Slit lamp): With its different types of illuminations and magnifications the slit lamp offers the observer a very detailed noninvasive thorough health assessment of the anterior segment.

Hyperemia assessment: The hyperemia or redness assessment is performed with a slit lamp with the usage of an artist-rendered or photographic grading scales (Efron et al. 2001).

Sodium Fluorescein: Fluorescein is a synthetic organic dye, utilized in combination with a slit lamp with blue light and a wratten 12 filter to examine corneal integrity. Loss or damaged epithelial cells will be visualized (called staining) and graded after severity. Fluorescein is most commonly dispensed in strips that are wetted with saline before instillation into the eye. Thus, the amount of dye used differs slightly between patients and this has the possibility to impact the reproducibility of the staining. It is also important to be aware of the molecular weight (high versus low) of these two variations of fluorescein as they will stain the anterior segment with different intensities. The grading of the intensity of the staining can vary between observers but is also dependent on which grading scale that was used for the assessment (Efron et al. 2001). Nevertheless, corneal fluorescein staining scores have been used in clinical trials of dry eye (Sall et al. 2000).

Rose bengal and lissamine green staining: In a slit lamp with white light, rose bengal will stain dead and damaged cells red and lissamine green will stain these cells green. Rose bengal is painful upon installation and, is therefore, seldom used today. Lissamine green is not painful upon installation and has become the most commonly used stain to aid in dry eye diagnosis. While rose bengal can stain cornea and conjunctiva, lissamine green staining is often observed in conjunctiva and at the lid margins of eyelid (Fig. 2).

Lid wiper epitheliopathy: The lid wiper is defined as the portion of the superior lid that wipes the ocular surface during blinking (Korb et al. 2002). This portion of the lid can be stained with fluorescein, rose bengal or lissamine green and the extent of the stain evaluated and graded (Korb et al. 2002, Korb et al. 2005, Korb et al. 2010). This test has shown to be useful in diagnosing patients who reported dry eye symptoms, and yet other tests were negative (Korb et al. 2010).

Impression cytology: This is a technique utilized to determine the cellular composition and/or cellular health of the conjunctiva. A piece of filter paper

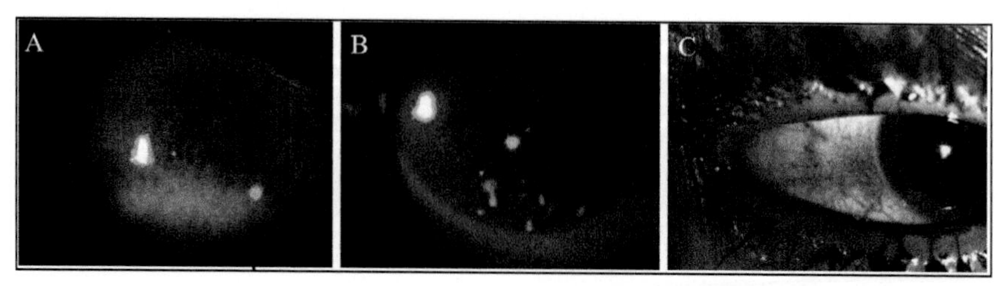

Figure 2 This figure shows vital dye staining of cornea and conjunctiva: (A-B) Two distinctive corneal fluorescein patterns seen in dry eye patients. (A) Severe inferior staining with diffusion and areas of coalescence of dye. (B) Filamentary keratitis. (C) Lissamine green staining showing intense diffuse exposure zone staining.

Color image of this figure appears in the color plate section at the end of the book.

is gently pressed against the conjunctival surface for a few seconds and then removed. Epithelial cells adhere to the filter paper during the application and removal process. After removal the filter paper can be fixed, stained and morphologically evaluated.

Impression cytology has been used to collect cells used for flow cytometry analysis since upregulation of MHC class II is correlated with severity of disease and also for evaluation of gene expression for inflammatory cytokines and matrix metalloproteinases (Baudouin et al. 2002, Brignole et al. 2000, de Paiva et al. 2009, Pisella et al. 2000).

GRADING SEVERITY

Dry eye disease is commonly divided into different levels, e.g. mild (occasional symptoms), moderate (symptoms more frequently) or severe (symptoms on a regular basis and often disabling). The different levels of the disease display a diverse degree of symptoms and outcomes of clinical tests. International Dry Eye Workshop and the Delphi Panel has therefore, suggested that the disease should be divided into categories depending on symptom severity and clinical test results to aid in diagnosis and treatment of the disease (Behrens et al. 2006, Dry Eye Workshop 2007a). The symptoms and expected clinical test results for each level of the disease are described and summarized in table 1:

- **Level 1:** Patients might report occasional symptoms mainly in environments stressful to the ocular surface, e.g. in dry office environments or while working in front of a computer. Test results will appear normal.

Table 1 Classification of dry eye severity according to signs and symptoms. Adapted from International Dry Eye WorkShop (Dry Eye Workshop 2007a). (mm = millimetres; min = minutes; s = seconds).

Severity	1	2	3	4
Discomfort	Mild (affected by environmental)	Moderate	Severe	Severe and disabling
Impact on vision	None	Sometimes	Frequent	Frequent and disabling
Conjunctival Hyperemia	None to mild	Mild if any	Moderate	Severe
Corneal staining	Zero to mild	Mild	Moderate often central	Severe and punctate
Meibomian gland dysfunction	Asymptomatic	Mild	Moderate	Severe
Tear break up time	10–30 s	≤ 10 s	5–9 s	< 5 s
Tear volume (Schirmer)	> 15 mm/5min	≤ 10 mm/5min	≤ 5 mm/5min	Immediate
Phenol red thread	>20 mm/15 s	15–20 mm/15 s	<15 mm/15 s	<15 mm/15 s

- **Level 2:** At this level of the disease symptoms are reported more frequently and on a more regular basis. Some test results might give an indication of dry eye disease but the values can also appear normal.
- **Level 3:** At this point of the disease progress the patients will report severe discomfort that frequently impacts their vision. Clinical test results will appear lower than normal values.
- **Level 4:** This is a severely dry eye. At this level the patient constantly suffers from dry eyes and the condition is often disabling. Test scores will appear abnormally low.

CONCLUSIONS

In summary, dry eye is a multifactorial disease that can be divided into several main and sub categories. The disease affects millions of people worldwide. Prevalence increases with age and women are more prone to develop the disease. Environmental factors, e.g. dry climate, computer work or contact lens wear increases the risk to develop the disease. Due to the low correlation between signs and symptoms a whole battery of tests are utilized to diagnose a dry eye and today there is little consensus on which combination of tests to use for an accurate diagnosis. Although research has improved our knowledge about dry eye over the last decades there is still a long way to go.

REFERENCES

Afonso, A.A., D. Monroy, M.E. Stern, W.J. Feuer, S.C. Tseng and S.C. Pflugfelder. 1999. Correlation of tear fluorescein clearance and Schirmer test scores with ocular irritation symptoms. Ophthalmology 106: 803–810.

Alamanos, Y., N. Tsifetaki, P.V. Voulgari, A.I. Venetsanopoulou, C. Siozos and A.A. Drosos. 2006. Epidemiology of primary Sjogren's syndrome in north-west Greece, 1982–2003 Rheumatology (Oxford) 45: 187–191.

Albietz, J.M., L.M. Lenton and S.G. McLennan. 2004. Chronic dry eye and regression after laser *in situ* keratomileusis for myopia. J. Cataract Refract. Surg. 30: 675–684.

Battat, L., A. Macri, D. Dursun and S.C. Pflugfelder. 2001. Effects of laser *in situ* keratomileusis on tear production, clearance and the ocular surface. Ophthalmology 108: 1230–1235.

Baudouin, C., F. Brignole, P.J. Pisella, M.S. De Jean and A. Goguel. 2002. Flow cytometric analysis of the inflammatory marker HLA DR in dry eye syndrome: results from 12 months of randomized treatment with topical cyclosporin A Adv. Exp. Med. Biol. 506: 761–769.

Behrens, A., J.J. Doyle, L. Stern, R.S. Chuck and P.J. McDonnell. 2006. Dysfunctional Tear Syndrome: A Delphi Approach to Treatment Recommendations. Cornea 25: 900–907.

Bergmanson, J.P.G. 2011. Clinical Ocular Anatomy and Physiology, 18th edition, Houston, TX, Texas. Eye Research and Technology Center.

Bowman, S.J., G.H. Ibrahim, G. Holmes, J. Hamburger and J.R. Ainsworth. 2004. Estimating the prevalence among Caucasian women of primary Sjogren's syndrome in two general practices in Birmingham, UK. Scand. J. Rheumatol. 33: 39–43.

Brignole, F., P.J. Pisella, M. Goldschild, J.M. De Saint, A. Goguel and C. Baudouin. 2000. Flow cytometric analysis of inflammatory markers in conjunctival epithelial cells of patients with dry eyes. Invest Ophthalmol. Vis. Sci. 41: 1356–1363.

Cheng, L., S.J. Muller and C.J. Radke. 2004. Wettability of silicone-hydrogel contact lenses in the presence of tear-film components. Curr. Eye Res. 28: 93–108.

de Paiva, C.S., S. Chotikavanich, S.B. Pangelinan, J.I. Pitcher, B. Fang, X. Zheng, P. Ma, W.J. Farley, K.S. Siemasko, J.Y. Niederkorn, M.E. Stern, Li D-Q and S.C. Pflugfelder. 2009. IL-17 disrupts corneal barrier following desiccating stress. Mucosal Immunology 2: 243–53.

de Paiva, C.S., Z. Chen, D.D. Koch, M.B. Hamill, F.K. Manuel, S.S. Hassan, K.R. Wilhelmus and S.C. Pflugfelder. 2006. The incidence and risk factors for developing dry eye after myopic LASIK. Am. J. Ophthalmol. 141: 438–445.

Dry Eye Workshop. 2007a. The definition and classification of dry eye disease: report of definition and classification subcommittee of the international dry eye workshop in: Report of the international dry eye workshop (DEWS) The ocular surface 5: 75–92.

Dry Eye Workshop. 2007b. The epidermiology of dry eye disease: Report of the epidermiology subcommittee of the international dry eye workshop in: Report of the international dry eye workshop (DEWS) The ocular surface 5: 93–107.

Efron, N., P.B. Morgan and S.S. Katsara. 2001. Validation of grading scales for contact lens complications. Ophthalmic Physiol. Opt. 21: 17–29.

Galor, A., W. Feuer, D.J. Lee, H. Florez, D. Carter, B. Pouyeh, W.J. Prunty and V.L. Perez. 2011. Prevalence and risk factors of dry eye syndrome in a United States veterans affairs population. Am. J. Ophthalmol. 152: 377–384.

Grus, F.H. and A.J. Augustin. 2001. High performance liquid chromatography analysis of tear protein patterns in diabetic and non-diabetic dry-eye patients. Eur. J. Ophthalmol. 11: 19–24.

Guillon, M. and C. Maissa. 2010. Tear film evaporation-effect of age and gender Cont. Lens Anterior Eye. 33: 171–175.

Gumus, K., C.H. Crockett and S.C. Pflugfelder. 2010. Anterior segment optical coherence tomography: a diagnostic instrument for conjunctivochalasis. Am. J. Ophthalmol. 150: 798–806.

Hay, E.M., E. Thomas, B. Pal, A. Hajeer, H. Chambers and A.J. Silman. 1998. Weak association between subjective symptoms or and objective testing for dry eyes and dry mouth: results from a population based study. Ann. Rheum. Dis. 57: 20–24.

Hom, M.M. and J.P. Shovlin. 2000. Manual of Contact Lens Prescribing and Fitting with CD-ROM, 2nd edition, Boston, MA, Butterworth Heinemann. USA.

Jones, D.T., D. Monroy, Z. Ji, S.S. Atherton and S.C. Pflugfelder. 1994. Sjogren's syndrome: cytokine and Epstein-Barr viral gene expression within the conjunctival epithelium. Invest Ophthalmol. Vis. Sci. 35: 3493–3504.

Jones, L.T. 1966. The lacrimal secretory system and its treatment Am. J. Ophthalmol. 62: 47–60.

Jonsson, R., P. Vogelsang, R. Volchenkov, A. Espinosa, M. Wahren-Herlenius and S. Appel. 2011. The complexity of Sjögren's syndrome: Novel aspects on pathogenesis. Immunol. Lett. 141: 1–9.

Kaiserman, I., N. Kaiserman, S. Nakar and S. Vinker. 2005. Dry eye in diabetic patients Am. J. Ophthalmol. 139: 498–503.

Koh, S., C. Tung, J. Aquavella, R. Yadav, J. Zavislan and G. Yoon. 2010. Simultaneous measurement of tear film dynamics using wavefront sensor and optical coherence tomography. Invest Ophthalmol. Vis. Sci. 51: 3441–3448.

Kojima, T., O.M. Ibrahim, T. Wakamatsu, A. Tsuyama, J. Ogawa, Y. Matsumoto, M. Dogru and K. Tsubota. 2011. The Impact of Contact Lens Wear and Visual Display Terminal Work on Ocular Surface and Tear Functions in Office Workers. Am. J. Ophthalmol. 152: 933–940.

Korb, D.R. 2000. Survey of preferred tests for diagnosis of the tear film and dry eye. Cornea 19: 483–486.

Korb, D.R., J.P. Herman, C.A. Blackie, R.C. Scaffidi, J.V. Greiner, J.M. Exford and V.M. Finnemore. 2010. Prevalence of lid wiper epitheliopathy in subjects with dry eye signs and symptoms. Cornea 29: 377–383.

Korb, D.R., J.P. Herman, J.V. Greiner, R.C. Scaffidi, V.M. Finnemore, J.M. Exford, C.A. Blackie and T. Douglass. 2005. Lid wiper epitheliopathy and dry eye symptoms. Eye Contact Lens 31: 2–8.

Korb, D.R., J.V. Greiner, J.P. Herman, E. Hebert, V.M. Finnemore, J.M. Exford, T. Glonek and M.C. Olson. 2002. Lid-wiper epitheliopathy and dry-eye symptoms in contact lens wearers. CLAO J. 28: 211–216.

Lee, A.J., J. Lee, S.M. Saw, G. Gazzard, D. Koh, D. Widjaja and D.T. Tan. 2002. Prevalence and risk factors associated with dry eye symptoms: a population based study in Indonesia. Br. J. Ophthalmol. 86: 1347–1351.

Lee, S.H. and S.C. Tseng. 1997. Rose bengal staining and cytologic characteristics associated with lipid tear deficiency. Am. J. Ophthalmol. 124: 736–750.

Lekhanont, K., D. Rojanaporn, R.S. Chuck and A. Vongthongsri. 2006. Prevalence of dry eye in Bangkok, Thailand. Cornea 25: 1162–1167.

Levinson, B.A., C.J. Rapuano, E.J. Cohen, K.M. Hammersmith, B.D. Ayres and P.R. Laibson. 2008. Referrals to the Wills Eye Institute Cornea Service after laser *in situ* keratomileusis: reasons for patient dissatisfaction. J. Cataract Refract Surg. 34: 32–39.

Lin, P.Y., S.Y. Tsai, C.Y. Cheng, J.H. Liu, P. Chou and W.M. Hsu. 2003. Prevalence of dry eye among an elderly Chinese population in Taiwan: the Shihpai Eye Study. Ophthalmology. 110: 1096–1101.

Macri, A., M. Rolando and S. Pflugfelder. 2000. A standardized visual scale for evaluation of tear fluorescein clearance. Ophthalmology 107: 1338–1343.

McCarty, C.A., A.K. Bansal, P.M. Livingston, Y.L. Stanislavsky and H.R. Taylor. 1998. The epidemiology of dry eye in Melbourne, Australia. Ophthalmology 105: 1114–1119.

Moore, J.E., G.T. Vasey, D.A. Dartt, V.E. McGilligan, S.D. Atkinson, C. Grills, P.J. Lamey, A. Leccisotti, D.G. Frazer and T.C. Moore. 2011. Effect of tear hyperosmolarity and signs of clinical ocular surface pathology upon conjunctival goblet cell function in the human ocular surface. Invest Ophthalmol. Vis. Sci. 52: 6174–6180.

Moss, S.E., R. Klein and B.E. Klein. 2008. Long-term incidence of dry eye in an older population. Optom. Vis. Sci. 85: 668–674.

Narayanan, S., W.L. Miller, T.C. Prager, J.A. Jackson, N.E. Leach, A.M. McDermott, M.T. Christensen and J.P. Bergmanson. 2005. The diagnosis and characteristics of moderate dry eye in non-contact lens wearers. Eye Contact Lens 31: 96–104.

Nichols, J.J. and P.E. King-Smith. 2003. Thickness of the pre- and post-contact lens tear film measured *in vivo* by interferometry. Invest Ophthalmol. Vis. Sci. 44: 68–77.

Nichols, J.J., G.L. Mitchell and P.E. King-Smith. 2005. Thinning rate of the precorneal and prelens tear films. Invest Ophthalmol. Vis. Sci. 46: 2353–2361.

Pflugfelder, S.C., A.J.W. Huang, P.T. Schuchovski, I.C. Pereira and S.C.G. Tseng. 1990. Conjunctival cytological features of primary Sjögren syndrome. Ophthalmology 97: 985–991.

Pflugfelder, S.C., D. Jones, Z. Ji, A. Afonso and D. Monroy. 1999. Altered cytokine balance in the tear fluid and conjunctiva of patients with Sjögren's syndrome keratoconjunctivitis sicca. Curr. Eye Res. 19: 201–211.

Pflugfelder, S.C., S.C.G. Tseng, K. Yoshino, D. Monroy, C. Felix and B.L. Reis. 1997. Correlation of goblet cell density and mucosal epithelial membrane mucin expression with rose bengal staining in patients with ocular irritation. Ophthalmology 104: 223–235.

Pflugfelder, S.C., S.C.G. Tseng, O. Sanabria, H. Kell, C.G. Garcia, C. Felix, W. Feuer and B.L. Reis. 1998. Evaluation of subjective assessments and objective diagnostic tests for diagnosing tear-film disorders known to cause ocular irritation. Cornea 17: 38–56.

Pisella, P.J., F. Brignole, C. Debbasch, P.A. Lozato, C. Creuzot-Garcher, J. Bara, P. Saiag, J.M. Warnet and C. Baudouin. 2000. Flow cytometric analysis of conjunctival epithelium in ocular rosacea and keratoconjunctivitis sicca. Ophthalmology 107: 1841–1849.

Prabhasawat, P. and S.C.G. Tseng. 1998. Frequent association of delayed tear clearance in ocular irritation Br. J. Ophthalmol. 182: 666–675.

Priori, R., E. Medda, F. Conti, E.A. Cassara, M.G. Sabbadini, C.M. Antonioli, R. Gerli, M.G. Danieli, R. Giacomelli, M. Pietrogrande, G. Valesini and M.A. Stazi. 2007. Risk factors for Sjögren's syndrome: a case-control study. Clin. Exp. Rheumatol. 25: 378–384.

Pult, H., C. Purslow and P.J. Murphy. 2011. The relationship between clinical signs and dry eye symptoms Eye 25: 502–510.

Rodriguez-Prats, J.L., I.M. Hamdi, A.E. Rodriguez, A. Galal and J.L. Alio. 2007. Effect of suction ring application during LASIK on goblet cell density. J. Refract. Surg. 23: 559–562.

Sall, K., O.D. Stevenson, T.K. Mundorf and B.L. Reis. 2000. Two multicenter, randomized studies of the efficacy and safety of cyclosporine ophthalmic emulsion in moderate to severe dry eye disease. CsA Phase 3 Study Group. Ophthalmology 107: 631–639.

Schaumberg, D.A., D.A. Sullivan, J.E. Buring and M.R. Dana. 2003. Prevalence of dry eye syndrome among US women. Am. J. Ophthalmol. 136: 318–326.

Schaumberg, D.A., R. Dana, J.E. Buring and D.A. Sullivan. 2009. Prevalence of dry eye disease among US men: estimates from the Physicians' Health Studies. Arch. Ophthalmol. 127: 763–768.

Schein, O.D., J.M. Tiesch, B. Munõz, K. Bandeen-Roche and S. West. 1997. Relation between signs and symptoms of dry eye in the elderly. A population-based perspective. Ophthalmology 104: 1395–1401.

Shoja, M.R. and M.R. Besharati. 2007. Dry eye after LASIK for myopia: Incidence and risk factors. Eur. J. Ophthalmol. 17: 1–6.

Situ, P., T.L. Simpson, L.W. Jones and D. Fonn. 2010. Effects of silicone hydrogel contact lens wear on ocular surface sensitivity to tactile, pneumatic mechanical and chemical stimulation. Invest Ophthalmol. Vis. Sci. 51: 6111–6117.

Solomon, A., D. Dursun, Z. Liu, Y. Xie, A. Macri and S.C. Pflugfelder. 2001. Pro- and anti-inflammatory forms of interleukin-1 in the tear fluid and conjunctiva of patients with dry-eye disease. Invest Ophthalmol. Vis. Sci. 42: 2283–2292.

Tomlinson, A., A.J. Bron, D.R. Korb, S. Amano, J.R. Paugh, E.I. Pearce, R. Yee, N. Yokoi, R. Arita and M. Dogru. 2011. The international workshop on meibomian gland dysfunction: report of the diagnosis subcommittee. Invest Ophthalmol. Vis. Sci. 52: 2006–2049.

Turner, A.W., C.J. Layton and A.J. Bron. 2005. Survey of eye practitioners' attitudes towards diagnostic tests and therapies for dry eye disease. Clin. Experiment Ophthalmol. 33: 351–355.

Uchino, M., D.A. Schaumberg, M. Dogru, Y. Uchino, K. Fukagawa, S. Shimmura, T. Satoh, T. Takebayashi and K. Tsubota. 2008. Prevalence of dry-eye disease among Japanese visual display terminal users. Ophthalmology 115: 1982–1988.

Uchino, M., M. Dogru, Y. Yagi, E. Goto, M. Tomita, T. Kon, M. Saiki, Y. Matsumoto, Y. Uchino, N. Yokoi, S. Kinoshita and K. Tsubota. 2006. The features of dry eye disease in a Japanese elderly population. Optom. Vis. Sci. 83: 797–802.

Uchino, M., Y. Nishiwaki, T. Michikawa, K. Shirakawa, E. Kuwahara, M. Yamada, M. Dogru, D.A. Schaumberg, T. Kawakita, T. Takebayashi and K. Tsubota. 2011. Prevalence and Risk Factors of Dry Eye Disease in Japan: Koumi Study. Ophthalmology 118: 2361–2367.

Vitali, C., S. Bombardieri, R. Jonsson, H.M. Moutsopoulos, E.L. Alexander, S.E. Carsons, T.E. Daniels, P.C. Fox, R.I. Fox, S.S. Kassan, S.R. Pillemer, N. Talal and M.H. Weisman. 2002. Classification criteria for Sjögren's syndrome: a revised version of the European criteria proposed by the American-European Consensus Group. Ann. Rheum. Dis. 61: 554–558.

Wilson, S.E. and R. Ambrosio. 2001. Laser *in situ* keratomileusis-induced neurotrophic epitheliopathy. Am. J. Ophthalmol. 132: 405–406.

Wittpenn, J.R., S.C. Tseng and A. Sommer. 1986. Detection of early xerophthalmia by impression cytology. Arch. Ophthalmol. 104: 237–239.

Young, G., J. Veys, N. Pritchard and S. Coleman. 2002. A multi-centre study of lapsed contact lens wearers. Ophthalmic Physiol. Opt. 22: 516–527.

Young, G., R.L. Chalmers, L. Napier, C. Hunt and J. Kern. 2011. Characterizing contact lens-related dryness symptoms in a cross-section of UK soft lens wearers. Cont. Lens Anterior. Eye 34: 64–70.

5

Anterior and Posterior Blepharitis

Taliana Freitas Bernardes[a,*] and *Adriana Alvim Bonfioli*[b]

SUMMARY

Blepharitis is a chronic inflammatory process of the eyelid margin. It is a common eye disorder throughout the world and can affect any age group. Many eye care professionals don't recognize the importance of the disease and its long-term consequences. The etiology of blepharitis is unknown and probably multifactorial. Common symptoms associated with blepharitis are burning sensation, irritation, tearing, photophobia, blurred vision, and red eyes. Clinical examination reveals the presence of scurf, telangiectatic vascular changes of the eyelid margin, inspissated meibomian glands, conjuntival hyperemia, punctuate keratopathy, corneal vascularization and ulceration. Basic treatment for blepharitis includes a long-term commitment to eyelid hygiene. Topical lubricants, antibiotics, corticosteroids and imunomodulators may be necessary in selected cases.

INTRODUCTION

The term blepharitis refers to the inflammation of the eyelid. It is a common eye disorder frequently misdiagnosed and untreated. Many eye care professionals don't recognize the importance of the disease and its irreversible long-term consequences. Some complications of chronic blepharitis are tear film instability and dry eye, lid morphologic changes and corneal abnormalities. The etiology of blepharitis is unknown and

Oculoplastics Department, Visio Clinica, Address Av. do Contorno 4747 suite 1705, Belo Horizonte, Minas Gerais, Brazil, Zip Code: CEP 30110-090.
[a]E-mail: taliana.freitas@gmail.com
[b]E-mail: dribonfioli@gmail.com
*Corresponding author

probably multifactorial (Dougherty and McCulley 1984, Oto et al. 1998). The treatment is prolonged and recurrences are common.

ANATOMY

The eyelid margin is divided in anterior and posterior portions by the grey line that represents the marginal region of the orbicularis muscle seen through the skin (Fig. 1). The Meibomian glands are arranged in parallel in a single row throughout the length of the tarsal plate. The number of separate glands in the upper lid varies from 25 to 40 and in the lower lid from 20 to 30 (Knop et al. 2011). The ducts of the Meibomian glands open onto the lid margin posterior to the grey line and anterior to the mucocutaneous junction. The secretion of the glands is controlled by neural and hormonal factors. The delivery of the secretion is related to the mechanical action of the lid muscles during blinking.

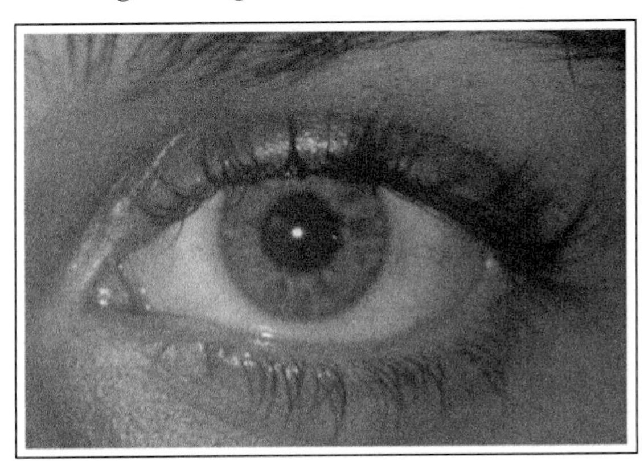

Figure 1 The eyelid margin is subdivided in anterior and posterior by the grey line.

Color image of this figure appears in the color plate section at the end of the book.

Meibomian secretion forms the superficial layer of the tear film and is responsible for stabilizing and protecting it against evaporation and outflow. It lubricates the lid margins avoiding maceration when in contact with the tear film, protects against contamination and provides a regular optical surface at the cornea.

TERMINOLOGY

The term **blepharitis** refers to inflammation of the lid as a whole (Fig. 2). **Marginal blepharitis** is inflammation of the lid margin and includes anterior and posterior blepharitis. **Anterior blepharitis** describes inflammation of the margin anterior to the grey line and **posterior blepharitis** describes

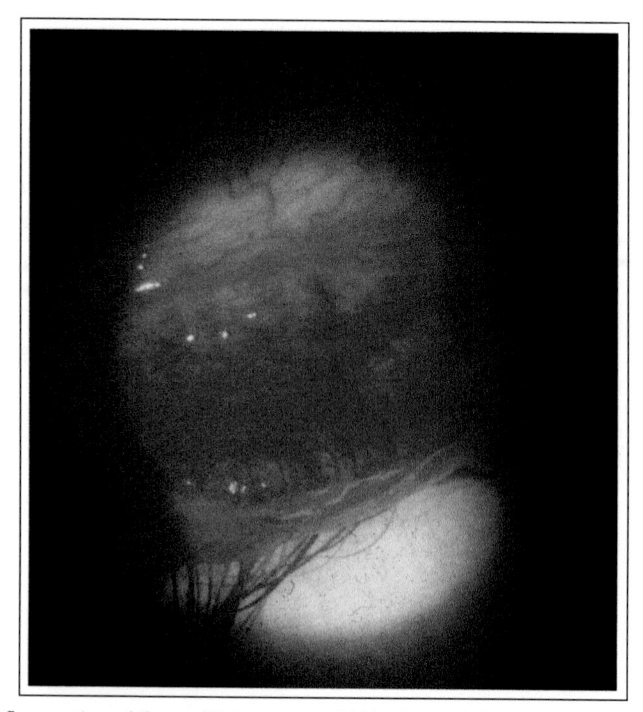

Figure 2 Inflammation of the eyelid (courtesy of Alfredo Bonfioli).

Color image of this figure appears in the color plate section at the end of the book.

inflammatory conditions of the posterior lid margin including **Meibomian gland dysfunction** (MGD).

Some authors use the terms posterior blepharitis and MGD as synonyms, which they are not. MGD is one of the causes of posterior blepharitis as well as infectious or allergic conjunctivitis, acne rosacea, and others (Nelson et al. 2011). There is also considerable overlap between anterior and posterior blepharitis.

MGD refers to functional abnormalities of the Meibomian glands associated with dry eye disease. The term meibomitis describe the type of MGD associated with inflammation of the Meibomian glands.

CLASSIFICATION

There are innumerous classifications of blepharitis based in anatomic, etiological and clinical criteria. The authors use the classification described in table 1 for its correlation with treatment protocols. Anterior blepharitis is usually infectious or seborrheic and posterior blepharitis (meibomitis) is commonly metabolic.

Table 1 This table summarize the Classification of chronic blepharitis.

Infectious blepharitis: bacterial, viral, fungal or parasitic.
Seborrheic blepharitis: associated with increased secretion of Meibomian lipids (hypersecretory).
Meibomitis: associated with inflammation of Meibomian glands (hypo or hypersecretory).

ETIOLOGY

Bacteria have been implicated in playing an important role in the pathogenesis of blepharitis (Dougherty and McCulley 1984, Oto et al. 1998). The conjunctival flora, in patients with blepharitis, has a greater number of bacteria than in normal individuals, specially *Staphylococcus epidermidis, Propionibacterium acnes, Corynebacterium sp.* and *Staphylococcus aureus* (Opitz and Tyler 2010). Bacterial lipase interferes with Meibomian secretion, increasing cholesterol concentration and favoring bacterial growth and proliferation (Dougherty and McCulley 1986, McCulley and Shine 2004). Toxic residues from bacteria, direct tissue invasion and immune mediated damage play an important role in the development of the disease.

The *Dermodex folliculorum* are small parasitic mites that live in hair follicles, sebaceous glands and meibomian glands. It is commonly seen at the face, neck, axillary and pubic regions. First seen by Henle and Berguer in 1841 and described in details by Simon in 1842 (Foton and Seys 1993), it was related to chronic blepharitis. It is still controversial and some authors believe *D. folliculorum* to be innocuous (Portelhinha and Cai 1983).

Seborrheic blepharitis is characterized by Meibomian gland hypersecretion. It is mostly associated to seborrheic dermatitis but also occurs in other conditions such as atopic disease and acne rosacea. The disorder is not associated with active inflammation of the Meibomian glands and there are no remarkable changes in its morphology. A hyperresponse of the glands to androgens is the most likely explanation for the seborrhea and the commonly associated acne. Topical prostaglandins analogues used in glaucoma treatment were implicated in blepharitis and chalazion formation acting directly to stimulate or alter Meibomian gland secretion (Cunniffe et al. 2010).

Meibomitis is characterized by inflammation of the Meibomian glands usually associated with systemis diseases like seborrheic dermatitis, atopic disease, acne rosacea and rarely psoriasis (Zhu and Tao 2011). It is hypersecretory when associated with excessive secretion of lipids by the Meibomian glands. Obliteration of the Meibomian ducts and orifices due to hyperkeratinization leads to hyposecretory meibomitis.

Acne rosacea is a chronic disease characterized by the presence of facial telangiectasias, erythema, papules, pustules, and hypertrophic sebaceous glands. After a few months it progresses to plaques and phymatous changes.

The rhinophima is the most advanced stage in the development of rosacea characterized by marked hypertrophic changes at the nose producing firm, reddish or purplish, lobulated masses. The exact etiology of rosacea is unknown and believed to be multifactorial. Over the years many causes have been suspected and not confirmed like genetic predisposition, dyspepsia with gastric hypochlorhydria, Helicobacter pylori infection, inflammatory bowel disease, seborrhea, Demodex folliculorum mites, endocrine disorders, vitamin deficiencies, microcirculatory disturbances, liver disease or psychogenic factors (Zengin et al. 1995). The disease usually starts around 30 to 40 years old and the signs and symptoms are frequent and occur early in the process of the disease. All suspects should be promptly referred to a dermatologist to confirm the diagnosis and identify the triggers.

Seborrheic dermatitis is a chronic condition characterized by recurrent erythema, edema, greasy scales, and yellowish crusts, predominantly in areas where there is a greater concentration of sebaceous glands. The etiology is unknown but in predisposed individuals the dermatitis may be triggered or exacerbated by certain factors such as climate change, emotional stress, fatigue or infections.

CLINICAL PICTURE

Blepharitis patients complain of burning sensation, irritation, tearing, photophobia, blurred vision, and red eyes. The symptoms are usually worse in the morning, because during sleep the inflamed lids are in close contact with the ocular surface. Also, without blinking, the lipid content of the tear film builds up, achieving a concentration 50% higher than in the rest of the day. A decrease in tear production occurs during the night, which is associated with the constant liberation of inflammatory mediators that induce the corneal damage.

Clinical examination reveals the presence of scurf, telangiectatic vascular changes of the eyelid margin, inspissated meibomian glands, conjuntival hyperemia, punctuate keratopathy, corneal vascularization and ulceration. Hard crusts around the base of the cilia (collarets) are typical of staphylococcal blepharitis. Seborrheic blepharitis is characterized by the presence of hyperaemia, lipid secretion and soft crusts at the lid margin and eyelashes. Patients with longstanding chronic blepharitis may present hypertrophy of the lid margin, scars, madarosis, trichiasis and poliosis (Fig. 3).

Posterior blepharitis has more pronounced symptoms with less clinical signs. Meibomian gland orifices may have little drops of lipid secretion (plugging or pouting) due to obstruction of the terminal ducts and extrusion of lipids and keratinized cell debris. Pressure on the tarsus releases thick and opaque toothpaste like lipid secretion. Tear film is oily and foamy. Chronic blepharitis may result in the obstruction of the gland ducts and structural changes. In such cases, there is a secondary Meibomian gland dysfunction that presents as a second peak of symptoms late in the day, caused by excessive evaporation of the tear film and dry eye. At the final stages, there is fibrosis and obliteration of the glands and reduction of the inflammation. The morning symptoms disappear and the evaporative symptoms late at the day tend to intensify (Gilbard et al. 1992).

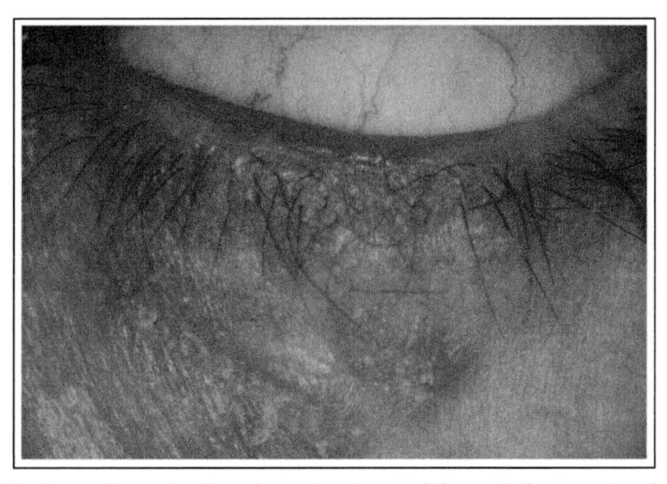

Figure 3 This figure shows the clinical examination: eyelid margin hyperemia, telangiectasy and ulceration, localized trichiasis and madarosis (cortesy of Alfredo Bonfioli).

Color image of this figure appears in the color plate section at the end of the book.

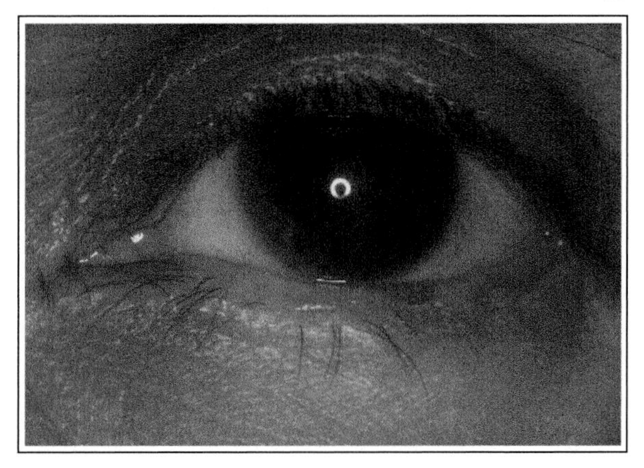

Figure 4 This figure shows the Discoid lupus (courtesy of Alfredo Bonfioli).

Color image of this figure appears in the color plate section at the end of the book.

DIFFERENTIAL

Most cases of discoid lupus affecting the eyelids present as blepharitis, with small infiltrated discoid plaques, edema, infiltration and inflammation of the margin (Feldman et al. 1995, Gloor et al. 1997) (Fig. 4).

Early sebaceous glands carcinoma present only subtle signs of malignancy and may be confused with less aggressive lesions. Meibomian gland carcinoma infiltrates the dermis and results in eyelid margin thickening. Sclerosing basal cell carcinoma (BAC) also infiltrates the dermis and may be mistaken for localized chronic blepharitis (Kanski 2007). Determining the clinical margins of the tumor is very difficult.

TREATMENT

Management of blepharitis varies greatly among health care providers. Patients' comprehension of the chronic nature of the condition is essential. Basic treatment for blepharitis includes a long-term commitment to eyelid hygiene. Topical lubricants, antibiotics, corticosteroids and immunomodulators may be necessary in selected cases.

Lid Hygiene

Properly instructing the patient on the correct methods of lid hygiene is the first step of the treatment. Frequently they develop their own processes and quickly abandon it as ineffective.

The application of warm compresses for a few minutes promotes melting of the altered lipids, evacuation and cleansing of the gland ducts. After the compresses the patient should apply pressure at the tarsus, moving from the medial to the lateral canthus, releasing gland obstruction and secretion. The margin is washed using mild shampoo (baby shampoo) diluted in water or commercial formulations. A cotton swab dipped in the solution is used to rub the base of the eyelashes, removing the crusts and toxic products.

Artificial Lubricants

Dry eye is a concomitant disease in many blepharitis patients. Increasing tear volume reduces hyperosmolarity and friction between the tarsal conjunctiva and the cornea. Lubricants also rinse toxins and debris and dilute inflammatory molecules. Preserved artificial tears may be used up to 6 times a day without toxicity.

Antibiotics

Several antibiotic agents have been used in blepharitis treatment in 2 to 4 week regimens. They may be available as an ointment, drops or oral presentations.

Topical fluoroquinolones provide excellent coverage against Gram positive and negative organisms. Concerns about bacterial resistance have limited its use for treating blepharitis. Macrolid antibiotics are effective against Gram positive and negative bacteria, atypical organisms and Chlamydia. They also have immunomodulatory and anti-inflammatory effects. Erythromycin is available as an ointment. Topical 1% Azitromycin is also available in some countries (Luchs 2010).

Systemic tetracycline has been used for the treatment of acne rosacea for decades. Doxycicline and minocycline are effective in lower doses and have fewer side effects. They have anti-inflammatory and antioxidative effects, and contribute to stabilize the tear film. The usual dosages are: tetracycline 250 mg 1 to 4 times a day, doxicycline 50 a 100 mg 1 to 2 times a day and minocycline 100 mg daily. The duration of treatment varies between 4 weeks and 3 months. The authors use doxycycline 100 mg, once a day for 10 days, followed by 50 mg once a day for 6 to 12 weeks.

Demodex Folliculorum

The role of D. folliculorum in the pathogenesis of blepharitis is controversial. Treatment involves rubbing the eyelid margins with cotton swabs dipped in ether (Aguiar and Berra 2001) or tea tree oil and shampoo (Portelhinha and Cai 1983).

Topical Steroids

Steroids may be used as ointment or drops, usually in combination with antibiotic agents. They are indicated when the patient presents with inflammation of the lid or the conjunctiva and should only be used in short term regimens (Geerling et al. 2011).

Tacrolimus

Tacrolimus 0.1% ointment two times a day has been used to treat psoriasiform blepharitis (Zhu and Tao 2011).

Cyclosporine

Topical cyclosporine is used to improve tear film production in patients with associated inflammatory dry eye disease such as rosacea or aqueous deficient dry eye. It may be used two times a day for 3 to 6 months.

Essential Fatty Acids

Increasing the intake of Omega-3 and 6 fatty acids (flaxseed oil supplements) is considered an adjunctive therapy for several forms of tear deficiencies. They have an anti-inflammatory effect and improve tear production and stability (Pinheiro et al. 2007). The authors use Flex seed oil 1g daily for 3 months to one year.

CONCLUSIONS

Blepharitis is a very common condition with variable signs and symptoms. The correct diagnosis and identification of the causes are essential for the treatment. Patient education and commitment are the basis for successfully managing the disease and preventing or minimizing the long-term damage to the eyelids and cornea.

ACKNOWLEDGMENT

I would like to give my sincere thanks to my teacher and guide, Adriana Alvim Bonfioli.

REFERENCES

Aguilar, A.J. and A. Berra. 2001. Chronic blepharitis. Arch. Alerg. Inmunol. Clin. 32: 98–100.
Bernardes, T.F. and A.A. Bonfioli. 2010. Blepharitis. Seminars in Ophthal. 25: 79–83.
Cunniffe, M.G., R. Mendel-Jiménez and M. González-Candial. 2010. Topical antiglaucoma treatment with prostaglandin analogues may precipitate meibomian gland disease. Ophthal. Plast. Reconstr. Surg. 27: 128–129.
Dougherty, J.M. and J.P. McCulley. 1984. Comparative bacteriology of chronic blepharitis. Br. J. Ophthalmol. 68: 524–528.
Dougherty, J.M. and J.P. McCulley. 1986. Bacterial lipases and chronic blepharitis. Invest Ophthalmol. Vis. Sci. 27: 486–491.
Feldman, G., A. Papa, E. Baroni and N. Amoros. 1995. Lupus eritematoso crónico de localización palpebral: respuesta a la talidomida en un caso. Arch. Argent. Dermatol. 45: 101–103.
Foton, F. and B. Seys. 1993. Density of Demodex folliculorum in rosacea: a case control study using standardized skin surface biopsy. British J. Dermatol. 128: 650–659.
Geerling, G., J. Tauber, C. Baudouin, E. Goto, Y. Matsumoto, T. O'Brien, M. Rolando, K. Tsubota and K.K. Nichols. 2011. The international workshop on meibomian gland dysfunction: Report of the subcommittee on management and treatment of meibomian gland disfunction. Invest Ophthalmol. Vis. Sci. 52: 2050–2064.

Gilbard, J.P., G.R. Cohen and J. Baum. 1992. Decrease tear osmolarity and absence of the inferior marginal tear strip after sleep. Cornea 11: 231–233.

Gloor, P., M. Kim, J.M. McNiff and D. Wolfley. 1997. Discoid lupus erythematosus presenting as asymmetric posterior blepharitis. Am. J. Ophthalmol. 124: 707–709.

Goldman, D.A. 2009. Treating blepharitis to maximize surgical success. Cataract Refractive Surgery Today 5: 61–63.

Groden, L.R., B. Murphy, J. Rodnite and G.I. Genvert. 1991. Lid flora in bleharitis. Cornea 10: 50–53.

Hom, M.M., J.R. Martinson, L.L. Knapp and J.R. Paugh. 1990. Prevalence of meibomian gland dysfunction. Optom. Vis. Sci. 67: 710–720.

Kanski, J.J. 2007. Clinical ophthalmology: A systematic approach. Windsor, UK.

Knop, E., N. Knop, T. Millar, H. Obata and D.A. Sullivan. 2011. The international workshop on meibomian gland dysfunction: Report of the subcommittee on anatomy, physiology, and pathophysiology of the meibomian gland. Invest Ophthalmol. Vis. Sci. 52: 1938–1978.

Luchs, J. 2010. Azithromycin in DuraSite for the treatment of blepharitis. Clin. Ophthalmol. 4: 681–688.

McCulley, J.P. and W.E. Shine. 2004. The lipid layer of tears: dependent on meibomian gland function. Exp. Eye Research 78: 361–365.

Nelson, J.D., J. Shimazaki, J.M. Benitez-del-Castillo, J.P. Craig, J.P. McCulley, S. Den and G.N. Foulks. 2011. The international workshop on meibomian gland dysfunction: Report of the definition and classification subcommittee. Invest Ophthalmol. Vis. Sci. 52: 1930–1937.

Opitz, D.L. and K.F. Tyler. 2010. Efficacy of Azithomycin 1% ophthalmic solution for treatment of ocular surface disease from posterior Blepharitis. Clin. Exp. Optom. 94: 200–206.

Oto, S., P. Aydin, N. Ciftcioglu and D. Durson. 1998. Slime production by coagulase-negative staphylococci isolated in chronic blepharitis. Eur. J. Ophthalmol. 8: 1–3.

Pinheiro, J.R. and M. Neuzimar. 2007. Uso oral do óleo de linhaça (Linum usitatissimum) no tratamento do olho seco de pacientes portadores da síndrome de Sjögren. Arq Bras Oftalmol. 70: 649–655.

Portellinha, W.M. and S. Cai. 1983. Avaliação clínica e laboratorial do uso de substância emoliente e detergente nas blefarites ciliares. Arq. Bras Oftalmol. 46: 134–137.

Saccà, S.C., A. Pascotto, G.M. Venturino, G. Prigione, A. Mastromarino, F. Baldi, C. Bilardi, V. Savarino, C. Brusati and A. Rebora. 2006. Prevalence and treatment of Helicobacter pylori in patients with blepharitis. Invest Ophthalmol. Vis. Sci. 47: 501–508.

Schaumberg, D.A., J.J Nichols, E.B. Papas, L. Tong, M. Uchino and K.K. Nichols. 2011. The international workshop on meibomian gland dysfunction: Report of the subcommittee on the epidemiology of, and associated risk factors for, MGD. Invest Ophthalmol. Vis. Sci. 52: 1994–2005.

Shimazaki, J., M. Sakata and K. Tsubota. 1995. Ocular surface changes and discomfort in patients with meibomian gland disfunction. Arch. Ophthalmol. 113: 1266–1270.

Skare, T.L. 2007. Reumatologia: princípios e prática. Rio de Janeiro. BR. pp. 52–54.

Tiffany, J.M. 1985. The role of meibomian secretion in the tears. Trans. Ophthalmol. Sac. 104: 396–401.

Zengin, N., H. Tol and K. Gunduz. 1995. Meibomian gland dysfunction and tear film abnormalities in rosacea. Cornea 14: 144–146.

Zhu, F. and J.P. Tao. 2011. Bilateral upper and lower eyelid severe Psoriasiform blepharitis: Case report and review of literature. Ophthal. Plast. Reconstr. Surg. 27 :138–139.

6

Keratitis

Flavia S.A. Pelegrino

SUMMARY

Keratitis is an inflammation of the cornea that can lead to visual impairment and even loss of vision. According to the World Health Organization (WHO), infectious diseases affecting the cornea are responsible for 25% of the causes of blindness worldwide. Even today many cases of keratitis are a challenge for ophthalmologists around the world. It may evolve slowly and be rapidly progressive with perforation of the cornea.

There are various types of keratitis, which may be related to cleaning of contact lenses, eye surgery, infections (attributed to the action of viruses, fungi, parasitic organisms and bacteria), trauma, use of certain medications (topical corticosteroids), associated with pre-existing ocular (atopic conjunctivitis, dry eye) and systemic disease (rheumatic and immunologic diseases), neurological (keratitis neurotrophic, toxic or nutritional) or unknown (Thygeson's keratitis) among other sub classifications. In general, keratitis may occur with symptoms that cannot distinguish the fungal, bacterial and viral causes, thus it would be necessary to take laboratory tests to confirm the presence of the causative agent.

Keratitis were classified as: ulcerative, non-ulcerative and may be present because of infectious or non-infectious causes. Its symptoms include pain, tearing, photophobia, decreased vision, conjunctival injection, purulent discharge, and eyelid secretions.

Early diagnosis and the correct treatment is the principal way to preserve vision. In advanced cases of the disease, when eye ulcers are

Ophthalmologic Clinic, 292 Querubim Uriel St., Zip code 13024-470, Campinas, São Paulo, Brazil; E-mail: flaviapelegrino@gmail.com present as a result of failure of the previous treatments, or loss of quality and quantity of vision, corneal transplant may be indicated.

INTRODUCTION

Corneal inflammation or keratitis is a significant cause of ocular morbidity around the world. Fortunately, the majority of the cases are successfully managed with medical therapy, but failure of therapy does occur, leading to devastating consequences of either losing vision or the eye.

Corneal inflammation may be ulcerative or non-ulcerative and may arise because of infectious or non-infectious causes. Non-ulcerative corneal inflammation may be confined to the epithelial layer or to the corneal stroma or may affect both.

In the pathogenesis of ulcerative keratitis, microorganisms such as bacteria, fungi, parasites (Acanthamoeba), or viruses play an important role. Approximately, 12.2% of all corneal transplantations are performed for active infectious keratitis. On the other hand, non-infectious ulcerative keratitis can be related to a variety of systemic or local causes, predominantly of autoimmune origin.

The cornea is a tissue highly specialized to refract and transmit light. Anatomically, the cornea consists of an outer stratified squamous non keratinized epithelium, an inner connective tissue stroma with keratocytes, and a single layer of hexagonal cells bordering the anterior chamber (Kanski 1994).

The course of healing of the corneal epithelium depends on the extent as well as intensity of the injury (such as chemical injury, bacterial infection, and surgical procedures).

INFECTIOUS ULCERATIVE KERATITIS

Bacterial Keratitis

Bacterial corneal ulceration is an ocular emergency due to the rapid progression of this corneal infection with the threat of loss of vision and potential corneal perforation. Prompt recognition, expedient evaluation, and rapid initiation of antibiotic therapy improve the patient's prognosis (Kumar et al. 2010). Knowledge of the various etiologic agents causing corneal ulceration and their various risk factors are important in order to select appropriate initial therapy (Smolin et al. 2005).

The majority of bacterial corneal ulcers occur in eyes with some predisposing risk factor such as, a corneal abrasion with contaminated matter, contact lens use, (Liesegang 1997) eyelid disease, and ocular surface disease. Corneal trauma with inoculation of organisms can occur from vegetable matter, industrial foreign bodies, makeup application, and administration of ocular medications (Torok et al. 1996). Corneas that have been compromised by conditions including prior herpetic infection, dry eye syndrome, bullous keratopathy, previous surgery, and atopic disease are susceptible to infection by opportunistic bacteria, most commonly the staphylococcal species.

Patients with AIDS (acquired immunodeficiency syndrome) from infection with human immunodeficiency virus (HIV) do not appear to be at increased risk of bacterial corneal ulcers but when they develop bacterial keratitis, it appears to take a more fulminate course than would otherwise be expected (Smolin et al. 2005).

Clinical Features: Symptoms and Signs of Corneal Ulcers

Decreased visual acuity, photophobia, pain, redness, swelling, and discharge. The severity depends on the pathogen organism, the condition of the host, and the delay in the patient being examined.

The precorneal tear film and meniscus in an actively infected ulcer are typically observed via slit lamp biomicroscopy examination containing numerous cells and debris, distinguishing it from a non infected corneal ulcer. The corneal epithelium is absent in areas of active infection. The magnitude of the corneal ulceration is a significant predictor of poor visual outcome after the infection resolves (Kanski 1994).

Corneal infection in the presence of an intact epithelium could occur and it is produced by organisms as *Neisseria gonorrhoeae* (Gram-negative cocci); *Corynebacterium diphtheria, Listeria sp.* (Gram-positive bacilli); and *Haemophilus sp.* (Gram-negative bacilli).

Other bacteria are capable of producing keratitis only after loss of corneal epithelial integrity associated with contact lens wear, ocular surface disease, which disrupts the defence mechanisms such as post herpetic corneal disease, bullous keratopathy, corneal exposure and dry eye. In the group of Gram-positive cocci bacteria, ulcers caused by *Streptococcus pneumoniae* usually follow trauma and if they are left untreated, perforation is common. *Beta Hemolytic Streptococcus* often causes severe corneal infections without any particular typical features. *Alpha Hemolytic Streptococcus* can cause indolent ulceration or crystalline keratopathy. *Staphylococcus aureus* causes a more severe corneal infiltration than *Staphylococcus epidermidis*, but *S. epidermidis* may be equally common as a cause of ulceration (Smolin et al. 2005). Long-standing staphylococcal ulcers have a tendency to bore deep into the cornea and may cause intrastromal abscesses that may lead to perforation. The most common course evolving infected ulcer by Gram-negative organism is corneal perforation and loss of the eyeball in a relatively short period of time if it is untreated. *Pseudomonas aeruginosa*

(Gram-negative bacilli) infection is associated with the use of contact lens (Liesegang 1997). After the immediate adhesion of organisms to the damaged epithelium, stromal invasion occurs rapidly within an hour. This leads to a descemetocele and eventual perforation within 2 to 5 d of the onset of infection. *Klebsiella sp., Escherichia coli, Proteus sp.* (Gram-negative bacilli) usually cause indolent ulcerations in previously compromised corneas.

Marginal keratitis (catarrhal ulcer) is a common condition caused by a hypersensitivity reaction to staphylococcal exotoxins. It is particularly common in patients suffering from chronic staphylococcal blepharitis (Kanski 1994). The examination of early cases shows the presence of sub epithelial infiltrate in 10, 2, 4, or 8 o'clock peripheral corneal position which is separated from the limbus by a clear zone of the cornea, and is accompanied by a breakdown of the overlying epithelium.

Parasitic Keratitis

Acanthamoeba Keratitis (AK)

Acanthamoeba histolyticum is a free-living protozoan found in air, soil and fresh or brackish waters that cause an infiltrative corneal ulceration. They exist in both active (trophozoite) and dormant (cystic) forms. The cystic form is able to survive for long periods under hostile environmental conditions, including chlorinated swimming pools, hot tubs and subfreezing temperatures in fresh water lakes. The cysts turn into trophozoites, under appropriate environmental conditions, which produce a variety of enzymes that aid in tissue penetration and destruction (Panjwani 2010). Those using contact lens are particularly at risk of acanthamoeba infection that may be misdiagnosed and confused with herpetic or fungal infection (Liesegang 1997).

Early diagnosis and appropriate therapy is a key to good prognosis. A provisional diagnosis of acanthamoeba keratitis (AK) can be made using the clinical features and confocal microscopy; although a definitive diagnosis requires culture, histology, or identification of the acanthamoeba deoxyribonucleic acid by polymerase chain reaction.

Clinical Features: Symptoms and Signs

The main symptoms of AK are foreign body sensation, photophobia, and severe pain (predominantly unilateral). Its signs are granular epithelial irregularity with punctate or dendriform changes and stromal infiltrates that can ensue and gradually coalesce into a crescentic or annular configuration. With progression, coalescent dense suppurate necrosis may occur and lead to ulceration and perforation. Corneal neovascularization is typically absent.

Nematodes Keratitis

Onchocerca volvulus is filarial nematode. The infection is transmitted to humans by the bite of a female blackfly of the genus simulium. Parasitic infections of the cornea may be due to direct ocular inoculation or occur during systemic infection.

Clinical Features: Symptoms and Signs

Its signs involve punctate keratitis, sclerosing keratitis and others. Punctate keratitis is the result of an inflammation reaction to degenerating microfilariae in the cornea. In most cases, the patient is asymptomatic with minimal visual impairment. Sclerosing keratitis is a stromal inflammation characterized by prolonged invasion of the cornea by microfilariae. Corneal opacification occurs as a result of stromal invasion by blood vessels (Smolin et al. 2005).

Fungal Keratitis

Most of the fungi implicated in keratitis may be called opportunistic organisms and are ubiquitous as plant pathogens or in the soil. The most common causes appear to be those most frequently encountered after an apparently negligible trauma (Kanski 1994).

Fungal keratitis should always be considered if the differential diagnosis of suppurated bacterial keratitis and herpetic stromal necrotic keratitis is suspected (Smolin et al. 2005). It primarily affects the corneal epithelium and stroma, although the endothelium and anterior chamber of the eye may be involved in more severe disease.

The clinical appearance of fungal keratitis varies with the infectious agent and stage of the disease. Filamentous fungal keratitis is usually caused by *Aspergillus* or *Fusarium spp.*, occurs in previously healthy individuals, most frequently secondary ocular trauma with organic matter (wood). Candida keratitis usually develops in association with pre-existing chronic corneal disease or in an immune-compromised or debilitated patient.

Clinical Features: Symptoms and Signs

There are two features that should lead to suspecting a fungal infection; stromal infiltrates with feathery, hyphate edges, and infiltrates that tend to be dry, grey, and somewhat elevated above the level of the corneal surface. Endothelial plaques or an anterior chamber reaction usually indicate a more severe infection with penetration of fungal elements into the anterior chamber. However, fungal keratitis may be an indolent infection.

Viral Keratitis

Herpes simplex virus (HSV) is a DNA virus with humans as the only host, and about 90% of the population is seropositive for HSV antibodies. Ocular herpes is a prevalent and recurrent eye infection. According to different clinical and immunological properties, HSV is subdivided in HSV-1 predominantly causes infection on the face, lips, and eyes; is usually acquired by kissing or coming into close with a person who either has a cold sore (herpes labials) or is shedding the virus asymptomatically and HSV-2 typically causes genital infection and is acquired venereally. It can be transmitted to the eye through infected genital secretions in neonates during passage through the birth canal (Holland et al. 1999).

Ocular herpes may be classified into two general groups: congenital and neonatal, primary and recurrent. Although the vast majority of all ocular herpes infections are caused by HSV-1, about 80% of neonatal cases are caused by HSV-2.

Primary HSV keratitis is often atypical. Initially, there may be just a nonspecific diffuse punctate keratitis and may be associated with generalized symptoms of a viral illness. The cause of the inflammation whether immunologic, infectious, or neurotrophic, must be determined.

Other causes of non-herpetic dendritic ulceration include: herpes zoster keratitis, a healing corneal abrasion, soft contact lens (pseudodendrite in the mid-peripheral cornea), and toxic keratopathies usually caused by excessive drop administration.

Clinical Features: Symptoms and Signs

There are four major categories of HSV keratitis:

1) Infectious epithelial keratitis; which is made up of cornea vesicles, dendritic ulcer, geographic ulcer, and marginal ulcer.
2) Neurotrophic keratopathy; which includes punctate epithelial erosions and neurotrophic ulcer.
3) Stromal keratitis; which is subdivided into necrotizing stromal keratitis and immune stromal keratitis.
4) Endothelitis; which has three clinical presentations: disciform, diffuse, and linear.

Herpes zoster ophthalmicus (HZO) is caused by human herpes virus 3 (HHV-3), the ocular damage may be direct as the result of cellular infiltration, or indirect by denervation (ophthalmic division of the trigeminal nerve) and ischaemia induced by vasculitis (Smolin et al. 2005). Approximately 15% of all cases of herpes zoster affect the ophthalmic division of the trigeminal nerve and is referred to as HZO. The ocular complications of

HZO are related with: viral spread, nerve damage, ischaemic vasculitis, and inflammatory granulomatous reaction (Sanjay et al. 2011). They can clinically be divided into three phases:

1) Acute phase: which resolves within 4 wk.
2) Chronic phase: which persists for years.
3) Relapsing phase: the lesions appear to be controlled, but can appear years later.

The management of herpes zoster involves a multidisciplinary approach aiming to reduce complications and morbidity. Patients with HZO are referred to ophthalmologists for prevention or treatment of its potential complications. Without prompt detection and treatment, HZO can lead to substantial visual disability

Clinical Features: Symptoms and Signs

The main signs could be the presence of corneal lesions including punctate epithelial keratitis (multiple peripheral lesions), micro-dendritic ulcers (unlike the dendritic ulcers of HSV, they are peripheral, broader, more plaque-like, and a more stellate than dendritic shape), nummular keratitis, and disciform keratitis. Other ocular complications include: uveitis, necrotizing retinitis, and cranial nerve palsies related with the eye.

Finally, the *Epstein-Barr virus* is a member of the family Herpesviridae and is a common cause of lifelong infection in humans. Keratitis may be unilateral or bilateral with onset in 1 to 4 wk after acute infectious mononucleosis (IM). The epithelial keratitis is self-limited, and rarely stromal keratitis may develop. Its main clinical features are irritation, watering of eyes, photofobia, and blurred vision.

NON INFECTIOUS ULCERATIVE KERATITIS

Non infectious ulcerative keratitis or interstitial keratitis refers to a non-ulcerative and non suppurated inflammation of the central corneal stroma, often with subsequent vascularization. It is a significant cause of visual impairment. It can be associated with bacterial pathogens, viral infection, and parasitic infestations, or also related to immunologic factors, and may arise without identifiable antecedent illness.

The corneal changes depend on the stage or duration of the disease. In general, the epithelium is intact, although it may often be oedematous. Depending on the severity, the entire process may remain localized in the periphery or move centrally. The diagnosis of the cause of interstitial keratitis must rely largely on the history, ocular as well systemic examination.

Thygeson's Superficial Punctuate Keratitis (TSPK)

It is rare, usually bilateral, but may be also unilateral with asymmetrical findings in almost all cases, non contagious. Chronic keratitis is characterized by remissions and exacerbations over years to decades (Kanski 1994). Although its exact etiology is controversial, TSPK has a genetic association with HLA-DR3, an antigen associated with immunogenic responses. A viral etiology is suspected but not proved.

Concerning permanent damage to the cornea and the potential to decreased visual acuity, TSPK can be considered as a benign condition and can remain for several years as a self-limited disease. The disease is often misdiagnosed and incorrectly treated.

Clinical Features: Symptoms and Signs

The clinical findings are superficial keratitis consisting of stellate, round or oval conglomerates of distinct granular, grey-white intraepithelial dots, which may be associated with a mild sub epithelial haze with minimal or no conjunctival involvement.

Mooren's Ulcer

It is a peripheral, progressive, painful ulcerative keratitis that could be caused by an ischaemic necrosis resulting from a vasculitis of the limbic vessels. Mooren's ulcer can only be diagnosed in the absence of an infectious or systemic cause and must be differentiated from other corneal abnormalities, such as Terrien's degeneration. The etiology of the Mooren's ulcer remains uncertain but it is often idiopathic (Sangwan et al. 1997). However, recent studies indicate that it is an autoimmune disease directed against a specific target molecule in the corneal stroma, probably triggered in genetically susceptible individuals by one of several possible provocateurs. There are two types of Mooren's ulcer:

1) A limited form, which is usually unilateral and mostly affects the elderly.
2) A progressive form, which is bilateral and typically affects relatively young individuals.

Clinical Features: Symptoms and Signs

This condition initially affects the peripheral of the cornea, may spread circumferentially and then centrally with no associated scleritis. The response to medical and surgical intervention is typically poor, and the visual outcome can be devastating.

Rosacea Keratitis

Rosacea is a chronic skin disease affecting up to 10% of the population. It includes various combinations of characteristic signs and symptoms in a centro-facial distribution. The National Rosacea Society (NRS) has classified rosacea into four subtypes (erythemato-telangiectatic, papulopustular, phymatous, and ocular) and one variant (lupoid or granulomatous) avoiding assumptions on pathogenesis and progression, its etiology is unknown, typically affects women between 30 and 50 yr.

Clinical Features: Symptoms and Signs

Keratitis occurs in about 50% of cases and may take several forms like punctate epitheliopathy, peripheral vascularization, and thinning (Smolin et al. 2005).

1) Punctate epitheliopathy: involving the inferior two-thirds of the cornea.
2) Peripheral vascularization: involving the infero-temporal and infero-nasal corneal quadrants, followed by central sub epithelial infiltrates.
3) Thinning occurs either by resolution of the infiltrates or by grosses ulceration. Perforation may occur as a result of severe peripheral or central melting.

Exposure Keratitis

It is caused by the desiccation of the corneal epithelium because of the inability of the lids to resurface the cornea with each blink. Small epithelial defects, corneal stroma ulceration, and neovascularization are present. Exposure keratitis can lead to infectious keratitis, corneal perforation, blindness and disfigurement. Chronic exposure of the cornea can occur following facial nerve paralysis, eyelid ectropion, eyelid cosmetic surgery, severe proptosis and scarring of the eyelids.

Use of artificial tears and taping shut of the eyelids are indicated to help this condition.

Systemic Collagen Vascular Disorders Keratitis

The presence of severe, persistent corneal infiltration, ulceration or thinning unexplained by coexistent ocular disease should look for for an associated systemic collagen vascular disease. The ocular lesions may precede the clinical manifestation of the systemic disease. Four main diseases should considered:

1) *Rheumatoid arthritis (RA).* It is the most common collagen vascular disorder that affects the peripheral cornea. The cornea and conjunctiva are frequently involved in RA. Peripheral ulcerative keratitis is a rare inflammatory disease (Squirrel et al. 1999) usually associated with RA, which may lead to rapid corneal keratolysis, perforation of the globe and visual failure (Smolin et al. 2005).
2) *Systemic lupus erythematosus (SLE).* The corneal manifestations of SLE are confined to the epithelium. Although, non-infiltrate and infiltrate marginal thinning, peripheral ulcerative keratitis and vascularization have been reported (Sivaraj et al. 2007). Sicca syndrome is quite common.
3) *Polyarteritis nodosa.* Ocular findings in the cornea are peripheral ulcerative keratitis, progressive, both centrally and circumferentially, associated with ocular pain and inflammation.
4) *Wegener's granulomatosis (WG).* It is a systemic inflammatory disease whose histopathologic features often include necrosis, granuloma formation, and vasculitis of small-to-medium-sized vessels. WG involves many interrelated pathogenic pathways that are genetic, cell-mediated, neutrophil-mediated, humoral, and environmental. Ophthalmologic involvement is an important cause of morbidity in WG patients, occurring in approximately half of these patients. The presence of unexplained orbital inflammatory disease, scleritis, peripheral ulcerative keratitis, cicatricial conjunctivitis, nasolacrimal duct stenosis, retinal vascular occlusion, or infrequently uveitis should consider the question of possible WG (Tarabishy et al. 2010).

NON-ULCERATIVE KERATITIS

Non-ulcerative or intersticial keratitis refers to a non-ulcerative and non suppurative inflammation of the central corneal stroma, often with subsequent vascularization. It is a significant cause of visual impairment and can be associated with bacterial pathogens; viral infection, parasitic infestations, congenital syphilis, tuberculosis and Cogan's syndrome or it may arise without identifiable antecedent illness (Schwartz et al. 1998).

The corneal changes depend on the stage or duration of the disease. In general, the epithelium is intact; although it may often be oedematous and no primary involvement of the endothelium is present. Depending on severity, the entire process may remain localized in the periphery or move centrally.

Diagnosis of the cause of interstitial keratitis must rely largely on the history, and ocular as well systemic examination. Two clinical conditions are highlighted:

1) *Nummular Keratitis:* characterized by fine granular deposits (just below the Bowman's membrane), which are surrounded by a halo of stromal haze. Herpes simplex virus (HSV) infection is the most common cause, followed by idiopathic etiology.
2) *Adenoviral Keratoconjunctivitis:* characterized by anterior stromal infiltrates that fade initially when treated with topical steroids, but recur if treatment is discontinued prematurely. They may persist for months and even years.

CONCLUSIONS

Keratitis is a threat to vision and the eye. It is a true ophthalmological emergency. Therefore, it is important to learn the difference between infectious and non infectious ulcerative keratitis.

Whenever the presence of corneal infection is suspected, a laboratory study and prompt indicative treatment should be established. Frequent monitoring is required, as well as adjusted treatment to ensure the best result.

REFERENCES

Holland, E.J. and G.S. Schwartz. 1999. Classification of herpes simplex virus keratitis. Cornea 18: 144–54.

Kanski, J.J. 1994. Clinical Ophthalmology Butterworth-Heinemann International Editions. 3th Edition.

Kumar, R.L., A. Cruzat and P. Hamrah. 2010. Current state of in vivo confocal microscopy in management of microbial keratitis. Semin Ophthalmol. 25: 166–70.

Liesegang, T.J. 1997. Contact lens-related microbial keratitis: Part I: Epidemiology. Cornea 16: 125–31.

Panjwani, N. 2010. Pathogenesis of acanthamoeba keratitis. Ocul. Surf. 8: 70–9.

Sangwan, V.S., P. Zafirakis and C.S. Foster. 1997. Mooren's ulcer: current concepts in management. Indian J. Ophthalmol. 45: 7–17.

Sanjay, S., P. Huang and R. Lavanya. 2011. Herpes zoster ophthalmicus. Curr. Treat Options Neurol. 13: 79–91.

Schwartz, G.S., A.R. Harrison and E.J. Holland. 1998. Etiology of Immunestromal (interstitial) keratitis. Cornea 17: 278–81.

Sivaraj, R.R., O.M. Durrani, A.K. Denniston, P.I. Murray and C. Gordon. 2007. Ocular manifestations of systemic lupus erythematosus. Rheumatology 46: 1757–62.

Smolin, G. and R.A. Thoft. 2005. The Cornea. Lippincott Williams & Wilkins, 4th Edition. UK.

Squirrell, D.M., J. Winfield and R.S. Amos. 1999. Peripheral ulcerative keratitis, corneal melt' and rheumatoid arthritis: A case series. Rheumatology 38: 1245–8.

Tarabishy, A.B., M. Schulte, G.N. Papaliodis and G.S. Hoffman. 2010. Wegener's granulomatosis: clinical manifestations, differential diagnosis, and management of ocular and systemic disease. Surv. Ophthalmol. 55: 429–44.

Torok, P.G. and T.H. Mader. 1996. Corneal abrasions: diagnosis and management Am. Fam. Physician 53: 2521–9.

7

Ocular Allergy

Laura Motterle

SUMMARY

Ocular allergy includes a range of pathological conditions of the ocular surface, from mild to sight-threatening entities. This chapter presents a brief review of the different clinical forms of allergic conjunctivitis.

INTRODUCTION

Allergy is a widespread disease and its prevalence keeps increasing worldwide, mostly in industrialized countries and in urbanizing areas. Allergic ocular hypersensitivity is one of the most frequent causes of complaint in the clinical practice, also in paediatric population.

The classification of allergic ocular diseases is still a matter of debate. However, traditionally it encompasses different groups (Fig. 1): seasonal/ chronic allergic conjunctivitis (SAC/PAC), vernal keratoconjunctivitis (VKC), atopic keratoconjunctivitis (AKC), and giant papillary conjunctivitis (GPC). Moreover contact dermatoconjunctivitis (CDC) may be included in this category.

SEASONAL/ PERENNIAL CONJUNCTIVITIS

Seasonal (SAC) and perennial (PAC) conjunctivitis are the most common forms of allergic ocular reaction to environmental allergens. Ocular symptoms are often associated with rhinitis. They are clinically significant not for their severity, but for their high incidence.

Ophthalmology Unit, Department of Neuroscience, University of Padova, Via Giustiniani 2, 35128, Padova, Italy; E-mail: laura.motterle@yahoo.it

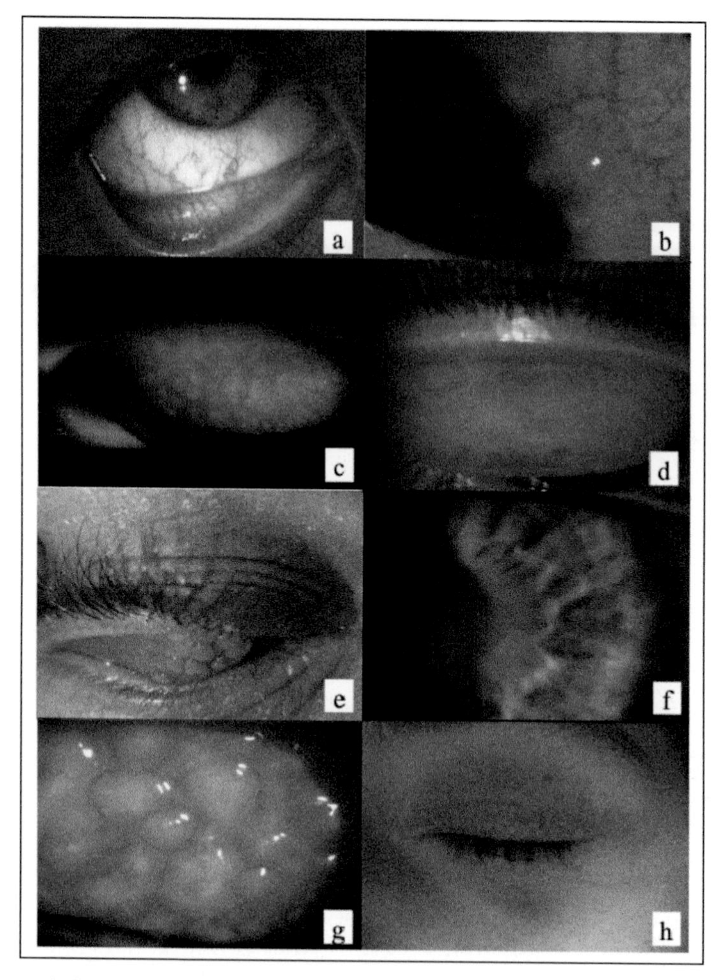

Figure 1 a) SAC, conjunctival hyperemia; b) VKC, Trantas' dots; c) VKC, giant papillae; d) VKC, reticular appearance of the conjunctiva on the upper tarsal plate; e) AKC, eczema of the eyelid and giant papillae on the upper tarsal plate; f) AKC, corneal ulcer; g) GPC, giant papillae and mucus discharge in a contact lens wearer; h) CDC, periocular contact dermatitis (Courtesy of Dr. A. Leonardi).

Color image of this figure appears in the color plate section at the end of the book.

Epidemiology

It has been estimated that at least 20% of population is affected by a form of ocular allergy and at least 80% are SAC/PAC. In 60% of cases it is associated with rhinitis. However prevalence data for allergic conjunctivitis are likely underreported (Rosario and Bielory 2011).

Clinical Features

SAC and PAC share the same signs, symptoms and pathogenesis, but differ in onset and course (Bielory 2008). SAC is an acute or sub-acute condition, whose manifestations are usually related to specific airborne seasonal allergens, mostly grass and tree pollens. Instead, PAC persists throughout the year with varying severity and periodical peaks. The allergens involved are mainly dust mites and animal dander.

The peculiar symptom is itching, accompanied with redness, tearing, mucous discharge, and burning. Usually both eyes are affected, although with different degrees of inflammation.

The conjunctiva looks hyperaemic, oedematous with papillary hypertrophy on the tarsal plate (Fig. 1a). In severe manifestations, also lid and periorbital oedema may develop. Cornea is never directly involved.

Differential diagnosis may include dry eye syndrome, infectious conjunctivitis and contact allergic conjunctivitis.

Pathogenesis

SAC/PAC are mediated by a typical Type 1 hypersensitivity mechanism; it is an immediate response that lasts clinically for 20–30 minutes.

After the sensitization, allergens bind to specific IgE on conjunctival mast cells, resulting in the release of histamine and other pharmacological mediators, as proteases, prostaglandins, leukotrienes and cytokines. All these factors initiate the recruitment of other inflammatory cells, such as eosinophils, neutrophils, basophils and T lymphocytes, which are responsible of the late-phase reaction and maintain local inflammation (Chigbu 2009). Specific IgE, histamine, eotaxine and ECP levels are increased in tears of SAC/PAC affected subjects.

The most potent mediator in this immunological mechanism is histamine, that acts through at least 4 different types of mediators (H1, H2, H3, H4). It is involved in the regulation of inflammatory cell migration, cellular activation, modulation of T-cell function, vascular permeability, smooth muscle contraction, and mucus secretion. It is responsible for itch and was found elevated in tears after conjunctival antigen challenge.

VERNAL KERATOCONJUNCTIVITIS

Vernal keratoconjunctivitis is a potentially severe inflammatory ocular condition that usually affects young children for many years, mostly in warm and hot regions of the world.

Epidemiology

VKC usually begins during infancy (mean age at the onset: 7.5 years) and tends to resolve with puberty. Much more rarely the disease occurs in adults. Among children males are more frequently affected than females with a ratio of approximately 3:1 (Leonardi et al. 2006).

The prevalence of VKC varies in different countries; it is higher in Mediterranean area, West and Central Africa, India and South America; however it is present also in North America, Japan, China and North Europe. Interestingly, immigration flows are changing this trend. This suggests that not only environmental and climate but also genetic factors play a role in the development of VKC.

Clinical Features

Symptoms are the same as in the other allergic conjunctivitis (itching, burning, foreign body sensation, tearing, mucus discharge, photophobia), but they are often so intense that they can impair daily activities. Despite the name, VKC is not a proper seasonal disease. It is usually a chronic remitting disorder which worsens in spring and summer, likely because of a major exposure to allergens and also to non-specific stimuli, as sun, heat, wind. Perennial cases are more frequent in subtropical and desert areas. Association with other atopic conditions, such as asthma, rhinitis and dermatitis, is observed in approximately a third of the patients.

VKC is classified in tarsal, limbal and mixed form, depending on the localization of inflammatory changes in the conjunctiva.

The tarsal form is characterized by important remodelling processes in the upper tarsal conjunctiva, which lead to the formation of giant papillae. These usually have a smooth polygonal shape, larger than 1 mm, with a vascular core, isolated or distributed along the entire tarsal conjunctiva in a cobblestone pattern (Fig. 1c). Giant papillae are surrounded by a sticky mucus discharge, which can take the form of a pseudomembrane (Maxwell-Lyon sign). The size of the papillae can even determine a ptosis, more evident in the active phases of the disease. Alternatively, a reticular appearance, secondary to a diffuse thickening and fibrosis, may be observed mainly in adults (Fig. 1d) (Leonardi 2002).

In the limbal form, a typical feature is the presence of gelatinous infiltrates, whitish-yellow chalky in appearance, single or multiple; they are sometimes calcified on the top, called Trantas'dots (Fig. 1b). These eosinophils aggregates are predominantly localized in the upper limbus, but they may be extended all around the cornea and associated with superficial corneal neovascularization. Limbus may appear thinned and opaque or marked by the presence of a pseudo-gerontoxon, caused by long-lasting altered permeability of limbal vasculature.

In the mixed tarsal/limbal form both clinical signs coexist to varying degrees.

Corneal involvement is a sign of severity of the disease. The progressive loss of corneal integrity leads to a punctuate epithelial keratopathy in mild cases, and to shield ulcers in the most severe manifestations. These typical aseptic ulcers are usually localized in the upper cornea and are more frequent in the tarsal form. A clinical grading of severity of the ulcer defines three levels: grade 1, clear ulcer that extends to the Bowman's membrane; grade 2, mild opaque ulcer that is partially occupied by inflammatory debris; grade 3, plaque ulcer that remains above the surrounding epithelium. Secondary microbial infection may rarely occur. These ulcers tend to heal leaving subepithelial opacities; whenever localized in the optic area, they can be sight-threatening.

Rarely chronic inflammation may lead to corneal ectasia and consequently different grades of astigmatism or even cheratoconus (Lodato Syndrome).

A practical classification in five grades and a correlation with treatment recommendations have been proposed (Bonini et al. 2007). Complications such as cataract and glaucoma are caused by excessive steroid treatment, not properly by the disease.

A differential diagnosis should include the other forms of allergic conjunctivitis and trachoma. Coexistence of VKC and trachoma should be considered, especially in some warm areas of the world.

Pathogenesis

The pathogenesis of VKC includes different mechanisms. Type I and type IV hypersensitivity reaction are involved, but they do not explain alone all the important conjunctival changes.

Conjunctival stroma and epithelium are infiltrated by an exaggerated amount of inflammatory cells, such as mast cells, eosinophils, basophils and neutrophils. High level of histamine, Th1 (IFN-γ) and Th2 cytokines (IL-3, IL-4, IL-5, IL-10, IL-13, GM-CSF) and other mediators are found in conjunctival tissue and tears of VKC patients (Leonardi 2002). Chemokines such as IL-8, MCP-1, RANTES and eotaxins are produced by cells of the immune system, but also by epithelial cells and fibroblasts. Fibrosis and tissue remodelling are related to an increased production of collagen type I and IV, that is linked to an over-expression of the TGF-β/Smad signalling pathway and modulated in conjunctival fibroblasts by histamine, IL-4, TGF-β1 and TNF-α (Leonardi et al. 2011). Altered expression of estrogen, progesterone, muscarinic and adrenergic receptors, NGF, substance P suggests a complex neuro-endocrine-immune regulation in the pathological mechanisms at the basis of VKC.

ATOPIC KERATOCONJUNCTIVITIS

Atopic keratoconjunctivitis is the most severe form of chronic allergic ocular disorder, affecting eyelids, conjunctiva and cornea. It is usually part of a dysregulated systemic immune condition, associated with atopic dermatitis and other allergic manifestations.

Epidemiology

The real prevalence of AKC may only be speculated: from 15% to 67.5% of subjects with atopic dermatitis have an ocular involvement and atopic dermatitis affects about 3–9% of the general population (Yanni and Barney 2008). Ocular manifestations usually begin in the second through fifth decade, although the onset of symptoms is reported from 7 up to 76 years of age. It is not clear if there is a gender prevalence. No racial or geographic predilection is described.

Clinical Features

AKC can manifest in different degrees of severity, from mild to blinding forms. It is usually bilateral and often asymmetrical.

A history of atopic dermatitis should be specifically questioned; in some cases the disease is evident only in childhood and there are no other pathological signs but in the ocular region in the adults.

Symptoms as itching, burning, tearing, mucus discharge, and photophobia vary in intensity and duration. It can afflict patients all year long or seasonally.

Typically anterior and/or posterior blepharitis impairs the eyelids; consequently, the tear film is often altered by the meibomian gland dysfunction. The skin of the lids may appear eczematous, thinning or thickening, lichenified and even macerated (Fig. 1e). *Staphilococcus aureus* infection is most likely to complicate the aspect of the skin.

Conjunctival inflammation may be localized both in the superior and inferior fornix and tarsal plate, accompanied by intense hyperemia and chemosis. In severe cases papillary hypertrophy and giant papillae in the upper tarsal area and gelatinous perilimbal infiltrate may develop, as in VKC. Occasionally the scarring process, due to long-lasting swelling, can determine a fornix foreshortening and even a symblepharon.

Cornea is injured in 75% of cases. Punctuate epithelial keratopathy and epithelial defects are mild signs, while corneal erosions, thinning, neovascularization, shield ulcers and plaque are worsening manifestations (Fig. 1f).

Predisposition to bacterial and herpetic infections is likely the result of a deficiency in the innate immunity. Keratoconus and cataract (often anterior polar or posterior subcapsular) are severe complications.

A clinical four-grade classification, based on principal signs and symptoms, was proposed with the intention of uniforming the terms related to AKC (Calonge and Herreras 2007).

Pathogenesis

AKC is a complex disease that probably involves both IgE and delayed hypersensitivity reaction, similarly to VKC.

Inflammatory cells, conjunctival epithelial cells and fibroblasts interconnect through the expression of different chemokines and adhesion molecules, amplifying the inflammatory processes.

Epithelium and "*substancia propia*" of the pathological conjunctiva are infiltrated by mast cells, eosinophils, lymphocites with an increased CD4/CD8 ratio, and show HLA-DR staining.

Th1-driven inflammatory response predominates, as proven by the increase of IL-2, IFN-γ and IL-12. Th2 response is also involved; a novel Th2 cytokine, IL-33, is hypothesized to play a central role in the allergic inflammatory cascade. Eosinophil expression of IL-4 and IL-8 is increased. Tear levels of eotaxin-1, a potent eosinophil chemoattractant, were found to correlate with disease severity. Moreover high serum concentration of ECP was found in AKC patients (Guglielmetti et al. 2010).

GIANT PAPILLARY CONJUNCTIVITIS

Giant papillary conjunctivitis was included in the spectrum of allergic ocular diseases when it was first described in 1974. However its pathogenesis has been revised and a major role is now attributed to the mechanical stimulus as trigger of the immunological reaction.

GPC affects mostly contact lens wearers, as well as patients with ocular prostheses, suture exposures, extruded scleral buckle, corneal foreign bodies, filtering blebs and cyanoacrylate tissue adhesives.

Epidemiology

Epidemiological studies focused mostly on contact lens wearers and only case reports consider other causes of GPC. Analysis in wide populations of contact lens wearers recorded about 15%–20% of GPC complications (Forister et al. 2009, Teo et al. 2011).

A higher prevalence of GPC in soft contact lenses than in rigid gas-permeable wearers is reported. The incidence of co-morbidity with other allergic manifestations varies between 12 and 26% in GPC patients. However, allergic subjects seem to have more severe complaints and clinical signs than non-allergic ones.

Clinical Features

Symptoms vary depending on the stage of GPC. A decreased lens tolerance is the general ailment referred by the patients; more specifically they complain of excessive lens movements, foreign body sensation, burning, itching (especially after lens removal), redness, mucus discharge, coated lenses and blurred or unstable vision.

GPC includes a range of progressive changes that are localized typically in the upper tarsal conjunctiva. In 90% of cases it is bilateral and quite symmetric. Although there may be a discrepancy between signs and symptoms, it can be classified in four stages. In the first stage (preclinical) symptoms are mild and the conjunctiva may appear quite normal or slightly hyperaemic. The second stage (mild) is characterized by lens awareness, itching and increased mucus production; there is a moderate injection and thickness of the upper tarsal conjunctiva, where a papillary reaction begins to be evident. At the third (moderate) level of severity, manifestations tend to exacerbate and lens coating causes fluctuating and cloudy vision. Conjunctival hyperaemia is marked and papillae are numerous and elevated; the top of the papillae can be eroded and stains with fluorescein or it can be whitish due to subconjunctival fibrosis. Stage four (severe) occurs when the subject cannot wear or even insert contact lenses. Mucus discharge is abundant and sticky, papillae are often larger than 1 mm and normal vasculature cannot be seen (Fig. 1g) (Donshik et al. 2008).

Symptoms and some clinical features mimic AKC and VKC, however in GPC the cornea is never involved and history of lens wearing or of exposure to foreign materials directs towards the diagnosis.

Pathogenesis

Two factors are supposed to be the cause of GPC: a chronic conjunctival micro-trauma and an immunological conjunctival susceptibility. Contact lenses design, polymer constitution, surface features, replacement intervals, wearing period and fitting characteristics can influence the development and the course of this disease (Stahl et al. 2004).

Histopathological analysis shows a great thickening of the conjunctiva with cells infiltration in the epithelium, constituted of eosinophils, basophils, plasma cells, macrophages and fibroblasts, T helper cells, memory T cells and epithelial cells expressing MHC molecules.

The mechanical stimulus as well as the coating on the lenses results in the release of neutrophilic chemotactic factors and other mediators that recruit Langerans cells and dendritic cells, enhancing antigen presentation to T cells and immune responsiveness. IgE, IgG and IgM levels are increased in tears and serum and C3 anaphilotoxin (C3a) is activated with subsequent release of vasoactive amines (Stapleton et al. 2003). The role of eosinophils in GPC pathogenesis is not clear; on the one hand eotaxin levels were found elevated in contact lens wearers (but not in ocular prosthesis wearers), on the other hand ECP serum level is not significantly increased in affected patients. Among cytokines, IL-6 levels are higher than in normal specimens; IL-6 and its soluble receptor IL-6sR are likely to be involved in stimulation of collagenase synthesis and therefore in the giant papillary proliferation (Donshik et al. 2008).

CONTACT DERMATOCONJUNCTIVITIS (CDC)

Contact ocular allergy is the consequence of an exaggerated immune response to small molecular weight chemicals which penetrate the skin of the eyelids and/or the conjunctiva. Topical drugs, preservatives, perfumes, cosmetics, and metals are common triggers.

Epidemiology

All eye drops and ointments may potentially cause a conjunctival contact reaction; among them, glaucoma medications, antibiotics, FANS, steroids, midriatics, local anesthetics, and contact lenses care solutions are frequently reported.

Eyelids contact allergic dermatitis is usually caused by cosmetics, nail lacquer, frames of eyeglasses containing plastics or metals (Novitskaya et al. 2011).

Clinical Features

Symptoms are characterized by itching, burning and foreign body sensation. Clinical signs may involve predominantly eyelids with oedema, hyperemia, and periocular dermatitis and/or conjunctiva with intense injection (Fig. 1h). They are usually bilateral. Skin prick or patch testing are sometimes useful to identify allergens.

Pathogenesis

The haptens do not have specific receptors; they enter the tissues for their lipofilic nature and bind to tissue proteins, that Langerhans' cells can process and present to T-lymphocytes. A delayed T cell-mediated hypersensitivity reaction develops 2–5 days after allergen exposure. The histopathological analysis reveals dendritic cells, basophils, eosinophils and Th1-lymphocytes (Chigbu 2009).

ACKNOWLEDGEMENTS

I would like to thank Dr. Andrea Leonardi (Ophthalmology Unit, Department of Neuroscience, University of Padova) for mentoring me and reviewing the contents of this chapter.

REFERENCES

Bielory, L. and M.H. Friedlaender. 2008. Allergic conjunctivitis. Immunol. Allergy Clin. North Am. 28: 43–58.

Bonini, S., M. Sacchetti, F. Mantelli and A. Lambiase. 2007. Clinical grading of vernal keratoconjunctivitis. Curr. Opin. Allergy ClinImmunol. 7: 436–441.

Calonge, M. and J.M. Herreras. 2007. Clinical grading of atopic keratoconjunctivitis. Curr. Opin. Allergy Clin. Immunol. 7: 442–445.

Chigbu, D.I. 2009. The pathophysiology of ocular allergy: a review. Contact Lens Anterior Eye 32: 3–15.

Donshik, P.C.,W.H. Ehler and M. Ballow. 2008. Giant papillary conjunctivitis. Immunol. Allergy Clin. North Am. 28: 83–103.

Forister, J.F., E.F. Forister, K.K. Yeung, P. Ye, M.Y. Chung, A. Tsui and B.A. Weissman. 2009. Prevalence of contact lens-related complications: UCLA contact lens study. Eye Contact Lens 35: 176–180.

Guglielmetti, S., J.K. Dart and V. Calder. 2010. Atopic keratoconjunctivitis and atopic dermatitis. Curr. Opin. Allergy Clin. Immunol. 10: 478–485.

Leonardi, A. 2002. Vernal keratoconjunctivitis: pathogenesis and treatment. Prog. Retin. Eye Res. 21: 319–339.

Leonardi, A., F. Busca, L. Motterle, F. Cavarzeran, I.A. Fregona, M. Plebani and A.G. Secchi. 2006. Case series of 406 vernal keratoconjunctivitis patients: a demographic and epidemiological study. Acta Ophthalmol. Scand. 84: 406–410.

Leonardi, A., A. Di Stefano, L. Motterle, B. Zavan, G. Abatangelo and P. Brun. 2011. Transforming growth factor-β/Smad-signalling pathway and conjunctival remodelling in vernal keratoconjunctivitis. Clin. Exp. Allergy 41: 52–60.

Novitskaya, E.S., S.J. Dean, J.P. Craig and A.B. Alexandroff. 2011. Current dilemmas and controversies in allergic contact dermatitis to ophthalmic medications. Clin. Dermatol. 29: 295–299.

Offiah, I. and V.L. Calder. 2009. Immune mechanisms in allergic eye diseases: what is new? Curr. Opin. Allergy ClinImmunol. 9: 477–481.

Rosario, N. and L. Bielory. 2011. Epidemiology of allergic conjunctivitis. Curr. Opin. Allergy Clin. Immunol. 11: 471–476.

Stahl, J.L. and N.P. Barney. 2004. Ocular allergic disease. Curr. Opin. Allergy Clin. Immunol. 4: 455–459.

Stapleton, F., S. Stretton, P.R. Sankaridurg, H. Chandoha and J. Shovlin. 2003. Hypersensitivity responses and contact lens wear. Contact Lens Anterior Eye 26: 57–69.

Teo, L., L. Lim, D.T. Tan, T.K. Chan, A. Jap and L.H. Ming. 2011. A survey of contact lens complications in Singapore. Eye Contact Lens 37: 16–19.

Yanni, J.M. and N.P. Barney. Ocular Allergy: clinical, Therapeutic and Drug Discovery Considerations. pp. 239–265. In: T. Yorio, A.F. Clark, M.B. Wax. [eds.]. 2008. Ocular therapeutics: eye on new discoveries. Elsevier, UK.

8

Cicatricial Conjunctivitis

Bety Yáñez Álvarez

SUMMARY

Cicatricial conjunctivitis comprises a group of disorders, characterized by scar formation in response to conjunctival inflammation affecting stromal layers of the conjunctiva. This is one of the ocular surface conditions more difficult to treat and currently represents a real challenge to preserve vision in these patients.

INTRODUCTION

Several infectous, physical agents and systemic mucocutaneuos autoimmune and not autoimmune diseases can produce conjunctival inflammation (Table 1). If any of these conditions affects conjunctival inner layers and tarsal tissue causing chronic inflammation, the healing response will be produce conjunctival fibrosis.

This chronic fibrosis, leads to the shortening of the conjunctival sac by the presence of symblepharon findings that characterize cicatricial conjunctivitis. Chronic inflammation can induce ocular surface disorders, abnormal position of the eyelids, dry eye and corneal neovascularization.

This conjunctival scarring may occur in some eyes as limited in time or chronically progressive. Chronic progressive type as presented in Stevens-Johnson syndrome/Toxic epidermal necrolysis, is leading to further damage the ocular surface and thus a higher risk of blindness caused by corneal opacity.

Recognize and diagnose the different causes of cicatricial conjunctivitis is important to the patient to receives the most appropriate medical

Department of Ophthalmology, Hospital Nacional Dos de Mayo, Cdra. 13 Av. Grau–Cercado de Lima, Lima, Peru; E-mail: byanez@hotmail.com

Table 1 This table summarize the main causes of cicatricial conjunctivitis.

Autoimmune conjunctivitis	Cicatricial pemphigoid Epidermolysis bullosa acquisita Linear IgA disease Dermatitis herpetiformis Bullous pemphigoid Paraneoplastic pemphigus Lichen planus pemphigoides
Allergic-atopic disorder	Atopic keratoconjunctivitis Stevens-Johnson syndrome/Toxic epidermal necrolysis
Systemic diseases	Sarcoidosis Progressive systemic sclerosis Sjögren's syndrome
Bacterial conjunctivitis	Trachoma Corynebacterium diphtheriae conjunctivitis Streptococcal conjunctivitis
Viral conjunctivitis	Adenoviral keratoconjunctivitis
Miscellaneous	Ocular rosacea Iatrogenic conjunctivitis Conjunctival trauma, chemical burns, or irradiation Porphyria cutanea tarda Erythroderma ichthyosiform congenital Self-induced cicatricial conjunctivitis

treatment because they may lead to severe side effects. In this chapter, three autoinmmune cicatricial conjunctivitis will be reviewed: the first one, caused by Stevens–Johnson syndrome (SJS), the second one caused by toxic epidermal necrolysis (TEN) and ocular mucous membrane pemphigoid (OMMP), and the last one a infectious cicatricial conjunctivitis secondary to Trachoma.

TOXIC EPIDERMAL NECROLYSIS AND STEVENS–JOHNSON SYNDROME

Definition

Stevens–Johnson syndrome (SJS) and toxic epidermal necrolysis (TEN), considered variants of a single disease, are severe acute-onset epidermolytic adverse mucocutaneous drug reactions, differing only by their extent of skin detachment (Harr and French 2010).

Epidemiology

SJS and TEN are rare diseases with an incidence of 1.9 cases of TEN per million inhabitants per year (Rzany et al. 1999). TEN and SJS have been observed worldwide and occur in all age groups but it is more common in elderly people.

The risk of SJS/TEN is major in HIV patients. Other factors associated to SJS/TEN are, herpes virus or *Mycoplasma pneumoniae*, radiotherapy, lupus erythematosus, and collagen vascular disease.

Clinical Features

SJS-TEN begins with a nonspecific flu-like illness for 1–3 days. Subsequently, a pruritic painful morbilliform eruption with initial lesions (poorly defined macules with darker centers) appears first on the face and upper trunk and may spread to the extremities that rapidly increase in number and size, with tendency to confluence and Nikolsky's sign.

Involvement of the buccal, ocular, genital mucosa occurs in more than 90% of patients but also internal organ involvement resulting in respiratory distress or profuse diarrhea. The mortality of SJS is approximately 5% and that of TEN 30–50% (Lebargy et al. 1997).

The reported incidence of ocular complications in SJS/TEN is 50–68% (Power et al. 1995). In the acute phase, SJS/TEN patients manifest severe conjunctivitis, conjunctival membrane or pseudomembrane formation or corneal erosion, and, in severe cases, to cicatrizing lesions, symblepharon, fornix foreshortening, and corneal ulceration (Chang et al. 2007) (Fig. 1).

In the chronic phase, corneal damage that can cause blindness is the most severe long-term complication for survivors of SJS/TEN. In a recent study (De Rojas et al. 2007) was reported the patterns of natural history of progression of ocular disease following onset of SJS/TEN and its classification in six types (Fig. 2).

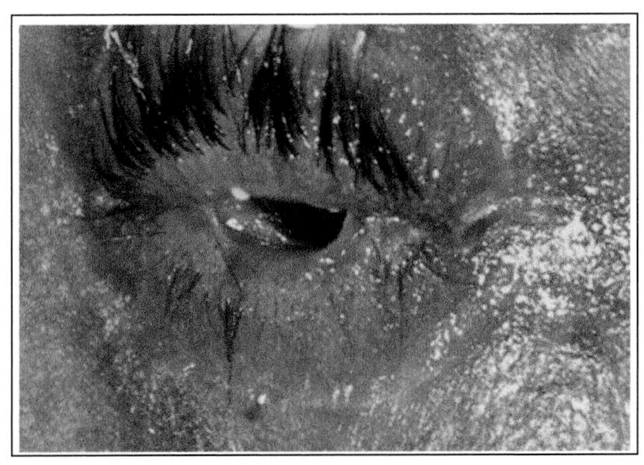

Figure 1 This figure shows the blepharoconjunctivitis and eyelid shortening in SJS.
Color image of this figure appears in the color plate section at the end of the book.

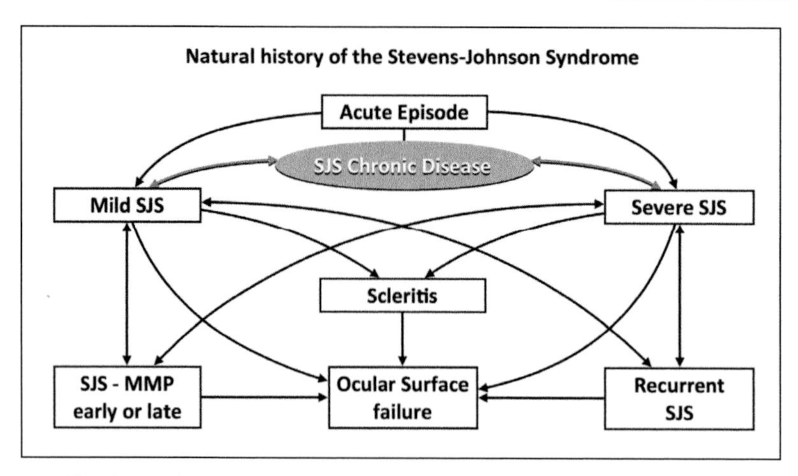

Figure 2 This figure shows the natural history of ocular disease in SJS/TEN.

Etiology and Pathogenesis

In Han and Thai Chinese population several studies confirmed the genetic susceptibility of individuals with HLA-B*1502 to carbamazepine (Chung et al. 2004). A strong association between HLA-B*5801 and SJS/TEN has been reported for allopurinol in Han Chinese, Japanese and Thai patients and in minor cases in patients of European origin.

Pathomechanism of SJS/TEN

To date CD8 T-cells as well as the cytolytic molecules FasL and granulysin are key players in the pathogenesis of SJS/TEN (Harr and French 2010).

Drugs

Drugs identified as having a high risk of inducing SJS/TEN are allopurinol, sulfamethoxazol/trimethoprim (SMX/TMP), sulfonamides, carbamazepine, antiepileptics, nevirapine and oxicam nonsteroidal inflammatory drugs (NSAID's) (Rzany et al. 1999).

Diagnosis

The acute prodromal systemic phase, with the history of drug ingestion followed by ocular and skin lesions is suggestive for the diagnosis of SJS/TEN. The Nikolsky sign is not specific for SJS/TEN.

Immunohistochemistry study showed deposits of immunoglobulins and complement components along the conjunctival epithelial basement membrane zone (BMZ).

Differential Diagnosis

Autoimmune blistering diseases are the principal differential diagnosis of SJS/TEN including linear IgA dermatosis and paraneoplastic pemphigus, pemphigus vulgaris and bullous pemphigoid, acute generalized exanthematous pustulosis (AGEP), disseminated fixed bullous drug eruption and staphyloccocal scalded skin syndrome (SSSS).

OCULAR MUCOUS MEMBRANE PEMPHIGOID

Definition

Ocular mucous membrane pemphigoid (OMMP), previously known as ocular cicatricial pemphigoid (OCP), is a blinding systemic autoimmune disease characterized by recurrent episodes of inflammation and progressive fibrosis of the conjunctiva and other mucosal surfaces (Thorne et al. 2004).

Etiology

Several epithelial BMZ components have been recognized by autoantibodies in the serum of OMMP patients, including BP230, BP180, LAD-1, laminin-332, type VII collagen, α6-integrin, and β4-integrin subunit (Chan et al. 1999, Oyama et al. 2006).

The presence of the HLA-DR4 allele substantially increases the risk of ocular disease (Zaltas et al. 1989). Furthermore, a prevalence of HLA-DQB1*0301 was first described in patients with OMMP (Ahmed et al. 1991).

Epidemiology

Its incidence is from 1: 12,000 to 1: 40,000, although it is believed that there would be underdiagnosed, especially in early stages of the disease (Nguyen and Foster 1996). Age standard of diagnosis is between 60 and 80 years, but has been reported in children. A predominance in females with 1,5 –2,2:1 a relationship.

Clinical Features

In OMMP, conjunctival lesions usually start in one eye but involve the other within a few years. The onset of symptoms (redness, burning, foreign body sensation, tearing, photophobia and blurred vision) may be a simple catarrhal conjunctivitis which lasts for years, with alternating periods of activity and remission.

Ocular complications of OMMP result from conjunctival erosions with subsequent scarring, symblepharon formation and progressive cicatrization with foreshortening of the fornices, anquiloblepharon (Elder and Collin 1996). The scarring of the conjunctiva also results in loss of goblet cells with subsequent dry eye and disruption of the corneal epithelium.

Eyelid malposition is also a common occurrence, with entropion and trichiasis resulting from the scarring process of the conjunctiva (Shiu and McNab 2005). All these changes result in chronic ocular surface disease contributing to corneal scarring and neovascularization (Fig. 3). Blindness and even loss of the globe due to perforation and endophthalmitis may occur (Nouri et al. 2001).

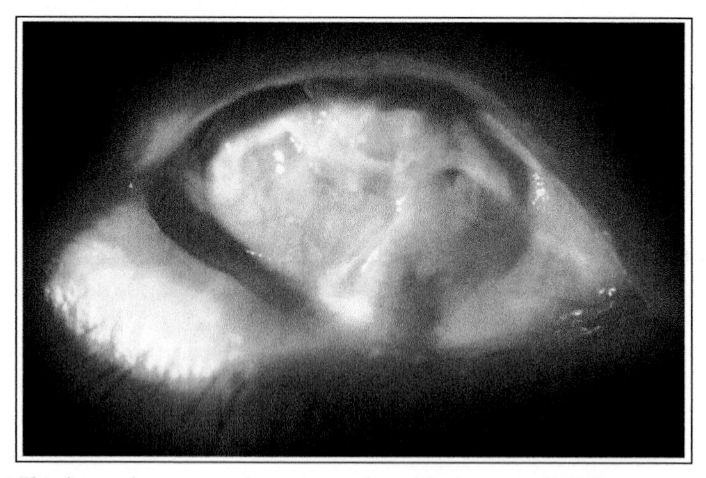

Figure 3 This figure shows corneal scarring and symblepharon in OMMP.

Color image of this figure appears in the color plate section at the end of the book.

Diagnosis

Direct immunofluorescence (DIF) investigation of a conjunctival biopsy is the most reliable diagnostic method that reveals linear deposition of IgG, C3, or less commonly IgA along the basement membrane zone in OMMP (Leonard et al. 1988). Only 80% of patients with OMMP have a positive conjunctival biopsy. In the case of a negative biopsy, the demonstration in the serum of auto-antibodies against BMZ antigens would be helpful in establishing the diagnosis (Higgins et al. 2006).

Differential Diagnosis

The following blistering diseases must be considered as differential diagnosis, bullous pemphigoid, epidermolysis bullosa acquisita, linear IgA bullous disease and bullous systemic lupus erythematosus.

In some patients who use different drugs (topical glaucoma medications, clonidine, practolol) has been reported a similar process limited only to the conjunctiva, called pseudo-ocular cicatricial pemphigoid (P-OCP).

TRACHOMA

Definition

It is a chronic keratoconjunctivitis caused by infection with *Chlamydia trachomatis* and is characterized by inflammatory changes in the conjunctiva in children with subsequent scarring, corneal opacity and blindness in adults.

Etiology

Endemic trachoma is caused by the four ocular serotypes of *C. trachomatis* (A, B, Ba, and C) (WHO 2003, Brunham et al. 1990).

C. trachomatis is probably transmitted between individuals by a variety of mechanisms, including: a) direct spread from eye to eye during close contact such as during play or sleep, b) spread of infected ocular or nasal secretions on fingers, c) indirect spread by fomites such as infected face-cloths, d) transmission by eye-seeking flies. No non-human reservoir of infection has been found, with flies only acting as passive vectors (Hu et al. 2010).

Epidemiology

Trachoma is a major cause of blindness in many less-developed countries, especially in poor, rural areas (Reskinoff et al. 2004). Blinding trachoma is believed to be endemic in over 50 countries, with the highest prevalence of active disease and trichiasis in Africa (Evans and Solomon 2011). Risk factors for trachoma are recognized associated to inadequate conditions of hygiene and poverty as water scarcity, limited access to latrines and crowded living conditions.

Active trachoma is predominantly seen in young children with equal prevalence in male and female, becoming less frequent and shorter in duration with increasing age (Dolin et al. 1998). Conjunctival scarring accumulates with age, usually becoming evident in the second or third decade of life (West et al. 1991).

Clinical Features

Active disease is characterized by recurrent episodes of chronic, follicular conjunctivitis. The World Health Organization (WHO) Simplified Trachoma Grading System (Table 2), which is used by trachoma control programs, subdivides active trachoma into two often coexisting clinical phenotypes: Trachoma Inflammation Follicular (TF) and Trachoma Inflammation Intense (TI) (Thylefors et al. 1987). Individuals are frequently asymptomatic or have only mild symptoms even if marked signs of inflammation are evident. If present, symptoms are similar to those associated with any chronic conjunctivitis: redness, discomfort, tearing, photophobia and scant mucopurulent discharge. Conjunctival follicles at the upper margin of the cornea leave shallow depressions after they resolve known as 'Herbert's pits' which, unlike follicles and papillae, are a pathognomonic sign of trachoma. In some eyes is also common to find severe follicular reaction on the conjunctiva of the everted upper lid, papillary hypertrophy that can obscure the deep tarsal vessels if severe enough, and upper cornea pannus in active disease.

Initially, conjunctival scarring is seen in the subtarsal conjunctiva, which can range from a few linear or stellate scars to thick, distorting bands of fibrosis. This conjunctival scarring could be strongly associated with non-chlamydial bacterial infection as shown by the results of a recent study (Hu et al. 2011).

Contraction of this scar tissue causes entropion and trichiasis. The most serious disease consequence from trachoma is blinding corneal opacification secondary a trichiasis. Other factors probably contribute for corneal damage as bacterial infection and chronic conjunctival inflammation (Evans and Solomon 2011).

Table 2 The simplified WHO system for the assessment of trachoma.

Grade	Description
TF	Trachomatous inflammation—Follicular: The presence of five or more follicles (>0.5 mm) in the upper tarsal conjunctiva
TI	Trachomatous inflammation—Intense: Pronounced inflammatory thickening of the tarsal conjunctiva that obscures more than half of the deep normal vessels
TS	Trachomatous scarring: The presence of scarring in the tarsal conjunctiva
TT	Trachomatous trichiasis: At least one lash rubs on the eyeball
CO	Corneal opacity: Easily visible corneal opacity over the pupil

Diagnosis

Diagnosis of trachoma is based on clinical signs. Various diagnostic tests have been used to detect *C. trachomatis*, but there is no 'Gold Standard' test.

Nucleic acid amplification tests, such as polymerase chain reaction (PCR) and quantitative real-time PCR (rt-PCR) are the current used for *C. trachomatis* detection (Solomon et al. 2003, Bailey et al. 1994).

Differential Diagnosis

Several conditions can produce a chronic follicular conjunctivitis, including conjunctivitis caused by viruses (e.g. adenovirus) and bacteria (e.g. *Staphylococcus aureus* and *Moraxella*).

Others causes of entropion and trichiasis that should be considered in the differential diagnosis, although most of these are relatively rare in trachoma endemic regions as cicatricial conjunctivitis caused by MMP, SJS, chemical injuries and drugs.

CONCLUSIONS

Cicatricial conjunctivitis is one of the ocular surface conditions more difficult to treat and represent a real challenge to eye care practitioners.

This chapter has summarized the most important conditions (Stevens–Johnson syndrome, toxic epidermal necrolysis, ocular mucous membrane pemphigoid and Trachoma) presenting the clinical feature, facilitating differential diagnosis and detailing the appropriate patient management to minimize the side effects to preserve patient's vision and their quality of live.

REFERENCES

Ahmed, A.R., C.S. Foster, M.M. Zaltas, G. Notani, Z. Awdeh, C.A. Alper and E.J. Yunis. 1991. Association of DQw7 (DQB1*0301) with ocular cicatricial pemphigoid. Proc. Natl. Acad. Sci. USA 88: 11579–11582.

Bailey, R.L., T.J. Hampton, L.J. Hayes, M.E. Ward, H.C. Whittle and D.C. Mabey. 1994. Polymerase chain reaction for the detection of ocular chlamydial infection in trachoma-endemic communities. J. Infect. Dis. 170: 709–712.

Brunham, R.C., M. Laga, J.N. Simonsen, D.W. Cameron, R. Peeling, J. McDowell, H. Pamba, J.O. Ndinya-Achola, G. Maitha and F.A. Plummer. 1990. The prevalence of Chlamydia trachomatis infection among mothers of children with trachoma. Am. J. Epidemiol. 132: 946–952.

Chan, R.Y., K. Bhol, N. Tesavihul, E. Letko, R.K. Simmons, C.S. Foster and A.R. Ahmed. 1999. The role of antibody to human β4-integrin in conjunctival basement membrane separation: possible *in vitro* model for ocular cicatricial pemphigoid. Invest Ophthalmol. Vis. Sci. 40: 2283–2290.

Chang, Y.S., F.C. Huang, S.H. Tseng, C.K. Hsu, C.L. Ho and H.M. Sheu. 2007. Erythema multiforme, Stevens–Johnson syndrome, and toxic epidermal necrolysis: acute ocular manifestations, causes, and management. Cornea 26: 123–129.

Chung, W.H., S.I. Hung, H.S. Hong, M.S. Hsih, L.C. Yang, H.C. Ho, J.Y. Wu and Y.T. Chen. 2004. Medical genetics: a marker for Stevens-Johnson syndrome. Nature 428: 486.

De Rojas, M.V., J.K.G. Dart and V.P.J. Saw. 2007. The natural history of Stevens–Johnson syndrome: patterns of chronic ocular disease and the role of systemic immunosuppressive therapy. Br. J. Ophthalmol. 91: 1048–1053.

Dolin, P.J., H. Faal, G.J. Johnson, J. Ajewole, A.A. Mohamed and P.S. Lee. 1998. Trachoma in The Gambia. Br. J. Ophthalmol. 82: 930–933.

Elder, M.J. and R. Collin. 1996. Anterior lamellar repositioning and grey line split for upper lid entropion in ocular cicatricial pemphigoid. Eye 10:439–442.

Evans, J.R. and A.W. Solomon. 2011. Antibiotics for trachoma. Cochrane Database Syst. Rev. 16; 3: CD001860.

Harr, T. and L.E. French. 2010. Toxic epidermal necrolysis and Stevens–Johnson syndrome. Orphanet. J. Rare Dis. 16; 5: 39.

Higgins, G.T., R.B. Allan, R. Hall, E.A. Field and S.B. Kaye. 2006. Development of ocular disease in patients with mucous membrane pemphigoid involving the oral mucosa. Br. J. Ophthalmol. 90(8): 964–967.

Hu, V.H., E.M. Harding-Esch, M.J. Burton, R.L. Bailey, J. Kadimpeul and D.C. Mabey. 2010. Epidemiology and control of trachoma: systematic review. Trop. Med. Int. Health 15: 673–691.

Hu, V.H., P. Massae, H.A. Weiss, C. Chevallier, J.J. Onyango, I.A. Afwamba, D.C.W. Mabey, R.L. Bailey and M.J. Burton. 2011. Bacterial infection in scarring trachoma. Invest Ophthalmol. Vis. Sci. 52: 2181–2186.

Lebargy, F., P. Wolkenstein, M. Gisselbrecht, F. Lange, J. Fleury-Feith, C. Delclaux, J. Roupie, E. Revuz and J.C. Roujeau. 1997. Pulmonary complications in toxic epidermal necrolysis: a prospective clinical study. Intensive Care Med. 23: 1237–1244.

Leonard, J.N., C.M. Hobday, G.P. Haffenden, C.E. Griffiths, A.V. Powles, P. Wright and L. Fry. 1988. Immunofluorescent studies in ocular cicatricial pemphigoid. Br. J. Dermatol. 118: 209–217.

Nguyen, Q.D. and C.S. Foster. 1996. Cicatricial pemphigoid: diagnosis and treatment. Int. Ophthalmol. Clin. 36: 41–60.

Nouri, M., H. Terada, E.C. Alfonso, C.S. Foster, M.L. Durand and C.H. Dohlman. 2001. Endophthalmitis after keratoprosthesis: incidence, bacterial causes, and risk factors. Arch. Ophthalmol. 119: 484–489.

Oyama, N., J.F. Setterfield, A.M. Powell, Y. Sakuma-Oyama, S. Albert, B.S. Bhogal, R.W. Vaughan, F. Kaneko, S.J. Challacombe and M.M. Black. 2006. Bullous pemphigoid Antigen II (BP 180) and its soluble extracellular domains are the mayor autoantigens in mucous membrane pemphigoid: the pathogenic relevance to HLA class II alleles and disease severity. Br. J. Dermatol. 154: 90–98.

Power, W.J., M. Ghoraishi, J. Merayo-Lloves, R.A. Neves and C.S. Foster. 1995. Analysis of the acute ophthalmic manifestations of the erythema multiforme/Stevens–Johnson syndrome/ toxic epidermal necrolysis disease spectrum. Ophthalmology 102: 1669–1676.

Reskinoff, S., D. Pascolini, D. Etya'ale, I. Kocur, R. Pararajasegaram, G.P. Pokharel and S.P. Mariotti. 2004. Global data on visual impairment in the year 2002. Bull World Health Organ. 82: 844–851.

Rzany, B., O. Correia, J.P. Kelly, L. Naldi, A. Auquier and R. Stern. 1999. Risk of Stevens–Johnson syndrome and toxic epidermal necrolysis during first weeks of antiepileptic therapy: a case-control study. Study Group of the International Case Control Study on Severe Cutaneous Adverse Reactions. Lancet 353: 2190–2194.

Shiu, M. and A.A. McNab. 2005. Cicatricial entropion and trichiasis in an urban Australian population. Clin. Experiment Ophthalmol. 3: 582–585.

Solomon, A.W., M.J. Holland, M.J. Burton, S.K. West, N.D. Alexander, A. Aguirre, P.A. Massae, H. Mkocha, B. Munoz, G.J. Johnson, R.W. Peeling, R.L. Bailey, A. Foster and D.C. Mabey. 2003. Strategies for control of trachoma: observational study with quantitative PCR. Lancet 362: 198–204.

Thylefors, B., C.R. Dawson, B.R. Jones, S.K. West and H.R. Taylor. 1987. A simple system for the assessment of trachoma and its complications. Bull World Health Organ 65: 477–483.

Thorne, J.E., G.J. Anhalt and D.A. Jabs. 2004. Mucous membrane pemphigoid and pseudopemphigoid. Ophthalmology 11: 45–52.

West, S.K., B. Munoz, V.M. Turner, B.B. Mmbaga and H.R. Taylor. 1991. The epidemiology of trachoma in central Tanzania. Int. J. Epidemiol. 20: 1088–1092.

World Health Organization (WHO). 2003. Report of the 2nd Global Scientific Meeting on Trachoma. Geneva, Switzerland, 25–27 August, 2003. WHO/PBD/GET 03. 1: 1–28.

Zaltas, M.M., R. Ahmed and C.S. Foster. 1989. Association of HLA-DR4 with ocular cicatricial pemphigoid. Curr. Eye Res. 8: 189–193.

9

Ocular Surface and Glaucoma

Rodrigo Martín Torres

SUMMARY

Glaucoma and Ocular Surface Disease (OSD) are pathologies with high prevalence around the world, which increase with age. Both require topical treatments for their management. However, detrimental effects of topical treatments for glaucoma could be developed by different mechanisms and could be produced by the drug itself or by preservatives. Moreover, a patient with glaucoma will need topical treatment for many years or decades. Adverse effects with glaucoma eye drops could be observed after long-term treatment or immediately after their first instillation. To decrease this problem, there are improved new eye drops formulations, some of them without preservatives, which are preferred for patients with OSD.

Glaucoma surgery could also negatively affect the ocular surfaces (OS) with toxicological effects of non-healing drugs as 5-Fluorouracil (5-FU) or Mitomycin C (MMC) and due to OS anatomical modification in filtration surgery (trabeculectomy). On the other hand, people with OSD who need glaucoma surgery, will have more incidents of surgical failure due to long time OS inflammation which will rapidly increase healing and scarring after surgery.

Patients with glaucoma need a careful OS examination, at the beginning and during their treatment, as scientific data shows that many of them will develop OSD throughout their lives, which can lead to worse topical treatment compliance (worse IOP control) and worse surgical outcome. Both pathologies glaucoma and OSD, if not appropriately managed could end in blindness.

Centro de Ojos Dr. Lódolo. Asociación Entrerriana de Oftalmología, Las Calandrias 4789 Barrio Las Acaias, Colonia Avellaneda, 3107, Entre Ríos–Argentine; E-mail: romator7@hotmail.com

INTRODUCTION

Glaucoma is an optic neuropathy with high prevalence around the world (Friedman et al. 2004). It is the second leading cause of blindness and is expected that 79.6 million people will be affected with glaucoma by 2020 (Quigley and Broman 2006).

Elevated intraocular pressure (IOP) is considered the greatest risk for its development, because of that, most treatments are indicated for decreasing IOP. Many patients undergo the instillation of eye drops once to twice a day, and sometimes it is necessary to combine two or more therapeutic drugs. If glaucoma can not be controlled by topical therapy, surgical treatments must be performed.

The ocular surface (OS) is the place where the eye drops will be absorbed. Moreover, in the OS different pathologies could be developed and "Dry Eye Syndrome" is one of the most frequently observed (McCarty et al. 1998). If the OS is damaged eye drops absorption could be altered, patients with glaucoma will suffer discomfort and their treatment compliance could be decreased. Without treatment, optic nerve damage will progress. However, long-term anti-glaucoma treatments are usually associated with some kind of OS toxicity, by the glaucoma drug itself or by other components of the formulations, as well as preservatives (Kalavala and Statham 2006). Because of that, new anti-glaucoma drops are seeking to decrease adverse effects and one of their targets is to improve the preservatives or to develop preservatives-free drops (Andres-Guerrero et al. 2011).

When glaucoma progress and pharmacological therapy are not enough, surgical treatment will be necessary. OS is important, because some surgical techniques could well disturb the OS anatomy and anti-metabolites as well as 5 Fluorouacil (5-FU) or Mitomycin C (MMC) could have harmful effects for the OS or for intra-ocular tissues (Shapiro et al. 1985, McDermott et al. 1994, Mietz et al. 1996, Georgopoulos et al. 2000, Vass et al. 2000, Dogru et al. 2003). Moreover, if a patient with glaucoma has an ocular surface disease (OSD) and needs surgery, he has more possibilities to develop post-operative complications and surgical failure. In this chapter, the relationship between OS and glaucoma will be described (Fig. 1), with the aim to review and actualize the knowledge and to provide tools to eye care practitioners to detect, understand and manage the problem of OS disease in glaucoma patients.

THE OCULAR SURFACE: "THE EYE DROPS ENTRY DOOR"

The lids, the eyelashes, the bulbar and tarsal conjunctiva, the limbus, the cornea, the principal and accessories lachrymal glands and finally the tear film, works together to care and to protect the corneal transparency (Rolando and Zierhut 2001).

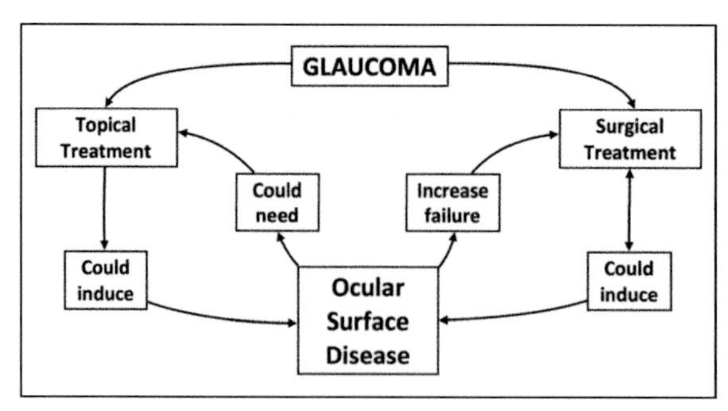

Figure 1 This figure shows the relationship between Glaucom and OSD.

This is an anatomical and functional unit called OS. Eye drops get in contact with this structure and drugs will be absorbed to release their pharmacological effect in the eye.

If the OS have problems, eye-drops absorption could be altered (increasing or decreasing their pharmacological or toxicological effects). Moreover, if the OS is healthy when the glaucoma treatment starts, a prolonged therapy will affect the OS and many patients could develop some kind of OSD, most frequently, the dry eye syndrome (Boudouin 1996, Asbell and Potapova 2005, Leung et al. 2008, Fechtner et al. 2010).

Also, most patients undergo the instillation of eye drops once or twice a day, and it is not unusual to find therapies that combine two or more drugs. If a patient needs to use a therapy which must be continued throughout their lifetime and this therapy causes discomfort, the treatment will probably be discontinued or induce a lack of patient compliance and glaucoma damage will progress. But, if the patient resists the pain, the redness and/or the itching, because "the doctor said" that it is important to prevent blindness and the drops are working "well" to control the glaucoma, OSD will progress and sometimes, severe visual impairment could well come by limbal stem cell deficiency and corneal transparency loss (Thorne et al. 2004).

GLAUCOMA DRUGS AND OS: WHERE IS THEIR RELATIONSHIP?

Glaucoma is a condition of the eye in which there is usually an elevation of the IOP that leads to progressive cupping and atrophy of the optic nerve head, deterioration of the visual fields and ultimately to blindness. Primary open-glaucoma is the most common type of glaucoma (Quigley 1996). Drugs used in the therapy include a variety of agents with different mechanism of action (Zimmerman 2000). The therapeutic goal is reducing the IOP (Rivera et al. 2008). Higher IOP provokes optic nerve damage and induce glaucomatous visual field loss. Reduction of the IOP may be accomplished by decreasing the rate of production or an increase of the rate of outflow (drainage) of aqueous humor.

Topical application is the most common route for glaucoma eye drops (Bartlell 2000). Most topical ocular preparations are commercially available as solutions or suspensions that are applied directly to the OS from the bottle, which serves as the eyedropper. Moreover in the drug itself (active agent) different inactive ingredients may be presented in ophthalmic products, as it shows the Table 1 (Bartlell 2000).

Table 1 This table summarize the inactive ingredients and their function.

Inactive ingredients	Function
Preservatives	Destroy or inhibit multiplication of microorganisms introduced into the product
Viscosity-Increasing Agents	Slow drainage of the product from the eye, increasing retention time of the active drug
Antioxidants	Prevent or delay deterioration of products by oxygen in the air
Wetting agents	Reduce surface tension, allowing drug solution to spread
Buffers	Help to maintain ophthalmic products in the range of pH 6 to 8
Tonicity agents	Helps ophthalmic products to be isotonic with the pre-ocular tear film

Glaucoma Eye Drops and OS: Preservative Versus Preservative-free

Compliance with any medical therapy is difficult, but it is easy to see why it is a major problem in glaucoma, where studies have shown that many patients do not take their eye medications conscientiously (Eamon et al. 2008). If a patient feels uncomfortable with the treatment, it will probably be suspended. Discomfort could be due to toxicity and adverse effects. These could be related by the drugs themselves or by the inactive ingredients, principally the preservatives (Mantelli et al. 2011).

Most eye drop formulations are used in multidose bottles that require the inclusion of an antimicrobial preservative in the solutions. A direct correlation has been reported between the presence of preservatives with symptoms and adverse effects experienced during antiglaucoma treatments (Boudouin and de Lunardo 1998, Jaenen et al. 2007, Kahook and Noecker 2008, Brasnu et al. 2008, Bai et al. 2010). Preservatives can cause epithelial toxicity, which get worse with time, as it happens with long-term treatments for glaucoma. Different *in vitro* and clinical studies have demonstrated their detrimental OS effect (Yee 2007, Becquet et al. 1998). Long-term and even 1-mon use of antiglaucoma drops with the cationic detergent benzalkonium chloride (BAK) induce subclinical inflammation of the conjunctiva with overexpression of human leukocyte antigen (HLA)-DE on epithelial cells (Baudouin et al. 2004, Rodrigues 2009).

BAK is the most commonly used preservative in topical ophthalmic formulations (0.025%–0.004%) and is especially toxic to the ocular surface cells (Debbasch et al. 2001, Cha et al. 2004). It could lead to the disruption of the hydrophobic barrier of the corneal epithelium. This is a side effect, however it permits an enhancement of their active compounds penetration and, subsequently, an improvement of the anti-glaucoma efficacy (Baudouin et al. 2010). BAK has been used as a preservative for more than 50 yr and meets the criteria required by both the United States Pharmacopoeia (USP) and the European Pharmacopoeia. When compared with new formulations, there are no statistical significant differences in rates of dry eye, ocular infection, or OSD (Schwartz et al. 2011).

However, the pharmacological industry is working to develop and to improve preservative-free formulations, supported by scientific data which shows that OS improve after their transition from BAK preserved to BAK-free treatments (Aihara et al. 2011). Nevertheless, preservative-free formulations are available in single-use vials or in special devices (such as the ABAK system) that, although still affordable, are more expensive. Another option is related to a reduction of the number of applications of the formulations, with substances as bio-adhesive polymers to enhance the viscosity of the ophthalmic vehicles, to reduce the drainage rate of the drugs and subsequently to increase their therapeutic efficacy (Andres-Guerrero 2011). Furthermore, some of the polymers are protective of the eye, and so their use is more appropriate for chronic treatments, such as the therapy for dry eye syndrome (Andres-Guerrero 2011).

There is an interesting study, which researched the conjunctival modifications in ocular hypertension and primary open-angle glaucoma, this could be added as another possible cause of ocular surface changes in glaucoma patients. The authors found the presence of epitheilal mycrocysts in hypertensive and glaucomatous eyes, and theorized that it could be related to one of the possible hydrodynamic pathways activated in hyperbaric ocular conditions instead of a side effect of antiglaucoma drugs or preservatives on conjuntvial epithelium (Ciancaglini et al. 2008).

Glaucoma Topical Treatment and Adverse Effects

The "perfect" glaucoma topical therapy must stop glaucoma progress, with no ocular side effects, with the lowest instillation frequency as possible. The present glaucoma therapeutic options are briefly described.

Prostaglandins Analogue

The prostaglandins analogues (Pgs) (latanoprost, bimatoprost, travatoprost, tafluprost) are very effective in decreasing IOP 25–30% by increasing uveal-scleral outflow (Zimmerman 2000). One of their main advantages over other treatments is that this drug needs to be instilled only once a day. This is very important for the patient's compliance.

However they were not recommended as the first line glaucoma therapy at the beginning of their ophthalmic use, because of their side effects (Zimmerman 2000), however they are now accepted worldwide and prescribed to begin a glaucoma treatment. Some of the problems with Pgs analogues are ocular redness (Fig. 2), stinging and burning, the possibility of the iris color changing (change from a light to darker iris), (Fig. 3) the growth of their eyelashes is effected (Feldman 2003).

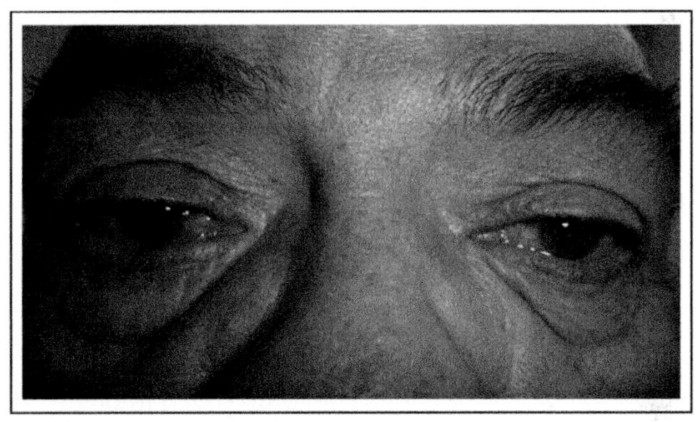

Figure 2 In this figure, a patient with pterigyum in his right eye shows a chronic redness in both eyes due to prostaglandins analogue treatment (Travatoprost).

Color image of this figure appears in the color plate section at the end of the book.

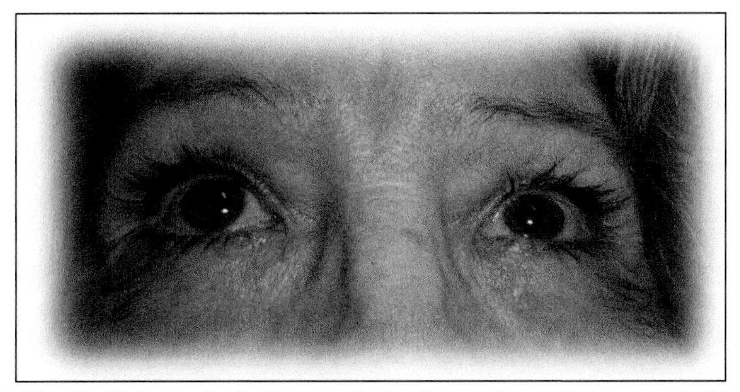

Figure 3 After the first month of using prostaglandins analogue (Latanoprost), this patient asked the ophthalmologist about the eyelashes... *principally the new eyelashes* in the skin, as it is possible to see in this figure at the bottom of the right and left.

Color image of this figure appears in the color plate section at the end of the book.

Moreover, there are possible complications in patients with ocular inflammation (uveitis or due to any surgical procedure) or with ocular infections. Most of the OSD (allergic keratoconjunctivitis, dry eye, blepharitis) are in part, an inflammatory disease, so in these patients Pgs analogues must be managed with care or another glaucoma treatment should be chosen.

Beta-blockers

Beta-blockers are a well known glaucoma treatment, which need to be instilled twice a day, but there are important contraindications and warnings in patients with asthma and/or cardiovascular diseases. Their IOP lowering effect could be increased in combination with another drug (such as dorzolamide, brinzolamide or brimonidine) in the same ophthalmic formulation. It needs to be instilled twice a day, however with chronic use, cases have been described of corneal "conjunctivalization" developing pseudo-pemphigoid disease (Foster and Sains de la Maza 2004, Thorne et al. 2004).

Alpha-2 Adrenergic Agonist

Alpha-2 adrenergic agonist (Apraclonidine, Brimonidine), are drugs more frequently associated with allergic reactions (approximately in 30% of patients) more with Apraclonidine hydrochloride than with Brimonidine tartrate. Usually, these agents were added to another glaucoma therapy, to increase their anti-glaucoma effect, and also need to be instilled twice a day. Atopic dermatitis could be associated to ocular allergy (Fig. 4). Patients with cardiovascular disease could suffer adverse effects.

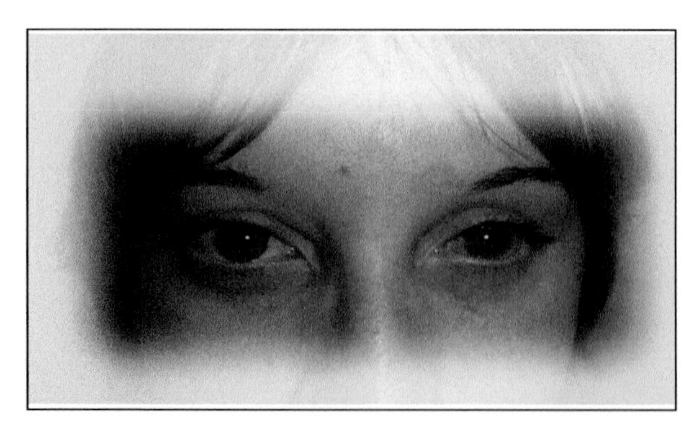

Figure 4 A female patient treated with Brimonidine in a fixed combination associated with Timolol, only in her left eye. Moreover the redness in the eye shows signs of atopic dermatitis at the periocular skin (eczematous lesions).

Color image of this figure appears in the color plate section at the end of the book.

Carbonic Anhydrase Inhibitors

Carbonic Anhydrase Inhibitors (dorzolamide, brinsolamide), principally decrease the aqueous humor production with some effect increasing its outflow. It needs to be instilled twice a day and usually in combination with another glaucoma drug (in the same or in a different eyedropper).

GLAUCOMA TREATMENT INDUCE OSD: MECHANISMS

A high prevalence (48.4–59%) of OSD symptoms were reported by patients using anti-glaucoma eye drops (Fechtner et al. 2010, Leung et al. 2008). Principally, the patients reported symptoms of dry eye and severe symptomatology on the OSDI questionnarire, however, the correlation with results of the OSD clinical test was poor (Leung et al. 2008). Because of this, it is important to understand the physiopathology between OSD induced by antiglaucoma eye drops.

An overexpression of two chemokine receptors in the conjunctival epithelium of glaucoma patients treated over a long period had been described (Baudouin et al. 2008), and the simultaneous overexpression of CCR4 and CCR5, suggesting that chronic use of topical treatments may stimulate both the Th1 and Th2 systems simultaneously. These findings suggest that inflammatory mechanisms combining allergy with toxicity are present together, ocurring simultaneously in the OS of glaucoma patients (Baudouin C et al. 2008).

OSD DUE TO GLAUCOMA MEDICATION: DIFFERENT CLINICAL PRESENTATION

The main OSD related with glaucoma medication are dry eye, allergy and Pseudo-pemphigoid.

Dry Eye

Chronic topical anti-glaucoma therapy has been shown to have detrimental effects on the ocular surface (Fechtner et al. 2010), which can include destabilization of the tear film and hyperosmolarity (International Dry Eye Worshop 2007). According to the 2007 International Dry Eye Workshop report, topical glaucoma medications and preservatives are responsible for a form of extrinsic evaporative dry eye induced by their pathologic effects on the ocular surface.

Also, the study "German Glaucoma and Dry Eye Register" in 2008 concluded that the type of glaucoma or eye drops played a pivotal role in dry eye occurrence and that the effect of dry eye could have also influenced the subjects´compliance with glaucoma medications (Erb et al. 2008). With this theory, the idea of "osmoprotection" in patients with glaucoma support that concurrent administration of artificial tears with antiglaucoma topical treatments in order to improve compliance, prognosis and possibly the efficacy of glaucoma surgery (Monaco et al. 2010).

Allergy

The typical clinical presentation of adverse allergy in a glaucoma patient is an eczematous lesion on the lids, swelling (Fig. 5), chemosis, intensive itching, and conjunctival hyperemia with papilar reaction. It appears in the first week when the patient starts the treatment. Normally, if the anti-glaucoma agent is discontinued, the signs and symptoms could be resolved, but sometimes it is necessary to add an anti-inflammatory and/or an anti-allergic topical agent. In these cases, it could be easy to detect the eye drop that led to the allergy as it was recently included.

In patients using two or more glaucoma eye drops, complaints could frequently be more progressive, after many years of treatment. At slit lamp, papilar reaction in the conjunctiva, keratitis punctata, tear film instability, are the most frequent signs observed. One option could be to suspend all treatment for two days, perform a new evaluation and restart the glaucoma treatment one by one, trying to detect which drug is inducing the allergy. Another possibility in the same case, is evaluate the eye drops (active and inactive ingredients) and only stop the instillation of one of the drugs. In accordance with the scientific data discussed above, if the patient is using Brimonidine twice daily and Latanoprost at night, keep treating the patient only with Latanoprost and evaluate if signs and symptoms resolve. Or, if the patient is treated with eye drops with preservatives, the possibility of changing the treatment to non-preservative eye drops could be considered.

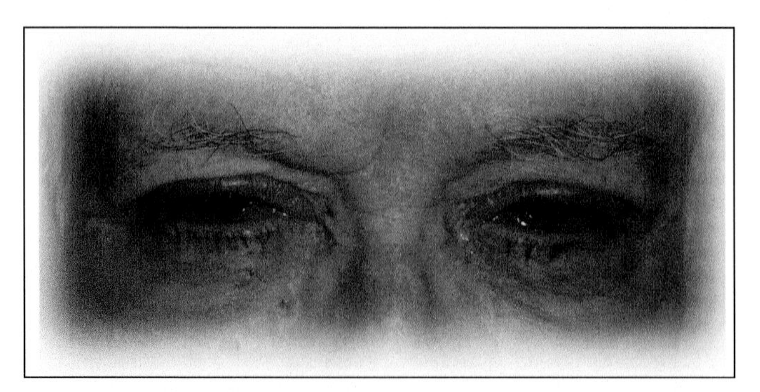

Figure 5 Patient with a typical bilateral allergic reaction with swelling of the lids, conjunctival redness, chemosis, and itching after starting the treatment with Brimonidine eye drops.

Color image of this figure appears in the color plate section at the end of the book.

Pseudo-pemphigoid

Pseudo-pemphigoid is one of the most severe long-term complications. Glaucoma medications are one of the most frequently presumed causes of this immunological disease (Thorne et al. 2004). The clinical characteristics of ocular mucous membrane pemphigoid and pseudo-pemphigoid are similar, therefore, immunohistologic evaluation of biopsied tissue is needed to confirm the diagnosis and perform the right treatment. The typical clinical course of this disease is characterized by slow progression from chronic conjunctivitis to sub-epithelial fibrosis, fornix foreshortening, symblepharon, and ankyloblepharon formation with OS keratinization and ultimately leads to blindness (Foster and Sains de la Maza 2004).

A patient with glaucoma and OS problems is always a challenge and no "simple" way exists to manage them. It is common to find that a patient with glaucoma using several kinds of eye drops (two different eye drops for glaucoma, one or two artificial tears, some anti-allergic eye drops, decongestants eye drops and even corticosteroids eye drops) for many years. While adding different eye drops (lubricants, anti-inflammatory, anti-allergic), practitioners are also including various potential allergic and toxicological causes of the initial problem. Moreover, many of these eye drops have the same preservative which one wants to avoid from glaucoma eye drops.

Moreover from the clinical picture, for many patients it could be relevant to establish the diagnosis by performing an OS laboratory test, as an example to detect IgE in tears (which led to the allergy) or to detect the lactoferrin levels and to correlate with tears osmolarity, which could be useful in dry eye diagnosis and management, as well as impression cytology (Merayo-Lloves et al. 2004).

Preserved eye drops are far more toxic than unpreserved eye drops, but both may induce alterations of the OS related to dose, frequency and duration of the treatment. These negative effects appear to be more easily elicited in predisposed patients, such as hyperreactive patients with underlying ocular allergy, elderly patients with dry eye and patients affected by severe ocular disease that need chronic therapy. However, some cases of OSD associated to glaucoma eye drops could be really serious and there are patients for whom none of the present glaucoma topical treatments are tolerated well. Moreover, if the glaucoma treatment was suspended, and the OS was improved, the problem of glaucoma would remain and a surgical treatment could be considered. From this premise, many ophthalmologists recommend glaucoma surgery from the beginning to avoid years and years of topical treatments, which will raise inflammatory cells and increase the possibility of post-operative complications (Sherwood et al. 1993), as will be discussed later.

GLAUCOMA SURGERY AND OCULAR SURFACE

There are different surgical techniques for glaucoma, however trabeculectomy is currently the most widely used around the world. An opening into the eye near the limbus is made beneath the conjunctiva and a partial-thickness scleral flap. The scleral flap offers some resistance to egress of aqueous, minimizing the chance of hypotony, and the resultant conjunctival bleb allows diffusion of the aqueous through the conjunctiva onto the surface of the eye (Katz et al. 1996). Basically, a fistula drains aqueous humor from the anterior chamber into the subconjunctival space (Katz et al. 1996). After surgery, if rapid healing and the scarring process develop it could lead to surgical failure (Chang et al. 2000). The success of trabeculectomy depends on the development of a functioning bleb. This process is influenced by the wound-healing response, being more intense in the early postoperative period, but continues indefinitely after surgery and results in increasing failure rate with longer follow-up (Katz et al. 1996).

The healthy OS and the surgical success are related by different aspects. One is linked to the antimitotic agents using intra and/or postoperative. Another is related with the OS anatomical change induced by the surgery itself. And finally the OSD of the patient could increase the percentage of surgical failures.

Antimitotic Drugs

Changes in the number of conjunctival fibroblasts and inflammatory cells were associated with increased risk of trabeculectomy failure (Broadaway et al. 1994). Also, there are changes in the aqueous composition, which could influence the scarring process, bleb appearance and its function over a long period (Freeman 2008). To avoid the wound-healing problem, many surgeons perform the application of mitomycin C (MMC) or 5-fluorouracil (5-Fu) intra-operative and/or post-operative (Skuta et al. 1992, Parrish 1992, The Fluorouracil Filtering Surgery Study Group 1996). Both drugs are useful in decreasing fibrosis and scarring and to increase the chances of successful filtration surgery. However, both drugs could produce OS toxicity and very serious problems. MMC and/or 5-FU could develop epithelial persistent defect, scleral thinning and atrophy with the risk of ocular perforation.

The MMC is an antibiotic derived from Streptomyces caespitosus. After intracellular enzymatic or spontaneous chemical reduction, MMC becomes a bifunctional or trifunctional alkylating agent, that is able to inhibit DNA synthesis and crosslink DNA between adenine and guanine at three different sites (Chabner et al. 2001). Although MMC acts primarily during the late G1 and S phases, it is non-cell cycle specific and rapidly dividing cells are preferentially sensitive to it effects (Chabner et al. 2001).

Moreover, MMC is considered a radiomimetic agent with long-term effects on tissues (McKelvie and Daniell 2001, Doagru et al. 2003). Experiments measuring intraocular levels have been performed and the presence of MMC in the anterior chamber was detected following trabeculectomy (Seah et al.1993, Kawase et al. 1992, Sarraf et al. 1993) and also following its topical application over intact corneal epithelium (Torres et al. 2006).

The 5-FU is a pyrimidine analogue, which has been shown to block mitosis of fibrocytes *in vitro* as *in vivo*. Their mechanisms of action are inhibiting synthesis of DNA. In some studies, there appeared to be some toxicity to the OS epithelium in the form of persistent epithelial defects. 5-FU has serious toxic effects on the ocular surface epithelium, which must be considered carefully when this drug is used in the treatment of glaucoma (The Fluorouracil Filtering Surgery Study Group 1996).

Sponge delivery variables and tissue levels have been studied for treatment with 5-FU (The Fluorouracil Filtering Surgery Study Group 1996, Wilkins et al. 2000) and the impact of different variables for MMC on intra-scleral concentrations was also assessed, the importance of standardizing all application parameters in order to obtain reproducible dosage and pharmacological effect was emphasized, avoiding their detrimental effect (Georgopoulos et al. 2000, Vas et al. 2000).

The carefully management of these drugs are relevant for the surgical success in glaucoma. Because of their worldwide-extended use, it is necessary to be alert of the short-term and long-term side effects. Basic research and long term clinical studies will help to develop safer wound-healing modulation drugs and improved surgical techniques.

Anatomical OS change after Trabeculectomy

Howover, if the bleb remains healthy, functioning and the IOP is controlled, another problem still exists: the foreign body sensation. The OS anatomo-unit is distorted after the surgery (Figs. 6 and 7).

The patient feels the bleb. Many patients complain of an unpleasant feeling under their upper eyelids for the rest of their lives. Blinking is disturbed, producing turbulence in the tear film that will affect the epithelial cells of the cornea. Sometimes, corneal Dellen could be produced. Keratitis becomes frequent and needs to be handle with care to avoid corneal infections. IOP is low, and the patient feels unwell. Lubricants therapy will improve this and could be necessary many times a day.

Also, a mild ptosis could be observed if only one eye was operated. A successful trabeculectoy is associated with a larger area and decreased vascularity of the bleb but not with diminished expression of the inflammatory marker by the ocular surface.

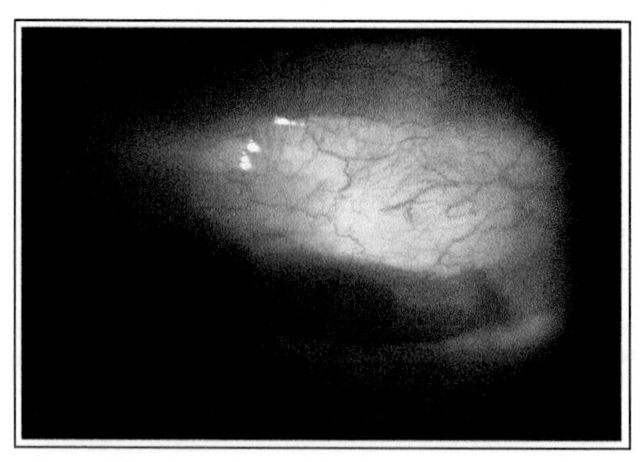

Figure 6 This figure shows a big functioning bleb. Complains about foreign body sensation was presented in this case.

Color image of this figure appears in the color plate section at the end of the book.

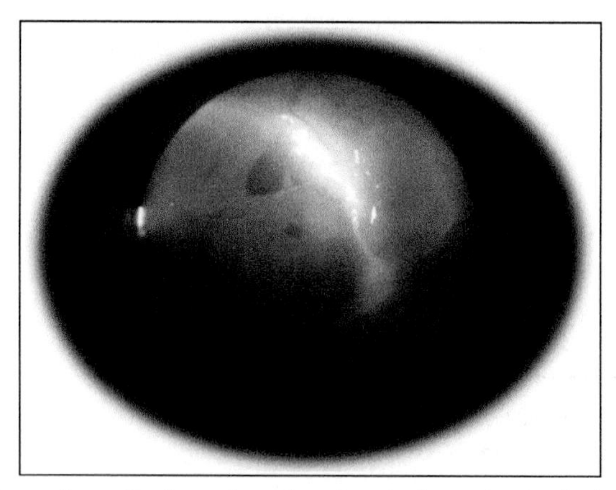

Figure 7 This figure shows the formation of conjunctival cyst; foreign body sensation in this case is obvious.

Color image of this figure appears in the color plate section at the end of the book.

Presence of subclinical inflammation in eyes without eye drops may result from the transcellular aqueous pathway towards the ocular surface, especially in functioning blebs with adjunctive mitomicyn-C.

Pre-existing OSD and Trabeculectomy

The wound-healing process is one of the most important factors in this kind of surgery: excessive healing and scarring of the bleb are associated with the surgical failure. In most of the OSD, acute or chronic inflammation appears and needs to be managed. If not, OS inflammatory markers will be increased which is associated with surgical failure (Cvenkel 2011).

But inflammatory markers are not only elevated in the dry eye syndrome or other OSD, moreover it could be elevated in glaucoma patients without OSD, and the severity is positively correlated with the duration and number of IOP-lowering medications used.

SUMMARY

This chapter reviewed the relationship between OSD and glaucoma. Both pathologies could happen in the same patient, both need topical treatments and sometimes surgical treatments. Topical treatments for glaucoma frequently cause OSD and it is common that many patients applying glaucoma eye drops need to also use ocular lubricants for the dry eye syndrome. Moreover, when filtration surgery is performed, different problems could arise and it is necessary to be careful of the short-term as well as of the long-term side effects. New topical pharmacological formulations and new surgical techniques for glaucoma treatments will decrease OS toxicity and adverse effects, improving the therapy for glaucoma and also the OS. Meanwhile, the OS in patients with glaucoma must be carefully and frequently checked.

ACKNOWLEDGMENTS

Faculty of Medicine, Buenos Aires University (UBA), Buenos Aires–Argentine. Dr. Hugo Dionisio Nano, from Clínica de Ojos Dr. Nano, San Miguel, Buenos Aires–Argentine. Dr. Jesús Merayo-Lloves, Instituto Oftalmológico Fernández-Vega, Oviedo–Spain. Prof. Carlos Pastor, Dra Margarita Calonge, Dr. José María Herreras, Instituto Universitario de Oftalmobiología Aplicada (IOBA), Universidad de Valladolid, Valladolid –Spain. Dr. Pablo Correas y Dr. Alfonso Manrique de Lara, from Clínica Oftalmológica Gran Canaria; Las Palmas de Gran Canaria, Gran Canaria –Spain. Prof. Jorge Alió, Vissum–Corporación Oftalmológica. Dr. Pablo Lódolo, from Centro de Ojos Dr. Lódolo, Paraná, Entre Ríos, Argetine and to the Asociación Entrerriana de Oftalmología (AEO).

REFERENCES

Aihara, M., S. Otani, J. Kozaki, K. Unoki, M. Takeuchi, K. Minami and K. Miyata. 2011. Long-term effect of BAK-free travatoprost on ocular surface and intraocular pressure in glaucoma patients after transition from Latanoprost. J. Glaucoma 26.

Asbell, P.A. and N. Potapova. 2005. Effects of topical antiglaucoma medications on the ocular surface. Ocul. Surf. 3: 27–40.

Bai, T., J.F. Huang and W. Wang. 2010. Short-term comparative study of the effects of preserved and unpreserved topical levofloxacin on the human ocular surface. Cutan. Ocul. Toxicol. 29: 247–253.

Bartlett, J.D. 2000. Dosage Forms and Routes of Administration. In: Ophthalmic Drugs Facts [eds.]. A Wolters Kluwer Company, St. Luis. United States of America, pp. 1–10.

Baudouin, C. 1996. Side effects of antiglaucomatous drugs on the ocular surface. Curr. Opini. Ophthalmol. 7: 80–6.

Baudouin, C. and C. de Lunardo. 1998. Short-term comparative study of topical 2% carteolol with and without benzalkonium chloride in healthy volunteers. Br. J. Opththalmol. 82: 39–42.

Baudouin, C., P. Hamard, H. Liang, C. Creuzot-Garcher, L. Bensoussan and F. Brignole. 2004. Conjunctival epithelial cell expression of interleukins and inflammatory markers in glacuoma patients treated over long term. Ophthtalmology 111: 2186–2192.

Baudouin, C., A. Labbe, H. Liang, A. Pauly and F. Brignole-Baudouin. 2010. Preservatives in the eyedrops: the good, the bad and the ugly. Prog. Retin. Eye Res. 29: 312–334.

Becquet, F., M. Goldschild, M.S. Moldovan, M. Ettaiche, P. Gastaud and C. Baudouin. 1998. Histopathological effects of topical ophthalmic preservatives on rat corneoconjunctival surface. Curr. Eye Res. 17: 419–25.

Brasnu, E., F. Brignole-Baudouin, L. Riancho, J.M. Guenoun, J.M. Warnet and C. Baudouin. 2008. *In vitro* effects of preservative-free tafluprost and preserved latanoprost, travatoprost and bimatoprost in a conjunctival epithelial cell line. Curr. Eye Res. 33: 303–312.

Broadaway, D.C., I. Grierson, C. O´Brian and R.A. Hitchings. 1994. Adverse effects of topical antiglaucoma medicacion. II The outcome of filtration surgery. Arch. Ophthalmol. 112: 1446–1454.

Chabner, B.A., D.P. Ryan and L. Paz-Ares. 2001. Antineoplasic agents. In: J.G. Hardman, L.E. Limbird, Goodman and A. Gilman, Editors, Goodman and Gilman's the Pharmacological Basis of Therapeutics [10th eds.]. McGraw Hill, New York, USA, pp. 1389–1459.

Cha, S.H., J.S. Lee, B.S. Oum and C.D. Kim. 2004. Corneal epithelial cellular dysfunction from benzalkonium chloride (BAC) *in vitro*. Clin. Exp. Ophthtalmol. 32: 180–184.

Chang, L., J.G. Crowston, M.F. Cordeiro, A.N. Akbar and P.T. Khaw. 2000. The role of the inmmune system in conjunctival wound healing after glaucoma surgery. Sur. Ophthalmol. 45: 49–68.

Ciancaglini, M., P. Carpineto, L. Agnifili, M. Nubile, V. Fasanella and L. Mastropasqua. 2008. Conjunctival modificatios in ocular hypertension ans primary open angle glaucoma: An *in vivo* confocal microscopy study. Invest Ophthalmol. Vis. Sci. 49: 3042–3048.

Cvenkel, B. and K.A. Kopitar. 2011. Correlation between filtering bleb morphology, expression of inflammatory marker hla-dr by ocular surface and outcome of trabeculectomy. J. Glaucoma 5.

Debbasch, C., F. Brignole, P.J. Pisella, J.M. Warner, P. Rat and C. Baudouin. 2001. Quaternary ammoniums ans other preservatives´ contribution in oxidative stress and apoptosis on Chang conjunctiva cells. Invest Ophthalmol. Vis. Sci. 42: 642–652.

Dogru, M., H. Ertyrk, J. Shimazaki, K. Tsubota and M. Gul. 2003. Tear function and ocular surface changes with mitomycin (MMC) treatment for primary corneal intraepithelial neoplasia. Cornea 22: 627–639.

Eamon, W., M.D. Leung, F.A. Medeiros and R.N. Weinreb. 2008. Prevalence of Ocular Surface Disease in Glaucoma Patients. J. Glaucoma 17: 350–355.

Erb, C., U. Gast and D. Schremmer. 2008. German register for glaucoma patients with dry eye: I: basic outcome with respect to dry eye. Graefes Arch. Clin. Exp. Ophthalmol. 246: 1593–601.

Fechtner, R.D., D.G. Godfrey, D. Budenz, J.A. Stewart, W.C. Stewart and M.C. Jasek. 2010. Prevalence of ocular surface complains in patients with glaucoma using topical intraocular pressure-lowering medications. Cornea 29: 618–21.

Feldman, R.M. 2003. Conjunctival hyperemia and the use of topical prostaglandins in glaucoma and ocular hypertension. J. Ocul. Pharmacol. Ther. 19: 23–35.

Foster, C.S. and M. Sains de la Maza. 2004. Ocular cicatricial pemphigoid review. Curr. Opin. Allergy Clin. Immunol. 4: 435–439.

Freeman, J. and D. Goddard. 2008. Elevated levels of transforming growth factor beta ans prostaglandin E2 in aqueous humor in patients undergoing filtration surgery for glaucoma. Can. J. Ophthalmol. 43: 370.

Friedman, D.S., R.C. Wolfs, B.J. O'Colmain, B.E. Klein, H.R. Taylor, S. West, M.C. Leske, P. Mithchell, N. Congdon and J. Kempen. 2004. Eye Disease Prevalence Research Group: Prevalence of open-angle glaucoma among adults in the United States. Arch. Ophthalmol. 122: 532–538.

Georgopoulos, M., C. Vass, I. El Menyawi, S. Radda, W. Graninger and R. Menapace. 2000. *In vitro* diffusion of Mitomycin-C into human sclera after episcleral application: impact of diffusion time. Exp. Eye Res. 71: 453–457.

International Dry Eye WorkShop (DEWS). 2007. Management and therapy of dry disease: report of the management and therapy subcommittee of the International Dry Eye WorkShop (2007). Ocul. Surf. 5: 163–78.

Jaenen, N., C. Baudouin, P. Pouliquen, G. Manni, A. Figueiredo and T. Zeyen. 2007. Ocular symptoms and signs with preserved and preservative-free glacuoma medications. Eur. J. Ophthalmol. 17: 341–349.

Kalavala, M. and B.N. Statham. 2006. Allergic contact dermatitis from timolol and dorzolamide eye drops. Contact Dermatitis 54: 345–350.

Kahook, M.Y. and R.J. Noecker. 2008. Comparison of corneal and conjunctival changes after dosing of travatoprost preserved with sofZia, latanoprost with 0.02% benzalkonium chloride and preservative-free artificial tears. Cornea 27: 339–343.

Katz, J.L., V.P. Costa and G.L. Spaeth. 1996. Filtration surgery. In: R. Ritch, M.B. Shields, T. Krupin. The Glaucomas [eds.]. Mosby. St. Louis, Missouri. USA, pp. 1661–1702.

Kawase, K., H. Matsushita, T. Yamamoto and Y. Kitazawa. 1992. Mitomycin concentration in rabbit and human ocular tissues after topical administration. Ophthalmology 99: 203–207.

Leung, E.W., F.A. Medeiros and R.N. Weinreb. 2008. Prevalence of ocular surface disease in glaucoma patients. J. Glaucoma 17: 350–5.

Mantelli, F., L. Tranchina, A. Lambiase and S. Bonini. 2011. Ocular surface damage by ophthalmic compounds. Curr. Opin. Allergy Clin. Immunol. 11: 464–470.

McCarty, C.A., A.K. Bansal, P.M. Livinston, Y.L. Stanislavsky and H.R. Taylor. 1998. The epidemiology of dry eye in Melbourne, Australia. Ophthalmology 105: 1114–1119.

McDermott, M.L., J. Wang and D.H. Shin. 1994. Mitomycin and the human corneal endothelium. Arch. Ophthalmol. 112: 533–537.

McKelvie, P.A. and M. Daniell. 2008. Impression cytology following mitomycin C therapy for ocular surface squamus neoplasia. Br. J. Ophthalmol. 85: 1115–1119.

Merayo-Lloves, J., R. Torres and A. Berra. 2004. Estudio de Superficie Ocular. In: J.M. Benítez del Castillo Sánchez, J.A. Durán de la Colina and M.T. Rodríguez Ares. Superficie Ocular. LXXX Ponencia Oficial de la Sociedad Española de Oftalmología Sociedad Española de Oftalmología. Spain, pp. 13–21.

Mietz, H., K. Addicks, W. Bloch and G.K. Krieglstein. 1996. Long-term intraocular toxic effects of topical mitomycin C in rabbits. J. Glaucoma 5: 325–333.

Monaco, G., V. Cacioppo, D. Consonni and T. Pasquale. 2010. Effects of osmoprotection on symptoms, ocular surface damage and tear film modifications caused by glaucoma therapy. Eur. J. Ophthalmol. 21: 243–50.

Parrish, R.K. 1992. Who should receive antimetabolites after filtering surgery? Arch. Ophthalmol. 110: 1069–1071.

Quigley, H.A. 1996. Number of people with glaucoma worldwide. Br. J. Ophthalmol. 80: 389–93.

Quigley, H.A. and A.T. Broman. 2006. The number of people with glaucoma worldwide in 2011 and 2010. Br. J. Ophthalmol. 58: 1131–1135.

Rivera, J.L., N.P. Bell and R.M. Feldman. 2008. Risk factors for primary open angle glaucoma progression: what we know and what we need to know. Cur. Opin. Ophthalmol. 19: 102–106.

Rodrigues, M. de L., D.P Felipe Costra, C.P. Soares, N.H. Deghaide, R. Duarte, F.S. Sakamoto, J.M. Furtado, J.S. Paula, E.A. Donadi and E.G. Soares. 2009. Immunohistochemical expression of HLA-DR in the conjunctiva of patients under topical prostaglandin analogs treatment. J. Glaucoma 18: 197–200.

Rolando, M. and M. Zierhut. 2001. The Ocular Surface and Tear Film and their Dysfunction in Dry Eye Disease. Surv. Ophthalmol. 45: S203–S210.

Sacu, S., G. Rainer, O. Findl, M. Georgopoulos and C. Vass. 2003. Correlation between the early morphological appearance of filtering blebs and outcome of trabeculectomy with mitomycin C. J. Glaucoma 12: 430–435.

Sarraf, D., R.D. Eezzuduemhoi, Q. Cheng, M.R. Wilson and D.A. Lee. 1993. Aqueous and vitreous concentration of mitomycin-C by topical administration after glaucoma filtration surgery in rabbits. Ophthalmology 100: 1574–1579.

Schwarts, G.F., S. Kotak, J. Mardekian and J.M. Fain. 2001. Incidence of new coding for dry eye and ocular infection in open-angle glaucoma and ocular hypertension patients treated with prostaglandin analogs: Retrospective analysis of three medical/pharmacy claims databases. BMC Ophthalmology 11: 14.

Seah, S.K., J.A. Prata, D.S. Minckler, R.T. Koda, G. Baerveldt, P.P. Lee and D.K. Heuer. 1993. Mitomycin-C concentration in human aqueous humour following trabeculectomy. Eye 7: 652–655.

Shapiro, M.S., R.A. Thoft, J. Friend, R.K. Parrish and M.G. Gressel. 1985. 5-Fluorouracil Toxicity to the Ocular Surface Epithelium. Invest. Ophthalmol. Vis. Sci. 26: 580–583.

Sherwood, M.B., C.S. Migdal, R.A. Hichtings, M. Sharir, T.J. Zimmerman and J.S. Schultz. 1993. Initial treatment of glaucoma: surgery or medications. Surv. Ophthalmol. 37: 293–305.

Skuta, G.L., C.C. Beeson, E.J. Higginbotham, P.R. Lichter, D.C. Musch, T.J. Bergstrom, T.B. Klein and F.Y. Falck. 1992. Intraoperative mitomycin versus postoperative 5-fluouracil in high-risk glaucoma filtering surgery. Ophthalmology 99: 438–444.

The Fluorouracil Filtering Surgery Study Group. 1996. Five-year follow-up of the Fluorouracil Filtering Surgery Study. Am. J. Ophthalmol. 121: 249–66.

Thorne, J.E., G.J. Anhalt and A.J. Douglas. 2004. Mucous Membrane Pemphigoid and Pseudopemphigoid. Ophthalmology 111: 45–52.

Torres, R.M., J. Merayo-Lloves, S.M. Daya, J.T. Blanco-Mezquita, M. Espinosa, M.J. Nozal, J.L. Bernal and J. Bernal. 2006. Presence of Mitomycin C in the anterior chamber alter application in PRK. J. Cataract. Refract. Surg. 32: 67–71.

Vass, C., M. Georgopoulos, I. el Menyawi, S. Radda and P. Nimmerrichter. 2000. Intrascleral concentration vs. depth profile of Mitomycin c after episcleral application: impact of applied concentration and volume of mitomycin-c solution. Exp. Eye Res. 70: 571–575.

Wand, M., R.N. Shaffer. Open-Angle Glaucoma. 2000. pp. 494–498. In: F.T. Fraunfelder and F.H. Roy. Current Ocular Therapy. W.B. Saunders Company. Philadelphia. United States of America.

Wilkins, M.R., N.L. Occleston, A. Kotecha, L. Waters and P.T. Khaw. 2000. Sponge delivery variables and tissue levels of 5-fluorouracil. Br. J. Ophthalmol. 84: 92–97.

Yee, R.W. 2007. The effect of drop vehicle on the efficacy and side effects of topical glaucoma therapy; a review. Curr. Opin. Ophthalmol. 18: 134–139.

Zimmerman, T.J. 2000. Agents for Glaucoma. In: Ophthalmic Drugs Facts [eds.]. A Wolters Kluwer Company, St. Luis. United States of America, pp. 173–176.

Section III: Care of Ocular Surface Disorders

10

Pathogenesis and Treatment of Dry Eye Syndrome

Cintia S. de Paiva

SUMMARY

Dry eye is a multifactorial condition that results in a dysfunctional lacrimal functional unit. Evidence suggests that inflammation plays is involved in the pathogenesis of the disease.

Changes in tear composition including increased cytokines, chemokines, metalloproteinases and the number of T cells in the conjunctiva are found in dry eye patients and in animal models. This inflammation is responsible in part for the irritation symptoms, ocular surface epithelial disease, and altered corneal epithelial barrier function in dry eye.

There are several anti-inflammatory therapies for dry eye that target one or more of the inflammatory mediators/pathways that have been identified and are discussed in detail.

INTRODUCTION

Dry eye disorders are very prevalent and are increasing with aging. It is the second most frequent cause for patients seeking eye care in US. With an increasing life-span and with baby-boomers getting older and older, there is an urgent need to better understand the mechanisms of meibomian gland disease and tear film disorders. Dry eye is the second most common problem of patients seeking eye care, and is characterized by eye irritation

Ocular Surface Center, Department of Ophthalmology, Cullen Eye Institute, Baylor College of Medicine, Houston, Texas, 6565 Fannin Street, NC 205-Houston, TX 77030;
E-mail: cintiadp@bcm.edu

symptoms, blurred and fluctuating vision, tear film instability, increased tear osmolarity and ocular surface epithelial disease (de Paiva 2003, Goto et al. 2002, Musch 1983, Pflugfelder et al. 1998). It is often a challenging clinical problem to identify because of its varying clinical presentation. Dry eye impacts quality of life by decreasing functional vision, i.e. the ability to perform daily activities such as reading, using a computer and driving (Miljanovic et al. 2007).

There is an increasing evidence that dry eye is an inflammatory disease. Disease or dysfunction of the tear secretory glands leads to changes in tear composition, such as hyperosmolarity that stimulate the production of inflammatory mediators on the ocular surface (Luo et al. 2004). Inflammation may in turn cause dysfunction or disappearance of cells responsible for tear secretion or retention (Niederkorn et al. 2006). Inflammation can also be initiated by chronic irritative stress (e.g. contact lenses) and systemic inflammatory/autoimmune disease (e.g. rheumatoid arthritis). Regardless of the initiating cause, a vicious cycle of inflammation may develop on the ocular surface in dry eye that leads to ocular surface disease.

Pathogenesis of Dry Eye: Role of Inflammation

The dry eye field is undergoing a transformation due to increased knowledge of molecular and cellular mechanisms of dry eye. Increased inflammatory cytokines, tear film osmolarity, metalloproteinases, chemokines and chemokines receptors inflammatory cascade and activation of immune cells have been implicated in the pathogenesis of dry eye.

Increased Inflammatory Cytokines

Increased production and activation of pro-inflammatory cytokines (interleukin [IL]-1 and tumor necrosis factor [TNF]-α) and proteolytic enzymes by stressed ocular surface and glandular epithelial cells, as well as by the inflammatory cells that infiltrate these tissues have been reported in dry eye (Lopez and Ubels 1991, 1993, Zhu et al. 2004).

Increased concentration of pro-inflammatory cytokines and chemokines in the tear fluid, such as IL-6, IL-1, and TNF-α has been extensively reported (Baudouin et al. 2005, de Paiva et al. 2007, Jones et al. 1994, Niederkorn et al. 2006, Pflugfelder et al. 1999, Rolando et al. 2005, Solomon et al. 2001, Stern et al. 2002, Yoon et al. 2007). In humans, significantly increased levels of IL-1α, IL-6, IL-8, TNF-α and transforming growth factor (TGF)-β1 RNA transcripts have been found in the conjunctival epithelium of Sjögren's syndrome, the most severe type of dry eye, compared to controls (Pflugfelder et al. 1999). However, the exact role of these cytokines in dry eye has not been fully elucidated.

Tear Film Osmolarity

Hyperosmolarity of the tear fluid has been recognized for decades as a common feature of all types of dry eye and it has been referred to as the "gold standard" for the diagnosis of dry eye (Farris 1994). It is also recognized as a pro-inflammatory stimulus (Noecker 2001, Tripathi 1992).

Exposure of cultured corneal epithelial cells to media of increasing sodium chloride concentration results in a concentration dependent increase in the production of the same pro-inflammatory factors that have been detected in the conjunctival epithelium and tear fluid of dry eye patients (i.e. IL-1, IL-8, TNF-α, and MMP-9) (Li et al. 2002, Nelson et al. 1994). *In vivo* and *in vitro*, hyperosmolarity was shown to stimulate the production of these inflammatory mediators by activating mitogen activated protein kinases (Annesley et al. 1981, Li et al. 2002, Luo et al. 2004).

Increased Metalloproteinases

Another pathologic change is an increased concentration and activity of matrix metalloproteinases (MMPs) in the tear fluid of dry eye patients (Afonso et al. 1999, Sobrin et al. 2000, Solomon et al. 2001). These enzymes, such as MMP-9, lyse a variety of different substrates including components of the corneal epithelial basement membrane and tight junction proteins (such as ZO-1 and occludin) that maintain corneal epithelial barrier function (de Paiva et al. 2006b, Pflugfelder et al. 2005). In a group of dry eye patients, we observed that tear MMP-9 activity levels increased as the severity of corneal disease progressed. Tear MMP-9 activity levels also correlated positively with corneal fluorescein staining scores and with low contrast visual acuity (Chotikavanich et al. 2009).

MMP-9 appears to play a physiological role in regulating corneal epithelial desquamation. In systemic vitamin A deficiency, there is reduced expression of MMP-9 and hyper stratification of the corneal epithelium. In contrast, the increased MMP-9 activity in keratoconjunctivitis sicca (KCS) is associated with deranged corneal epithelial barrier function (increased fluorescein permeability), increased corneal epithelial desquamation (punctate epithelial erosions) and corneal surface irregularity (de Paiva et al. 2006b, Luo et al. 2004, Pflugfelder et al. 2005).

Chemokines and Chemokines Receptors

Chemokines are small peptides that can induce recruitment of nearby responsive cells. They are important in sites of inflammation because they can amplify the cascade by attracting more and more inflammatory cells. In human cystic fibrosis patients, the tear levels of the chemokine macrophage inflammatory protein 1α (MIP-1α) were significantly higher when compared with healthy controls and correlated with dry eye findings in these patients (Mrugacz et al. 2007).

Inflammatory Cascade and Immune System

The increase in soluble and cellular inflammatory mediators in the tear fluid, conjunctiva and lacrimal glands initiates an inflammatory cascade on the ocular surface, evidenced by increased expression of immune activation and adhesion molecules (HLA-DR and ICAM-1) by the conjunctival epithelium. These molecules function to attract and retain inflammatory cells in the conjunctiva.

Upregulation of expression of HLA-DR and ICAM-1 has been reported in a variety of dry eye diseases, such as ocular rosacea, non-Sjögren's syndrome (SS) aqueous tear deficient and SS patients, but the highest expression of HLA-DR was found in the conjunctival epithelium of SS patients (Brignole et al. 2000, Tsubota et al. 1999). Increased HLA-DR antigen expression by the conjunctival epithelium detected by flow cytometry has been observed as a universal feature of dry eye (Baudouin et al. 2002, Brignole et al. 2000, Pisella et al. 2000).

Role of T Cells

Sjögren's syndrome (SS) is characterized by; dry eyes, dry mouth, vasculitis and neurologic disease. The cardinal manifestation of SS is dryness, resulting from exocrine gland dysfunction. At the cellular level, the involved exocrine tissues (including the lacrimal gland) are infiltrated with lymphocytes, monocytes and plasma cells (Carsons 2001). T cell infiltration of the conjunctiva has been observed in both SS and non-SS KCS (Pflugfelder et al. 1990, Raphael et al. 1988).

There is increased evidence that $CD4^+$ T cells are involved in the pathogenesis of dry eye. In an animal model, increased infiltration of $CD4^+$ T cells in the goblet cell rich area was accompanied by increased expression of IFN-γ, goblet cells loss and conjunctival metaplasia (de Paiva et al. 2007). Inflammation in the lacrimal glands, cornea, and conjunctiva, resulting in decreased tear production and conjunctival goblet cell loss, was transferable from wild type mice subjected to experimental dry eye to T-cell–deficient nude mice that have not been exposed to desiccating stress (Niederkorn et al. 2006). Using a similar approach, $CD4^+$ T cells adoptively transfer fromanimals deficient in the autoimmune regulator (aire) gene to immunodeficient recipient micecaused advanced ocular surface keratinization (Li et al. 2008).

MANAGEMENT AND ANTI-INFLAMMATORY THERAPY IN DRY EYE

Artificial Tears and Punctual Plugs

Dry eye disease is chronic and its management requires a full evaluation of the subject. Careful examination of the individual with special attention to systemic medications has been proved to ameliorate symptoms. In some cases, investigation of autoimmune diseases is indicated.

Traditionally, the first line of therapy for dry eye has been the use of artificial tears, to replace the decreased tears. The rationale here is to provide temporary relief of symptoms. There are several types of artificial tears in the market with different compounds in their formulation. Artificial tears are useful in controlling symptoms in mild patients, with the caveat that most of their effect is transient. Artificial tears also face the challenge of blurring vision and causing discomfort or ocular damage themselves due to frequent administration. For most severe patients, where administration of eye drops occurs more than four times a day, preservative-free artificial tears are indicated (Noecker 2001).

Punctual plugs have been used with success in mild patients. An increasing understanding of the pathogenesis of dry eye has shown that punctual plugs can worse symptoms of dry eye in some cases where their own tear film is saturated with inflammatory cytokines and matrix metalloproteinases.

Rationale for Anti-inflammatory Therapy in Dry Eye

Although suspected for decades that inflammation is implicated in the pathogenesis, it wasn't until the approval of cyclosporine A for treatment of dry eye that the role of inflammation began to receive serious attention. The advent of cyclosporine changed the way physicians, patients and pharmaceutical companies viewed dry eye: it became a treatable disease, where therapy can make a significant improvement in quality of life of affected patients. It changed radically the modalities of treatment available for dry eye disease.

Corticosteroids

Corticosteroids are potent anti-inflammatory agents that are routinely used to control inflammation in many organs. Corticosteroids have multiple mechanisms of action. They work through traditional glucocorticoid receptor mediated pathways to directly regulate gene expression and they also work through non-receptor pathways to interfere with transcriptional regulators of pro-inflammatory genes, such as NF-kB. Among their multiple biological activities, corticosteroids inhibit inflammatory cytokine and chemokine production, decrease the synthesis of matrix metalloproteinases and lipid mediators of inflammation (e.g. prostaglandins), decrease expression of cell adhesion molecules (e.g. ICAM-1) and stimulate lymphocyte apoptosis (Aksoy et al. 1999, Brunner et al. 2001, Dursun et al. 2001, Hashimoto et al. 2000, Liden et al. 2000, Solomon et al. 2000a). They have been reported to decrease the production of a number of inflammatory cytokines (IL-1, IL-6, IL-8, TNF-α, GM-CSF) and MMP-9 by the corneal epithelium (Dialilian 2001).

Corticosteroids have been successfully used to treat the corneal epithelial disease in dry eye (Marsh and Pflugfelder 1999, Pflugfelder et al. 2004, Wachtel et al. 1999) in several clinical studies (Marsh and Pflugfelder 1999). This therapy was effective even for patients suffering with severe KCS who had no improvement from maximum aqueous enhancement therapies. In a prospective, randomized clinical trial, topical treatment of dry eye patients with non-preserved methylprednisolone and punctual plugs significantly decreased the severity of ocular irritation symptoms and corneal fluorescein staining compared to the group that received punctual occlusion alone (Sainz de la Maza et al. 2000). A randomized, double-masked, placebo-controlled study of loteprednol etabonate showed that a subset of patients with the most severe inflammatory signs at entry, treated topically with loteprednol, showed a significantly greater decrease in central corneal fluorescein staining scores when compared to its vehicle. In another open-label randomized study, patients with KCS that received fluorometholone plus artificial tear substitutes experienced lower symptom severity scores, fluorescein and rose bengal staining than either patients receiving either artificial tear substitute alone, or artificial tear substitute plus flurbiprofen (Avunduk et al. 2003).

Short-termuses of concomitant topical corticosteroid and Cyclosporine have been shown to reduce stinging and improve compliance with the use of cyclosporine drops (Byun et al. 2009, Sheppard 2011).

Taken together, these studies indicate that topical corticosteroids are an important tool in the management of dry eye. While no steroid related complications were observed in these short-term clinical trials, there is the potential for toxicity with long-term use, such as increase on intraocular pressure and cataracts. This may limit the use of more potent steroids for chronic therapy of dry eye. The risk-benefit ratio may be better with "soft steroids" such as fluorometholone and loteprednol etabonate that have less intraocular activity and a lower likelihood of raising intraocular pressure.

Tetracyclines and their Derivatives

The tetracyclines have anti-inflammatory as well as antibacterial properties that may make them useful for the management of chronic inflammatory diseases. These agents decrease the activity of collagenase, phospholipase A2, and several matrix metalloproteinases. They also decrease the production of IL-1α and TNF-α in a wide range of tissues, including the corneal epithelium (Li et al. 2001, 2006, Solomon et al. 2000b).

Tetracyclines have been used extensively to treat rosacea (Dursun 2001, Frucht-Pery et al. 1989, Jansen and Plewig 1997, Macdonald and Feiwel 1972). Rosacea, including its ocular manifestations, is an inflammatory disorder, occurring mainly in adults with peak severity in the third and fourth decades. It is characterized by vasomotor instability (flushing) of the face, neck, and upper chest. Chronic facial inflammation may lead to persistent facial erythema and telangiectasia formation, and, ultimately, significant deformity of the central face. Tetracycline derivatives (e.g. minocycline, doxycycline) have been recommended as treatment options for chronic blepharitis because of their high concentration in tissues, low renal clearance, long half-life, high level of binding to serum proteins, and decreased risk of photosensitization (Hoeprich and Warshauer 1974).

Doxycycline was discovered in the early 1960's as a semi-synthetic long-acting tetracycline derivative useful as a bacterial ribosome inhibitor in a wide variety of microbes. In sub-antimicrobial doses, it is also an effective primary treatment for rosacea, sterile corneal ulceration, and effective adjunctive treatment for adult periodontitis (Akpek et al. 1997, Caton et al. 2000, Seedor et al. 1987). Doxycycline has been shown to effectively inhibit MMP-9 in a wide variety of mouse and human cells including prostate epithelium, epidermal keratinocytes, and the aortic endothelium (Hanemaaijer et al. 1997, Hanemaaijer et al. 1998, Lokeshwar 1999, Qin et al. 2006). A phase 3 FDA clinical trial is under way to evaluate the efficacy of a topical novel preparation of doxycycline in dry eye patients.

Because of the improvement observed in small clinical trials of patients with meibomianitis (Frucht-Pery et al. 1989, Ta et al. 2003), the American Academy of Ophthalmology recommends the chronic use of either doxycycline or tetracycline for the management meibomianitis (McCulley 1982). Larger randomized placebo-controlled trials assessing symptom improvement rather than surrogate markers are needed to clarify the role of this antibiotic in blepharitis treatment (Voils et al. 2005). Several studies have described the beneficial effects of minocycline and other tetracycline derivatives (e.g. doxycycline) in the treatment of chronic blepharitis (Dougherty et al. 1991, Gulbenkian 1980, Hoeprich and Warshauer 1974, Shine 2003). Studies have shown significant changes in the aqueous tear parameters, such as tear volume and tear flow following treatment that with tetracycline derivatives (e.g. minocycline). One study also demonstrated a decrease in the aqueous tear production that occurred along with clinical improvement (Aronowicz et al. 2006).

Cyclosporine

Cyclosporine A (CsA) is a lipophilic cyclic undecapeptide isolated from the fungus *Hypocladium inflatum gams* (Matsuda and Koyasu 2000). It was first introduced for clinical use in 1983, as an immunosuppressant drug to prevent organ rejection after transplant. The immunomodulatory effect of the drug has proved to be beneficial for treatment of a broad group of diseases that have in common an underlying inflammatory response in their physiopathology (psoriasis, rheumatoid arthritis, ulcerative colitis, etc.).

The potential of CsA for treating dry eye disease was initially recognized in dogs that develop spontaneous KCS (Kaswan 1989). Since the initial description, the therapeutic efficacy of CsA for human KCS has been well documented, with evidence ranging from several small single center randomized double-masked clinical trials to several large multicenter randomized, double-masked clinical trials (Gunduz and Ozdemir 1994, Laibovitz et al. 1993, Stevenson 2000).

Two independent phase 3 clinical trials compared twice daily treatment with 0.05% or 0.1% CsA or vehicle in 877 patients with moderate to severe dry eye disease (Sall et al. 2000, Stevenson 2000). When the results of the two Phase 3 trials were combined for statistical analysis, patients treated with CsA, 0.05% or 0.1%, showed significantly ($P \leq 0.05$) greater improvement in two objective signs of dry eye disease (corneal fluorescein staining and anesthetized Schirmer test values) than those treated with vehicle. An increased Schirmer test score was observed in 59% of patients treated with CsA, with 15% of patients having an increase of 10 mm or more. In contrast, only 4% of vehicle-treated patients had this magnitude of change in their Schirmer test scores ($P < 0.0001$). CsA 0.05% treatment also produced significantly greater improvements ($P < 0.05$) in three subjective measures of dry eye disease (blurred vision symptoms, need for concomitant artificial tears, and the global response to treatment). No dose-response effect was noted. Both doses of CSA exhibited an excellent safety profile with no significant systemic or ocular adverse events, except for transient burning symptoms after instillation in 17% of patients. Burning was reported in 7% of patients receiving the vehicle. No CsA was detected in the blood of patients treated with topical CsA for 12 months. Clinical improvement that was observed in these trials from CsA was accompanied by improvement in other disease parameters. Treated eyes had an approximately 200% increase in conjunctival goblet cell density (Kunert 2002). Furthermore, there was decreased expression of immune activation markers (i.e. HLA-

DR), apoptosis markers (i.e. Fas), and the inflammatory cytokine IL-6 by the conjunctival epithelial cells (Brignole et al. 2001, Turner et al. 2000). The numbers of CD3, CD4, and CD8-positive T lymphocytes in the conjunctiva decreased in cyclosporine-treated eyes, while vehicle-treated eyes showed an increased number of cells expressing these markers (Kunert 2002). Following treatment with 0.05% cyclosporine, there was a significant decrease in the number of cells expressing the lymphocyte activation markers CD11a and HLA-DR, indicating less activation of lymphocytes compared with vehicle-treated eyes.

In December 2002, US Food and Drug Administration approved CsA 0.05% ophthalmic emulsion for treatment of dry eye disease and it has been a record number of prescriptions in the USA and worldwide.

Essential Fatty Acids

Essential fatty acids are necessary for complete health and they cannot be synthesized by vertebrates and must be obtained from dietary sources. Among the essential fatty acids are 18 carbon omega-6 and omega-3 fatty acids. In the typical western diet, 20–25 times more omega-6 than omega-3 fatty acids are consumed. Omega-6 fatty acids are precursors for arachidonic acid and certain pro-inflammatory lipid mediators (PGE_2 and LTB4). In contrast, certain omega-3 fatty acids (e.g. EPA found in fish oil) inhibit the synthesis of these lipid mediators as well as block production of IL-1 and TNF-α (Endres et al. 1989, James 2000). A beneficial clinical effect of fish oil omega-3 fatty acids on rheumatoid arthritis has been observed in several double-masked placebo-controlled clinical trials (James and Cleland 1997, Kremer 2000). In a prospective placebo-controlled clinical trial of the essential fatty acids linoleic acid and gamma-linoleic acid administered orally twice daily produced significant improvement in ocular irritation symptoms (Macsai 2008) and ocular surface lissamine green staining (Barabino et al. 2003).

Anti-inflammatory Therapy in Mice: Lessons from Animal Models

The use of mice in dry eye research has increased steadily over the last twenty years. Despite size disparity, the similarities in the anatomic structure of human and mouse eye greatly outweigh the differences. Also, the immunological system of both human and mice share more similarities than disparities, which make mice an ideal candidate to study immunological responses. A good accumulation of data of inflammation and dry eye disease came from animal models of SS. Several candidate compounds for dry eye treatment have been first evaluated in mice for efficacy in treating corneal epithelial diseases and improving inflammatory markers before moving to clinical trials. A prototype of this category is the Resolvin EX (de Paiva et al. 2011), a immune response mediator derived via lipoxygenation from the essential dietary omega-3 polyunsaturated fatty acids, eicosapentaenoic acid and docosahexaenoic acid which is now in phase II of clinical trials.

The use of the experimental dry eye model induced by environmental stress and lacrimal gland blockade has provided a substantial body of evidence for inflammation and autoimmunity in dry eye. C57BL/6 mice under desiccating stress had goblet cell loss and corneal barrier disruption mimicking the human disease (de Paiva et al. 2006a, 2007, Dursun et al. 2002) and they also showed increased expression of inflammatory cytokines (IL-1α, IL-6, and TNF-α) transcripts in the corneal epithelium and conjunctiva; increased concentrations of MIP-1α, MIP-1β, monokine induced by interferon-gamma, and interferon-gamma-inducible protein (IP)-10 proteinsin the corneal epithelium and conjunctiva (Corrales et al. 2006, Yoon et al. 2007).

Previous reported studies using experimental dry eye model demonstrated that doxycycline was efficacious in decreasing gelatinolytic activity in the ocular surface epithelia, as well as decreasing levels of MMP-9 mRNA transcripts, and preventing experimental dry eye-induced increase in inflammatory cytokines IL-1 and TNF-α (de Paiva et al. 2006b). Doxycycline also improved corneal surface regularity and improved corneal barrier function (de Paiva et al. 2006a). At the cellular level, doxycycline preserved apical epithelial cell area and the tight-junction protein occludin, resulting in a decreased number of desquamating epithelial cells from the surface of the cornea (Beardsely et al. 2008, de Paiva et al. 2006a). These findings were also confirmed in cultured human corneal epithelial cells treated with doxycycline subjected to osmotic stress that increases their production of MMP-9 (Pflugfelder et al. 2005).

The corticosteroid methylprednisolone was noted to preserve corneal epithelial smoothness and barrier function in an experimental murine model of dry eye (de Paiva et al. 2006a). This was attributed to its ability to maintain the integrity of corneal epithelial tight junctions and decrease desquamation of apical corneal epithelial cells (de Paiva et al. 2006a). A concurrent study showed that methylprednisolone prevented an increase in MMP-9 protein in the corneal epithelium, as well as gelatinase activity in the corneal epithelium in response to experimental dry eye (de Paiva et al. 2006b).

Compounds that inhibit leukocyte migration into the ocular surface tissues such as integrin α4β1 integrin, chemokine receptor 2 (CCR2), or that used another topical therapies such as topical epigallocatechin gallate or alpha-linolenic acid significantly decreased corneal fluorescein staining compared with both vehicle and untreated controls and decreased in CD11b

(+) cell number, expression of corneal IL-1α and TNF-α, and conjunctival TNF-α (Ecoiffier et al. 2008, Goyal 2011, Goyal et al. 2009, Lee et al. 2011, Rashid et al. 2008).

Cyclosporine A has been shown to increase the density of goblet cells, while decreasing apoptosis of the corneal epithelium and preventing desiccating stress-induced increase of IL-17A and IFN-γ while reducing the number of $CD4^{+}$T cells infiltrating the conjunctiva. It also retained the intra-epithelial lymphocytes population in the same area (de Paiva et al. 2010, Pangelinan et al. 2008, Strong et al. 2005, Sun and Wang 2006).

Therapies directed to block directly or indirectly T helper cytokines, IL-17A and IFN-γ, have recently shown to be efficacious in murine dry eye models and show great promise for future human studies (Chauhan et al. 2009, Chen et al. 2011, de Paiva et al. 2009, Zhang et al. 2011).

Autologous Serum Drops

Autologous blood serum drops are another anti-inflammatory therapy because serum contains high concentration of essential tear components that have been shown to stabilize the tear film and improve ocular surface disease (Kojima et al. 2005, Rocha et al. 2000, Tsubota et al. 1999, Tsubota et al. 1999). They are prepared from the patients' own serum and then further diluted in saline. The challenges in using autologous serum drops (frequent drawing of blood, preservation of sterility and appropriated facilities willing to prepare them) prevent the broad use of this anti-inflammatory therapy.

CONCLUSION

Dry eye is a chronic disease that can profoundly affect quality of life. There are many available options currently in the market but only approved drug specifically designed to treat inflammation in dry eye syndromes, cyclosporine A. A lot of current knowledge and support for anti-inflammatory therapy comes from animal studies.

REFERENCES

Afonso, A.A., L. Sobrin, D.C. Monroy, M. Selzer, B. Lokeshwar and S.C. Pflugfelder. 1999. Tear fluid gelatinase B activity correlates with IL-1alpha concentration and fluorescein clearance in ocular rosacea. Invest Ophthalmol. Vis. Sci. 40: 2506–12.

Akpek, E.K., A. Merchant, V. Pinar and C.S. Foster. 1997. Ocular rosacea: patient characteristics and follow-up. Ophthalmology 104: 1863–7.

Aksoy, M.O., X. Li, M. Borenstein, Y. Yi and S.G. Kelsen. 1999. Effects of topical corticosteroids on inflammatory mediator-induced eicosanoid release by human airway epithelial cells. J. Allergy Clin. Immunol. 103: 1081–91.

Annesley, W.H. Jr., J.J. Augsburger and J.L. Shakin. 1981. Ten year follow-up of photocoagulated central serous choroidopathy. Trans. Am. Ophthalmol. Soc. 79: 335–46.

Aronowicz, J.D., W.E. Shine, D. Oral, J.M. Vargas and J.P. McCulley. 2006. Short term oral minocycline treatment of meibomianitis. Br. J. Ophthalmol. 90: 856–60.

Avunduk, A.M., M.C. Avunduk, E.D. Varnell and H.E. Kaufman. 2003. The comparison of efficacies of topical corticosteroids and nonsteroidal anti-inflammatory drops on dry eye patients: a clinical and immunocytochemical study. Am. J. Ophthalmol. 136: 593–602.

Barabino, S., M. Rolando, P. Camicione, G. Ravera, S. Zanardi, S. Giuffrida and G. Calabria. 2003. Systemic linoleic and gamma-linolenic acid therapy in dry eye syndrome with an inflammatory component. Cornea 22: 97–101.

Baudouin, C., F. Brignole, P.J. Pisella, M.S. De Jean and A. Goguel. 2002. Flow cytometric analysis of the inflammatory marker HLA DR in dry eye syndrome: results from 12 months of randomized treatment with topical cyclosporin A. Adv. Exp. Med. Biol. 506: 761–9.

Baudouin, C., H. Liang, D. Bremond-Gignac, P. Hamard, R. Hreiche, C. Creuzot-Garcher, J.M. Warnet and F. Brignole-Baudouin. 2005. CCR 4 and CCR 5 expression in conjunctival specimens as differential markers of T (H)1/T (H)2 in ocular surface disorders. J. Allergy Clin. Immunol. 116: 614–9.

Beardsely, R.M., C.S. de Paiva, D.F. Power and S.C. Pflugfelder. 2008. Desiccating stress decreases apical corneal epithelial cell size-modulation by the metalloproteinase inhibitor doxycycline. Cornea 27: 935–40.

Brignole, F., P.J. Pisella, J.M. De Saint, M. Goldschild, A. Goguel and C. Baudouin. 2001. Flow cytometric analysis of inflammatory markers in KCS: 6-month treatment with topical cyclosporin A. Invest Ophthalmol. Vis. Sci. 42: 90–5.

Brignole, F., P.J. Pisella, M. Goldschild, J.M. De Saint, A. Goguel and C. Baudouin. 2000. Flow cytometric analysis of inflammatory markers in conjunctival epithelial cells of patients with dry eyes. Invest Ophthalmol. Vis. Sci. 41: 1356–63.

Brunner, T., D. Arnold, C. Wasem, S. Herren and C. Frutschi. 2001. Regulation of cell death and survival in intestinal intraepithelial lymphocytes. Cell Death. Differ. 8: 706–14.

Byun, Y.J., T.I. Kim, S.M. Kwon, K.Y. Seo, S.W. Kim, E.K. Kim and W.C. Park. 2009. Efficacy of combined 0.05% cyclosporine and 1% methylprednisolone treatment for chronic dry eye. Cornea 31.

Carsons, S. 2001. A review and update of Sjögren's syndrome: manifestations, diagnosis, and treatment. Am. J. Manag. Care 7: S433–S443.

Caton, J.G., S.G. Ciancio, T.M. Blieden, M. Bradshaw, R.J. Crout, A.F. Hefti, J.M. Massaro, A.M. Polson, J. Thomas and C. Walker. 2000. Treatment with subantimicrobial dose doxycycline improves the efficacy of scaling and root planing in patients with adult periodontitis. J. Periodontol. 71: 521–32.

Chauhan, S.K., A.J. El, T. Ecoiffier, S. Goyal, Q. Zhang, D.R. Saban and R. Dana. 2009. Autoimmunity in dry eye is due to resistance of Th17 to Treg suppression. J. Immunol. 182: 1247–52.

Chen, Y., S.K. Chauhan, D.R. Saban, Z. Sadrai, A. Okanobo and R. Dana. 2011. Interferon-{gamma}-secreting NK cells promote induction of dry eye disease. J. Leukoc. Biol. 89: 965–72.

Chotikavanich, S., C.S. de Paiva, D.Q. Li, J.J. Chen, F. Bian, W.J. Farley and S.C. Pflugfelder. 2009. Production and activity of matrix metalloproteinase-9 on the ocular surface increase in dysfunctional tear syndrome. Invest Ophthalmol. Vis. Sci. 50: 3203–9.

Corrales, R.M., M.E. Stern, C.S. de Paiva, J. Welch, D.Q. Li and S.C. Pflugfelder. 2006. Desiccating stress stimulates expression of matrix metalloproteinases by the corneal epithelium. Invest Ophthalmol. Vis. Sci. 47: 3293–302.

de Paiva, C.S., S. Chotikavanich, S.B. Pangelinan, J.I. Pitcher, B. Fang, X. Zheng, P. Ma, W.J. Farley, K.S. Siemasko, J.Y. Niederkorn, M.E. Stern, D-Q Li and S.C. Pflugfelder. 2009. IL-17 disrupts corneal barrier following desiccating stress. Mucosal Immunology 2: 243–53.

de Paiva, C.S., R.M. Corrales, A.L. Villarreal, W. Farley, D.Q. Li, M.E. Stern and S.C. Pflugfelder. 2006a. Apical corneal barrier disruption in experimental murine dry eye is abrogated by methylprednisolone and doxycycline. Invest Ophthalmol.Vis. Sci. 47: 2847–56.

de Paiva, C.S., R.M. Corrales, A.L. Villarreal, W.J. Farley, D.Q. Li, M.E. Stern and S.C. Pflugfelder. 2006b. Corticosteroid and doxycycline suppress MMP-9 and inflammatory cytokine expression, MAPK activation in the corneal epithelium in experimental dry eye. Exp. Eye Res. 83: 526–35.

de Paiva, C.S., J.L. Lindsey and S.C. Pflugfelder. 2003. Assessing the severity of keratitis sicca with videokeratoscopic indices. Ophthalmology 110: 1102–9.

de Paiva, C.S., J.K. Raince, A.J. McClellan, K.P. Shanmugam, S.B. Pangelinan, E.A. Volpe, R.M. Corrales, W.J. Farley, D.B. Corry, D.Q. Li and S.C. Pflugfelder. 2010. Homeostatic control of conjunctival mucosal goblet cells by NKT-derived IL-13. Mucosal. Immunol. 4: 397–408.

de Paiva, C.S., E. Schwartz, P. Gjörstrup and S.C. Pflugfelder. 2011. Resolvin E1 (RX-10001) reduces corneal epithelial barrier disruption and protects against goblet cell loss in a murine model of dry eye. Cornea (in press).

de Paiva, C.S., A.L. Villarreal, R.M. Corrales, H.T. Rahman, V.Y. Chang, W.J. Farley, M.E. Stern, J.Y. Niederkorn, D.Q. Li and S.C. Pflugfelder. 2007. Dry Eye-Induced Conjunctival Epithelial Squamous Metaplasia Is Modulated by Interferon-{gamma}. Invest Ophthalmol. Vis. Sci. 48: 2553–60.

Dialilian, A. 2001. Effects of dexamethasone and cyclosporine A on the production of cytokines by human corneal epithelial cells and fibroblasts. ARVO Abstracts, 2001.

Dougherty, J.M., J.P. McCulley, R.E. Silvany and D.R. Meyer. 1991. The role of tetracycline in chronic blepharitis. Inhibition of lipase production in staphylococci. Invest Ophthalmol. Vis. Sci. 32: 2970–5.

Dursun, D., M.C. Kim, A. Solomon and S.C. Pflugfelder. 2001. Treatment of recalcitrant recurrent corneal erosions with inhibitors of matrix metalloproteinase-9, doxycycline and corticosteroids. Am. J. Ophthalmol. 132: 8–13.

Dursun, D., A.M. Piniella and S.C. Pflugfelder. 2001. Pseudokeratoconus caused by rosacea. Cornea 20: 668–9.

Dursun, D., M. Wang, D. Monroy, D.Q. Li, B.L. Lokeshwar, M.E. Stern and S.C. Pflugfelder. 2002. A mouse model of keratoconjunctivitis sicca. Invest Ophthalmol. Vis. Sci. 43: 632–8.

Ecoiffier, T., A.J. El, S. Rashid, D. Schaumberg and R. Dana. 2008. Modulation of integrin alpha4beta1 (VLA-4) in dry eye disease. Arch. Ophthalmol. 126: 1695–9.

Endres, S., R. Ghorbani, V.E. Kelley, K. Georgilis, G. Lonnemann, J.W. Van der Meer, J.G. Cannon, T.S. Rogers, M.S. Klempner and P.C. Weber. 1989. The effect of dietary supplementation with n-3 polyunsaturated fatty acids on the synthesis of interleukin-1 and tumor necrosis factor by mononuclear cells. N. Engl. J. Med. 320: 265–71.

Farris, R.L. 1994. Tear osmolarity a new gold standard? Adv. Exp. Med. Biol. 350: 495–503.

Frucht-Pery, J., A.S. Chayet, S.T. Feldman, S. Lin and S.I. Brown. 1989. The effect of doxycycline on ocular rosacea. Am. J. Ophthalmol. 107: 434–5.

Goto, E., Y. Yagi, Y. Matsumoto and K. Tsubota. 2002. Impaired functional visual acuity of dry eye patients. Am. J. Ophthalmol. 133: 181–6.

Goyal, S., S.K. Chauhan and R. Dana. 2011. Blockade of prolymphangiogenic vascular endothelial growth factor C in dry eye disease. Arch. Ophthalmol. 12.

Goyal, S., S.K. Chauhan, Q. Zhang and R. Dana. 2009. Amelioration of murine dry eye disease by topical antagonist to chemokine receptor 2. Arch. Ophthalmol. 127: 882–7.

Gulbenkian, A.,J. Myers and D. Fries. 1980. Hamster flank organ hydrolase and lipase activity. J. Invest Dermatol. 75: 289–92.

Gunduz, K. and O. Ozdemir. 1994. Topical cyclosporin treatment of keratoconjunctivitis sicca in secondary Sjögren's syndrome. Acta Ophthalmol. (Copenh) 72: 438–42.

Hanemaaijer, R., T. Sorsa, Y.T. Konttinen, Y. Ding, M. Sutinen, H. Visser, V.W. van Hinsbergh, T. Helaakoski, T. Kainulainen, H. Ronka, H. Tschesche and T. Salo. 1997. Matrix metalloproteinase-8 is expressed in rheumatoid synovial fibroblasts and endothelial cells. Regulation by tumor necrosis factor—alpha and doxycycline. J. Biol. Chem. 272: 31504–9.

Hanemaaijer, R., H. Visser, P. Koolwijk, T. Sorsa, T. Salo, L.M. Golub and V.W. van Hinsbergh. 1998. Inhibition of MMP synthesis by doxycycline and chemically modified tetracyclines (CMTs) in human endothelial cells. Adv. Dent. Res. 12: 114–8.

Hashimoto, S., Y. Gon, K. Matsumoto, I. Takeshita, S. Maruoka and T. Horie. 2000. Inhalant corticosteroids inhibit hyperosmolarity-induced, and cooling and rewarming-induced interleukin-8 and RANTES production by human bronchial epithelial cells. Am. J. Respir. Crit. Care Med. 162: 1075–80.

Hoeprich, P.D. and D.M. Warshauer. 1974. Entry of four tetracyclines into saliva and tears. Antimicrob. Agents Chemother. 5: 330–6.

James, M.J. and L.G. Cleland. 1997. Dietary n-3 fatty acids and therapy for rheumatoid arthritis. Semin.Arthritis Rheum 27: 85–97.

James, M.J., R.A. Gibson and L.G. Cleland. 2000. Dietary polyunsaturated fatty acids and inflammatory mediator production. Am. J. Clin. Nutr. 71: 343S–8S.

Jansen, T. and G. Plewig. 1997. Rosacea: classification and treatment. J. R. Soc. Med. 90: 144–50.

Jones, D.T., D. Monroy, Z. Ji, S.S. Atherton and S.C. Pflugfelder. 1994. Sjögren's syndrome: cytokine and Epstein-Barr viral gene expression within the conjunctival epithelium. Invest Ophthalmol. Vis. Sci. 35: 3493–504.

Kaswan, R.L., M.A. Salisbury and D.A. Ward. 1989. Spontaneous canine keratoconjunctivitis sicca. A useful model for human keratoconjunctivitis sicca: treatment with cyclosporine eye drops. Arch. Ophthalmol. 107: 1210–6.

Kojima, T., R. Ishida, M. Dogru, E. Goto, Y. Matsumoto, M. Kaido and K. Tsubota. 2005. The effect of autologous serum eyedrops in the treatment of severe dry eye disease: a prospective randomized case-control study. Am. J. Ophthalmol. 139: 242–6.

Kremer, J.M. 2000. n-3 fatty acid supplements in rheumatoid arthritis. Am. J. Clin. Nutr. 71: 349S–51S.

Kunert, K.S., A.S. Tisdale and I.K. Gipson. 2002. Goblet cell numbers and epithelial proliferation in the conjunctiva of patients with dry eye syndrome treated with cyclosporine. Arch. Ophthalmol. 120: 330–7.

Laibovitz, R.A., S. Solch, K. Andriano, M. O'Connell and M.H. Silverman. 1993. Pilot trial of cyclosporine 1% ophthalmic ointment in the treatment of keratoconjunctivitis sicca. Cornea 12: 315–23.

Lee, H.S., S.K. Chauhan, A. Okanobo, N. Nallasamy and R. Dana. 2011. Therapeutic efficacy of topical epigallocatechin gallate in murine dry eye. Cornea 30: 1465–72.

Li D-Q, Z. Chen, X.J. Song, W. Farley and S.C. Pflugfelder. 2002. Hyperosmolarity Stimulates Production of MMP-9, IL-1ß and TNF- by Human Corneal Epithelial Cells Via a c-Jun NH2- terminal kinase pathway. Invest Ophthalmol. Vis. Sci. 43: 1981.

Li, D.Q., L. Luo, Z. Chen, H.S. Kim, X.J. Song and S.C. Pflugfelder. 2006. JNK and ERK MAP kinases mediate induction of IL-1beta, TNF-alpha and IL-8 following hyperosmolar stress in human limbal epithelial cells. Exp. Eye Res. 82: 588–96.

Li, S., K. Nikulina, J. DeVoss, A.J. Wu, E.C. Strauss, M.S. Anderson and N.A. McNamara. 2008. Small proline-rich protein 1B (SPRR1B) is a biomarker for squamous metaplasia in dry eye disease. Invest Ophthalmol. Vis. Sci. 49: 34–41.

Li, Y., K. Kuang, B. Yerxa, Q. Wen, H. Rosskothen and J. Fischbarg. 2001. Rabbit conjunctival epithelium transports fluid, and P2Y2 (2) receptor agonists stimulate Cl (–) and fluid secretion. Am. J. Physiol. Cell Physiol. 281: C595–C602.

Liden, J., I. Rafter, M. Truss, J.A. Gustafsson and S. Okret. 2000. Glucocorticoid effects on NF-kappaB binding in the transcription of the ICAM-1 gene. Biochem. Biophys. Res. Commun. 273: 1008–14.

Lokeshwar, B.L. 1999. MMP inhibition in prostate cancer. Ann. N.Y. Acad. Sci. 878: 271–89.

Lopez, B.D. and J.L. Ubels. 1991. Quantitative evaluation of the corneal epithelial barrier: effect of artificial tears and preservatives. Curr. Eye Res. 10: 645–56.

Lopez, B.D. and J.L. Ubels. 1993. Artificial tear composition and promotion of recovery of the damaged corneal epithelium. Cornea 12: 115–20.

Luo, L., D.Q. Li, A. Doshi, W. Farley, R.M. Corrales and S.C. Pflugfelder. 2004. Experimental dry eye stimulates production of inflammatory cytokines and MMP-9 and activates MAPK signaling pathways on the ocular surface. Invest Ophthalmol. Vis. Sci. 45: 4293–301.

Macdonald, A. and M. Feiwel. 1972. Perioral dermatitis: aetiology and treatment with tetracycline. Br. J. Dermatol. 87: 315–9.

Macsai, M.S. 2008. The role of omega-3 dietary supplementation in blepharitis and meibomian gland dysfunction (an AOS thesis). Trans. Am. Ophthalmol. Soc. 106: 336–56.

Marsh, P. and S.C. Pflugfelder. 1999. Topical nonpreserved methylprednisolone therapy for keratoconjunctivitis sicca in Sjögren syndrome. Ophthalmology 106: 811–6.

Matsuda, S. and S. Koyasu. 2000. Mechanisms of action of cyclosporine. Immunopharmacology 47: 119–25.

McCulley, J.P., J.M. Dougherty and D.G. Deneau. 1982. Classification of chronic blepharitis. Ophthalmology 89: 1173–80.

Miljanovic, B., R. Dana, D.A. Sullivan and D.A. Schaumberg. 2007. Impact of dry eye syndrome on vision-related quality of life. Am. J. Ophthalmol. 143: 409–15.

Mrugacz, M., B. Zelazowska, A. Bakunowicz-Lazarczyk, M. Kaczmarski and J. Wysocka. 2007. Elevated tear fluid levels of MIP-1alpha in patients with cystic fibrosis. J. Interferon Cytokine Res. 27: 491–5.

Musch, D.C., A. Sugar and R.F. Meyer. 1983. Demographic and predisposing factors in corneal ulceration. Arch. Ophthalmol. 101: 1545–8.

Nelson, J.D., M.M. Drake, J.T. Brewer, Jr. and M. Tuley. 1994. Evaluation of a physiological tear substitute in patients with keratoconjunctivitis sicca. Adv. Exp. Med. Biol. 350: 453–7.

Niederkorn, J.Y., M.E. Stern, S.C. Pflugfelder, C.S. de Paiva, R.M. Corrales, J. Gao and K. Siemasko. 2006. Desiccating Stress Induces T Cell-Mediated Sjögren's Syndrome-Like Lacrimal Keratoconjunctivitis. J. Immunol. 176: 3950–7.

Noecker, R. 2001. Effects of common ophthalmic preservatives on ocular health. Adv. Ther. 18: 205–15.

Pangelinan, S.B., C.S. de Paiva, R. Singh, K. Agusala, J.W. Farley, M.E. Stern and Pflugfelder, S.C. 2008. Topical Cyclosporine Emulsion Modulates Immune Response in Experimental Dry Eye. Invest Ophthalmol. Vis. Sci. 49: 440.

Pflugfelder, S.C., W. Farley, L. Luo, L.Z. Chen, C.S. de Paiva, L.C. Olmos, D.Q. Li and M.E. Fini. 2005. Matrix metalloproteinase-9 knockout confers resistance to corneal epithelial barrier disruption in experimental dry eye. Am. J. Pathol. 166: 61–71.

Pflugfelder, S.C., A.J.W. Huang, P.T. Schuchovski, I.C. Pereira and S.C.G. Tseng. 1990. Conjunctival cytological features of primary Sjögren syndrome. Ophthalmology 97: 985–91.

Pflugfelder, S.C., D. Jones, Z. Ji, A. Afonso and D. Monroy. 1999. Altered cytokine balance in the tear fluid and conjunctiva of patients with Sjögren's syndrome keratoconjunctivitis sicca. Curr. Eye Res. 19: 201–11.

Pflugfelder, S.C., S.L. Maskin, B. Anderson, J. Chodosh, E.J. Holland, C.S. de Paiva, S.P. Bartels, T. Micuda, H.M. Proskin and R. Vogel. 2004. A randomized, double-masked, placebo-controlled, multicenter comparison of loteprednol etabonate ophthalmic suspension, 0.5%, and placebo for treatment of keratoconjunctivitis sicca in patients with delayed tear clearance. Am. J. Ophthalmol. 138: 444–57.

Pflugfelder, S.C., S.C.G. Tseng, O. Sanabria, H. Kell, C.G. Garcia, C. Felix, W. Feuer and B.L. Reis. 1998. Evaluation of subjective assessments and objective diagnostic tests for diagnosing tear-film disorders known to cause ocular irritation. Cornea 17: 38–56.

Pisella, P.J., F. Brignole, C. Debbasch, P.A. Lozato, C. Creuzot-Garcher, J. Bara, P. Saiag, J.M. Warnet and C. Baudouin. 2000. Flow cytometric analysis of conjunctival epithelium in ocular rosacea and keratoconjunctivitis sicca. Ophthalmology 107: 1841–9.

Qin, X., M.A. Corriere, L.M. Matrisian and R.J. Guzman. 2006. Matrix metalloproteinase inhibition attenuates aortic calcification. Arterioscler. Thromb. Vasc. Biol. 26: 1510–6.

Raphael, M., S. Bellefqih, J.C. Piette, H.P. Le, P. Debre and G. Chomette. 1988. Conjunctival biopsy in Sjögren's syndrome: correlations between histological and immunohistochemical features. Histopathology 13: 191–202.

Rashid, S., Y. Jin, T. Ecoiffier, S. Barabino, D.A. Schaumberg and M.R. Dana. 2008. Topical omega-3 and omega-6 fatty acids for treatment of dry eye. Arch. Ophthalmol. 126: 219–25.

Rocha, E.M., F.S. Pelegrino, C.S. de Paiva, A.C. Vigorito and C.A. de Souza. 2000. GVHD dry eyes treated with autologous serum tears. Bone Marrow Transplant 25: 1101–3.

Rolando, M., S. Barabino, C. Mingari, S. Moretti, S. Giuffrida and G. Calabria. 2005. Distribution of Conjunctival HLA-DR Expression and the Pathogenesis of Damage in Early Dry Eyes. Cornea 24: 951–4.

Sainz De La Maza Serra, C.C. and O. Kabbani. 2000. Nonpreserved topical steroids and lacrimal punctal occlusion for severe keratoconjunctivitis sicca. Arch. Soc. Esp. Oftalmol. 75: 751–6.

Sall, K., Stevenson, O.D., T.K. Mundorf and B.L. Reis. 2000. Two multicenter, randomized studies of the efficacy and safety of cyclosporine ophthalmic emulsion in moderate to severe dry eye disease. CsA Phase 3 Study Group. Ophthalmology 107: 631–9.

Seedor, J.A., H.D. Perry, T.F. McNamara, L.M. Golub, D.F. Buxton and D.S. Guthrie. 1987. Systemic tetracycline treatment of alkali-induced corneal ulceration in rabbits. Arch. Ophthalmol. 105: 268–71.

Sheppard, J.D., S.V. Scoper and S. Samudre. 2011. Topical loteprednol pretreatment reduces cyclosporine stinging in chronic dry eye disease. J. Ocul. Pharmacol. Ther. 27: 23–7.

Shine, W.E., J.P. McCulley and A.G. Pandya. 2003. Minocycline effect on meibomian gland lipids in meibomianitis patients. Exp. Eye Res. 76: 417–20.

Sobrin, L., Z. Liu, D.C. Monroy, A. Solomon, M.G. Selzer, B.L. Lokeshwar and S.C. Pflugfelder. 2000. Regulation of MMP-9 activity in human tear fluid and corneal epithelial culture supernatant. Invest Ophthalmol. Vis. Sci. 41: 1703–9.

Solomon, A., D. Dursun, Z. Liu, Y. Xie, A. Macri and S.C. Pflugfelder. 2001. Pro- and anti-inflammatory forms of interleukin-1 in the tear fluid and conjunctiva of patients with dry-eye disease. Invest Ophthalmol. Vis. Sci. 42: 2283–92.

Solomon, A., M. Rosenblatt, D. Li, D. Monroy, Z. Ji, B.L. Lokeshwar and S.C. Pflugfelder. 2000a. Doxycycline inhibition of interleukin-1 in the corneal epithelium. Am. J. Ophthalmol. 130: 688.

Solomon, A., M. Rosenblatt, D.Q. Li, Z. Liu, D. Monroy, Z. Ji, B.L. Lokeshwar and S.C. Pflugfelder. 2000b. Doxycycline inhibits the interleukin-1 system in the corneal epithelium. Invest Ophthalmol. Vis. Sci. 41: 2544–57.

Stern, M.E., J. Gao, T.A. Schwalb, M. Ngo, D.D. Tieu, C.C. Chan, B.L. Reis, S.M. Whitcup, D. Thompson and J.A. Smith. 2002. Conjunctival T-cell subpopulations in Sjögren's and non-Sjogren's patients with dry eye. Invest Ophthalmol. Vis. Sci. 43: 2609–14.

Stevenson, D., J. Tauber and B.L. Reis. 2000. Efficacy and safety of cyclosporin A ophthalmic emulsion in the treatment of moderate-to-severe dry eye disease: a dose-ranging, randomized trial. The Cyclosporin A Phase 2 Study Group. Ophthalmology 107: 967–74.

Strong, B., W. Farley, M.E. Stern and S.C. Pflugfelder. 2005. Topical Cyclosporine Inhibits Conjunctival Epithelial Apoptosis in Experimental Murine Keratoconjunctivitis Sicca. Cornea 24: 80–5.

Sun, J. and J. Wang. 2006. Cyclosporine inhibits apoptosis in experimental murine xerophthalamia conjunctival epithelium. J. Huazhong. Univ. Sci. Technolog. Med. Sci. 26: 469–71.

Ta, C.N., W.E. Shine, J.P. McCulley, A. Pandya, W. Trattler and J.W. Norbury. 2003. Effects of minocycline on the ocular flora of patients with acne rosacea or seborrheic blepharitis. Cornea 22: 545–8.

Tripathi, B.J., R.C. Tripathi and S.P. Kolli. 1992. Cytotoxicity of ophthalmic preservatives on human corneal epithelium. Lens Eye Toxic. Res. 9: 361–75.

Tsubota, K., K. Fukagawa, T. Fujihara, S. Shimmura, I. Saito, K. Saito and T. Takeuchi. 1999. Regulation of human leukocyte antigen expression in human conjunctival epithelium. Invest Ophthalmol. Vis. Sci. 40: 28–34.

Tsubota, K., E. Goto, H. Fujita, M. Ono, H. Inoue, I. Saito and S. Shimmura. 1999. Treatment of dry eye by autologous serum application in Sjögren's syndrome. Br. J. Ophthalmol. 83: 390–5.

Tsubota, K., E. Goto, S. Shimmura and J. Shimazaki. 1999. Treatment of persistent corneal epithelial defect by autologous serum application. Ophthalmology 106: 1984–9.

Turner, K., S.C. Pflugfelder, Z. Ji, W.J. Feuer, M. Stern and B.L. Reis. 2000. Interleukin-6 levels in the conjunctival epithelium of patients with dry eye disease treated with cyclosporine ophthalmic emulsion. Cornea 19: 492–6.

Voils, S.A., M.E. Evans, M.T. Lane, R.H. Schosser and R.P. Rapp. 2005. Use of macrolides and tetracyclines for chronic inflammatory diseases. Ann. Pharmacother. 39: 86–94.

Wachtel, M., K. Frei, E. Ehler, A. Fontana, K. Winterhalter and S.M. Gloor. 1999. Occludin proteolysis and increased permeability in endothelial cells through tyrosine phosphatase inhibition. J. Cell Sci. 112: 4347–56.

Yoon, K.C., C.S. de Paiva, H. Qi, Z. Chen, W.J. Farley, D.Q. Li and S.C. Pflugfelder. 2007. Expression of th-1 chemokines and chemokine receptors on the ocular surface of C57BL/6 mice: effects of desiccating stress. Invest Ophthalmol. Vis. Sci. 48: 2561–9.

Yoon, K.C., I.Y. Jeong, Y.G. Park and S.Y. Yang. 2007. Interleukin-6 and tumor necrosis factor-alpha levels in tears of patients with dry eye syndrome. Cornea 26:431–7.

Zhang, X., W. Chen, C.S. de Paiva, E.A. Volpe, N.B. Gandhi, W.J. Farley, D.Q. Li, J.Y. Niederkorn, M.E. Stern and S.C. Pflugfelder. 2011. Desiccating Stress Induces CD4 (+) T-Cell-Mediated Sjögren's Syndrome-Like Corneal Epithelial Apoptosis via Activation of the Extrinsic Apoptotic Pathway by Interferon-gamma. Am. J. Pathol. 179: 1807–14.

Zhu, X., S. Topouzis, L.F. Liang and R.L. Stotish. 2004. Myostatin signaling through Smad2, Smad3 and Smad4 is regulated by the inhibitory Smad7 by a negative feedback mechanism. Cytokine 26: 262–72.

11

Management of Keratitis

Flavia S.A. Pelegrino

SUMMARY

The treatment of infectious and noninfectious ulcerative keratitis, and non-ulcerative keratitis were summarized. Improper management can lead to marked loss of vision. The contact lens wear is a major risk factor for infectious keratitis. Acanthamoeba and fungal keratitis are the most expensive forms of infectious keratitis to treat. Noninvasive methods and molecular techniques have improved diagnosis of infectious keratitis. Fortified topical antibiotics and fluoroquinolones are still the mainstay of bacterial keratitis therapy. Voriconazole and new routes of administration of conventional antifungals appear promising for fungal keratitis. Antivirals and amelioration of host inflammatory response are promising for viral keratitis; the host response is also crucial in pathogenesis of Pseudomonas aeruginosa keratitis. Trauma-induced bacterial and fungal keratitis and contact lens-associated keratitis are preventable entities. Improved modalities of diagnosis and treatment have improved the outcome of infectious keratitis, but therapy of acanthamoebal, fungal and P. aeruginosa keratitis is still a challenge. Effective strategies must neutralize potential risk factors and counter host response over-activity without impairing killing of infecting microorganisms. Management of factor risks (include trauma, eyelid abnormalities, such as ectropion, entropion, tear film abnormalities and neurological deficiencies (trigeminal nerve paralysis or facial nerve paralysis) improve visual prognosis. In some such cases, surgery is required. Systemic and local immunosuppressive therapy in noninfectious ulcerative keratitis control corneal disease and prevents complications such as corneal perforation, endophthalmitis, and lost of the eye.

Ophthalmologic Clinic, 292 Querubim Uriel St., Zip code 13024-470, Campinas, São Paulo, Brazil; E-mail: flaviapelegrino@gmail.com

INTRODUCTION

The vast majority of corneal ulcers occur when additional systemic or local factors disrupt the corneal defense mechanisms. Local factors include conditions that alter the lids (such as entropion, ectropion, and seventh nerve palsy), disorders of lacrimation (such as dry eye, dacryocystitis), conjunctival disorders (such as pemphigoid, vernalconjunctivitis, Stevens-Johnson), corneal diseases (such as neurotrophic disease, bullous keratopathy, herpetic ulceration), and systemic factors such as alcoholism, coma, diabetes, nutritional or immune deficiency.

The prevention should be the treatment or elimination bacterial keratitis risk factors in ulcer patients as part of the management of the ulcer and to decrease the risk for subsequent ulcers. Risk management may include minimizing or eliminating extended-wear contact lens use, treating dry eyes or lid abnormalities, and using eye protection to prevent corneal foreign bodies.

INFECTIOUS ULCERATIVE KERATITIS

Bacterial Keratitis

Clinical evaluation of corneal disease with slit-lamp biomicroscopy to identified clinical sign of corneal ulcers should be the first step to diagnosis. Next, scraping for laboratory analysis with either a Kimura spatula or the bent tip of a 21-gauge hyper-dermic needle under slit-lamp after the instillation of a topical anesthetic. The material should be placed on to glass slides for Gram staining and on to culture media plate (Blood agar, Thyoglycolate broth, Chocolate agar). Antimicrobial agents should be used as soon as preliminary investigations have been completed. Two of the most common bacteria to cause microbial keratitis are *Pseudomonas aeruginosa* and *Staphylococcus aureus.* Antibiotic therapy to treat keratitis caused by these bacteria is either monotherapy with a fluoroquinolone (ciprofloxacin, ofloxacin—both second-generation fluoroquinolones, or gatifloxacin, moxifloxacin—fourth-generation fluorquinolones) or combination therapy with fortified aminoglycoside (gentamicin). Ciprofloxacin is the most potent of the fluoroquinolones, and has extended broad-spectrum activity with excellent *in vitro* activity against *Pseudomonas aeruginosa, Chlamydia trachomatis, Haemophilus influenzae, Neisseria gonorrhoeae, Staphylococcus aureus, Staphylococcus epidermidis, Enterobacter sp.* The fourth-generation fluorquinolones demonstrate increased susceptibility for *Staphylococcus aureus* isolates that are resistant to second-generation fluorquinolones. (Kowalski et al. 2003).

Table 1 This table summarize the main doses of topical and subconjunctival antibiotics for corneal ulcers treatment.

Antibiotic	Topical dose	Subconjunctival dose
Erytromycin	5 mg/gm ointment	100mg/0.5ml
Ciprofloxacin	3 mg/ml	
Vancomicin	50 mg/ml	25 mg/0.25ml
Amikacin sulfate	50 mg/ml	25 mg/0.5ml
Gentamicin	10–20 mg/ml	20 mg/0.5ml
Tobramycin	10–20 mg/ml	20 mg/0.5ml
Moxifloxacin	5 mg/ml	
Gatifloxacin	3 mg/ml	
Ofloxacin	3 mg/ml	

Ciprofloxacin 0.3% topical ophthalmic solution is extremely well tolerate with no significant corneal and conjunctival epithelial toxicity as opposed to topical fortified gentamicin 0.2%. Tobramycin and gentamicin ophthalmic solutions are associated with punctate epithelial keratitis and pseudomembranous conjunctivitis, both of which are reversible without sequels.

There is no difference in the efficacy of fourth-generation fluorquinolone in patients with bacterial keratitis compared with patients treated with ofloxacin (0.3%) or fortified tobramycin (1.33%) and cephazolin (5%) (Shah et al. 2010).

The treatment of Chlamydial epithelial keratitis is oral tetracycline, 1.0 to 1.5 mg daily for 4 weeks and topical tetracycline or erythromycin 2 times daily for 5 days and repeated monthly for 6 months.

The use of topical corticosteroids in microbial keratitis may promote bacterial growth, suppress corneal repair and induces corneal perforation.

Parasitic Keratitis: Acanthamoeba Keratitis (AK)

Early diagnosis and wide epithelial debridement of Acanthamoeba keratitis plays a crucial role in successful medical treatment. Recommended therapy would include the cationic antiseptic agents, chlorhexidine or polyhexamethylene biguanide (PHMB), in combination with propamidine isethionate and neomycin antibiotic. Surgical intervention should be avoided until a medical cure has been achieved. Penetrating keratoplasty should be deferred if at all possible until a medical cure has been achieved (Lindquist 1998). Ultraviolet-A light and riboflavin therapy have been shown a therapeutic option in non-responsive cases with topical agents (Khan et al. 2011). Recalcitrant chronic Acanthamoeba stromal keratitis may be treated with extended systemic voriconazole 200 mg twice daily with good preservation of vision (Tu et al. 2010).

Fungal Keratitis

The fungal etiology should be suspected in cases of keratitis that do not respond to antibacterial agents, especially in cases of vegetative trauma or extended wear contact lens usage. These cases should be scraped and sent for KOH or Gomorimethenamine silver stains as well as culture on Saboraud'agar. Scraping positive for fungal elements or the appearance for fungal growth by 36 to 48 hours. After culture should prompt the initiation of antifungal therapy.

The only commercially available antifungal drug in the United States of America is Natamycin (Pimaricin 5% suspension). However, various other drugs can be compounded into eye drops (by compounding pharmacies) or subconjunctival doses are effective. The most commonly used drugs are Voriconazole (1%), Amphotericin B (0.15%), Fluconazole (1%), and Miconazole (1%). Typically, the topical antifungals are given every hour initially. The duration of treatment is from 3 to 4 weeks on average. Antiseptics such as Chlorhexidine (0.2%) and Povidone iodine (5%) may be used, but are not as effective. Topical cycloplegic agents may be used. Systemic antifungal medications have been advocated as adjunctive therapy in severe cases, especially ulcers with anterior chamber reaction, but there have been no controlled studies showing a clear benefit of adding systemic antifungals. Therapeutic keratoplasty should be performed if after treatment the keratitis continues to progress.

Viral Keratitis

Punctate epithelial keratitis is the hallmark of viral infections. The cure rate is in order of 95% with treatment, which should promote rapid healing with minimal adverse effects. The initial treatment is with antiviral drugs (drops or ointment) for 14 days. By day 4, the lesion should start to diminish in size and, by day 10, it should have healed.

There are several antiviral agents with proven efficacy in the treatment of human herpetic keratitis (Ohashi 1997). These drugs are antimetabolites and include idoxuridine or IDU, trifluorothymidine, vidarabine or Ara-A, acyclovir.

The therapy for herpes zoster ophthalmicus (HZO) is systemic acyclovir (800 mg 5 times daily for 10 days) (Sanjay et al. 2011). The use of systemic and topical corticosteroids appears to be indicating to non-immunocompromised patients suffering the vasculitic complications of HZO such as; severe scleritis, episcleritis, corneal immune disease, and uveitis. Persistent

Table 2 Usual concentration and frequency of use of available antiviral drugs.

Drug	Concentration	Frequency
Trifluridine (Viroptic) drops	1.0%	Every 1–2h by day (total of 9 doses) for 14 days
Acyclovir (Zovirax) ointment	3.0%	5 times daily for 14–21 days
Idoxuridine (Stoxil) ointment	0.5%	5 times daily for 14 days
Idoxuridine (Stoxil, Herplex, Dendrid) drops	0.1%	Every 4h by day, every 2h at night for 14 days
Vidarabine (Vira-A) ointment	3.0%	5 times daily for 14 days

vasculitis and neuritis may result in chronic ocular complications, the most important of which are neurotrophic keratitis, mucus plaque keratitis, and lipid degeneration of corneal scars. Post-herpetic complications, especially post-herpetic neuralgia, are observed in well over half of patients with HZO. The chronic pain of post-herpetic neuralgia is treated locally with cold compresses and lidocaine 5% cream.

A vaccine (varicella-zoster virus) is suggested for patients who are at risk of developing herpes zoster and has been shown to boost immunity against herpes zoster virus in older patients.

NONINFECTIOUS ULCERATIVE KERATITIS

Mooren's Ulcer

Limited form, which is usually unilateral, may be treated with: topical corticosteroids (used at hourly intervals), topical cycloplegics, therapeutic soft contact lens, or patching of the eyelids. Systemic therapy immunosuppressive (cyclosporine, steroids) and with cytotoxic drugs (cyclophosphamide, azathioprine, or metotrexate) may be required in patients unresponsive to conventional treatment and in bilateral progressive disease. Surgical excision of 4 mm of the perilimbal conjunctiva adjacent to the ulcer and application of cryotherapy to the peri-limbal conjunctiva in the area adjacent to the ulceration may be effective in some cases resistant to systemic therapy. Amniotic membrane transplantation may be a useful treatment for selected patients with Mooren's ulcer especially where systemic immunosuppressive drugs are unavailable (Ngan et al. 2011).

Rosacea Keratitis

Many short- and long-term treatments are available to alleviate the symptoms and effects of rosacea. Topical therapy with topical corticosteroids, cyclosporine 0.05% ophthalmic emulsion, and systemic therapy with oral antibiotics (Tetracycline 250 mg 4 times daily for 1 month followed by 250 mg daily for at least 6 months or Doxycicline 100 mg daily for 1 month, after that 50 mg daily for 1 month, then 50 mg every other day for 1 month) are successful in alleviating the ocular lesions (Oltz et al. 2011).

Systemic therapy with corticosteroids, non-steroidal anti-inflammatory agents, and immunosuppressive may be used. Conditions associate as such as dry eye (Donnenfel et al. 2009), microbial keratitis, and scleritis must be identified and treated.

Systemic Collagen Vascular Disorders Keratitis

The treatment to four main diseases should considered:

1) *Rheumatoid arthritis (RA).* Local therapy with topical corticosteroids, systemic therapy with corticosteroids, nonsteroidal anti-inflammatory agents, and immune-suppressive are indicated. Conditions associates such as; dry eye, microbial keratitis, and scleritis must be identified and treated.
2) *Systemic lupus erythematosus (SLE).* Treatment of the systemic disease, as well as local therapy for the keratitis and tear deficiency (topical artificial tears, therapeutic soft contact lenses, punctual occlusion, etc.) and topical corticosteroid are necessary.
3) *Polyarteritis Nodosa.* Adequate systemic therapy with steroids and/or cytotoxic drugs is necessary. Local therapy is not indicated.
4) *Wegener's Granulomatose (WG).* Topical corticosteroid is ineffective, systemic treatment with combination of steroids and cyclophosphamide may be beneficial (Tarabishy AB et al. 2010).

NON-ULCERATIVE KERATITIS

Cogan's Syndrome

It is non-syphilitic interstitial keratitis and vestibule-auditory symptoms shows bilateral interstitial keratitis as most common inflammatory ophthalmologic condition. The major disease-related morbidities were due to vestibule-auditory disease and only infrequently due to systemic manifestations such as vasculitis, with or without aortitis.

It is necessary to be careful before institution of protracted courses of high-dose corticosteroids and/or chemotherapy for patients without pronounced systemic disease or severe eye disease unmanageable by topical or periocular corticosteroids alone (Haynes et al. 1980). A novel therapeutic treatment with anti-TNF-alpha has seen an option in clinical course that suggests progression of the disease (Fricker et al. 2007).

Luetic Interstitial Keratitis

This disease usually occurs between the ages of 5 and 25 years with acute bilateral pain and severe blurring of vision. Diagnosis is centered on a high level of clinical suspicion and includes treponemal specific and non-treponemal serologic tests. All patients with newly diagnosed syphilis should be tested for co-infection with human immunodeficiency virus, as the risk factors are similar for both diseases. The preferred treatment for all stages of syphilis remains parenteral penicillin; topical corticosteroids and cycloplegics. With proper diagnosis and prompt antibiotic treatment, the majority of cases of syphilis can result in a cure (Kiss et al. 2005).

CONCLUSION

The diagnosis and appropriate treatment of corneal infections requires isolate and identify causative agents. However, the clinical history directed and organized can provide much information and valuable hints for diagnostic, avoiding problems with the treatment.

The ophthalmologists should get the history of current illness, ocular history (including the medication used), general medical history (including systemic and allergic medications), and family history.

REFERENCES

Donnenfeld, E. and S.C. Pflugfelder. 2009. Topical ophthalmic cyclosporine: pharmacology and clinical uses. Surv. Ophthalmol. 54: 321–38.

Fricker, M., A. Baumann, F. Wermelinger, P.M. Villiger and A. Helbling. 2007. A novel therapeutic option in Cogan diseases? TNF-alpha blockers. Rheumatol. Int. 27: 493–5.

Haynes, B.F., M.I. Kaiser-Kupfer, P. Mason and A.S. Fauci. 1980. Cogan syndrome: Studies in thirteen patients, long-term follow-up, and a review of the literature. Medicine (Baltimore). 59: 426–41.

Kiss, S., F.M. Damico and L.H. Young. 2005. Ocular manifestations and treatment of syphilis. Semin Ophthalmol. 20: 161–7.

Khan, Y.A., R.T. Kashiwabuchi, S.A. Martins, J.M. Castro-Combs, S. Kalyani, P. Stanley, D. Flikier and A. Behrens. 2011. Riboflavin and ultraviolet light a therapy as an adjuvant treatment for medically refractive Acanthamoeba keratitis: report of 3 cases. Ophthalmology 118: 324–31.

Kowalski, R.P., D.K. Dhaliwal, L.M. Karenchak, E.G. Romanowski, F.S. Mah, D.C. Ritterband and Y.J. Gordon. 2003. Gatifloxacin and moxifloxacin: an *in vitro* susceptibility comparison to levofloxacin, ciprofloxacin, and ofloxacin using bacterial keratitis isolates. Am. J. Ophthalmol. 136: 500–5.

Lindquist, T.D. 1998. Treatment of Acanthamoeba keratitis. Cornea 17: 11–6.

Ngan, N.D. and H.T. Chau. 2011. Amniotic membrane transplantation for Mooren's ulcer. Clin. Experiment Ophthalmol. 39: 386–92.

Ohashi, Y. 1997. Treatment of herpetic keratitis with acyclovir: benefits and problems. Ophthalmologica 211: 29–32.

Oltz, M. and J. Check. 2011. Rosacea and its ocular manifestations. Optometry 82: 92–103.

Sanjay, S., P. Huang and R. Lavanya. 2011. Herpes zoster ophthalmicus. Curr. Treat Options Neurol. 13: 79–91.

Shah, V.M., R. Tandon, G. Satpathy, N. Nayak, B. Chawla, T. Agarwal, N. Sharma, J.S. Titiyal and R.B. Vajpayee. 2010. Randomized clinical study comparative evaluation of fourth-generation fluoroquinolone with the combination of fortified antibiotics in the treatment of bacterial corneal ulcers. Cornea 29: 751–7.

Tarabishy, A.B., M. Schulte, G.N. Papaliodis and G.S. Hoffman. 2010. Wegener's granulomatosis: clinical manifestations, differential diagnosis, and management of ocular and systemic disease. Surv. Ophthalmol. 55: 429–44.

Tu, E.Y., C.E. Joslin and M.E. Shoff. 2010. Successful treatment of chronic stromal acanthamoeba keratitis with oral voriconazole monotherapy. Cornea 29: 1066–8.

12

Management of Ocular Allergy

Laura Motterle

SUMMARY

This chapter presents a survey on anti-allergic drugs and specific suggestions for each type of ocular allergy. While SAC/PAC and GPC are usually well controlled by the therapy, severe forms of VKC and AKC would need further investigations and new medications.

INTRODUCTION

The conjunctiva is an active immunological tissue, constantly exposed to the environment, which gets inflamed easily in predisposed subjects. A number of non-pharmacological measures and different drugs are available for treatment of ocular allergic manifestations.

NON-PHARMACOLOGICAL MEASURES

The first step in prevention of local allergic reaction is allergen avoidance, if possible, or at least minimizing exposure. For this reason the identification of susceptible antigens is necessary. Although in many cases of allergic conjunctivitis particular allergen sensitivities are not detected, specific IgE blood analysis, Prick-test and/or conjunctival provocation test (CPT) may give further indications.

Environmental control is the most basic and cost effective action. Several measures are particularly recommended in case of dust mites allergy: removing or covering their reservoirs (carpets, curtains, mattresses,

Ophthalmology Unit, Department of Neuroscience, University of Padova, Via Giustiniani 2, 35128, Padova, Italy; E-mail: laura.motterle@yahoo.it pillows, soft toys), reducing temperature and humidity, cleanness (with water temperature over 131°F/55°C) and use of acaricides and HEPA air filters. When the sources of allergens are animals, like cats, dogs, rabbits and other furry pets, they should be banned from the home or at least from the bedrooms. Likewise, plants and food should be avoided, if they are known to induce allergy.

Another beneficial generic procedure is ocular lubrication with artificial tears, to dilute allergens and inflammatory mediators from the ocular surface. Cold compresses will give some symptoms relief.

Patients should be instructed not to rub their eyes, because this may cause mechanical mast cell degranulation and worsening of symptoms.

THERAPY OPTIONS

For the treatment of allergic conjunctivitis both topical and systemic medications are available. Topical drugs include different categories: mast cell stabilizers, antihistamines, nonsteroidal anti-inflammatory drugs (NSAID), corticosteroids, and immunosuppressive agents.

Mast cell stabilizers prevent mast cell degranulation, by reducing intracellular calcium-dependent histamine secretion and lowering the release of pro-inflammatory mediators. They should be considered as protective drugs. They may take within 24 to 48 hours to improve symptoms. Cromolyn sodium 4.0%, lodoxamide tromethamine 0.1%, pemirolast potassium 0.1%, nedocromil sodium 2.0% and N acetyl-aspartyl-glutamic acid (NAAGA) 6% are included in this group.

Antihistamines are the major pharmacological group used to control allergic symptoms. They act mainly by preventing the binding of histamine to histamine-receptors (H-R). They have evolved particularly over the last decade, reaching longer duration, better selectivity, and multiple mechanisms of action. Thus new formulations result more efficacious on both early and late phases of allergic responses (Abelson et al. 2011).

The first-generation antihistamines, as antazoline and pheniramine, were surpassed by longer lasting and better tolerated agents, as levocabastine and emedastine. Both of them display their activity not only as H1-receptors antagonist. Levocabastine also reduces the expression of adhesion molecules on conjunctival epithelium and the eosinophils infiltration. Emedastine is able to antagonize histamine-stimulated vascular permeability in the conjunctiva.

Dual-acting agents, which act both as H1-antagonists and as mast cell stabilizers in variable degrees, provide a prompt relief and exert a preventative function. Olopatadine, ketotifen, azelastine, epinastine are included in this class of medications. Olopatadine hydrochloride is also able to reduce the expression of cell adhesion molecules and inflammatory markers. Ketotifen fumarate inhibits inflammatory mediator release from mast cells, basophils, and eosinophils, chemotaxis and degranulation of eosinophils, and leukotriene activity. Azelastine hydrochloride decreases expression of ICAM-1, reduces eosinophil chemotaxis, and inhibits platelet-activating factor. Epinastine hydrochloride has affinity for the H1, H2, $\alpha1$, $\alpha2$, and 5-HT2 receptors.

A new dual-action formulation, called alcaftadine, was approved recently. Besides its mast cell-stabilization effect and H1-R and H2-R antagonism, it has been shown to antagonize H4-R *in vitro* (Ono and Lane 2011).

Main systemic side effects of antihistamines are drowsiness and dryness of the eyes, nose, and mouth. It has to be considered in case of dry eye. Some preparations contain also a vasoconstrictor (such as naphazoline hydrochloride), whose prolonged use should be avoided, as this can result in a reactive hyperaemia.

Topical nonsteroidal anti-inflammatory drugs (NSAID), ketorolac, indomethacin and diclofenac, may be indicated in mild cases of ocular allergy for itching relief; in fact they are capable of reducing tear levels of (PG) E2, a pruritogenic prostaglandin.

Topical corticosteroids are highly effective drugs for the treatment of allergic inflammation, but their use needs to be monitored by the ophthalmologist, as they are associated with the well-known side effects, such as glaucoma, cataract formation, and increased risk of infections of the ocular surface. Generally, low absorption steroids, such as clobetasone, fluorometholone and desonide, show a lower incidence of intraocular pressure rise and lens opacity. They penetrate the cornea but are inactivated quickly in the anterior chamber before they reach the trabecular meshwork or the lens receptors. More potent steroids are prednisolone, betamethasone and dexamethasone that are useful in most severe conjunctivitis. Loteprednol etabonate seems to be effective in allergic conjunctivitis and is rarely associated with a significant rise in intraocular pressure.

Cyclosporine A (CsA), a fungal antimetabolite, exerts multiple anti-inflammatory functions. It inhibits interleukin-2 dependent activation of lymphocytes, and interferes with antigen processing and presentation of antigen to the uncommitted T lymphocytes. Moreover it has been shown *in vitro* to reduce collagen production and to induce apoptosis of conjunctival fibroblasts from VKC patients (Leonardi et al. 2001). As an ophthalmic emulsion, CsA can be applied safely also for long periods of time. No systemic absorption has been detected and local side effects are limited (burning at the instillation, rare mild corneal epitheliopathy).

Tacrolimus is a strong immunosuppressant (10 to 100 times more potent than CsA) that inhibits the proliferative response of lymphocytes to antigen stimulation on a variety of T cell associated immune reactions.

Oral antihistamines may be associated to the topical treatment. Second-generation antihistamines (cetirizine hydrochloride, desloratadine, fexofenadine hydrochloride, loratadine, alcaftadine) should be preferred, for their generally limited adverse effects (Bielory 2005).

Subcutaneous or sublingual immunotherapy may be beneficial to some extent, although it seems more effective in relief of nasal symptoms than ocular manifestations.

Efficacy of treatments is primarily based on signs and symptoms relief. However, if more objective information is needed, as in clinical trials and follow-up, lacrimal or brush cytology may be useful. A promising alternative technique is the confocal scanning laser microscopy; it is an *in vivo* non-invasive technique that resembles histological analysis of conjunctiva and allows an assessment of inflammatory cell density and tissues changes (Wakamatsu et al. 2009).

Investigation is focused on the identification of new targets, novel drug delivery systems and genetic therapy (Mishra et al. 2011).

Regulation of histamine release is still the main goal of treatment. The research points to new medications directed against H4-R, the last discovered histamine receptor, and to spleen tyrosine kinase (Syk) inhibitors. Syk regulates the phosphorylation of enzymes (such as phopholipase-C, phosphotidylinositol-3 kinase and protein kinase), which regulate the release of histamine.

New anti-lymphocyte drugs may be considered also for the treatment of the allergic inflammation: a novel drug, mycophenolic acid that inhibits T and B lymphocyte replication, and inhibitors of Janus protein kinase-3 (JAK-3) that is involved in the activation and proliferation of T-cells.

Toll-like receptors may be another target of new compounds. They play a role in the activation of the innate immunity and seem to regulate Th1/Th2 lymphocyte balance in the hypersensitivity reaction.

Monoclonal antibodies against different inflammatory mediators were developed and may have a potential therapeutic role in ocular allergy: anti-IgE (omalizumab), anti-eotaxin 1, anti-CD-20 (rituximab), anti-TNFα (infliximab), anti-CD2 (alefacept).

SEASONAL/PERENNIAL CONJUNCTIVITIS

Most of the patients with SAC/PAC are treated using either topical mast cell stabilizers or antihistamines. They both show similar efficacy in relieving symptoms, only a limited evidence of a faster action of antihistamines (Owen et al. 2004). Dual-acting drugs offer an advantage in regulating the inflammation through a low dosage. Systemic antihistamine may be added in case of extraocular allergic involvement, since rhinitis is frequently associated to the ocular symptoms.

Topical FANS are an adjuvant in the therapy, but they are not beneficial if taken alone.

Low absorption steroids, as loteprednol, clobetasone, fluorometholone and desonide, may be prescribed at need for abating the ailments.

VERNAL KERATOCONJUNCTIVITIS

VKC management is based on the severity of the disease (Sacchetti et al. 2010). However, generic preventative measures should be suggested to all patients, such as avoiding non specific stimuli (heat, sun, wind, salt water) and specific allergen exposure, using hats with visors, wrap-around sunglasses and swimming goggles, frequent hand, face and hair washing.

The pharmacological treatment may be prolonged for many years; it is seasonal or almost continuous in case of year-round symptoms. Since most of the VKC patients are children, parents' instruction and collaboration is essential for compliance.

The first-line drugs are antihistamines and mast cell stabilizers; in mild cases they are effective, and might be associated with oral antihistamines. The combination of drugs with different mechanisms of action may help reaching a better control of the inflammation. Thus, drugs with mast cell stabilizing effects, such as lodoxamide and NAAGA, or dual acting drugs, such as olopatadine, azelastine and ketotifen, may be associated with medication, such as emedastine or levocabastine, that also reduce IL-1, IL-6 and IL-8 production besides the H1-antagonism.

Topical steroids ensure a prompt relief of signs and symptoms, but a protracted administration is contraindicated due to their side effects. Low absorption steroids, as loteprednol, clobetasone, fluorometholone and desonide, are sometimes beneficial, but prednisolone, dexamethasone and betamethasone are more potent. They can be prescribed in pulsed short cycles at need, without suspending antihistamine therapy.

In most severe forms of VKC, when patients develop a steroid-dependence or have no advantage from standard therapy, CsA ophthalmic emulsion (from 0.5% to 2% in castor or olive oil, BSS or artificial tears) is a very effective treatment. It can be applied from one to four times daily depending on the phases of the disease (Leonardi 2002). Commercial 0.1% and 0.05% ophthalmic solutions, where available, seems effective as well (Ebihara et al. 2009, Ozcan et al. 2007). In addition, the use of tacrolimus eye-drops or ointment has been proposed in refractory cases (Kheirkhah et al. 2011).

Occasionally, in worst cases, a short-term immunosuppression with systemic corticosteroids is appropriate. Ocular infections are uncommon.

In the presence of corneal ulcer, steroid eye-drops or ointment with preventative antibiotic coverage and eye patching are necessary and a strict follow-up is required. In case, amniotic membrane patching can enhance re-epithelization.

Surgical removal of corneal plaques is sometimes useful. Giant papillae excision is convenient in the event of relevant mechanical pseudo-ptosis.

ATOPIC KERATOCONJUNCTIVITIS

AKC is the most severe chronic form of allergic disease of the ocular surface; both allergic and proper non-allergic signs and symptoms and complications have to be monitored.

Environmental and preventative measures are necessary, although far from sufficient. As AKC is not only an ocular disorder, a systemic therapeutic approach is recommended. Antihistamine oral therapy is the first-step treatment; it is quite safe and can be taken all-year round with limited side effects. However, systemic steroids are often necessary to obtain a significant control of the inflammation. In case of refractive AKC and as an alternative to steroids, a T-cell targeted approach with low dose oral cyclosporine A or tacrolimus (FK-506) usually guarantees good results (Cornish et al. 2010, Stumpf et al. 2006). Alternatively or in addition to steroids or to cyclosporine, other immunosuppressive agents, like azathioprine and mycophenolate mofetil, are used in severe atopic dermatitis. The trend for future treatment focuses on biological therapies; omalizumab, a humanised monoclonal IgE antibody that complexes with free circulating IgE, demonstrated its efficacy in asthma control and has been proposed also for AKC (Williams et al. 2005).

Ocular therapy includes antihistamines and mast cell stabilizers, that are able to reduce the ocular swelling only partially, and topical steroids, that usually ensure a rapid palliation of symptoms. A number of steroid-sparing medications were assessed, such as cyclosporine eye-drops and tacrolimus ointment or eye-drops (FK-506) (Ebihara et al. 2009, Attas-Fox et al. 2008).

Topical or systemic antibiotics are often necessary to treat infectious blepharitis or microbial keratitis, as well as antiviral has to be administered when a herpetic infection occurs. In case of recurrent HSV keratitis, oral acyclovir 400 mg twice daily is prescribed for a long period as prophylaxis against relapses (Guglielmetti et al. 2010).

Surgery may be necessary in case of complications. Cataract may develop also in younger patients; typically in AKC it is an anterior polar cataract, while posterior capsular opacity is secondary to a prolonged steroid treatment. Glaucoma in steroid-dependent subjects sometimes required a filtering surgery, in some cases a drainage implants. The outcome is quite poor, due to the long-lasting inflammation of the conjunctiva. As in VKC, removal of corneal plaques and giant papillae excision is sometimes useful.

GIANT PAPILLARY CONJUNCTIVITIS

GPC is determined by continuous or excessive exposure of ocular surface to contact lenses or other foreign material in subjects with conjunctival oversensitivity; thus two measures should be considered for its control. The first point is to decrease or totally remove discomforting items from the eye; the second aim is to modulate the conjunctival immune response.

The best results are achieved in case of refitting with new contact lenses after a period of discontinuation for 3 to 4 weeks. Since lenses with large diameter and excessive edge lift are usually more traumatic, the change of contact lens material or design is the most beneficial action; rigid gas-permeable or glyceryl methylmethacrylate lenses are usually more tolerated. Also frequent replacement or the use of disposable daily wear contact lenses may be suggested (Donshik et al. 2008).

A revision in cleaning manage can resolve about 50% of contact lens intolerance, as coating decrease reduces antigen exposure of the conjunctiva. Coating of contact lens begins half an hour after insertion and accumulates as time passes. It tends to occur mostly in lenses with higher water content. Daily cleaning with hydrogen peroxide or unpreserved saline solution and enzymatic cleaning once or twice a week should be encouraged. However, even the best cleaning procedure remove only 75% of the coating.

The pharmacological control of inflammation is not necessary in preclinical and mild stages of GPC. Topical mast cell stabilizers, such as cromolyn sodium, nedocromil sodium or lodoxamide, induce symptoms relief, reducing itching and mucus production. Non-steroidal and steroidal anti-inflammatory agents are recommended, depending on the severity of conjunctival swelling. Since there is no risk of topical side effects in case of GPC related to ocular prosthesis, steroids can be administered as first treatment.

CONTACT DERMATOCONJUNCTIVITIS

Treatment consists of avoiding contacts with the allergen. The causing agent can be often identified by an accurate clinical history; otherwise skin prick and patch testing may be assessed.

In most of the cases the therapy is not necessary, but the healing process may be facilitated using mild topical corticosteroids for a short time. Antihistamines may be associated to provide itch relief.

CONCLUSIONS

In conclusion, different medications are available for treatment of allergic ocular inflammation and usually they are promptly effective and safe. However, in the most severe forms of allergic conjunctivitis the therapy tends to be long and frustrating, particularly for paediatric patients. New drugs are being developed with the aim of modifying the course of such diseases.

ACKNOWLEDGMENT

I would like to thank Dr. Andrea Leonardi (Ophthalmology Unit, Department of Neuroscience, University of Padova) for mentoring me and reviewing the contents of this chapter.

REFERENCES

Abelson, M.B., J.T. McLaughlin and P.J. Gomes. 2011. Antihistamines in ocular allergy: are they all created equal? Curr. Allergy Asthma Rep. 11: 205–211.

Attas-Fox, L., Y. Barkana, V. Iskhakov, S. Rayvich, Y. Gerber, Y. Morad, I. Avni and D. Zadok. 2008. Topical tacrolimus 0.03% ointment for intractable allergic conjunctivitis: an open-label pilot study. Curr. Eye Res. 33: 545–549.

Bielory, L., K.W. Lien and S. Bigelsen. 2005. Efficacy and tolerability of newer antihistamines in the treatment of allergic conjunctivitis. Drugs 65: 215–228.

Cornish, K.S., M.E. Gregor and K. Ramaes. 2010. Systemic cyclosporin A in severe atopic keratoconjunctivitis. Eur. J. Ophthalmol. 20: 844–851.

Donshik, P.C., W.H. Ehler and M. Ballow. 2008. Giant papillary conjunctivitis. Immunol. Allergy Clin. North Am. 28: 83–103.

Ebihara, N., Y. Ohashi, E. Uchio, S. Okamoto, N. Kumagai, J. Shoji, E. Takamura, Y. Nakagawa, K. Nanba, A. Fukushima and H. Fujishima. 2009. A large prospective observational study of novel cyclosporine 0.1% aqueous ophthalmic solution in the treatment of severe allergic conjunctivitis. J. Ocul. Pharmacol. Ther. 25: 365–372.

Guglielmetti, S., J.K. Dart and V. Calder. 2010. Atopic keratoconjunctivitis and atopic dermatitis. Curr. Opin. Allergy Clin. Immunol. 10: 478–485.

Kheirkhah, A., M.K. Zavareh, F. Farzbod, M. Mahbo and M.J. Behrouz. 2011. Topical 0.005% tacrolimus eye drop for refractory vernal keratoconjunctivitis. Eye 25: 872–880.

Leonardi, A. 2002. Vernal keratoconjunctivitis: pathogenesis and treatment. Progr. Retin. Eye Res. 21: 319–339.

Mishra, G.P., V. Tamboli, J. Jwala and A.K. Mitra. 2011. Recent patents and emerging therapeutics in the treatment of allergic conjunctivitis. Recent Pat Inflamm. Allergy Drug Discov. 5: 26–36.

Ono, S.J. and K. Lane. 2011. Comparison of effects of alcaftadine and olopatadine on conjunctival epithelium and eosinophil recruitment in a murine model of allergic conjunctivitis. Drug Des. Devel. Ther. 5: 77–84.

Owen, C.G., A. Shah, K. Henshaw, L. Smeeth and A. Sheikh. 2004. Topical treatments for seasonal allergic conjunctivitis: systematic review and meta-analysis of efficacy and effectiveness. Br. J. Gen. Pract. 54: 451–456.

Ozcan, A.A., T.R. Ersoz and E. Dulger. 2007. Management of severe allergic conjunctivitis with topical cyclosporin a 0.05% eye drops. Cornea 26: 1035–1038.

Sacchetti, M., A. Lambiase, F. Mantelli, V. Deligianni, A. Leonardi and S. Bonini. 2010. Tailored approach to the treatment of vernal keratoconjunctivitis. Ophthalmology 117: 1294–1299.

Stumpf, T., N. Luqmani, P. Sumich, S. Cook and D. Tole. 2006. Systemic tacrolimus in the treatment of severe atopic keratoconjunctivitis. Cornea 25: 1147–1149.

Wakamatsu, T.H., N. Okada, T. Kojima, Y. Matsumoto, O.M. Ibrahim, M. Dogru, E.S. Adan, K. Fukagawa, C. Katakami, K. Tsubota, J. Shimazaki and H. Fujishima. 2009. Evaluation of conjunctival inflammatory status by confocal scanning laser microscopy and conjunctival brush cytology in patients with atopic keratoconjunctivitis (AKC). Molecular Vision 15: 1611–1619.

Williams, P.B. and J.D. Sheppard. 2005. Omalizumab: a future innovation for treatment of severe ocular allergy? Expert Opin. Biol. Ther. 5: 1603–1609.

13

Management of Cicatricial Conjunctivitis

Bety Yáñez Álvarez

SUMMARY

The management of cicatricial conjunctivitis depends on the underlying ocular or systemic disease. In the case of infectious conjunctivitis such as trachoma or those produced by bacterias, antibiotic treatment is specific for each case.

In this chapter management of the most frequent cicatricial conjunctivitis; Stevens–Johnson syndrome, Ocular Mucous Membrane Pemphigoid, and Trachoma are discussed.

INTRODUCTION

The treatment of autoimmune cicatricial conjunctivitis is a challenge; controlling conjunctival inflammation is necessary in many cases with the use of immunosuppresive drugs. Many of them have severe side systemic effects.

It is extremely important that the ophthalmologist recognizes in chronic progressive conjunctival inflammation the toxicity induced by drugs and reduces this inflammation in order to prevent further ocular surface damage, especially in the corneal epithelium. To treat severe dry eye free tear substitutes must be indicated, topic autologous serum, cyclosporine eye drops to 0.05% and conservative environmental measures.

Department of Ophthalmology, Hospital Nacional Dos de Mayo, Cdra. 13 Av. Grau–Cercado de Lima, Lima, Peru; E-mail: byanez@hotmail.com

As surgical intervention (cataract or conjunctival oculoplastic surgery) may provoke disease exacerbation, these procedures only will be performed in patients in whom conjunctival inflammation is fully controlled by medical therapy.

Finally, in the end-stage inflammatory ocular disease the osteo-odonto-keratoprosthesis (OOKP) has better results in eyes with the presence of severe tear deficiency, often accompanied by chronic conjunctival inflammation, cicatrization and limbal stem cell failure.

TOXIC EPIDERMAL NECROLYSIS AND STEVENS–JOHNSON SYNDROME

Treatment

Stevens–Johnson syndrome (SJS/TEN) is a life threatening condition and therefore supportive care is an essential part of the therapeutic approach and prompt identification and withdrawal of the culprit drug(s). Meticulous skin and mucous membrane care is important in the acute phase. Amniotic membrane coverage of the ocular surface in its entirety coupled with the use of intensive short-term topical corticosteroids is associated with the preservation of good visual acuity and an intact ocular surface.

To date, a specific therapy for SJS/TEN that has shown efficacy in controlled clinical trials unfortunately does not exist. Several treatments have been reported to be beneficial, systemic steroids, high-dose intravenous immunoglobulin therapy (Trent et al. 2007) plasmapheresis (Yamada et al. 1998), Ciclosporin (CsA) (Valeyrie-Allanore et al. 2010) and systemic humanized monoclonal antibodies (daclizumab) (Fiorelli et al. 2010).

The management of the chronic stage of SJS/TEN requires recognition of the toxicity induced by drugs and reduces conjunctival inflammation. This may be related to trichiasis or severe dry eye and a recurrent endogenous inflammation. It is in these cases that the use of systemic immunosuppression with corticosteroids and/or steroid-sparing agents must be considered. To treat severe dry eye, free tear substitutes, topic autologous serum, cyclosporine eye drops to 0.05% and conservative environmental measures (humidifiers, moist chamber spectacles) must be indicated.

If patients have conjunctival inflammation or evidence of progressive conjunctival shrinkage, cataract surgery or oculoplastic surgery on the conjunctiva should be avoided until the disease is controlled.

Corneal transplantation in SJS/TEN patients with severe ocular surface disease is associated with a poor prognosis. Persistent epithelial defects occurring after penetrating or lamellar keratoplasty often progress to corneal melting and perforation. Transplanted limbal stem cells or keratoepithelioplasty in these chronically inflamed eyes often elicit graft rejection and loss of donor epithelial cells, resulting in progressive conjunctivalization, scarring, and visual loss (Kenyon and Tseng 1989, Thoft 1984). Other new ocular surface reconstructive procedures such as amniotic membrane and cultivated epithelial transplantation have yielded promising results for the treatment of SJS/TEN (Tseng et al. 1998, Koizumi et al. 2001).

In the end-stage inflammatory ocular disease the osteo-odonto-keratoprosthesis (OOKP) has proved to be the only keratoprosthesis (KPro) surgery with reasonable long-term anatomical and functional success rates for this group of patients (Falcinelli et al. 2005, Liu et al. 2008). This is a modified keratoprosthesis that has better results in eyes with the presence of severe tear deficiency, often accompanied by chronic conjunctival inflammation, cicatrization and limbal stem cell failure. These conditions include Stevens–Johnson syndrome, mucous membrane pemphigoid and severe chemical injury.

Invented by Strampelli, (Strampelli 1963) and later modified by Falcinelli, the OOKP uses the patient's own tooth and alveolar bone to support an optical cylinder (Fig. 1). A buccal mucous membrane covering provides vascular support and physical and microbiological protection.

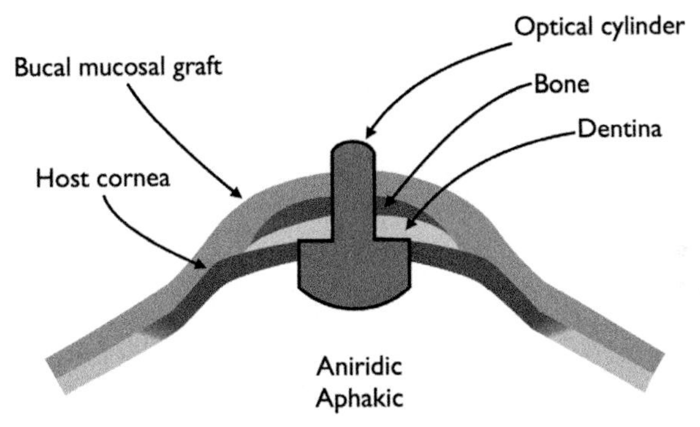

Figure 1 This figure shows the schematic of osteo-odonto-keratoprosthesis (OOKP) (Adaptation from Liu et al. 2008).

Color image of this figure appears in the color plate section at the end of the book.

OCULAR MUCOUS MEMBRANE PEMPHIGOID (OMMP)

Treatment

The extent and progression of ocular disease as well as treatment options can be monitored according to the staging system of Foster (Foster 1986). Ocular lubricants such as nonpreserved artificial tears, ointment, and gels are useful in improving surface lubrication. Concomitant blepharitis should be recognized and treated appropriately; it can contribute to eyelid scarring with trichiasis, and cause conjunctival and corneal inflammation as well.

Topical corticosteroids are ineffective in controlling the progression of disease and should be avoided for treatment of acute relapses. Intralesional corticosteroid injections may produce short-term remission, but lack long-term benefits and may even induce cataract formation. Both topical and subconjunctival applications of mitomycin C have been described in the treatment of advanced OMMP (Celis-Sanchez et al. 2002).

Systemic treatment is necessary for patients with intermediate stage of disease. Prednisone is the most useful drug, because of its potent anti-inflammatory and immunosuppressive effects, and rapid effects can be seen at a dose of 1 mg/kg daily (Nguyen and Foster 1996). For mild disease without rapid progression, dapsone may be initiated at 25 to 50 mg per day, increasing monthly from 25 to 50 mg until clinical remission is achieved or until the maximum tolerated dose is reached (usually 200 mg per day) (Miserocchi et al. 2002). Azathioprine 1–2 mg/kg daily and mycophenolate mofetil 2–3 g/daily also are effective (Bialasiewicz et al. 1994, Megahed et al. 2001). More aggressive treatment with cyclophosphamide 1–2 mg/kg daily alone or in association with prednisone 1 mg/kg daily is required in severe ocular mucous membrane pemphigoid or primary progressive course of disease (Kirtschig et al. 2002). The successful use of intravenous immunoglobulins (IVIG) has been proved in OMMP although not in controlled trials. Other immunomodulatory approaches includes inhibitors of pro-inflammatory cytokines, such as tumor necrosis factor-α, etanercept and infliximab (Heffernan and Bentley 2006). The advantages of the treatment with rituximab, the monoclonal anti-CD20 have been evaluated.

Because surgical intervention may provoke disease exacerbation, it should only be performed in patients in whom OMMP is fully controlled by medical therapy. Entropion with secondary trichiasis may be treated with cryotherapy, marginal rotation of the eyelid, or lid splitting with repositioning of the lash-bearing portion of the lid margin (Elder and Collin 1996). Surgical reconstruction of the fornices may be accomplished with mucous membrane grafting or amniotic membrane grafting, possibly augmented with topical or subconjunctival mitomycin (Tseng et al. 2005).

Corneal transplant can restore the clarity of the visual axis scarred from OMMP, but stem cell deficiency, dry eye, lid malposition, and conjunctival inflammation can contribute to graft failure or rejection (Tsubota et al. 1996). Limbal stem cell deficiency may be addressed with limbal stem cell grafting prior to corneal transplant to increase the success of the corneal graft. Placement of a keratoprosthesis or osteo-odonto-keratoprosthesis also offers an alternative for patients with severe visual disability due to end-stage scarring.

TRACHOMA

Treatment

With the goal of eliminating blinding trachoma by the year 2020, the Global Alliance for the Elimination of Blinding Trachoma (GET2020) was formed in 1998, including the WHO, trachoma endemic countries and organizations working in the field. Control activities focus on the implementation of the SAFE strategy, **S**urgery for trichiasis, **A**ntibiotics for infection, **F**acial cleanliness (hygiene promotion) and **E**nvironmental improvements, to reduce transmission of the organism (WHO 2003).

For mild trichiasis with a few peripheral lashes in the absence of significant entropion, epilation of the eyelashes may be a reasonable alternative to surgery. Several chirurgic procedures (Merbs et al. 2005, Bowman et al. 2000) are in routine use by trachoma control programs. These generally involve a full thickness incision through the tarsal plate combined with several everting sutures to turn the distal part of the eyelid outwards. Others techniques described are the bilamellar tarsal rotation (BLTR), and the posterior lamellar tarsal rotation (PLTR), including the Trabut procedure.

One of the major problems is a high post-surgery trichiasis recurrence rate, ranging from about 20% in the first 2 yr to 60% after 3 yr (Merbs et al. 2005, Bowman et al. 2000).

The WHO recommends two antibiotic treatment regimes: either 1% tetracycline eye ointment twice daily for 6 wk or a single oral dose of azithromycin. Azithromycin for trachoma control is not currently recommended for children under 6 mon or pregnant women, and therefore tetracycline ointment is the treatment of choice for these groups of patients (Cochereau et al. 2007).

While oral azithromycin would seem to be a safe option, a potential alternative is azithromycin eye drops. A clinical trial of short duration azithromycin eye drops found that at 2 mon, the cure rate and safety of topical 1.5% azithromycin twice a day for 3 d has a similar efficacy as a single oral 20 mg/kg dose of azithromycin for the treatment of active trachoma in children was not inferior to oral azithromycin (Huguet et al. 2010).

CONCLUSIONS

The management of chronic progressive conjunctival inflammation is a challenge to the ophthalmologist, combining topical and systemic treatment and controlling the toxicity induced by these treatments and side systemic effects. In many cases a treatment combination is necessary to reduce the conjunctival inflammation.

REFERENCES

Bialasiewicz, A.A., W. Förster, H. Radig, K.B. Hüttenbrink, S. Grewe and H. Busse. 1994. Syngeneic transplantation of nasal mucosa and azathioprine medication for therapy of cicatricial ocular mucous membrane pemphigoid. Study of 9 patients with 11 eyes. Ophthalmologe 91: 244–250.

Bowman, R.J., O.S. Soma, N. Alexander, P. Milligan, J. Rowley, H. Faal, A. Foster, R.L. Bailey and G.J. Johnson. 2000. Should trichiasis surgery be offered in the village? A community randomized trial of village vs. health centre-based surgery. Trop. Med. Int. Health 5: 528–533.

Celis-Sanchez, J., N. Lopez, M. Garcia, F. Gonzalez and M.J. Ortiz. 2002. Subconjunctival mitomycin C for the treatment of ocular cicatricial pemphigoid. Arch. Soc. Esp. Oftalmol. 77: 501–506.

Cochereau, I., P. Goldschmidt, A. Goepoqui, T. Afghani, L. Delval, P. Pouliquen, T. Bourcier and P.Y. Robert. 2007. Efficacy and safety of short duration azithromycin eye drops versus azithromycin single oral dose for the treatment of trachoma in children: a randomized, controlled, double-masked clinical trial. Br. J. Ophthalmol. 91: 667–672.

Elder, M.J. and R. Collin. 1996. Anterior lamellar repositioning and grey line split for upper lid entropion in ocular cicatricial pemphigoid. Eye 10: 439–442.

Falcinelli, G., B. Falsini, M. Taloni, P. Colliardo and G. Falcinelli. 2005. Modified osteo-odonto-keratoprosthesis for treatment of corneal blindness: long-term anatomical and functional outcomes in 181 cases. Arch. Ophthalmol. 123: 1319–1329.

Fiorelli, V.M., P.E. Dantas, A.T. Jackson and M.C. Nishiwaki-Dantas. 2010. Systemic monoclonal antibody therapy (daclizumab) in the treatment of cicatrizing conjunctivitis in stevens-jhonson syndrome, refractory to conventional therapy. Curr. Eye Res. 35: 1057–1062.

Foster, C.S. 1986. Cicatricial pemphigoid. Am. Ophthalmol. 84: 527–663.

Heffernan, M.P. and D.D. Bentley. 2006. Successful treatment of mucous membrane pemphigoid with infliximab. Arch. Dermatol. 142: 1268–1270.

Huguet, P., L. Bella, E.M. Einterz, P. Goldschmidt and P. Bensaid. 2010. Mass treatment of trachoma with azithromycin 1.5% eye drops in the Republic of Camerron: feasibility, tolerance and effectiveness. Br. J. Ophthalmol. 94: 157–160.

Kenyon, K.R. and S.C. Tseng. 1989. Limbal autograft transplantation for ocular surface disorders. Ophthalmology 96: 709–722.

Kirtschig, G., D. Murrell, F. Wojnarowska and N. Khumalo. 2002. Interventions for mucous membrane pemphigoid/cicatricial pemphigoid and epidermolysis bullosa acquisita: a systematic literature review. Arch. Dermatol. 138: 380–384.

Koizumi, N., T. Inatomi, T. Suzuki T, C. Sotozono and S. Kinoshita. 2001. Cultivated corneal epithelial stem cell transplantation in ocular surface disorders. Ophthalmology 108: 1569–1574.

Liu, C., S. Okera, R. Tandon, J. Herold, C. Hull and S. Thorp. 2008. Visual rehabilitation in end-stage inflammatory ocular surface disease with the osteo-odonto-keratoprosthesis: results from the UK. Br. J. Ophthalmol. 92: 1211–1217.

Megahed, M., S. Schmiedeberg, J. Becker and T. Ruzicka. 2001. Treatment of cicatricial pemphigoid with mycophenolate mofetil as a steroid-sparing agent. J. Am. Acad. Dermatol. 45: 256–259.

Merbs, S.L., S.K. West and E.S. West. 2005. Pattern of recurrence of trachomatous trichiasis after surgery surgical technique as an explanation. Ophthalmology 112: 705–709.

Miserocchi, E., S. Baltatzis, M.R. Roque, A.R. Ahmed and C.S. Foster. 2002. The effect of treatment and its related side effects in patients with severe ocular cicatricial pemphigoid. Ophthalmology 109: 111–118.

Nguyen, Q.D. and C.S. Foster. 1996. Cicatricial pemphigoid: diagnosis and treatment. Int. Ophthalmol. Clin. 36: 41–60.

Strampelli, B. 1963. Keratoprosthesis with osteodental tissue. Am. J. Ophthalmol. 89: 1029–1939.

Thoft, R.A. 1984. Keratoepithelioplasty. Am. J. Ophthalmol. 97: 1–6.
Trent, J.T., M. Fangchao, F. Kerdel, S. Fie, L.E. French, P. Romanelli and R.S. Kirsner. 2007. Dose of intravenous immunoglobulin and patient survival in SJS and toxic epidermal necrolysis. Expert Review of Dermatology 2: 299–303.
Tseng, S.C., M.A. Di Pascuale, D.T. Liu and A. Baradaran-Rafii. 2005. Intraoperative mitomycin C and amniotic membrane transplantation for fornix reconstruction in severe cicatricial ocular surface diseases. Ophthalmology 112: 896–903.
Tseng, S.C., P. Prabhasawat, K. Barton, T. Gray and D. Meller. 1998. Amniotic membrane transplantation with or without limbal allografts for corneal surface reconstruction in patients with limbal stem cell deficiency. Arch. Ophthalmol. 116: 431– 441.
Tsubota, K., Y. Satake, M. Ohyama, I. Toda, Y. Takano, M. Ono, N. Shinozaki and J. Shimazaki. 1996. Surgical reconstruction of the ocular surface in advanced ocular cicatricial pemphigoid and Stevens–Johnson syndrome. Am. J. Ophthalmol. 122: 38–52.
Valeyrie-Allanore, L., P. Wolkenstein, L. Brochard, N. Ortonne, B. Maitre, J. Revuz, M. Bagot and J. Roujeau. 2010. Open trial of ciclosporin treatment for Stevens–Johnson syndrome and toxic epidermal necrolysis. Br. J. Dermatol. 163: 847–853.
World Health Organization (WHO). 2003. Report of the 2nd Global Scientific Meeting on Trachoma. Geneva, Switzerland, 25–27 August, 2003. WHO/PBD/GET 03. 1: 1–28.
Yamada, H., K. Takamori, H. Yaguchi and H. Ogawa. 1998. A study of the efficacy of plasmapheresis for the treatment of drug induced toxic epidermal necrolysis. Ther. Apher. 2: 153–156.

14

Management Options for Limbal Stem Cell Deficiency

Hernán Martínez-Osorio,[a,*] María F. de la Paz and Rafael I. Barraquer

SUMMARY

The maintenance of corneal epithelium relies on the presence of stem cells located at the limbus. Damage to this population of sufficient severity causes limbal stem cell deficiency (LSCD), a condition characterized by the inability to maintain a normal corneal epithelium leading to progressive neovascularization and conjunctivalization, chronic inflammation and scarring. The management of LSCD comprises treating the underlying disease as well as general measures to improve the ocular surface environment. However, once a severe LSCD is established, the restoration of a physiological corneal surface requires the contribution of a new population able to substitute for the lost limbal stem cells. In unilateral cases, the unaffected eye may provide these, classically in the form of a keratolimbal autologous transplant (KLAU). Bilateral cases, however, require allogeneic keratolimbal grafts (KLAL) which, have a high risk of rejection and, consequently, limited long term results. The progress of cell culture makes possible to use *ex vivo* amplified tissue from either cultivated limbal epithelium (CLET) or another autologous origin such as oral mucosa epithelium (COMET). Nevertheless, patients with corneal stromal scarring and loss of transparency will require a lamellar or penetrating corneal graft for visual rehabilitation once the surface has been stabilized. Cases with multiple graft failures and/or a hostile ocular surface environment may

Intitut Universitari Barraquer, C/Muntaner 314 08021, Barcelona, Spain;
[a]E-mail: hernanophth@hotmail.com
*Corresponding author

necessitate a keratoprosthesis, either all-alloplastic as the Boston types (BKP) or—especially in the context of severe dry eye and keratinization of the ocular surface—of the autologous biological-haptic type such as osteo-odonto-keratoprosthesis (OOKP) and tibial osteo-keratoprosthesis (T-OKP).

INTRODUCTION

The Stem Cells of the Ocular Surface

At least two major epithelial cell lines have been recognized on the ocular surface: corneal and conjunctival epithelia (Wei et al. 1993, 1996). The stem cells originating the corneal epithelium have been identified at the corneosclero-conjunctival junction of the limbus (Schermer et al. 1986) thus referred as "limbal stem cells". Conjunctival epithelial stem cells have been identified at the mucocutaneous junction of the lid margins (Liu et al. 2007, Wirtschafter et al. 1997, 1999), the conjunctival fornix (Budak et al. 2005, Lavker et al. 1998, Pellegrini et al. 1999, Wei et al. 1995), the tarsal conjunctiva (Chen et al. 2003), and the bulbar conjunctiva (Nagasaki and Zhao 2005, Pellegrini et al. 1999) including the limbus (Pe'er et al. 1996). It is currently accepted that the conjunctival stem cells are dispersed all over the conjunctiva. Moreover, oligopotent stem cells capable of generating either conjunctival or corneal phenotypes depending on the environment have been found dispersed throughout the entire ocular surface including the cornea (Majo et al. 2008).

The concept of stem cells originated half a century ago, most prominently from the pioneering work of E. Donnall Thomas on bone marrow transplantation for which he received the Nobel Prize in 1990 (Bensinger et al. 1990, Berenson et al. 1990). By definition, stem cells are those capable of self-renewal, giving rise to new cells, of development into different cell types and of functional restoration of cell populations. Stem cells tend to be undifferentiated or poorly differentiated—smaller in size and with less cytoplasm—with slow cell cycle, low mitotic activity and high longevity. Due to these properties, a stem cell´s capacity for self-renewal is unlimited and has indefinite proliferative ability. Anomalous changes in stem cell's replication may lead to cloning of abnormal and dysfunctional cells having a high proliferative capacity, thereby giving rise to neoplasia.

In order to retain their low differentiation and proliferative properties, stem cells require a suitable anatomical niche. These sites tend to be protected from external aggression, having rich innervation and vascularization locally or nearby as well as other sheltering features such as a surrounding pigmentation-defensive from radiation. In the ocular surface, such an environment is found at the palisades of Vogt of the corneoscleral limbus (Hall and Watt 1989, Miller et al. 1993, Morrison et al. 1997).

Differentiation of Limbal Stem Cells

Stem cells may divide symmetrically or asymmetrically (Lajtha 1979, Morrison et al. 1997). During an asymmetric division, one of the cells remains in the niche as a stem cell, while the other, called a transient amplifying cell, is destined to undergo further division and differentiation into several types of mature cells (Morrison et al. 1997). This differentiation process implies that the proliferative potential of a transient amplifying cell becomes less than that of the original stem cell (Potten and Loeffler 1990). Further on, the transient amplifying cells evolve into post-mitotic cells and, finally, into terminally differentiated cells, which thereafter become incapable of cellular division (Lehrer et al. 1998).

The histological organization of the corneal epithelium is consistent with the vertical turnover scheme of squamous epithelia: the basal cells give rise to the progressively more superficial layers, which ultimately desquamate. The role of a centripetal cell dynamics in the maintenance of the corneal epithelium had been proposed as early as 1951 (Friedenwald 1951) and later formulated as the "XYZ hypothesis" (Thoft and Friend 1983), pointing at the limbus as the source or this centripetal long-term renewal. Stem cells for the corneal epithelium have been identified within the crypts of the limbal palisades of Vogt (Cotsarelis et al. 1989, Davanger and Evensen 1971, Schermer et al. 1986), the transient amplifying cells in the limbus and the basal layer of the peripheral cornea, and the post-mitotic and terminally differentiated cells at the central basal and the remaining layers of the corneal epithelium (Dua and Azuara-Blanco 2000). However, the exclusivity of the limbus as the sole location of corneal epithelial stem cells has been more recently challenged (Majo et al. 2008).

The Spectrum of Limbal Stem Cell Deficiencies

Corneal diseases are one of the most common causes of blindness in the world, especially in countries with less developed economies (Garg et al. 2005). While corneal grafting may be a valid therapeutic option for patients with stromal opacities or endothelial failure, those with extensive lesions of the ocular surface generally have poor results unless a healthy corneal epithelium is restored first. This has been explained by these conditions having in common a critical loss of the epithelial stem cells at the limbus, leading to what has been named as "limbal stem cell deficiency" (LSCD) (Espana et al. 2004, Holland 1996, Holland and Schwartz 2000, Maruyama-Hosoi et al. 2006, Solomon et al. 2002).

LSCD may originate from any process disrupting or destroying the normal limbal architecture and/or the epithelial stem cells and their niche, thereby causing a deficiency in the natural turnover of the corneal epithelium. Depending on the particular situation, this will lead to a number of manifestations including an aberrant and irregular corneal epithelium, recurrent or persistent epithelial defects (PED), progressive corneal neovascularization and invasion by conjunctiva-like tissue (conjunctivalization), chronic inflammation and scarring. Conjunctivalization can be confirmed by the presence of goblet cells and the expression of specific conjunctival cytokeratins and mucins.

LSCD may be unilateral or bilateral, reversible or (more typically) irreversible. The possible etiologies basically include trauma, inflammatory and degenerative diseases. Cases related to contact lens abuse (Barraquer et al. 2004) (Fig. 1) and post-surgical, especially after multiple surgeries involving the ocular surface—i.e. recurrent pterygia or ocular surface neoplasia—may be partial and reversible, although not always (Jenkins et al. 1993).

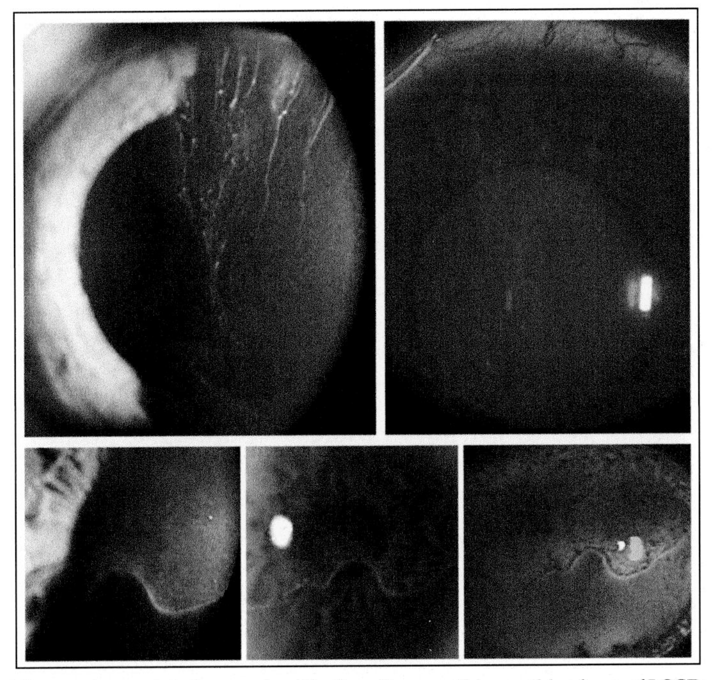

Figure 1 Contact lens-related corneal epitheliopathy, possibly a milder form of LSCD. Above, a 28 year old soft contact lens user since 9 year before and symptomatic for 1 year, showing a linear or "fimbriae" irregular epithelial pattern in a superior triangular area originating from the limbus. Below, a 48 year old patient, contact lens user for 28 year, and with reduced vision since 2 months before, with an avascular layer of altered epithelium as an "advancing wave" over the whole superior half of the cornea, clearly demarcated by line from the inferior normal-appearing zone. Both improved with the suspension of contact lens use and unpreserved lubricants.

Color image of this figure appears in the color plate section at the end of the book.

Apart from surgical trauma and contact lens chronic microtrauma, the causes may include other factors as chronic ischemia of the limbus, toxicity due to antimetabolites (5FU, mitomycin C, etc.), cryotherapy, radiation therapy, or even the chronic use of topical glaucoma drugs (D'Aversa et al. 1997).

The most common cause of irreversible LSCD is severe chemical or—less frequently—thermal injuries. While these are often unilateral, some bilateral involvement is not rare ranging from partial to complete. Inflammatory and degenerative causes of LSCD are relatively rare but pose the greatest challenges. These include autoimmune diseases involving the ocular surface such as Stevens–Johnson (SJS) and Lyell's syndromes (toxic epidermal necrolysis), mucous membrane or ocular cicatricial pemphigoid (OCP) and graft-versus-host disease (GVHD). Other causes of chronic conjunctivitis or keratoconjunctivitis that can lead to LSCD are severe trachoma, atopy and rosacea.

Degenerative congenital entities such as aniridia keratopathy, ectodermal dysplasia, autosomal dominant keratitis and KID syndrome, share with the former group a chronic progressive nature, which tends to evolve to bilateral and irreversible LSCD (Holland and Schwartz 2000, Barraquer et al. 2004).

TREATMENT OPTIONS FOR LIMBAL STEM CELL DEFICIENCY

General Management

The management of LSCD comprises that of the causing disease when applicable, as well as general measures to improve the ocular surface environment, including the lacrimal and palpebral functions, treating inflammation and promoting epithelial growth. Special consideration should be made with dry eye. The complete medical armamentarium from unpreserved lubricants to autologous blood derivatives, mucolytic and antiprotease agents, as well as immunomodulators, may have to be applied depending on the condition. As a rule, LSCD cases should be treated for dry eye one step ahead of the severity levels recommended by Delphi panel, and always including the treatment of meibomian gland dysfunction (Behrens et al. 2006). In patients with milder LSCD as in post-surgical or contact lens-related epitheliopathy, these measures may suffice together with the avoidance of the noxious stimulus.

Oculoplastic surgery will be a common prerequisite in cases with corneal exposure, irregularity of the lid margins, trichiasis/distichiasis causing abrasion of the ocular surface and/or symblepharon. The latter will usually involve fornix reconstruction using mucous (i.e. oral mucosa) or amniotic membrane grafts. Before any surgical procedure for restoring LSCD is attempted, it is essential that the ocular surface environment is optimized and all inflammation has subsided. This may imply a long period, even after clinical resolution, as the expression of high levels of HLA-DR has been found up to 24 months after chemical burns (Gicquel et al. 2007). Moreover, patients with autoimmune inflammatory disease and those undergoing allogeneic limbal transplant will require a long period of immunosuppression—when not for life (Holland and Schwartz 2000). Dry eye therapy will generally have to be maintained as well after LSCD surgical procedures.

Sequential Conjunctival Epitheliectomy

The surgical options to treat LSCD most often consist in the transfer to the damaged cornea of viable epithelial stem cells. However, in cases with only a partial LSCD, the remaining limbal stem cell population outside the involved area may suffice to restore the normal corneal epithelium, provided the abnormal tissue is removed. This may the reason why some traditional treatments have been effective, as simply scraping an aberrant epithelium growing over the cornea and/or the application of 1% silver nitrate (AgNO3) on the abnormal limbus in cases of superior limbic keratoconjunctivitis and advancing wave-like keratopathy (D'Aversa et al. 1997).

Sequential conjunctival epitheliectomy (SCE), as described by Harminder Dua, follows a similar rationale. The conjunctivalization of only a sector of the cornea would indicate that the LSCD is partial and a local breach in the "limbal barrier" has permitted the invasion of this area by the abnormal tissue. Therefore, if this is tissue removed, the normal corneal and limbal epithelium from the unaffected adjacent sectors could be able to grow over the denuded area restoring the normal corneal surface (Dua 1998).

The SCE technique involves gently scraping the invading conjunctival-like epithelium with a dry or slightly alcohol-soaked cellulose spear, and/or a crescent (Desmarres) blade. In order to incline the balance in favor of the corneal phenotype in the ensuing competition for repopulating the denuded area, the conjunctival scrapings may have to be repeated every 24–48 hours until normal corneal re-epithelialization is attained, which usually takes 5 to 10 days (Fig. 2). SCE also can help to reduce the conjunctival invasion after an autologous or allogeneic limbal transplantation.

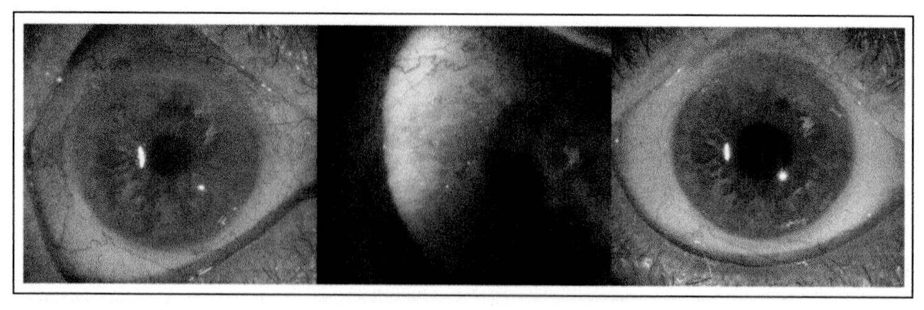

Figure 2 This figure shows a partial LSCD due to by ocular cicatricial pemphigoid treated with sequential conjunctival epitheliectomy (SCE). Left: Preoperative appearance with sector conjunctivalization of the superior cornea. Center: Appearance immediately after a superior SCE. Right: Postoperative appearance of the same eye 1 month later, showing the repopulation of the superior cornea with normal-appearing epithelium.

Color image of this figure appears in the color plate section at the end of the book.

Keratolimbal and Conjunctival Limbal Autologous Transplantation

The concept of limbal or—as currently termed—keratolimbal autologous transplantation (KLAU) was first described in the form of sectorial grafts from the contralateral eye by José I. Barraquer, during the 1st World Cornea Congress of 1964, and then by Benedetto Strampelli as an annular 360° limbal graft (King and Mc Tigue 1965, Strampelli and Restivo Manfridi 1966). However, these procedures remained anecdotal (Barraquer and Rutllán 1982) until popularized by Kenyon and Tseng during the last decades of the past century (Kenyon and Tseng 1989).

Currently KLAU has become the standard procedure for unilateral irreversible LSCD and in partial cases where SCE has failed. The KLAU technique involves in the first place a complete removal of all abnormal fibrovascular tissue or pannus growing over the cornea (superficial keratectomy), followed by the transplantation of one or two thin lamellar, 90° to 120° sector grafts taken from the unaffected upper and lower limbus of the contralateral eye (Fig. 3) (Holland and Schwartz 2004).

In order to preserve and include the limbal epithelial crypts, it is recommended to retain at least one mm of donor conjunctiva from the limbus section (Dua et al. 2005). Once the grafts are sutured in place over the recipient limbus the ocular surface is protected by a therapeutic contact lens, an amniotic membrane patch, or a temporary tarsorraphy to promote re-epithelialization.

While the original technique included a thin layer of corneoscleral tissue as a substratum for the limbus (Kenyon and Tseng 1989), some surgeons use "conjunctival limbal" grafts not including corneoscleral tissue (Meallet et al. 2003, Santos et al. 2005). It is not clear whether these differences in

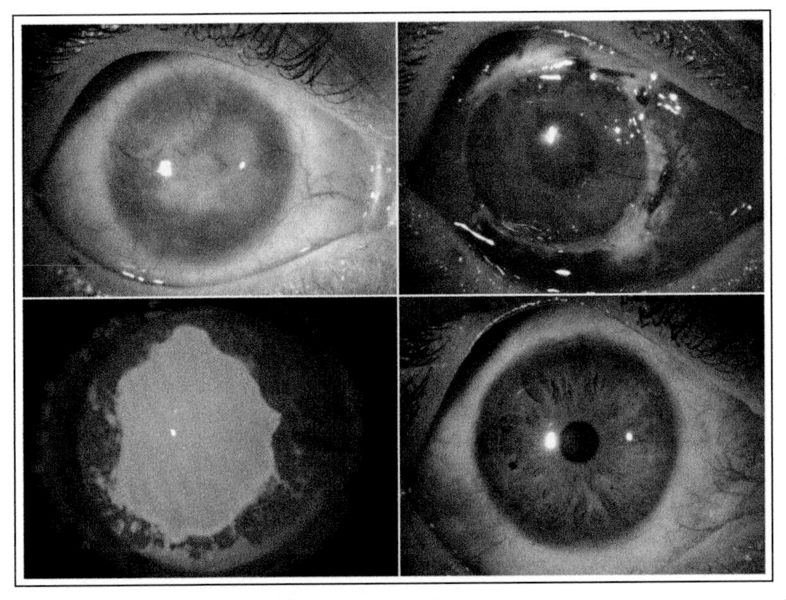

Figure 3 KLAU in a case of unilateral alkali injury. Top left: Initial appearance 7 year after sustaining a lime burn, with counting fingers vision. Top right: Immediate postoperative appearance after 2-sector KLAU from the fellow normal eye. Bottom left: Early re-epithelialization of the host cornea, at 3 days post-op. Bottom right: result at 18 month, with stable ocular surface and visual acuity of 20/50 with 2 diopters of mixed astigmatism, improving to 20/25 with contact lens. At 4 years, it reached 20/20.

Color image of this figure appears in the color plate section at the end of the book.

technique may influence the outcomes, which are generally excellent (see below), making KLAU the benchmark for this type of surgery. This is no surprise, as an autograft does not face the risk of rejection. However, the presence of chronic inflammatory disease or severe dry eye—which is rare in these unilateral cases—could jeopardize the outcome even in the absence of rejection. On the other hand, the newer tissue culture techniques offer a less invasive alternative, minimizing the amount of required donor tissue.

Keratolimbal Allogeneic Transplantation

Bilateral total LSCD, as is the common result in autoimmune inflammatory and degenerative diseases or trauma severe enough to completely destroy the limbal stem cells in both eyes, preclude the repopulation of the cornea from autologous limbal stem cells. The use of an allogeneic keratolimbal transplant (KLAL) requiring long term systemic immunosuppression was, until recently, almost the only option for these cases (Espana et al. 2004, Holland and Schwartz 2000).

KLAL can be obtained from a HLA-matched living relative of the patient in the form of sector grafts, following a technique similar to that described for KLAU (Fig. 4).

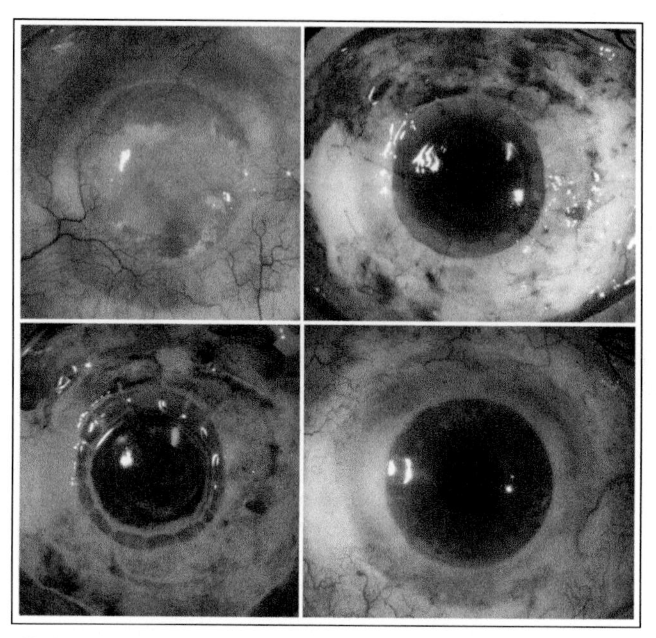

Figure 4 Total bilateral LSCD after trauma and multiple ocular surgeries (above, left), treated with 2-sector KLAL from his HLA-identical sister, combined with a simultaneous PK. Immediate (above, right), early result showing re-epithelialization (below, left), and result after 4 years (below, right) with a stable ocular surface and a clear central graft without immunosuppressive therapy, despite peripheral neovascularization and lipid deposition.

Color image of this figure appears in the color plate section at the end of the book.

Nevertheless, more frequently the tissue will come from unrelated cadaveric origin—usually of unknown histocompatibility—as an annular 360° keratolimbal graft. This maximizes the amount of viable limbal stem cells transferred, not limited by the risk of side effects as in the case of a living donor.

As is a general requirement for allogeneic donors, those for KLAL must be screened for transmissible diseases including HIV, Hepatitis B and C, HTLV, syphilis and prion-related disease. A vigilant management of inflammation, both preoperative and postoperative is also mandatory in KLAL recipients.

As an allograft in a vascularized environment, KLAL are subject with frequent immune rejection. This can take the form of an acute rejection with vasculitis and necrosis of the graft, or as a chronic and protracted, little-symptomatic rejection process. Even with sustained immunosuppression, a progressive loss of the transplanted limbal stem cells explains the limited long-term results of KLAL (see below). This, combined with the difficulty in obtaining suitable fresh limbal tissue for transplantation, the inherent risk of viral disease transmission, and the side effects of long-term systemic immunosuppression, stimulated the search for alternative treatments for bilateral irreversible LSCD.

Cultivated Limbal Epithelial Transplantation

Tissue engineering is a concept in regenerative medicine entailing the production of substitute tissue from cells of the body and other biological materials to repair a damaged or destroyed tissue. This often involves using *ex vivo* expansion of stem cells or transient amplifying cells and their differentiation into a predestined normal cell population. The transplanted tissue ideally must be able to replace, as well as regenerate, the affected tissue restoring its physiological function. Over the past decade, *ex vivo* expansion of limbal cells and their transplantation (CLET) has become a new option to treat LSCD (Daya et al. 2005, Koizumi et al. 2001, Pellegrini et al. 1997, Rama et al. 2001, 2010, Sangwan et al. 2003, 2006, 2011, Schwab et al. 2000, Shimazaki et al. 2002, Shortt et al. 2008, 2010, Tsai et al. 2000).

In tissue engineering, as well as in stem cell expansion, the interaction between the cell population and the substrate or scaffolding material is a critical factor. This substrate must be as similar as possible to the extracellular matrix where the original cells are found, so as to permit cell migration, proliferation, differentiation and maintenance of its physiological functions. The cell-substrate complex must be stable enough to allow transportation, manipulation and grafting onto the patient. It was the group of Graziella Pellegrini in 1997; who first reported on the successful isolation and *ex vivo* expansion of limbal cells for LSCD (Pellegrini et al. 1997). These cells were grown over a 3T3 fibroblast substrate, similar to those used for dermal keratinocyte culture (Lindberg et al. 1993, Rheinwald and Green 1975, Sun and Green 1977). Since then, several groups from different continents have published alternative techniques for the *ex vivo* expansion of limbal cells, most of which use amniotic membrane as a substrate (Grueterich et al. 2003, Koizumi et al. 2001, Sangwan et al. 2003, Schwab et al. 2000, Shimazaki et al. 2002, Tsai et al. 2000). Still others have also described the use of a fibrin matrix (Rama et al. 2001, 2010) or a contact lens (Di Girolamo et al. 2009) as a substrate.

A live stem cell must be able to repopulate/reproduce functionally once grafted, and this capability will determine success or failure of the entire *ex vivo* expansion process. At present, characterization of epithelial stem cells is done by indirect methods using biological markers. Most of these markers have been extrapolated from hematopoietic stem cells, although our knowledge about their sensitivity and specificity in other tissues is limited. Using these markers, it has been reported that only 0.5% of the limbal epithelium are stem cells (Wolosin et al. 2004). Nevertheless, because of the severity of the LSCD conditions treated and the relative low risks of these external and little-invasive procedures, CLET was tested directly in human patients (Rama et al. 2001). The capability of the CLET grafted cells of functional repopulation/reproduction, as well as the expression in the grown cells of already established markers for stem cells characterization, was thus proven. The development animal models of LSCD and confirming the functionality CLET—in immune deficient rats with specific genetic abnormalities of the ocular surface epithelium—have been carried out only following the human application of CLET (Liu et al. 2003, Yoshida et al. 2006).

For CLET, a small biopsy of 1–2 mm width is taken from a superior keratolimbal area of the unaffected eye. The biopsy may then transport to the laboratory and finally cultured for 2–4 weeks until a confluent epithelial layer is obtained. Because of the small size of the biopsy, occasionally the primary culture is not successful. When this occurs, a further biopsy can be obtained.

However, minimizing the harvested tissue makes CLET safer than KLAU for the donor eye. While in most cases of KLAU the donor eye does not suffer appreciable side effects, partial LSCD has been described following the larger limbal biopsy required for KLAU in a previously normal appearing donor eye in a chronic contact lens user (Jenkins et al. 1993). Moreover CLET could be applied to bilateral cases of LSCD with partial sparing of one eye, including cases not amenable to KLAU for the real risk of decompensating the donor eye ocular surface. However, in cases with complete 360° bilateral LSCD there is no healthy autologous limbus available, and the use of allogeneic limbus for CLET meets the same rejection problems as in KLAL. This stimulated further research on alternative autologous donor tissues.

Cultivated Oral Mucosa Epithelial Transplantation

A variety of tissues have been tested as possible epithelial stem cell donors for LSCD alternative to the limbus, including oral mucosa epithelium (Inatomi et al. 2005, 2006a, b, Nakamura et al. 2004, 2011, Nishida et al. 2004), conjunctival epithelium (Kawasaki et al. 2006, Ono et al. 2007, Sangwan et al. 2003, Tanioka et al. 2006), adipose tissue (Arnalich-Montiel et al. 2008), hair follicles (Blazejewska et al. 2009), and dental pulp (Gomes et al. 2010). Moreover, the detection of mesenchymal stem cells in the corneal stroma (Choong et al. 2007, Du et al. 2005, Thill et al. 2007), in the limbus (Polisetty et al. 2008) and in the conjunctiva has brought out a number of implications regarding possible new approaches to surgical reconstruction of the ocular surface (Nadri et al. 2008a, b).

Autologous oral mucosa had been one of the tissues most frequently used for the reconstruction of the ocular surface, especially as a substitute for conjunctiva in symblepharon and recurrent pterygium surgery (Barraquer and Charoenrook 2009). Recently, oral mucosa graft as a surrogate for limbus, combined with amniotic membrane transplantation, has been found effective in 7 cases of total LSCD (Liu et al. 2011).

In 2004 cultivated autologous oral mucosal epithelial transplantation (COMET) was first introduced by a Japanese group for the treatment of total bilateral LSCD (Nishida et al. 2004). The major advantage of this technique is avoiding the need for a prolonged immunosuppressive and/or topical steroid therapy, along with their added risks, respectively systemic and local as secondary infection and glaucoma. COMET has been applied successfully to patients with severe ocular burns, inflammatory diseases as SJS and OCP, and congenital aniridia keratopathy. The morphological appearance of a COMET epithelium is similar to that of a CLET. The transplanted oral epithelium retains its transparency, although the resulting visual acuities are somewhat lower than those attainable with KLAU or CLET. This is due to a higher amount of light scattering from the oral epithelium, which tends to retain a superficial punctate staining pattern with fluorescein. On the other hand, COMET can be used for the simultaneous reconstruction of the conjunctival fornixes whenever required, and it has been shown to inhibit significantly the postoperative symblepharon formation during the long-term follow-up (Nakamura et al. 2011).

The use of a highly hydrated therapeutic soft contact lens is recommended in the early postoperative period after a CLET or COMET, with instillation of artificial eye drops every 15 minutes to 2 hours. Providing the adequate protection for the epithelial cells from dryness and mechanical stress appears to be a crucial aspect for the clinical success of these procedures (Nakamura et al. 2011, Shortt et al. 2010).

Amniotic Membrane Transplantation

The virtues of amniotic membrane (AM) as an ocular surface reconstructive material—a substrate that physically facilitates the growth of epithelia and provides a number of anti-inflammatory and growth-promoting factors—are well established. AM can be used as a bandage patch, combined with other types of surface reconstructive surgery, or as a substratum for the limbal stem cells graft (Meallet et al. 2003). In the first case, the AM is applied with the epithelial side down; in the second, its epithelium should not be peeled off and the membrane always placed with the epithelial side up. The epithelial growth-promoting properties of AM may suffice in cases with partial LSCD, and AM as a permanent graft has been shown to restore the entire corneal surface in eyes with as little as 30°–60° of intact limbus (Tseng et al. 2004). However, AM is not a source of viable limbal stem cells, and therefore its role in total LSCD can only be ancillary.

Corneal Transplantation (keratoplasty)

Because LSCD is often accompanied by severe corneal stromal opacity and/or corneal endothelial dysfunction, a significant number of patients will also require a lamellar (LK) or penetrating keratoplasty (PK) for visual rehabilitation (Baradaran-Rafii et al. 2010). A large diameter LK including the limbus (Vajpayee et al. 2000) or a penetrating limbo-keratoplasty (Reinhard et al. 2004) have been proposed as alternatives including the equivalent of a KLAL together with the central cornea. However, simultaneous keratoplasty and limbal transplant—either classical or cultivated—tend to be associated to a higher number of complications. For example, opportunistic infections have been reported in PK combined with KLAL and occasionally with CLET or COMET (Nakamura et al. 2011, Shortt et al. 2010). This risk is enhanced by immunosuppression, as required in these combined procedures, especially in the presence of an ocular surface not yet stabilized. Better results have been obtained when keratoplasty is delayed for several months after epithelium-repopulating procedures (Basu et al 2011, Nakamura et al. 2011, Sangwan et al. 2011).

Artificial Cornea (keratoprosthesis)

Despite the substantial progress in the field of corneal epithelial repopulation, not all eyes with severe LSCD are suitable for regenerative strategies. In patients with multiple transplant failures, and especially with severe dry eye or keratinization of the ocular surface, an artificial cornea or keratoprosthesis may be the only remaining option. The idea of replacing the cornea with an artificial device dates back to Pellier de Quensy in 1771 and was first tried in a human eye by Weber in 1855. This consisted in a glass cylinder with anterior and posterior annular flanges, and stayed in place for 6 months before being expulsed (Temprano 1991).

Strampelli was the first in recognizing that the major difficulty with keratoprosthesis stems from the fact that they represent a mesoprosthesis: they occupy the boundary between the external and internal media, as opposed to an exoprosthesis, i.e. an artificial limb, or an endoprosthesis, i.e. an artificial heart valve. As long as the materials of a mesoprosthesis do not "heal" with the surrounding tissues, the surface epithelium in charge of maintaining the barrier between the external and internal media will keep trying to either cover the prosthesis or to grow under and externalize it. Moreover, the failure of the surface epithelium in growing over the prosthesis tends to trigger enzyme activation, leading to melting of the tissue around the prosthesis and hence to failure due to a leak, infection and/or expulsion. In 1963, Strampelli devised a possible solution to this problem by adding to the alloplastic optical cylinder an osteodental plate that would act as haptic, actually healing to the ocular surface tissues. A leak-proof union between the plastic cylinder and the osteodental haptic was ensured by their "mineral" (chemically stable) properties and the use of cyanoacrylate glue—whose long-term stability under strong stresses had been proven by the extensive experience in dentistry. This is the rationale underlying osteo-odonto-keratoprosthesis (OOKP), as developed by Strampelli and later modified and popularized by the Falcinelli and by Temprano (Falcinelli et al. 2005, Strampelli 1963, Temprano 1991).

The concept of OOKP—and its very successful clinical record (see below)—establishes a major division between all alloplastic keratoprosthesis and those using an autologous biological haptic. Today, one of the most popular among the former is the Boston keratoprosthesis (BKP), developed by Claes Dohlman during the last decade of the past century (Dohlman and Doane 1994). The BKP is a nut and bolt type of artificial cornea, a concept that can be traced back to 1951 (Györffy 1951). The device is inserted through a central trephination in a corneal donor disk and held in place by a posterior plastic plate that acts as a nut secured by means of a titanium washer. The corneal disk is then sutured to the recipient as in a standard PK. Type I BKP is designed to end at the anatomic ocular surface and requires the use of a therapeutic contact lens and antibiotic drops for life. Type II BKP has a longer optical cylinder for transpalpebral placement in cases with the most severe keratinized dry eye (Pujari et al. 2011). Type I BKP has been proven a successful alternative for patients with multiple corneal allograft failures and in cases with poor prognosis for a conventional allograft as in the pediatric group (Yaghouti et al. 2001). Although BKP has also been used to treat total bilateral LSCD with end-stage dry eye in cases of SJS or OCP the long-term results have been less satisfactory (Sayegh et al. 2008). Currently a normal Schirmer test is recommended as a prerequisite for type I BKP (Pujari et al. 2011, Sejpal et al. 2011).

The ability of autologous tissues to heal over the ocular surface would explain the success of the autologous biologic-haptic types such as OOKP and its variants (Falcinelli et al. 2005, Liu et al. 2005, 2008, Temprano 1991), compared to the all-alloplastic prostheses like BKP, and especially in the very difficult cases with severe dry eye, ocular surface keratinization, and persisting inflammatory disease. OOKP uses as haptic for the optical cylinder an osteodental plate obtained from one of the patient's tooth, including material from its root and the alveolar bone as well as the osteodental ligament. Large monoradicular teeth such as canines are best suited for OOKP. A full thickness oral mucosa graft is placed over the ocular surface, so the prosthesis will be sandwiched between the cornea and this mucosa layer, pierced by the anterior tip of the optical cylinder. Its posterior end enters the globe by a close-fitting central trephination in the cornea. The OOKP procedure is performed in two stages, three to four months apart, which allows for soft tissue to grow around and revascularize the osteodental plate. This serves to improve the viability of the plate and to exclude infection. Because of the significant number of edentulous patients among the OOKP candidates, José Temprano at the Barraquer Institute of Barcelona (Spain), developed since 1975 a tibial osteo-keratoprosthesis (T-OKP), using as haptic a tibial bone disc from the same patient (Temprano 1991).

As the cylinder extends deep into the anterior chamber, both the OOKP/T-OKP and the BKP require the patient to be aphakic or pseudophakic, and the power of the optic has to be calculated accordingly. To reduce the risk of formation of retrocorneal fibrous membrane, aphakia may be preferable to pseudophakia. The main limitation of OOKP and T-OKP compared to BKP is the smaller visual field of the former. While this could be improved by modifying the design of the optical cylinder, combining both types is possible, as some bilateral patients formerly treated successfully with OOKP in one eye have tolerated well a BKP in the second eye (Fig. 5).

In eyes with severe LSCD and opaque corneas, neither the fundus status nor pupillary reflexes can be assessed. Ocular electrophysiology testing such as electroretinogram (ERG) and visual evoked potentials (VEP) are helpful to estimate visual recovery potential and establish post-surgical prognosis in these patients. Eyes with a normal ERG before undergoing OOKP or T-OKP achieved better visual outcomes than those with subnormal, abnormal, or absent ERG.

Patients with preoperative normal or mildly subnormal flash VEP demonstrated greater visual improvement than those in which VEP was clearly abnormal. When comparing both eyes of the same patient, VEP was useful in detecting the eye with better prognosis (de Araujo et al. 2011).

LONG-TERM RESULTS OF THE SURGICAL TREATMENTS FOR LSCD

A number of outcomes have been proposed to determine the success or failure of surgical LSCD procedures. These include the degree of corneal transparency, corneal vascularization, conjunctivalization, inflammation, epithelial defect, photophobia, pain, and improvement of visual acuity (Shortt et al. 2008). However, the definition of these is variable or imprecise in many instances. More recently, impression cytology, confocal microscopy

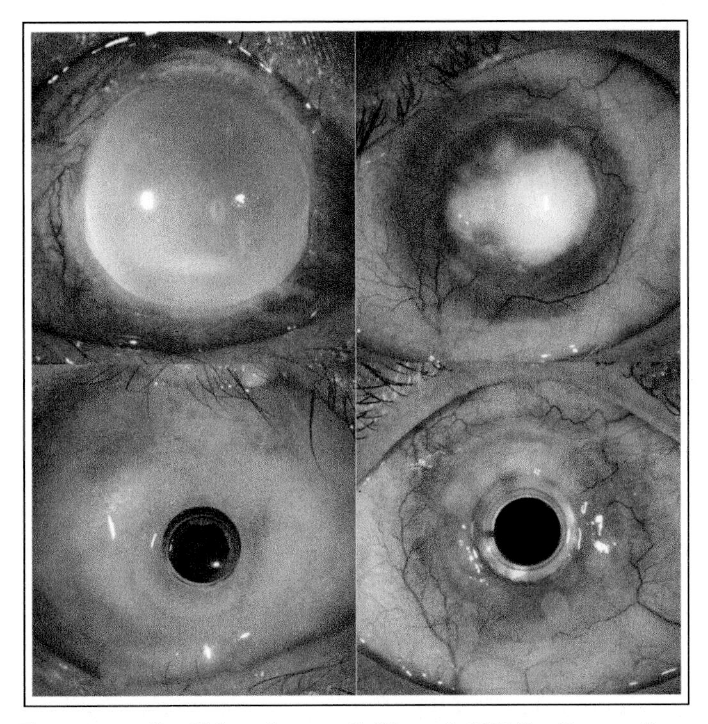

Figure 5 This patient suffered bilateral severe alkali burns in 1985. Top: Preoperative appearance of the right eye (at left) after 3 years and of the left eye (at right), 13 years after the accident with a failed penetrating keratoplasty, severe vascularization and secondary lipid keratopathy. Bottom (respectively): Results 13 years after OOKP in the right eye and 3 years after BKP in the left eye.

Color image of this figure appears in the color plate section at the end of the book.

and questionnaires have been suggested as they may provide objective parameters of success (Shortt et al. 2010).

Keratolimbal Autologous (KLAU) Transplantation and Cultivated Limbal Epithelial Transplantation (CLET)

Excellent results have been obtained with KLAU, with 100% survival over a long-term follow-up in some series (Kenyon and Tseng 1989, Miri et al. 2010). Nevertheless, treating any active inflammation and insuring an optimal ocular surface environment remains mandatory. Failure of KLAU was reported in 2 of 5 patients with bilateral chronic contact lens associated epitheliopathy (Jenkins et al. 1993) and in 2 of 10 cases of conjunctival limbal autografts (Santos et al. 2005) for several types of unilateral total LSCD.

A recent review found that 581 eyes worldwide had received, between 1999 and 2010, either autologous or allogeneic CLET for unilateral or bilateral LSCD. Of these eyes, CLET alone improved vision in 16% to 54%—depending on the particular series—and another 18% to 38% (27.5% in average) additionally required one or more PK for visual restoration. Corneal allograft survival rate in these series was between 58% and 90% (Baylis et al. 2011, Di Iorio et al. 2010, Shortt et al. 2010).

A multicentric study of 88 eyes with chemical injuries showed autologous CLET to be successful in the long term in 73% of cases (Sangwan et al. 2006). Comparing culture methods, the same study found that co-culture technique has better clinical outcomes than the traditional keratinocyte-based cultures. More recently, the same group found that CLET using xeno-free explant culture technique was successful in ocular surface restoration in 71% of the 200 recipient eyes (Sangwan et al. 2011). This success rate is in the range of those found by similar studies: 68% of 107 eyes (Rama et al. 2010) to 80% of 166 eyes (Di Iorio et al. 2010). They also recommend performing CLET at least 3 months after the initial injury, when inflammation is completely under control.

A recent study comparing the timing of PK associated to autologous CLET found that delaying PK for 6 weeks has better long term outcomes than a single-stage procedure (Basu et al. 2011). Most patients in this series (78.7%) had LSCD from alkali burns and included 40% of children. Overall, ocular surface stabilization at year 4 was achieved in 74.5% of eyes, with visual acuity better than 20/40 in 71.4% of eyes with clear grafts. Kaplan-Meier survival rate at 1 year was 80 ± 6% (median survival, 4 years) of the 35 grafts using 2-stage procedure versus 25 ± 13% (median survival, 6 months) of the 12 single-stage cases. Recurrence of LSCD was more common after single-stage (58.3%) than 2-stage (14.3%). The results of the delayed PK after CLET are also far better than those performed in eyes with LSCD without prior limbal transplantation (33%–46%), or in eyes with other high risk factors for corneal graft failure (32%–67%). In order to prevent amblyopia in children, it is recommended to perform PK as soon as the ocular surface stabilizes after CLET (Basu et al. 2011).

Different studies have analyzed the restoration of the corneal epithelium after CLET. In patients operated of LK or PK after CLET, phenotypic analysis of the corneal button epithelium did not show expression of goblet cells markers as Muc5ac, Muc4 or Muc1, or conjunctival epithelial markers as cytokeratin (CK) 19 or CK13, except in failed cases. In successful cases, the epithelium exhibited corneal phenotype by expression of CK3 and CK12 and even of DeltaNp63alpha, a marker for high proliferative potential (Colabelli Gisoldi et al. 2010, Kawashima et al. 2007, Pauklin et al. 2009). Apart from the CK profile, corneal phenotype can be demonstrated by impression cytology and confocal microscopy (Shortt et al. 2008).

Keratolimbal Allografts (KLAL) and Allogeneic CLET

As a rule, the long-term results of KLAL and of allogeneic CLET have been rather disappointing. Although some cases of KLAL have maintained a healthy corneal epithelium for years, many patients have required in the long run one or multiple repeated limbal grafts. Some of the published reports have considered as "success" the cumulative duration of all procedures including regrafts. By year 5, anatomical success rates of KLAL vary between 23.7 % and 51%, with preservation or improvement of vision between 34.9% and 44.6% (Solomon et al. 2002, Tsubota et al. 1995). A comparative study shows that living-related KLAL fare better than cadaveric KLAL; long term success rate was 89% for the former and 33% the latter, while vision improved in 78% of patients in both groups (Miri et al. 2010).

The limited survival of KLAL has been explained because transplanted cells die gradually because of immunologic rejection and a hostile environment. A study of the long-term outcomes after KLAL identified factors leading to the loss of donor cells and failure—usually after 3 to 4 years—including chronic allograft rejection, infrequent blinking, blink-related microtrauma, conjunctival inflammation, increased IOP, severe dry eye, lagophthalmos, pathogenic symblepharon, and prior KLAL or PK failure (Liang et al. 2009).

The published studies on the origin of the epithelial cells that repopulate the ocular surface after limbal allotransplantation also point to a limited survival of the donor cells. No donor derived DNA could be detected in 5 recipient corneas between 3 and 5 years after KLAL (Henderson et al. 2001), or 9 months in 7 recipient corneas after cadaveric donor-CLET (Daya et al. 2005). However, a third study detected only donor DNA in one eye 3.5 years after KLAL, and a mixed profile of donor and host DNA in two other cases (Djalilian et al. 2005). While these variable results may reflect differences in the DNA detection techniques, the latter evidence supports the use of systemic immunosuppressive therapy.

So far, no studies have examined whether tissue matching improves the survival of KLAL or allogeneic CLET. However, this could be inferred from a series of 48 patients with total LSCD who underwent penetrating limbo-keratoplasty. At year 5th, 65% grafts with 0–1 HLA mismatches were transparent in contrast than 14% of unmatched grafts (Reinhard et al. 2004). A case of allogeneic CLET from the same HLA-identical living-related donor who had previously donated bone marrow—and despite having induced a GVHD that led to total bilateral LSCD—successfully reconstructed the ocular surface for 31 months without systemic immunosuppression (Meller et al. 2009). Our experience in a case using tissue from HLA-identical living-related donor (Fig. 4) also suggests that tissue matching may improve the survival of KLAL grafts.

While published reports using systemic cyclosporine A in allogeneic CLET span only 6 months (Daya et al. 2005, Shortt et al. 2008), longer term systemic immunosuppression is recommended for KLAL and high risk PK patients when using tacrolimus—until at least 18 months from the last rejection episode (Miri et al. 2010), or mycophenolate mofetil—for up to 2 years in penetrating limbo-keratoplasty for total bilateral LSCD (Reinhard et al. 2004). Using combined therapy with mycophenolate mofetil and tacrolimus, KLAL rejection rate was 17%, which is significantly lower than in other series (Liang et al. 2009).

Cultivated Oral Mucosa Transplantation (COMET)

The results of COMET for bilateral total LSCD have been reported in 59 patients since 2004. At year 3rd, visual acuity (VA) had improved in 95% of 19 eyes followed for a mean of 55 months (Nakamura et al. 2011). However, the mean best-corrected VA was of about 20/40 only. While 74% of these cases received also amniotic membrane transplantation, 37% developed PED postoperatively; 71% of them suffered from SJS and 90% had also received mitomycin C as adjunctive therapy to prevent scarring after fornix reconstruction, performed simultaneously to COMET. Ocular hypertension was observed in 3 eyes (16%).

In another series of 40 patients with a mean follow-up of 25.4 months (Satake et al. 2011) COMET was followed by "immediate" epithelialization in all patients. However, the ocular surface stability declined during the first 6 months leading to PED. At year 2nd and 3rd, the acquisition of a stable ocular surface was 59.0% and 53.1%, respectively. The different phenotype of oral mucosa compared to corneal epithelium would explain this tendency to PED, and all patients had peripheral corneal vascularization and stromal opacity that affected the final visual acuity.

To improve the results of CLET and COMET we need to progress in several aspects, as determining the actual number of stem cells present in CLET and COMET cell sheets, a better knowledge of the behavior of transplanted cells and the role of the adequate limbal "niche" in the survival of these. Beyond the provision of new limbal cells, future advances may require the reconstruction of these "niches" and possibly the replacement of other components of the ocular surface necessary for maintaining its physiology (Baradaran-Rafii et al. 2010).

Autologous Biological-haptic Keratoprostheses (OOKP, T-OKP)

Of the several types of artificial cornea that have been used in cases with severe LSCD, the longest record corresponds to OOKP. The group of Falcinelli reported 85% (of 181 cases) anatomical survival after 18 years, and that of Hille 100% survival at 8 years (25 cases) (Falcinelli et al. 2005, Hille et al. 2006). At our institution, the anatomical results for OOKP performed in 145 eyes—excluding the second eye in bilateral cases—between 1974 and 2005 were of 68% Kaplan-Meier survival at 8th years and 59% at 18 years post-op (de la Paz et al. 2008). While these lower survival rates might reflect the fact that 37% of our patients, live abroad—which could explain a suboptimal follow-up—the overall very long-term success of OOKP in theses series is quite remarkable.

The published functional results of OOKP follow a similar pattern, with more optimistic figures in the Falcinelli series compared to ours. They report a mean VA of 0.69 (Snellen decimal scale) at the last follow-up after 18 years. We found "maximal VA"—defined as the best at any point after the procedure—between 0.8 and 1.0 in 48% of eyes, with an additional 24% between 0.3 and 0.7. However, Kaplan-Meier functional survival defined as VA $\geq$ 0.05 was of 50% after 5 years and of 39% after 10 years of follow-up, leveling to around 20% after more than 30 years (de la Paz et al. 2008). Using a "VA by Time" index—which averages the best corrected VA at a particular time after the procedure—we found an increase to 0.33 at 2nd year followed by a slow decline to 0.28 at 6 years (Michael et al. 2008). Even these relatively modest results are remarkable considering the initial condition of these patients.

An analysis of the factors influencing the outcomes showed that the anatomic survival of OOKP at 5 years was similar for all diagnosis except for OCP with already the poorest results. After 10 years, those with thermal and chemical burns fared better than those with SJS or trachoma. Surprisingly, the functional survival (VA $\geq$ 0.05 decimal) at both 5 and 10 years was better for SJS compared to burns and OCP. Postoperative complications are common after any keratoprosthesis procedure and OOKP does not escape this rule. The most frequent in our series was extrusion of the prosthesis (28%), followed by retinal detachment (16%), uncontrolled glaucoma (11%), infection (9%), retroprosthetic membrane (5%) and vitreous hemorrhage (3%). The effect of these complications on the outcomes is variable, with as a rule greater impact on the functional than on the anatomic survival. For example, glaucoma is a major cause of final decrease in VA, although these cases had the highest rate of anatomical success (90% at 10 years). The influence of age is not very clear; while those above 70 years had the best 10-year anatomical survival, there is a trend to better functional survival in the younger patients. This could be due to the age-related progression of either the chronic ocular (glaucoma and retinal/macular problems) or systemic conditions (de la Paz et al. 2008). The frequent postoperative complications of OOKP are confirmed in other series including one recent from India (Iyer et al. 2010).

The results of T-OKP cover shorter follow-ups that those of OOKP. At 1.5 years, the anatomical survival (Kaplan-Meier) of 80% in our T-OKP series of 82 eyes is similar to the 3 of 4 cases reported by Hille et al. At five years it was of 65.1%, and 48% at 10 years for T-OKP (vs. 73.7% and 60% respectively for OOKP). However, these differences did not reach statistical difference. Regarding functional survival (VA ≥ 0.05 decimal) it was not statistically different from that of OOKP for the first 2 years, but decreased to 29% and 17% at 5 and 10 years respectively (vs. 50% and 39% for OOKP), which is significantly worse. Nevertheless, the results according to diagnosis varied greatly in some instances, with worse anatomical and functional survival for T-OKP versus OOKP in SJS, and the reverse in the case of OCP, at least for the first 5 years (de la Paz et al. 2008). The "VA by Time" index was somewhat lower in T-OKP versus OOKP (0.28 vs. 0.33 at 2 years), but with coinciding confidence intervals, which suggests no statistical difference. On the other hand, these differences may reflect some selection bias, since the T-OKP included more OCP (older) patients and the OOKP series more with chemical injuries (younger) and trachoma (Michael et al. 2008).

Alloplastic Keratoprostheses (Boston keratoprosthesis types I and II)

The first report on long-term results of BKP included 63 patient eyes operated by Dohlman between 1990 and 1997 and followed for a minimum of 21 months. One prosthesis extruded spontaneously and another 10 were removed due to complications (82.5% anatomical success). Functional success—defined as VA ≥ 0.1 decimal—was best for graft failure and non-cicatrizing conditions (83% and 68% at 2 and 5 years respectively), then for OCP (72% and 43%) and chemical burns (64% and 25%), and clearly worse for SJS (33% and 0%) (Yaghouti et al. 2001). A further study from the same institution, on 15 patients of SJS operated between 2000 and 2005 with either BKP types I or II found better results, with no cases of extrusion or endophthalmitis and maintenance of VA ≥ 0.1 over 2.5 ± 2.0 years. However, the authors concluded that the outcomes of BKP in SJS are still substantially less favorable than in non-autoimmune diseases (Sayegh et al. 2008). A more recent report focused on Type II BKP included 29 eyes, mostly with OCP (51.7%) or SJS/Lyell (41.4%). VA ≥ 0.1 was achieved in 79.3% and persisted in 46.2% of the 13 eyes followed for more than 5 years. Anatomical success was 58.6% at 5 years without extrusion or replacement. The authors conclude that type II BKP is a viable option for corneal blindness from severe autoimmune ocular surface disease (Pujari et al. 2011).

The other major published study on BKP type I included 23 eyes with various conditions causing LSCD—7 cases of chemical injury and 6 of SJS, among others—and compared the outcomes with those receiving a BKP for indications other than LSCD. Anatomical (retention) failure was higher in the LSCD group (0.148/eye-year) versus the non-LSCD group (0.114/eye-year). However, when SJS patients were excluded from the former, the failure rate dropped (0.056/eye-year) under that of the non-LSCD group. This confirms the poor prognosis of BKP type I in cases of SJS. On the other hand, functional results were better in the LSCD group, with percentages of VA ≥ 20/50 (69%, 88% and 67%, respectively at 1, 2, and 3 years), superior to those of the non-LSCD group. The most common postoperative complications in the LSCD group were PED formation (56.5% of eyes) and sterile corneal necrosis (30%), whereas retroprosthetic membrane formation (46%) was the most common postoperative complication non-LSCD group, followed by PED formation (23%). The development of a PED was found to be a significant risk factor for sterile corneal stromal necrosis and infectious keratitis. These results support the use of the BKP type I in managing bilateral, non-immune-mediated LSCD (Sejpal et al. 2011).

ALGORITHM FOR THE MANAGEMENT OF LSCD

The selection of one particular option in the surgical management of LSCD will depend on the unique characteristics of each patient, as well as the personal experience and preferences of the surgeon. To propose a detailed algorithm for the treatment of these cases would inevitably incur into an exercise of simplification of this very complex set of pathological conditions. Nevertheless, the available data already enables the design of a general set of rules that can be helpful in orienting this selection process (Fig. 6).

After the condition of the ocular surface environment has been improved as possible with the use of the applicable general nonsurgical measures (see above) and the LSCD has proven to be irreversible, a surgical indication is established. Whenever present, the correction of symblepharon, trichiasis or other oculoplastic conditions with negative impact on the ocular surface constitutes a prerequisite to the LSCD procedures proper. Starting from this point, the first dichotomy is established between partial and total LSCD.

Cases with partial LSCD of either non-inflammatory or inflammatory etiology can be treated as a first instance with SCE, and only when this fails—or is likely to fail due to the extensiveness of the damage or other associated factors as a persisting inflammation, etc.—a one-sector KLAU or KLAL will be indicated. All of these procedures can be combined with an AM patch or graft, which may enhance the success rates (Meallet et al. 2003, Tseng et al. 2004).

The cases with total (360°) LSCD can be divided in three groups: (1) unilateral total LSCD; generally of non-inflammatory, non-progressive etiology, most commonly after chemical injuries or other trauma, (2) bilateral total LSCD of the same group of etiologies as the former, and (3) bilateral

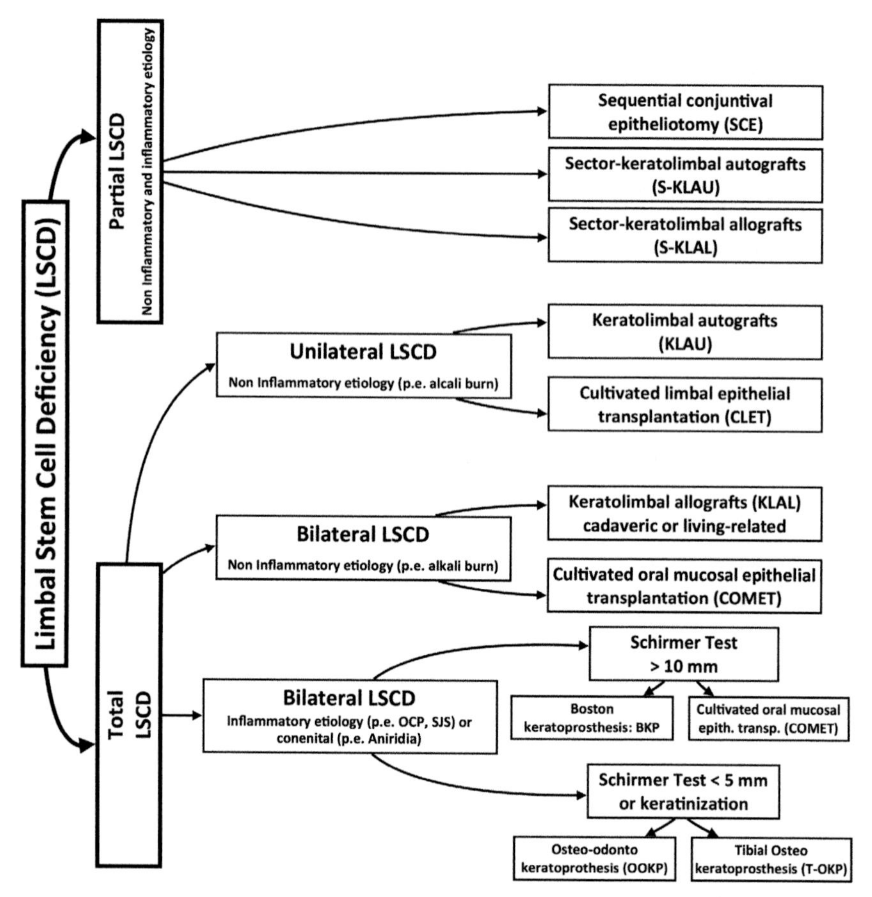

Figure 6 Basic algorithm for the surgical treatment of limbal stem cells deficiency (LSCD). SCE: Sequential conjunctival epitheliectomy. S-KLAU: Sectorial keratolimbal autograft. S-KLAL: Sectorial keratolimbal allograft. CLET: Cultivated limbal epithelial transplantation. COMET: Cultivated oral mucosal epithelial transplantation. BKP: Boston keratoprosthesis. OOKP: Osteo-odonto-keratoprosthesis. T-OKP: Tibial osteo-keratoprosthesis. OCP: Ocular cicatricial pemphigoid. SJS: Stevens–Johnson syndrome.

total LSCD caused by inflammatory diseases such as SJS and OCP or other chronic progressive conditions as in congenital aniridia keratopathy.

In unilateral total LSCD, the standard procedure has classically been a 2-sector KLAU, which has been proved to be a very effective and safe procedure. However, the lower impact on the donor cornea associated with CLET makes it a better option where the involved technologies have become available. There are to date no controlled trials comparing CLET and KLAU, however the success appears to be similar, at least in alkali burns (Tuft and Shortt 2009). The higher complexity and cost of CLET compared to KLAU is a factor to be considered, and the latter remains a valid option in cases with no involvement of the contralateral eye. If the latter has been partially affected and there is doubt about its suitability as a donor—risking the iatrogenic induction of LSCD—CLET is clearly preferred.

In bilateral total LSCD of non-progressive etiologies, KLAL is the classical alternative. This is usually an annular 360° graft when the donor is cadaveric and 2 sector grafts when it comes from a living relative. A large diameter corneal graft including the limbus, either lamellar (Vajpayee et al. 2000) or penetrating with a limbal lamellar flange (Reinhard et al. 2004) have also been used as a variant of KLAL. However, KLAL have a limited survival due to immunologic rejection and the hostile environment. In this group, tissue-engineering techniques would in principle have the same advantages as compared to KLAU. In bilateral cases with partial sparing of one eye, autologous CLET may still be possible as long as the small limbus biopsy can be taken safely from the less affected eye. Nevertheless, CLET would have to be allogeneic in cases of total bilateral LSCD, facing the same rejection problems as KLAL. Autologous COMET offers a fairly good compromise, possibly better for these patients.

In cases with OCP, SJS and other inflammatory autoimmune diseases, as well as those with aniridia and other progressive degenerative disorders, the associated LSCD tends to become total and bilateral. The chronic inflammatory or otherwise progressive nature of the causing condition and/or the presence of a severe dry eye implies a very poor prognosis for most epithelial reconstructive techniques. A keratoprosthesis becomes the best option for most of these eyes. If the lacrimal function is relatively preserved (Schirmer test > 10 mm), COMET—or even an oral mucosa graft plus AM—may have chances of success, provided the condition of the ocular surface environment is optimized, applying the available measures for dry eye and to control inflammation. Allogeneic CLET from HLA-identical living-related donor may have better long-term results (Meller et al. 2009). However, HLA-matched donors may not be available in many cases. On the other hand, these patients are good candidates for the all-alloplastic keratoprosthesis such as the BKP type I, which has the advantage of being technically simpler.

Finally, when the dry eye is severe (Schirmer test < 5 mm) and/or there is ocular surface keratinization, the prognosis of all tissue engineering techniques is poor. The long-term results of BKP type I have also been limited, especially in SJS (Sejpal et al. 2011). While the results of BKP type II have been more encouraging in this group (Pujari et al. 2011), those of the autologous biological-haptic keratoprosthesis as OOKP or T-OKP appear so far to be superior (Falcinelli et al. 2005; de la Paz et al. 2010).

CONCLUSION

LSCD pose some of the most difficult challenges to the cornea and anterior segment specialist. The advances in the understanding of the underlying pathology and contributing factors, as well as in the medical and surgical treatments available—especially in the highly technological domains of tissue engineering and artificial cornea—are opening new and promising perspectives for the long-term restoration of vision in these patients.

While a set of basic rules for selecting the surgical options can already be established, elucidating which is best for each patient remains a complex task. To advance in this direction, the need for further hard evidence as provided by prospective comparative trials is warranted.

REFERENCES

Arnalich-Montiel, F., S. Pastor, A. Blazquez-Martinez, J. Fernandez-Delgado, M. Nistal, J.L. Alio and M.P. De Miguel. 2008. Adipose-derived stem cells are a source for cell therapy of the corneal stroma. Stem Cells 26: 570–579.

Baradaran-Rafii, A., M. Ebrahimi, M.R. Kanavi, E. Taghi-Abadi, N. Aghdami, M. Eslani, P. Bakhtiari, B. Einollahi, H. Baharvand and M.A. Javadi. 2010. Midterm outcomes of autologous cultivated limbal stem cell transplantation with or without penetrating keratoplasty. Cornea 29: 502–509.

Barraquer, J. and J. Rutllán. 1982. Atlas de Microcirugía de la Córnea. Ediciones Scriba, S.A., Barcelona, Spain.

Barraquer, R.I., M. de Toledo and E. Torres. 2004. Distrofias y Degeneraciones Corneales. Atlas y Texto. Espaxs Publicaciones Médicas, S.A., Barcelona, Spain. pp. 160–162.

Barraquer, R.I. and V.K. Charoenrook. 2009. Management of pterygium–complications and recurrence. In: S. Boyd and L. Wu. [eds.]. Management of Complications in Ophthalmic Surgery. Jaypee–Highlights, Panama, Rep. of Panama, pp. 93–104.

Basu, S., A. Mohamed, S. Chaurasia, K. Sejpal, G.K. Vemuganti and V.S. Sangwan. 2011. Clinical Outcomes of Penetrating Keratoplasty after Autologous Cultivated Limbal Epithelial Transplantation for Ocular Surface Burns. Am. J. Ophthalmol. 17.

Baylis, O., F. Figueiredo, C. Henein, M. Lako and S. Ahmad. 2011. 13 years of cultured limbal epithelial cell therapy: a review of the outcomes. J. Cell Biochem. 112: 993–1002.

Behrens, A., J.J. Doyle, L. Stern, R.S. Chuck, P.J. McDonnell, D.T. Azar, H.S. Dua, M. Hom, P.M. Karpecki, P.R. Laibson, M.A. Lemp, D.M. Meisler, J.M. Del Castillo, T.P. O'Brien, S.C. Pflugfelder, M. Rolando, O.D. Schein, B. Seitz, S.C. Tseng, Setten G. Van, S.E. Wilson and S.C. Yiu. 2006. Dysfunctional tear syndrome: a Delphi approach to treatment recommendations. Cornea 25: 900–907.

Bensinger, W.I., R.J. Berenson, R.G. Andrews, D.F. Kalamasz, R.S. Hill, I.D. Bernstein, J.G. Lopez, C.D. Buckner and E.D. Thomas. 1990. Positive selection of hematopoietic progenitors from marrow and peripheral blood for transplantation. J. Clin. Apher. 5: 74–76.

Berenson, R.J., W.I. Bensinger, R. Hill, R.G. Andrews, J. Garcia-Lopez, D.F. Kalamasz, B.J. Still, C.D. Buckner, I.D. Bernstein and E.D. Thomas. 1990. Stem cell selection-clinical experience. Prog. Clin. Biol. Res. 333: 403–410.

Blazejewska, E.A., U. Schlotzer-Schrehardt, M. Zenkel, B. Bachmann, E. Chankiewitz, C. Jacobi and F.E. Kruse. 2009. Corneal limbal microenvironment can induce transdifferentiation of hair follicle stem cells into corneal epithelial-like cells. Stem Cells 27: 642–652.

Budak, M.T., O.S. Alpdogan, M. Zhou, R.M. Lavker, M.A. Akinci and J.M. Wolosin. 2005. Ocular surface epithelia contain ABCG2-dependent side population cells exhibiting features associated with stem cells. J. Cell Sci. 118: 1715–1724.

Chen, W., M. Ishikawa, K. Yamaki and S. Sakuragi. 2003. Wistar rat palpebral conjunctiva contains more slow-cycling stem cells that have larger proliferative capacity: implication for conjunctival epithelial homeostasis. Jpn. J. Ophthalmol. 47: 119–128.

Choong, P.F., P.L. Mok, S.K. Cheong and K.Y. Then. 2007. Mesenchymal stromal cell-like characteristics of corneal keratocytes. Cytotherapy. 9: 252–258.

Colabelli Gisoldi, R.A., A. Pocobelli, C.M. Villani, D. Amato and G. Pellegrini. 2010. Evaluation of molecular markers in corneal regeneration by means of autologous cultures of limbal cells and keratoplasty. Cornea 29: 715–722.

Cotsarelis, G., S.Z. Cheng, G. Dong, T.T. Sun and R.M. Lavker. 1989. Existence of slow-cycling limbal epithelial basal cells that can be preferentially stimulated to proliferate: implications on epithelial stem cells. Cell 57: 201–209.

D'Aversa G., J.L. Luchs, M.J. Fox, P.S. Rosenbaum and I.J. Udell. 1997. Advancing wave-like epitheliopathy. Clinical features and treatment. Ophthalmology 104: 962–969.

Davanger, M. and A. Evensen. 1971. Role of the pericorneal papillary structure in renewal of corneal epithelium. Nature 229: 560–561.

Daya, S.M., A. Watson, J.R. Sharpe, O. Giledi, A. Rowe, R. Martin and S.E. James. 2005. Outcomes and DNA analysis of *ex vivo* expanded stem cell allograft for ocular surface reconstruction. Ophthalmology 112: 470–477.

de Araujo, A.L., V. Charoenrook, M.F. de la Paz, J. Temprano, R.I. Barraquer and R. Michael. 2011. The role of visual evoked potential and electroretinography in the preoperative assessment of osteo-keratoprosthesis or osteo-odonto-keratoprosthesis surgery. Acta Ophthalmol. (in press).

de la Paz, M.F., J.A. De Toledo, V. Charoenrook, S. Sel, J. Temprano, R.I. Barraquer and R. Michael. 2011. Impact of clinical factors on the long-term functional and anatomic outcomes of osteo-odonto-keratoprosthesis and tibial bone keratoprosthesis. Am. J. Ophthalmol. 151: 829–839.

Di Girolamo, N., M. Bosch, K. Zamora, M.T. Coroneo, D. Wakefield and S.L. Watson. 2009. A contact lens-based technique for expansion and transplantation of autologous epithelial progenitors for ocular surface reconstruction. Transplantation 87: 1571–1578.

Di Iorio, E., S. Ferrari, A. Fasolo, E. Bohm, D. Ponzin and V. Barbaro. 2010. Techniques for culture and assessment of limbal stem cell grafts. Ocul. Surf. 8: 146–153.

Djalilian, A.R., S.P. Mahesh, C.A. Koch, R.B. Nussenblatt, D. Shen, Z. Zhuang, E.J. Holland and C.C. Chan. 2005. Survival of donor epithelial cells after limbal stem cell transplantation. Invest Ophthalmol. Vis. Sci. 46: 803–807.

Dohlman, C.H. and M. Doane. 1994. Keratoprosthesis in end-stage dry eye. Adv. Exp. Med. Biol. 350: 561–564.

Du, Y., M.L. Funderburgh, M.M. Mann, N. SundarRaj and J.L. Funderburgh. 2005. Multipotent stem cells in human corneal stroma. Stem Cells 23: 1266–1275.

Dua, H.S. 1998. The conjunctiva in corneal epithelial wound healing. Br. J. Ophthalmol. 82: 1407–1411.

Dua, H.S. and A. Azuara-Blanco. 2000. Limbal stem cells of the corneal epithelium. Surv. Ophthalmol. 44: 415–425.

Dua, H.S., V.A. Shanmuganathan, A.O. Powell-Richards, P.J. Tighe and A. Joseph. 2005. Limbal epithelial crypts: a novel anatomical structure and a putative limbal stem cell niche. Br. J. Ophthalmol. 89: 529–532.

Espana, E.M., M.D. Pascuale, M. Grueterich, A. Solomon and S.C. Tseng. 2004. Keratolimbal allograft in corneal reconstruction. Eye 18: 406–417.

Falcinelli, G., B. Falsini, M. Taloni, P. Colliardo and G. Falcinelli. 2005. Modified osteo-odonto-keratoprosthesis for treatment of corneal blindness: long-term anatomical and functional outcomes in 181 cases. Arch. Ophthalmol. 123: 1319–1329.

Friedenwald, J.S. 1951. Growth pressure and metaplasia of conjunctival and corneal epithelium. Doc. Ophthalmol. 5–6: 184–192.

Garg, P., P.V. Krishna, A.K. Stratis and U. Gopinathan. 2005. The value of corneal transplantation in reducing blindness. Eye 19: 1106–1114.

Gicquel, J.J., R. Navarre, M.E. Langman, A. Coulon, S. Balayre, S. Milin, M. Mercie, A. Rossignol, A. Barra, P.M. Levillain, J.M. Gombert and P. Dighiero. 2007. The use of impression cytology in the follow-up of severe ocular burns. Br. J. Ophthalmol. 91: 1160–1164.

Gomes, J., A. Monteiro, B. Geraldes, G.B. Melo, R.L. Smith, M. Silva, N.F. Lizier, A. Kerkis, H. Cerruti and I. Kerkis. 2010. Corneal reconstruction with tissue-engineered cell sheets composed of human immature dental pulp stem cells. Invest Ophthalmol. Vis. Sci. 51: 1408–1414.

Grueterich, M., E.M. Espana and S.C. Tseng. 2003. *Ex vivo* expansion of limbal epithelial stem cells: amniotic membrane serving as a stem cell niche. Surv. Ophthalmol. 48: 631–646.

Györffy I. 1951. Acrylic corneal implant in keratoplasty. Am. J. Ophthalmol. 34: 757–758.

Hall, P.A. and F.M. Watt. 1989. Stem cells: the generation and maintenance of cellular diversity. Development 106: 619–633.

Henderson, T.R., D.J. Coster and K.A. Williams. 2001. The long-term outcome of limbal allografts: the search for surviving cells. Br. J. Ophthalmol. 85: 604–609.

Hille, K., A. Hille and K.W. Ruprecht. 2006. Medium term results in keratoprostheses with biocompatible and biological haptic. Graefes Arch. Clin. Exp. Ophthalmol. 244: 696–704.

Holland, E.J. 1996. Epithelial transplantation for the management of severe ocular surface disease. Trans. Am. Ophthalmol. Soc. 94: 677–743.

Holland, E.J. and G.S. Schwartz. 2000. Changing concepts in the management of severe ocular surface disease over twenty-five years. Cornea 19: 688–698.

Holland, E.J. 2004. The Paton lecture: Ocular surface transplantation: 10 years' experience. Cornea 23: 425–431.

Inatomi, T., T. Nakamura, N. Koizumi, C. Sotozono and S. Kinoshita. 2005. Current concepts and challenges in ocular surface reconstruction using cultivated mucosal epithelial transplantation. Cornea 24: S32–S38.

Inatomi, T., T. Nakamura, N. Koizumi, C. Sotozono, N. Yokoi and S. Kinoshita. 2006a. Midterm results on ocular surface reconstruction using cultivated autologous oral mucosal epithelial transplantation. Am. J. Ophthalmol. 141: 267–275.

Inatomi, T., T. Nakamura, M. Kojyo, N. Koizumi, C. Sotozono and S. Kinoshita. 2006b. Ocular surface reconstruction with combination of cultivated autologous oral mucosal epithelial transplantation and penetrating keratoplasty. Am. J. Ophthalmol. 142: 757–764.

Iyer, G., V.S. Pillai, B. Srinivasan, G. Falcinelli, P. Padmanabhan, S. Guruswami and G. Falcinelli. 2010. Modified osteo-odonto keratoprosthesis—the Indian experience—results of the first 50 cases. Cornea 29: 771–776.

Jenkins, C., S. Tuft, C. Liu and R. Buckley. 1993. Limbal transplantation in the management of chronic contact-lens-associated epitheliopathy. Eye 7: 629–633.

Kawasaki, S., H. Tanioka, K. Yamasaki, N. Yokoi, A. Komuro and S. Kinoshita. 2006. Clusters of corneal epithelial cells reside ectopically in human conjunctival epithelium. Invest Ophthalmol. Vis. Sci. 47: 1359–1367.

Kawashima, M., T. Kawakita, Y. Satake, K. Higa and J. Shimazaki. 2007. Phenotypic study after cultivated limbal epithelial transplantation for limbal stem cell deficiency. Arch. Ophthalmol. 125: 1337–1344.

Kenyon, K.R. and S.C. Tseng. 1989. Limbal autograft transplantation for ocular surface disorders. Ophthalmology 96: 709–722.

King, J.H. and Mc Tigue J.W. [eds.]. 1964. The Cornea. World Congress. Butterwoths, Baltimore, USA. p. 354.

Koizumi, N., T. Inatomi, T. Suzuki, C. Sotozono and S. Kinoshita. 2001. Cultivated corneal epithelial stem cell transplantation in ocular surface disorders. Ophthalmology 108: 1569–1574.

Lajtha, L.G. 1979. Stem cell concepts. Differentiation 14: 23–34.
Lavker, R.M., Z.G. Wei and T.T. Sun. 1998. Phorbol ester preferentially stimulates mouse fornical conjunctival and limbal epithelial cells to proliferate *in vivo*. Invest Ophthalmol. Vis. Sci. 39: 301–307.
Lehrer, M.S., T.T. Sun and R.M. Lavker. 1998. Strategies of epithelial repair: modulation of stem cell and transit amplifying cell proliferation. J. Cell Sci. 111: 2867–2875.
Liang, L., H. Sheha and S.C. Tseng. 2009. Long-term outcomes of keratolimbal allograft for total limbal stem cell deficiency using combined immunosuppressive agents and correction of ocular surface deficits. Arch. Ophthalmol. 127: 1428–1434.
Lindberg, K., M.E. Brown, H.V. Chaves, K.R. Kenyon and J.G. Rheinwald. 1993. *In vitro* propagation of human ocular surface epithelial cells for transplantation. Invest Ophthalmol. Vis. Sci. 34: 2672–2679.
Liu, C., B. Paul, R. Tandon, E. Lee, K. Fong, I. Mavrikakis, J. Herold, S. Thorp, P. Brittain, I. Francis, C. Ferrett, C. Hull, A. Lloyd, D. Green, V. Franklin, B. Tighe, M. Fukuda and S. Hamada. 2005. The osteo-odonto-keratoprosthesis (OOKP). Semin. Ophthalmol. 20: 113–128.
Liu, C., S. Okera, R. Tandon, J. Herold, C. Hull and S. Thorp. 2008. Visual rehabilitation in end-stage inflammatory ocular surface disease with the osteo-odonto-keratoprosthesis: results from the UK. Br. J. Ophthalmol. 92: 1211–1217.
Liu, C.Y., D.E. Birk, J.R. Hassell, B. Kane and W.W. Kao. 2003. Keratocan-deficient mice display alterations in corneal structure. J. Biol. Chem. 278: 21672–21677.
Liu, J., H. Sheha, Y. Fu, M Giegengack and S.C. Tseng. 2011. Oral mucosal graft with amniotic membrane transplantation for total limbal stem cell deficiency. Am. J. Ophthalmol. 152: 739–747.
Liu, S., J. Li, D.T. Tan and R.W. Beuerman. 2007. The eyelid margin: a transitional zone for 2 epithelial phenotypes. Arch. Ophthalmol. 125: 523–532.
Majo, F., A. Rochat, M. Nicolas, G.A. Jaoude and Y. Barrandon. 2008. Oligopotent stem cells are distributed throughout the mammalian ocular surface. Nature 456: 250–254.
Maruyama-Hosoi, F., J. Shimazaki, S. Shimmura and K. Tsubota. 2006. Changes observed in keratolimbal allograft. Cornea 25: 377–382.
Meallet, M.A., E.M. Espana, M. Grueterich, S.E. Ti, E. Goto and S.C. Tseng. 2003. Amniotic membrane transplantation with conjunctival limbal autograft for total limbal stem cell deficiency. Ophthalmology 110: 1585–1592.
Meller, D., T. Fuchsluger, M. Pauklin and K.P. Steuhl. 2009. Ocular surface reconstruction in graft-versus-host disease with HLA-identical living-related allogeneic cultivated limbal epithelium after hematopoietic stem cell transplantation from the same donor. Cornea 28:233–236.
Michael, R., V. Charoenrook, M.F. de la Paz, W. Hitzl, J. Temprano and R.I. Barraquer. 2008. Long-term functional and anatomical results of osteo- and osteoodonto-keratoprosthesis. Graefes Arch. Clin. Exp. Ophthalmol. 246: 1133–1137.
Miller, S.J., T.T. Sun and R.M. Lavker. 1993. Hair follicles, stem cells and skin cancer. J. Invest Dermatol. 100: 288S–294S.
Miri, A., B. Al-Deiri and H.S. Dua. 2010. Long-term outcomes of autolimbal and allolimbal transplants. Ophthalmology 117: 1207–1213.
Morrison, S.J., N.M. Shah and D.J. Anderson. 1997. Regulatory mechanisms in stem cell biology. Cell 88: 287–298.
Nadri, S., M. Soleimani, J. Kiani, A. Atashi and R. Izadpanah. 2008a. Multipotent mesenchymal stem cells from adult human eye conjunctiva stromal cells. Differentiation 76: 223–231.
Nadri, S., M. Soleimani, Z. Mobarra and S. Amini. 2008b. Expression of dopamine-associated genes on conjunctiva stromal-derived human mesenchymal stem cells. Biochem. Biophys. Res. Commun. 377: 423–428.
Nagasaki, T. and J. Zhao. 2005. Uniform distribution of epithelial stem cells in the bulbar conjunctiva. Invest Ophthalmol. Vis. Sci. 46: 126–132.

Nakamura, T., T. Inatomi, C. Sotozono, T. Amemiya, N. Kanamura and S. Kinoshita. 2004. Transplantation of cultivated autologous oral mucosal epithelial cells in patients with severe ocular surface disorders. Br. J. Ophthalmol. 88: 1280–1284.

Nakamura, T., K. Takeda, T. Inatomi, C. Sotozono and S. Kinoshita. 2011. Long-term results of autologous cultivated oral mucosal epithelial transplantation in the scar phase of severe ocular surface disorders. Br. J. Ophthalmol. 95: 942–946.

Nishida, K., M. Yamato, Y. Hayashida, K. Watanabe, K. Yamamoto, E. Adachi, S. Nagai, A. Kikuchi, N. Maeda, H. Watanabe, T. Okano and Y. Tano. 2004. Corneal reconstruction with tissue-engineered cell sheets composed of autologous oral mucosal epithelium. N. Engl. J. Med. 351: 1187–1196.

Ono, K., S. Yokoo, T. Mimura, T. Usui, K. Miyata, M. Araie, S. Yamagami and S. Amano. 2007. Autologous transplantation of conjunctival epithelial cells cultured on amniotic membrane in a rabbit model. Mol. Vis. 13: 1138–1143.

Pauklin, M., V. Kakkassery, K.P. Steuhl and D. Meller. 2009. Expression of membrane-associated mucins in limbal stem cell deficiency and after transplantation of cultivated limbal epithelium. Curr. Eye Res. 34: 221–230.

Pe'er, J., G. Zajicek, H. Greifner and M. Kogan. 1996. Streaming conjunctiva. Anat. Rec. 245: 36–40.

Pellegrini, G., C.E. Traverso, A.T. Franzi, M. Zingirian, R. Cancedda and M. De Luca. 1997. Long-term restoration of damaged corneal surfaces with autologous cultivated corneal epithelium. Lancet 349: 990–993.

Pellegrini, G., O. Golisano, P. Paterna, A. Lambiase, S. Bonini, P. Rama and M. De Luca. 1999. Location and clonal analysis of stem cells and their differentiated progeny in the human ocular surface. J. Cell Biol. 145: 769–782.

Polisetty, N., A. Fatima, S.L. Madhira, V.S. Sangwan and G.K. Vemuganti. 2008. Mesenchymal cells from limbal stroma of human eye. Mol. Vis. 14: 431–442.

Potten, C.S. and M. Loeffler. 1990. Stem cells: attributes, cycles, spirals, pitfalls and uncertainties. Lessons for and from the crypt. Development 110: 1001–1020.

Pujari, S., S. Siddique, C.H. Dohlman and J. Chodosh. 2011. The Boston Keratoprosthesis Type II: The Massachusetts Eye and Ear Infirmary Experience. Cornea (in press).

Rama, P., S. Bonini, A. Lambiase, O. Golisano, P. Paterna, Luca M. De and G. Pellegrini. 2001. Autologous fibrin-cultured limbal stem cells permanently restore the corneal surface of patients with total limbal stem cell deficiency. Transplantation 72: 1478–1485.

Rama, P., S. Matuska, G. Paganoni, A. Spinelli, M. De Luca and G. Pellegrini. 2010. Limbal stem-cell therapy and long-term corneal regeneration. N. Engl. J. Med. 363: 147–155.

Reinhard, T., H. Spelsberg, L. Henke, T. Kontopoulos, J. Enczmann, P. Wernet, P. Berschick, R. Sundmacher and D. Bohringer. 2004. Long-term results of allogeneic penetrating limbo-keratoplasty in total limbal stem cell deficiency. Ophthalmology 111: 775–782.

Rheinwald, J. G. and H. Green. 1975. Serial cultivation of strains of human epidermal keratinocytes: the formation of keratinizing colonies from single cells. Cell 6: 331–343.

Sangwan, V.S., G.K. Vemuganti, S. Singh and D. Balasubramanian. 2003. Successful reconstruction of damaged ocular outer surface in humans using limbal and conjunctival stem cell culture methods. Biosci. Rep. 23: 169–174.

Sangwan, V.S., H.P. Matalia, G.K. Vemuganti, A. Fatima, G. Ifthekar, S. Singh, R. Nutheti and G.N. Rao. 2006. Clinical outcome of autologous cultivated limbal epithelium transplantation. Indian J. Ophthalmol. 54: 29–34.

Sangwan, V.S., S. Basu, G.K. Vemuganti, K. Sejpal, S.V. Subramaniam, S. Bandyopadhyay, S. Krishnaiah, S. Gaddipati, S. Tiwari and D. Balasubramanian. 2011. Clinical outcomes of xeno-free autologous cultivated limbal epithelial transplantation: a 10-year study. Br. J. Ophthalmol. 95: 1525–1529.

Santos, M.S., J.A. Gomes, A.L. Hofling-Lima, L.V. Rizzo, A.C. Romano and R. Belfort. 2005. Survival analysis of conjunctival limbal grafts and amniotic membrane transplantation in eyes with total limbal stem cell deficiency. Am. J. Ophthalmol. 140: 223–230.

Satake, Y., K. Higa, K. Tsubota and J. Shimazaki. 2011. Long-term outcome of cultivated oral mucosal epithelial sheet transplantation in treatment of total limbal stem cell deficiency. Ophthalmology 118: 1524–1530.

Sayegh, R.R., L.P. Ang, C.S. Foster and C.H. Dohlman. 2008. The Boston keratoprosthesis in Stevens–Johnson syndrome. Am. J. Ophthalmol. 145: 438–444.

Schermer, A., S. Galvin and T.T. Sun. 1986. Differentiation-related expression of a major 64K corneal keratin *in vivo* and in culture suggests limbal location of corneal epithelial stem cells. J. Cell Biol. 103: 49–62.

Schwab, I.R., M. Reyes and R.R. Isseroff. 2000. Successful transplantation of bioengineered tissue replacements in patients with ocular surface disease. Cornea 19: 421–426.

Sejpal, K., F. Yu and A.J. Aldave. 2011. The Boston keratoprosthesis in the management of corneal limbal stem cell deficiency. Cornea 30: 1187–1194.

Shimazaki, J., M. Aiba, E. Goto, N. Kato, S. Shimmura and K. Tsubota. 2002. Transplantation of human limbal epithelium cultivated on amniotic membrane for the treatment of severe ocular surface disorders. Ophthalmology 109: 1285–1290.

Shortt, A.J., G.A. Secker, M.S. Rajan, G. Meligonis, J.K. Dart, S.J. Tuft and J.T. Daniels. 2008. *Ex vivo* expansion and transplantation of limbal epithelial stem cells. Ophthalmology 115: 1989–1997.

Shortt, A.J., S.J. Tuft and J.T. Daniels. 2010. *Ex vivo* cultured limbal epithelial transplantation. A clinical perspective. Ocul. Surf. 8: 80–90.

Solomon, A., P. Ellies, D.F. Anderson, A. Touhami, M. Grueterich, E.M. Espana, S.E. Ti, E. Goto, W.J. Feuer and S.C. Tseng. 2002. Long-term outcome of keratolimbal allograft with or without penetrating keratoplasty for total limbal stem cell deficiency. Ophthalmology 109: 1159–1166.

Strampelli B. 1963. Osteo-odontokeratoprosthesis. Ann. Ottalmol. Clin. Ocul. 89: 1039–1044.

Strampelli B. and M.L. Restivo. 1966. Total keratectomy in leukomatous eye associated with autograft of a keratoconjunctival ring removed from the controlateral normal eye. Ann. Ottalmol. Clin. Ocul. 92: 778–786.

Sun, T.T. and H. Green. 1977. Cultured epithelial cells of cornea, conjunctiva and skin: absence of marked intrinsic divergence of their differentiated states. Nature 269: 489–493.

Tanioka, H., S. Kawasaki, K. Yamasaki, L.P. Ang, N. Koizumi, T. Nakamura, N. Yokoi, A. Komuro, T. Inatomi and S. Kinoshita. 2006. Establishment of a cultivated human conjunctival epithelium as an alternative tissue source for autologous corneal epithelial transplantation. Invest Ophthalmol. Vis. Sci. 47: 3820–3827.

Temprano J. 1991. Queratoplastias y queratoprótesis. LXVII Ponencia de la Sociedad Española de Oftalmología. Art Book 90, S.L., Barcelona.

Thill, M., K. Schlagner, S. Altenahr, S. Ergun, R.G. Faragher, N. Kilic, J. Bednarz, G. Vohwinkel, X. Rogiers, D.K. Hossfeld, G. Richard and U.M. Gehling. 2007. A novel population of repair cells identified in the stroma of the human cornea. Stem Cells Dev. 16: 733–745.

Thoft, R.A. and J. Friend. 1983. The X, Y, Z hypothesis of corneal epithelial maintenance. Invest Ophthalmol. Vis. Sci. 24: 1442–1443.

Tsai, R.J., L.M. Li and J.K. Chen. 2000. Reconstruction of damaged corneas by transplantation of autologous limbal epithelial cells. N. Engl. J. Med. 343: 86–93.

Tseng, S.C., E.M. Espana, T. Kawakita, M.A. Di Pascuale, W. Li, H. He, T.S. Liu, T.H. Cho, Y.Y. Gao, L.K. Yeh and C.Y. Liu. 2004. How does amniotic membrane work? Ocul. Surf. 2: 177–187.

Tsubota, K., I. Toda, H. Saito, N. Shinozaki and J. Shimazaki. 1995. Reconstruction of the corneal epithelium by limbal allograft transplantation for severe ocular surface disorders. Ophthalmology 102: 1486–1496.

Tuft, S.J. and A.J. Shortt. 2009. Surgical rehabilitation following severe ocular burns. Eye 23: 1966–1971.

Vajpayee, R.B., S. Thomas, N. Sharma, T. Dada and G.C. Tabin. 2000. Large-diameter lamellar keratoplasty in severe ocular alkali burns: A technique of stem cell transplantation. Ophthalmology 107: 1765–1768.

Wei, Z.G., G. Cotsarelis, T.T. Sun and R.M. Lavker. 1995. Label-retaining cells are preferentially located in fornical epithelium: implications on conjunctival epithelial homeostasis. Invest Ophthalmol. Vis. Sci. 36: 236–246.

Wei, Z.G., R.L. Wu, R.M. Lavker and T.T. Sun. 1993. In vitro growth and differentiation of rabbit bulbar, fornix and palpebral conjunctival epithelia. Implications on conjunctival epithelial transdifferentiation and stem cells. Invest Ophthalmol. Vis. Sci. 34: 1814–1828.

Wei, Z.G., T.T. Sun and R.M. Lavker. 1996. Rabbit conjunctival and corneal epithelial cells belong to two separate lineages. Invest Ophthalmol. Vis. Sci. 37: 523–533.

Wirtschafter, J.D., J.M. Ketcham, R.J. Weinstock, T. Tabesh and L.K. McLoon. 1999. Mucocutaneous junction as the major source of replacement palpebral conjunctival epithelial cells. Invest Ophthalmol. Vis. Sci. 40: 3138–3146.

Wirtschafter, J.D., L.K. McLoon, J.M. Ketcham, R.J. Weinstock and J.C. Cheung. 1997. Palpebral conjunctival transient amplifying cells originate at the mucocutaneous junction and their progeny migrate toward the fornix. Trans. Am. Ophthalmol. Soc. 95: 417–429.

Wolosin, J.M., M.T. Budak and M.A. Akinci. 2004. Ocular surface epithelial and stem cell development. Int. J. Dev. Biol. 48: 981–991.

Yaghouti, F., M. Nouri, J.C. Abad, W.J. Power, M.G. Doane and C.H. Dohlman. 2001. Keratoprosthesis: preoperative prognostic categories. Cornea 20: 19–23.

Yoshida, S., S. Shimmura, T. Kawakita, H. Miyashita, S. Den, J. Shimazaki and K. Tsubota. 2006. Cytokeratin 15 can be used to identify the limbal phenotype in normal and diseased ocular surfaces. Invest Ophthalmol. Vis. Sci. 47: 4780–4786.

15

Ocular Surface Reconstruction

David J. Galarreta,[1,2,a,*] *Ester Carreño,*[2] *Alejandro Portero,*[2] *Belen Carrasco*[1] and *Jose M. Herreras*[1,2]

SUMMARY

The management of ocular surface disease has benefited from different and several new surgical methods that amplify the wide spectrum of ocular surface reconstruction procedures.

Advances in microsurgery, reintroduction of amniotic membrane transplantation, developments of new materials, new perspectives in pharmacological immunomodulation and the critical understanding of the location and role of the limbal stem cells represent the highlights of the last decades.

These advancements have led to improvements in success rates in the management of patients with classically poor prognosis. However, corneal graft is still the last step in visual rehabilitation after ocular surface reconstruction, and represents the main and the most extended technique among these procedures, but even in this case, new sight is possible with the improvements in lamellar surgery.

INTRODUCTION

The ocular surface is rich in complexity and functionality; consisting of the eyelids, tear film, conjunctiva, and cornea. Severe ocular surface disease

[1]Hospital Clínico Universitario Valladolid—University of Valladolid, Paseo Ramón y Cajal 3, 47011 Valladolid, Spain.
[a]E-mail: davidgalarreta@hotmail.com
[2]Instituto Universitario de Oftalmobiología Aplicada (IOBA-Eye Institute)—University of Valladolid, Paseo de Belén, 17—Campus Miguel Delibes, 47011 Valladolid, Spain.
*Corresponding author

(OSD) may affect every single component of the functional unit. The wider the affected area, the more difficult is the treatment and the reconstruction procedures. The steps in the treatment must be orientated to provide better conditions to the cornea. To achieve this homeostasis is vital for the maintenance of good corneal epithelium, which is needed to keep corneal clarity and the optical function.

Better understanding of the ocular surface, especially through the knowledge that the corneal limbus is the source of limbal stem cells for the replacement of normal corneal epithelium, has changed the way to focus reconstruction procedures.

EVOLVING CONCEPTS OF OCULAR SURFACE RECONSTRUCTION

Ocular surface rehabilitation (OSR) algorithm can be divided in three main steps (Table 1):

1. Stabilization: Stabilization of conditions that can trigger the failure of the procedure, mainly represented by glaucoma, dryness, eyelid dysfunction and inflammation.
2. Ocular surface reconstruction procedures: Transplantation of the diseased conjunctiva or limbus from donors or tissue-engineered grafts.
3. Corneal graft: Penetrating or lamellar grafts for the final visual rehabilitation.

Stabilization Procedures

A specific sequential paradigm to stabilize the ocular surface is mandatory to ensure the best opportunity for graft survival. Several points must be considered to this stepwise approach.

Glaucoma

The presence and severity of glaucoma must be elucidated. Aggressive and a surgical approach with placement of a tube shunt in patients on more than one topical glaucoma medication is performed to avoid multiple medications that are toxic to the future transplanted tissue.

Eyelid Disease

The status of the eyelids and lashes can trigger the failure of the ocular surface graft. Surgical correction of pre-existing exposure, lagophthalmos, and eyelids or eyelashes malposition need an aggressive management before any transplantation.

Table 1 This table summarizes the algorithm of ocular surface rehabilitation in ocular surface disease. Modified from Daya et al. 2011 and Holland and Schwartz 2004.

1. Stabilization procedures	
Problem	**Treatment**
Glaucoma	Tube shunt in > 1 medication
Eyelid disease	Exposure surgery. Misdirected lashes treatment
Dryness	Lubrication preservative free
Inflammation	Topical anti-inflammatories. Systemic immunomodulation
Structural failures	Ophthalmic adhesives
Ocular surface disease complications	Amniotic membrane transplantation (AMT)
2. Ocular surface reconstruction procedures	
Problem	**Treatment**
Unilateral or bilateral Partial limbal deficiency	Sequential Sectoral Conjunctival Epitheliectomy (SSCE)
Unilateral conjunctival disease	Conjunctival limbal autograft (CAU) Ex vivo cultivated conjunctival autograft (EVCAU)
Bilateral conjunctival disease	Cadaveric, living relative or not conjunctival allograft (c, lr or nlr-CAL) *Ex vivo* cultivated cadaveric, living-related or not conjunctival allograft. (EVc, lr, lnr-CAL)
Unilateral limbal deficiency with minimal to moderate conjunctival disease	Keratolimbal autograft (KLAU) *Ex vivo* cultivated limbal autograft (EVLAU)
Bilateral limbal deficiency with minimal to moderate conjunctival disease	Keratolimbal allograft (KLAL) *Ex vivo* cultivated cadaveric, living-related or not limbal allograft. (EVc, lr, lnr-LAL)
Unilateral limbal deficiency with severe conjunctival disease	Combined conjunctival-keratolimbal autograft (CLAU)
Bilateral limbal deficiency with moderate to severe conjunctival disease	Cadaveric, Living-relative or not conjunctival limbal allograft (c, lr or nlr-CLAL)
Severe disease of palpebral conjunctiva and fornix	Oral, nasal, peritoneum or intestinal mucosal autograft (O, N, P or IMAU)
Unilateral or bilateral Partial limbal deficiency	Sequential Sectoral Conjunctival Epitheliectomy (SSCE)
3. Keratoplasties	
Problem	**Treatment**
Whole cornea disease	Penetrating keratoplasty (PK)
Superficial stromal or Bowman disease	Superficial anterior lamellar keratoplasty (s-ALK)
Deep or whole stromal disease	Deep anterior lamellar keratoplasty (DALK)
Endothelial disease	Descemet stripping automated endothelial keratoplasty (DSAEK) Descemet membrane endothelial keratoplasty (DMEK)
Non viable keratoplasty	Boston Keratoprosthesis in wet eye Osteo-odonto and tibial bone keratoprosthesis in dry eye

Dryness

The right lubrication of the ocular surface is essential for viability of ocular surface reconstruction procedures, however are not sufficient to resolve the ocular surface disease and inflammation must be treated as well. The single most critical advance in the treatment of dryness came with the elimination of preservatives, especially benzalkonium chloride, from lubricants. The abscence of preservatives is more important than the particular polymeric agent used in ocular lubricants.

Inflammation

Preoperative inflammation must be aggressively managed before any ocular surface graft. Transplantation performed in an inflamed ocular surface always shows poorer prognosis. If it is necessary, topical or systemic immunomodulation allows a greater chance for success.

Ophthalmic Adhesives

Ophthalmic adhesives can offer an intermediate step when tissue is not available for ocular surface reconstruction. Tissue adhesives have been used in ophthalmology in different ways like a structural filler or support instead of donor tissue, as an adhesive to avoid sutures and as a barrier to facilitate epithelial healing and prevent collagen breakdown (Chan and Boisjoly 2004).

Tissue adhesives can be subdivided into synthetic adhesives (e.g. cyanoacrylate derivatives) and biologic adhesives (e.g. fibrin-based adhesives).

Cyanoacrylate derivatives are synthetic and nonbiodegradable, and may induce an inflammatory foreign body reaction, including neovascularization and tissue necrosis (Kaufman et al. 2003). In ophthalmology, cyanoacrylate derivatives have mainly been used in the management of progressive corneal thinning and small, uncomplicated corneal perforations of non infectious causes (Fig. 1). Most of the uses of adhesives are only a temporizing or adjunctive measure previous to a keratoplasty or conjunctival flap, however some cases may heal completely with the glue (Sharma et al. 2003). The use of antibiotic prophylaxis is recommended to avoid infectious infiltrates widely reported (Cavanaugh and Gottsch 1991).

Fibrin-based adhesives simulate the final stage of the coagulation cascade, creating a biocompatible fibrin matrix similar to a plasma clot. They are biologic and biodegradable and consequently can be used under a superficial covering layer (e.g. conjunctiva, amniotic membrane). The

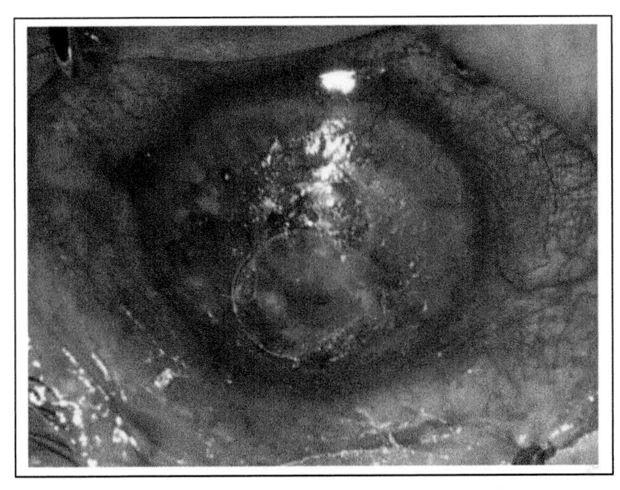

Figure 1 This figure shows the surgical view of a paracentral corneal perforation due to a herpetic keratitis sealed with cyanoacrylate and a plastic shell before the contact lens fitting. Note the air bubble centred due to supine position.

Color image of this figure appears in the color plate section at the end of the book.

degradation does not lead to toxic metabolites but occurs much more quickly with possible desirable or undesirable effects.

Additionally, they have useful applications such as to minimize the use of sutures, improve hemostasis of the conjunctiva and underlying tissues, and promote adhesion of a mucosal graft or amniotic membrane (Kaufman et al. 2003).

Limitations of fibrin glues still remain, because this blood product uses human thrombin and still carries a risk of infection from a contaminated donor pool (Atrah 1994, Cederholm-Williams 1994, Radosevich et al. 1997).

Amniotic Membrane Transplantation

The amniotic membrane (AM) constitutes the inner wall of the fetal membranes, and is a single layer of epithelium with an underlying stroma rich in extra cellular matrix and collagens. As a fetal tissue it is considered to be immunologically inert and to have several physiologic properties, including induction of adhesion and migration of epithelial cells to reduce inflammatory, cicatricial, and angiogenic reactions (Dua et al. 2004).

Kim and Tseng reintroduced the AM in 1995 for ocular surface diseases (Kim and Tseng 1995). Encouraging results led to a large and maybe over expressed use in ophthalmology. Indications need to be clarified and compared in clinical trials to classical methods for a good evidence-based medicine to evaluate its true potential.

In ophthalmology, AM has been used as a graft to replace the damage stromal matrix with its epithelium up when it is expected to become covered by host conjunctival or corneal epithelium; as a patch with its epithelium down, which facilitates trapping of inflammatory cells in the stroma, reducing inflammation; and in a combination of both.

Mechanism of action. The underlying physiologic mechanisms through which AM confers beneficial effects on the reconstruction of the ocular surface have not been completely elucidated yet (Table 2).

Table 2 This table summarizes the principal mechanisms of action of amniotic membrane (AM).

Mechanisms of Action of Amniotic Membrane
Prolongs life span and maintains clonogenicity of epithelial progenitor cells.
AM expresses epidermal growth factor, nerve growth factor, hepatocyte growth factor, and keratinocyte growth factor.
Facilitates the migration of epithelial cells, reinforces the adhesion of the basal epithelium, promotes cellular differentiation, and prevents cellular apoptosis.
Promotes goblet cell differentiation when combined with conjunctival fibroblasts.
Inhibitory effects on various proteases, including decreased expression of MMP-1 and 2.
Decreases expression of IL-1, IL-2, IFN-γ and TNF-α.
Suppresses TGF-β signalling system and myofibroblast differentiation of normal fibroblasts.
Decreases expression of VEGF.

MMP: matrix metalloproteinase. VEGF: vascular endothelial growth factor. TNF: tumour necrosis factor. IFN: interferon.

Evidence indicates that AM modulates the level of various cytokines, growth factors, enzymes, and receptors (Li et al. 2005). The three main favourable properties of the AM are anti-inflammatory, antiangiogenic and antiscarring effects.

Clinical indications of AM. The clinical indications of conjunctival and corneal diseases are summarized in Table 3.

Amniotic membrane as a graft for conjunctival surface reconstruction. Amniotic membrane transplantation can be used to reconstruct the conjunctival surface as an alternative to conjunctival graft after removal of large lesions and to restore normal stroma and provide a healthy basement membrane to promote epithelial proliferation and differentiation (Tseng et al. 1997).

Amniotic membrane transplantation as a graft for corneal surface reconstruction. Studies on the efficacy of AM to provide tectonic support in impending or recent corneal perforation have yielded different and controversial results. However, it is an excellent method when a corneal graft is not available or in some perforations, especially those related to vasculitic peripheral

Table 3 Mean clinical indications of amniotic membrane transplantation.

Function of AM	Conjunctival diseases	Corneal diseases
Graft	Pterigium. Bulbar reconstruction Symblepharonlysis Conjunctivochalasis Bleb leakage or revision	Persistent epithelial defect Partial limbal stem cell deficiency Stevens–Johnson syndrome Chemical burns. Band keratopathy
Patch	Acute chemical burns Stevens–Johnson syndrome	Persistent epithelial defects Acute chemical burns Preventing scar after PRK or PTK Refractory inflammatory ulcerative keratitis
Carrier in tissue engineered grafts	EVCAU	EVLAU

AM: Amniotic membrane, PRK: Photorefractive keratectomy, PTK: Phototherapeutic keratectomy, EVCAU: *Ex vivo* cultivated conjunctival autograft, EVLAU: *Ex vivo* cultivated limbal autograft.

ulcerative keratitis where the corneal graft has bad results (Lee and Tseng 1997, Kruse et al. 1999, Chen et al. 2000).

Amniotic membrane as a patch. Persistent epithelial defect is an extended indication for AM, providing excellent help in cases resistant to classical treatments (Fig. 2).

One of the most promising indications of AM is in the treatment of acute chemical and thermal injury of the ocular surface. Primary care of the acute phase will eventually determine the amount of ensuing damage.

Early epithelialization and inflammation control can prevent later complications associated to chronic phases like scarring, symblepharon formation or total stem cell loss.

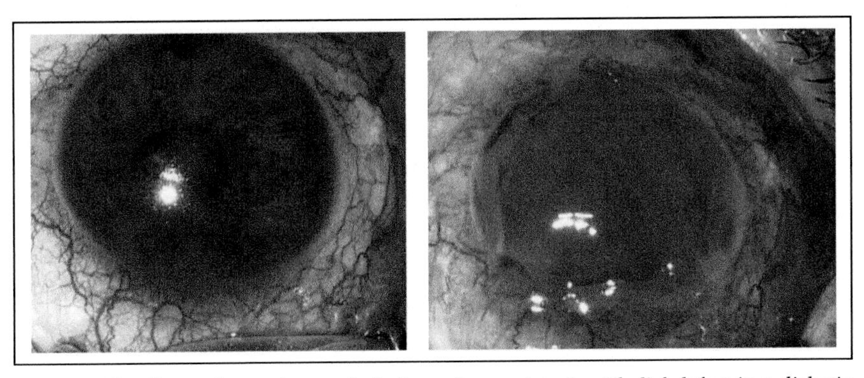

Figure 2 This figure shows the surgical view of a persistent epithelial defect in a diabetic patient after a pars plana vitrectomy before and after the amniotic membrane transplantation with a 10/0 nylon buried running suture technique.

Color image of this figure appears in the color plate section at the end of the book.

AM within 2 weeks of injury suppressed inflammation and promoted early epithelialization in mild to moderate disease states, however in severe burns, limbal stem cell (LSC) deficiency is not prevented and additional LSC transplantation is required, but the ocular surface condition is much better (Meller et al. 2000).

Amniotic membrane as a carrier for transplantation of expanded limbal stem cells ex vivo. This new indication to use the AM as a carrier for limbal grafts will be mentioned below.

Limitations. AM is a substrate transplantation and an excellent carrier for anti-inflammatory molecules, but can not treat by itself complete stem cells deficiency, severe dryness, neurotrophic states or ischemias that prevent the host tissue to supply the ingredients for ocular surface restoration.

Ocular Surface Reconstruction Procedures

The limbus, or corneoscleral junction, is the point at which the cornea becomes continuous with the sclera. Limbal epithelial stem cells (LESC) are believed to reside in the basal layer of this area. LESCs share common features with other adult somatic stem cells including small size and high nuclear to cytoplasmic ratio. However, no reliable and specific marker for LESC exists yet, but multiple negative and positive markers have been identified (Schlötzer-Schrehardt and Kruse 2005). LESC depend on specialized niches to maintain them in an undifferentiated state and regulate their function. Greater knowledge regarding stem cell behaviour, surgical advances and the tissue engineering technique has led a new approach in OSD.

The clinical signs of limbal stem cell deficiency (LSCD) include conjunctivalization of the cornea (the hallmark sign), vascularization, fibrovascular pannus, and often persistent epithelial defects and scarring (Dua et al. 2003).

LSCD can be classified on unilateral or bilateral, partial or total and congenital or acquired deficiency. A number of well-established surgical approaches, depending on several factors, have been adopted to deliver LESCs with the goal of restoring normal ocular surface function (Table 1).

Sequential Sectoral Conjunctival Epitheliectomy

The objective of sequential sectoral conjunctival epitheliectomy (SSCE) is removing conjunctival epithelium covering an area of the cornea or limbus and preventing from crossing the limbus until the denuded surface is covered by corneal epithelium derived cells (Dua 1998).

It is well established that large ocular surface epithelial defects, involving the cornea heal by centripetal migration of epithelial cells from remaining intact corneal epithelium and by circumferential migration of limbal epithelial cells along the limbus, arising from the two ends of the remaining intact limbal epithelium (Dua and Forrester 1990). However, often, the centripetally migrating healing conjunctival epithelium reaches the limbus and crosses it to cover the cornea. This conjunctival epithelial sheet contact inhibits the circumferentially migrating limbal epithelium, and the area of limbus covered by it without limbal stem cells (Dua and Forrester 1990). SSCE is performed to prevent conjunctival epithelium from reaching the limbus. It is effectively held back until repopulation of the limbus from surviving limbal cells is achieved. Similarly, if conjunctivalization has already occurred, the epithelium over the affected area of the cornea and limbus is removed to allow healing from healthy limbus-derived cells.

The same principles will apply following resection of limbal lesions and other surgical procedures performed during ocular surface reconstruction. The epithelial defect generally resolves over several days. Because this procedure does not repair the actual limbal stem cells, some abnormal conjunctival epithelial cells will also appear on the periphery. If the symptoms have not improved enough after one treatment, the procedure can be repeated (Dua et al. 2010).

Kerato-conjunctival Limbal Grafts

Limbal tissue is used to manage limbal deficiencies. Moderate or severe diseases of conjunctiva can condition the use of conjunctival tissue as well (Table 1). The technique uses grafts of bulbar conjunctiva that extend approximately 0.5 mm onto the clear cornea centrally, thus containing limbal cells. The area harvested is subsequently placed onto the already cleared area of limbal deficiency. The origin of the graft can be autologous in cases of unilateral disease, obtained from the contralateral eye or an allograft coming from cadaveric, living relative or living not relative donors. Allografts must be treated with topical and systemic immunomodulators to keep the graft viability. Long-term survival depends on the plasticity of host´s stem cells coming from the bone marrow to colonize the niche. However, active inflammation and severe OSD due to systemic factors can make the graft fail (Dua et al. 2010).

Ex vivo Expansion of Stem Cells

This procedure uses autologous or allogenic expanded stem cells grown on human amniotic membrane to restore the ocular surface. The limbal specimen is obtained from the other healthy eye (autologous) in unilateral cases, or from allogenic origin (from cadaver donor or living relative or non-relative donor) in bilateral disease (Dua et al. 2010). The piece of limbal tissue is subsequently cultivated and expanded on amniotic membrane *in vitro* always according to the guidelines for good manufacturing practice (GMP) (Bosse et al. 1997), with strictly regulated procedures and high quality control tests to manipulate stem cells as "medicinal products". The transplantation is performed on the denuded corneal surface at the disordered eye, after superficial keratectomy to remove fibrovascular ingrowth (Fig. 3). This procedure improves the ocular surface condition, but in many cases an additional corneal graft is needed for visual rehabilitation. The cost and technical difficulties currently limit the availability of this procedure.

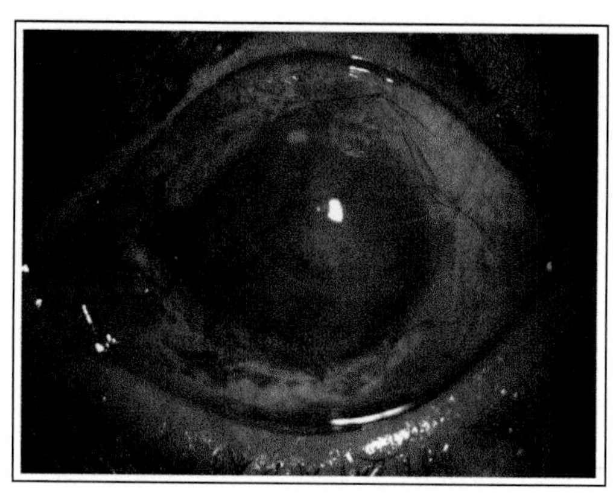

Figure 3 This figure shows the *ex vivo* expanded limbal graft carried in amniotic membrane covered by a great diameter therapeutic contact lens.

Color image of this figure appears in the color plate section at the end of the book.

Keratoplasties

After ocular surface stabilization, consideration of a subsequent keratoplasty can be entertained. Different options can be performed. Significant stromal scarring with good endothelial function allows an anterior lamellar keratoplasty.

A posterior lamellar transplantation is possible after endothelial failure with a preserved normal stroma. In patients with stromal and endothelial disease, penetrating keratoplasty is required for visual rehabilitation. Finally, some special conditions demand keratoprosthesis as the last solution.

Penetrating Keratoplasty

Penetrating keratoplasty (PK) has been the surgical treatment of choice for visual rehabilitation in corneal disease. PK involves full-thickness replacement of the cornea, and is a highly successful procedure and is still the most common type of corneal transplant surgery performed worldwide today (Krachmer et al. 2010).

Surgical technique has been widely described and in experienced groups has excellent results. Development of topical steroid eye drops (to prevent and treat corneal graft rejection) led to greatly enhanced PK graft survival rates and antiviral treatment improved as well the results in herpetic keratitis.

However, problems like astigmatism and poor long-term survival rates led to considering other procedures. The cornea is anatomically a multi-layered structure, and disease may only affect selective layers, hence individual lamellar surgical replacement of only the pathological corneal layers, while retaining unaffected layers represents the new challenge.

Lamellar Keratoplasties

Lamellar keratoplasty is a surgical method for partial transplantation of donor corneal tissue after dissection of the equivalent part of the diseased host corneas. Refinements in microsurgical technique have improved the problems that appeared in the first years of the technique. However, visual quality of vision was often impaired by scarring within host-recipient stroma interface, or by incongruence of graft and host bed. Lamellars has distinct advantages over PK (Table 4).

However, the risk of rejection still remains but with lower incidence and is limited to the grafted tissue. These characteristics condition new patterns of rejection that the ophthalmologist should learn to recognize.

Lamellar keratoplasties can be subdivided in anterior (superficial and deep) and posterior.

Table 4 Main advantages of lamellar surgery over penetrating keratoplasty (PK).

Main advantages of lamellar surgery over PK
Fewer intraocular complications.
No risk of endothelial rejection in anterior lamellar, or epithelial and stromal rejection in posterior lamellar.
No need for long-term immunomodulation and therefore a decreased risk of infection, glaucoma, and cataract in the anterior lamellar.
Superior wound strength.
Fewer rigid criteria for donor corneal tissue selection depending on harvested lamella.
Better prognosis if a second graft is needed.
Less amount of astigmatism in posterior lamellar.
Duplication of tissue (Split techniques).

Superficial anterior lamellar keratoplasty (s-ALK). For superficial stromal disorders a microkeratome or femtosecond laser assisted lamellar resection from 100 to 400 μm stromal depth followed by transplantation of a donor lamella of the same dimensions onto the recipient bed can be performed. The automated or laser assisted nature of the dissection results in smoother and enhanced high quality as compared to manual dissection. The major disadvantage of this procedure is the possibility of sub-optimal visual outcomes due to interface related problems. Conditions such as anterior stromal scarring following infectious keratitis or corneal stromal dystrophies affecting superficial layers are the main indications of this procedure (Shousha et al. 2011).

Deep anterior lamellar keratoplasty (DALK). DALK can be performed in most diseases treated by PK in which there is a healthy corneal endothelium. The principle of this technique is to remove all recipient stromal tissue to Descemet's membrane (DM) and place a donor cornea over the bed after DM and endothelium removal. The main parameter for good visual function after DALK is the thickness of residual recipient stromal bed (Fig. 4) (Ardjomand et al. 2007). The drawback of DALK is that deep dissection is time consuming and relatively difficult with a steep learning curve.

Two main techniques are performed to get this aim. Melles reported a technique for effective tissue dissection using air injection in the anterior chamber. The difference in optical index between cornea and anterior chamber filled with air makes the dissection easier (Melles et al. 1999). Anwar described the "big bubble technique" consisting in forceful intra-stromal air injection to split DM away from the posterior stroma. Clean

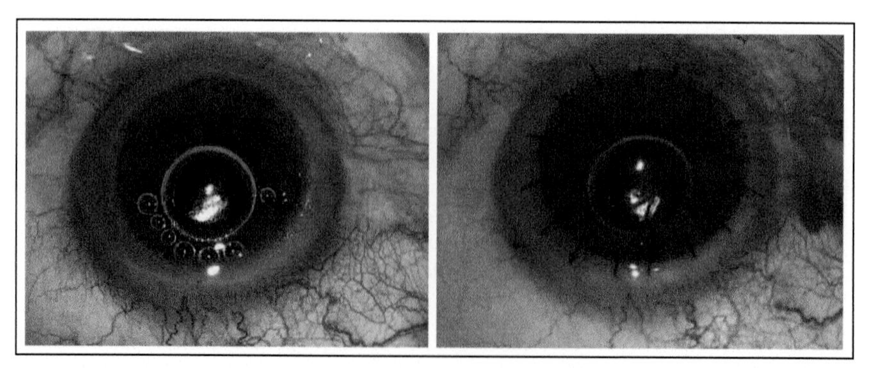

Figure 4 Left: Surgical view of the bared Descemet membrane after deep lamellar dissection. Small bubbles of air show the integrity of the anterior chamber. Right: Graft sutured with 10/0 nylon interrupted suture technique. Coloured graft is due to the use of trypan blue in the harvest of Descemet-endothelium complex for a DMEK with the split technique.

Color image of this figure appears in the color plate section at the end of the book.

separation between posterior stroma and DM due to the forceful injection results in a higher quality of vision (Anwar and Teichmann 2002). However, irregular astigmatism is still the main problem after a successful surgical procedure and this event significantly influences the visual outcome.

DALK must be the first option in severe OSD with a healthy endothelium to avoid endothelial rejection and because sometimes these diseases require an additional graft to keep the optical axis clear.

Posterior keratoplasty. Corneal conditions such as; aphakic and pseudophakic bullous keratopathy, graft failure, and Fuchs dystrophy may cause epithelial and stromal oedema due to endothelial cell dysfunction.

Corneal transplantation has advanced rapidly during the last decade in these conditions with a shift from full-thickness PK procedures to partial-thickness posterior lamellar graft. Surgical procedures have evolved from deep lamellar endothelial keratoplasty (DLEK) to Descemet stripping endothelial keratoplasty (DSEK), or Descemet stripping automated endothelial keratoplasty (DSAEK) to Descemet membrane endothelial keratoplasty (DMEK). The last two are the most commonly applied posterior corneal lamellar procedures. DSAEK consists in removing the diseased host endothelium and DM through a small incision similar to a cataract surgery and replacing the recipient's diseased endothelium and DM with a donor endothelium/DM/thin lenticle of posterior stroma of between 50 and 250 μm depending on the donor harvesting technique used. The thickness of the stromal lamella injected in DSAEK seems to influence the final visual acuity by alteration of the posterior curvature of the cornea. The thinner the graft, the better are the functional results. Thus, the invention of DMEK for the sole replacement of the endothelium–DM layer intends to circumvent potential problems associated with DSAEK (Fig. 5). DMEK shows spectacular results in visual outcome but donor preparation and intraocular manipulation limits the widespread use of this cheaper technique (Dapena et al. 2009).

However, these techniques can only be performed in a few cases of severe OSD due to the common condition of the corneal stroma that requires PK.

Globally, one of the most important advantages of the lamellar procedures is the concept of using one donor cornea for the treatment of multiple patients (split technique) to cope with the shortage of donor corneas. Here, a good-quality donor cornea can be divided into three parts to obtain the peripheral corneoscleral rim with viable limbal stem cells for cadaveric or *ex vivo* expanded limbal stem-cell transplantation; a whole or partial anterior lamellar disc for DALK or superficial lamellar keratoplasty; and a posterior lamellar disc for DMEK or DSAEK (Heindl et al. 2011).

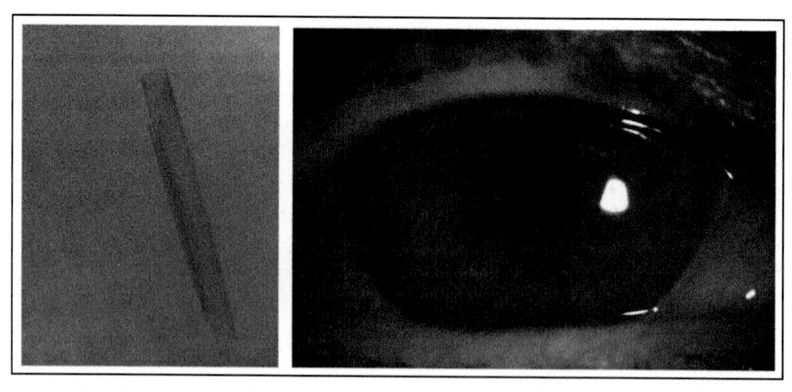

Figure 5 Left: This figure shows the endothelium-Descemet membrane roll after harvesting from a donor cornea coloured by trypan blue. Right: Clinical photo 6 mon after DMEK. Note small-pigmented fibrotic peripheral folds in the limit of the graft with a transparent cornea.

Color image of this figure appears in the color plate section at the end of the book.

Keratoprosthesis

The history of keratoprosthesis began with Pellier de Quengsy in 1789, who, first proposed that an artificial cornea could be implanted in place of a natural cornea opacified by disease or infection (Pellier de Quengsy 1789). Given the limitations of corneal transplantation, the need for a safe and effective alternative to corneal allograft is obvious. Candidates for keratoprosthesis implantation can be classified into three main groups: autoimmunity-related corneal opacity and ulceration, chemical injury, and corneal allograft failure.

Although many keratoprosthesis designs have been proposed in recent years, two devices that are most commonly implanted: the osteo-odonto or tibial bone keratoprosthesis (OO or TBKP) and the Boston keratoprosthesis (KPro™) (Fig. 6).

Each of them has its strengths and weaknesses in specific clinical conditions, and therefore the indications for each device are different. The main characteristics of these devices are described in Table 5 (Krachmer et al. 2010).

CONCLUSION

The management of the ocular surface disease has benefited from different and several new advances in microsurgery, reintroduction of amniotic membrane transplantation, developments of new materials, new perspectives in pharmacological immunomodulation and the critical understanding of the location and role of the limbal stem cells.

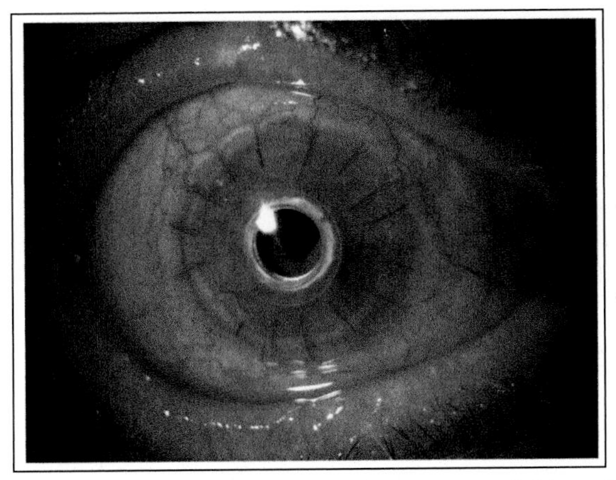

Figure 6 This figure shows the Boston Keratoprosthesis in an aphakic patient with three previous failed penetrating keratoplasties.

Color image of this figure appears in the color plate section at the end of the book.

Table 5 Mean indications and characteristics of keratoprosthesis. OOKP: Osteo-odonto keratoprosthesis. TBKP: Tibial bone keratoprosthesis. KPro: Boston Keratoprosthesis.

Type of keratoprosthesis	Indications	Main advantages	Main disadvantages
OOKP **TBKP**	Chemical injury Severe autoimmune disorder	Ability to be retained despite conditions of severe dry eye	The most invasive and technically difficult Insertion requires two surgeries Worst cosmesis
KPro	Corneal allograft failure Chemical injury Severe autoimmune disorder	Type I, performed in one stage Type I, technically similar to standard corneal graft	Type II, insertion requires two surgeries

However, corneal graft is still the last step in visual rehabilitation after ocular surface reconstruction, and represents the main and the most extended technique among these procedures, but even in this case, new sight has come with the improvements in lamellar surgery.

ACKNOWLEDGMENTS

We would like to thank all the members of the Ocular Surface Units at Hospital Clinico Universitario de Valladolid and Instituto Universitario de Oftalmobiología Aplicada (IOBA Eye Institute).

REFERENCES

Anwar, M. and K.D. Teichmann. 2002. Big-bubble technique to bare Descemet's membrane in anterior lamellar keratoplasty. J. Cataract Refract Surg. 28: 398–403.

Ardjomand, N., S. Hau, J.C. Mcalister, C. Bunce, D. Galarreta, S.J. Tuft and D.F.P. Larkin. 2007. Quality of Vision and Graft Thickness in Deep Anterior Lamellar and Penetrating Corneal Allografts. Am. J. Ophthalmol. 143: 228–235.

Atrah, H.I. 1994. Fibrin glue. BMJ. 308: 933–4.

Bosse, R., M. Singhofer-Wowra, F. Rosenthal and G. Schulz. 1997. Good manufacturing practice production of human stem cells for somatic cell and gene therapy. Stem Cells 15: 275–80.

Cavanaugh, T.B. and J.D. Gottsch. 1991. Infectious keratitis and cyanoacrylate adhesive. Am. J. Ophthalmol. 111: 466–72.

Cederholm-Williams, S.A. 1994. Fibrin glue. Br. Med. J. 308: 1570.

Chan, S.M. and H. Boisjoly. 2004. Advances in the use of adhesives in ophthalmology. Curr. Opin. Ophthalmol. 15: 305–10.

Chen, H.J., R.T. Pires and S.C. Tseng. 2000. Amniotic membrane transplantation for severe neurotrophic corneal ulcers. Br. J. Ophthalmol. 84: 826–33.

Dapena. I., L. Ham and G.R. Melles. 2009. Endothelial keratoplasty: DSEK/DSAEK or DMEK the thinner the better? Curr. Opin. Ophthalmol. 20: 299–307.

Daya, S.M., C.C. Chan and E.J. Holland. 2011. Cornea Society Nomenclature for Ocular Surface Rehabilitative Procedures. Cornea 30: 1115–1119.

Dua, H.S. 1998. The conjunctiva in corneal epithelial wound healing. Br. J. Ophthalmol. 82: 1407–11.

Dua, H.S. and J.V. Forrester. 1990. The corneoscleral limbus in human corneal epithelial wound healing. Am. J. Ophthalmol. 110: 646–56.

Dua, H.S., A. Joseph, V.A. Shanmuganathan and R.E. Jones. 2003. Stem cell differentiation and the effects of deficiency. Eye 17: 877–85.

Dua, H.S., A. Miri and D.G. Said. 2010. Contemporary limbal stem cell transplantation—a review. Clin. Experiment Ophthalmol. 38: 104–17.

Dua, H.S., J.A. Gomes, A.J. King and V.S. Maharajan. 2004. The amniotic membrane in ophthalmology. Surv. Ophthalmol. 49: 51–77.

Gomes, J.A., A. Romano, M.S. Santos and H.S. Dua. 2005. Amniotic membrane use in ophthalmology. Curr. Opin. Ophthalmol. 16: 233–40.

Heindl, L.M., S. Riss, B.O. Bachmann, K. Laaser, F.E. Kruse and C. Cursiefen. 2011. Split cornea transplantation for 2 recipients: a new strategy to reduce corneal tissue cost and shortage. Ophthalmology 118: 294–301.

Holland, E.J. and G.S. Schwartz 2004. The Paton lecture: Ocular surface trasplantation: 10 years´ experience. Cornea 23: 425–31.

Kaufman, H.E., M.S. Insler, H.A. Ibrahim-Elzembely and S.C. Kaufman. 2003. Human fibrin tissue adhesive for sutureless lamellar keratoplasty and scleral patch adhesion: a pilot study. Ophthalmology 110: 2168–72.

Kim, J.C. and S.C. Tseng. 1995. Transplantation of preserved human amniotic membrane for surface reconstruction in severely damaged rabbit corneas. Cornea. 14: 473–84.

Krachmer, J.H., M.J. Mannis and E.J. Holland. 2010. Cornea. St. Louis, Mo., London. Mosby. UK.

Kruse, F.E., K. Rohrschneider and H.E. Volcker. 1999. Multilayer amniotic membrane transplantation for reconstruction of deep corneal ulcers. Ophthalmology 106: 1504–10.

Lee, S.H. and S.C. Tseng. 1997. Amniotic membrane transplantation for persistent epithelial defects with ulceration. Am. J. Ophthalmol. 123: 303–12.

Li, H., J.Y. Niederkorn, S. Neelam, E. Mayhew, R.A. Word, J.P. McCulley and H. Alizadeh. 2005. Immunosuppressive factors secreted by human amniotic epithelial cells. Invest Ophthalmol. Vis. Sci. 46: 900–7.

Meller, D., R.T. Pires, R.J. Mack, F. Figueiredo, A. Heiligenhaus, W.C. Park, P. Prabhasawat, T. John, S.D. McLeod, K.P. Steuhl and S.C. Tseng. 2000. Amniotic membrane transplantation for acute chemical or thermal burns. Ophthalmology 107: 980–9.

Melles, G.R., F.J. Rietveld, W.H. Beekhuis and P.S. Binder. 1999. A technique to visualize corneal incision and lamellar dissection depth during surgery. Cornea 18: 80–6.

Pellier de Quengsy, G. 1789. Précis ou cours d'operations sur la chirurgie des yeux. Montpellier: Rigaut, Roullet. Paris, Didot, Mequignon 94.

Radosevich, M., H.I. Goubran and T. Burnouf. 1997. Fibrin sealant: scientific rationale, production methods, properties, and current clinical use. Vox Sang 72: 133–43.

Sharma, A., R. Kaur, S. Kumar, P. Gupta, S. Pandav, B. Patnaik and A. Gupta. 2003. Fibrin glue versus N-butyl-2-cyanoacrylate in corneal perforations. Ophthalmology 110: 291–8.

Shousha, M.A., S.H. Yoo, G.D. Kymionis, T. Ide, W. Feuer, C.L. Karp, T.P. O'Brien, W.W. Culbertson and E. Alfonso. 2011. Long-term results of femtosecond laser-assisted sutureless anterior lamellar keratoplasty. Ophthalmology 118: 315–23.

Schlötzer-Schrehardt, U. and F.E. Kruse. 2005. Identification and characterization of limbal stem cells. Exp. Eye Res. 81: 247–64.

Tseng, S.C., P. Prabhasawat and S.H. Lee. 1997. Amniotic membrane transplantation for conjunctival surface reconstruction. Am. J. Ophthalmol. 124: 765–74.

Section IV: Ocular Surface and Contact Lens

16

Contact Lenses Definitions

Raul Martin Herranz,[a,*] *Guadalupe Rodriguez Zarzuelo* and *Victoria de Juan Herraez*

SUMMARY

This chapter provides a description of some main aspects related to contact lens wear; including common types of contact lenses, the materials used to manufacture these and their subsequent properties; and some new applications of contact lenses.

INTRODUCTION

There are different materials with which to manufacture contact lenses. Depending on the physical proprieties of the material (and the manufacturing process), different types of contact lenses are available to correct refractive errors (myopia, hyperopia, astigmatism and presbyopia) and to manage other ocular conditions (specially irregular astigmatism and others). In these special cases, contact lenses are often the only option to improve the patient's vision.

The variety of lenses available, materials, and designs permit the wearer and the practitioner a wider range of options to choose from in order to satisfy the wearer's needs.

Instituto Universitario de Oftalmobiología Aplicada (IOBA-Eye Institute), Optometry Research Group, School of Optometry, University of Valladolid. Paseo de Belén, 17-Campus Miguel Delibes, 47011 Valladolid, Spain.
[a]E-mail: raul@ioba.med.uva.es
*Corresponding author

CONTACT LENS WEAR

More than 100 million people wear contact lenses in the world; the main contact lens indication is the correction of ametropia (myopia, hyperopia and astigmatism). However, contact lenses can also be fitted with other intent, to improve the visual function, especially in patients with irregular corneae likes keratoconus, after complicated corneal surgery or in ocular injuries. Other different uses of contact lens are discussed in chapter 18.

In general terms, contact lenses are prescribed with different replacement schedules and different wearing schedules (Dillehay and Allee 2000).

According to replacement frequency, lenses may be classified as: *disposable lenses* (intended for single use), *frequent-replacement lenses* (when lenses are removed from the eye, are cleaned and are reused for a limited time, then disposed). Many manufacturers and practitioners recommend a specific replacement time, for example daily disposable (for one only use), weekly, fortnightly, monthly, three-, six-monthly or other depending the wearer characteristics, lens material and other factors.

On the other hand, according to the wearing schedule of the lenses; there are four main categories: *daily wear* (when the lenses are worn during the day and they are removed before sleeping, without overnight wear), *extended wear* (when the lenses are worn during the day and during sleep, for periods no longer than six consecutive nights before their removal), *continuous wear* (when the lenses are worn for up to 30 consecutive nights without removal from the eye), and *flexible wear* (when the lenses are mainly for daily wear with an occasional overnight use or during sleep, for example 2–3 nights per week or occasional nap with the lens in place).

Contact lenses must always be fitted and prescribed by a practitioner after an eye examination to determine if the patient is suitable for contact lens wear in order to minimize the risk of future contact lens complications. If a person is deemed suitable to wear contact lenses, the practitioner will recommend the most appropriate lens material, design, wearing and replacement schedule, care regimen and a follow-up schedule.

Also, every contact lens wearer needs to know how to handle the lenses for insertion and removal, how to clean and disinfect the lenses and the lens case correctly. The patient should be educated on the limitations of contact lens wear, the problems that can occur through non-compliance, and on the signs and symptoms of infection and inflammation of the eye.

CONTACT LENS MATERIALS

The procedure to identify the materials to hard and soft contact lenses is described in the international BS EN ISO11539:1999 (Gasson and Morris 2003). Each material is classified by a six-part code: with a prefix, a stem, a series suffix, the oxygen permeability (Dk) range and the code of surface modification (Table 1).

For example the material Enflufocon B III 3 (Boston EO, Polymer Technology) used the USAN prefix "Enflu", the stem focon indicates a hard material (<10% of water), the USAN B indicates the second formulation of this polymer, the Group suffix III, indicates that this material contains silicon and fluorine and finally the number 3 said that it has a Dk in the range of 31 to 60 ISO units.

Table 1 This table summarize the material classification (modify from Gasson and Morris 2003). Dk = the oxygen permeability.

Prefix	This is the first part of the material name administered by United States Adopted Names (USAN). For example, material Lotrafilcon A has the USAN code *"lotra"*.		
Stem	This is de last part of the material name and used; *falcon* for soft contact lenses (hydrogels with >10% of water content by mass) and *focon* for rigid lenses materials (<10% of water content).		
Series Suffix	The USAN also provides a series suffix with a capital letter added to the stem to indicate the revision level of the chemical formula of each material. For example, capital letter A means the original formulation, B the second and so on. This can be omitted if there is one formulation only.		
Group Suffix	The group suffix expressed with roman numbers different properties if is a rigid or soft contact lens:		
		Rigid Lenses	**Soft Lenses**
	I	Without silicon or fluorine	<50% of H_2O, non-ionic
	II	With silicon but without fluorine	≥50% of H_2O, non-ionic
	III	With both silicon and fluorine	<50% of H_2O, ionic
	IV	With fluorine but without silicon	≥50% of H_2O, ionic
Dk Range	This is a numerical code which identifies the permeability to the oxygen in determinates ranges defined to contact lens wear. Dk units are: (cm^2/s) *[ml O_2/ml * hPa)]. Dk value is 0 if <1Dk units; 1 between 1 to 15 units; 2 for 16 to 30 units; 3 for 31 to 60 units; 4 to 61 to 100 units; 5 to 101 to150 units; 6 to 151 to 200 units and new categories in increments of 50 Dk units.		
Modification Code	To inform that the surface of the lens have been modified with different chemical characteristics from the original material a lower case m is used.		

CONTACT LENSES WITHOUT WATER

There is a large group of polymers used to manufacture contact lenses that do not incorporate water in their final state. These materials are called *rigid or hard lenses* (non gas-permeable lenses manufactured with polymethyl methacrylate-PMMA) and also *rigid or hard gas permeable lenses* (lenses manufactured with oxygen permeable polymers), which are most popular because there are a wide range of oxygen permeability materials available nowadays.

Polymethyl Methacrylate (PMMA) Contact Lenses

Polymethyl methacrylate (PMMA) was the first material used for the manufacture of hard contact lenses to replace the original glass scleral designs in the 1940s.

The PMMA contact lenses were used for over 40 years but have declined in popularity due to its gas impemeability and hypoxic effects. However, some contact lens wearers could benefit from it's good surface wettability and ease of painting upon for cosmetic use.

Hard Gas-permeable Contact Lenses

The hard gas-permeable contact lens is the generic name of a large variety of contact lenses manufactured with many different materials that do not contain water.

The main materials used in hard gas-permeable contact lenses are the *cellulose acetate butyrate (CAB)* with good wettability (Shiobara et al. 1989, Gasson and Morris 2003) and performance but low Dk, some corneal adhesion problems with variations in corneal curvature (Briceño-Garbi 1984) and high lens flexure (Harris et al. 1982); *the silicon acrylates* that are copolymers in varying proportions of acrylate (PMMA) with good performance in daily wear, low to medium Dk and good dimensional stability but protein deposit and breakage problems (Gasson and Morris 2003); *the fluorosilicon acrylates,* sometimes called fluorocarbons, improve the lens wettability and deposit resistance (Shiobara et al. 1989, Gasson and Morris 2003) with high Dk values that permit longer wear. However, this material is brittle and requires careful patient manipulation. *Fluoropolymers* incorporate more than 50% by weight of fluorine without silicon and obtain a good surface wettability with high Dk but several problems related with the lens flexure (Gasson and Morris 2003). For these reasons, fluopolymers have a limited application today.

Silicone Contact Lenses

Silicone-rubber based flexible contact lenses (silicone-elastomeric lenses) were used for therapeutic, aphakic and paediatric applications for many years (Gurland 1979). These lenses offer a high Dk value and durability, but present important limitations that limit their clinical practice, with hydrophobic lens surfaces that facilitate the lipid deposition (Huth and Wagner 1981) and risks of binding to the cornea (Rae and Huff 1991).

CONTACT LENSES WITH WATER

There is a wide range of water-incorporating polymers used to manufacture contact lenses. Many of them are derived polymer of hydroxyethyl methacrylate (HEMA) but others have a different chemical formulation (silicone hydrogels). In all of these materials the water content is expressed to describe the percentage of water that each polymer takes under specified conditions.

Hydrogel (Soft) Contact Lenses

Hydrogel or soft lenses have been produced with water contains from 18% to 85% water and many of these lenses still currently use HEMA (hydroxyethyl methacrylate) which is the first hydrogel material used for soft contact lens manufacture in the 1960's (Wichterle and Lim 1960).

The main advantages of the soft contact lens are related with the initial comfort versus that of hard contact lenses wear. Also, these soft lenses have great reproducibility and are easier to manufacture that permit cheaper lenses useful to disposable wear with frequent replacement from daily, weekly, fortnightly, monthly or other depending the wearer characteristics and practitioner's recommendations.

On the other hand, conventional hydrogel soft contact lenses present some disadvantages mainly related with their relatively low Dk values that induces corneal oedema (Bonnano 2001, Holden and Mertz 1984, Martin et al. 2007), corneal vascularization in long-term wearers (Holden et al. 1985, Weissman and Chan 1996) especially with high contact lens powers (Martin 2007b) and other complications (Suchecki 2003) in a small percentage of soft contact lenses wearers (Chalmers et al. 2010).

Another disadvantage with hydrogels is the dehydratation of the lens (specially in high water content lenses) that induces signs and symptoms of dryness; this is one of the most common reasons for discontinuation of contact lens use (Richdale 2007).

Silicone Hydrogel Contact Lenses

Silicone hydrogel material represents a new era in contact lens technology, because it combines silicone rubber with hydrogel monomers (Gasson and Morris 2003) to provide unique properties. These lenses first appeared commercially in 1998 and since then have become a popular lens choice.

The name of these materials maintains the suffix-*filcon* following the USAN codification. The silicone component permits high levels of oxygen permeability and the hydrogel still provides confortable wear. However, because silicone is a hydrophobic material the lens manufacture requires a 'plasma' surface treatment to render biocompatibility and wettable surface (Lai and Friends 1997).

The main advantage of these lenses is a high Dk value that permits consideration for extended wear and still allows an easy adaptation process similar to conventional soft lenses (Martin and Alonso 2010). However, these lenses demonstrate some complications related with the higher modulus (the modulus of a material describes how well it resists deformation) that can induce superior epithelial arcuate lesions (SEAL), epithelial toxicity with determinate preservative solutions (specially polyaminopropyl biguanide) (Jones et al. 2002), and formation of mucin balls (Miller et al. 2003).

SPECIALIST USES OF CONTACT LENSES

While the main reason to prescribe contact lenses is to correct myopia, hyperopia, astigmatism or presbyopia there are different clinical situations that requires the use of contact lens to improve the visual function further than that which is obtained with conventional spectacles (Rabinowitz 1998).

These include patients with irregular cornea due to corneal pathologies like keratoconus, Pellucid marginal degeneration and others (Rabinowitz 1998, Zadnik et al. 1998). In these cases a special contact lens may permit an improvement of the vision, special rigid gas-permeable lenses would be the first option to patient management (Rabinowitz 1998, Zadnik et al. 1998, McMonnies 2005).

Also, patients with irregular cornea after ocular surgery (corneal refractive surgery like PRK or LASIK, corneal transplantation, etc.) or after severe eye injury should beneficiate of the use of contact lenses to improve their vision (Martin and Rodriguez 2005, Martin et al. 2011), especially with hard gas-permeable contact lens.

The therapeutic uses of contact lenses are described in chapter 18 with more detail.

Recently, different contact lens sensors (Leonardi 2004, Sanchez et al. 2011) have been proposed to non-invasive measurement and continuous assessment of the intraocular pressure that could permit in a near future news contact lens application to eye care.

CONCLUSIONS

Contact lenses are a safe and common method used to correct the refractive errors with millions of wearers worldwide. However, different polymers with different properties and indications are available and all should be considered. The correct combination between material, cleaning solutions, replacement frequency, wearing schedule and appropriate aftercare is required to choose the most appropriate lens for each wearer and a safe wearing system with minimum ocular surface disturbances.

REFERENCES

Bonnano, J.A. 2001. Effects of contact lens-induced hypoxia on the physiology of the corneal endothelium. Optom. Vis. Sci. 78: 783–790.

Briceño-Garbi, E.A. 1984. Variations in corneal curvature and refractive error in CAB gas-permeable contact lens wearers. J. Am. Optom. Assoc. 55: 217–9.

Chalmers, R.L., L. Keay, B. Long, P. Bergenske, T. Giles and M.A. Bullimore. 2010. Risk factors for contact lens complications in US clinical practices. Optom. Vis. Sci. 87: 725–35.

Dillehay, S.M. and V. Allee. 2000. Material selection. In: Bennet E.S. and V. Allee. [eds.]. Clinical manual of contact lenses. Lippicontt Williams & Wilkins, Philadelpia. USA, pp. 239–258.

Gasson, A. and J. Morris. 2003. The contact lens manual. A practical guide to fitting. Butterworth-Heinemann, London. UK

Gurland, J.E. 1979. Use of silicone lenses in infants and children. Ophthalmology. 86: 1599–1604.

Harris, M.G., K.E. Sweeney, S. Rocchi and D. Pettit. 1982. Flexure and residual astigmatism with cellulose acetate buterate (CAB) contact lenses on toric corneas. Am. J. Optom. Physiol Opt. 59: 858–62.

Holden, B.A. and G.W. Mertz. 1984. Critical oxygen levels to avoid corneal edema for daily and extended wear contact lenses. Invest Ophthalmol. Vis. Sci. 25: 1161–1167.

Holden, B.A., D.F. Sweeney, A. Vannas, K.T. Nilsson and N. Efron. 1985. Effects of long-term extended contact lens wear on the human cornea. Invest Ophthalmol. Vis. Sci. 26: 1489–1501.

Huth, S. and H. Wagner. 1981. Identification and removal of deposits on polydimethylsiloxane silicone elastomer lenses. Int. Contact Lens Clin. 19–26.

Jones, L., N. MacDougall and L.G. Sorbara. 2002. Asymptomatic corneal staining associated with the use of balafilcon silicone-hydrogel contact lenses disinfected with a polyaminopropyl biguanide-preserved care regimen. Optom. Vis. Sci. 79: 753–61.

Lai, Y.C. and G.D. Friends. 1997. Surface wettability enhancement of silicone hydrogel lenses by processing with polar plastic molds. J. Biomed. Mater Res. 35: 349–56.

Leonardi, M., P. Leuenberger, D. Bertrand, A. Bertsch and P. Renaud. 2004. First steps toward noninvasive intraocular pressure monitoring with a sensing contact lens. Invest Ophthalmol. Vis. Sci. 45: 3113–3117

McMonnies, C.W. 2005. The biomechanics of keratoconus and rigid contact lenses. Eye Contact Lens 31: 80–92

Martin, R. 2007. Corneal conjunctivalization in long standing contact lenses wearers. Clin. Exp. Optom. 90: 26–30.

Martin, R. and E. Alonso. 2010. Comparison of the number of visits and diagnostic lenses required to fit RGP, conventional hydrogel and silicone hydrogel contact lenses. J. Optom. 3: 169–174.

Martin, R. and G. Rodriguez. 2005. Reverse geometry contact lens fitting after corneal refractive surgery. J. Refract. Surg. 21: 753:756

Martin, R., V. de Juan, G. Rodríguez, R. Cuadrado and I. Fernandez. 2007. Measurement of corneal swelling variations without removal of the contact lens during extended wear. Invest Ophthalmol. Vis. Sci. 48: 3043–50.

Martin, R., G. Rodriguez and V. de Juan. 2011. Contact lens correction of regular and irregular astigmatism. In: Goggin M [ed.]. Astigmatism. In Tech, Rijeka. Croatia, pp. 1–24 (in press).

Miller, T.J., E.B. Papas, J. Ozkan, I. Jalbert and M. Ball. 2003. Clinical appearance and microscopic analysis of mucin balls associated with contact lens wear. Cornea 22: 740–745.
Rabinowitz, Y. 1998. Keratoconus. Surv. Ophthalmol. 42: 297–319.
Rae, S., J. Huff. 1991. Studies on initiation of silicone elastomer lens adhesion *in vitro*: binding before the indentation ring. CLAO J. 17: 181–186.
Richdale, K., L.T. Sinnott, E. Skadahl and J.J. Nichols. 2007. Frequency of and factors associated with contact lens dissatisfaction and discontinuation. Cornea 26: 168–74.
Sánchez, I., V. Laukhin, A. Moya, R. Martin, F. Ussa, E. Laukhina, A. Guimera, R. Villa, C. Rovira, J. Aguiló, J. Veciana and J.C. Pastor. 2011. Prototype of a nanostructured sensing contact lens for noninvasive intraocular pressure monitoring. Invest Ophthalmol. Vis. Sci. 52: 8310–5.
Shiobara, M., C.M. Schnider, A. Back and B.A. Holden. 1989. Guide to the clinical assessment of on-eye wettability of rigid gas-permeable lenses. Optom. Vis. Sci. 66: 202–6.
Suchecki, J.K., P. Donshik and W.H. Ehlers. 2003. Contact lens complications. Ophthalmol. Clin. North Am. 16: 471–84.
Weissman, B.A. and W.K. Chan. 1996. Corneal pannus associated with contact lens wear. Am. J. Ophthalmol. 121: 540–546.
Wichterle, O. and D. Lim. 1960. Hydrophilic gels for biological use. Nature 85: 117–118.
Zadnik, K., J.T. Barr, T.B. Edrington, D.B. Everett, M. Jameson, T.T. McMahon, J.A. Shin, J.L. Sterling, H. Wagner and M.O. Gordon. 1998. Baseline findings in the collaborative longitudinal evaluation of keratoconus (CLEK) study. Invest Ophthalmol. Vis. Sci. 39: 2537–2546.

17

Contact Lens Induced Ocular Surface Alterations

Helmer Schweizer[1,a,2,*] and *Inma Perez-Gomez*[1,b]

SUMMARY

There is a large history of reported changes of the anterior ocular surface related with contact lens wear. Most of them are temporary and, while impacting the comfort and not being pleasant for the wearers, are not vision or eye health threatening. Frequent replacement, here especially daily disposable contact lenses, as well as the availability of many silicon hydrogel materials and different lens care products give eye care practitioners new tools in how to deal with them. Beside the technical aspects, it is also the wearer's behavior, hygiene and compliance with the care regimen and replacement schedule that contribute to the incidence, frequency and severity of such alterations.

This chapter looks at some potential alterations, how they present themselves, what may be the (potential) causes, their clinical and practical significance and how they could, should or must be attended to, best be prevented.

INTRODUCTION

Eye care professionals (ECP) have observed and noted changes in the anterior ocular surface since the first insertion of a contact lens (CL) in

[1]c/o Alcon SA, Avenue Louis-CasaÏ 58, CH 1216 Cointrin-Geneva, Switzerland.
[a]E-mail: helmer.schweizer@cibavision.com
[b]E-mail: inma.perez@cibavision.com
[2]Universities of Novi Sad, Serbia and Velika Gorica, Croatia.
*Corresponding author

1888. These came along with visible signs and reported symptoms. The careful study of those helped to make improvements of the CLs and led to on-going innovation ever since. The initial learning curve was, no surprise, very steep.

Breakthrough innovations, like new materials (PMMA vs. glass, rigid gas permeable (RGPs) vs. PMMA, RPGs vs. soft (HEMA or hydrogel (Hy)) CLs or silicone-hydrogels (SiHy) CLs vs. Hy CLs etc.), advances in the geometries, the introduction of planned vs. open end replacement [eventually leading to the introduction of daily disposable (DD) CL] etc. did not only make some of the past learning obsolete, but launched new curves, as is always the case with the entry of disruptive technologies.

The knowledge base, the observational skills and the instruments (from naked eye observation to loupes, to bio-microscopes (a.k.a. slit lamps), to confocal microscopes or (S)OCTs [(spectral) ocular coherence tomographers], from simple keratometers to topographers etc.) developed either independent (spill over from general ophthalmology or optics) or in parallel with the CLs (materials, geometries and manufacturing). Better understanding of the causes, signs and symptoms also led to an increase of observable, yet symptom free alterations (like keratocyte density, polymorphism and -megethism). An advancement in one area (e.g. instruments) was sometimes the foundation or even a prerequisite for a big step forward of another one (e.g. knowledge), while at other times these were more in parallel. The goal was always to better understand the cause for the observed signs, their link to the reported symptoms and to provide a (re-)solution to them. In other words: to make CLs safer to wear and more comfortable while using.

It is worth noting, that each progress was also followed by a higher demand, i.e. longer daily usage (hours per day), more days per week, or even the notion or desire to be able to sleep in the CLs. An ever increasing wearer base and the higher demands also elevated the likelihood to (still) observe signs, symptoms and alterations, sometimes a bit less, sometimes even more.

Wearers as well as ECP often forget the increase in demand that they place on the CLs when they express some disappointment about the "new" CLs' performance. Example: Increases in Dk/t (oxygen transmissibility) was considered to allow going from daily wear (DW) to extended wear (EW), which is a huge requirement to place on the CLs performance, if no difference on eye health is expected to happen. Some alterations are more pronounced or frequent in more demanding conditions, like EW or even continuous wear (CW), and less so in DW or occasional wear (OW).

An ideal CL corrects vision, changes and/or enhances the eye color, provides the benefits of a bandage or that of a therapeutic application. It would always be comfortable to wear, would not do any harm and have no side effects, like anterior surface alterations. All this, despite the fact that it is basically a "foreign body" applied into the tear film in front of the cornea, the bulbous and lid conjunctiva.

A CL does, however, just by its sheer presence, always "interact" with its environment (air, tear film, and anterior ocular surfaces). This interaction may result in alterations of those surfaces. These are usually undesired, but may as well be so. The change of the cornea's shape by affecting its epithelium is, in the case of an Ortho-K CL, an example for the latter.

Today's well designed, made and selected ("fitted") standardized or custom made CLs can be comfortably worn by a large number of (individually very different) people for many hours, even days and nights in a row, i.e. even during sleep. Some, like the rigid gas permeable ones, may require an adaption period. Being comfortable does, however, not mean that the CLs do not alter or has not already altered the ocular surface.

SIGNS AND SYMPTOMS

Visible and measureable, thus objective signs, and reported subjective symptoms often accompany, and are part of ocular surface alterations. Wearers may or may not complain about these symptoms (bad vision, pain, photophobia, uncomfortable CLs, etc.), and/or signs (redness, discharge, white spots on the cornea, etc.).

A symptom is defined as (Merrian-Webster 2011) "a subjective evidence of disease or physical disturbance, broadly something that indicates the presence of bodily disorder" and sign is defined as (Merriam-Webster 2011) as "an objective evidence of plant or animal disease."

Asymptomatic alterations may remain undetected by the wearer (no symptoms and often also no obvious signs), until observed and recognized by a skilled ECP (with the help of one or various instruments, most noteworthy, the bio-microscope, at the time of the fitting or a regular after care visit). This adds to the importance of the initial fitting and regular follow-ups.

The presence of symptoms and signs (i.e. the CL not being comfortable, the eye being red and the (perceived) vision being low, etc.) does not always mean that permanent, severe and undesirable alterations are present. Severe symptoms and signs, such as high levels of pain, photophobia and redness may "force" the wearer to (at least temporarily) cease CL use. Severe and persisting cases of signs and/or symptoms may also result in an unscheduled visit to an ECP, who in turn can then identify and address (hopefully eliminate) the cause of them, while the condition is still "active".

While signs and symptoms usually go together, a wearer may also report symptoms (subjective evidence), but there may be no signs (objective evidence). In some rare cases, the symptoms may then be of a more psychological (imagined) nature and truly lack a physical basis (sign). Vice versa, there may be signs, but no symptoms (like in an asymptomatic infiltrative keratitis (AIK), an epithelial scar, neovascularization and others). These signs may then only be detected during the (re-) fitting or after care visits.

The manifestation of signs can range from being almost non-existent to severe. Various grading scales, like the CCLRU or the Efron ones, have been developed to facilitate a standardized, meaningful capturing and recording of the signs. These scales provide commonly agreed reference points that allow an easier comparison of cases and exchange between practitioners and academics, as well as better student education and training. There are no standard grading scales for symptoms. Their subjective nature and more quantitative description make this much more difficult, if not impossible.

In summary, the presence of alterations, their magnitude and importance can only be recognized during the fitting, unscheduled or scheduled after-care visits. The latter should therefore take place on a regular basis and in short enough frequencies. Symptoms provide guidance to what signs the ECP may/shall look for in the CL/eye examination.

PRESENTATION OF OCULAR SURFACE ALTERATIONS

The sign or symptoms causing interaction between the CL and the anterior ocular surface can be direct, i.e. directly caused by the CL itself (e.g. by the edge of CL mechanically interacting with the corneal or conjunctival epithelium) or indirect (e.g. the global allergic reaction of the lid conjunctiva due to allergenic deposits on the CL surface or the lens care solution that surrounded the CL when it was applied).

The signs and or symptoms can be temporary (transient) or permanent. Permanent ones are usually the most undesired. Temporary ones that happen very infrequent and do so at low grades (see below) may be tolerated (not accepted, though), but should still be monitored. They can be an early warning signal for the development of permanent alterations. Temporary ones can slip detection (and required attention), even when they happen in a relatively high frequency, as they may not be present at the actual time of the after care visit. The ECP should therefore ask at each visit about episode(s) or experience(s) of signs and symptoms related the eye, the vision or the use of the CL since the last time.

Alterations may be mild, moderate or severe in appearance (signs). Grading scales facilitate the capturing and recording of the signs. These commonly agreed on reference points allow for an easier documentation (status and development), comparison of cases etc.

Alterations may represent a low, moderate or high potential threat to the eye's health, the vision and thus require an appropriate level of attention/(re-) action. Such action ranges from immediate and with all force to only taking note of and put them under future observation or monitoring.

Wearers often do not realize (gradually developing) alterations, even or especially if they are of the permanent type (like neovascularization). They may think that those were always there, especially when the alterations happened, at least in early stages, without symptoms and easily visible signs (e.g. new limbal blood vessels are very small and the low level of the "all time" or occasional high(er) levels of limbal redness is considered "normal", i.e. nothing to worry about (except the not so nice appearance). Others are quite visible with the naked eye or a normal loupe, e.g. a CLPU, or cause enough pain/photophobia to force one to see an ECP.

Signs may happen and develop slowly and over time, e.g. neovascularization, or rapidly, e.g. CLARE. They may also be there for a short duration (temporary) or longer time (persistent, chronic up to permanent). The time of the onset and the duration can again provide insight about the potential causes and what could be done in order to address, prevent them in the future.

WHY ARE THERE OCULAR SURFACE ALTERATIONS AT ALL?

ECPs aim to avoid any, but certainly severe and permanent alterations induced by CL wear. They do so through a careful fitting process, i.e. the optimal selection of the CL type material, design, wearing modality and replacement frequency, as well as the lens care product(s)/system (LCP) for each individual wearer. This is amended by a thorough instruction on the application and removal, hygiene, care and "what to do if". All of this is repeated at the regular after care visits. Despite an initially optimal "fit" etc., basic limitations in the CL performance—do not forget: they are essentially and still a foreign body to the eye—combine with variations and/or changes in the wearing conditions (temperature, humidity, air flow, etc.), day to day personal condition (tiredness, stress, body hydration status, work/tasks to be performed and others), overall health, seasonal allergies, medication, days per week and hours per day of wear, and last but not the least compliance (wearing regimen for example, sleep with inadequate CLs, insufficient care efforts, a switch of the LCP(s) without consultation and/or replacement schedule) can and will influence the CL fit or performance, and thus may well result in alterations. Even when all or most of the above mentioned elements are kept the same, the eye itself over time and the tear film may change (quality and/or quantity), along with the overall aging of the body.

The comparison of all data collected during the fitting process (baseline) and the carefully conducted after care visits help the well trained ECP (observation skills and knowledge) to detect undesired alterations. The determination of the potential cause and its severity/importance guide drive the decision if and how to manage it/them. Not all changes are, however, undesired (see Ortho-Keratology above).

TYPES AND APPEARANCE OF OCULAR SURFACE ALTERATIONS

Alterations can be observed at all ocular surfaces (cornea, bulbous conjunctiva, sclera, lid conjunctiva/epithelium and the lid margins). The main alterations described in this chapter are: corneal epithelial thinning; corneal epithelial "weakening"; superior epithelial arcuate lesion (SEAL); mucin balls and dimple veiling; deposits on the CL and CL induced papillary conjunctivitis (CLPC); microcysts and vacuoles; diffuse punctate staining; symptomatic (IK) and asymptomatic infiltrative keratitis (AIK); CL induced peripheral ulcer (CLPU); CL related acute red eye (CLARE); keratocyte density; stromal pH change; corneal swelling, stromal opacities, striae and folds; endothelial blebs, polymorphism and polymegethism; limbal hyperemia and corneal vascularization; and conjunctival folds.

Corneal Epithelial Thinning

The landmark Gothenburg study (Holden 1985a) was one of the first to describe in depth the physiological effects of soft CL EW (5 yr) on the cornea. The corneal epithelium of the CL wearers was significantly thinner than that of the non CL wearing, fellow eyes. Many of the epithelial changes, among them thinning, reversed to non CL wear level, after 1 mon of ceasing use. Measuring epithelial thickness is not easy. Modern instruments, like OCTs, make this task much simpler. Ren et al. published that the epithelial thinning can be seen as a short term reaction to over night CL wear (Ren et al. 2002). It is most pronounced in 30 nights EW of high Dk RGPs and low Dk (Hy) than in high Dk (SiHy) 6 nights EW or high Dk (SiHy) 30 nights EW. Wearers of high Dk soft CLs also showed greater adaptive recovery over time.

Stapleton presented at the 2004 BLCA congress on the "Effect of long term CL wear on the corneal epithelium" (Stapleton 2004). She and her colleagues Jalabert and Sweeney looked at two groups of CL wearers (14 EW SiHy (mean experience 5±1 yr) and 23 Hy ones (13±4 yr). They also compared these two groups to an age matched 18 non CL wearing control group. One of their findings is that the topographical location (central and four peripheral locations) did not affect the corneal thickness observations.

The epithelium was thinnest in the Hy group (46±10 μm), followed by the SiHy one (54±14 μm) and the control group (58±9 μm). The difference was significant between all groups. The Hy group also showed reduced basal epithelial cell regularity. This was not seen in the SiHy and the control group. The transparency of the basal cell layer was not affected in both CL groups. In another presentation on a similar study, Jalabert et al. did, however, not find the central Dk to be a predictor for the epithelium thinning and suggest that there may also be other factors (Jalabert personal communication 2005). Among those could be the fit and thus the mechanical pressure of the CL on the epithelium. Central Dk/t (CL parameter) or average Dk/t may also be a better predictor than Dk (a material parameter).

In summary, CL wear results in a short term epithelial thinning. This is reversed over a year. For soft CL, the effect is stronger when there is not sufficient oxygen supply to the cornea, i.e. low Dk/t. That high Dk RGPs show an even stronger effect than the low Dk soft lenses may also indicate a mechanical component. The adaptive recovery for all CL types is very positive with regards to eye health. By fitting CL with a high Dk/t and not only high Dk (e.g. high Dk, but thicker RGP lens), the ECP can reduce the short term effect and accelerate the adaptive recovery.

Corneal Epithelial "Weakening"

The epithelium is the cornea's first line of self-defense. Microbes have difficulties to penetrate into or even through a healthy epithelium. Microplicae and micorvilli on its surface are involved in anchoring and stabilizing the tear film's mucous layer. The wing cells extend around the hemispherical tops of the basal cells (vertical attachment). The periphery of the wing cells looks like a "velcro". This allows for very tight junctions between the cells (horizontal) and to the layers below and above (vertical). It is this tightness that prevents micro-organisms to penetrate into or go through a healthy epithelium (Evans 2010). When the epithelium is "weakened", i.e. traumatized, chemically impacted by preservatives from LCP (solutions) or a not well functioning metabolism [mainly lack of oxygen (Ichiyama 1999)], the micro-organism can easily penetrate towards the basal lamina, the next line of defense. Ladage et al. (Ladage et al. 2003) found that SiHy CL showed less reduction in the proliferation rate of the epithelial basal call layer (–33.8%) versus the Hy ones (–40.8%), which means that oxygen supply is important. The epithelium's top cells-layer is continuously replaced by cells shedding off at the surface and being replaced by those that migrate from the limbus to the center and up. To remain healthy, the epithelium's integrity and metabolism (cell itself, cell layers and renewal rate) needs to be maintained as best as possible.

In summary, it is important to ensure sufficient oxygen supply to the cornea and thus its epithelium. In doing so, the health status is best maintained and the self defense is strongest. Next to oxygen, one should avoid weakening or disrupting the corneal epithelium by mechanical effects, i.e. the fit of the CL or the finger nails during manipulation. Foreign bodies that are trapped under the CL may also cause such trauma. LCPs can affect the epithelium integrity as well. Staining the cornea reveals the type of the alteration of the epithelium, its depth and extent. Any grade of three or higher on the CCLRU grading scale should trigger a more detailed search for the potential cause(s) and a possible resolution.

Superior Epithelial Arcuate Lesion (SEAL)

When mentioning superior epithelial arcuate lesion (SEAL), many practitioners immediately associate this condition with a) SiHy CLs and b) material rigidity. In 1987, Hine et al. reported about the aetiology of SEALs (Hine et al. 1987), Jones et al. also describe them as being "relatively common complications in soft CL wear" and to be "found in all soft CL types" (Jones et al. 1995). These 2 to 5 mm long and 0.1 to 0.3 mm wide splits of the epithelium can be found in the superior and peripheral cornea (within 1–3 mm of the limbus and between 10 and two o'clock) and run parallel to the limbus. They can present themselves combined with symptoms (irritations, burning, itching and photophobia) and an associated reduced wearing time or asymptomatic. They may be bi- or monocular. In the case with irritations, superior bulbar and palpebral conjunctival redness may be present, too. Their existence may be attributed to the combination of the CL material, design, dehydration characteristics, hypoxia, and solution used. The pressure that the upper lid exercises on the CL when it moves over it during the blink and the CL movement during the blink over the cornea may be the mechanics and the friction required to break/split the epithelium.

The condition usually resolves within a few days (three to four) of non CL wear. Frequent use of lubricants/artificial tears may be advisable during that time. Some ECPs may also recommend the prophylactic use of antibiotics. Thirty seven percent of the cases will have a reoccurrence when using the same CL and or LCP. It may therefore be better to think about an alternative CL (design, fit etc.) or LCP.

CL design and the choice of base curve are more important than the material's rigidity (often mentioned as being a major driver). The case of changes made by one manufacturer to its high modulus SiHy CL that showed a much higher incidence of SEALs than the other SiHy CL launched at the same time (all attributed it to the modulus) is proof to this. The change of the back surface design from a tri-curve to an aspheric one (better alignment with the eye's topography and no junctions between the curves) and also offering a steeper base curve (less, yet sufficient movement on steeper corneas or those with a low numeric eccentricity) strongly reduced the incidence rate of SEALs with this CL to a very, very low number. Furthermore, Lakkis reported on two incidents (one in the test and one in the control eye) of SEALs in a 50- patients, six months DW contra-lateral study of low rigidity SiHy CLs (Lakkis 2011). A switch of the LCP to a preservative free one (see weakening of the epithelium) or a DD CL (no care) may also be considered.

In summary, SEALs have been observed and reported since the 1980s. Their cause is multi-factorial. CL material rigidity (modulus) may be an important factor, but an appropriate combination of CL design (Stapelton 2006) and choice of base curve (best fit) addresses this factor's potential impact. A SEAL is a chronic condition. With asymptomatic patients, it may not be seen at all or in active stage at the after care visits, while traces thereof may still be present. A few days of non-wear (and the use of lubricants) usually resolve the condition. To prevent the high reoccurrence rate, the ECP may change the CL fit (base curve), design, material or the LCP.

Mucin Balls and Dimple Veiling

Mucin balls (Fig. 1) are generally translucent and can either be scattered or clumped in the superior corneal quadrant. They are round, vary in size (from 10–200 μm), but may appear flattened under the CL. There may be as many as 50 of them (Fonn 2000).

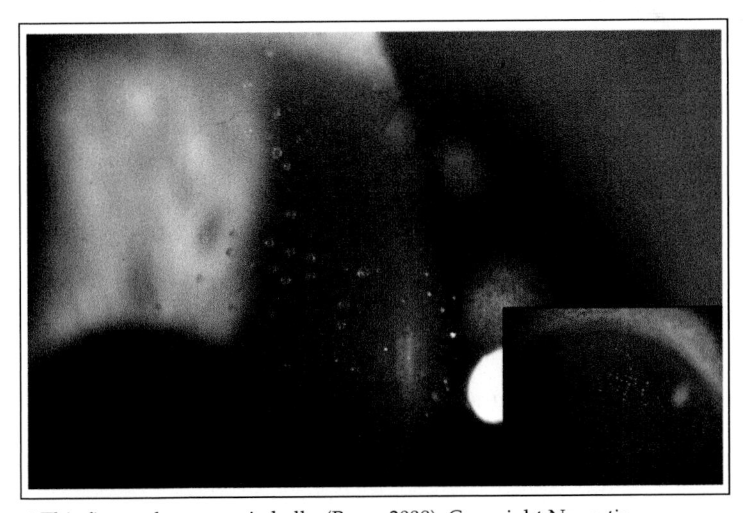

Figure 1 This figure shows mucin balls. (Bruce 2008), Copyright Novartis.

Color image of this figure appears in the color plate section at the end of the book.

That they do not move with the CL and disappear (blinked away), when the CL is removed makes it easier to distinguish them versus microcysts and vacuoles. When staining with fluorescein, the impressions that they made on the epithelium may still be observed, for a short time.

The occurrence of the mucin balls is patient-specific. There are usually no symptoms, and no effect on the high contrast visual acuity.

Similar to the SEALs, the occurrence of mucin balls was already described before the launch of SiHy CLs (Polse 1979, Fleming et al. 1994), but received attention only thereafter (Sweeney 2000). Efron published a two-part paper with a very thorough review and attempted to differentiate various types of mucin balls versus microcysts, vacuoles and dimple veiling (Efron 2004a). In case of presence, they can be observed minutes after CL insertion. According to Naduvilath, subjects that display mucin balls, have a 1.6 times higher probability of developing CLPUs (Naduvilath 2003).

The increased incidence and appearance of mucin balls was also attributed to the SiHy's higher rigidity, their surface treatment and thus friction when moving on the eye (Fleming 1994). Again, similar to the SEALs, the same modifications in the CL design (from tri-curve to an aspheric back surface) of one of the early SiHys and the additional offering of a steeper base curve considerably lowered the incidence. Fleiszig's work showed that it is beneficial to maintain an intact mucous layer, in order to have the lowest attachment of bacteria to the corneal epithelium (Fleiszig 1994).

Air bubbles that are trapped under a CL also leave impressions that, similar to those from mucin balls, fill with tear film and stain with fluorescein. This condition is called dimple veiling. It is much more common in rigid CL wear, rarely with soft CLs (Fleiszig et al. 1994).

In summary, mucin balls may be visible in individual patients and in combination with individual CLs, shortly after insertion. The likely cause is a combination of the patient's corneal topography, the CL design, parameter (base curve, i.e. fit) and material. While asymptomatic, the ECP may still aim to re-fit these patients with different CL (parameter, design material, etc.) as a non-disrupted mucous layer and no epithelial impressions are, if achievable, preferred.

Deposits on the CL and CL Induced Papillary Conjunctivitis (CLPC)

Deposit formation on the anterior and posterior surface of a CL was and still is very common. These deposits can come from the tear film (endogenic) or from the hands (exogenic, during manipulation), make up or its remover etc. Some CL materials are more prone than others [e.g. ionic vs. non ionic ones (Minarik 1989)], as well as to certain types of deposits (proteins (Emch 2009, Fleiszig 1994), lipids (Bontempo 1994), mucins, carbohydrates, etc.).

The amount also depends on the water content, the CL age and the LCP used (Zhao 2009).

Before the advent of frequent replacement CL, wearers invested a lot of effort and money in trying slow down build up by removing as many deposits as possible, either by daily cleaning, i.e. manual rubbing with a surface cleaner and rinsing and weekly, bi-weekly use of enzymatic protein removers.

Jelly bumps (Fig. 2) are (at least at a late stage) easily visible, white spots of heavy deposits on the CL surfaces. They can cause mechanical irritations as well contribute to toxic effects by accumulating LCP preservatives within them.

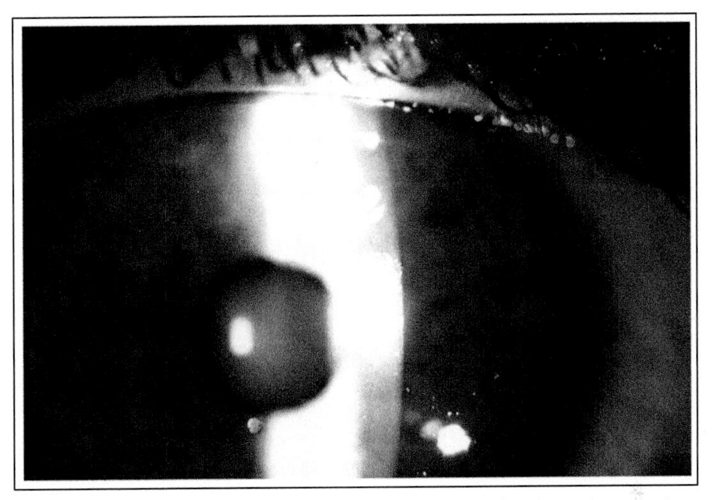

Figure 2 This figure shows Jelly bumps (Bruce 2008), Copyright Novartis.
Color image of this figure appears in the color plate section at the end of the book.

Another important aspect is the potential denaturing of the protein deposits, which can make them allergens and thus contribute, next to mechanical factors, to the development of a condition called "contact lens induced papillary conjunctivitis" (CLPC), an inflammatory reaction of the whole superior palpebral conjunctiva [which is a continuous extension of the bulbar conjunctiva under the upper eye lid (Stapelton 2006)].

Abelson and Leung dispute the classification of CLPC/GPC (Fig. 3) as being of allergic nature (Abelson 2009). They do not fully distinguish between the "localized" (see later) and the "general: CLPC".

The proteins in the deposits are specially a potential cause for an allergic reaction. It may, however, be misleading to make a quick decision and assume that CL that have 79±12% (SiHy) of the lysozme deposits being denatured are worse than Hy ones with only 40±11%.

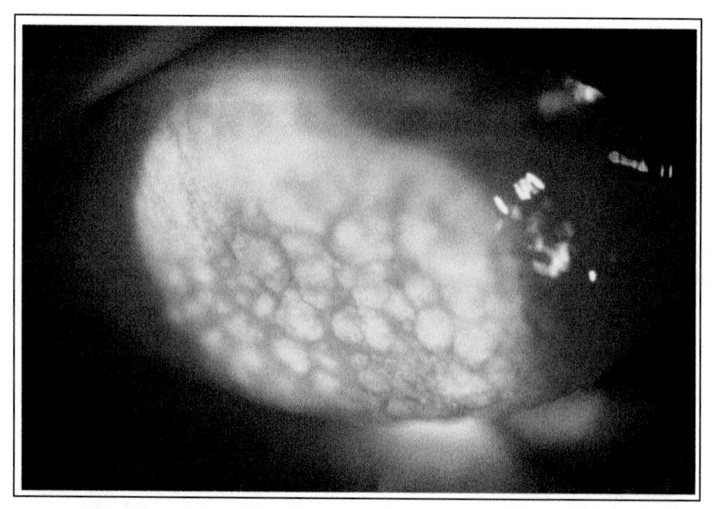

Figure 3 This figure shows CPG (Bruce 2008), Copyright Novartis.
Color image of this figure appears in the color plate section at the end of the book.

It is important to look at the absolute amount of the protein deposits and calculate. Seventy nine percent of 0.8±0.4 μg is far less than the 40% of 1577±320 μg (Senchyna 2002).

General CLPC is also sometimes described as GPC (gigant papillary conjunctivitis). Symptoms are pain, itching, foreign body sensation and so forth. The lid may also swell and its surface temperature increases (redness). Once a general CLPC condition develops, it often reoccurs, even after a break in CL wear, requiring the use of antihistamines and no more visible conjunctivitis before the restart. The old CL should be discarded and wear only be restarted with fresh, new ones. Also changing to shorter frequency replacement CL should be considered, or those that are known for attracting fewer deposits. The most radical change and the one with a high probability for success is the use of DD CL. It takes the LCP aspect out of the equation. The use of preservative free peroxide systems may be considered and preferred.

Localized CLPC is usually more in the center of the upper lid conjunctiva and closer to the lid border. It is believed, to be caused by mechanical irritation, i.e. the lid moving over the CL edge during blinks (humans blink about 14.000 times a day) and is not of allergic nature. Too flat fitting CL (edge lifts too much off the surface), rather thick CL edges or the edge's apex being too high, may also cause the irritation. A change in CL design or type (e.g. from rigid to soft) should be considered.

Zhao (Zhao et al. 2009) found no or only a weak correlation of cholesterol deposits on SiHy CLs and clinical variables like inflammatory keratitis (IK), asymptomatic IK (AIK), asymptomatic infiltrates (AI), corneal erosions (CE) and CL induced acute red eye (CLARE). LCP induced corneal staining [solutions induced corneal staining (SICS), contrary to CL induced conjunctival staining (CLICS)] was significantly correlated with cholesterol or protein deposits (four CL types and four multipurpose LCPs (MPS) reviewed). Protein deposits also significantly correlate with "on-eye" performance, like conjunctival staining, CL front surface wetting and fit tightness. There was no correlation between protein or deposits and clinical variables. SICS with signs of grade 3 and higher require clinical action (Efron 1996a).

A special type of deposits is the "horizontal band". Incomplete blinking causes it. It can be made visible by applying fluorescein and observing the eye while blinking. Incomplete blinks (Fig. 4) leave a band of heavier fluorescence in between the lids, after the blink.

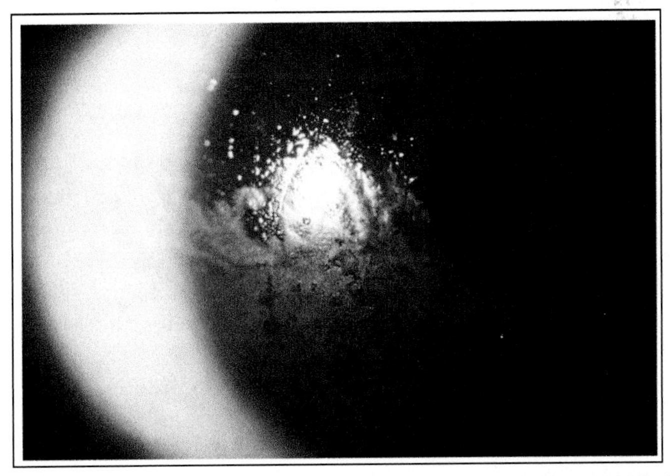

Figure 4 This figure shows the CL deposits with incomplete blink (Bruce 2008), Copyright Novartis.

Color image of this figure appears in the color plate section at the end of the book.

This represents the tear film being not completely pushed downwards, the eye's lids therefore do not close fully and the eye may not show full Bell's phenomena either. Deposits within the tear film and on the CL surface pushed into this band (and left there) accumulate and attach more and more to the CL surface(s) (lack of the "wiper" going fully over the CL surface). This band can be (at least partially) in front of the pupil and may impact vision (symptom).

In summary, deposit formation is inevitable in CL wear. It is part of the "aging by use" of the CL. More frequent and especially planned replacement resulted in less deposit formation. It is therefore unusual today, to use either special surface cleaners or protein removing tablets, other than with open end use of CL. A good daily rub and rinse with a MPS or the use of peroxide LCP can be very beneficial, too. One of the most difficult conditions to handle conditions caused by deposits is CLPC. Successful re-start of CL wear is often only possible by switching the wearer to DD CL and to avoid the use of LCP. Lowest surface friction and the most wettable CL should be considered, too. Localized CLPC requires a careful look at the fit and avoiding a fit that is too flat. In high water Hy, fitting flat is a precaution to not end up with too tight CL, because they steepen when dehydrating during wear. A different edge or overall design may also work better for an individual patient. Some CL materials are more prone to certain deposits and thus may be less adequate for an individual wearer. The choice of LCP and the wearer's compliance with the care regimen can also influence the amount of observed deposits. When seeing band-like deposits, check for incomplete blink, make the wearer aware and instruct him to willingly, completely blink several times at certain intervals, like every quarter of an hour.

Microcysts and Vacuoles

Microcysts might be confused with vacuoles or mucin balls. Ruben et al. first reported about microcysts (Ruben et al. 1976). They suspected a microcyst to be a sign of chronic changes in the epithelium. Zantos and Holden [used the term "microvesicles" (Zantos and Holden 1978)] and Josephson (Josephson 1979) confirmed this. With the number of papers being published since, it is fair to say that today's knowledge can be considered to be quite complete (Fig. 5). Holden found that up to 10 microcysts could be seen in non CL wearers (Holden 1987). The prevalence is higher in EW than in DW and in lower Dk versus higher Dk CL (Efron 1996b). The correlation with the oxygen supply therefore revived the interest in microcysts when really high Dk SiHy CLs were launched.

One can observe microcysts even at low magnifications (15x), in the central and para-central cornea. They appear as small scattered gray opaque dots in focal illumination. With indirect retro-illumination, they are transparent refractile inclusions. Their shape is generally spherical or ovoid. To properly differentiate them from vacuoles etc. (that take on a similar appearance) requires a careful observation with higher magnification (40x). Microcysts show a reversed illumination (due to the higher refractive index (Bron 1973) of the assumed accumulated dead cell's material (Bergmanson 1987) within the microcyst). Vacuoles and dimple veiling show an unreversed one (Zantos 1983). Microcysts and vacuoles remain visible after CL removal, where as mucin balls are no longer (indentions may fill with fluorescein). The probability to also see superficial punctate

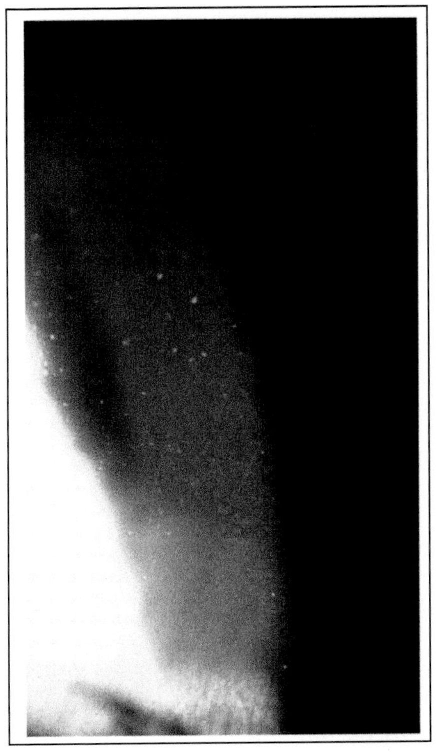

Figure 5 This figure shows microcysts and vacuoles (Bruce 2008), Copyright Novartis.
Color image of this figure appears in the color plate section at the end of the book.

staining (anterior epithelial surface cells breaking open) increases with the number of microcysts, as they emerge from the deeper layers and rub out of the epithelium. Their size can vary from 15 to 50 μm. The large ones would therefore have the same diameter as the whole epithelium thickness, which is why one may normally see only those with diameters of up to 20 μm (Efron 1996b). Based on the numbers observed, Efron established the grading scale showed in Table 1.

Vision is only affected when the number is very high. Microcysts can already start to develop after 1 wk of EW of Hy CL. Keay and colleagues found that significant numbers only appear after 3 mon, where as "*... the number of microcysts remained within the normal range at all visits ...*" for a lotrafilcon A SiHy CL worn continuously for 30 nights (Keay et al. 2001). The same paper reports that the number of microcysts drops back to normal within 6 mon after cessation of CL wear. It does so, however, only after initial increase (spike, also named "rebound effect") after 1 wk, going back to base line after 1 mon, before further decline (Holden 1985b). Keay and

Table 1 Microcystic grading scale (Efron 1996b).

Grade	Number of microcysts	Clinical interpretation
0	0	No microcysts visible
1	1–4	This number of microcysts is commonly observed in wearers of all types of contact lenses and in a normal non contact lens wearer
2	5–30	This represents a mild microcystic response that warrants monitoring
3	31–100	This represents a moderate microcystic response that warrants monitoring and generally requires clinical intervention. Slight staining may be observed, vision may be reduced slightly, and the patient may be experiencing discomfort
4	> 100	This represents a severe microcystic response that demands immediate clinical intervention. Moderate staining will be observed, vision may be reduced slightly, and the patient may be experiencing discomfort

colleagues observed a similar rebound effect when switching wearers after 12 mon of 7 d of Hy EW to 30 nights SiHy CW (Keay et al. 2001).

Vacuoles often occur concurrently with microcysts (Zantos 1983). The number is usually low (10 per eye), but the incidence of vacuoles when wearing Hy CL on an EW basis, is high (32%) (Zantos 1978). Vacuoles are similar in size than microcysts (20 to 50 µm), but larger on average and are filled with gas or fluid of lower refractive index then the surrounding; hence do not show reversed illumination (Josephson 1988). There is no report about any difference in their incidence in SiHy, compared to Hy wear.

In summary, differential diagnosis between microcysts, vacuoles, mucin balls and dimple veiling is very important. The reversed illumination characteristic distinguishes microcysts mainly from the others. While a small number of microcysts can also be observed in a normal, non CL wearing population, seeing microcysts in CL wearers is a sign of oxygen deprivation. They may be non compliant and sleep in Hy CLs or the cornea may have a generally high oxygen demand. Moving them to SiHy CLs is strongly advised. The fitting of SiHy CLs from the start on is also recommended.

Diffuse Punctate Staining

Staining the eye with fluorescein and/or other stains, like bengal rose or lyssamine green is an important part of the initial fitting process and each after care visit. The CCLRU grading scale divides the cornea in five approximately equal sectors, a central (circle) one (1), the areas to the left (2), to the right (3), above (4) and below (5) of the central one. A grading scale amends this; that refers to the extent, depth and type of the staining. Grades for each go from 0 to 4 score.

Central staining and snowflake like inferior staining that can be seen when (thin) Hy CL dehydrate, as well as diffuse corneal and conjunctival staining can sometimes be seen in DW patients. McNally and colleagues described the appearance of central staining, observed with extremely thin Hy CL and compared two different designs of the same material, as well as the effect of the water content (McNally et al. 1987). They found that the higher the water content, the earlier was the onset, the greater was the maximum level and the more wearers were affected by central staining, which peaked after 2 to 3 hr and waned thereafter. A parallel surface design (central thickness was 10% to 20% higher then in the –3.00 one, which, in turn, had a two to three times thicker mid-periphery than the parallel one) showed peripheral staining (mainly superior), where as the –3.00 D power design had central staining. The staining of the parallel design was much later and in lower grades. They concluded that other factors besides dehydration also came into play.

A snowflake like pattern was in the past observed with ultra thin and high water Hy CL and attributed to localized CL dehydration (Holden 1986a). They are seen far less since the advent of the mid water content and relatively thicker planned replacement CL and are almost non existent in low water content and high Dk SiHy CL, due to improved dehydration characteristics (Jones 2002).

Solution toxicity has long been a quest with Hy and even rigid CL wear. Some preservatives (e.g. thiomersal, benzalkonimum chloride (BAK) (Schepper 1993) and chlorhexidine) or even complete formulations of lens care solutions were found to cause more staining (Fig. 6) than others and were consequently discontinued or even legally banned (thiomersal). While usually asymptomatic, such staining has been connected to inflammatory events (Jalabert 2006).

The development and use of single step hydrogen peroxide solutions and multipurpose solutions (MPS) put an end to this. It resurfaced with the launch of the SiHy CLs. Jones was among the first to report about differences in staining of certain SiHy CLs and LCP combinations (Jones 2003).

A one step hydrogen peroxide solution has since emerged as the one causing the least staining and has become the gold standard in this respect.

In summary, different types of diffuse corneal staining may be observed in CL and non CL wearers. Next to lack of oxygen supply, they may be related to the dehydration of (thin) Hy CLs. Lower water content SiHy CLs/materials show less dehydration. Another possible cause for staining is the LCP used to care for the CL. Certain combinations of CL and LCP seem to yield more staining than others. A one step hydrogen peroxide LCP may be the first choice to resolve such a condition. The alternative option is the use of DD SiHy CL.

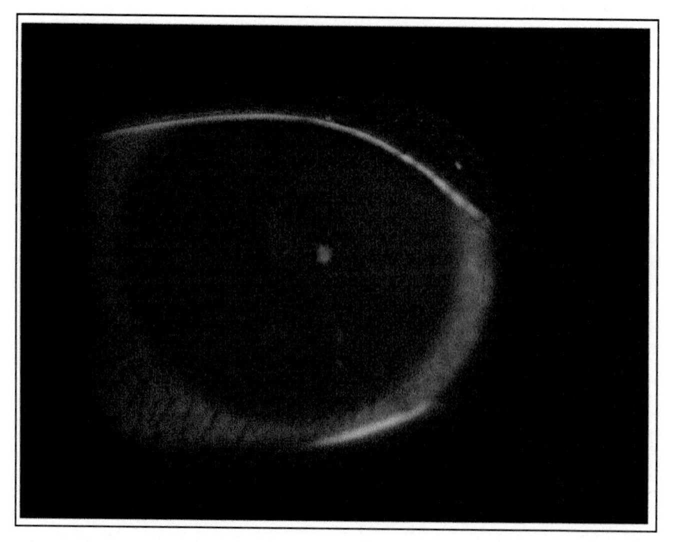

Figure 6 This figure shows a solution toxicity keratitis (Bruce 2008), Copyright Novartis.

Color image of this figure appears in the color plate section at the end of the book.

Symptomatic (IK) and Asymptomatic Infiltrative Keratitis (AIK)

"Contact Lens Acute Red Eye (CLARE), Contact Lens Induced Peripheral Ulcer (CPLU), Infiltrative Keratitis (IK) and Asymptomatic Infiltrative Keratitis (AIK) are probable inflammatory responses of the cornea and conjunctiva caused by the presence of bacteria colonizing the CL and are not considered to be infections of the cornea or conjunctiva surface" (Wilcox 2000). Microbial Keratitis (MK) is the only event that is considered infectious, i.e. showing rapid bacteria growth in tissues. Other corneal infiltrative events that are not classified as MK are IK and AIK, CLARE and CLPU. Contrary to CLARE, there is no association of sleeping with CL with IK. Its onset is usually later in the day (Sankaridurg 1999a). Watery discharge and mild to moderate limbal and bulbar conjunctival hyperaemia can be observed. IK may be found to be related to corneal trauma/corneal erosions, possibly combined with foreign body traces or matter under the CL. The infiltrative response is usually under these erosions or traces. Another type is a diffuse, dim infiltration in the mid-periphery to periphery. These infiltrations are in the anterior stroma and come from the limbus. Neither the posterior stroma, nor the endothelium is involved. Non trauma cases often have bacteria involvement (on the CL and in the CL case). The epithelium above the infiltrate can show punctate or strong defects. The most common accompanying symptoms are redness, irritation and excessive tear flow (watering) (Holden 2000). There is normally no effect on vision.

IK usually resolves simply on discontinuation of CL wear and does not require treatment, not even a prophylactic one, except in cases with severe trauma. It may, however, take more than a week. The use of artificial tears may be welcomed by the wearer. Some opacity may be visible after resolution. With a low probability for recurrence (0.07) (Holden 2000), the prognosis for successfully resuming CL wear is very good.

AIK is considered a non significant adverse event. It is a mild corneal infiltrative event that can be seen in EW and in DW of Hy CL, as well as in non CL wear. Being asymptomatic, it may be observed at a regular after care visit or even go undetected. The signs are similar to but more faint as those of an IK, as is the management (Holden 2000). The infiltrates in AIK, if any, are small and focal.

In summary, IK is a general category for what is not classified as an MK, CLPU or CLARE. It can be in combination with trauma or just a diffuse staining (likely bacteria related, similar to CLARE). Due to the symptoms and the possibility of the involvement of a larger corneal epithelium disruption, one has to interrupt CL wear. The condition disappears after a week to 20 d, maybe leaving a small opacity. Cases of AIK can be seen at after care visits, as well as in non CL wearers. The bacterial involvement makes hygiene and compliance with the LCP regimens an important talking point in the discussion with IK and AIK patients. The use of (SiHy) DD may also be discussed.

Contact Lens Peripheral Ulcer (CLPU)

While contact lens peripheral ulcers are mostly observed in association with EW, Grant reports some events in DW, too (Grant 1988). There is almost always a unilateral single lesion. The onset is usually acute, but individual wearers may present with or without symptoms, or with a healed lesion. In an acute stage, CLPU is characterized by strong bulbar and limbal hyperaemia, "pointing" towards the peripheral corneal location of the ulcer. The ulcer itself is a yellowish-white, dense, well defined (circular shaped) and focal, corneal infiltrate with a diameter ranging from 0.2 to 1.2 mm (Fig. 7). The depth of the infiltrate is always into the anterior stroma and shows a complete loss of the epithelium, above the infiltrate. Fluorescein will therefore rapidly diffuse into the stroma. Additional fine, diffuse infiltration extends from the limbus and surrounds the ulcer. It resolves to an anterior stromal scar (which can last for up to 6 mon plus).

This can also be seen at after care visits, in symptomless cases (the epithelium may be complete or only shows punctate staining). Symptoms are mainly redness, pain or soreness, irritation, photophobia and watering (Holden 2000). Vision is usually not affected.

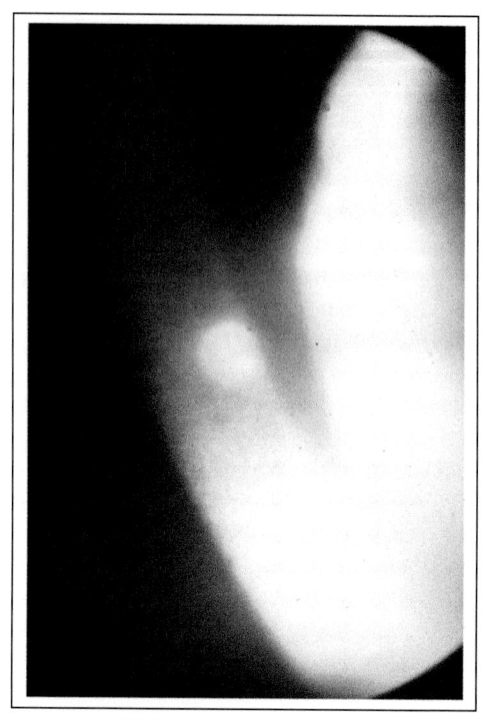

Figure 7 This figure shows a CLPU (Bruce 2008), Copyright Novartis.
Color image of this figure appears in the color plate section at the end of the book.

Acute CLPU requires a temporary intermittence of CL wear until resolution of the infiltrate, ulcer. Like CLARE and IK, CLPUs usually resolve with cessation of CL wear. The use of artificial tears may help and also supports infection prevention. The scar may persist more than 6 mon. One has to be careful to not mix up early stages of MK with CLPU. CLPU symptoms decrease after CL removal, while they continue to increase in severity in MK, necessitating medical treatment. MK also shows and includes lid oedema, discharge, general limbal and bulbar redness, large and irregular focal infiltrates, and satellite lesions etc. CLPU are also present in the use of SiHy CL and their cause may therefore be more trauma-related rather than the results of oxygen deprivation alone. A healthy epithelium may, however, be less prone to trauma and more capable of repairing itself.

In summary, CLPU may be a painful and unpleasant experience for the CL wearer, but is not vision threatening, if the CL is immediately removed and hygiene is maintained in the healing phase. The typical round scars can also be seen at after care visits and may reflect a past case of asymptomatic CLPU. A healthy and strong epithelium (high Dk SiHy CL), together with a well fitted CL that shows good wettability and deposit resistance (less mechanical irritation or CL binding), lower the potential of epithelial trauma.

Contact Lens Acute Red Eye (CLARE)

Contact lens acute red eye (CLARE) was first described in 1978, by Holden and Zantos (at that time called the "red eye reaction") (Holden and Zantos 2000). CLARE is (almost) always observed during EW (Holden and Zantos 2000, Fonn 2005). It is a corneal infiltrative event, characterized by its sudden onset. Wearers wake up in the middle of their sleep, with symptoms of eye irritations or pain, redness and watery eyes, or the symptoms are noticed shortly after normal awakening. CLARE is usually unilateral. The signs are very noticeable bulbar and conjunctival hyperaemia, and watery discharge. These may be joined with low levels of conjunctival chemosis. Most characteristics are, however, mild, diffuse infiltration of the peripheral to mid-peripheral cornea with groups of slight focal infiltrates, without separation between the infiltrates and the limbus (the infiltrates appear to flow in from the blood vessels). The infiltrates are limited to the anterior stroma and may occupy a small sector of the limbus or go all around it (360°). Sankaridurg found an average of 240±113° (Sankaridurg 1999a). Vision is usually not affected. CL harvested from CLARE affected eyes have a significant load of Gram-negative bacteria, few, if any, Gram-positive ones (Sankaridurg 1999b, Sankaridurg 1996). Medical intervention is normally not needed during the mandatory disruption of CL wear. The resolution is without sequela and leaves a clear cornea.

In summary, CLARE is an unpleasant experience, especially from the point of view of the pain and appearance. It resolves without trace and does not require treatment, other than no CL wear until clearance. Its occurrence can best avoided by proper hygiene, compliance (care regimen, replacement etc.), and only sleeping with adequate CLs, i.e. SiHy ones. Moving to (SiHy) DD CLs may also be a good choice.

Keratocyte Density

Keratocytes are located in the corneal stroma, between the collagen fibers and proteoglycans. Their role is to actively contribute to maintain corneal health, transparency and structural stability. They do so, by regulating the collagen fibril size and the spacing between the proteoglycan matrix. They communicate via a network of gap junctions within the stroma (Watsky 1995).

Using PMMA lenses and electron microscopy, Bergmanson and Chu found degenerating and dying keratocytes in primates (Bergmanson and Chu 1982). Using confocal microscopy, Kallinikos and Efron found a decrease in anterior and posterior keratocyte density in low Dk Hy CL wear (control) compared to eyes that wore SiHy or hyper Dk RGP CLs (test) under anoxic conditions (Kallinikos and Efron 2004). They did not find any difference in corneal thickness between the control and test CLs and between the two different test CLs. They also studied the mechanical influence and concluded that mechanical forces may also play a role in the "dysgenesis or apoptosis" of keratocytes.

In summary, the lack of oxygen supply though the CL and mechanical pressure lead to undesired changes in the keratocytes. It is therefore important to always achieve a best possible fit and use the best oxygen transmissible CLs.

Stromal pH Change

In 1987, Bonnano and Polse reported about corneal acidosis in CL wear and as a consequence of hypoxia and increase in CO_2 (Bonnano and Polse 1987). In 1991 and 1992, they and co-workers followed up and found changes in the stromal pH after CL wear (Rivera et al. 1991, Polse et al. 1992), and, most importantly a direct correlation between stromal pH and CL Dk/t. They concluded that a minimum Dk/t value of 16 would be required to avoid a drop in stromal pH for DW and a value of 80 for EW, which was not available at the time. They further concluded that "stromal acidosis lowers the rate at which the cornea recovers from an oedematous load", which lowers corneal hydration control (CHC), which in turn may induce endothelial polymegethism (Fig. 8). CHC is further and increasingly impacted by corneal pH values lower than 7.3. In 1999, Harvitt and Bonanno revised the minimal Dk/t numbers to 35 for DW and 125 for over night wear (Harvitt and Bonanno 1999).

Meng et al. stipulated after a one hour closed eye experiment that the new SiHy CLs "may eliminate corneal acidosis and reduced epithelial barrier functions that accompany closed eye CL wear with lower Dk materials" (Meng et al. 2000).

In summary, today's SiHy CL provides sufficient oxygen transmissibility for open eye condition, not only centrally, but across the whole CL and in a wide range of powers. Many Hy CL do not even pass the central DW criteria. The EW threshold of Dk=125 is surpassed by some CL (at least centrally), but higher Dk materials are still needed to match this criterion across the whole CL in a much wider range of powers.

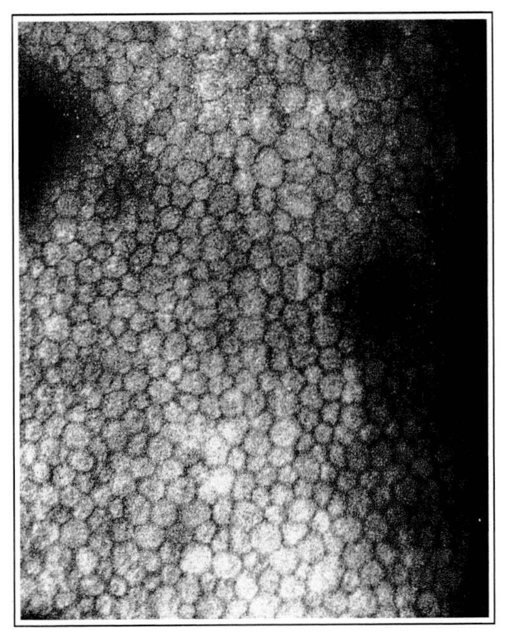

Figure 8 This figure shows endothelial polymegethism (Bruce 2008), Copyright Novartis.

Corneal Swelling, Stromal Opacities, Striae and Folds

Corneal swelling is a visible consequence of corneal oedema, hypoxia. Low levels do not show signs and need to be measured. Levels of more than 5% result in increasing opacity and the development of striae and folds. La Hood and Grant found that one striae became visible with a corneal swelling of 5%.

They identified a link between the number of striae and folds, and the level of corneal oedema (La Hood and Grant 1990). Five striae were equivalent with 8% swelling and a first fold was visible with a mean swelling of 8%. Five folds and ten striae were seen at 11% swelling. Striae are in the central stroma, whereas folds are in the posterior part.

One of the studies cited the most in CL research is the 1984 publication of Holden and Merz (Holden and Merz 1984). It concluded that a minimum Dk/t of 24 units is needed for DW and 87 for EW to see no more than the physiological swelling of 4% (assumed to be the overnight swelling without a CL). After Harvitt and Bonnano posted the 35 and 125 values (see above, pH), Holden et al. revisited their results and also found 35 and 125, when lowering the threshold from 4% to 3%–3.2% (Barr 2005). In a summary report, Fonn said in 2005: *"... that silicone hydrogel lens wear significantly reduces overnight central corneal swelling to about the same amount (app. three percent) as no lens wear"* (Fonn 2005).

In summary, corneal swelling is a sign of hypoxia. It is higher in closed eye condition. Various studies reported between 1.5% and 5.5% swelling without CL wear, (majority showed 3.0% to 3.5%) (Zantos 1977). High Dk/t SiHy CLs yield only a small increase of central swelling compared to non CL wear, while many Hy CLs do not even meet the 35 Dk/t threshold.

Today, one should only see striae and folds in very sensitive, oxygen demanding patients or when (non compliant) wearers sleep in CL that do not have adequate Dk/t. These wearers must strongly be advised to upgrade to SiHy CL. There is still room for higher Dk materials, in order to account for whole CL Dk/t and changes of available oxygen in different altitudes.

Endothelial Blebs, Polymorphism and Polymegethism

Zantos and Holden first mentioned the possibility that the endothelium may also show signs of hypoxia (Zantos and Holden 1977). They noticed a change in the endothelial within a short time after commencing soft CL wear. They called the observed black (non reflecting areas) "blebs" (Fig. 9). It was associated with a marked increase of the cell separation. Schoessler and Woloschak then added the observation of polymegethism (Fig. 8) to the belbs (Schoessler and Woloschak 1981, Schoessler 1983).

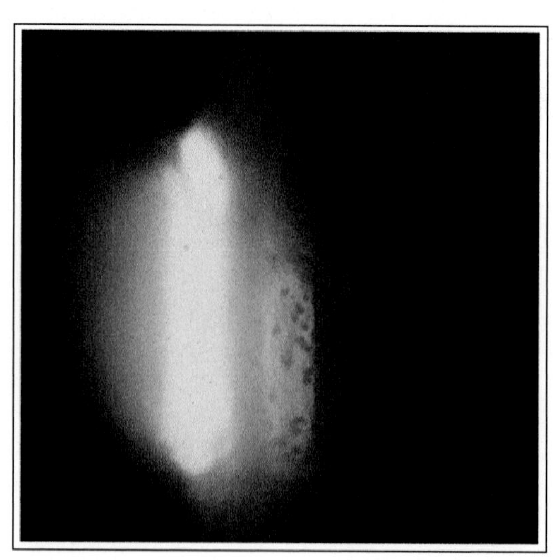

Figure 9 This figure shows endothelial blebs (Bruce 2008), Copyright Novartis.

Color image of this figure appears in the color plate section at the end of the book.

Efron published a very comprehensive paper on CL induced endothelial blebs (Efron 1996c). Williams found a correlation to hypoxia, as soft CL (at that time only low Dk) showed more blebs than RGP and thicker CL of the same material showed more than thinner ones (lower Dk/t) (Williams 1986a).

Holden and Williams identified an adaptation process, with the numbers of blebs decreasing after the initial 8 d of EW (Williams and Williams 1986b). Inagaki et al. looked at three different (by Dk/t) RGP and soft CLs after 20 min of closed eye (Inagaki et al. 2003). The control (non CL wearing eye) showed 0.4% area of blebs. Hyper Dk/t rigid and soft CLs showed only 1.7% and 1.2% area, respectively. The medium and low Dk/t CLs showed up to 5 times more. The response disappeared in Hpyer Dk/t CLs after only 5 min, whereas it took up to 30 min in the low Dk/t ones.

In summary, when fitting high Dk/t CLs, the inevitable (innocuous) bleb response can be reduced to very low levels, but not eliminated. Polymorphism and -megethism are also endothelial conditions. Contrary to blebs, they do not appear after short term wear, but are rather a sign of long term oxygen deprivation.

In 1996, Efron published a review of CL induced polymegethism (Efron 1996d). Polymegethism describes a deviation from the normal approximately same size and polymorphism a deviation from their regular hexagonal shape. The two are often observed together. Cell size is usually expressed as the endothelial cell density (ECD), the number of cells per mm^2. The coefficient of variation (COV) refers to the variation in apparent cell size. It is calculated by "dividing the standard deviation of the cell areas by the arithmetic mean area of all cells in that field". Due to the body's aging process, ECD decreases from about "4000 at birth to 2200 at age 80 years". COV increases in that period. This implies that observations need to take the subject's age into consideration. Changes are a mean degree of change in excess of that expected for a given age. Corneal polymegethism may be a sign for "corneal exhaustion" and a potential upcoming intolerance to CLs that cannot be explained by other factors.

To assess the impact and capabilities of the modern materials, Doughty et al. refitted Hy wearers (average 5.5 yr of wear) with high Dk/t SiHy CLs (Doughty et al. 2006). After only 6 mon, they observed a slight (while not significant) increase in mean endothelial cell area (358 to 363 μm^2) and a slight decrease in the mean COV (30.2% to 29.1%). Mean ECD also decreased slightly (from 2821 to 2774 cells/mm^2). There was a strong correlation between the ECD and a not significant decrease in central corneal thickness (CCT). As CCT decreased, the apparent ECD did decrease as well. The percentage of normal six-sided ("hexagonal") cells increased slightly (58.3 to 60.1%).

They concluded that "... *it remains to be established, however, whether these changes can be directly attributed to oxygen effects (reduced hypoxia and hypercapnia altering the endothelial cells) or to a mechanical effect (in which changes in corneal thickness result in a reorganization of the corneal endothelium)*" (Doughty et al. 2006).

In summary, hypoxia is seen as the main cause for polymegethism and polymorphism that exceeds the expected, age related level in CL wearers. Hence, one should consider to fit all patients with high oxygen transmissible CL, in order to prevent the condition to appear. There are early indications that refitting wearers of low Dk/t CL with high Dk/t CL may result in an improvement, at least a stabilization of the status.

Limbal Hyperaemia and Corneal Vascularization

Asymptomatic limbal hyperaemia (Fig. 10) that is not associated with an inflammatory event or a trauma is a sign of corneal hypoxia (Stapelton 2006). Grades three and four (CCLRU) can easily be seen by the wearer and used to be a common sign at the end of the day when wearing Hy CL. ECP should look, but more importantly ask for such observations, usually referred to as reddish eye, the frequency and severity, at each after care visit. Chronic limbal hyperaemia is a stimulus for corneal vascularization. SiHy materials have shown to be able to reduce limbal hyperaemia (Papas 1994) and should therefore be considered as CL material of first choice, in a prophylactic approach, rather then a reactive one. Corneal vascularization is a permanent condition. The blood vessels may empty and thus not be

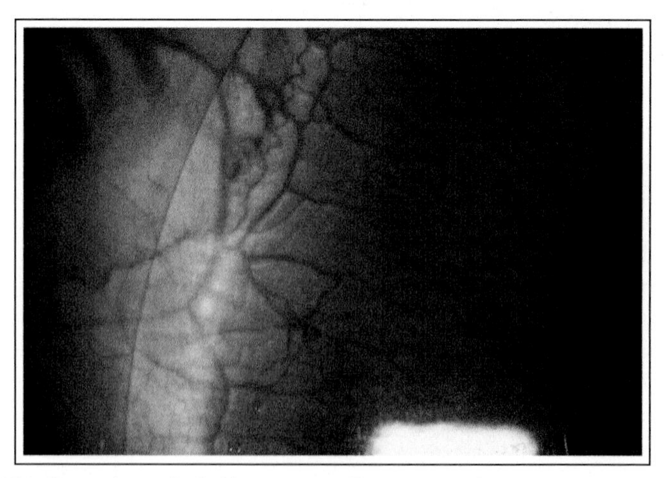

Figure 10 This figure shows limbal hyperaemia (Bruce 2008), Copyright Novartis.

Color image of this figure appears in the color plate section at the end of the book.

easily visible anymore, but can rapidly refill—and may even resume growth (McMonnies 1983). Limbal hyperaemia and vascularization occurs more often in EW of Hy CL than in DW (Holden 1986).

High central Dk/t values may make an ECP feel good, however as most CL wearers are myopes, and minus CLs have the thickest part at the edge of the front optical zone. Depending on the material's refractive index, the CL thickness may be two to threefold in the center, even for powers as low as –4.00 D.

In summary, limbal hyperaemia can appear infrequently or frequently and vary in severity even for a given wearer. Chronic limbal hyperaemia leads to corneal vascularization, a non reversible condition. It should therefore be avoided. High oxygen transmissible materials (SiHy) are nowadays available with a wide variety of features and parameters. ECP can ease their minds by always fitting a high Dk/t CL. They can thus act prophylactic and account for the regular naps that the wearer takes with CL in place, not to speak about the non compliant, notorious sleepers in Hy CLs wearer, who often deny doing so and refuse to accept the idea of any potential long term damages.

Conjunctival Folds

The observation and grading of lid parallel conjunctiva folds (LIPCOF), as a predictor for successes in CL wear, were introduced by Sickenberger (Sickenberg 2000). Grades two and three showed a correlation with reduced comfort, dryness sensation etc. This analysis has, however, not yet become a part of the standard eye examination procedures on a European or global scale.

In 2005, Lofstrom and Kruse wrote about signs that they had seen in the conjunctiva and called it the conjunctival epithelial flap (Loftrom and Kruse 2005). They associated this with CL induced conjunctival staining (CLICS). Their observation (mainly with EW) showed a shuffling or budging of conjunctival tissue by the CL edge, even to the point where it was "digging" into the tissue. Different CL edge designs (knifepoint, chisel and round) of different overall CL designs and materials resulted in more or less of the flaps and CLICS. It may be interesting to compare different edge designs of CL with the same overall design and made of the same material. OCT technology has since facilitated such observations and helped in designing better CL.

In summary, LIPCOF seem to be only relevant if seen in grade two or higher. If used as a contraindication, one may incorrectly exclude the potentially successful CL wearer. ECP may, upon the observation of higher grade LIPCOF choose the best fitting, low water SiHy CL, to lower potential issues caused by dehydration of high water Hy CL.

In case of the observation of CLICS or conjunctival epithelial folds, one may change to a CL with a different design, material and thus a different overall fit.

CONCLUSIONS

With the many signs and symptoms listed above, the reader may think that CL wear is loaded with issues and problems. One may even question why people still wear CL. It is therefore important to put things in perspective. Today, there are more than 100 million CL wearers and the number still increases, despite the substitute (non reversible) offer of refractive surgery and the complementary offer of spectacles. All the progress on the knowledge and understanding of the signs and symptoms has led and will lead to the launch of improved CL (materials, geometries, etc.), systems (frequent replacement, disposables, DD, CW, etc.) and advances in the fitting (preventive) and the ability to manage them. In some cases, the taking away of one major overlaying sign, like SiHy materials have done to hypoxia, allows to better understand and address others, previously secondary ones.

Knowledge is the key to success. Once the signs and symptoms are more understood, the potential cause(s) can be better isolated and addressed in their management. Today, ECP have a whole arsenal of possible actions at hand. But beside all the technical aspects, a lot still depends on the individual wearer's behavior and compliance. Thus, ECP do not only have to be experts in the use of the instruments, the anamnestic and diagnosis, the selection of the management approach, but also in communication, in making the wearers understand their vital role for a life-long successful use of CLs. The fitting process is only the start of a continuous dialogue (at the after care visits or ad hoc) between the wearer and the ECP. The ECP can and should always introduce new technology that may help to solve existing issues, but maybe even more prevent them to surface at all.

REFERENCES

Abelson, M.B. and S. Leung. 2008. The Naked Eye—The Truth About GPC. Review of cornea and contact lenses 9: 16.

Barr, J., D. Fonn, J.A. Bonanno, D. Cavanagh, R. Hill, B.A. Holden, E. Papas, D.F. Sweeney and G. Wilson. 2005. Elemental need: achieving normoxia; measuring oxygen uptake; sustaining limbal health; O2 measurements and needs. Contact Lens Spectrum, (Supp.) August 2005. 1–15.

Bergmanson, J. and L.W. Chu. 1982. Corneal response to rigid contact lens wear. Br. J. Ophthalmol. 66: 667–675.

Bergmanson, J.P. 1987. Histopathological analysis of the corneal epithelium after contact lens wear. J. Am. Optom. Ass. 58: 812.

Bonnano, J. and K.A. Polse. 1987. Corneal acidosis during contact lens wear. Effects of hypoxia and CO2. Invest. Ophthalmol. Vis. Sci. 28: 1514–1520.

Bron, A.J. and R.C. Tripathi. 1973. Cystic disorders of the epithelial epithelium I: Clincal aspects. Br. J. Ophthalmol. 57: 361.

Bruce, A.S. 2008. A guide to clinical contact lens management, CIBA Vision Corporation, USA.

Doughty, M.J., B.M. Aakre, A.E. Ystenaes and E. Svarverud. 2006. Short-term adaptation of the human corneal endothelium to continuous wear of silicone hydrogel (lotrafilcon A) contact lenses after daily hydrogel lens wear. Optom. Vis. Sci. 82: 473–480.

Efron, N. 1996a. Contact lens-induced corneal staining. 1996. Optician 212: 18–26.

Efron, N. 1996b. Contact lens induced epithelial microcysts. Optician. Vol. 211 : 24–28.

Efron, N. 1996c. Contact lens induced endothelial blebs. Optician 212: 34–7.

Efron, N. 1996d. Contact lens induced polymegethism. Optician 212: 20–8.

Efron, N. 2004a. Mucin balls—Part 1 and Mucin balls—Part 2. Optician 227: 18–21.

Emch, A.J. and J.J. Nichols. 2009. Protein identified from care solution extractions of silicone hydrogels. Optom. Vis. Sci. 82: 123–31.

Evans, D. and S. Fleiszig. 2010. Paper at BCLA conference May 2010 Birmingham. United Kingdom.

Fleiszig, S.M., T.S. Zadi and G.B. Pier. 1994. Modulation of pseudomonas aeruginosa adherence to the corneal surface by mucous. Infect Immunol. 62. 1799–1804.

Fleming, C., R. Austen and S. Davies. 1994. Pre-corneal deposits during soft contact lens wear. Optom. Vis. Sci. 71: 152–153.

Fonn, D., N. Pritchard and K. Dumbleton. 2000. Factors affecting the success of silicone hydrogels. pp. 214 - 234. In D.F. Sweeney [ed.]. Silicone Hydrogels—The rebirth of continuous wear contact lenses. Butterworth-Heinemann. Oxford, United Kingdom.

Fonn, D. and A. Sivak. 2005. Benefits and challenges of high-Dk/t materials. Contact Lens Spectrum. 2: 28–31.

Grant, T., M. Chong and C. Vajdic. 1988. Contact lens induced peripheral ulcers during hydrogel contact lens wear. J. Cont. Lens Ass. Ophthalmol. 24: 145–151.

Harvitt, D.M. and J.A. Bonanno. 1999. Re-evaluation of the oxygen diffusion model for predicting minimum contact lens Dk/t values needed to avoid corneal anoxia. Optom Vis. Sci. 76: 713–9.

Hine, N., A. Back and B.A. Holden. 1987. Aetiology of arcuate epithelial lesions induced by hydrogels. Brit Contact Lens Ass. Trans. Ann. Clinical Conference 47–50.

Holden, B.A. and G.W. Merz. 1984. Critical oxygen level to avoid corneal edema for daily and extended wear contact lenses. Invest. Ophthalmol. Vis. Sci. 25: 1161–7.

Holden, B.A., D.F. Sweeney and A. Vannas. 1985a. Effects of long-term extended contact lens wear on the human cornea Invest. Ophthalmology Vis. Sci. 26: 1489–501.

Holden, B.A. and A. Vannas. 1985b. Effects of longterm extended contact lens wear on the human cornea. Ophthalmol. Vis. Sci. 26: 1489–1501.

Holden, B.A., D.F. Sweeney and R.G. Seger. 1986a. Epithelial erosions caused by thin high water content lenses. Clin. Exp. Optom. 69: 103–107.

Holden, B.A. and D.F. Sweeney and H. Swarbrick. 1986b. The vascular response to long-term extended contact lens wear. Clin. Exp. Optom. 69: 112–9.

Holden, B.A., T. Grant, M. Kotow, C. Schnider and D.F. Sweeney. 1987. Epithelial microcysts with daily and extended wear of hydrogel and rigid gas permeable contact lenses. Invest Ophthalmol. Vis. Sci. 28: 372

Holden, B.A., P.R. Sankaridurg and I. Jalabert. 2000. Adverse events and infections: which ones and how many? Pp 151. In: Sweeney. D.F. [ed.]. Silicone Hydrogels—The rebirth of continuous wear contact lenses. Butterworth-Heinemann. Oxford. UK.

Ichiyama, H., N. Yokoi and A. Nishizawa. 1999. Fluorophotometric assessment of rabbit cornealepithelial barrier function after rigid contact lens wear. Cornea 18: 87–91.

Inagaki, Y., A. Akahori, K. Sugimoto, A. Kozai, S. Mitsunaga and H. Hamano. 2003. Comparison of corneal endothelial bleb formation and disappearance process between rigid gas-permeable and soft contact lenses in three classes of Dk/l. Eye Contact Lens 29: 234–7.
Jalabert, I. 2005. The effect of long term wear of soft lenses of low and high oxygen transmissibility on the corneal epithelium AAO conference.
Jalabert, I., N. Carnt and N. Naduvilath. 2006. The relationship between solution toxicity, corneal inflammation and ocular comfort in sot contact lens daily wear. 2006 ARVO e-abstract.
Josephson, J.E. 1979. Coalescing microcysts after long term use of extended wear lenses. Int. Contact Lens Clin. 6: 24.
Josephson, J.E., G. Zantos and B. Chaffey. 1988. Differentiation of corneal complications observed in contact lens wearers. J. Am. Optom. Ass. 59: 679–85.
Jones, L. and D. Jones. 1995. Photofile: Part Three: Superior epithelial arcuate lesions. Optician 208: 32–3.
Jones, L., C. May, L. Nazar and T. Simpson. 2002. *In vitro* evaluation of the dehydration characteristics of silicone hydrogel and conventional hydrogel contact lens materials. Contact Lens Anterior Eye 25: 147–156.
Jones, L., K. Dumbleton, S. Bayer, D. Fonn and B. Long. 2003. Corneal staining associated with silicone-hydrogel materials used on a daily wear basis with RENU and AOSEPT care regimens. Program book AAO Int. congress.
Kallinikos, P. and N. Efron. 2004. On the etiology of keratocyte loss during contact lens wear. Invest Opthal. Vis. Sci. 45: 3011–3020.
Keay, L., I. Jalabert, D.F. Sweeney and B.A. Holden. 2001. Microcysts: Clinical significance and differential diagnosis. Optometry 72: 452–460.
Ladage, P.M., D.H. Ren and W.M. Pentroll. 2003. Effects of eyelid closure and disposable and silicone hydrogel extended wear contact lens wear on rabbit corneal epithelial proliferation. Invest. Ophthalmol. Vis. Sci. 44: 1843–9.
La Hood, D. and T. Grant. 1990. Striae and folds as indicators of corneal edema. Optom. Vis. Sci. 67: 196.
Lakkis, C. and P. Ther. 2011. Clinical performance of silver salt-infused silicone hydrogel contact lenses over six-months of daily wear. AAO presentation.
Loftrom, T. and A. Kruse. 2005. A conjunctival response to silicone hydrogel lens wear. Contact Lens Spectrum 9: 42–4.
McMonnies, C.W. 1983. Contact lens induced corneal vascularisation. Int. Contact Lens Clin. 10: 12–15.
McNally, J., R. Chalmers and R. Payor. 1987. Corneal epithelial disruption with extremely thin hydrogel lenses. Clinical Experimental Optometry 70: 106–10.
Meng, C.L., A.D. Graham, K.A. Polse, N.A. Mcnamara and T.G. Tieu. 2000. The effects of one -hour wear of high –Dk soft contact lenses on corneal pH and epithelial permeability. CLAO. 26: 130–3.
Merrian-Webster. 2011. Online Dictionary http://www.websters-online-dictionary.org/Visited on September 02. 2011. 17:00.
Minarik, I. and J. Rapp. 1989. Protein deposits on individual hydrophilic contact lenses; effects of water and ionicity. CLAO J. 15: 185–8.
Naduvilath, T.J. 2003. Modeling of risk factors associated with soft contact lens related corneal infiltractive events. PhD Thesis (Newcastle, University of Newcastle).
Papas, E., C. Flemming, R. Austen and B.A. Holden. 1994. High Dk soft contact lenses reduce the limbal vascular response. Optom. Vis. Sci. 71: 14.
Polse, KA. 1979. Tear flow under hydrogel contact lenses. Invest. Ophthalmol. Vis. Sci. 38: 201.
Polse, K.A., R. Rivera, C. Gan, J. Bonanno and S. Cohen. 1992. Contact lens wear affects corneal pH: Implications and new directions for contact lens reserach. J. Br. Contact Lens Ass. 15: 171–7.

Ren, D.H., K. Yamamoto, and P.M. Ladage. 2002. Adaptive effects of 30-night wear of hyper O2 permeable contact lenses on bacterial binding and corneal epithelium. A 1-year clinical trial. Ophthalmology 109: 27–40.
Rivera, R.K., C.M. Gan, K. Polse, J.A. Bonnano and I. Fatt. 1993. Contact lenses affect corneal stromal pH. Optom. Vis. Sci. 70: 991–997.
Ruben, M., N. Brown, N. Lobascher, J. Chaston and J. Morris. 1976. Clinical manifestation secondary to soft contact lens wear. Brit. J. Ophthalmol. 60: 529.
Sankaridurg, P.R., M. Wilcox, S. Sharma, U. Gopinathan, D. Janakiraman, S. Hickson, N. Vuppala, D.F. Sweeney, G.N. Rao and B.A. Holden. 1996. Haemophilus influenza adherent to contact lenses is associated with the production of acute ocular inflammation. J. Clin. Mircobiol. 34: 2426–2431.
Sankaridurg, P.R. 1999a. Corneal infiltrative conditions with EW of disposable hydrogels. PhD thesis. University of South Wales. Sydney, Australia.
Sankaridurg, P.R., S. Sharma, M. Wilcox, D.F. Sweeney, T.J. Naduvilath, B.A. Holden and G.N. Rao. 1999b. Colonization of hydrogel lenses with Streptoccocus pneumoniae: Risk of development of corneal infiltrates. Cornea 18: 289–295.
Schepper, J. and M.S. Vaughan. 1993. A new *in vitro* method for assessing the potential toxicity of soft contact lens solutions. CLAO J. 19: 54–7.
Schoessler, J.P. and M.J. Woloschk. 1981. Corneal endothelium in veteran PMMA contac lens wearers. Int. Contact Lens Clin. 8: 19.
Schoessler, J.P. 1983. Corneal endothelial ploymegethism associated with extended wear. Int. Contact Lens Clin. 10: 144.
Senchyna, M., M.A. Glasier, N. Thorgood, L. Jones and K. Dumbleton. 2002. Assessing the degree of denatured lysozyme deposited on conventional and silicone-hydrogel contact lenses. AAO conference.
Sickenberg, W., H. Pult and B. Sickenberger. 2000. LIPCOF and contact lens wearers—A new tool to forecast subjective dryness and degree of comfort of contact lens wearers. Contactologia. 22: 74–9.
Stapleton, F. 2004. BCLA conference book abstract. UK.
Stapelton, F., S. Streeton, E. Papas, C. Skotnitsky and D.F. Sweeney. 2006. Silicone hydrogel contact lenses and the ocular surface. The Ocular Surface 4: 24–43.
Sweeney, D.F., L. Keay, I. Jalabert, P.R. Sankaridurg and B.A. Holden. 2000. Clinical performance of silicone hydrogel lenses. In: D.F. Sweeney [ed.]. Silicone Hydrogels—The rebirth of continuous wear contact lenses. Butterworth-Heinemann. Oxford. UK, pp. 128–32.
Watsky, M.A. 1995. Keratocyte gap junctional communication in normal and wounded rabbit corneas and human corneas. Invest. Ophtalmol. Vis. Sci. 36: 2568–2576.
Wilcox, M., R. Padmaja, P.R. Sankaridurg, J. Lan, D. Pearce, A. Thakur, H. Zhu, L. Keay and F. Stapleton. Inflammation and infection and the effects of the closed eye, pp. 64. In: D.F. Sweeney (ed.). 2000. Silicone Hydrogels—The rebirth of continuous wear contact lenses. Butterworth-Heinemann, Oxford, UK.
Williams, L. 1986a. Transient endothelial changes *in vivo* human corneas. PhD thesis. University of New South Wales, Sydney, Australia.
Williams, L. and B.A. Holden. 1986b. The bleb response of the endothelium decreases with extend wear of contact lenses. Clin. Exp. Optom. 69: 90.
Zantos, S.G. and B.A. Holden. 1977. Transient endothelial changes soon after wearing soft contact lenses. Am. J. Optom. Physiol. Opt. 54: 856.
Zantos, S.G. and B.A. Holden. 1978. Ocular changes associated with continuous wear of soft contact lenses. Aust. J. Optom. 61: 418.
Zantos, S.G. 1983. Cystic formations in the corneal epithelium during extended wear of contact lenses. Int. Contact Lens Clin. 10: 128.
Zhao, Z., N.A. Carnet, Y. Aliwarga, X. Wei, T. Naduvilath, Q. Garret, J. Korth and M.D. Wilcox. 2009. Care regimen and lens material influence on silicone hydrogel contact lens deposition. Optom. Vis. Sci. 86: 251–9.

18

Therapeutic use of Contact Lenses

Victoria de Juan Herraez,[1,a,*] *Stephanie Campbell*,[2] *Guadalupe Rodriguez Zarzuelo*[1] and *Raul Martin Herranz*[1]

SUMMARY

This chapter provides a description of how contact lenses may be used for therapeutic purposes. The aim may be to improve visual function (patients with irregular astigmatism, with existing pathology or high anisometropia), to change the corneal surface shape in order to improve the vision without aids, to promote healing of a damaged and painful corneal epithelium, or to manage post surgical complications such as aqueous humour leakage.

THERAPEUTIC CONTACT LENS DEFINITION

A therapeutic contact lens is used in an attempt to restore visual function following pathology or trauma to the eye, or it may serve as a bandage lens, relieving pain and allowing the eye to heal. A therapeutic contact lens may also incorporate a refractive design in order to correct ametropia, but this is not always necessary.

[1]Instituto Universitario de Oftalmobiología Aplicada (IOBA-Eye Institute) Optometry Research Group, School of Optometry, University of Valladolid, Paseo de Belén, 17-Campus Miguel Delibes, 47011 Valladolid, Spain.
[a]E-mail: victoria@ioba.med.uva.es

[2]School of Optometry and Vision Sciences. Cardiff University, MaindyRd. CF24 4LU Cardiff Wales, UK; E-mail: campbells3@cardiff.ac.uk
*Corresponding author

INTRODUCTION

Bandage lenses have become more diverse in recent years due to developments in contact lens materials and designs, especially with increased oxygen permeability in silicone hydrogels (Kanpolat and Uçakhan 2003), and the advances in ocular surface understanding and the development of corneal refractive surgery.

Therapeutic contact lenses can be both rigid or soft in nature, they may also be 'hybrid' in design where a HEMA soft lens skirt surrounds an RGP centre. Sometimes, an RGP lens is placed over a thin daily lens to provide a cushioning effect, whilst maintaining the clarity and refractive nature of the overlying hard lens. This is known as a 'piggyback' lens.

In fact, almost any type of lens or lens combination may be used in a therapeutic setting. This demonstrates the diverse and dynamic nature of the fitting 'experiment' used to compliment ophthalmological treatment. Some main fitting indications are explained below.

Contact Lenses to Improve Visual Function

A disrupted corneal surface as produced in keratoconus or ocular surface injury or disease often results in irregular astigmatism or high ametropia, which cannot be corrected adequately with spectacles or regular soft contact lenses alone. A spherical RGP lens to bridge, or vault over these defects allow the eyes natural tear film to 'fill in' and mask these irregularities. This means that the front surface of the contact lens now becomes the eye's new primary refractive surface, and it's regular shape allows the light to refract more efficiently, thus often improving vision to a great extent. Sometimes glasses are prescribed to be worn also, to allow residual astigmatism, or anisometropia to be corrected for, or a near addition to be prescribed.

Spherical back surface RGP lenses working in this way have the great benefit of hiding the effects of the frequently changing astigmatism of the corneal surface under the lens. When fitting, due to the new ocular surface produced by the lens, it is important to approximate the RGP prescription required using retinoscopy and subjective techniques from first principles as the final result often differs greatly from the refractive error expected.

Large diameter contact lenses which are supported by the sclera rather than resting on the cornea are indicated here, the stability of the lens and the visual improvement is often noticeably better with 14 mm or 18mm varieties, for example. Anecdotally, this seems to work very well for severe keratoconus where the cone is highly prominent and the lens usually unstable. Due to it's ability to bridge the cornea, great amounts of astigmatism can be attempted to be corrected with the use of a spherical lens and the tear lens only.

There is evidence to show an altered corneal nerve distribution around the cone, perhaps contributing to this discomfort (Al-Aqaba et al. 2011). These patients are inherently atopic, and suffer from allergic conditions such as excema, hayfever and asthma. This means wearing times of lenses are often limited and the patient may require topical medication to reduce inflammation of the tarsal conjunctiva and a good spare pair of glasses for occasions when contact lens wear is difficult.

Soft lenses have been produced using a more rigid polymer that allows some displacement of tears to correct irregular astigmatism, as above. The lenses are designed to be thicker with a higher water content, larger in diameter and so more stable to incorporate the residual astigmatism as required into each lens and are lathe cut (Kerasoft, Ultravision, Leighton Buzzard, UK). This balance between RGP clarity of vision and the comfort of a soft lens may be useful in some cases.

Injury may result in astigmatism in the healing cornea as the collagen fibres contract to bind together again. Corneal scars often heal flat, and should this be central, then the healed cornea can result in a flatter central zone, steepening toward the periphery. The lenses that complement this unusual corneal shape are called reverse-geometry lenses and may result in good visual acuity where conventional RGP designs do not provide an acceptable fit or vision (Martin and de Juan 2007).

As scars heal flat, the position of scalpel insertion in cataract surgery is often considered to correct preoperative astigmatism, and take advantage of this healing effect. Likewise, ocular surgery where the cornea has been cut or sutured, may result in irregularities and affect refraction. Corneal graft surgery is a great example of where ophthalmologists and optometrists must work together in order to work out which sutures are to be removed in order to aid balanced healing and result in the most spherical prescription so that emmetropia is guided and a therapeutic lens may be avoided.

Contact Lenses to Heal and Relieve Pain

This is often the primary indication for fitting a therapeutic lens. It is because of this use that these therapeutic type of lenses are often referred to as 'bandage' lenses (Fig. 1). As above, many lenses may be used to relieve this purpose but when a short term relief of pain is required for a healing epithelium, a soft hydrogel is the lens of choice. An average-commercially available hydrogel has been shown to provide adequate fit and comfort in 80% of cases (Rubinstein 1995).

Pain is generally caused by exposed nerve endings underlying a damaged epithelium, and the nerves make contact with the air and are severely irritated by the forces of the eyelid upon blinking. Conversely, it is the rough surface of the cornea in band keratopathy can cause the tarsal

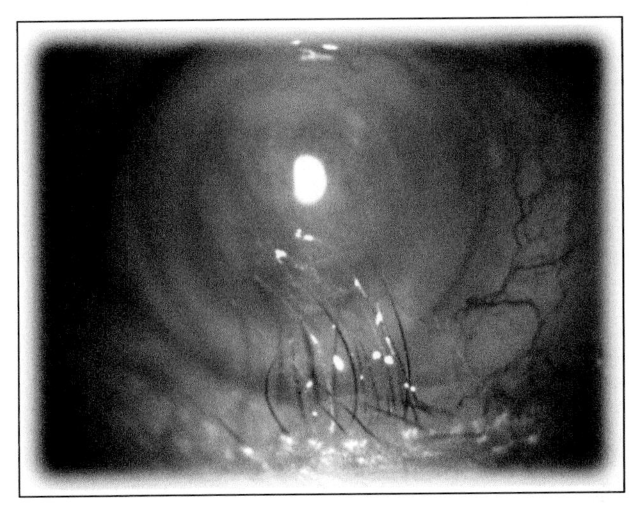

Figure 1 This figure shows the use of a soft bandage contact lens to protect the cornea in a patient with trichiasis in the inferior eyelid.

Colour image of this figure appears in the color plate section at the end of the book.

conjunctiva pain (Erlich 2001). Placing a contact lens over the corneal surface spreads the eyelid closure force over a larger area and protects both surfaces as necessary.

Often in an epithelial erosion, the epithelium heals a lot during the night, and when the patient wakes, the new unstable epithelium sticks to the tarsal conjunctiva and is ripped off when the eyes are first opened or rubbed. A bandage contact lens allows the superficial corneal layers to heal under the protection of this lens, preventing sleeptime adhesion and allowing uninterrupted time for anchorage and a stable attachment of epithelium to the basal lamina attaching the epithelium to the stroma more permanently.

It is well accepted that silicon hydrogel materials provide a much better oxygen transmissibility than regular hydrogels. Perhaps due to the natural stiffness of the silicon hydrogel, it is thought that standard hydrogel materials provide better comfort initially, but it has been demonstrated that particularly over longer periods of time, and especially with overnight wear, silicon hydrogel contact lenses provide improved comfort over conventional materials (Martin et al. 2007).

Continuous wear is often a requirement of a therapeutic lens, and while overnight wear is a great risk factor for corneal infiltrates and infection, such events are less clinically severe with silicon materials than standard hydrogel materials when soft lenses are compared (Efron et al. 2010). This may be due to the presumed biocompatibility of a higher oxygen environment in the pre lens tear film, and the oxygen levels available to the respiring healing epithelial cells.

Evidence shows that large diameter RGPs may also be useful to relieve pain and protect the eye in conditions such as Stevens–Johnson syndrome, ocular cicatricial pemphigoid, exposure keratitis, superior limbal keratoconjunctivitis and filamentary keratitis (Romero-Rangel et al. 2000), providing more comfortable wear than conventional RGP. (DeNaeyer 2008).

Post-surgical Contact Lenses

Leaking blebs may occur following trabeculectomy surgery resulting in incomplete closure of the wound; if leakage is not severe enough to require a stitch then a large diameter soft contact lens may be used to help flatten and protect the healing area. These are usually reviewed in 2–3 days post surgery, whilst monitoring symptoms of discomfort and blur, which may indicate hypotony. Specialist soft contact lenses with "sectoral management" are now available which allow for a flatter fitting edge in a position of the practitioner's choice in order to accommodate ocular surface anomalies such as very large blebs whilst maintaining an optimal fit without edge fluting.

LASIK (laser-assisted *in situ* keratomileusis) flattens the central cornea under a flap of lifted epithelium. A therapeutic contact lens may be used in order to maintain the "flap" in its preferred position while superficial attachment takes place. LASEK (laser assisted sub-epithelial keratectomy) is a procedure that removes the whole epithelium before laser-ablation of the anterior stroma takes place. With the basal lamina having been removed, which usually acts as the adhesion of the epithelium and stroma, regrowth may be slower and require the protection of the cushioning bandage lens as described above.

Contact Lenses in Allergic Eye Disease

It has been demonstrated that patients suffering from hypersensitivity reactions may benefit from lenses even when the eye is in an irritated state. The likely reason for this is that a large diameter soft lens prevents tarsal conjunctiva inflammatory response like papillae from rubbing on the cornea and creating further discomfort. When fitted with minimal movement and a stable lens fit, this will limit the frictional irritation of the lens (Quah et al. 2006).

Where refractive correction is necessary, daily disposable contact lenses prevent the gradual insidious hypersensitivity reaction that can result from preservatives in a cleaning solution becoming stuck in the contact lens matrix. Daily disposable contact lenses have been linked to the prevention of allergic reaction to airborne antigens such as pollen (Wolffsohn 2011). It may be the physical barrier to the external environment and also the effects of the lubricating agent provided by the surface treatments of modern hydrogel lenses.

It is important that any topical medication is carefully considered and that any continuous wear modality is closely monitored in clinic so that any further serious hypersensitive reaction is picked up at an early stage. Where topical steroids are used in conjunction with a therapeutic lens, it is vital that the patient understands the additional risks of microbial keratitis and what the signs and symptoms of an infection are in addition to what they are already experiencing. Steroids carry a high risk with therapeutic contact lens wear as the eye's immune system is compromised, especially in a damaged eye more open to pathogens.

Contact Lenses for Drug Delivery

The concept of drug delivery system through a hydrophilic contact lens was introduced in the 1960's (Wichterle 1960). There are different approaches to drug delivery systems through a contact lens (Xinming et al. 2008). The main idea is that the drug will be 'soaked up' by the contact lens and released over a longer period of time and so have more contact time with the cornea than conventional drug delivery by eye drops would allow. This allows more drugs to be used by the eye and hopefully with a more stable sustained effect rather than peaks in drug activity. This is particularly important in the treatment of glaucoma (Lavik et al. 2011), to remove the reliance on patient compliance of eye drop usage for effective treatment.

The popularity of corneal refractive surgery has developed methods to avoid damage to the epithelium and minimise pain. There still exists certain techniques where this is unavoidable and PRK (photorefractive keratectemy) is one. Patients are required to use pain-relieving drops, including anaesthetics for some time after surgery. While a bandage lens as above will protect the anterior eye from the external environment, therapeutic lenses have also demonstrated their ability to act ac a drug delivery device. Peng (Peng et al. 2012) have demonstrated the ability of a silicon hydrogel lens in conjunction with vitamin E to provide a slow release of anaesthetic drug over 2 days. This is importance for patient comfort, eye healing and to avoid excessive amount of drug present on the cornea at any time.

The efficacy of delivery of the drug from a contact lens will depend on the relative molecule size and interaction between the drug and the lens polymer, and so it is not possible to design a single contact lens to serve all drugs (Ciolino et al. 2011).

Contact Lenses to Prevent Anisometropia

There are situations in which healthy eyes require contact lenses also. Lenses incorporating refractive power may be considered therapeutic because they are necessary to maintain optimal visual function, and it may not be possible wear glasses in order to achieve this. For example, very young babies who have undergone removal of congenital cataract are not often fitted with an intraocular lens (IOL) until 6 weeks of age in order to allow the eye to mature in size pre-operatively.

Contact lens used until this stage to correct the large hypermetropic error will allow normal visual stimulation, and hence allow normal development of vision. Another example may be a situation of high anisometropia acquired in adulthood, particularly following surgery. Contact lenses can be used to balance the refractive power between the eyes required in spectacles and so reduce anisokonic effects and make vision more comfortable, and encourage binocularity.

Contact Lens to Change Corneal Shape

Orthokeratology is a method of correcting refractive errors using a series of rigid contact lenses worn during the sleeping hours to change the shape of the cornea overnight, thus altering the refractive error. The premise is that progressive flattening of the cornea through the use of a series of flatter fitting lenses will reduce myopia temporarily.

Painted Contact Lenses

Painted contact lenses have a variety of uses: cosmetic, therapeutic and handling. Although some lenses are commercially available with predetermined colors and powers, others are custom made for each patient (Figs. 2 and 3).

There are several indications for painted lens use:

- Aniridia or polycoria, resulting either congenitally or surgically, can result in severe visual discomfort, photophobia, because of the lack of pupil aperture (aniridia) or multiple retinal images (from polycoria). If corrected using a painted lens with just one appropriately selected pupil, visual ability is often much improved. A large diameter soft lens is typical, which provides stability and minimal movement on the eye.
- Occlusion of an eye may be of benefit in cases of intractable or monocular diplopia, and also in orthoptic treatment in an attempt to correct dense amblyopia in the fellow eye.

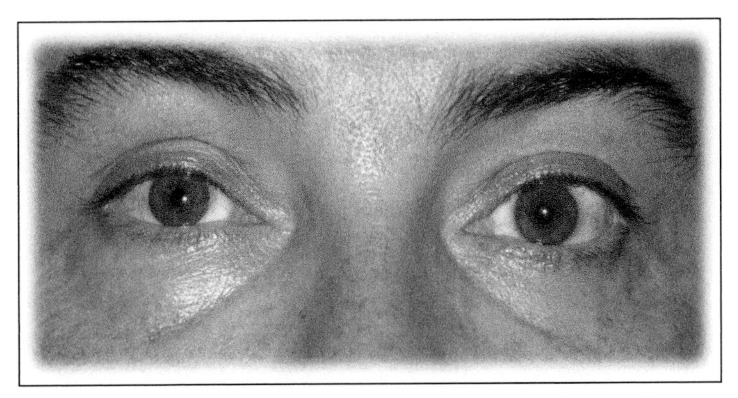

Figure 2 This figure shows the cosmetic effect of a painted contact lens in right eye.
Colour image of this figure appears in the color plate section at the end of the book.

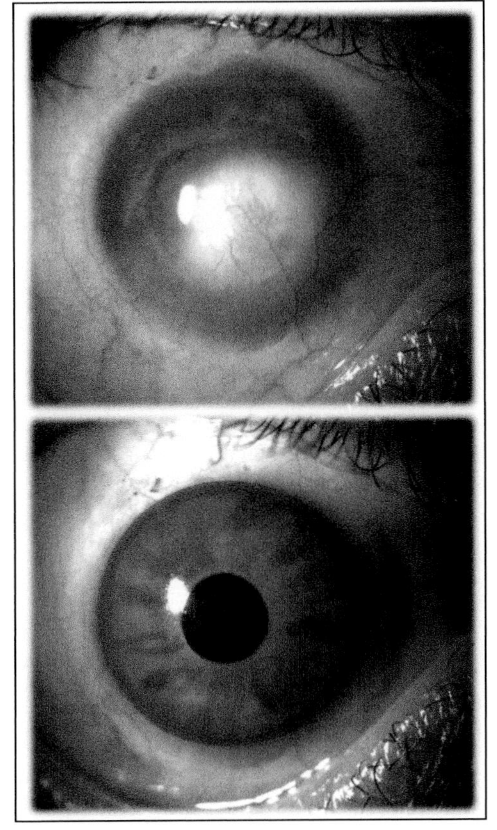

Figure 3 This figure shows the cosmetic contact lens effect with low biomicroscopic magnification.
Colour image of this figure appears in the color plate section at the end of the book.

Coloured lenses are sometimes used monocularly in patients suffering from colour deficiency to help them distinguish colour hues in daily life. It is important to note that the use of these lenses is not permitted for use during colour vision testing for job requirements.

The main use of coloured lenses is for cosmetic purpose-replicating the appearance of the healthy eye upon the surface of an eye which has been opaque, squinting, or altered in shape.

CONCLUSIONS

Therapeutic lenses are diverse in nature and in fitting approach. Lenses are often used off-license and there are few limitations or guidelines; but in many cases these contact lenses help in pain management, improving aesthetic effect and increase patient's quality of live.

REFERENCES

Al-Aqaba, M., L. Faraj, U. Fares, A.M. Otri and H.S. Dua. 2011. The morphologic characteristics of corneal nerves in advanced keratoconus as evaluated by acetylcholinesterase technique. AJO 152: 374–376.

Bendoriene, J. and U. Vogt. 2006. Therapeutic use of silicone hydrogel contact lenses in children. Eye Contact Lens 32: 104–8.

Cheng-Chun, P., M.T. Burke and A. Chauhan. 2012. Transport of topical anesthetics in vitamin e loaded silicone hydrogel contact lenses. Langmuir. 28: 1478–87.

Ciolino, J.B., S.P. Hudson, A.N. Mobbs, T.R. Hoare, N.G. Iwata, G.R. Fink and D.S. Kohane. 2011. A prototype antifungal contact lens. Invest Ophthalmol. Vis. Sci. 52: 6286–91.

DeNaeyer, G.W. 2008. Exploring the therapeutic applications of contact lenses. Optometric Management 12 (www.optometric.com).

Efron, N., P.B. Morgan, M.A. Hill, M.K. Raynor and A.B. Tullo. 2010. Incidence and morbidity of hospital-presenting corneal infiltrative events associated with contact lens wear. Clin. Exp. Optom. 88: 232–239.

Lavik, E, M.H. Kuehn and Y.H. Kwon. 2011. Novel drug delivery systems for glaucoma. Eye 25: 578–86.

Martin, R. and V. de Juan. 2007. Reverse geometry contact lens fitting in corneal scar caused by perforating corneal injuries. Cont. Lens Anterior Eye 30: 67–70.

Martin, R., V. de Juan, G. Rodriguez, S. Martin and S. Fonseca . 2007. Initial comfort of lotrafilcon A silicone hydrogel contact lenses versus etafilcon A contact lenses for extended wear. Cont. Lens Anterior Eye 30: 23–28.

Quah, S.A., C. Hemmerdinger, S. Nicholson and S.B. Kaye. 2006. Treatment of refractory vernal ulcers with large-diameter bandage contact lenses. Eye Contact Lens 32: 245–7.

Romero-Rangel, T., P. Stavrou, J. Cotter, P. Rosenthal, S Baltatzis and C.S. Foster. 2000. Gas-permeable scleral contact lens therapy in ocular surface disease. Am. J. Ophthalmol. 130: 25–32.

Rubinstein, M.P. 1995. Disposable Contact lenses as therapeutic devices. J. Br. Contact Lens Assoc. 18: 95–7.

Wichterle, O.L.D. 1960. Hydrophilic gels for biological use. Nature 185: 117–118.

Wolffsohn, J.S. and J.C. Emberlin. 2001. Role of contact lenses in relieving ocular allergy. Contact Lens Anterior Eye 3: 169–172.

Xinming, L., C. Yingde, A.W. Lloyd, S.V. Mikhalovsky, S.R. Sandeman, C.A. Howel and L. Liewen. 2008. Polymeric hydrogels for novel contact lens-based ophthalmic drug delivery systems: a review. Contact Lens Anterior Eye 31: 57–64.

19

The Tear Film Interaction with Contact Lenses

Montani Giancarlo,[1,*] *Arima Valentina*[2] and *Maruccio Giuseppe*[2,3]

SUMMARY

Contact lens (CLs) fitting causes modifications in the tear film structure with potential negative effects on patient comfort and tolerability. This condition is known as contact lens related dry eye (CLRDE) and its effects lead to a large percentage of dropouts of CLs wearers. In order to reduce this condition to a minimum, it is essential that the eye care professionals (ECPs) chooses the best combination of CLs, rewetting drops, and care systems that best match the tear film structure and the visual needs of the CLs wearer. The first part of this chapter will present the effects that CLs wear has on tear film structure and on its components, while the second part will describe the CLRDE management using the products in the field of CLs, rewetting drops and contact lens care products.

INTRODUCTION

A 1995 report from the National Eye Industry indicated that CLRDE is a major sub-classification of dry eye disease, and various visual activities, including computer terminal work or wearing CLs in low humidity

[1]Centro di Ricerche in Contattologia, Università del Salento, Via per Arnesano, Lecce, 73100 Italy; E-mail montani.gc@libero.it

[2]NNL Institute Nanoscience-CNR, Via per Arnesano, Lecce, 73100 Italy.

[3]Dipartimento di Matematica e Fisica, Università del Salento, Via per Arnesano, Lecce, 73100 Italy.
*Corresponding author

environmental conditions, can intensify this condition (Lemp 1995). In various studies it has been noted that up to 75% of contact lens (CL) wearers present symptoms of dryness (Sindt 2007, Begley et al. 2001, Fonn et al. 1999, Pritchard and Fonn 1995, Nichols and Sinnott 2006). A non CL wearer who has moderately dry eyes would probably not have symptoms until he begins to wear CL. The pathogenesis of CLRDE is complex and multifactorial (Sindt 2007, Asbell and Ucakhan 2006) and can be summarized in Fig. 1.

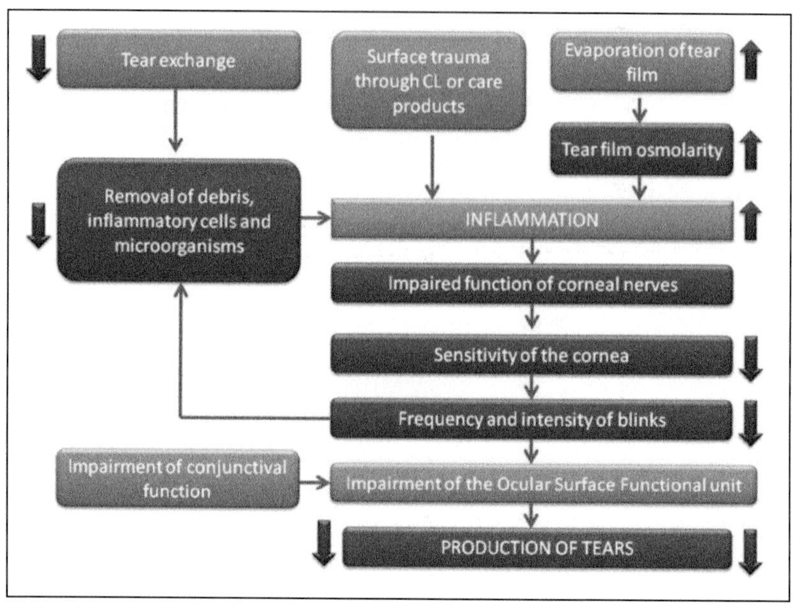

Figure 1 Possible pathogenetic mechanism of CLRDE (modified from Asbell and Ucakhan 2006).

CLRDE Induced Contact Lens Dropout

A challenge for all ECPs is to reduce to a minimum the number of wearers who discontinue use of their CLs. Numerous studies have been conducted to quantify the percentage of wearers who every year discontinue CLs revealing values from 5% to 34% (Begley et al. 2001, Fonn et al. 1999, Pritchard and Fonn 1995, Nichols and Sinnott 2006). The average dropout in America was 17%, in Asia 31%, and in Europe/Middle East/Africa 30.4% (Rumpakis 2010). Pritchard et al. found a dropout rate in Canada of 34% (Pritchard et al. 1999) and Richdale et al. found a dropout rate of 24% in USA (Richdale et al. 2007). Considering that there are more than 125 million CL wearers in the world, it is easy to understand the significant economic impact on the CLs market due to dropout.

The reasons for dropout have been studied in various works and the leading cause for discontinuation across all regions of the world has been found to be the reduction of comfort. Young found discomfort as the major cause of dropout (51%), and the most common symptom of discomfort was dryness (40%) (Young 2002). Weed cited the main reasons for discontinuation as discomfort (27%), dryness (16%) and red eyes (11%) (Weed 1993). Pritchard listed the primary reasons for CLs abandonment as discomfort (40% for soft CLs and 58% for hard CLs) and dryness (10% for soft CLs and 11% for hard CLs) (Pritchard 1999). Richdale showed that ocular symptoms such as discomfort and dryness were the leading causes of patient dissatisfaction and discontinuation (Richdale 2007). A research conducted in six clinical sites in North America showed that the symptoms of ocular dryness and discomfort are relatively common among CLs wearers, and that they get worse toward the end of the day (Begley et al. 2001). These findings suggest that ECPs should examine patients who wear CLs toward the end of the day to match symptomatic patients better. The Dry Eye Work Shop Study (DEWS) also reported that the primary reasons for CLs intolerance were discomfort and dryness (Smith 2007).

CONTACT LENS INTERACTION WITH TEAR FILM

Contact Lens and Tear Film: The Pre and Post-lens Film

The tears, responsible for eye cleaning and lubrification, form a three layer film (with lipid, aqueous and mucous layers) on the cornea that appears relatively thin compared with the CL thickness. As a result, the shear mass of any CL may disrupt normal tear physiology by thinning and break-up of the tear film, (Holly 1981, Kline et al. 1975, Patel 1987) by interrupting tear film reformation, (Holly 1981, Tomlinson 1992) by rupturing the lipid layer (Guillon in Holly 1986, Hamano 1981) with consequent increases in tear film evaporation, (Tomlinson et al. 1982) and by causing, in the case of soft CLs, per-evaporation of fluid from the corneal tissue (Holden 1986).

As soon as a CL is inserted, reflex tearing may cause a transient hypo-osmolar tear film and a thick pre-lens tear film. However, during the initial settling period following CL insertion, the pre-lens tear film thins over the center of the CL (Nichols et al. 2004, Wang et al. 2003a, Nichols et al. 2003). Measurements of pre-lens tear film, thinning time (PLTFTT), and non-invasive break up time (NIBUT) may help in predicting dryness symptomatology.

The average published thickness of the pre-lens tear film measured by interferometric techniques is about 3 µm (King-Smith et al. 2004), in agreement with other results obtained using different methodologies (Creech et al. 1998, King-Smith et al. 2004, Guillon in Holly 1986). Guillon

(Guillon et al. 1993) optically exploited the composition of the pre-lens film by using wide diffusive lighting of Tearscope and analyzed tear components by high-performance liquid chromatography; they observed that the lipid layer shows meshwork or flow patterns. Open meshwork indicates the presence of a thin lipid layer; on the contrary, close meshwork or flow indicates a thicker, more stable lipid layer. Occasionally, an amorphous pattern was seen that indicates a thick, very stable lipid layer. The aqueous layer was only visible when the lipid layer was thin (open meshwork). The observation of blue and red fringes indicated a thin aqueous layer of about 2–3 μm that tended to dry. Faint patterns correspond to a thicker (3 μm) and more stable aqueous layer. When the aqueous layer was not visible, the lipid layer was calculated to be >90 nm and appeared as either an amorphous pattern or a colored fringe pattern. Under these conditions, no conclusion was reached about the thickness of the aqueous layer or its stability, but it was assumed to be adequate.

There are a number of studies about the influence of CLs parameters such as material (Wang et al. 2003a), thickness (Guillon in Holly 1986) and water content of the CLs (Young et al. 1991, Faber et al. 1991) on the thickness of the pre-lens tear film.

Although some studies support the view that the thickness of the pre-lens tear film is independent on CLs type (Wang et al. 2003b), it is noteworthy that the rate of thinning is higher for silicone than for hydrogel CLs (Nichols et al. 2005). Additionally there are some evidences that thick hydrogel CLs are associated with a thicker pre-lens tear film compared to thin hydrogel CLs (Guillon in Holly 1986). The effect of CLs water content on the pre-lens tear film is somewhat controversial, since some studies report that high water content hydrogel CLs are associated with a thicker pre-lens tear film (Young et al. 1991) and other studies state that water content and lens type are not relevant to pre-lens tear stability (Faber et al. 1991).

For examining the PLTFTT, it is possible to use a TearScope Plus (Keeler Ltd., Windsor, UK) or a biomicroscopy at 40x magnification putting a bright, white illumination at the temporal section of the CL. Subjects were instructed to blink once and hold their eyes open until instructed to blink again. Specular reflection was performed, and the initial pre-lens tear film typically appeared as white. The measurement of the time of appearance of a colored interference fringe within the reflection indicated the PLTFTT. Biomicroscopy was found to be less efficient due to the limited scrutinized area, the evaporation from heat generated by the slit lamp bulb, and epiphora secondary to photophobia. A PLTFTT less than 3 s was demonstrated to be strongly correlated to CLRDE symptoms measured with Ocular Surface Disease Index (OSDI) (Hom and Bruce 2009).

To the thinning of pre-lens layer, a break-up may follow and an evaporative-dehydration process begins that draws water through the lens and out of the post-lens tear film toward the environment, leading to corneal staining (Fonn 2007). Blinking is the natural process that partially rehydrates the eye for deposition of a new pre-lens film on the lens surface; after some time a periodic steady state of hydration is reached (Nichols et al. 2005).

To evaluate the capacity of the pre-lens tear film to maintain a moist surface of the lens between eye blinks, the relationship between the PLTFTT and the inter blink interval (IBI) can be measured. If the result of the relationship is ≥ 1, then the pre-lens lacrimal film is stable from one blink to another, maintaining an uniform lens surface wetness; on the contrary, if the results indicate a relationship lower than 1, the tear film breaks before the blink and causes a dry zone on the lens surface.

Some experiments (Guillon et al. 2002) demonstrated that the tear composition influences the NIBUT. A low level of phospholipids resulted in NIBUTs of about 4.5 s; levels higher than 0.06mg/ml were associated to significantly higher NIBUTs (about 6.5 s). Additionally, higher cholesterol esters levels were detected in CL wearers with dryness symptomatology.

However, the pre and post lens tear films originated by lens insertion are thinner than the original three layer tear film and result in a different composition because chemical/physic interactions of the original tear solution occur at the lens surface. Eyes appear less hydrated and inflammation may easily occur.

Contact Lens Interaction with Tear Film Components

After lens insertion, the tear film is strongly destabilized and interactions of the lens material with the water of the aqueous medium, with proteins and lipids dissolved in the aqueous and lipid layers occur. Deposits on lens strongly compromise the rigid lens optical performance; on the contrary, for hydrogel lenses adequate blinking and hydration are very crucial factors for observing clear optical images.

Lens interaction with water is a process mainly dependent on the lens surface and less on the bulk; water lost through lens material dehydration is relatively minor compared with water evaporated from the anterior lens surface (Cedarstaff et al. 1983).

However, the hydration-dehydration mechanism is dependent on the lens material. For hydroxyethyl methacrylate (HEMA) lens (soft or hydrogel lenses), the surface-hydration mechanism was explained considering the chemical composition and spatial arrangement of the lens polymer. It is known that the HEMA complex polymeric matrices contain hydrophobic ends oriented inside the lens matrix and hydrophilic tails exposed on the lens surface.

Dehydration causes a polymer reorganization called hydrophobization with hydrophilic ends moving inward toward the center of the material matrix to seek moisture, and the hydrophobic ends outward toward the dehydrated surface. As a consequence of this process, water is repelled from the lens surface. Clinically, the hydrophobization of the lens or ocular surface leads to a decreased tear break-up time and soaking a CL in saline doesn't reverse the process (Ketelson et al. 2005).

For HEMA hydrogel soft CL it was observed that bulk water content changes significantly only within the first 5–10 min and fluctuates very little throughout the wearing period (Efron et al. 1987). The amount of initial water fluctuation was found to be not dependent on the water content of the lens material and to be strongly related to the lens thickness, osmolarity of the storage solution and the temperature of the lens upon insertion. Water content reduction is associated to a decreased oxygen transmissibility and lens diameter after a 6 hr wearing period (Tranoudis et al. 2004a).

Silicone hydrogel CLs have low water content (24–36%) and high oxygen transmissibility, and are subject to less bulk dehydration over time (Morgan et al. 2003, Fornasiero et al. 2005).

It is noteworthy that the hydration mechanism is influenced by environmental humidity; it was demonstrated that CL comfort decreases markedly with the arrival of the winter season or dry indoor heat and that 100% humidity for 30 min improves the pre-lens lipid layer stability of hydrogel lens wearers (Korb 1994).

The interaction between proteins and lens is explained by lens dehydration that is caused by the lens exposed to the hydrophobic regions, which can easily interact with hydrophobic portion of proteins. Lens parameters such as water content, pore size, age of the lens and surface charge (ionicity) influence the rate of deposit adsorption.

Intrinsic protein features related to surface charge and size might also affect the deposit. At the lens interface, proteins denature, thus stimulating the immunological response (Michaud et al. 2002, Senchyna et al. 2004).

The effect of protein deposition is very critical for HEMA CLs. Small, positively charged proteins like lysozyme are easily adsorbed on negatively charged materials with relatively large pore size. High levels of lysozyme deposition were measured on conventional hydroxyethyl methacrylate/methacrylic acid (HEMA/MAA) lens of different water content and ionicity. Higher water CLs (and hence, larger-pored polymers) are reported to absorb more proteins, such as lysozyme, into their structure.

Lysozyme deposition is less efficient for hydroxyethyl methacrylate/glycerol methacrylate (HEMA/GMA) lens; this indicates that carboxymethylation is probably a more significant factor in protein spoliation than water content alone (Maldonado-Codina et al. 2004). In addition, lysozyme was found to interact not only with the lens surface but to penetrate readily into negatively charged HEMA materials (Lord et al. 2006). In many cases, diffusive penetration of lysozyme may be promoted by an increase in effective pore size due to an increase of charge density (Garrett et al. 2000).

Interestingly, protein deposition is reduced on silicone hydrogel materials, due to their relatively hydrophobic nature and small pore size (Senchyna et al. 2004). Only small amounts of denatured lysozyme were observed on silicone CLs.

Lipids are strongly attracted by hydrophobic areas of the lens surface. Both HEMA and silicone CLs have such areas that result in lipid deposition. However, silicone lenses show a higher affinity for lipids than hydrogel CLs (Lorentz et al. 2007).

Daily and extended modalities of wear in low oxygen transmission lenses create a state of hypoxia in the cornea. Silicone hydrogel CLs seem to alleviate corneal hypoxia (Stapleton et al. 2006) and determine an improvement in dryness and ocular comfort (Brennan et al. 2002, Long et al. 2006, Woods et al. 2002, Dillehay et al. 2005, Chalmers et al. 2005, Schafer et al. 2003).

Oxygen transmissibility in HEMA-based lenses depends on water content (high water content means higher oxygen transmission); however the tear film break-up always induces eye dehydration with consequent substantial reduction of the oxygen transmission through HEMA-based CLs (Tranoudis et al. 2004b).

The CL's interaction with water, protein, and lipids has consequences on the tear film composition. It was reported that the level of some tear components increases early in the adaptation process following lens insertion and then returns to baseline levels following adaptation. No large differences in protein absorption are reported for soft or rigid CLs wearers except for Serum IgG levels that are higher in soft CLs wearers than rigid gas permeable lens wearers (Sengör et al. 1990). In the presence of pathological changes induced by CL wear, tear fluid plasmin was found to be elevated (Virtanen et al. 1994).

THE MANAGEMENT OF CONTACT LENS RELATED DRY EYE (CLRDE)

The management of CLRDE may involve treatment of various aspects at the same time, starting with the management of any blepharitis and

Meibomian gland dysfunction (MGD) and then passing to the appropriate CL, rewetting drops and care system.

Contact Lens Fitting

The correct treatment of CLRDE typically starts with the substitution of the type of CL utilized. The type of lens chosen should be based on the principle that the lens dehydrates less and maintains a stable surface wetness. To reduce to a minimum the dehydration with hydrogel materials, it is advised to use low hydrations non-ionic polymers in respect to those of high hydration, particularly if they would belong to the ionic group (Efron and Young 1988, Pritchard and Fonn 1995, Efron et al. 1987).

ECPs should however consider a few exceptions to this rule, the polymer inside their matrix incorporates certain specific molecules. For example, Omafilcon A (Group II), a biomimetic material, contains a synthetic molecule analogue of the natural phospholipid phosphatidylcholine, named phosphoricoline, that is able to create or mimic the biological surface. This material has a higher resistance on eye dehydration (Lemp 1999, Hall et al. 1999) and lens deposits (Young et al. 1997a), reduced corneal staining (Young et al. 1997b, Lemp 1999) with an increased PLTFTT (Lemp 1999).

Another material that has been demonstrated to be effective in the management of CLRDE is Hioxifilcon A (Group II), a copolymer containing glycerol methylmethacrylate (GMA) which demonstrates the chemical property of binding water molecules tightly with a slower dehydration and a quicker rehydration (Businger 2000). In head to head studies, symptoms and ocular surface staining associated with CLRDE can be significantly alleviated by the use of Hioxofilcon A or Omafilcon A lenses, which yielded similar findings (Riley et al. 2005).

There also exists a series of lenses that release humidifying substances, which due to their nature are suited for daily disposable use. The daily disposable lenses made in Nelfincon A contain two components: polyvinyl alcohol (PVA), and polyethylene glycol (PEG) into the matrix of the polymer during polymerization and are released during the use of the lens. These lenses are immersed in a solution with PEG and hydroxypropyl methylcellulose (HPMC) is inserted in the solution in which the lens is immersed also in association with HPMC.

Another daily disposable lens that contains a wetting agent in the polymer matrix is Etafilcon A, which has the addition of PVP that demonstrates an elevated wetting capability (analogous to the mucin layer, resulting in a better lubrication) and reduces lens dehydration. Another lens has the polymerization of the Filcon IV with hyaluronic acid, which is released from the lens at the ocular surface level guaranteeing lubrication and lens wettability. The material Filcon IV is also used for the production of a daily disposable lens that includes, in the polymerization phase, the AquaActract biomimetic additive which functions to maintain a longer hydration of the lens.

The use of CLs made in silicone hydrogel (SH) materials has been proposed to manage CLRDE (Chalmers et al. 2008, Schafer et al. 2007, Fonn 2007). SH lenses when compared with conventional lenses appear to dehydrate less *in vitro* (Jones et al. 2002) and *in vivo* (Morgan and Efron 2003). Further, it has been demonstrated that these materials form less bonds with protein deposits (Jones et al. 2003, Senchyna et al. 2004).

The presence of silicone in these materials makes the lens more hydrophobic in respect to hydrogel materials with a higher affinity for lipid deposits (Jones et al. 2003). In order to reduce to a minimum this phenomenon, the SH polymers have evolved starting with those that have a surface plasma treatment (Balafilcon A, Lotrafilcon A, Lotrafilcon B, Asmofilcon A), passing then to materials that incorporate internally wetting agents based on PVP (such as; Galyfilcon A, Senofilcon A and Narafilcon A), and finally to materials that do not require either plasma treatments or internal wetting agents (i.e. Comfilcon A, Filcon II). Even though not all studies have demonstrated a significant difference in patient comfort associated with the use of hydrogel conventional lenses with a low hydration in respect to silicon hydrogel lenses (Coles-Brennan 2006, Fonn and Dumbleton 2003), the majority of them indicate that SH lenses are associated with better comfort (Brennan et al. 2002, Long and McNally 2006, Dumbleton et al. 2006, Chalmers et al. 2002, Chalmers et al. 2005).

Rewetting Agents

The rewetting drops have varied tasks to perform: rehydrate the CLs, re-equilibrate the tear film, and act as a lubricant. Thus it is essential that after the drop instillation its effect must last as long as possible. Despite their ingredients, rewetting drops tend to have a short ocular residence time, draining through the patient's nasolacrimal duct quickly after instillation, with the remainder being quickly absorbed by the cornea, conjunctiva, and nasal mucosa with at least 90% loss for each instillation (Tonge et al. 2001). For this reason, rewetting drops generally must be re-instilled frequently throughout the day to be effective. Several investigations and clinical trials have demonstrated the toxicity of preservatives and their allergic potential found in the multiuse rewetting drops (Asbell and Ucakhan 2006, Baudouin 2001, Schaefer et al. 1994).

With hydrophilic CLs the risk increases because the lenses can absorb the preservative on instillation and then release these potentially toxic substances onto the ocular surface over a prolonged period of time. For this reason it is important that these drops are "non-preserved". Their different levels of viscosity and their various rehologic characteristics, osmolarity and pH categorize rewetting drops. The substances used for lubrication include:

Cellulose Esters: Methylcellulose, hydrossyetilcellulose, hydrosypropymethyl-cellulose (HPMC) and carbomethylcellulose. Carbomethylcellulose has demonstrated cytoprotection since it reduces the bioactivity of disinfecting agents in multipurpose solutions (MPSs), thus protecting the eye from ocular surface insult if used prior to lens insertion (Vehige et al. 2003, Szczotka-Flynn 2006).

Mucopolysaccarides: Hyaluronic acid is a natural compound widely used in CL for its hydration qualities, and muco-mimetic and muco- adhesive properties. Hyaluronic acid in solution strongly interacts with the mucin layer of the tear film; this interaction allows hyaluronic acid to be trapped on the ocular surface for a long period (Szczotka-Flynn 2006). Further, the muco-mimetic property attracts and maintains water molecules, reduces evaporation (Nakamura et al. 1993), and increases tear film stability (Mengher et al. 1986).

Finally, hyaluronic acid can stimulate the process of corneal repair (Aragona et al. 2002, Nishida et al. 1991, Inoue and Katakami 1993). Other mucopolysaccarides are TSPs (Tamarind Seed Xyloglucan) which can be found isolated or in conjunction to hyaluronic acid (Uccello-Barretta et al. 2010). Their expanded structure, similar to MUC1 of the epithelial glycocalyx, and their absence of charge favors the formation of bonds between the mucin and the mucous-water layer, allowing the adhesion to the superficial epithelial cell membrane. TSP has visco-elastic and muco-adhesive properties similar to hyaluronic acid (Burgalassi et al. 2000).

Synthetic Polymers: Carbopol (polyacrylic acid), PVA (polyvinyl alcohol), povidone (polyvinylpyrrolidone), poloxamer (polyoxyethylene-polyoxypropylene) are water-soluble compounds with a good superficial tension and epithelial absorption property. They are able to stabilize the tear film by lowering the superficial tension. Carbopol in respect to other synthetic polymers presents a longer time of residence on the cornea and conjunctival surfaces. PVA, despite its low viscosity, remains attached for a long period of time to the epithelium and it is a good mucous-mimetic; finally, poloxamer has a good tension-active and wetting capacity.

Synthetic Mucomimetic: HpGuar is a synthetic polysaccharide that, in contact with the tears, forms a film that bonds to epithelial glycocalyx. It is noted for its lubrication properties, which support the mucous layer in its wetting function.

In-eye Cleaners and Rewetting Agents

To maintain a stable pre-lens tear film it is important that lenses remain clean during the time of use. Deposit formation could create an imperfect CL surface over which tears cannot spread correctly. A lens surface that is not wetted enough promotes deposition and relative denaturation. For these reasons, it is important that lens surface remains wet enough during the fitting time. To reach this goal there are some products available that reduce protein buildup, rewet and clean lenses during wear.

They also contain surfactant agents with both a hydrophilic and a hydrophobic end in their structure. When surfactants are used to clean the lens the hydrophobic ends cluster around the debris to form micelles and the hydrophilic ends react with water to remove the micelles. When they are used like wetting agents, the hydrophobic ends interact with the dry hydrophobic lens surface exposing the hydrophilic ends at the lens surface allowing the lens to regain hydrophilic nature (Szczotka-Flynn 2006).

To obtain a stable pre-lens tear film, some products are available that reduce protein buildup, rewet and clean lenses during wear. One such product is a PHMB-preserved solution that uses tromethamine as the emulsifier/buffer, HPMC as the lubricant, tyloxapol as a surfactant and EDTA as a chelating agent. The recommended dosage is one to two drops up four times a day. Another product is a Polyquad-preserved solution that uses Tetronic 1304 as its surfactant/wetting agent and RLM-100 (lauryl ether carboxylic acid) as its surfactant. The recommended usage is two drops as needed during the day for minor irritations and two drops four times each day to prevent protein buildup.

Contact Lens Care Solutions

Some multipurpose solutions (MPSs) for soft lens disinfection have been modified in their composition in order to reduce surface tension, improve wettability, increase viscosity and improve CL comfort. Similar to rewetting drops, HPMC, and hyaluronic acid have also been added to release hydrating fluid throughout the day and to protect the eye from dryness and irritation.

Some MPSs use a combination of components to protect the lenses from drying out. For example, one product uses a combination of Dexpant-5, an ingredient found in dry eye products, and Sorbitol, an ingredient that attracts moisture. Another product contains Tetronic 1304, a surfactant that retains moisture and also, protects the lens from future protein buildup and C9-ED3A (nonanolyethyleneddiaminetriacetic acid), which is a surface wetting agent. A new MPS contains a wetting agent named HydraGlyde Moisture Matrix [EOBO-poly (oxyethylene)-poly (oxy-butylene)], a synthetic block copolymer that improves the hydrophilic properties of SH lenses. It also has an extremely high affinity for the hydrophobic areas of all soft CLs that are not wetted by the tear film.

CONCLUSIONS

CLRDE is the most important cause of CLs dropout with a significant economic impact on the CLs market. For this reason, it is important for the ECPs to know the interaction of CLs wear on tear film as well as the characteristics of CLs products in the market that are useful to minimize CLRDE effects.

REFERENCES

Aragona, P. and V. Papa. 2001. Long term treatment with sodium hyaluronate-containing artificial tears reduces ocular surface damage in patients with dry eye. Br. J. Ophthalmol. 86: 181–4.

Asbell, P.A. and O.O. Ucakhan. 2006. Dry Eye and Contact Lenses. In: P.A. Asbell and M.A. Lemp [eds.]. Dry Eye Disease. Thieme, New York, USA, pp. 114–131.

Baudouin, C. 2001. The pathology of dry eye. Surv. Ophthalmol. 45: S211–S220

Begley, C.G., B. Caffery, K.K. Nichols and R. Chalmers. 2001. Responses of contact lens wearers to a dry eye survey. Optom. Vis. Sci. 77: 40–6

Brennan, N.A., M.L. Coles, T.L. Comstock and B. Levy. 2002. A 1-year prospective clinical trial of Balafilcon A (PureVision) silicone-hydrogel contact lenses used on a 30-day continuous wear schedule. Ophthalmology 109: 1172–7.

Burgalassi, S., L. Raimondi, R. Pirisino, G. Banchelli, E. Boldrini and M.F. Saettone. 2000. Effect of xyloglucan (tamarind seed polysaccharide) on conjunctival cell adhesion to laminin and on corneal epithelium wound healing. Eur. J. Ophthalmol. 10: 71–76.

Bursinger, U. 2000. A new material on the block of frequent replacement lenses. C L Spectrum 15: 14–7.

Cedarstaff, T. and A. Tomlinson. 1983. A comparative study of tear evaporation rates and water content of soft contact lenses. Am. J. Optom. Physiol. Opt. 60: 167–74.

Chalmers, R., B. Long, S. Dillehay and C. Begley. 2008. Improving contact-lens related dryness symptoms with silicone hydrogel lenses. Optom. Vis. Sci. 85: 778–784.

Chalmers, R.L., J.J. McNally, C.D. McKenney and S.R. Robirds. 2002. The role of dryness symptoms in discontinuation of wear and unscheduled lens removals in extended wear of silicone hydrogel lenses. Invest Ophthalmol. Vis. Sci. 43: E-abstract 3088.

Chalmers, R.L., S. Dillehay, B. Long, J.T. Barr, P. Bergenske, P. Donshik, G. Secor and J. Yoakum. 2005. Impact of previous extended and daily wear schedules on signs and symptoms with high Dk lotrafilcon A lenses. Optom. Vis. Sci. 82: 549–54.

Coles-Brennan, C., N.A. Brennan, H.R. Connor and R.G. McIlroy. 2006. Do silicone-hydrogels really solve end-of-day comfort problems? Invest Ophthalmol. Vis. Sci. 47: E-abstract 106.

Creech, J.L., L. Do, I. Fatt and C.J. Radke. 1998. *In vivo* tear-film thickness determination and implications for tear-film stability. Curr. Eye Res. 17: 1058–66.

Dillehay, S., B. Long and J. Barr. 2005. Summary of 3 year in-practice trial in the US with lotrafilcon A silicone hydrogel soft contact lenses. Optom. Vis. Sci. 82: E-abstract 050080.

Dumbleton, K., N. Keir, A. Moezzi, Y. Feng, L. Jones and D. Fonn. 2006. Objective and subjective responses in patients refitted to daily-wear silicone hydrogel contact lenses. Optom. Vis. Sci. 83: 758–68.
Efron, N. and G. Young. 1988. Dehydration of hydrogen contact lenses *in vitro* and *in vivo*. Ophthalm. Physiol. Opt. 8: 253–256.
Efron, N. and P.B. Morgan. 1999. Hydrogel contact lens dehydration and oxygen transmissibility. CLAO J. 25: 148–51.
Efron, N., N.A. Brennan, A.S. Bruce, D.I. Duldig and N.J. Russo. 1987. Dehydration of hydrogel lenses under normal wearing conditions. CLAO J. 13: 152–6.
Faber, E., T.R. Golding, R. Lowe and N.A. Brennan. 1991. Effect of hydrogel lens wear on tear film stability. Optom. Vis. Sci. 68: 380–4.
Fonn D. 2007. Targeting contact lens-induced dryness and discomfort: what properties will make lenses more comfortable? Optom. Vis. Sci. 84: 279–85.
Fonn, D. and K. Dumbleton. 2003. Dryness and discomfort with silicone hydrogel contact lenses. Eye Contact Lens 29: S101–S104.
Fonn, D., P. Situ and T. Simpson. 1999. Hydrogel dehydration and subjective comfort and dryness ratings in symptomatic and asymptomatic contact lens wearers. Optom. Vis. Sci. 76: 770–4.
Fornasiero, F., F. Krull, J.M. Prausnitz and C.J. Radke. 2005. Steady-state diffusion of water through soft contact lens materials. Biomaterials 26: 5704–16.
Garrett, Q., B. Laycock and R.W. Garrett. 2000. Hydrogel lens monomer constituents modulate protein sorption. Invest Ophthalmol. Vis. Sci. 41: 1687–95.
Guillon, J.P. and M. Guillon. 1993. Tear film examination of the contact lens patient. Optician 206: 21–29.
Guillon, M., C. Maissa, K. Girard-Claudon and P. Cooper. 2002. Influence of the tear film composition on tear film structure and symptomatology of soft contact lens wearers. In Lacrimal gland, tear film, and dry eye syndromes 3: basic science and clinical relevance, PTS A&B Book Series: Advances in experimental medicine and biology. USA.
Hall, B., S. Jones, G. Young and S. Coleman. 1999. The on-eye dehydration of Proclear Compatible Lenses. CLAO J. 25: 233–237.
Hamano, H. 1981. The change of pre-corneal tear film by the application of contact lenses. Contact Lens 7: 205–9.
Holden, B.A., D.F. Sweeney and R.G. Seger. 1986. Epithelial erosions caused by thin high water content lenses. Clin. Exp. Optom. 69: 103–7.
Holly, F.J. 1981. Tear film physiology and contact lens wear. II: contact lens-tear film interaction. Am. J. Optom. Physiol. Opt. 58: 331–41.
Holly, F.J. 1986. The preocular tear film in health, disease, and contact lens wear. Lubbock, Texas: Dry Eye Institute. USA.
Holly, F.J. and B.S. Hong. 1982. Biochemical and surface characteristics of human tear proteins. Am. J. Optom. Physl. Opt. 59: 43–50.
Hom, M.M. and A.S. Bruce. 2009. Prelens tear stability: relationship to symptoms of dryness. Optometry 80: 181–184.
Inoue, M. and C. Katakami. 1993. The effect of hyaluronic acid on corneal epithelial cell proliferation. Invest. Ophthalmol. Vis. Sci. 34: 2313–5.
Jones, L. and C. May. 2002. *In vitro* evaluation of the dehydration characteristics of silicone-hydrogel and conventional hydrogel contact lens materials. Contact Lens and Anterior Eye 25: 147–156.
Jones, L. and M. Senchyna. 2003. Lysozyme and lipid deposition on silicone hydrogel contact lens materials. Eye Contact Lens 29: S75–S79.
Ketelson, H.A., D.L. Meadows and R.P. Stone. 2005. Dynamic wettability properties of a soft contact lens hydrogel. Colloids Surf. B Biointerfaces 40: 1–9.
King-Smith, P.E., B.A. Fink, R.M. Hill, K.W. Koelling and J.M. Tiffany. 2004. The thickness of the tear film. Curr. Eye Res. 29: 357–68.

Kline, L.N. and T.J. De Luca. 1975. Effect of gels lens wear on the pre-corneal tear film. ICLC. 1: 56–9.
Kopecek, J. 2009. J. Polym. Sci. Pol. Chem. 47: 5929–5946.
Korb, D.R. 1994. Tear film-contact lens interactions. In Advances in experimental medicine and biology. 350: 403–10.
Lemp, M.A. 1995. Report of the National Eye Institute/Industry workshop on Clinical Trials in Dry Eyes. CLAO J. 21: 221–232.
Lemp, M.A. 1999. Omafilcon A (Proclear) soft contact lenses in a dry-eye population. CLAO J. 25: 40–47.
Long, B. and J. McNally. 2006. The clinical performance of a silicone hydrogel lens for daily wear in an Asian population. Eye Contact Lens 32: 65–71.
Lord, M., M.H. Stenzel, A. Simmons and B.K. Milthorpe. 2006. The effect of charged groups on protein interactions with poly (HEMA) hydrogels. Biomaterials 27: 567–75.
Lorentz, H. and L. Jones. 2007. Lipid deposition on hydrogel contact lenses: how history can help us today. Optom. Vis. Sci. 84: 286–95.
Maldonado-Codina, C. and N. Efron. 2004. Impact of manufacturing technology and material composition on the clinical performance of hydrogel lenses. Optom. Vis. Sci. 81: 442–54.
Mengher, L.S. and K.S. Pandher. 1986. Effect of Sodium hyaluronate (0.1%) on break-up time (NIBUT) in patients with dry eyes. Br. J. Ophthalmol. 70: 422–7.
Michaud, L. and C.J. Giasson. 2002. Overwear of contact lenses: increased severity of clinical signs as a function of protein adsorption. Optom. Vis. Sci. 79: 184–92.
Morgan, P.B. and N. Efron. 2003. *In vivo* dehydration of silicone hydrogel contact lenses. Eye Contact Lens 29: 173–6.
Morgan, P.B. and N. Efron. 2003. *In vivo* dehydration of silicone hydrogel contact lenses. Eye Contact Lens 29: 173–176.
Nakamura, M., M. Hikida and T. Nakano. 1993. Characterization of water retentive properties of hyaluronan. Cornea 12: 433–6.
Nichols, J.J. and P. King-Smith. 2003. Thickness of the pre- and post-contact lens tear film measured *in vivo* by interferometry. Invest. Ophthalmol. Vis. Sci. 44: 68–77.
Nichols, K.K., J.J. Nichols and G.L. Mitchell. 2004. The lack of association between signs and symptoms in patients with dry eye disease. Cornea 23: 762–770.
Nishida, T., M. Nakamura and H. Mishima. 1991. Hyaluronan stimulates corneal epithelial migration. Exp. Eye Res. 53: 753–8.
Patel, S. 1987. Constancy of the front surface desiccation times for Igel 67 lenses *in vivo*. Am. J. Optom. Physiol. Opt. 64: 167–71.
Pritchard, N. and D. Fonn. 1995. Dehydration, lens movement and dryness ratings of hydrogel contact lenses. Ophthalmic Physiol. Opt. 15: 281–286.
Pritchard, N., D. Fonn and D. Brazeau. 1999. Discontinuation of contact lens wear: a survey. Int. Contact Lens Clin. 26: 157–62.
Pult, H., P.J. Murphy and C. Purslow. 2010. Clide-index: a novel method to diagnose and measure contact lens induced dry eye. Contact Lens and Anterior Eye 33: 256–300.
Richdale, K., L.T. Sinnott, E. Skadahl and J.J. Nichols. 2007. Frequency of and factors associated with contact lens dissatisfaction and discontinuation. Cornea 26: 168–74.
Rumpakis, J. 2010. New data on contact lens dropouts: an international perspective. Rew. of Optom. 1: 1–4.
Schaefer, K., M.A. George and M.B. Abelson. 1994. A scanning electron microscopic composition of the effects of two preservative- free artificial tear solutions on the corneal epithelium as compared to posphate buffered saline and a 0,02% benzalkonium chloride control. Adv. Exp. Med. Biol. 350: 459–464.
Schafer, J., G.L. Mitchell and R.L. Chalmers. 2007. The stability of dryness symptoms after refitting with silicone hydrogel contact lenses over 3 years. Eye Contact Lens. 33: 247–252.

Schafer, L., J. Barr and C. Mack. 2003. A characterization of dryness symptoms with silicone hydrogel lenses. Optom. Vis. Sci. 80: 187.

Senchyna, M., L. Jones, D. Louie, C. May, L. Forbes and M.A. Glasier. 2004. Quantitative and conformational characterization of lysozyme deposited on balafilcon and etafilcon contact lens materials. Curr. Eye Res. 28: 25–36.

Sengör, T., S. Gürgül, S. Ögretmenoglu, A. Kiliç and H. Erker. 1990. Tear immunology in contact lens wearers. Contactologia. 12E: 43–45.

Sindt, C. 2007. Contact lens strategies for the patient with dry eye. Ocular Surf. 5: 294–307.

Smith, J.A. 2007. Research in dry eye: report of the Research Subcommittee of the International Dry Eye WorkShop. Ocul. Surf. 5: 179–93.

Stapleton, F., S. Stretton and E.B. Papas. 2006. Silicone hydrogel contact lenses and the ocular surface. Ocul. Surf. 4: 24–43.

Sullivan, B.D., D. Whitmer and K.K. Nichols. 2010. An objective approach to dry eye disease severity. Invest. Ophthalmol. Vis. Sci. 51: 6125–6130.

Szczotka-Flynn, L.B. 2006. Chemical properties of contact lens rewetters. A review on hyaluronic acid as a contemporary ingredient in contact lens rewetters. CL Spectrum. 21: 40–45.

Tomlinson, A. 1992. Complications of contact lens wear. St. Louis, Mosby. USA.

Tomlinson, A. and T.H. Cedarstaff. 1982. Tear evaporation from the human eye: effects of contact lens wear. J BCLA. 5: 141–50.

Tonge, S., B. Tighe, V. Franklin and A. Bright. 2001. Contact lens care, Part 6: Comfort drops, artificial tears and dry-eye therapies. Optician 222: 27–32.

Tranoudis, I. and N. Efron. 2004a. Parameter stability of soft contact lenses made from different materials. Cont. Lens Anterior Eye 27: 115–31.

Tranoudis, I. and N. Efron. 2004b. In-eye performance of soft contact lenses made from different materials. Cont. Lens Anterior Eye 27: 133–48.

Uccello-Barretta, G., S. Nazzi, Y. Zambito, G. Di Colo, F. Balzano and M. Sansò. 2010. Synergistic interaction between TS-polysaccharide and hyaluronic acid: Implications in the formulation of eye drops. Int. J. of Pharmaceutics 395: 122–131.

Vehige, J.G., P.A. Simmons, C. Anger, R. Graham, L. Tran and N. Brady. 2003. Cytoprotective properties of carboxymethyl cellulose (CMC) when used prior to wearing contact lenses treated with cationic disinfecting agents. Eye Contact Lens 29: 177–180.

Virtanen, T., N. Honkanen, M. Harkonen, A. Tarkkanen and T. Tervo. 1994. Elevation of tear fluid plasmin activity of contact lens wearers studied with a rapid fluorometric assay. Cornea 13: 210–213.

Wang, J., D. Fonn, T. Simpson, L. Sorbara, R. Kort and L. Jones. 2003a. Topographical thickness of the epithelium and total cornea after overnight wear of reverse–geometry rigid contact lenses for myopia reduction. Invest Ophthalmol. Vis. Sci. 44: 4742–4746.

Wang, J., D. Fonn, T.L. Simpson and L. Jones. 2003b. Pre-corneal and pre and postlens tear film thickness measured indirectly with optical coherence tomography. Invest Ophthalmol. Vis. Sci. 44: 2524–2528.

Woods, J., N. Brennan and A. Jaworski. 2002. Ocular signs and symptoms in patients completing 3 years with silicone hydrogel contact lenses in 30-day continuous wear. Contact Lens Anterior Eye 25: 198–9.

Young, G. 2004. Why one million contact lens wearers dropped out? Contact Lens Ant. Eye 27: 83–5.

Young, G. and N. Efron. 1991. Characteristics of the pre-lens tear film during hydrogel lens wear. Ophthalmic. Physiol. Opt. 11: 53–8.

Young, G., J. Veys and S. Coleman. 2002. A multicentre study of lapsed contact lens wearers. Opht. Physiol. Opt. 22: 516–527.

Young, G., R. Bowers and B. Hall. 1997a. Clinical comparison of Omafilcon A with four control materials. CLAO J. 23: 249–258.

Young, G., R. Bowers, B. Hall and M. Port. 1997b. Six-month clinical evaluation of a biomimetic hydrogel contact lens. CLAO J. 23: 226-35.

Contact Lens Care Solutions and Ocular Surface

José Manuel González-Méijome, Miguel F. Ribeiro* and *Daniela Lopes-Ferreira*

SUMMARY

Safe wear of non-daily disposable contact lenses require care systems to ensure disinfection and cleaning to allow the lens to be reused after a period of non lens wear (typically overnight). To do so, modern care systems, also known as multipurpose solutions (MPS) or multipurpose disinfecting solutions (MPDS) try to mimic the environment of the ocular surface regarding pH, osmolality, surface tension or viscosity while introducing other singular properties as a wide-spectrum antimicrobial activity and high cleaning performance. This delicate equilibrium of properties is challenging and sometimes results in adverse events that have drawn the attention of the CL industry and scientific community to these systems with an increased intensity during the past decade.

Even with disposable lenses, safety is still a concern and presently we are in front of a new paradigm of solutions development, which must provide reinforced disinfecting capabilities and at the same time improve the compatibility with the ocular surface. Indeed, disinfection efficacy against different new strains of microorganisms and compatibility with the ocular surface have been two aspects widely covered in the recent scientific literature, topics that will also be addressed in this chapter.

Clinical & Experimental Optometry Research Lab (CEORLab), University of Minho, Campus de Gualtar 4710-057, Braga–Portugal; E-mail: jgmeijome@fisica.uminho.pt
*Corresponding author

INTRODUCTION

Care solutions are essential for contact lens (CL) safety and durability. They were first intended for cleaning and disinfecting non-planned replacement lenses, commonly using products with poor compatibility with the ocular surface, such as hydrogen peroxide, chlorhexidine or benzalconium chloride. In spite of their efficacy against microorganisms, they were somewhat inconvenient, at least in their earlier formulations particularly when used with hydrophilic lenses that absorb such components into their matrix.

Care systems were initially complex requiring that the wearer must use different products; nowadays the care systems have evolved into complex formulations that meet all the requirements for a care system in a single package in order to make their use easier and more convenient to the CL wearer and to promote compliance. However, this combination of properties into the same package requires an optimal balance between the disinfecting and cleaning activity and the respect for the ocular surface homeostasis.

Today, development of effective care systems is a challenging task once they have to be effective for cleaning and disinfection, and at the same time respect the physiology of the ocular surface, be convenient for the wearer and when possible improve the comfort of the CL during the whole day. These are the requirements for current MPDS that represent over 80% of the all care systems prescribed (Morgan et al. 2010). With the advent of disposable CL cleaning capability has become less important and industry makes efforts to provide MPDS with new features that address one of the main unsolved issues for contact lens wearers: all day comfortable wear. As a result, current MPDS are the result of complex chemical and biological engineering that must have a critical combination of properties such as pH, osmolality, viscosity maintaining high standards of disinfection and cleaning efficacy.

REQUIREMENTS AND COMPOSITION OF CURRENT CARE SOLUTIONS

The basic requirements for a CL care system are to clean tear and environmental products deposited on the surface of the lens, to remove the materials that had been strongly adhered to the CL material, to disinfect the lens by reducing the populations of microorganisms to safety levels, to warrant that the CL material remains fully hydrated, and desirably, to improve the CL wetting properties to extend wearing comfort during the whole day.

Table 1 shows some examples of soft CL MPDS marketed and their composition. Composition of MPDS is critical because of the potential

Table 1 Composition of different marketed MPDS. Information collected from different sources (Dalton et al. 2008, Santodomingo-Rubido et al. 2006).

Care System	Manufacturer	Preservative/Biocidal Agents	Buffer	Surfactant	Others
Soft Wear Saline	CIBA Vision	Hydrogen peroxide (0.006%)	Sodium borate		Sodium chloride; sodium borate; boric acid; sodium perborate; phosphonic acid
SoloCare Aqua	CIBA Vision	PHMB (0.0001%)		Poloxamer 407	Sodium phosphate; tromethamine; sorbitol; EDTA (0.025%)
Menicare Soft	Menicon	PHMB (0.0001%)	2-amino-2-methyl-1,3-propanediol (AMPD)	Macrogolglycerol hydroxystearate 60	Glycolic acid, glycine, propylene glycol, EDTA
Sensitive Eyes MPS	Bausch & Lomb	PHMB (0.00005%)	Boric acid		Edatate disodium; poloxamine; boric acid; sodium chloride
ReNu MultiPlus	Bausch & Lomb	PHMB (0.0001%)	Boric acid		Sodium borate, hidroxyalkilphosphonate
Opti-Free RepleniSH	Alcon	Polyquaternium-1 (0.001%) Myristamidopropyl dimethylamine or ALDOX (0.0005%)		Poloxamine (Tetronic 1304)	
Opti-Free Express	Alcon	Polyquad (0.001%) Myristamidopropyl dimethylamine or ALDOX (0.0005%) EDTA (0.05%)	Boric acid	Poloxamine (Tetronic 1304)	
Biotrue	Bausch & Lomb	Polyaminopropyl biguanide (0.00013%) Polyquaternium-1 (0.0001%)	Sodium borate, boric acid, socium chloride	Poloxamine Sulfobetaine	Hyaluronan
Complete RevitaLens	AMO-Abbott	Alexadine (0.00016%) Polyquaternium-1 (0.0003%)	Boric acid, sodium borate decahydrate, sodium chloride, trisodium citrate dehydrate	Tetronic 904	

OptiFree EverMoist/ PureMoist	Alcon	Polyquaternium-1 (0.001%) + Aldox (0.0006%)	Boric acid	Poloxamine (Tetronic 1304)	Sorbitol, citrate, EDTA (0.05%), HydraGlyde (EOBO-41, poly-oxyethylene-poloxybutylene) Aminomethylpropanol (AMP-95)
Sinergi MPDS	Sauflon	Oxipol™ Oxychlorite complex (sodium chlorite and hydrogen peroxide)	Phosphate	Poloxamer	Hydroxypropylmethylcellulose (HPMC), polyvinyl pyrrolidone (PVP)
Clear Care*	CIBA Vision	Hydrogen peroxide (3%)			Platinum disc for neutralization
AOSept*	CIBA Vision	Hydrogen peroxide (3%)			Platinum disc for neutralization
Ultracare*	Abbott	Hydrogen peroxide (3%)			Catalase tablets for neutralization

PHMB: polyhexamethylene biguanide; EDTA: ethylenediamine tetra-acetic acid; HPMC: hydroxypropyl methylcellulose. *hydrogen peroxide in its neutral form.

impact on the interaction with the contact lenses and with the ocular surface. Similar compositions can be described for rigid gas permeable lenses (RGP). Eventually, in the case of RGP materials, due to their inability to absorb the active ingredients, other preservative or biocidal agents as derivates of chlorhexidine or even benzalkonium chloride (BAK), no longer incorporated to soft CL care systems, can be used. Other major difference with soft CL systems is that RGP MPDS use to have higher viscosity to improve the comfort of the CL by increasing the wetting capabilities of the surface and cushioning the contact between the RGP CL and the ocular surface.

PROPERTIES OF CARE SYSTEMS

Considering the intimal relationship of care solutions with the contact lens material and the ocular surface, the properties of the care systems are critical for a good performance.

Some of the most relevant properties include: pH, osmolality, surface tension, and viscosity. The relevance of these properties, from the clinical point of view range from the impact on insertion comfort to the dynamic interaction of the lens with the ocular surface (cornea, bulbar conjunctiva and tarsal conjunctiva). In brief, a solution with pH or osmolality out of the physiological range (Table 2), could result in discomfort at insertion; an excessive surface tension will result in decreased spread of the solution over the CL or the ocular surface; too low viscosity will decrease the lubricating effect of the tears in the interaction of the lens with the ocular surface; too high viscosity will result in an improved lubricating effect but eventually

Table 2 This table summarizes the results of Dalton et al. regarding the physical and chemical properties of different contact lens care systems (Dalton et al. 2008).

Care System	pH	Osmolality (mOsm/Kg)	Surface Tension (mN/m)	Viscosity (cP @ 34°C)
Average for Normal Tears	6.2–9.0	305	40–46	1.5–5.0
Soft Wear Saline	6.97	303.5	67.9	0.70
Solo Care Aqua	7.23	310	35.1	0.81
Sensitive Eyes MPS	7.29	286.7	38.0	0.83
ReNu MultiPlus	7.38	286.2	36.3	0.84
Complete Moisture Plus	7.28	304.3	40.5	1.92
Opti-Free RepleniSH	7.88	277.2	29.7	0.71
Opti-Free Express	7.82	225.0	31.2	0.76
Clear Care*	6.76	293.7	42.9	0.69
AOSept*	6.66	290.7	70.3	0.69
Ultracare*	7.18	329.0	43.2	0.86

*hydrogen peroxide in its neutral form.

could result in poor visual due to irregularities in the tear film or even lens binding to the ocular surface. Dalton et al. evaluated these properties in 10 care systems including 7 MPDS solutions and 3 neutralized hydrogen peroxide solutions (Dalton et al. 2008). The main results of this study are summarized in Table 2 along with the normal range for the normal tears.

Although the normal pH of tears has been defined between 6.2 and 9.0, care systems must be formulated to maintain the pH within a narrower range between 6.6 and 7.8, which is also known as the comfort zone (Carney and Hill 1976). As seen in table 2, most of the solutions do respect these desired limits, although presenting significant differences between them.

Beyond their intrinsic properties, it is also interesting to evaluate the potential changes of such parameters overtime. Previous studies have demonstrated that properties such as pH can remain within the normal physiological range during a significant period of time in spite of some fluctuations (Lopez-Alemany and Montes 1998). Recent studies have addressed the changes in pH and osmolality over a month for different products and particularly when the lens cases are subjected to different cleaning, rinsing and drying procedures (Abengozar-Vela et al. 2011). Some cleaning and drying procedures of the storage lens cases can affect osmolality (Abengozar-Vela et al. 2011) such that air-drying, today recommended by several manufacturers, can increase significantly the osmolality of the solution where the lenses are immersed overnight if a tissue-wiping step is not added to the process.

DISINFECTING EFFICACY

Disinfecting efficacy of CL care systems is one of the most critical properties and it is measured as the reduction of viable organisms or colony forming units (CFU's) during the recommended disinfection time or during a certain period of time as set by FDA and ISO standards. The usual number of organisms inoculated is in the order of 10^6 units or CFU's. Typically, a care system should be challenged against several organisms, including gram-negative and gram-positive bacteria, mold and yeast. The results are expressed in terms of logarithm units of reduction such that 1 log unit corresponds to the elimination or neutralization of 90% of the organisms initially inoculated (10^5 CFU's remaining); similarly 2 log units are equivalent to a double reduction of 90% (10^4 CFU's remaining), etc. After a 5 log reduction, only 10 CFU's will remain and after a 6 log reduction, only 1 CFU will remain. Standards require at least a 3 log reduction for bacteria (gram-positive and gram-negative) and 1 log reduction for mold and yeast.

Figure 1 presents the requirements set by FDA and ISO required for a given care system although these regulations are under review to incorporate more microorganisms.

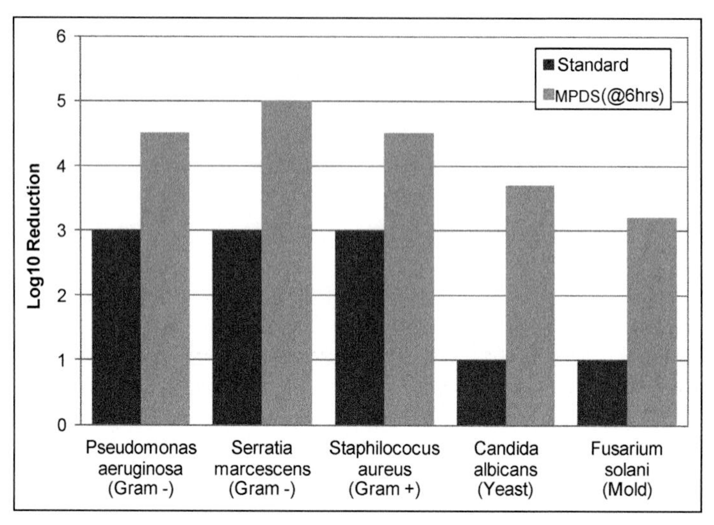

Figure 1 Standards of reduction in CFU viability and performance of a given MPDS against gram-positive and gram-negative bacteria, yeast and mold and the example of efficady of a MPDS after 6 hours.

Indeed, most CL care systems and particularly those new systems being developed are being challenged against a much wider spectrum of microorganisms, including those responsible for serious public health concerns as antibiotic-resistant microorganisms and clinical isolates (extracted from contaminated cases or infected eyes instead of using only strains from laboratory cultures) in order to provide stronger evidences of the efficacy of the MPDS against "real world microorganisms". This has been the case in a study published by Zhu et al. (Zhu et al. 2007) where up to 10 gram-positive bacteria, 8 gram-negative bacteria and 3 fungi (2 yeast and 1 mold) were evaluated. Other study conducted by Kilvington et al. (Kilvington et al. 2010) on a new MPDS also evaluated different microorganisms beyond those recommended by FDA and ISO standards. *Acanthamoeba castellanii*, one of the most virulent and challenging ocular microorganism, either in the trophozoite and cyst forms were evaluated, providing quite different results among different solutions, particularly in the cystic form where some solutions were ineffective to achieve a 3 log reduction; such results reinforce the need for tighter standards for MPDS evaluation. Further, these authors evaluated the efficacy of different solutions after evaporation of the solution, a condition that could derive from poor compliance from the patient and that could act as a risk factor for ocular infection due to reduced biocidal activity of the solution and/or to increased resistance of the microorganisms in such environment.

INTERACTIONS OF CARE SYSTEMS WITH THE OCULAR SURFACE

The components of the CL solutions can be related to several adverse reactions in the ocular surface by mediating a toxic reaction due to prolonged contact once the products adhered to the lens material leak to the ocular surface. Several studies have tried to relate these clinical events presenting in the form of ocular surface damage to the presence of certain components in the care systems. The disinfecting or biocidal agents or their interaction with other components included in the formulation have been pointed as responsible for such reactions.

Polyhexamethylene biguanide (PHMB) also known as polyhexanide, Dymed, polyhexidine and polyaminopropyl biguanide has been frequently related to solution induced corneal staining (SICS). However, the laboratory research was unable to establish an *in vitro* relationship with the *in vivo* clinical findings. On the other hand, polyquaternium-1 also known, as Polyquad is other frequently used disinfecting agent also involved in such laboratory tests whose results are described in further detail below and will be summarized in Table 3.

A study from Santodomingo-Rubido et al. reported that different concentrations of certain ingredients commonly found in commercial CL solutions could be related with higher or lower citotoxicity of experimental cell lines from animal models (mouse lung cells) (Santodomingo-Rubido et al. 2006). Interestingly, the authors found that while some MPDS solutions were invariably associated to high or low levels of citotoxicity at any concentration between 1.25 to 10% of MPDS, the citotoxicity activity was largely dependent on concentration for other MPDS which was attributed to the biocidal agent included or to the combination of this agent with other ingredients such as buffer ingredients. Was also interesting to observe how certain surfactants such as Poloxamer 237 or Poloxamer 407 were invariably related to low and high citotoxicity irrespective of the concentration used, while citotoxicity of other surfactant Tetronic 1107 was dependent on concentration. Finally, cytotoxicity of most isotonic and buffer agents was found to be highly dependent on concentration (Santodomingo-Rubido et al. 2006).

Other authors used to look at these issues in different ways using different laboratory assays. Chuang et al. used ApopTag Fluorescein Apoptosis assay to detect cell death after exposure to different MPDS, fluorescein permeability to investigate the corneal epithelial barrier function and tight junction proteins zonula occludens (ZO)-1 and occludin were

Table 3 This table summarizes the results of different studies addressing the citotoxicity of several MPDS.

Authors (year)	**Solutions evaluated**	**Main outcomes**
Santodomingo-Rubido et al. (2006)	Complete Moisture Plus* MeniCare Soft Opti-Free Express ReNu Multiplus Renu MoistureLoc* SoloCare Aqua	*Citotoxicity Activity (at medium concentration of 5%)* Complete < MeniCare < SoloCare < OptiFree < ReNu MultiPlus = ReNu MoistureLoc
Chuang et al. (2008)	Complete Easy Rub OptiFree Express OptiFree RepleniSH ReNu Multiplus	*Apoptosis Rates* Complete < ReNu < OptiFree RepleniSH < OptiFree Express
Imayasu et al. (2008)	MPS1: PHMB MPS2: PHMB + Poloxamine + Boric Acid MPS3: Alexidine + Poloxamine + Boric Acid MPS4: Polyquad + Poloxamine + Boric Acid	*Transepithelial Electrical Resistance Culture in Collagen Membrane (@120 min.)* MPS1 < MPS3 < MPS2 < MPS4 Combination of poloxamine and boric acid as potential cause of tigh junction damage
Mowrey-McKee et al. (2002)	AOSept (neutral) Complete Comfort Plus OptiFree Express ReNu ReNu Multiplus SoloCare Soft	*Different Assays* Only OptiFree Express was considered citotoxic for trypan blue uptake test and cell regrowth (absent) following exposure to test material. ReNu products showed citotoxicity on quantification of viable cells

* Currently not marketed.

evaluated using immunofluorescent staining (Chuang et al. 2008). Imayasu et al also evaluated the tight junction integrity of the corneal epithelial cells with ZO-1 (tight junction-related protein) labeling under laser confocal microscopy.

To investigate the changes of ultrastructure in tight junctions of human corneal epithelial cells, the same authors observed an ultrathin cross-section of the cell on collagen membrane using transmission electron microscopy. (Imayasu et al. 2008) Mowrey-McKee et al., evaluated citotoxicity of CL care systems using different modifications of the United States Pharmacopeia (USP) elution test required by U.S. Food and Drug Administration (FDA) and global standards organizations such as the International Standards Organization (ISO) for *in vitro* cytotoxicity testing of lens care solutions (Mowrey-McKee et al. 2002). Results of these and other evaluations are summarized in table 3. Despite all these studies found significant differences between the citotoxicity of different MPDS, the different experimental conditions and techniques used limit the ability to compare and summarize the results from all these studies. Moreover, their results obtained *"in vitro"* did not relate much with the clinical observations such that, care systems reported as being safe in the clinical setting revealed worse behaviour in the laboratory assays and vice-versa.

CHALLENGES FOR CURRENT MPDS DEVELOPMENT

After the outbreaks of MPDS-related infections during the last 5 years (Khor et al. 2006, Patel and Hammersmith 2008, Tu and Joslin 2010), there has been a great deal of discussion regarding the efficacy of these systems and the sufficiency of current FDA and ISO regulations regarding the tests that new systems must undergo for approval.

This issue is aggravated by the known lack of compliance of many CL wearers with the correct guidelines of use of their CL and care systems. Such behaviours could eventually result in an increased risk of adverse reactions. Among some of the modifiable patient compliance-related behaviours, the following have been identified as increased risk factors (relative risk, RR) for CL associated infections: inadequate hand-washing (RR=4.5 (Stapleton et al. 2007)), inadequate case cleaning (RR=4.0 (Houang et al. 2001)), failure to use correct disinfecting solution (RR=21.8 to 55.9; (Houang et al. 2001, Radford et al. 1998), failure to rub and rinse lenses (RR=3.5 (Radford et al. 1995)) and topping off solution (RR=2.5 (Saw et al. 2007)), among others (Morgan et al. 2011).

As a result, some new MPDS have been developed to meet criteria beyond the current regulatory standards. As such, new organisms as different strains of bacteria (i.e. metacilyn resistant *Staphylococcus aureus* also known as MRSA) or amoeba such as *Acanthamoeba* either in its trophozoite or cystic forms have became usual in the evaluation of biocidal activity of new care systems as previously described in this chapter (Kilvington et al. 2010; Zhu et al. 2007).

ROLE OF CARE SOLUTIONS IN IMPROVING CL's CLINICAL PERFORMANCE

When it comes to the relationship between adverse events in the ocular surface and the potential impact of MPDS on their aetiology, it is hard to establish a direct relationship. However, it is commonly accepted that the interaction between some MPDS ingredients and the ocular surface, mediated or not thought particular interactions of those with the CL material, can have a negative impact in the tolerance of CL with the ocular surface (Andrasko et al. 2006, Carnt et al. 2007, Garofalo et al. 2005).

Nowadays, care systems no longer pretend only to perform well in terms of cleaning and disinfecting efficacy but play also an important role in CL performance by improving physical comfort at insertion and through the whole period of wear.

Indeed, several studies pointed out the role of care systems in preventing or recovering from ocular surface damage related to CL as is the case of superficial punctuate corneal staining and SICS. SICS identification require that at least 4 out of 5 regions of the cornea are affected by this superficial staining (Fig. 2).

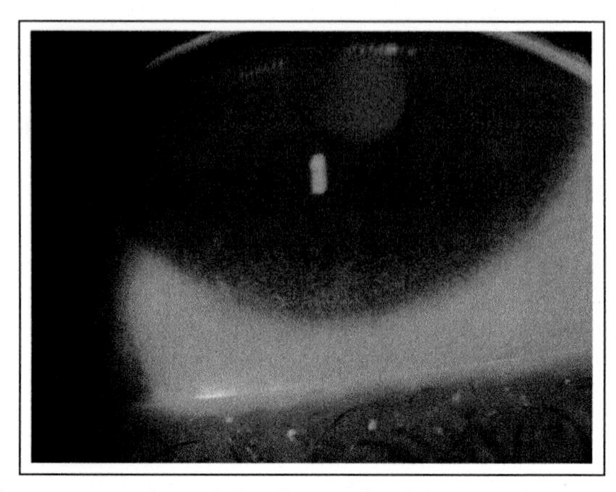

Figure 2 Typical appearance of a peripheral corneal staining covering 4 quadrants (superior one not seen in this photograph).

Color image of this figure appears in the color plate section at the end of the book.

Current knowledge advocates for the role of care systems in reducing SICS events. Hydrogen peroxide has been proposed as an effective strategy in case of corneal staining potentially related to multipurpose solution (Carnt et al. 2007, Papas et al. 2007). Additionally, the pre-application of carboxymethylcellulose (CMC) lubricating and wetting drops (artificial tears) has demonstrated to significantly reduce the incidence of corneal staining in conventional hydrogel CL (Coles et al. 2004) and silicone hydrogel soft CL wearers (Paugh et al. 2007).

Another field of interest is the introduction of wetting agents and other active ingredients into new MPDS formulations to improve the hydration retention of CL materials and improve the long-term comfort. With this purpose, new solutions such as Biotrue (Bausch & Lomb) incorporate derivates of hyaluronic acid to make the lens material hydration to last and the company claims up to 20 hours of hyaluronan retention in the matrix of different hydrogel and silicone hydrogel CL with the potential benefit of improved end-of-day comfort (Scheuer et al. 2010). An additional purpose of this formulation is to maintain native tear proteins in their natural state instead of denaturating onto the CL material to keep the antimicrobial activity of lysozime and preventing adverse reactions on the ocular surface

(Barniak et al. 2010). Although the benefits of these new formulations on lens dehydration and whole day lens comfort will need to be further demonstrated by independent studies, the active implication of the CL care systems to improve CL comfort over time present is a promissory strategy to improve the whole CL wearing experience and eventually decrease the drop-out rate among CL wearers as most of these discontinuations seem to be directly related with dryness and discomfort symptoms (Pritchard et al. 1999).

Finally, the modes in which care systems are used seem to be critical to improve their performance in several aspects. Although in the recent past MPDS use to include a non-rub claim, it has been proved that the inclusion of rub and rinse steps improves significantly the performance of MPDS solutions in terms of disinfection efficacy (Zhu et al. 2011). Rub and rinse step in addition to other compliance recommendations such as tissue-wipe and air-dry demonstrated to be very effective in eliminating microorganisms and biofilms from CL storage cases (Wu et al. 2010, Wu et al. 2011).

CONCLUSIONS

Advances in current MPDS formulations allowed the incorporation of highly efficient disinfection systems while keeping high levels of ocular surface biocompatibility. These facts, along with their potential role to improve contact lens comfort and tolerance overtime make MPDS key factors in the future success of CL industry.

DISCLOSURE

JM González-Méijome is or has been during the last 5 years consultant for several companies with interests in contact lens care systems. The author does not have any financial interest in those care systems or brands.

REFERENCES

Abengozar-Vela, A., F.J. Pinto, J.M. Gonzalez-Meijome, M. Rallo, C. Seres, M. Calonge and M.J. Gonzalez-Garcia. 2011. Contact Lens Case Cleaning Procedures Affect Storage Solution pH and Osmolality. Optom. Vis. Sci. 88: 1414–21.

Andrasko, G.J., K.A. Ryen, R.J. Garofalo and J.M. Lemp. 2006. Compatibility of Silicone Hydrogel Lenses With Multi-Purpose Solutions. ARVO Meeting Abstracts 47: 2392.

Barniak, V.L., S.E. Burke and S. Venkatesh. 2010. Comparative evaluation of multi-purpose solutions in the stabilization of tear lysozyme. Cont. Lens Anterior. Eye 33: S7–11.

Carney, L.G. and R.M. Hill. 1976. Human tear pH. Diurnal variations. Arch. Ophthalmol. 94: 821–824.

Carnt, N.A., M.D.P. Willcox, V.E. Evans, T.J. Naduvilath, D. Tilia, E.B. Papas, D.F. Sweeney and B.A. Holden. 2007. Corneal staining with various contact lens solution-silicone hydrogel lens combinations and its significance: The IER Matrix Study. Contact Lens Spectrum 22: 38–43.

Chuang, E.Y., D.Q. Li, F. Bian, X. Zheng and S.C. Pflugfelder. 2008. Effects of contact lens multipurpose solutions on human corneal epithelial survival and barrier function. Eye Contact Lens 34: 281–286.

Coles, M.L., N.A. Brennan, V. Shuley, J. Woods, C. Prior, J.G. Vehige and P.A. Simmons. 2004. The influence of lens conditioning on signs and symptoms with new hydrogel contact lenses. Clin. Exp. Optom. 87: 367–371.

Dalton, K., L.N. Subbaraman, R. Rogers and L. Jones. 2008. Physical properties of soft contact lens solutions. Optom. Vis. Sci. 85: 122–128.

Garofalo, R. J., N. Dassanayake, C. Carey, J. Stein, R. Stone and R. David. 2005. Corneal staining and subjective symptoms with multipurpose solutions as a function of time. Eye Contact Lens 31: 166–174.

Houang, E., D. Lam, D. Fan and D. Seal. 2001. Microbial keratitis in Hong Kong: relationship to climate, environment and contact-lens disinfection. Trans. R. Soc. Trop. Med. Hyg. 95: 361–367.

Imayasu, M., A. Shiraishi, Y. Ohashi, S. Shimada and H.D. Cavanagh. 2008. Effects of multipurpose solutions on corneal epithelial tight junctions. Eye Contact Lens 34: 50–55.

Khor, W. B., T. Aung, S.M. Saw, T.Y. Wong, P.A. Tambyah, A.L. Tan, R. Beuerman, L. Lim, W.K. Chan, W.J. Heng, J. Lim, R.S. Loh, S.B. Lee and D.T. Tan. 2006. An outbreak of Fusarium keratitis associated with contact lens wear in Singapore. JAMA 295: 2867–2873.

Kilvington, S., L. Huang, E. Kao and C.H. Powell. 2010. Development of a new contact lens multipurpose solution: Comparative analysis of microbiological, biological and clinical performance. J. Optom. 3: 134–142.

Lopez-Alemany, A. and R. Montes Mico. 1998. pH of multipurpose contact lens solutions over time. Cont.Lens Anterior Eye 21: 7–10.

Morgan, P. B., C.A. Woods, I.G. Tranoudis, M. Helland, N. Efron, R. Knajian, C.N. Grupcheva, D. Jones, K.-O. Tan, A. Pesinova, O. Ravn, J. Santodomingo, E. Vodniaznszky, N. Erdinest, H.I. Hreinsson, G. Montani, M. Itol, J. Bendoriene, E. van der Worp, J. Hsiano, G. Phillips, J.M. Gonzalez-Méijome, S. Radu, V.Belousov and J.J. Nichols. 2010. International contact lens prescribing in 2009. Contact Lens Spectrum 30–36.

Morgan, P.B., N.Efron, H. Toshida and J.J. Nichols. 2011. An international analysis of contact lens compliance. Cont. Lens Anterior. Eye 34: 223–228.

Mowrey-McKee, M., A. Sills and A. Wright. 2002. Comparative cytotoxicity potential of soft contact lens care regimens. CLAO J. 28: 160–164.

Papas, E.B., N. Carnt, M.D. Willcox and B.A. Holden. 2007. Complications associated with care product use during silicone daily wear of hydrogel contact lens. Eye Contact Lens 33: 392–393.

Patel, A. and K. Hammersmith. 2008. Contact lens-related microbial keratitis: recent outbreaks. Curr. Opin. Ophthalmol 19: 302–306.

Paugh, J.R., H.J. Marsden, T.B. Edrington, P.N. Deland, P.A. Simmons and J.G. Vehige. 2007. A pre-application drop containing carboxymethylcellulose can reduce multipurpose solution-induced corneal staining. Optom.Vis. Sci. 84: 65–71.

Pritchard, N., D. Fonn and D. Brazeau. 1999. Discontinuation of contact lens wear: a survey. Int. Contact Lens Clin. 26: 157–162.

Radford, C.F., A.S.Bacon, J.K. Dart and D.C. Minassian. 1995. Risk factors for acanthamoeba keratitis in contact lens users: a case-control study. BMJ 310: 1567–1570.

Radford, C.F., D.C. Minassian and J.K. Dart. 1998. Disposable contact lens use as a risk factor for microbial keratitis. Br. J. Ophthalmol. 82: 1272–1275.

Santodomingo-Rubido, J., O. Mori and S.Kawaminami. 2006. Cytotoxicity and antimicrobial activity of six multipurpose soft contact lens disinfecting solutions. Ophthalmic Physiol. Opt. 26: 476–482.

Saw, S. M., P.L. Ooi, D.T. Tan, W.B. Khor, C.W. Fong, J. Lim, H.Y. Cajucom-Uy, D. Heng, S.K. Chew, T. Aung, A.L. Tan, C.L. Chan, S. Ting, P.A. Tambyah and T.Y. Wong. 2007. Risk factors for contact lens-related fusarium keratitis: a case-control study in Singapore. Arch. Ophthalmol. 125: 611–617.

Scheuer, C.A., K.M. Fridman, V.L. Barniak, S.E. Burke and S. Venkatesh. 2010. Retention of conditioning agent hyaluronan on hydrogel contact lenses. Cont. Lens Anterior. Eye 33: S2–S6.

Stapleton, F., L. Keay, I. Jalbert and N. Cole. 2007. The epidemiology of contact lens related infiltrates. Optom. Vis. Sci. 84: 257–272.

Tu, E.Y. and C.E. Joslin. 2010. Recent outbreaks of atypical contact lens-related keratitis: what have we learned? Am. J. Ophthalmol. 150: 602–608.

Wu, Y.T., H. Zhu, M. Willcox and F. Stapleton. 2010. Impact of air-drying lens cases in various locations and positions. Optom. Vis. Sci. 87: 465–468.

Wu, Y.T., H. Zhu, M. Willcox and F. Stapleton. 2011. The effectiveness of various cleaning regimens and current guidelines in contact lens case biofilm removal. Invest Ophthalmol. Vis. Sci. 52: 5287–5292.

Zhu, H., A. Ding, M. Bandara, M.D. Willcox and F. Stapleton. 2007. Broad spectrum of antibacterial activity of a new multipurpose disinfecting solution. Eye Contact Lens 33: 278–283.

Zhu, H., M.B. Bandara, A.K. Vijay, S. Masoudi, D. Wu and M.D. Willcox. 2011. Importance of rub and rinse in use of multipurpose contact lens solution. Optom. Vis. Sci. 88: 967–972.

Index

C

D

U

V

W

X

Z

Color Plate Section

Chapter 1

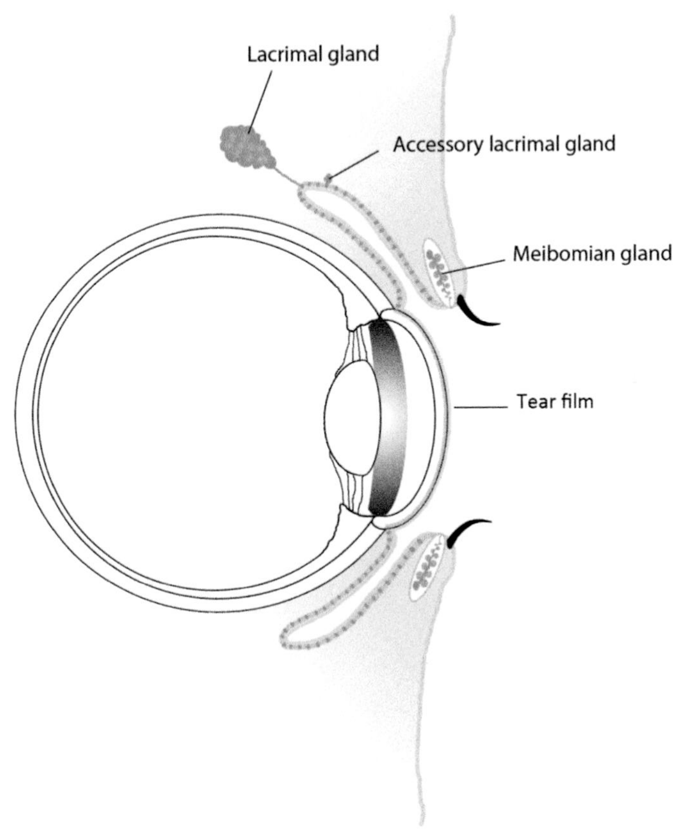

Figure 1 This figure shows a cross-sectional view of the ocular surface system with the continuous epithelium highlighted in pink and the tear film in blue.

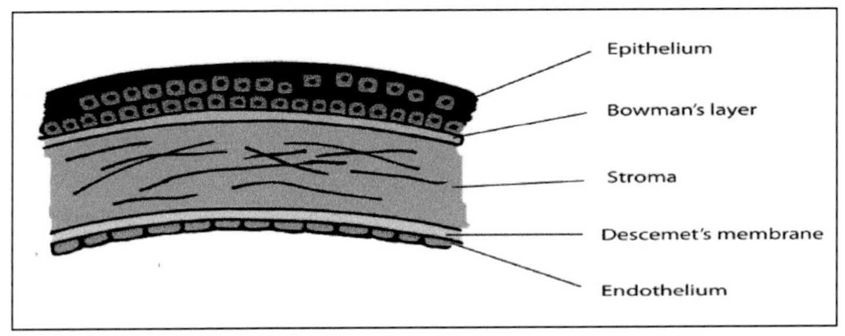

Figure 2 This figure shows a schematic illustration of the different layers of the cornea (not drawn to scale).

Chapter 4

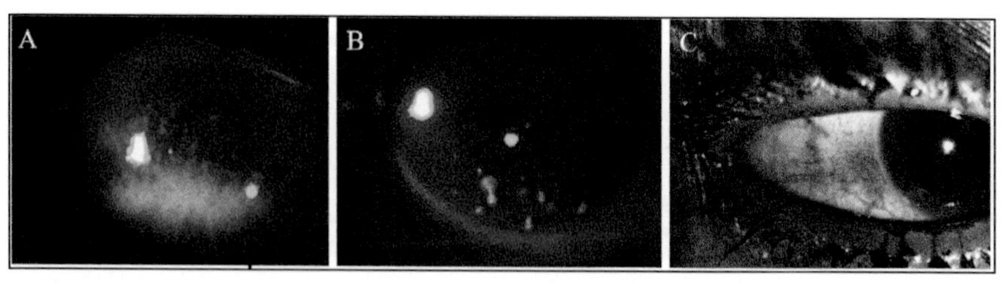

Figure 2 This figure shows vital dye staining of cornea and conjunctiva: (A-B) Two distinctive corneal fluorescein patterns seen in dry eye patients. (A) Severe inferior staining with diffusion and areas of coalescence of dye. (B) Filamentary keratitis. (C) Lissamine green staining showing intense diffuse exposure zone staining.

Chapter 5

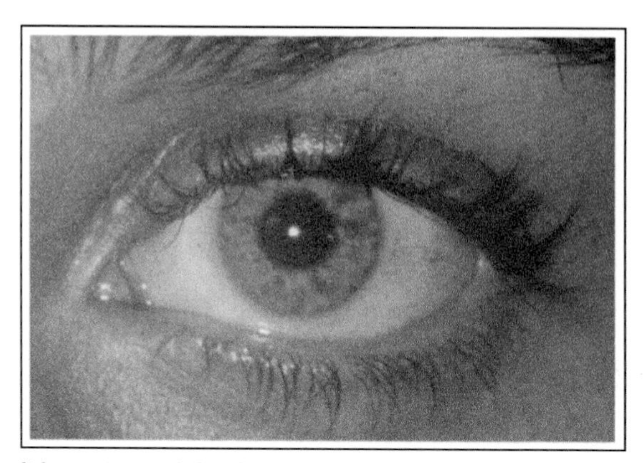

Figure 1 The eyelid margin is subdivided in anterior and posterior by the grey line.

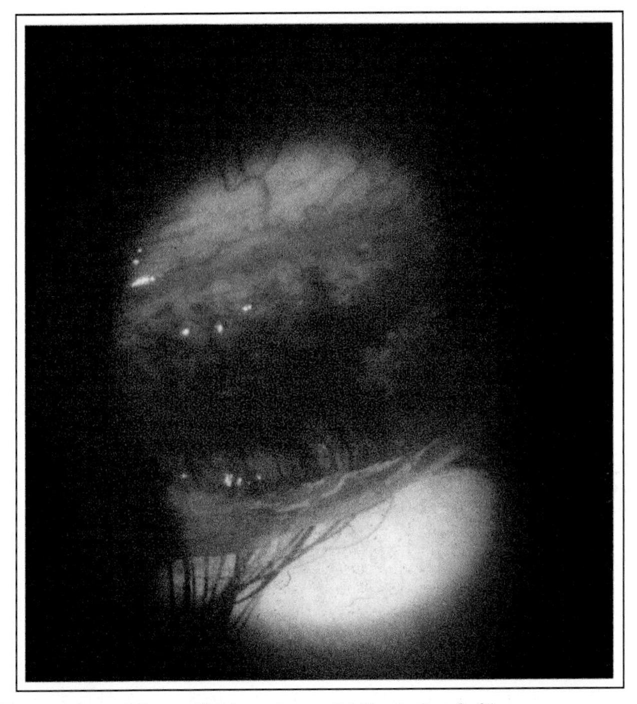

Figure 2 Inflammation of the eyelid (courtesy of Alfredo Bonfioli).

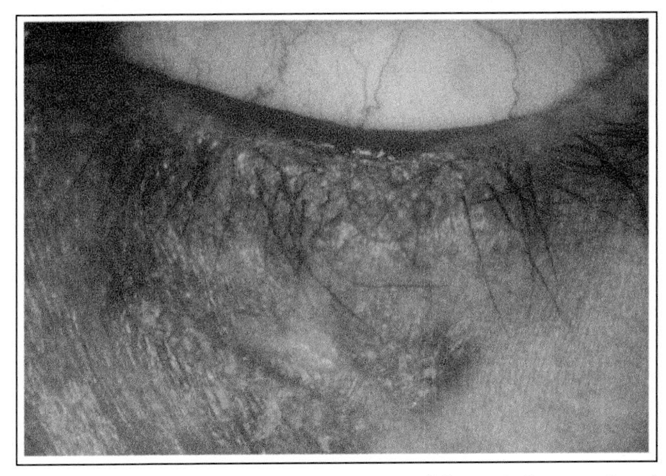

Figure 3 This figure shows the clinical examination: eyelid margin hyperemia, telangiectasy and ulceration, localized trichiasis and madarosis (cortesy of Alfredo Bonfioli).

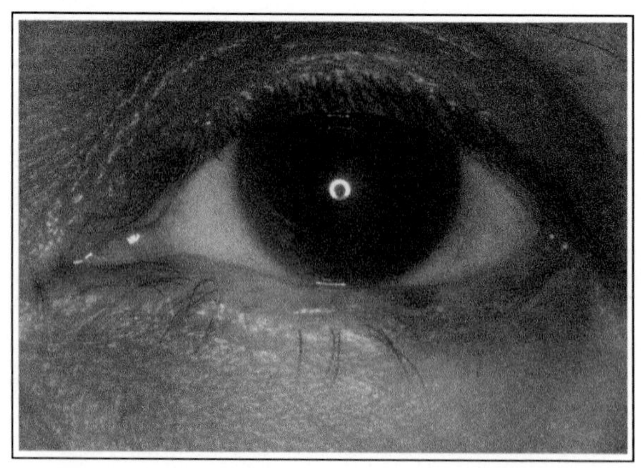

Figure 4 This figure shows the Discoid lupus (courtesy of Alfredo Bonfioli).

Chapter 7

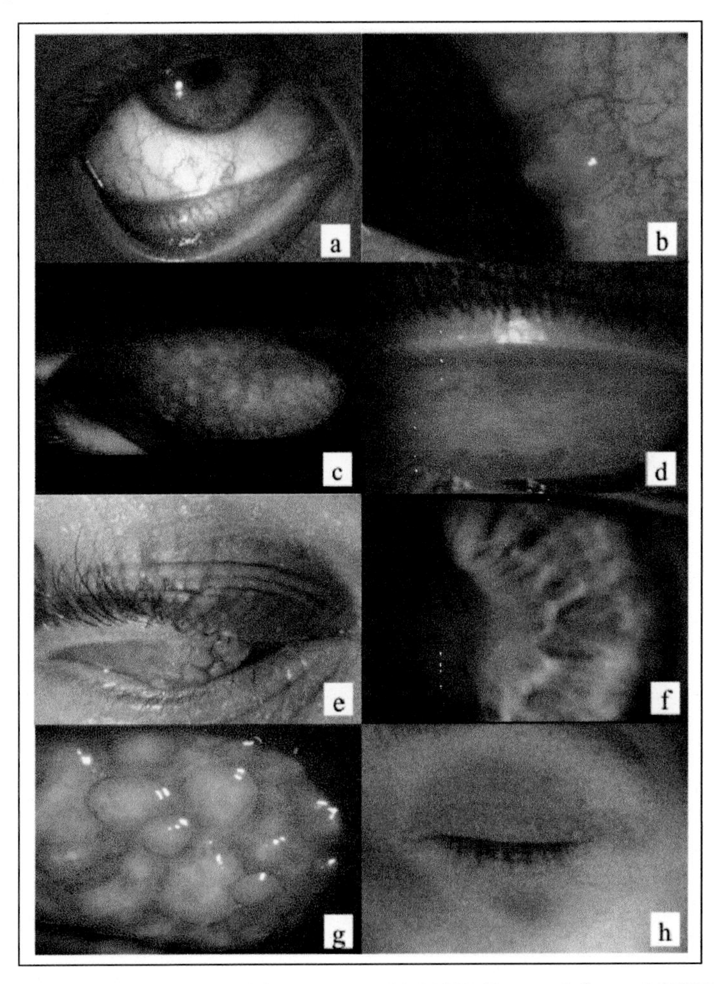

Figure 1 a) SAC, conjunctival hyperemia; b) VKC, Trantas' dots; c) VKC, giant papillae; d) VKC, reticular appearance of the conjunctiva on the upper tarsal plate; e) AKC, eczema of the eyelid and giant papillae on the upper tarsal plate; f) AKC, corneal ulcer; g) GPC, giant papillae and mucus discharge in a contact lens wearer; h) CDC, periocular contact dermatitis (Courtesy of Dr. A. Leonardi).

Chapter 8

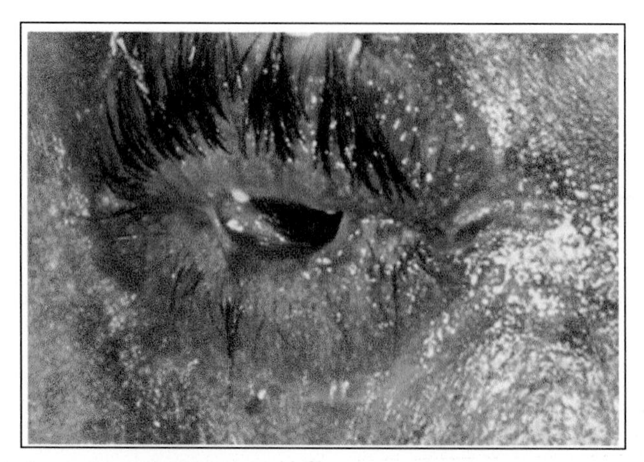

Figure 1 This figure shows the blepharoconjunctivitis and eyelid shortening in SJS.

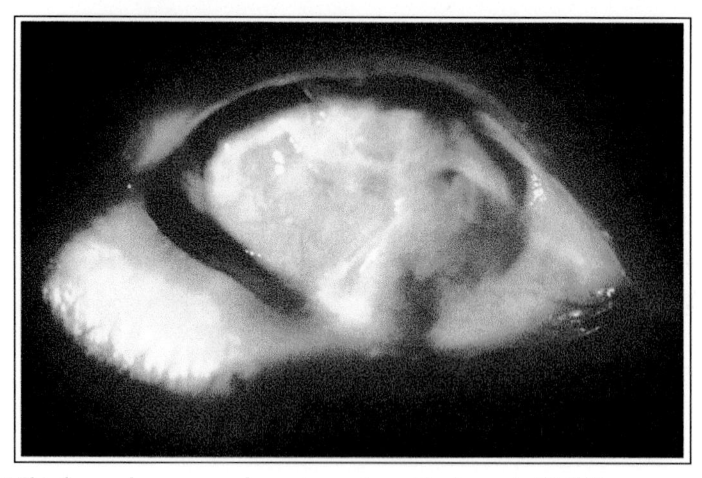

Figure 3 This figure shows corneal scarring and symblepharon in OMMP.

Chapter 9

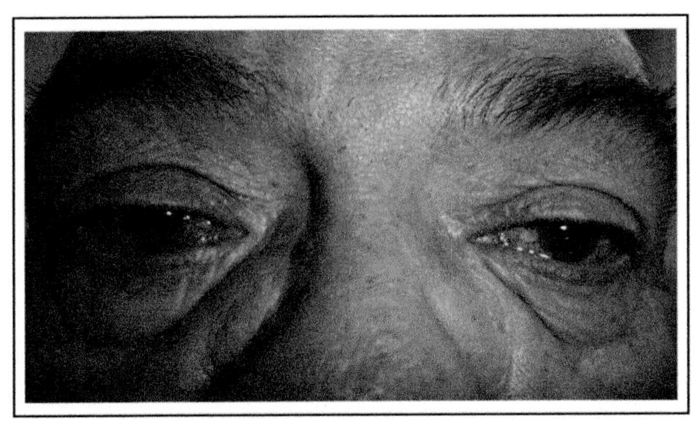

Figure 2 In this figure, a patient with pterigyum in his right eye shows a chronic redness in both eyes due to prostaglandins analogue treatment (Travatoprost).

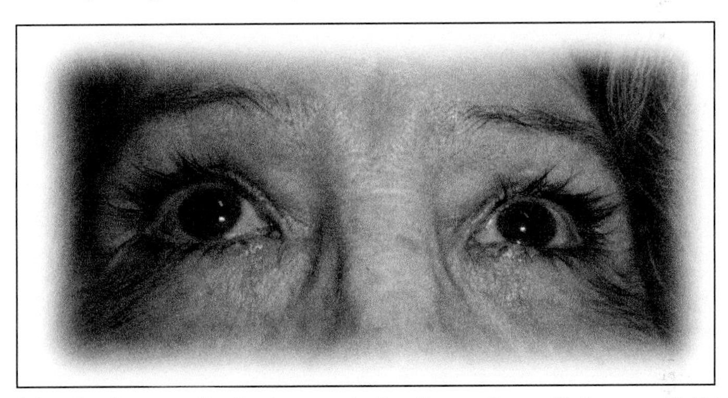

Figure 3 After the first month of using prostaglandins analogue (Latanoprost), this patient asked the ophthalmologist about the eyelashes... *principally the new eyelashes* in the skin, as it is possible to see in this figure at the bottom of the right and left.

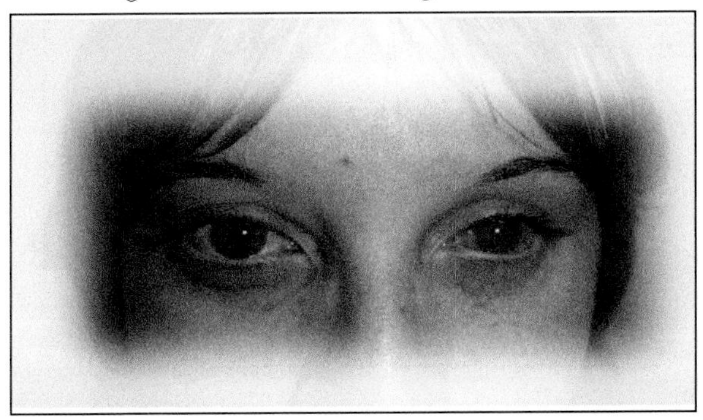

Figure 4 A female patient treated with Brimonidine in a fixed combination associated with Timolol, only in her left eye. Moreover the redness in the eye shows signs of atopic dermatitis at the periocular skin (eczematous lesions).

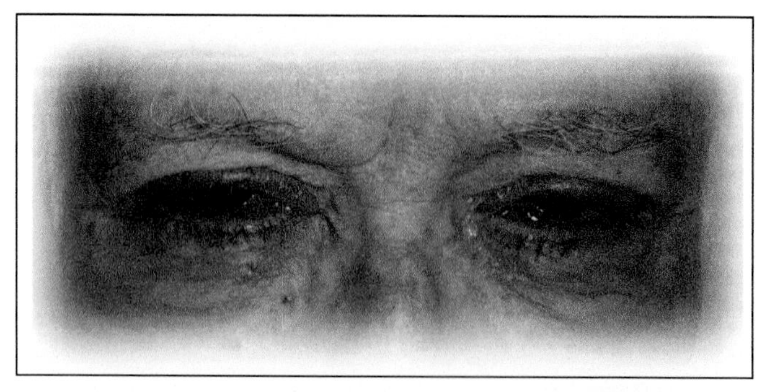

Figure 5 Patient with a typical bilateral allergic reaction with swelling of the lids, conjunctival redness, chemosis, and itching after starting the treatment with Brimonidine eyedrops.

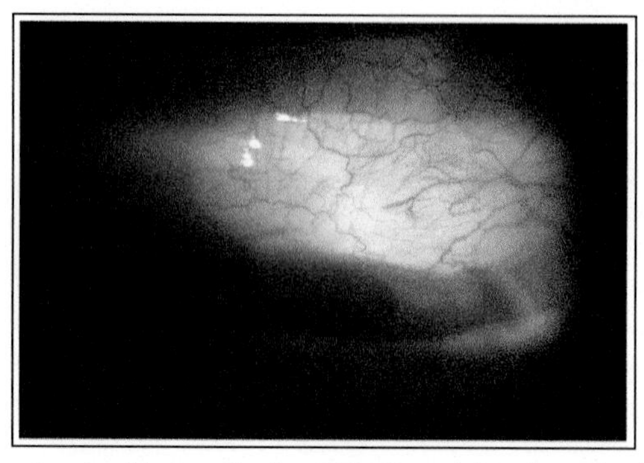

Figure 6 This figure shows a big functioning bleb. Complains about foreing body sensation was presented in this case.

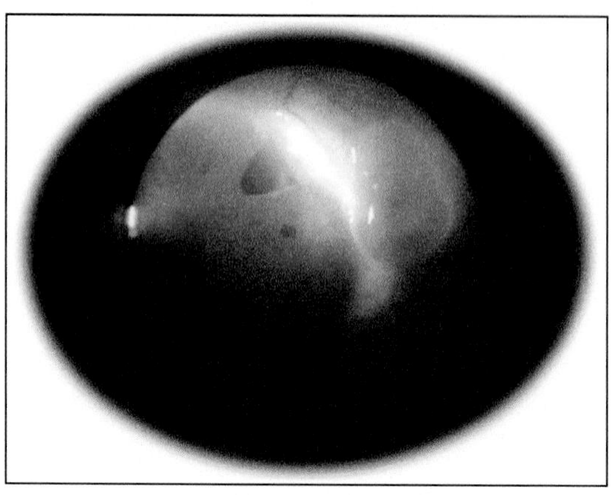

Figure 7 This figure shows the formation of conjunctival cyst; foreing body sensation in this case is obvious.

Chapter 13

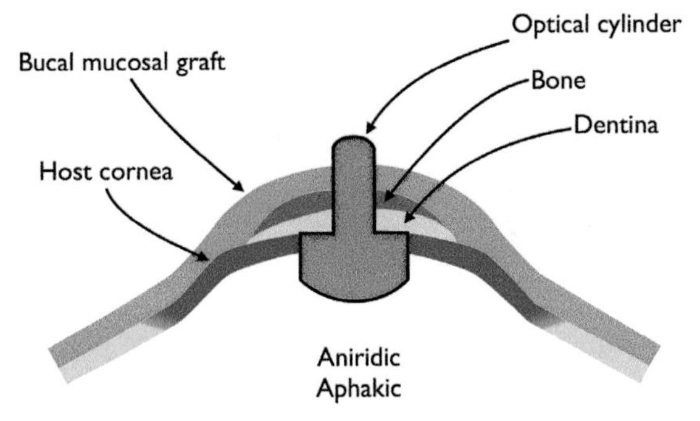

Figure 1 This figure shows the schematic of osteo-odonto-keratoprosthesis (OOKP) (Adaptation from Liu et al. 2008).

Chapter 14

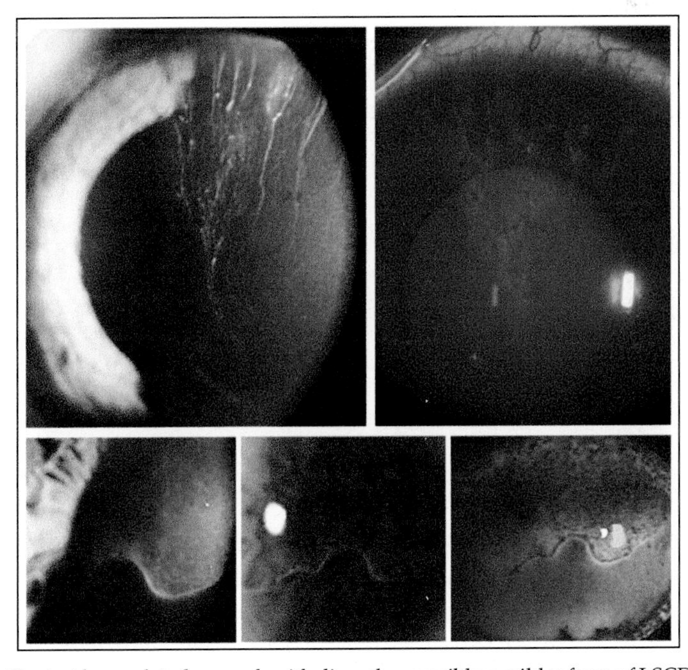

Figure 1 Contact lens-related corneal epitheliopathy, possibly a milder form of LSCD. Above, a 28 year old soft contact lens user since 9 year before and symptomatic for 1 year, showing a linear or "fimbriae" irregular epithelial pattern in a superior triangular area originating from the limbus. Below, a 48 year old patient, contact lens user for 28 year, and with reduced vision since 2 months before, with an avascular layer of altered epithelium as an "advancing wave" over the whole superior half of the cornea, clearly demarcated by line from the inferior normal-appearing zone. Both improved with the suspension of contact lens use and unpreserved lubricants.

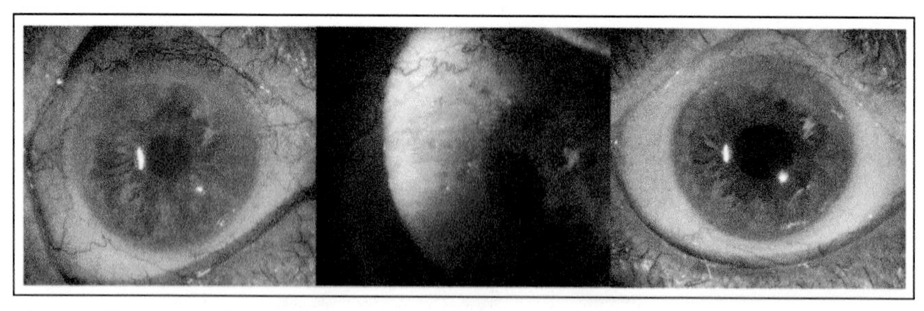

Figure 2 This figure shows a partial LSCD due to by ocular cicatricial pemphigoid treated with sequential conjunctival epitheliectomy (SCE). Left: Preoperative appearance with sector conjunctivalization of the superior cornea. Center: Appearance immediately after a superior SCE. Right: Postoperative appearance of the same eye 1 month later, showing the repopulation of the superior cornea with normal-appearing epithelium.

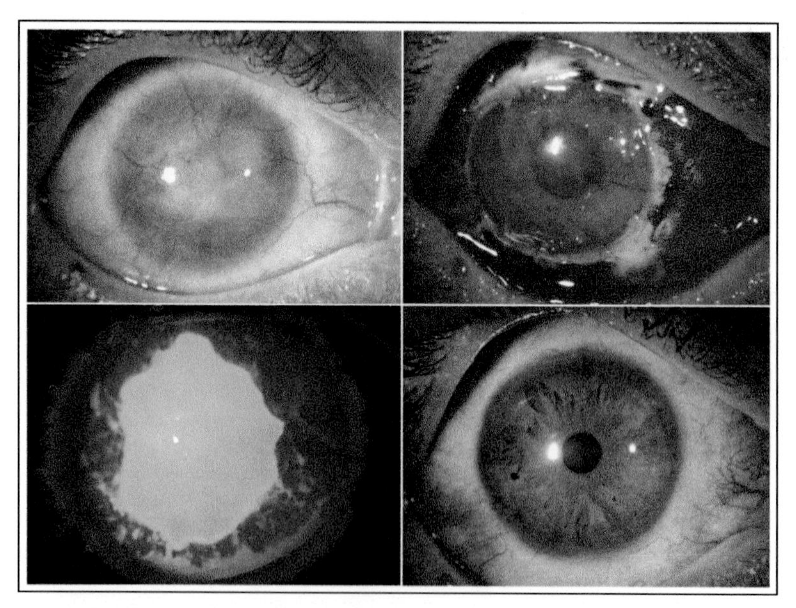

Figure 3 KLAU in a case of unilateral alkali injury. Top left: Initial appearance 7 year after sustaining a lime burn, with counting fingers vision. Top right: Immediate postoperative appearance after 2-sector KLAU from the fellow normal eye. Bottom left: Early re-epithelialization of the host cornea, at 3 days post-op. Bottom right: result at 18 month, with stable ocular surface and visual acuity of 20/50 with 2 diopters of mixed astigmatism, improving to 20/25 with contact lens. At 4 years, it reached 20/20.

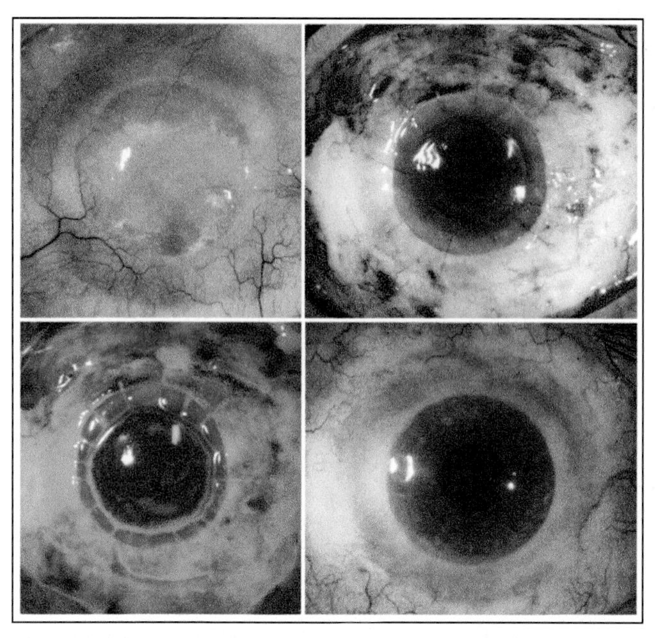

Figure 4 Total bilateral LSCD after trauma and multiple ocular surgeries (above, left), treated with 2-sector KLAL from his HLA-identical sister, combined with a simultaneous PK. Immediate (above, right), early result showing re-epithelialization (below, left), and result after 4 years (below, right) with a stable ocular surface and a clear central graft without immunosuppressive therapy, despite peripheral neovascularization and lipid deposition.

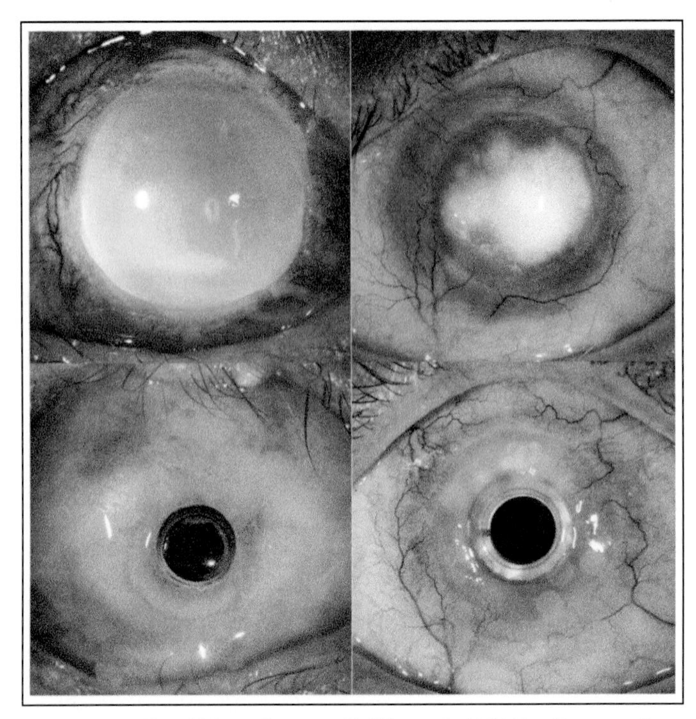

Figure 5 This patient suffered bilateral severe alkali burns in 1985. Top: Preoperative appearance of the right eye (at left) after 3 years and of the left eye (at right), 13 years after the accident with a failed penetrating keratoplasty, severe vascularization and secondary lipid keratopathy. Bottom (respectively): Results 13 years after OOKP in the right eye and 3 years after BKP in the left eye.

Chapter 15

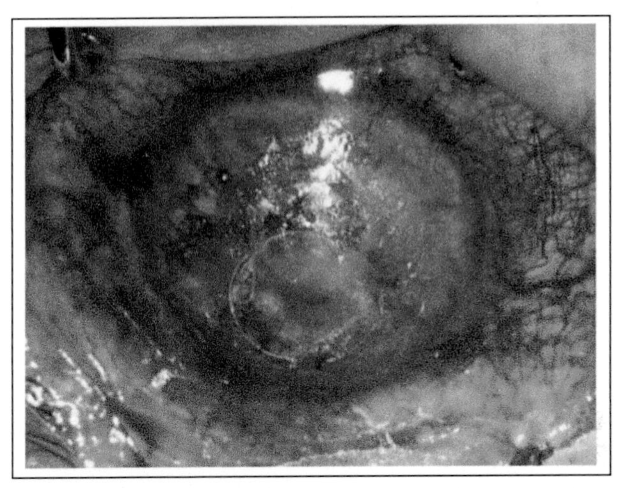

Figure 1 This figure shows the surgical view of a paracentral corneal perforation due to a herpetic keratitis sealed with cyanoacrylate and a plastic shell before the contact lens fitting. Note the air bubble centred due to supine position.

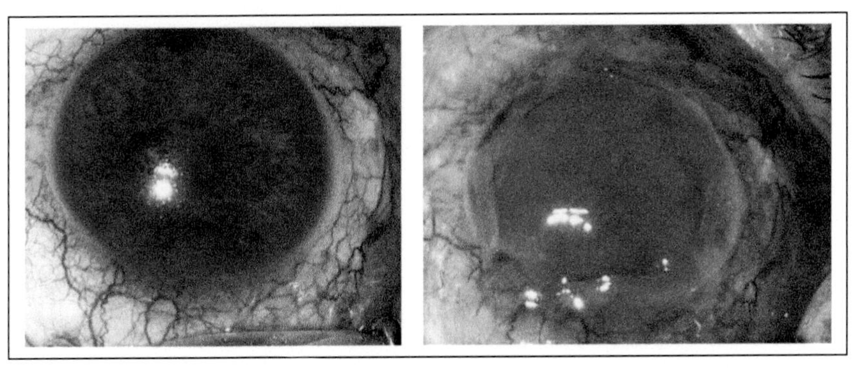

Figure 2 This figure shows the surgical view of a persistent epithelial defect in a diabetic patient after a pars plana vitrectomy before and after the amniotic membrane transplantation with a 10/0 nylon buried running suture technique.

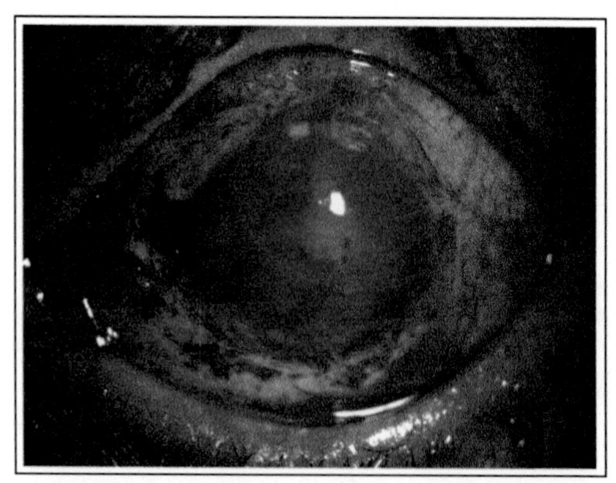

Figure 3 This figure shows the *ex vivo* expanded limbal graft carried in amniotic membrane covered by a great diameter therapeutic contact lens.

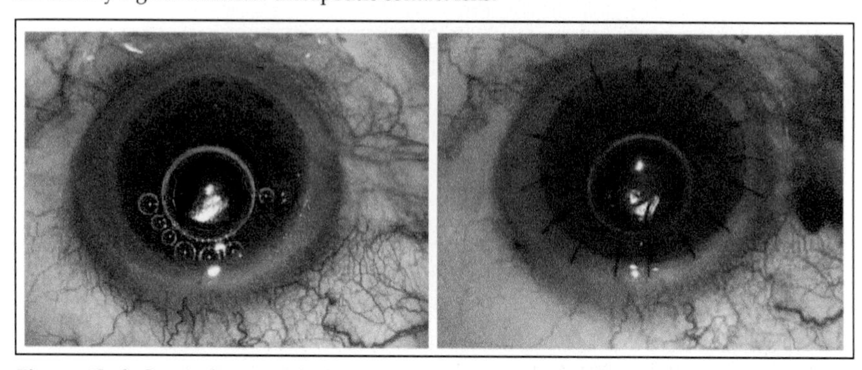

Figure 4 Left: Surgical view of the bared Descemet membrane after deep lamellar dissection. Small bubbles of air show the integrity of the anterior chamber. Right: Graft sutured with 10/0 nylon interrupted suture technique. Coloured graft is due to the use of trypan blue in the harvest of Descemet-endothelium complex for a DMEK with the split technique.

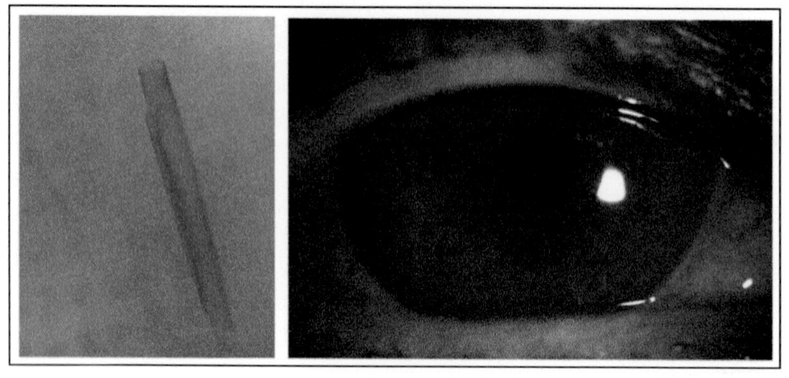

Figure 5 Left: This figure shows the endothelium-Descemet membrane roll after harvesting from a donor cornea coloured by trypan blue. Right: Clinical photo 6 mon after DMEK. Note small-pigmented fibrotic peripheral folds in the limit of the graft with a transparent cornea.

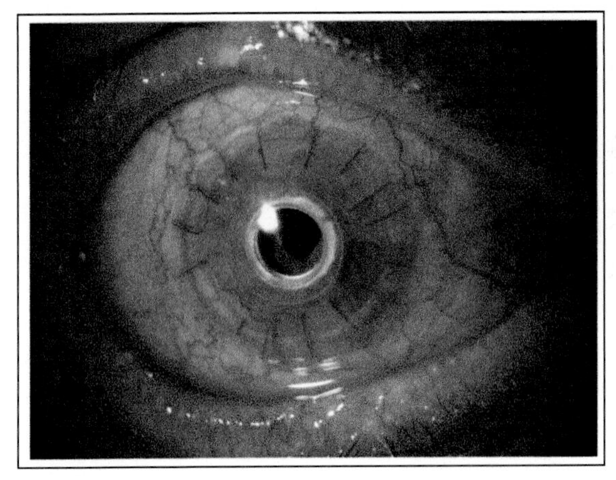

Figure 6 This figure shows the Boston Keratoprosthesis in an aphakic patient with three previous failed penetrating keratoplasties.

Chapter 17

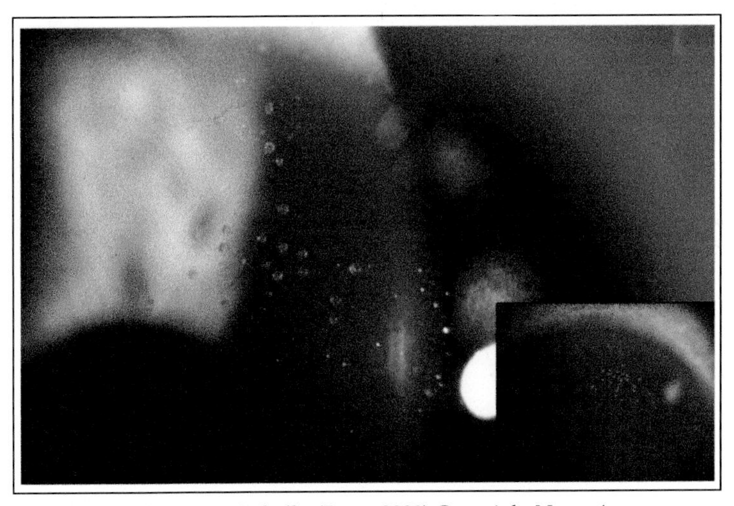

Figure 1 This figure shows mucin balls. (Bruce 2008) Copyright Novartis.

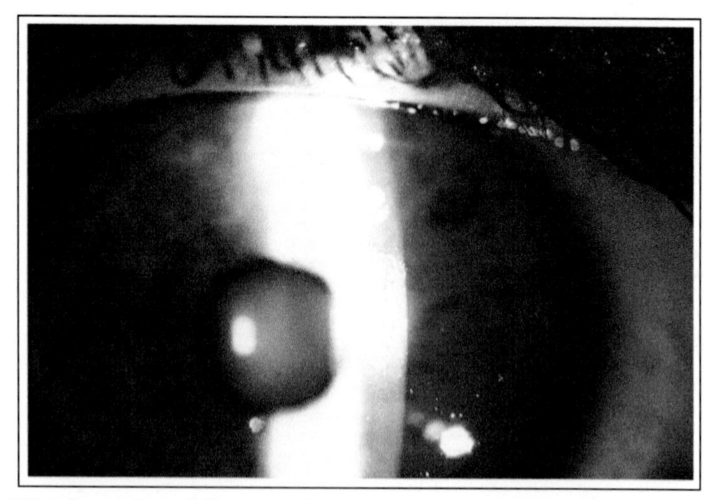

Figure 2 This figure shows Jelly bumps (Bruce 2008) Copyright Novartis.

Figure 3 This figure shows CPG (Bruce 2008) Copyright Novartis.

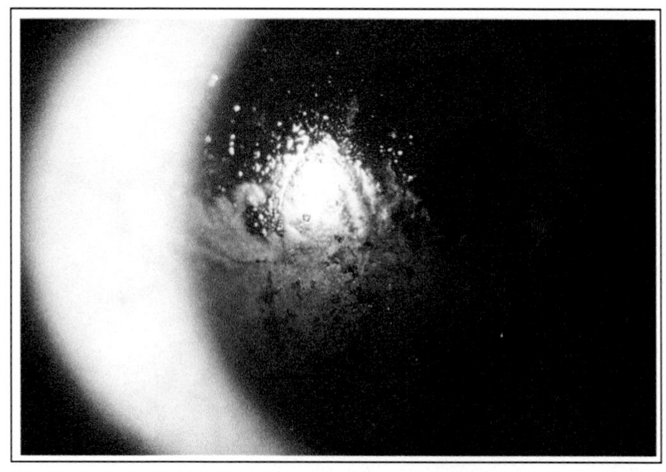

Figure 4 This figure shows the CL deposits with incomplete blink (Bruce 2008) Copyright Novartis.

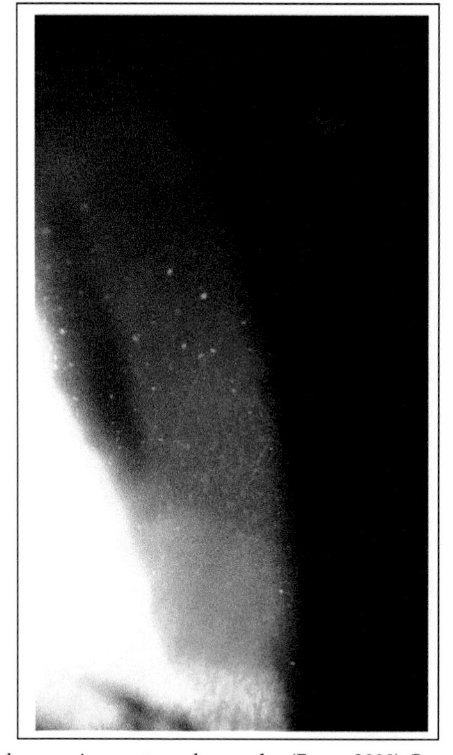

Figure 5 This figure shows microcysts and vacuoles (Bruce 2008) Copyright Novartis.

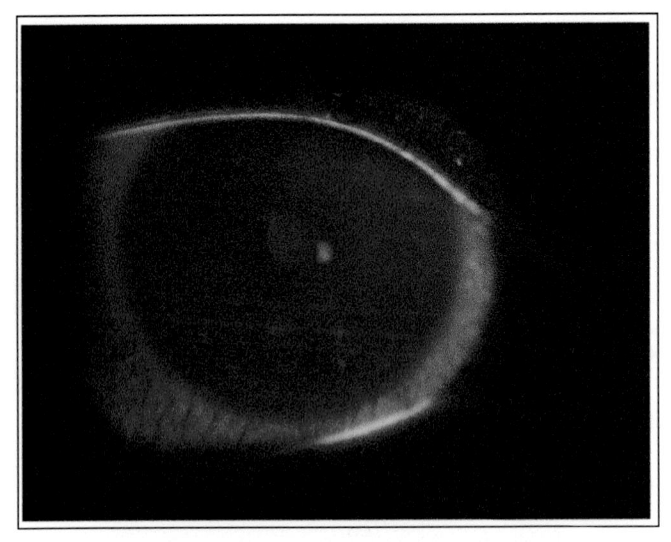

Figure 6 This figure shows a solution toxicity keratitis (Bruce 2008), Copyright Novartis.

Figure 7 This figure shows a CLPU (Bruce 2008), Copyright Novartis.

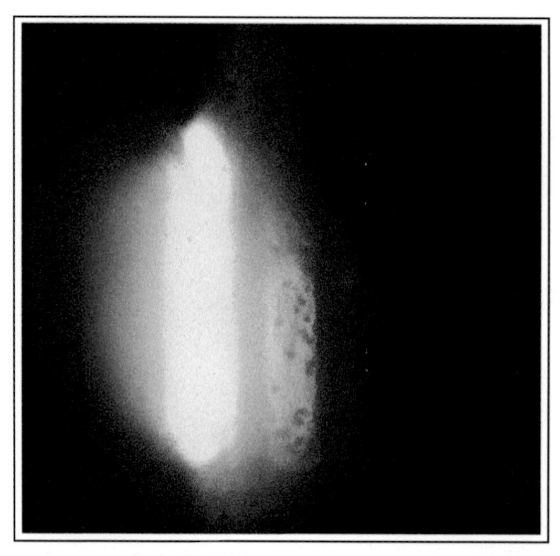

Figure 9 This figure shows endothelial blebs (Bruce 2008), Copyright Novartis.

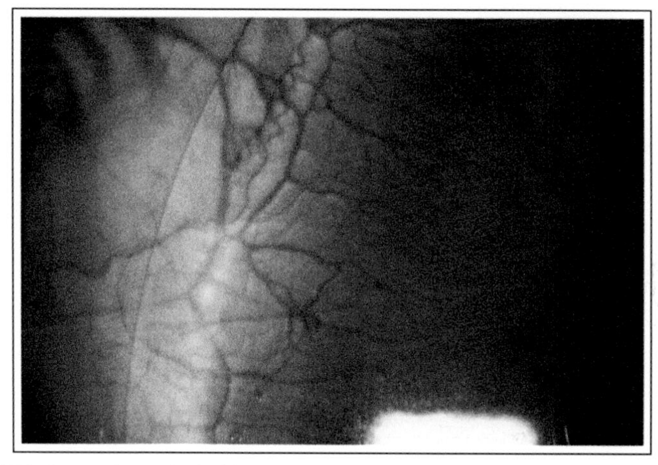

Figure 10 This figure shows limbal hyperaemia (Bruce 2008), Copyright Novartis.

Chapter 18

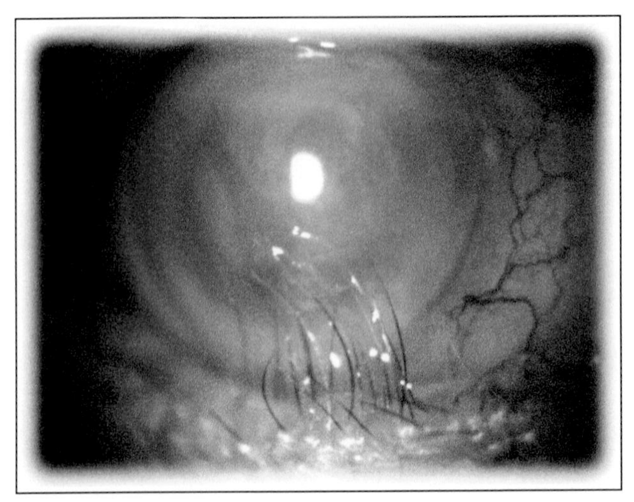

Figure 1 This figure shows the use of a soft bandage contact lens to protect the cornea in a patient with trichiasis in the inferior eyelid.

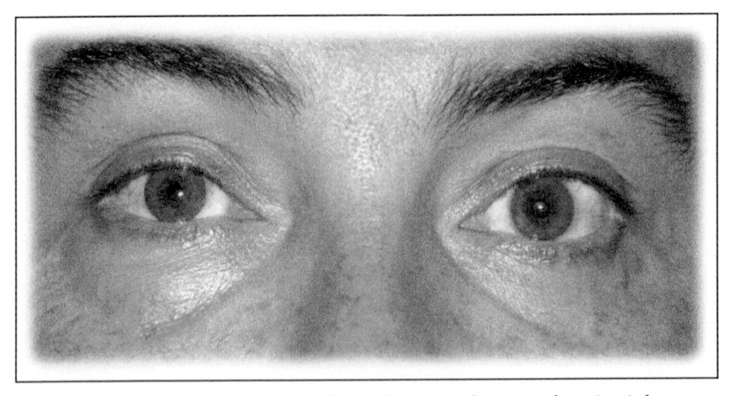

Figure 2 This figure shows the cosmetic effect of a painted contact lens in right eye.

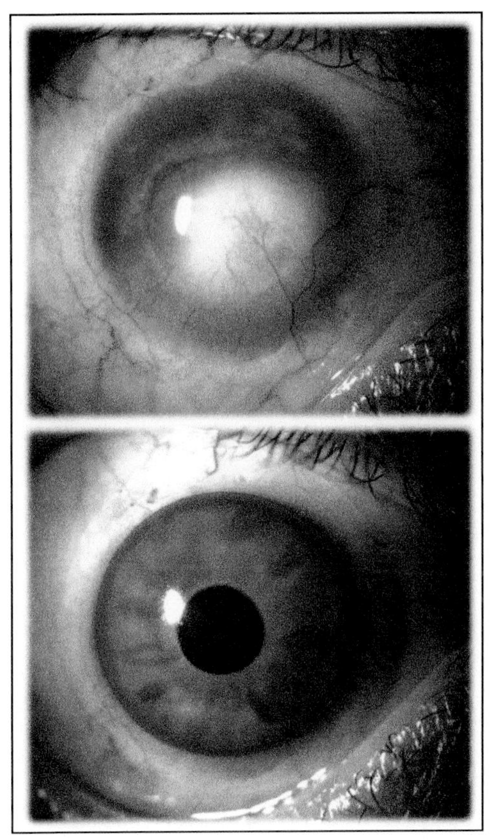

Figure 3 This figure shows the cosmetic contact lens effect with low biomicroscopic magnification.

Chapter 20

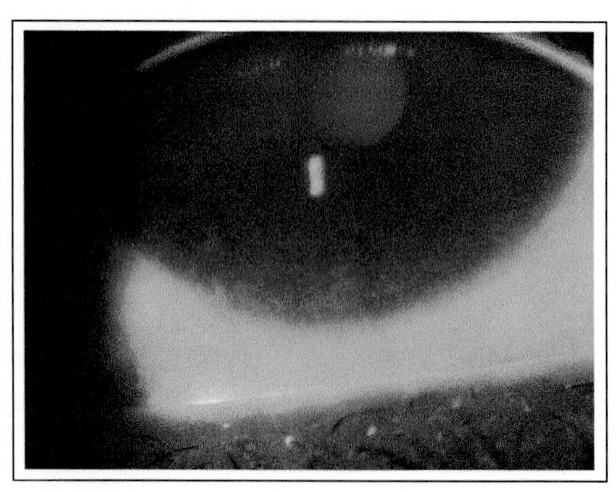

Figure 2 Typical appearance of a peripheral corneal staining covering 4 quadrants (superior one not seen in this photograph).

Agile Model-Based Development Using UML-RSDS

Kevin Lano
Department of Informatics
King's College London, London
United Kingdom

CRC Press is an imprint of the
Taylor & Francis Group, an **informa** business
A SCIENCE PUBLISHERS BOOK

CRC Press
Taylor & Francis Group
6000 Broken Sound Parkway NW, Suite 300
Boca Raton, FL 33487-2742

First issued in paperback 2020

ISBN-13: 978-1-4987-5222-0 (pbk)
ISBN-13: 978-0-367-78285-6 (pbk)

Visit the Taylor & Francis Web site at
http://www.taylorandfrancis.com

and the CRC Press Web site at
http://www.crcpress.com

Preface

Model-based development is moving into the mainstream of software engineering, and it is an approach that all software developers should know about as one option to use in solving software problems. In this book we give a practical introduction to model-based development, model transformations and agile development, using a UML and Java-based toolset, UML-RSDS. Guidance on applying model-based development in a range of domains is provided, and many examples are given to illustrate the UML-RSDS process and techniques. The book is suitable both for professional software engineers and for postgraduate and undergraduate teaching.

I would like to acknowledge the contributions of my research team members and colleagues who have helped in the development of the UML-RSDS tools and method: Kelly Androutsopoulos, Pauline Kan, Shekoufeh Kolahdouz-Rahimi, Sobhan Yassipour-Tehrani, Hessa Alfraihi, David Clark, Howard Haughton, Tom Maibaum, Iman Poernomo, Jeffery Terrell and Steffen Zschaler. The support of Imperial College London, King's College London and the EPSRC is also acknowledged.

I dedicate this book to my wife, Olga.

Contents

Chapter 1

Introduction

In this chapter we introduce model-based development and UML-RSDS, and discuss the context of software development which motivates the use of such methods and tools.

1.1 Model-based development using UML-RSDS

Model-based development (MBD) is an approach which aims to improve the practice of software development by (i) enabling systems to be defined in terms closer to their requirements, abstracted from and independent of particular implementation platforms, and (ii) by automating development steps, including the writing of executable code.

A large number of MBD approaches, tools and case studies now exist, but industrial uptake of MBD has been restricted by the complexity and imprecision of modelling languages such as UML, and by the apparent resource overheads without benefit of many existing MBD methods and tools [3, 5, 6, 7, 8].

UML-RSDS[1] has been designed as a lightweight and agile MBD approach which can be applied across a wide range of application areas. We have taken account of criticisms of existing MBD approaches and tools, and given emphasis on the aspects needed to support practical use such as:

- *Lightweight method and tools*: usable as an aid for rapidly developing parts of a software system, to the degree which developers

[1] 'Rigorous Specification, Design and Synthesis', although 'Rapid Specification, Design and Synthesis' would also be appropriate.

find useful. It does not require a radical change in practice or the adoption of a new complete development process, or the use of MBD for all aspects of a system.

- *Independent* of other MBD methods or environments, such as Eclipse/EMF.
- *Non-specialist*: UML-RSDS uses only a core subset of UML class diagram and use case notations, which are the most widely-known parts of UML.
- *Agile*: incremental changes to systems can be made rapidly via their specifications, and the changes propagated to code.
- *Precise*: specifications in UML-RSDS have a precise semantics, which enables reliable code production in multiple languages.

The benefits of our MBD approach are:

- Reduction in coding cost and time.
- The ability to model an application domain, to define a DSL (domain specific language) for a domain, and to define custom code generators for the domain.
- Reducing the gap between specification and code, so that the consequences of requirements and specification choices can be identified at an early development stage.
- The ability to optimize systems at the platform-independent modelling level, to avoiding divergence between code and models caused by manual optimization of generated code.
- The ability to formally analyse DSLs and systems at the specification stage.

These capabilities potentially reduce direct software development costs, and costs caused by errors during development and errors persisting in delivered products. Both time-to-market and product quality are potentially improved.

Figure 1.1 shows the software production process which is followed using UML-RSDS: specifications are defined using UML class diagrams and use cases, these can be analysed for their internal quality and correctness, and then platform-independent designs are synthesised (these use a pseudocode notation for a subset of UML Activity diagram notation). From these designs executable code in a particular object-oriented programming language (currently, Java, C# or C++) can then be automatically synthesised.

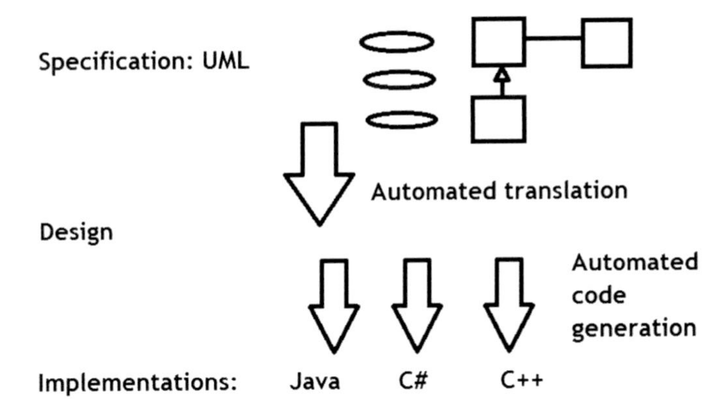

Figure 1.1: UML-RSDS software production process

Unlike many other MBD approaches, which involve the management of multiple models of a system, UML-RSDS specifies systems using a single integrated Use Case and Class Diagram model. This simplifies the specification and design processes and aligns the approach to Agile development practices which are also based on maintaining a single system model (usually the executable code).

Some typical uses of UML-RSDS could be:

- Modelling the business logic of an application and automatically generating the code for this, in Java, C# or C++.
- Modelling and code generation of a component within a larger system, the remainder of which could be developed using traditional methods.
- Defining a model transformation, such as a migration, refinement, model analysis, model comparison, integration or refactoring, including the definition of custom code generators.

UML-RSDS is also useful for teaching UML, OCL, and MBD and agile development principles and techniques, and has been used on undergraduate and Masters courses for several years. In this book we include guidance on the use of the tools to support such courses. An important property of MBD tools is that the code they produce should be *correct*, *reliable* and *efficient*: the code should accurately implement the specification, and should contain no unintended functionality or behaviour. To meet this requirement, the UML-RSDS code generators have been developed to use a proven code-generation strategy which both ensures correctness and efficiency.

In the following chapters, we illustrate development in UML-RSDS using a range of examples (Chapter 2), describe in detail the UML notations used (Chapters 3, 4 and 5) and the process of design synthesis adopted (Chapter 6). Chapter 7 describes how model transformations can be expressed in UML-RSDS. An illustrative example of UML-RSDS development is given in Chapter 8.

Chapter 9 describes design patterns and refactorings that are supported by UML-RSDS. Chapter 10 explains how UML-RSDS systems can be composed. Chapters 11, 12 and 13 describe how migration, refinement and refactoring transformations can be defined in UML-RSDS. Chapters 14 and 15 describe how bidirectional, incremental and exploratory transformations can be defined in UML-RSDS. Chapters 16, 17 and 18 describe in detail the development process in UML-RSDS, following an agile MBD approach. The development of specialised forms of system is covered in Chapters 19 (reactive systems) and 20 (enterprise information systems).

Chapter 21 describes how the tools can be used to support UML, MBD and agile development courses, and gives examples of case studies and problems that can be used in teaching.

The Appendix gives technical details of the UML-RSDS notations and tool architecture.

1.2 The 'software crisis'

The worldwide demand for software applications has been growing at an increasing pace as computer and information technology is used pervasively in more and more areas: in mobile devices, apps, embedded systems in vehicles, cloud computing, health informatics, finance, enterprise information systems, web services and web applications, and so forth. Both the pace of change, driven by new technologies, and the complexity of systems are increasing. However, the production of software remains a primarily manual process, depending upon the programming skills of individual developers and the effectiveness of development teams. This labour-intensive process is becoming too slow and inefficient to provide the software required by the wide range of organisations which utilize information systems as a central part of their business and operations. The quality standards for software are also becoming higher as more and more business-critical and safety-critical functionalities are taken on by software applications.

These issues have become increasingly evident because of high-profile and highly expensive software project failures, such as the multi-billion pound costs of the integrated NHS IT project in the UK [4].

New practices such as agile development and model-based development have been introduced into the software industry in an attempt to improve productivity and quality. Agile development tries to optimize the manual production of software by using short cycles of development, and by improving the organization of development teams and their interaction with customers. This approach remains dependent on developer skills, and subject to the limits which hand-crafting of software systems places on the rate of software production. Considerable time and resources are also consumed by the extensive testing and verification used in agile development.

Model-based development, and especially model-driven development (MDD) attempts to automate as much of the software development process as possible, and to raise the level of abstraction of development so that manually-intensive activities are focussed upon the construction and analysis of *models*, free from implementation details, instead of upon executable code. *Model transformations* are an essential part of MBD and MDD, providing automated refactoring of models and automated production of code from models, and many other operations on models. Transformations introduce the possibility of producing, semi-automatically, many different platform-specific versions of a system from a single platform-independent high-level specification of a system.

MBD certainly has considerable potential for industrializing software development. However, problems remain with most of the current MBD approaches:

- The development process may be heavyweight and inflexible, involving multiple models – such as the several varieties of behaviour models in UML – which need to be correlated and maintained together.

- Support tools may be highly complex and not be interoperable with each other, requiring transformation 'bridges' from tool to tool, which increases the workload and possibilities for introducing errors into development.

- The model-based viewpoint conflicts with the code-centered focus of traditional development and of traditional computer science education.

For these reasons we have aimed to make UML-RSDS a lightweight and agile MBD approach which requires minimum development machinery, and which provides simple tool support for MBD, interoperable with external tools for analysis and verification. We have also endeavoured to make the tools compatible with traditional practices, so that developers can use UML-RSDS in combination with conventional coding. We report on experiences of use of UML-RSDS for education and in industry in Chapter 21.

1.3 Model-based development concepts

MBD is founded, naturally enough, on the concept of *models*. The dictionary definition of a model is:

> Something that represents another thing, either as a physical object that is usually smaller than the real object, or as a simple description that can be used in calculations. (Cambridge Advanced Dictionary, 2015)

Models are representations of actual artifacts (cars, bridges, buildings, executable software systems, and even living organisms [1]) which are used to analyse, simulate and document the artifacts. The representations can be physical scale models or prototypes, diagrams or mathematical models, or computer simulations.

In this book we are concerned with models of real-world data and entities, and with models of processing on these data and entities, to be implemented in software. The models will be either visual representations in a subset of the Unified Modelling Language (UML) [2] or textual representations (in pseudocode). These models can be used in discussions between the stakeholders of a software system (customers, users, clients) and developers, to agree on the scope and functionality of the system, and to precisely express the system requirements. Models can also be checked for validity (that they are internally consistent and satisfy other properties). In UML-RSDS, models serve as a description of the system design and can be used to generate the design and executable implementations of the system in multiple programming languages, with minimal human intervention.

A key principle of UML-RSDS is that *"The specification is the system"*, and that development work should be focussed upon defining a correct specification, instead of upon low-level coding.

For example, Fig. 1.2 shows part of a UML class diagram model for the data typically found in a university teaching administration system: such a system will record what are the courses on offer, who teaches the courses, and which students are attending which courses. This could be an initial model of the data of the system, and should be enriched with further attributes (such as the year of a student, and the level of a course) and classes (such as a representation of degree programmes). Even in this outline form it already conveys some information about the data, for example, that each course has a single lecturer and that each student belongs to a single department, whilst a department has sets *students* of *Student*s, *courses* of *Course*s and *staff* of *Lecturer*s. The UML-RSDS tools can check that this model is a valid class diagram, and generate outline Java, C# and C++ code from the diagram.

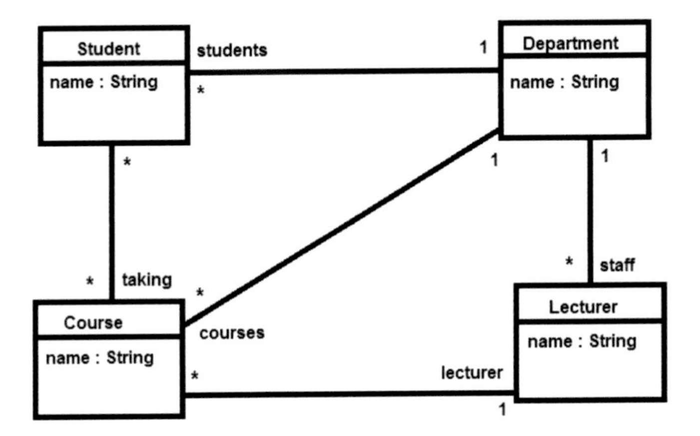

Figure 1.2: Class diagram example

The generated Java code for the Course class is:

```
class Course
{ private String name = "";
  private Lecturer lecturer;

  public Course(Lecturer lecturerx)
  { lecturer = lecturerx; }

  public String getname()
  { return name; }

  public void setname(String namex)
  { name = namex; }

   ... other update and query operations ...
}
```

The model abstracts from code details (for example, the visibilities *private* and *public* of Java instance variables and operations) and presents a logical view of the data manipulated and stored by the system. Thus it is potentially easier for stakeholders to understand than the code. The model also has a precise mathematical semantics (as we discuss in Chapter 18), enabling formal verification and analysis of the system if required.

Model-based Development is a term for those software development approaches where models are an essential part of the development process. UML-RSDS could be termed a *Model-driven Development* (MDD) approach, which are those MBD approaches where models are the primary artifact used in development, and code production is mainly automated. Another, closely related, MDD approach is the Model-driven Architecture (MDA), an OMG standard (www.omg.org/mda) for MBD using the UML. MDA is based on the idea of separating Platform Independent Models (PIMs) from Platform-specific Models (PSMs) of a system: a developer should express the logical business data and rules in a platform-independent manner using UML notations such as class diagrams, then, for the required implementation platforms, map these PIMs to PSMs for the specific platforms. From the PSMs code can be generated automatically (Fig. 1.3).

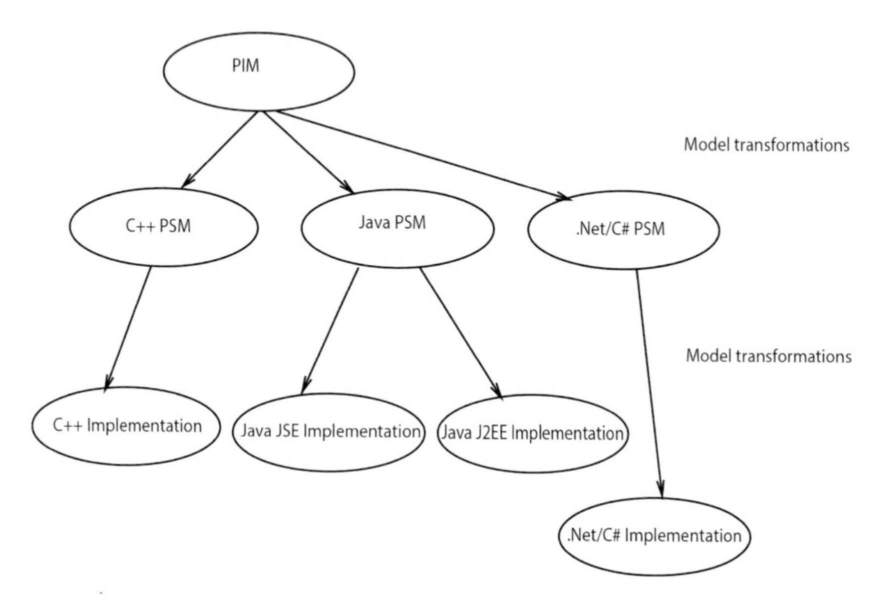

Figure 1.3: MDA process

Compared to UML-RSDS, MDA is broader in scope: the UML-RSDS specification in Fig. 1.1 is only independent of platform across the range of Java-like languages: a single PSM representation (the design) is used for all these languages. The PIM to PSM step in MDA may only be semi-automated, whilst the specification to design step in UML-RSDS is highly automated, and developers do not need to manually construct a PSM.

Summary

In this chapter we have introduced the concepts of model-based development and agile development. We have described UML-RSDS, and we gave the motivation for the UML-RSDS approach.

References

[1] OpenWorm project, www.artificialbrains.com/openworm, 2015.

[2] M. Fowler and K. Scott, *UML Distilled: Applying the Standard Object Modeling Language*, Addison-Wesley, 2003.

[3] J. Hutchinson, J. Whittle, M. Rouncefield and S. Kristoffersen, *Empirical Assessment of MDE in Industry*, ICSE 11, ACM, 2011.

[4] The Independent, *NHS pulls the plug on its £11bn IT system*, 3 August 2011.

[5] M. Petre, *"No shit" or "Oh, shit!": responses to observations on the use of UML in professional practice*, Softw Syst Model, 13: 1225–1235, 2014.

[6] B. Selic, *What will it take? A view on adoption of model-based methods in practice*, Software systems modeling, 11: 513–526, 2012.

[7] A. Vallecillo, *On the Industrial Adoption of Model Driven Engineering*, International Journal of Information Systems and Software Engineering for Big Companies, Vol. 1, No. 1, pp. 52–68, 2014.

[8] J. Whittle, J. Hutchinson and M. Roucefield, *The state of practice in Model-driven Engineering*, IEEE Software, pp. 79–85, May/June 2014.

Chapter 2

Overview of Development Using UML-RSDS

In this chapter we give an overview of the UML-RSDS development approach by means of some simple examples illustrating the different ways in which it can be used in software development: to develop components in a larger system, to develop self-contained applications, or to define model transformations.

2.1 Component development: statistical correlation calculator

As part of a system development it may be necessary to create some utility components which provide services to other parts of the system, but which are otherwise independent of the system. These components could therefore be reused in other systems.

In this section we describe such a component which carries out a numerical computation of the *Pearson correlation coefficient* between two sets of values: a measure of how linearly dependent one set of values is on the other. This computation could be used in a wide range of data analysis applications. Figure 2.1 shows the class diagram of this component.

This diagram means that each *CorrelationCalculator* has an associated sequence *datapoints* of *DataPoint* objects, each of which has an x and y coordinate in a graph (for example, as in Fig. 2.2). The x coordinate values of the data points represents one set of values, to be compared with another set, represented by the y coordinate values.

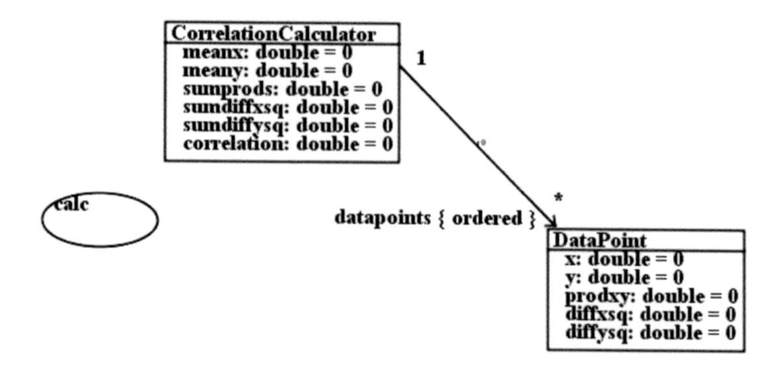

Figure 2.1: Correlation calculator class diagram

There are other auxiliary data items which are used to store properties of the dataset as a whole, such as the mean values *meanx* and *meany* of the x and y coordinates of data points. The Pearson correlation of the x and y values in this example is 0.91, indicating that there is a strong positive linear correlation between the values.

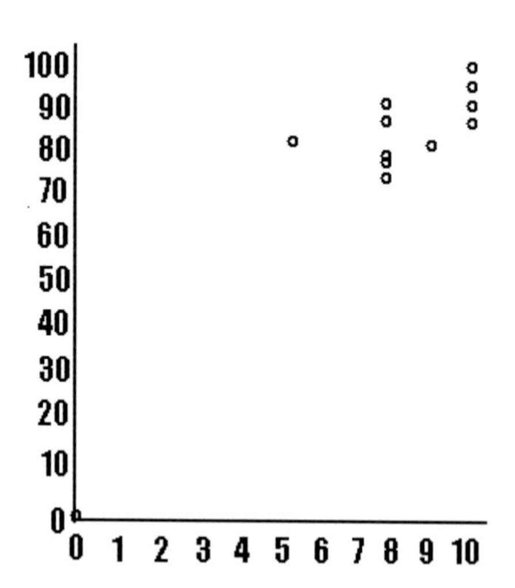

Figure 2.2: Correlation example

The calculator component has (initially at least) just one use case, *calc*, which can be described informally as:

Compute the Pearson correlation coefficient

$$P = \frac{\Sigma_d (x_d - \overline{x})(y_d - \overline{y})}{\sqrt{(\Sigma_d (x_d - \overline{x})^2 \Sigma_d (y_d - \overline{y})^2)}}$$

of a non-empty (at least 2 points) series of real-valued data points (x_d, y_d), where at least 2 points have different x_d values, and at least 2 have different y_d values.

Use cases express the functional capabilities or services which a system/component provides. This informal use case can be specified by a UML use case *calc*, represented as an oval in Fig. 2.1. In UML-RSDS use cases are shown together with the class diagram of the data that they operate upon,[1] and the relations between the use cases and classes can also be displayed on the diagram.

UML-RSDS use cases can be given a *precondition* or assumption: a logical condition that should be satisfied before the use case can be invoked. For the correlation calculator, the precondition is the constraint that there are two or more data points, not all of which are vertically aligned or horizontally aligned:

$$
\begin{array}{l}
datapoints.size > 1 \ \& \\
datapoints{\rightarrow}exists(d1 \mid datapoints{\rightarrow}exists(d2 \mid d1.x \neq d2.x)) \ \& \\
datapoints{\rightarrow}exists(d1 \mid datapoints{\rightarrow}exists(d2 \mid d1.y \neq d2.y))
\end{array}
$$

on *CorrelationCalculator*.

The functionality of a use case is defined by a series of *postcondition* constraints, which describe the logical processing steps that the use case should carry out in order to achieve its complete functionality. The conjunction of these postconditions also defines the final state achieved at termination of the use case. For *calc* we can define the following series (1), (2), (3), (4) of postcondition constraints on *CorrelationCalculator*. (1) is:

$$
\begin{array}{l}
meanx = datapoints.x{\rightarrow}sum()/datapoints.size \ \& \\
meany = datapoints.y{\rightarrow}sum()/datapoints.size
\end{array}
$$

This first constraint computes the mean x and y values, $\overline{x}$ and $\overline{y}$.

$$
\begin{array}{l}
(2): \\
\quad d : datapoints \ \Rightarrow \\
\qquad\qquad d.prodxy = (d.x - meanx) * (d.y - meany) \ \& \\
\qquad\qquad d.diffxsq = (d.x - meanx) * (d.x - meanx) \ \& \\
\qquad\qquad d.diffysq = (d.y - meany) * (d.y - meany)
\end{array}
$$

This constraint computes, for each datapoint d, the values of $(x_d - \overline{x})(y_d - \overline{y})$, $(x_d - \overline{x})^2$ and $(y_d - \overline{y})^2$. A constraint $P \Rightarrow Q$ has the intuitive meaning "if P is true, Q should also be true". P is called the *antecedent* or *condition* of the constraint, Q is its *succedent*. d here is an additional

[1]In standard UML they are shown in separate diagrams.

quantified variable – the constraint is applied for each possible value of d in *datapoints*, for each *CorrelationCalculator*.

$$(3):$$
$$sumprods = datapoints.prodxy{\rightarrow}sum()\ \&$$
$$sumdiffxsq = datapoints.diffxsq{\rightarrow}sum()\ \&$$
$$sumdiffysq = datapoints.diffysq{\rightarrow}sum()$$

This constraint computes the main terms of the correlation calculation.

$$(4):$$
$$sumdiffxsq > 0\ \&\ sumdiffysq > 0\ \Rightarrow$$
$$correlation = sumprods/(sumdiffxsq * sumdiffysq){\rightarrow}sqrt()$$

This constraint computes the final correlation result, if it is well-defined.

The postcondition constraints define the successive steps of the correlation calculation, computing the data averages and standard deviations. The final result is placed in the *correlation* attribute. At termination of the use case, *all of the postconditions are true*, so that *correlation* expresses the correct correlation value according to the original formula.

This example illustrates a key principle of UML-RSDS:

> *A postcondition constraint P means "Make P true" when interpreted as a specification of system behaviour.*

The implementation of *calc* must make all the postconditions true, and hence solve the problem. The order of the postconditions is not logically significant, but provides a guide to the design synthesis and code generation process. The ordering of the postconditions will normally result in synthesised code which sequentially executes the corresponding code fragments. The constraints should be ordered so that data is defined before it is used – as with attributes such as *meanx* and *prodxy* in the above example.

The precondition of the *calc* use case can be derived from the *definedness* conditions of its postcondition constraints (Chapter 18), in this case that both *sumdiffxsq* and *sumdiffysq* are positive. In turn, this means that at least one *d.diffxsq* and at least one *d.diffysq* term are non-zero, which is ensured by the formalised use case precondition.

From this specification, a platform-independent design can be generated using the Generate Design option on the UML-RSDS Synthesis menu (Fig. 2.3). Language-specific code can then be synthesised in Java, C# or C++ for incorporation of this component in a number of different systems: an API to the component is provided by the Controller class in the generated code.

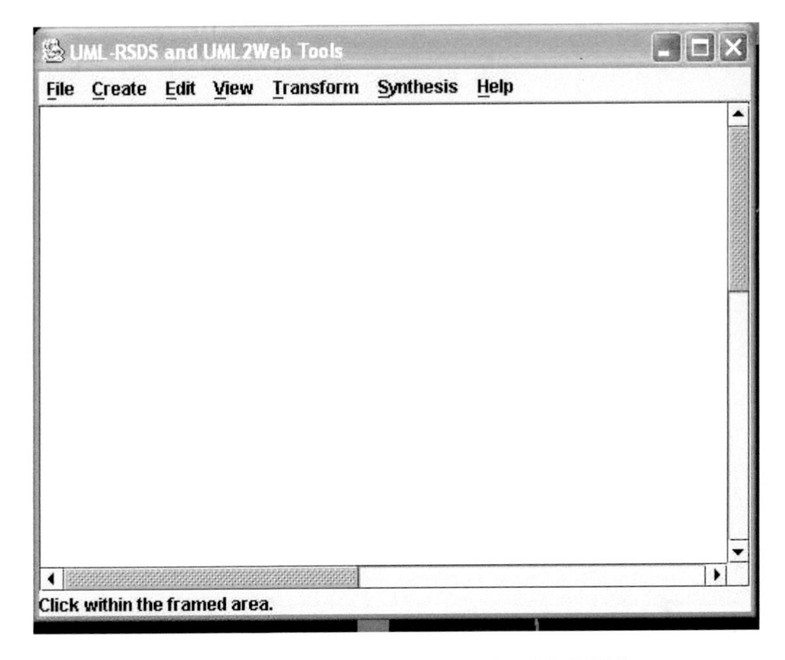

Figure 2.3: Main GUI frame of UML-RSDS

Table 2.1 shows the efficiency of the calculator on different input sizes, using the Java 4, Java 6 and C# generated code. Since the constraints can each be implemented by a bounded loop (Chapter 6), a linear increase in execution time with input size is expected, and this is observed. The Java 6 implementation seems generally to be the most efficient.

Table 2.1: Test results for correlation calculator

Input size	*Java 4*	*Java 6*	*C#*
1000 data points	16ms	16ms	5ms
10,000 data points	47ms	26ms	21ms
50,000 data points	78ms	51ms	59ms
100,000 data points	141ms	63ms	105ms

2.2 Application development: Sudoku solver

This system tries to solve a partially completed Sudoku puzzle, and identifies if the puzzle is unsolvable. For simplicity, we consider 4-by-4 puzzles, but the approach should work (in principle) for any size of puzzles. The board is divided into four columns, four rows and four 2-by-2 corner subparts, each column, row and subpart must be completely filled with the numbers 1, 2, 3, 4, without duplicates, to solve the puzzle. The following shows a partially completed board, where 0 is used to mark an unfilled square.

```
2 1 0 0
0 0 0 0
0 0 0 0
0 0 0 3
```

The following strategies could be used to fill in the blank squares:

- (R1): if a blank square is restricted by the Sudoku rules to only have one possible value *val*, then set its value to *val*.
- (R2): if there is a choice of two or more values for a blank square, make a random choice amongst these and put on the square.
- (R3): if there is no possible value that can be used for a blank square, abandon the solution attempt.

R1 should be applied as many times as possible before applying R2. This is an effective strategy in many cases, although it does not provide any facility for backtracking and undoing choices made by R2. We will consider a backtracking approach in Chapter 15.

The data of a Sudoku puzzle can be formalised as a class diagram (Fig. 2.4), and the above informal rules expressed as postconditions of use cases *solve* (for R1 and R2) and *check* (for R3) operating on this class diagram.

A query operation of *Square* is defined to return the set of possible values for the square:

```
Square::
query possibleValues() : Set(int)
post:
  result = Integer.subrange(1,4) - parts.elements.value
```

Integer.subrange(a,b) is the sequence Sequence{a, ..., b} of integers from a up to b, inclusive. The result of *possibleValues* is the set of values in 1..4 which are not already on some square in the board parts

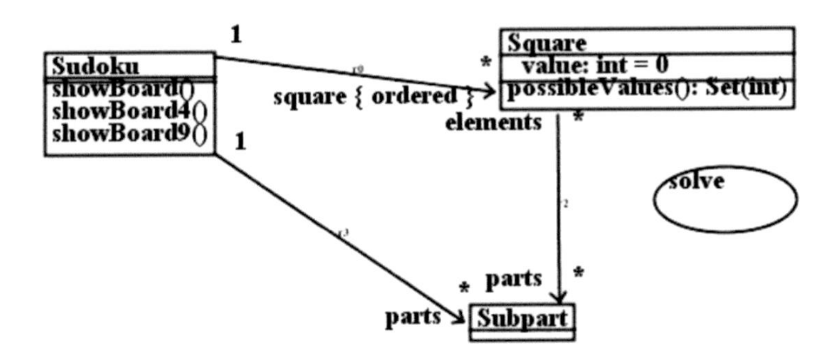

Figure 2.4: Sudoku class diagram

in which the square itself occurs (there are always three such parts: one row, one column and one of the four 2-by-2 corner regions).

An operation *showBoard* of the *Sudoku* class displays the board state:

```
Sudoku::
showBoard()
pre: true
post:
  ( square[1].value + " " + square[2].value + " " +
    square[3].value + " " + square[4].value )->display() &
  ( square[5].value + " " + square[6].value + " " +
    square[7].value + " " + square[8].value )->display() &
  ( square[9].value + " " + square[10].value + " " +
    square[11].value + " " + square[12].value )->display() &
  ( square[13].value + " " + square[14].value + " " +
    square[15].value + " " + square[16].value )->display() &
  ""->display()
```

R1 can then be formalised by a constraint (C1) iterating over the *Sudoku* class:

```
sq : square & sq.value = 0 &
v = sq.possibleValues() & v.size = 1  =>
                    sq.value = v->min() & self.showBoard()
```

This can be read as "for all squares sq on the board, if the square is blank, and the set v of possible values has a single element, set the square value to that value". v here is a *Let-variable*, a read-only identifier which holds the value of its defining expression, *sq.possibleValues*(), so that this expression only needs to be evaluated once in the constraint. The board is also displayed to the console after the update.

$R2$ can be formalised as constraint $C2$:

```
sq : square & sq.value = 0 &
v = sq.possibleValues() & v.size > 0  =>
                                sq.value = v->min() & self.showBoard()
```

This is very similar to $C1$, but handles the case where there is more than one possible value for *sq*.

Finally, $R3$ is formalised by $C3$:

```
sq : square & sq.value = 0 &
v = sq.possibleValues() & v.size = 0  =>
      ("No possible value for square: " + sq)->display()
      & self.showBoard()
```

$C1$ is our first example of a constraint which both reads and writes the same data (the values of squares) and which therefore requires a *fixed-point iteration* implementation: while the assumption

```
sq : square & sq.value = 0 & v = sq.possibleValues() &
v.size = 1
```

remains true for some *sq* : *square*, the conclusion

```
    sq.value = v->min() & self.showBoard()
```

is performed. At termination of $C1$, the assumption is false for all squares, and so no square has a unique possible value, i.e.:

```
sq : square & sq.value = 0 & v = sq.possibleValues()  =>
v.size /= 1
```

for all *sq*.

This example also illustrates the key principle:

> *A postcondition constraint P means "Make P true" when interpreted as a specification of system behaviour.*

In this case, the principle means that the implementation of a constraint such as $C1$ must continue to apply the constraint to elements of the model (Sudoku squares) until it is true for all elements: while there is any square with a unique possible value, the constraint implementation will continue to execute, to try to make the conclusion of $C1$ true for such squares. At termination, the constraint $C1$ is therefore established for all squares, because there is no square for which the antecedent/condition of $C1$ is true.

Logically, the succedent of $C1$ contradicts the antecedent (because $value = 0$ in the antecedent, but is set to a non-zero value in the succedent). So the constraint has the form

$$\forall x : S \cdot (P \Rightarrow not(P))$$

The only way to establish $P \Rightarrow not(P)$ is to establish $not(P)$. Thus a specification using a constraint such as $C1$ does have a precise meaning, even though it may seem contradictory at first sight.

Such constraints are pervasively used for refactoring transformations and active systems such as game-playing programs, problem solvers and other software 'agents' because they directly express requirements of the form "If there is an unwanted situation s, then make changes to remove or correct s":

$$\textit{Unwanted situation s} \quad \Rightarrow \quad \textit{Make changes to remove/correct s}$$

Here, the unwanted situation is a blank square on the board.

$C2$ is very similar to $C1$. Because $C2$ writes data which $C1$ reads, they are inter-dependent (a choice made by $C2$ will typically enable some new applications of $C1$) and so (if they are defined in the same use case) they will be grouped together into a single fixed-point iteration with the behaviour $(C1*;\ C2)*$, i.e.: "Apply $C1$ repeatedly as many times as possible, then apply $C2$, and repeat this process as many times as possible". This behaviour only terminates when no square satisfies either the assumption of $C1$ or $C2$, this means that either all squares are filled, or that some blank square has no possible value. $C3$ will show which squares, if any, satisfy the latter case. $C3$ itself does not update any data that it reads (or indeed any data at all), so can be given a bounded (for loop) iteration, as with the constraints of the correlation calculator. All of these design choices are performed automatically by the "Generate Design" option of the tools. More details on this process are given in Chapter 6.

An example execution of this system is shown below:

```
2 1 0 4   <- applied C1 to choose 4
0 0 0 0
0 0 0 0
0 0 0 3

2 1 3 4   <- applied C1 to choose 3
0 0 0 0
0 0 0 0
0 0 0 3
```

```
2 1 3 4
3 4 0 0   <- applied C2 to choose 3; C1 to choose 4
0 0 0 0
0 0 0 3

2 1 3 4
3 4 0 0
0 0 0 0
0 2 0 3   <- applied C1 to choose 2

2 1 3 4
3 4 0 0
0 3 0 0   <- applied C1 to choose 3
0 2 0 3

2 1 3 4
3 4 1 2   <- applied C2 to choose 1; C1 to choose 2
0 3 0 0
0 2 0 3

2 1 3 4
3 4 1 2
0 3 0 1   <- applied C1 to choose 1
0 2 0 3

2 1 3 4
3 4 1 2
4 3 0 1   <- applied C1 to choose 4
0 2 0 3

2 1 3 4
3 4 1 2
4 3 2 1   <- applied C1 to choose 2
0 2 0 3

2 1 3 4
3 4 1 2
4 3 2 1
1 2 0 3   <- applied C1 to choose 1

2 1 3 4
3 4 1 2
4 3 2 1
1 2 4 3   <- applied C1 to choose 4
```

The start of the second row with 3 is an application of C2, as is the choice of 1 in the third position of this row. Otherwise, square fillings are the result of C1.

An initial formalisation of R1 as a constraint could have been (C0):

```
sq : square & sq.value=0 & sq.possibleValues()->size()=1 =>
          sq.value = sq.possibleValues()->min() &
          self.showBoard()
```

Here the function *possibleValues*() is evaluated twice, which is a waste of computational effort, and could slow the solver significantly for larger versions of Sudoku. Instead, we can avoid the duplicated computation by introducing the new variable v, setting this to *sq.possibleValues*() and then using v in place of the operation call in the two places in C0 where the operation result is read.

This illustrates a further important UML-RSDS principle:

> *Improve the efficiency of a system at the specification level where possible, whilst keeping a clear and platform-independent specification style.*

In the above case the use of the let-variable v slightly increases the syntactic complexity of the constraint, but may improve its clarity because it makes explicit that the same value is being used throughout the constraint. This is an application of the Remove Duplicated Expression Evaluations optimization pattern, we give more details of relevant design patterns in Chapter 9.

2.3 Transformation development: class diagram refactoring

This is an example of an *update-in-place* model transformation, which carries out a *refactoring* of a class diagram to improve its quality. The aim of the transformation is to remove situations of apparently duplicated attributes in different classes from the diagram, for example, if all subclasses (more than 1) of a given class all have an attribute with identical name and type, then these copies can be replaced by a single attribute in the superclass. Figure 2.5 shows the metamodel for the source and target language of the transformation. This is a small subset of the UML 2.0 class diagram language. It will be assumed that in the initial model to be refactored:

- No two classes have the same name.
- No two types have the same name.
- The owned attributes of each class have distinct names within the class, and do not have common names with the attributes of any superclass.

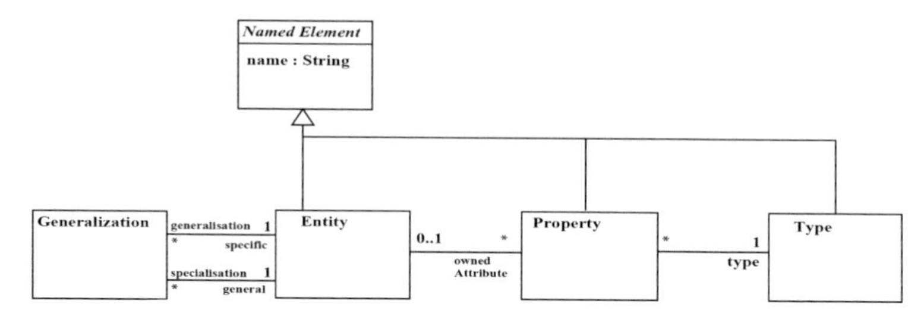

Figure 2.5: Basic class diagram metamodel

- There is no multiple inheritance, i.e., the multiplicity of *generalisation* is restricted to 0..1.

These properties must also be preserved by the transformation.

The informal transformation steps are the following:

(1) **Pull up common attributes of direct subclasses:** If the set of all direct subclasses $g = c.specialisation.specific$ of a class c has two or more elements, and all classes in g have an owned attribute with the same name n and type t, add an attribute of this name and type to c, and remove the copies from each element of g. (Fig. 2.6). This is the "Pull up attribute" refactoring of [2].

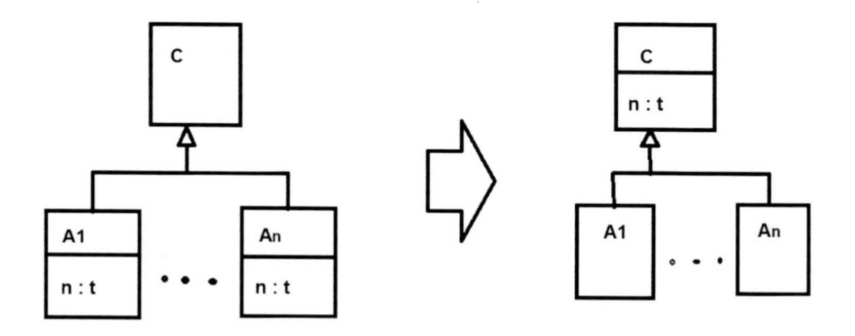

Figure 2.6: Rule 1

(2) **Create subclass for duplicated attributes:** If a class c has two or more direct subclasses g, $g = c.specialisation.specific$, and there is a subset $g1$ of g, of size at least 2, all the elements of $g1$ have an owned attribute with the same name n and type t, but there are elements of $g - g1$ without such an attribute, introduce a new class $c1$ as a subclass of c. $c1$ should also be set as a direct superclass of all those classes in g which own a copy of the cloned attribute. (In order to minimise the number of new classes introduced, the

largest set of subclasses of c which all contain a copy of the same attribute should be chosen). Add an attribute of name n and type t to $c1$ and remove the copies from each of its direct subclasses. (Fig. 2.7).

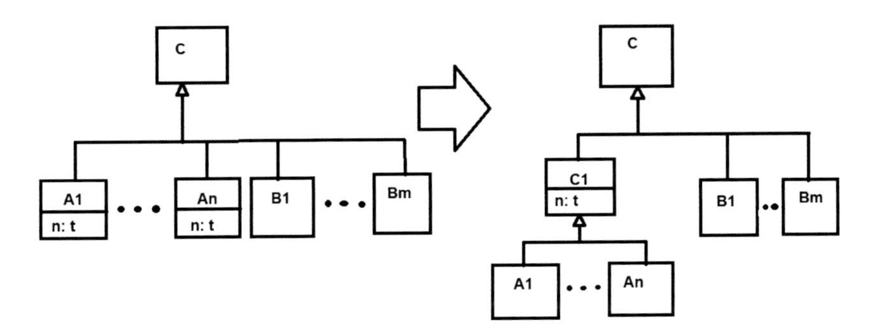

Figure 2.7: Rule 2

This is the "Extract superclass" refactoring of [2].

(3) Create root class for duplicated attributes: If there are two or more root classes all of which have an owned attribute with the same name n and type t, create a new root class c. Make c the direct superclass of all root classes with such an attribute, and add an attribute of name n and type t to c and remove the copies from each of the direct subclasses. (Fig. 2.8).

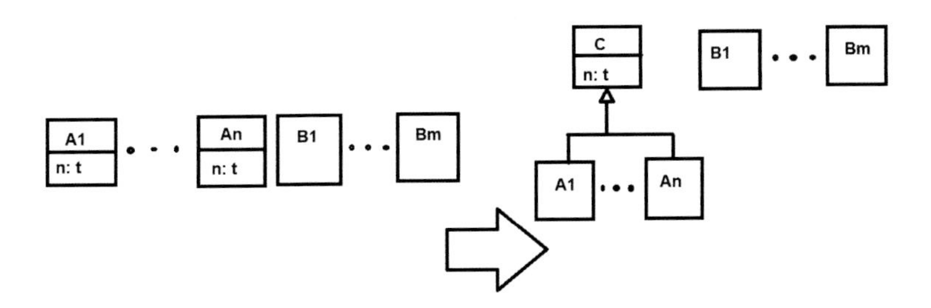

Figure 2.8: Rule 3

We will just consider the first scenario here, for simplicity. In common with most refactorings, this can be viewed as a logical assertion:

$$\textit{Unwanted situation s in model} \quad \Rightarrow \quad \textit{Modify model to remove s}$$

In this case the unwanted situation is that there is a duplicated attribute in every subclass of a class, and the modification is to promote one attribute copy to the superclass and to delete all the other copies.

A possible transformation rule that formally expresses scenario 1 is then:

$$
\begin{array}{l}
(C1): \\
a : \mathit{specialisation.specific.ownedAttribute}\ \& \\
\mathit{specialisation.size} > 1\ \& \\
\mathit{specialisation.specific}{\rightarrow}\mathit{forAll}(\\
\quad \mathit{ownedAttribute}{\rightarrow}\mathit{exists}(b \mid b.\mathit{name} = a.\mathit{name}\ \&\ b.\mathit{type} = a.\mathit{type})) \quad \Rightarrow \\
\qquad a : \mathit{ownedAttribute}\ \& \\
\qquad \mathit{specialisation.specific.ownedAttribute}{\rightarrow}\mathit{select}(\\
\qquad\qquad\qquad \mathit{name} = a.\mathit{name}){\rightarrow}\mathit{isDeleted}()
\end{array}
$$

This rule operates on instances of *Entity*. An instance (*self*) of *Entity*, and instance a of *Property* match the constraint test if: (i) a is in the set of attributes of all direct subclasses of *self*, (ii) there is more than one direct subclass of *self*, and (iii) every direct subclass of *self* has an attribute with the same name and type as a.

The conclusion of the constraint specifies that (i) the property a is moved up to the superclass *self* (by adding it to *self.ownedAttribute* we implicitly remove a from its current class, because an attribute can belong to at most one class according to Fig. 2.5), (ii) all other attributes with name $a.name$ are deleted from all direct subclasses of *self*.

$s{\rightarrow}\mathit{isDeleted}()$ is a built-in operator of UML-RSDS, which deletes the object or set of objects s from their model, removing them from all the entity types and association ends in which they occur.

As with the Sudoku solver, applying the constraint $(C1)$ to particular elements in a model may change the application of the constraint to other elements – for example, promoting the attribute $x : T$ from subclasses C and D of class B in Fig. 2.9 up to B will enable a further application of the constraint to B and its sibling class E. This means that the rule has to be applied repeatedly until no further change takes

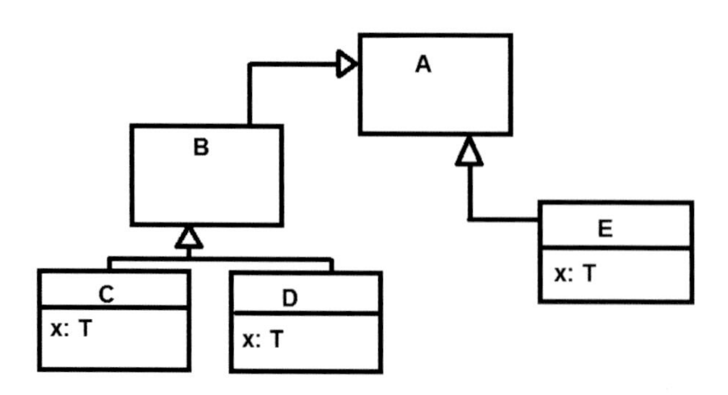

Figure 2.9: Rule 1 example

place – which means until the rule cannot be applied, and in that case no situation in the diagram matches the rule condition. This also means that all cases of the quality problem (duplicated attributes in sibling classes) have been removed from the diagram. The specification is very concise, and corresponds quite closely with the informal requirement for the rule. However, several questions should be asked about the correctness of such fixed-point rules:

- Does the execution terminate?
- Does the rule application preserve any necessary invariant properties of the model – such as single inheritance in our case?
- Does it preserve the semantic meaning of the model?

For this example it is possible to show that these properties are true: the transformation process terminates because each rule application reduces the number of *Property* instances by at least one – so there can only be finitely many rule applications. The rule does not create classes, types or generalisations, and does not change class nor type names, so the invariants concerning class name uniqueness, type name uniqueness and single inheritance are clearly preserved. Attributes cannot have names in common with any ancestor class attribute, so promoting an attribute to a superclass cannot create a name clash in the superclass, and the rule maintains the name uniqueness property (since copies of the attribute in the subclasses are deleted). Finally, regarding the semantic meaning of the class diagram, this can be considered to consist of the possible objects that could be instantiated from the concrete (leaf) classes of the diagram, characterised by their class name and attribute names and types. The collection of such object templates is preserved by the transformation, which simply relocates the place at which attributes are declared.

2.4 Integrating UML-RSDS applications with external applications

The scope of UML-RSDS application development for general software applications is primarily the business tier or business-logic tier within a three- or more-tier architecture (Fig. 2.10).

This tier contains the core business functionality of the application, and additionally contains object-oriented code representations of the business data. There should be no technology-specific code for particular UI or data storage technologies, instead such functionalities should be invoked via interface components. In UML-RSDS such components can be declared as classes with the stereotype *external* or *externalApp*

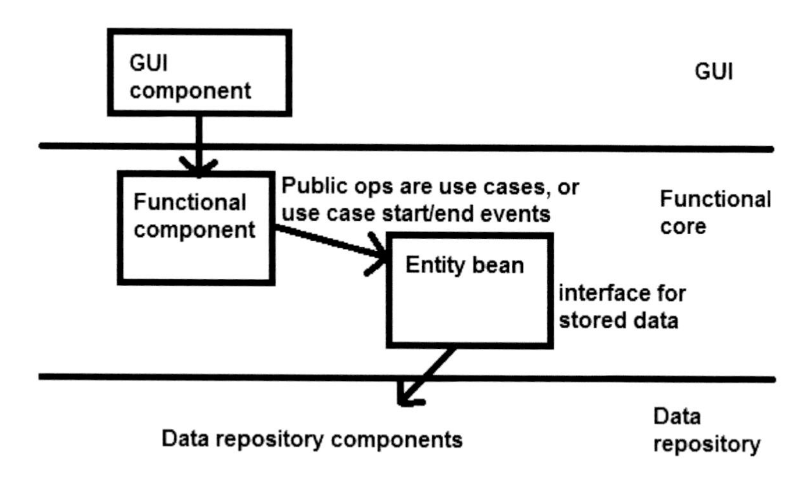

Figure 2.10: UML-RSDS systems architecture

(Chapter 10). The operations offered by these components should be listed in the class (only the operations needed by the current system need to be listed). The operations should have parameter types that are valid in UML-RSDS (boolean, int, long, double, String, class type names). The actual code of the components would be given externally, e.g., by hand-written Java classes or by separate UML-RSDS developments. For example, in the case of the correlation calculator, it may be required to read the datapoints from a spreadsheet CSV file, with one datapoint on each row of the spreadsheet. Such input could be the responsibility of an interface component *SpreadsheetReader* which has operations $getFirstRow() : Sequence(String)$ and $getNextRow() : Sequence(String)$. For each row a datapoint would be created, for use by the main *calc* use case.

For model transformations, on the other hand, the simple file input and output facilities provided within synthesised transformations, and the simple synthesised GUI, may be sufficient, without the need for additional external software (Chapter 7).

2.5 What is different about UML-RSDS?

Most MBD tools and methods have originated from academic research, and often are highly specialised and sophisticated. The Xmodeller tool from Xactium [1] was an example of this type of product. Such tools give many advanced capabilities for software development – to those developers and organisations who are able to use them. The barriers to their use include the cost of training, and the inability to combine the tools with existing development practice, in particular with manual coding practices. Tool costs, and dependence upon small research-based companies for tool support and upgrades, are other negative factors which have discouraged potential users from adopting them.

We attempt to overcome some of these problems with UML-RSDS by keeping the notations needed and the development steps as simple as possible: only (subsets of) UML class diagram, use case and OCL notations are needed. Apart from various checks on specifications, the main process is design and code generation. Rather than having separated UML diagrams and multiple system models, which then need to be correlated and maintained consistently, we combine use case and class diagrams together in a single diagram. This combined model is adequate for the specification of the data and behaviour of many software systems, and it gives a clear representation of the purpose of the class diagram in terms of the required services of the system. The generated code can be easily used by and integrated with manually-produced code. In contrast to sophisticated MBD tools, we have found that even second year undergraduate students are able to learn and apply UML-RSDS successfully within a short time. UML-RSDS can be used as part of component-based development (CBD) or service-oriented architecture (SOA) approaches, and with either an agile or plan-based development process.

Another agile MBD approach is Executable UML (xUML) [6]. This differs from UML-RSDS in several ways:

- xUML models systems using class diagrams, use case diagrams, domains, sequence diagrams, class collaboration diagrams, state machines and an action language.
- Developers must model the properties of the implementation platform.
- A concurrent execution model is assumed.

The approach is a relatively 'heavyweight' model-based development method, involving the construction of multiple inter-related models. In contrast, UML-RSDS uses only class diagrams, OCL and use cases, in a single integrated model. It does not require implementation platform modelling, and it has a sequential execution model.

Use cases are the main unit of incremental development in agile methods such as Scrum and eXtreme Programming (XP): the product backlog and iteration backlog are both expressed as prioritised lists of use cases (or "stories" as they are termed in Scrum). Therefore it seems appropriate to make use cases a central part of the specification models. Use cases can represent services in a SOA, or the operations of a component in CBD. The functionality of use cases is in turn defined using the data and operations of classes in the class diagram. State machines can be used in UML-RSDS to model the intended life histories of objects, and detailed behaviour of operations, but are optional. UML-RSDS has a higher level of abstraction than xUML: use case functionality is defined by a series of logical constraints, rather than actions in an action language (a design level of description). Alternative designs can be generated from the same specifications. Some other agile MBD approaches also use xUML-style action languages [4] and suffer from the costs of too-detailed low-level coding which these languages require, and which seems to negate the benefits (apart from platform-independence) of MBD. On the other hand, UML-RSDS is more restrictive than xUML, since the implementation platforms supported are restricted to C++, Java (two variants) and C#, whilst xUML supports a wider range of platforms. New code generators for UML-RSDS (e.g., to ANSI C) can be developed using UML-RSDS itself, but this involves a significant amount of work (Section 12.2 illustrates this procedure).

Summary

In this chapter we have illustrated the use of UML-RSDS on a range of examples, and we described the general UML-RSDS development process and compared UML-RSDS to other MBD approaches.

References

[1] T. Clark and P.-A. Muller, *Exploiting model driven technology: a tale of two startups*, Software Systems Modelling, 11: 481–493, 2012.

[2] M. Fowler, K. Beck, J. Brant, W. Opdyke and D. Roberts, *Refactoring: Improving the Design of Existing Code*, Addison-Wesley, 1999.

[3] S. Kolahdouz-Rahimi, K. Lano, S. Pillay, J. Troya and P. Van Gorp, *Evaluation of Model Transformation Approaches for Model Refactoring*, Science of Computer Programming, 2013, http://dx.doi.org/10.1016/j.scico.2013.07.013.

[4] I. Lazar, B. Parv, S. Monogna, I.-G. Czibula and C.-L. Lazar, *An Agile MDA Approach for Executable UML Structured Activities*, Studia Univ. Babes-Bolyai, Informatica, Vol. LII, No. 2, 2007.

[5] R. Matinnejad, *Agile Model Driven Development: an intelligent compromise*, 9th International Conference on Software Engineering Research, Management and Applications, pp. 197–202, 2011.

[6] S. Mellor and M. Balcer, *Executable UML: A Foundation for Model-driven Architectures*, Addison-Wesley, Boston, 2002.

[7] M.B. Nakicenovic, *An Agile Driven Architecture Modernization to a Model-Driven Development Solution*, International Journal on Advances in Software, Vol. 5, Nos. 3, 4, pp. 308–322, 2012.

Chapter 3

Class Diagrams

Class diagrams are the central specification notation of UML and of UML-RSDS. This chapter describes how class diagrams can be defined in UML-RSDS and used to specify system data.

3.1 Class diagram concepts

A class diagram specifies the data of a system (and possibly data external to a system which it needs to be aware of). The diagram is a graphical representation of:

Classes or *entity types* – such as *Student* or *Lecturer* in Fig. 3.1. These denote the collection of all student or lecturer instances currently existing in the system (perhaps stored as objects in a Java executable, or as rows in a database table, for example). *Abstract* classes do not have instances of their own, instead their instances actually belong to the *concrete* subclasses of the class, at the leaves of the inheritance hierarchy beneath it.

Attributes of classes denote the features of instances of the class, and the type of the value of these features. For example, in Fig. 3.1, every student has a name, which is a string.

Operations of a class represent computed properties of instances of the class (in the case of *query*, *instance scope* operations), or computed properties independent of particular instances (*query*, *static* operations), or represent updates to the state of instances. Query operations must have return types, an example is $possibleValues() : Set(Integer)$ in Fig. 2.4.

Associations between classes (or between a class and itself) denote the links which exist between instances of one class and another. For example, in Fig. 3.1, each department has an associated set of students (the students registered in the department).

Inheritances between classes, representing that one class (at the source of the inheritance arrow) is a specialisation of another (at the target of the arrow). For example, the inheritance of *NamedElement* by *Entity* in Fig. 2.5. In UML-RSDS it is a rule that superclasses must be abstract. Classes may have multiple subclasses, but at most one superclass (single inheritance).

Class diagrams define the structure and inter-relationships of data managed by the system, and hence underpin the definition of the functionality of the system.

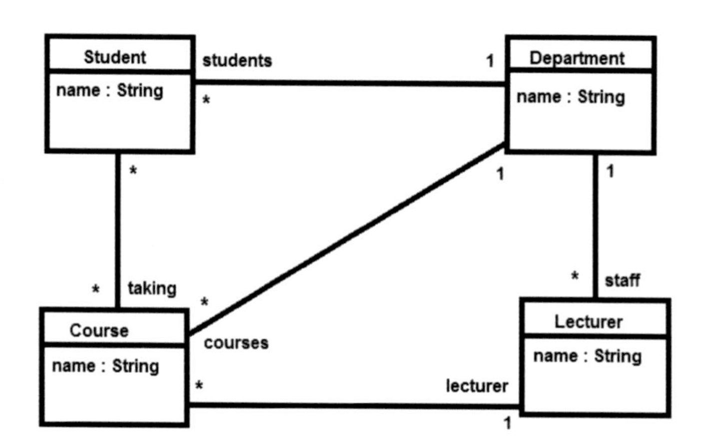

Figure 3.1: Class diagram example

3.2 Class diagram notations

The following class diagram notations are taken directly from UML 2.0:

- Classes, shown as rectangles, with a centered name, usually with an initial capital letter. Abstract classes are shown with italic font names, otherwise standard font is used.
- Enumerated types, shown as rectangles, with a centered name, and with the individual elements (values) of the enumeration listed in the second compartment of the enumeration box. An example is *Status* in Fig. 15.4. The name of the enumerated type can then be used as a type for attributes (in Java, C# and C++

enumeration type elements are typically implemented as small unsigned integers).

- Attributes of primitive (int, long, double, boolean, or enumeration) type or of String type. These are shown in the next compartment of the class box. Static (class scope) attributes are shown underlined. Attributes cannot be abstract. If an attribute is a primary key/identity attribute for the class (distinct objects must have different values for the attribute), this is indicated by the constraint {*identity*} beside the attribute. Initial values are also shown after the declaration.

- Operations, with optional input parameters and return types, are shown in the next compartment of the class box, beneath the attributes. Italic font denotes an abstract operation, underlining denotes a static operation. The declaration

  ```
  op(x : T) : RT
  ```

  denotes an operation with input parameter x of type T, and with result of type RT. A query (non-updating) operation has the constraint {*query*} written beside it in UML class diagrams. Query operations must have a return type.

- Associations, denoted by lines between classes (including between a class and itself, in the case of a *reflexive* association). Each end of the line can have a rolename, and both ends should have a multiplicity. In UML-RSDS the default is a unidirectional association (navigated from the source class – at the end without the rolename – to the target class – the end with the rolename), but bidirectional associations with rolenames at both ends can be defined. Multiplicities have the standard meanings (Table 3.1). Other multiplicities can be expressed by using constraints on the size of the collection denoted by the rolename. An association end of * or 0..1 multiplicity can be ordered, this is denoted by the constraint {*ordered*} at the end.

- Inheritance is shown by an arrow with a triangle head, pointing to the superclass.

Table 3.1: Multiplicity symbols in UML

Multiplicity	*Meaning*
*	Any finite number (including 0) of objects at this end of the association can be associated with one object at the other end
1	Exactly one object at this end of the association is associated with any object at the other end
0..1	At most one object at this end of the association is associated with any object at the other end

Some less common but still possible modelling elements are:

- Association classes: denoted by an association line with a dashed line to a class box.
- Qualified associations: denoted with a small box at the end with the qualifier index variable, and the variable is written in the box. In UML-RSDS we assume that the index is always String-valued.
- Interfaces: denoted as classes but with the stereotype ≪ *interface* ≫ written at the top.
- Interface inheritance: inheritance of an interface by another interface or by a class.
- Aggregation: an association with a whole-part meaning, the association end at the 'whole' class end is marked with a filled diamond.

To illustrate the construction of class diagrams, the following simple example system will be modelled:

- There is data on the courses, students and lecturers within each department of a college.
- Students are registered for a set of courses, and each lecturer teaches a set of courses.
- Each student, course, department and lecturer has a name, which is a string.

Figure 3.1 shows the class diagram which models this data.

3.2.1 The class diagram editor

The class diagram editor is the main panel of the UML-RSDS tools. The menu options are, in brief:

File menu: options *Set name* to set the package name of the class diagram model, *Recent* to load a recent class diagram (in the file *output/mm.txt*), *Load data* to load a class diagram in text format, *Load Ecore* to load a class diagram *output/mm.ecore* in Ecore XML format, *Save* to save the current model in text format (into *output/mm.txt*), *Save as*, to save the class diagram in a number of formats, including *Save data* to save in text format, *Save Ecore* to save as an Ecore metamodel definition (in *output/My.ecore*) and *Save model* to save as an instance model (in *output/model.txt*) of the UML-RSDS metamodel. *Print* and *Exit* options are also on the *File* menu. A *Convert* menu allows XMI data in XML files (by default, *output/xsi.txt*) to be converted to model data format (in *output/model.txt*).

Create menu: options

Class – create an entity type, including interfaces and abstract classes

Type – create an enumerated type

Association – create an association between two classes or between a class and itself

Inheritance – draw an inheritance arrow from the subclass to the superclass (the superclass must be an abstract class or an interface)

Association Class – create an association class

Invariant – introduce a constraint for a class or for the system.

Edit menu – options to *Move*, *Delete* and *Modify* visual elements. *Modify* is the default action and clicking on a class, use case or association (at the association label, usually in the centre of the association line) invokes this option.

View menu – options to view the invariants, types, operations, etc., of the system.

Transform menu – options to apply refactoring, refinement or design pattern transformations to a class diagram or system.

It is possible to load a class diagram in parts: if a class (with the same name) is defined both in the current displayed model and in a loaded file, then the union of the features and stereotypes of the two definitions is taken.

3.2.2 Drawing a class diagram

To create our example system, we create the four classes *Student*, *Course*, *Lecturer*, *Department*. Classes are created by selecting the *Create Class* option, entering the name, and clicking on the location for the class. Figure 3.2 shows the class definition dialog. Options in the *stereotypes* field include *abstract* for abstract classes, and *interface* for interfaces. *leaf* denotes that the class cannot have subclasses. *persistent* denotes that the class represents persistent data (e.g., data repository data in an enterprise information system). In this example the default *none* option is used.

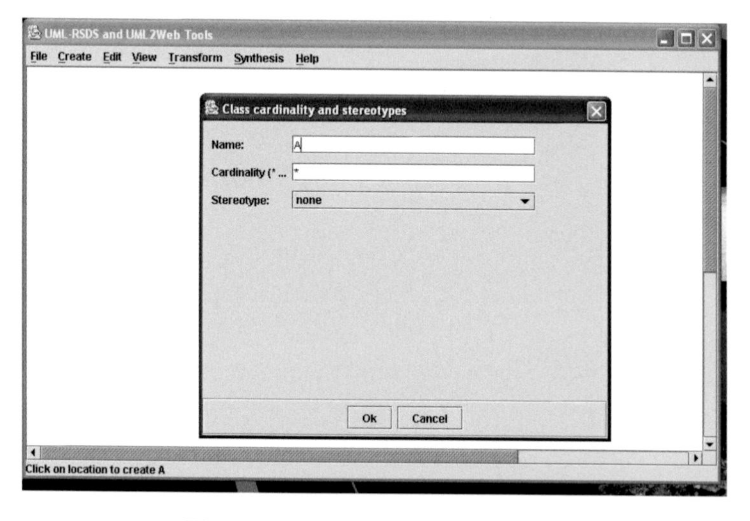

Figure 3.2: Class definition dialog

Class names (by convention) should begin with a capital letter, they can then consist of any further letters or digits. The classes are then edited by clicking on them or by using the *Edit* menu *Modify* option to add attributes *name* : *String* to each entity. The attribute dialog (Fig. 3.3) is used to define attributes. Only the name and type fields are mandatory. Attributes cannot be of an entity type or of a collection type: associations should be used to model data in these cases. The Scope field indicates if the attribute is static (class scope) or not (instance scope). The Uniqueness field indicates if the attribute is a primary key/identity attribute or not.

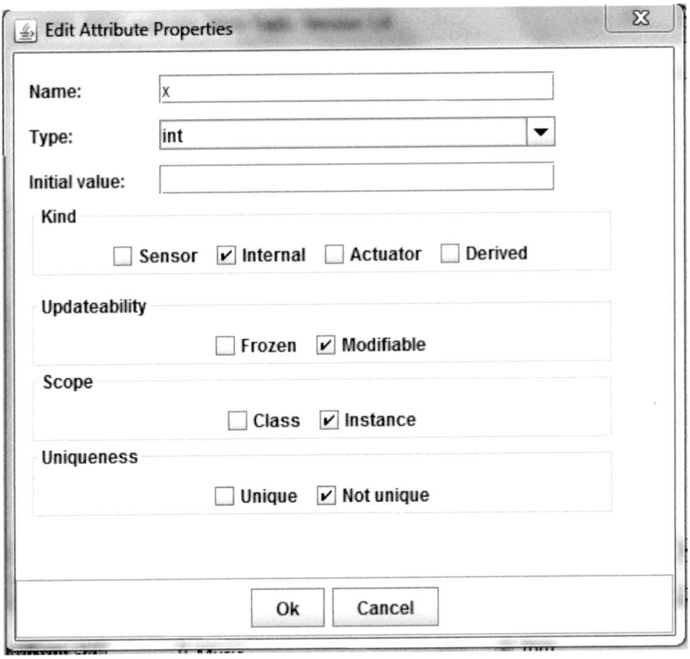

Figure 3.3: Attribute definition dialog

Associations are defined using the dialog of Fig. 3.4. Associations are created by selecting *Create Association* and dragging the mouse from the source class to the target class. Waypoints can be created by pressing the space key at the point (provided the class diagram editor panel has focus – click on the editor panel to ensure this). The *Role*2 field must be filled in – this is the end to which the association is directed. *Role*2 will become a feature of the source class. *Role*1 may be left blank, it is filled in for bidirectional associations, and then it will become a feature of the target class.

Association stereotypes include *implicit* (for associations calculated on demand using a predicate, rather than being stored), *source* (for associations belonging to the source metamodel of a model transformation), etc. The default *none* can be used in most cases. The construction process for our example should produce the class diagram of Fig. 3.1.

3.2.3 Operations

Operations can be added to classes using the *Modify* option (or by clicking on the class in the editor), and the edit class dialog (Fig. 3.5) and the operation definition dialog (Fig. 3.6).

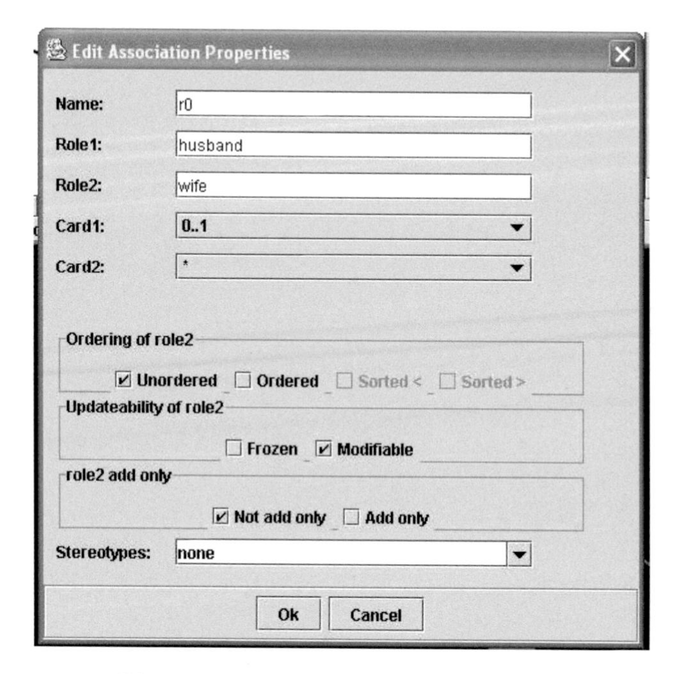

Figure 3.4: Association definition dialog

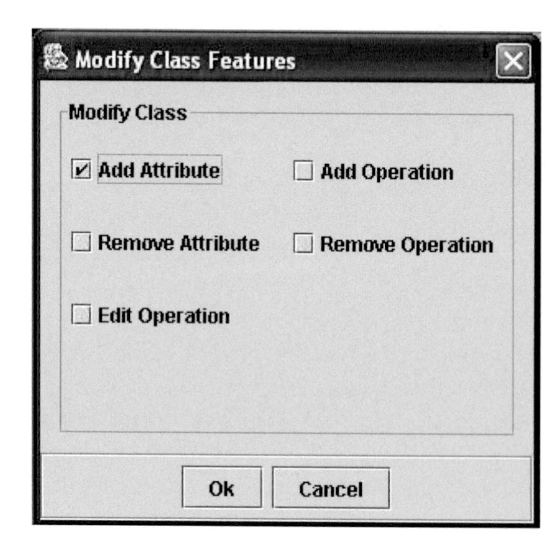

Figure 3.5: Edit class dialog

In the operation dialog both the precondition and postcondition predicate should be entered, `"true"` can be entered to denote no constraint.

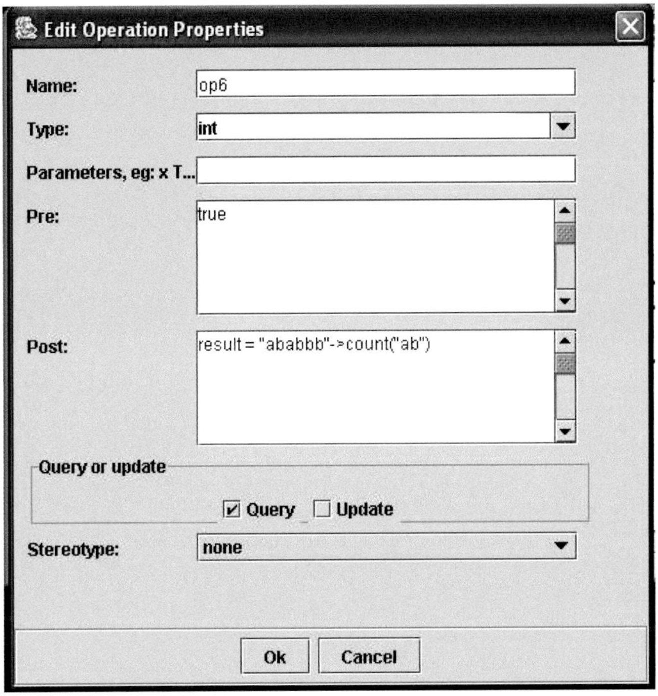

Figure 3.6: Operation definition dialog

Query operations must have a result type, and some equation *result* = *e* as the last conjunct in each conditional case in their postcondition (as in Fig. 3.6 and the factorial example below). Update operations normally have no return type, but this is possible. Input parameters and their types are entered in a list, for example:

```
x int y int
```

for an operation with two integer input parameters.

Operations can be defined recursively, for example:

```
static query factorial(i : int) : int
pre: i >= 0
post:
  (i <= 1  =>  result = 1) &
  (i > 1  =>  result = i*factorial(i-1))
```

Operations can be called by the notation *obj*.*op*(*params*) as usual. Such calls are expressions, in the case of query operations, and statements in the case of update operations. If *op* is a query operation then this expression can be used within other expressions as a value of its declared return type. Update operations should not be used as values or in contexts (such as conditional tests or constraint antecedents) where a pure value expression is expected. Static query operations with at least one parameter, or instance scope query operations with no parameter, can be stereotyped as *cached*, which means that caching is used for their implementation: the results of operation applications are stored in order to avoid recomputation of results. This can make a substantial difference to efficiency in the case of recursive operations. Only operations owned by a class (not an interface or use case) can be cached. For example, for the function defined as:

```
static query f(x : int) : int
pre: x >= 0
post:
  ( x = 0  =>  result = 1 ) &
  ( x = 1  =>  result = 2 ) &
  ( x > 1  =>  result = f(x-1) + f(x-2) )
```

there are 331,160,281 calls to f when evaluating $f(40)$ without caching, and an execution time of 7.5s. With caching there are only 41 calls, and an execution time of 10ms. However, the user should ensure that cached behaviour is really what is intended – if the operation result could legitimately change if the operation is invoked at different times with the same argument values then this stereotype should not be used.

3.2.4 Rules and restrictions

UML-RSDS class diagrams are valid UML 2.0 class diagrams, but there are the following additional language rules in UML-RSDS:

Single inheritance: a class can have at most one direct superclass. However, it may inherit directly from multiple interfaces.

No concrete superclasses: All superclasses must be abstract, and all leaf classes must be concrete, in a completed system specification.

These restrictions are intended to improve the quality of the models and the quality of generated code. Both multiple inheritance and concrete superclasses introduce ambiguous semantics into models and should be avoided by modellers.

The following UML class diagram features are not supported, again because of their complex semantics, which may lead developers into producing erroneous models:

- n-ary associations for $n > 2$
- nested classes
- visibility annotations

- subset relationships between associations
- constraints such as disjointness on families of inheritances.

3.2.5 Code synthesis from class diagrams

Outline code in Java, C# or C++ can be produced from UML-RSDS class diagrams, by the Generate X options on the Synthesis menu. The architecture of the implemented system has a standard structure (Fig. 3.7): a simple user interface component (GUI.java in a Java implementation) invokes operations of a Singleton class (Controller) which has operations corresponding to the use cases of the system, and which acts as a single point of access to the system functionality. There are also classes corresponding to each class of the class diagram, and their operations are invoked by the controller class.

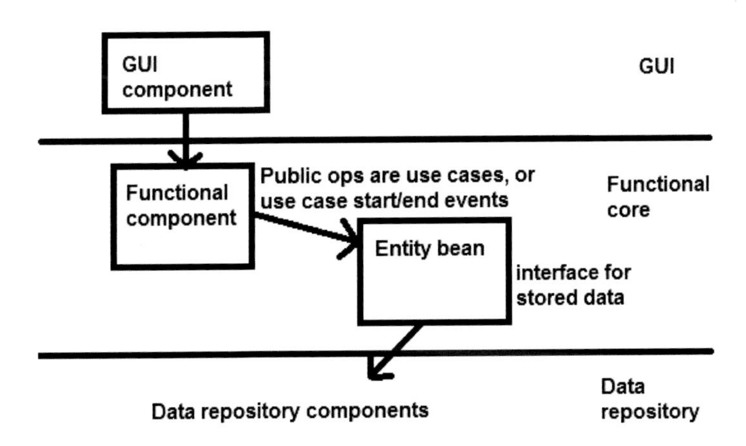

Figure 3.7: UML-RSDS systems architecture

Java code can be generated from a class diagram model by the option Generate Java on the Synthesis menu. There are two options: Java 4 and Java 6. The main difference is that the Java 6 code uses HashSet to represent unordered association ends, and ArrayList to represent ordered association ends, whilst the Java 4 code uses Vector for both. The Java 6 version can be more efficient for systems involving unidirectional unordered associations.

For our example model, the generated code consists of Java classes for each class of Fig. 3.1, in a file *Controller.java*, together with the main class, *Controller*, of the system. A file *SystemTypes.java* contains an interface with utility operations (for computing OCL expressions), and *ControllerInterface.java* contains an interface for *Controller*. These files are in the same Java package *pack* if this is specified as the system name. A file *GUI.java* defines a simple user interface for the system, with buttons for each use case (an example is shown in Fig. 3.13). The file *Controller.java* is also displayed to the user (Fig. 3.8).

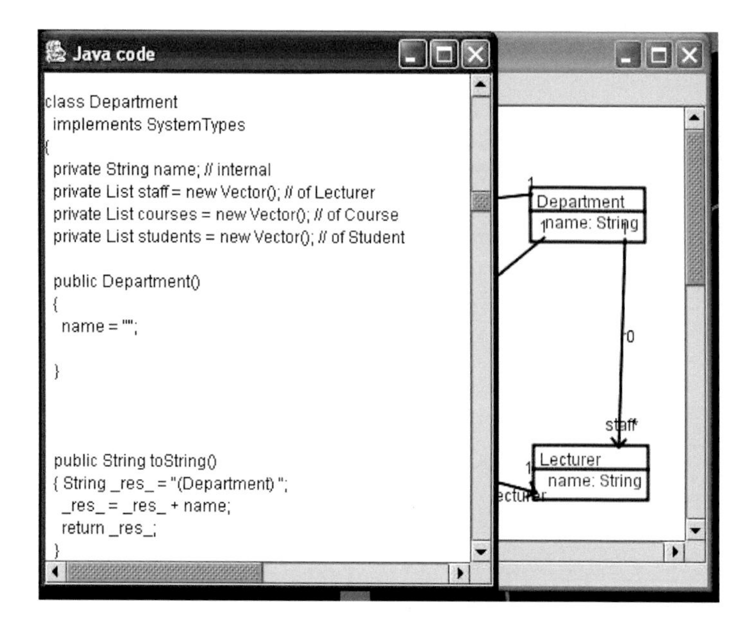

Figure 3.8: Display of Java

The Java code can be used directly in other applications, a standard interface of operations is provided for each class C. In the case of the Java 4 generated code, this is:

- A no-argument constructor $C()$ which sets attributes and roles to the default values of their types.
- A constructor $C(T\ attx, ..., D\ rolex)$ which takes input values for each *final* or *frozen* instance attribute and for each 1-multiplicity role.
- An operation $setatt(T\ attx)$ for each modifiable instance attribute $att : T$
- An operation $setrole(D\ rolex)$ for each modifiable instance role $role : D$.
- An operation $setAllf(List\ objs, T\ val)$ to set $ob.f$ to val for each $ob : objs$ for each modifiable feature $f : T$.

For sequence-valued roles f there is an operation $setAllf(List\ objs, int\ ind, D\ fx)$ to set $ob.f[ind]$ to fx for $ob : objs$.

- Operations $setrole(List\ rolex)$, $addrole(D\ rolexx)$, $removerole(D\ rolexx)$, $addAllrole(List\ objs, D\ rolexx)$, $removeAllrole(List\ objs, D\ rolexx)$ for each modifiable many-valued role $role : Set(D)$ or $role : Sequence(D)$. The *removerole* and *removeAllrole* operations are omitted if the role is *addOnly*. An operation $setrole(int\ i, D\ rolex)$ is included for sequence-valued roles, to set the i-th element of *role* to *rolex*.
- An operation $getrole() : D$ for a 1-multiplicity role $role : D$.
- An operation $getatt() : T$ for an attribute $att : T$.
- An operation $getrole() : List$ for a many-valued role *role*.
- A static operation $getAllf(List\ objs) : List$ which returns the set (i.e., with duplicate values removed) of all $ob.f$ values for $ob : objs$, for any data feature f of the class.
- A static operation $getAllOrderedf(List\ objs) : List$ which returns the sequence/bag of all $ob.f$ values for $ob : objs$, for any data feature f of the class (with duplicate values preserved). This corresponds to OCL $objs \rightarrow collectNested(f)$.
- For interfaces I, the above static operations are instead placed in an inner class *IOps* of I.
- There are also specialised operations for manipulating elements of qualified associations using qualifier index values.

Both sets and sequences are represented by the Java List type, and are implemented as Vectors. The Java 6 code instead uses Collection as a general Collection type, and HashSet and ArrayList for sets and sequences, respectively. One limitation compared to standard UML is that multiple inheritance is not fully supported (except for implementation in C++). Only multiple interfaces can be inherited by a single class, not multiple classes. However, unlike Java, such interfaces can contain instance features, which can be inherited by multiple classes. If this is used to simulate multiple inheritance, then before synthesising Java, the Refactoring transformation *Push down abstract features* should be used to copy such features down to all subclasses, thus achieving the effect of multiple class inheritance.

Controller.java contains a public set of operations to modify the objects of a system:

- An operation $setatt(E\ ex, T\ attx)$ for each modifiable instance attribute $att : T$ of class E.
- An operation $setrole(E\ ex, D\ rolex)$ for each modifiable instance role $role : D$ of E.
- Operations $op(E\ ex, pardec)$ and $AllEop(List\ exs, pardec)$ for each update operation $op(pardec)$ of class E.
- Operations $List\ AllEop(List\ exs, pardec)$ for each query operation $rT\ op(pardec)$ of class E.
- Operations $addrole(E\ ex, D\ rolexx)$, $removerole(E\ ex, D\ rolexx)$ for each modifiable many-valued role $role : Set(D)$ or $role : Sequence(D)$ of class E. $removerole$ is omitted if the role is $addOnly$.
- Operations $killE(E\ ex)$ and $killE(List\ exs)$ to remove E instances ex and collections exs of E instances from the model.
- An operation $public\ rT\ uc(pardec)$ for each general use case uc of the system.

Note that $objs.op(e)$ for a collection $objs$ and query operation op is interpreted by the Controller operation $AllEop(objs, e)$, and always returns a sequence-typed result (representing a bag of values) even if $objs$ is set-valued. Generation of C# follows the same structure as for Java, with classes E.cs for each UML class E, Controller.cs and SystemTypes.cs being produced. C# ArrayList is used in place of Java List. If a system name is specified, this name is used as the C# namespace name of the generated code. For C++, class declarations are placed in a file Controller.h, and class code is placed in Controller.cpp. The template class *set* is used to represent UML-RSDS sets, and *vector* is used to represent UML-RSDS sequences.

3.2.6 Bi-directional associations

Associations with an inverse can be specified by filling in the 'Role 1' field in the association creation dialog (as in Fig. 3.4). If the association is drawn from class E1 to class E2, then an inverse association from E2 to E1 will be created also. The inverse association is treated as a feature of E2. The generated Java code will automatically maintain the two associations as mutual inverses: updates to the forward association will generally cause updates to the inverse association also, and vice-versa. This maintenance code is in the Controller operations such as addrole2(E1 e1x, E2 e2x) and killE2(E2 e2x).

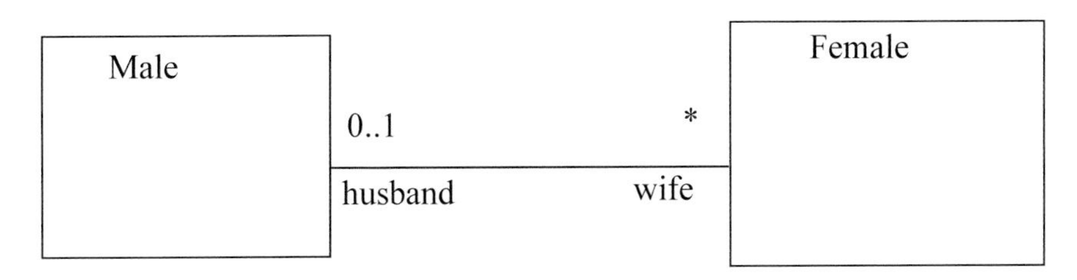

Figure 3.9: Bidirectional association example

In Fig. 3.9, adding a w : *Female* instance to $m.wife$ also implicitly adds m to $w.husband$, and so is only permitted if the resulting size of $w.husband$ is no more than 1. Likewise, removal of an object from one end of a bidirectional association implies removal of an object at the other end also. Table 3.2 gives details of the updates which are automatically provided in generated code by UML-RSDS, where there is an association between distinct classes A and B, with unordered roles ar and br at these ends of the association. These updates aim to maintain the invariant

$$A{\rightarrow}forAll(a \mid B{\rightarrow}forAll(b \mid \\ (b : a.br \Rightarrow a : b.ar) \ \& \ (a : b.ar \Rightarrow b : a.br)))$$

for a many-many bidirectional association, or

$$A{\rightarrow}forAll(a \mid B{\rightarrow}forAll(b \mid \\ (b = a.br \Rightarrow a : b.ar) \ \& \ (a : b.ar \Rightarrow b = a.br)))$$

for a many-one bidirectional association.

The simplest case is many-many associations, particularly *—*. However, setting the br end of this association to a set bs: $ax.br = bs$ needs the opposite end ar to be updated to (i) remove ax from $bx.ar$ if bx was previously linked to ax and will be unlinked by the assignment because $bx \notin bs$; (ii) add ax to $bx.ar$ if bx was previously unlinked to ax and will be linked by the assignment because $bx \in bs$. Assignments to many-multiplicity ends of bidirectional associations may therefore be computationally expensive. Null values may arise in the case of bidirectional 1—* associations modified by $\notin$ or $=$ at the many end, and in the case of 1—1 associations, and this may cause a failure of semantic correctness unless care is taken to also assign valid objects in place of the null links.

In some cases, the existence of an inverse association may enable more efficient code to be produced. However, it also closely semantically links the classes A and B, so that these must be contained in the same code module. In general, bidirectional associations should only be introduced if they are necessary for the system being developed,

Table 3.2: Derived association updates

ar end	*br end*	*br update*	*ar update*
$*$ or 0..1	$*$ or 0..1	$bx : ax.br$ $bx \notin ax.br$ $ax.br = bs$	$ax : bx.ar$ $ax \notin bx.ar$ $(ax.br@\text{pre}-bs)\rightarrow forAll(bx \mid ax \notin bx.ar)$ $(bs - ax.br@\text{pre})\rightarrow forAll(bx \mid ax : bx.ar)$
$*$ or 0..1	1	$ax.br = bx$ $bx \neq ax.br@\text{pre}$	$ax \notin ax.br@\text{pre}.ar$ $ax : bx.ar$
1	$*$ or 0..1	$bx : ax.br$ $bx \notin ax.br$ $ax.br = bs$	$bx \notin bx.ar@\text{pre}.br$ $bx.ar = ax$ $bx.ar = null$ $bx.ar = null$ for $bx : ax.br@\text{pre}-bs$ $bx.ar = ax$ and $bx \notin bx.ar@\text{pre}.br$ for $bx : bs - ax.br@\text{pre}$
1	1	$ax.br = bx$	$ax.br@\text{pre}.ar = null$ $bx.ar@\text{pre}.br = null$ $bx.ar = ax$

i.e., navigation in both directions along the association is required for the functionality of the system. If one end of a bidirectional association is {*addOnly*}, then the other end is also implicitly {*addOnly*} (for a many-valued opposite end), or {*frozen*}, for a single-valued opposite end (because a change to the opposite end of object x requires removal of x from any existing forward role sets).

3.2.7 Qualified associations

A qualified association *br* from class A to class B represents a String-indexed map from each A object to individual B objects or to sets of B objects (depending on if the multiplicity at the B end is 1 or not). The multiplicity at the A end is assumed always to be *, and this end is un-named. Qualified associations cannot be static. Figure 3.10 shows an example of a qualified association, which represents a situation where the students of a department are uniquely identifiable by an id value. The qualified association aims to provide efficient lookup of objects by the index (in Java, the Map type directly supports qualified associations, as does the map template type in C++).

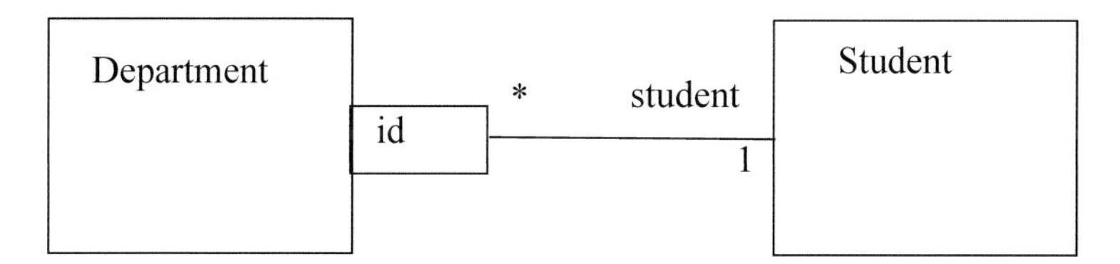

Figure 3.10: Qualified association example

For a qualified association from class A to class B, the Java code generated for A will represent br as a Java Map, and include operations to modify and read br, based on the index values. For many-valued br, the operations are:

- An operation $setbr(String\ ind, List\ rolex)$ to set $br[ind]$ to a set $rolex$ of B objects.
- Operations $addbr(String\ ind, B\ rolexx)$, $removebr(String\ ind, B\ rolexx)$, to add/remove $rolexx$ from $br[ind]$.
- An operation $getbr(String\ ind) : B$ to return $br[ind]$.

Corresponding operations are also defined in the Controller class. Analogous representations are used in C# (the Hashtable class) and C++ (the map template class). Qualified associations should not be used in source or target languages for model transformations, because their values cannot be stored in/read from model data text files. However, they can be used as auxiliary data to support transformation processing.

3.2.8 Association classes

An association class is a class C which is also an association between classes A and B (normally a many-many association). Objects of C are also links $a \mapsto b$ of pairs of related A and B objects, and their attribute values define property values specific to individual links. For example, a salary specific to an employment link between a company and an employee (Fig. 3.11).

In translation to Java, C#, C++ or B AMN, association classes are represented by the two implicit many-one associations $a : C \to A$ and $b : C \to B$. UML does not permit C to be the same class as A or B, we additionally require A and B to be distinct, and unrelated to each other or to C by inheritance. Instances of C should not be created directly, but only as a result of linking an A object to a B object by the association. Likewise, deleting C objects is a consequence of unlinking

A and B objects. The set of C objects linked to a particular $ax : A$ is $C \rightarrow select(a = ax)$, likewise for the C's linked to some B object. The values of association classes cannot be stored in/read from model data text files.

Figure 3.11 shows an example of an association class, which represents a situation where there are companies and employees, linked by employments. Each employee may have several employments, each with an individual salary. An employee's total salary would then be expressed as:

$$Employment \rightarrow select(employee = e) \rightarrow collect(salary) \rightarrow sum()$$

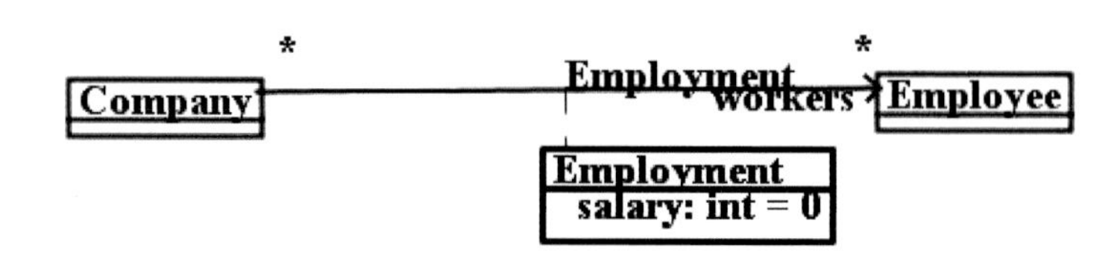

Figure 3.11: Association class example

3.2.9 *Aggregation*

An aggregation is a specific kind of association which has the informal meaning of a whole-part relation. For example, a Sudoku board has a whole-part relation to the board squares (Fig. 3.12). The parts cannot exist without the whole, nor be shared between two wholes.

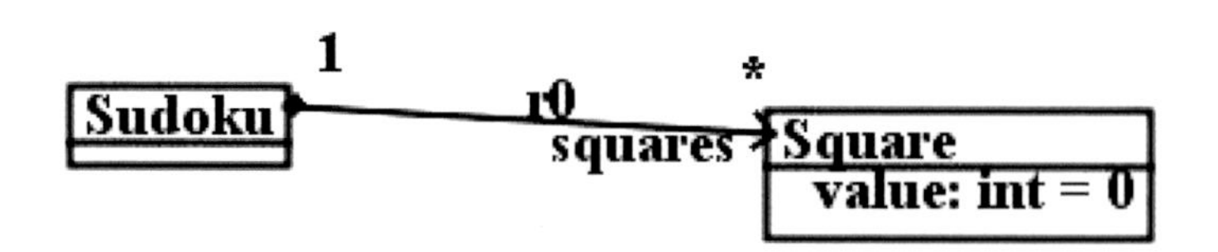

Figure 3.12: Aggregation example

The form of aggregation we use in UML-RSDS is the strong concept of the parts being owned by the whole, known as *composition*. An aggregation should have 1 or 0..1 multiplicity at the 'whole' end (the end with the black diamond), and * multiplicity at the 'part' end. The semantics of aggregations is the same as for associations except that aggregations cause *deletion propagation* from the whole to the parts: when a whole object is deleted, so are all its attached parts, in the same execution step. If there is an aggregation br from class A to class

B, then any call of *ax→isDeleted*() for an instance $ax : A$ also results in a call *ax.br→isDeleted*(), even if the association is optional at the A end.

Aggregation, qualified associations and association classes are specialised modelling mechanisms, which should only be used if needed in a particular application.

3.3 Models and metamodels

Model-based development is centered on the use of application models, such as class diagrams of a system, instead of on the direct use of application code. As we have seen in this chapter, a class diagram such as Fig. 3.1 can be used to generate all the structural and book-keeping code of a system. Only the application-specific functionalities are missing – these can be specified as use cases in UML-RSDS, as described in Chapter 5. Models are not necessarily represented as diagrams, but could consist of text, and be stored as text files in a specific format (such as XML). UML-RSDS class diagrams are stored in text format in a file *mm.txt* in the */output* subdirectory: the *Save* option on the *File* menu saves the diagram to this file, and the *Recent* option loads the model in *mm.txt* into the class diagram editor. It is possible to edit the class diagram data in *mm.txt* using a text editor, although care must be taken to use the same format for the data.

Class diagrams have *instances*, or *instance models*, in the same way that classes have individual objects as their instances. An instance model m for a class diagram L consists of finite collections of objects (instances of E) for each class (entity type) E of L (the collections may be empty, but all objects of m must belong to some concrete class of L). An object *obj* in m should have an attribute value for each declared attribute of its class in L (including inherited attributes). Links between objects are recorded in m, and represent the instances of associations in L. An instance model corresponds to a possible set of program objects and object feature values that may exist at some time point during the execution of a program that implements the class diagram. The programs synthesised by UML-RSDS from class diagrams can be initialised with a particular instance model (stored in *output/in.txt*) by means of the option *loadModel* of the generated GUI (for Java4 generated code). They can also store the current program state in *output/out.txt* using the option *saveModel*. Figure 3.13 shows an example of the generated GUI.

The text format for instance models in UML-RSDS is very simple, consisting of lines of text of three possible forms:

```
e : T
```

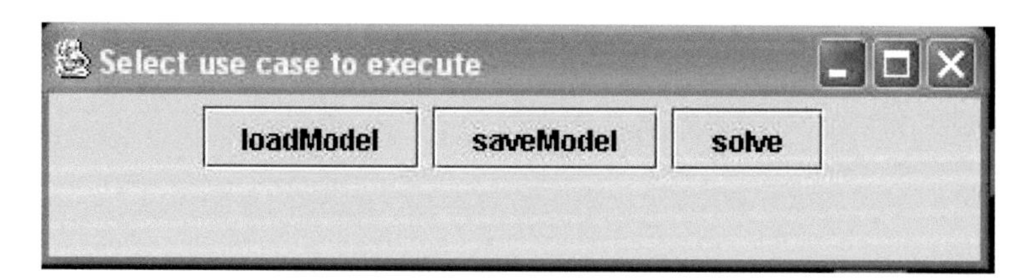

Figure 3.13: GUI example

declaring that the identifier e is an instance of concrete class T,

```
e.f = v
```

declaring that the attribute or 1-multiplicity role f of e has value v, and

```
obj : e.role
```

declaring that *obj* is a member of the many-valued (0..1 or *-multiplicity) role *role* of e. For sequence-valued roles, the order in which these membership assertions occur in the file is the order of elements in *role*. Spaces must be present around the = and : symbols in the lines of the file.

For example, we could have the example instance model

```
d1 : Department
d1.name = "Mathematics"
l1 : Lecturer
l1.name = "David"
l1 : d1.staff
c1 : Course
c1.name = "Algebra1"
c1 : d1.courses
c1.lecturer = l1
```

for the Department class diagram of Fig. 3.1. Here there is a single instance of the *Department* class, a single instance of *Lecturer*, and a single link between these instances. There is an instance of *Course*, linked to the department and lecturer instances. The declaration of an object (such as $d1$: *Department*) must precede any other line that uses the object.

Class diagrams have two separate but related roles in MBD: (i) they can describe the data and structure of particular *applications*, as (executable) specifications of these applications – Fig. 3.1 is an example of such a class diagram; (ii) they can describe *languages* (such as English, Russian, Java or a modelling language such as UML) – Fig. 2.5 is an example of such a class diagram. A class diagram at the language level is often called a *metamodel* because its classes (or *entity types*) have instances at the (application) model level. In turn, an application model has instances at the instance model (program execution) level. Figure 3.14 shows the relationships between the different model levels. Both the language level and application level models can be represented using class diagram notation, and are stored in file mm.txt by the UML-RSDS tools. Instance-level models are represented as text files (in model.txt, in.txt, out.txt, etc.).

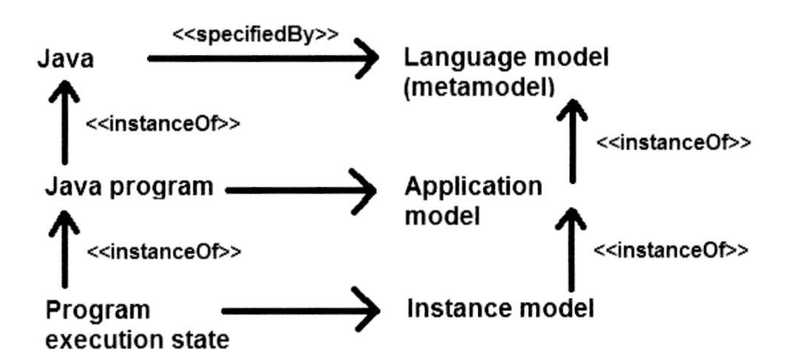

Figure 3.14: Modelling levels

Our application-level class diagram (Fig. 3.1) can be partly represented as an instance of the language-level class diagram (Fig. 2.5) as follows:

```
e1 : Entity
e1.name = "Student"
e2 : Entity
e2.name = "Department"
e3 : Entity
e3.name = "Course"
e4 : Entity
e4.name = "Lecturer"
p1 : Property
p1.name = "name"
p2 : Property
p2.name = "name"
p3 : Property
p3.name = "name"
p4 : Property
p4.name = "name"
p1 : e1.ownedAttribute
p2 : e2.ownedAttribute
p3 : e3.ownedAttribute
p4 : e4.ownedAttribute
t1 : Type
t1.name = "String"
```

```
p1.type = t1
p2.type = t1
p3.type = t1
p4.type = t1
```

The full UML-RSDS class diagram language is given in Fig. B.2.

The ability to represent application models as instances of metamodels enables a developer to apply transformations (such as class diagram refactoring, Section 2.3) to the application models: such transformations can operate on application models in the same way that conventional programs can operate on instance models. In the case of the department application model, the restructuring rule 3 can be applied to rationalise the class diagram structure: the *name* : *String* attribute declared in the four separate root classes *Department*, *Student*, *Lecturer*, *Course*, can be replaced by a single attribute in a new superclass of these classes.

As far as development in UML-RSDS is concerned, there is no practical difference between specifications concerning application models or languages: in both cases the specifications are expressed in terms of the entity types and features of the model, and the model itself is visually represented as a class diagram. In the case of application-level specifications, the implementation of the UML-RSDS specification operates on instance models (in text format), which represent instances of the application model. In the case of language-level specifications, the implementation operates on instance models which represent application models that are instances of the language (as in the above example of the teaching class diagram as an instance of UML).

3.4 Domain-specific languages

A domain-specific language (DSL) is a notation (graphical, textual or a mix of the two), together with a formal language definition, intended to represent the concepts and individual elements within particular application domains.

For example, a notation to describe distributed systems consisting of physical computing devices, connections between devices, and software running on the devices, could be used to draw diagrams such as that in Fig. 3.15. This visual notation is referred to as the *concrete syntax* of the DSL, in the same way that the textual forms of Java statements and declarations is the concrete syntax of Java.

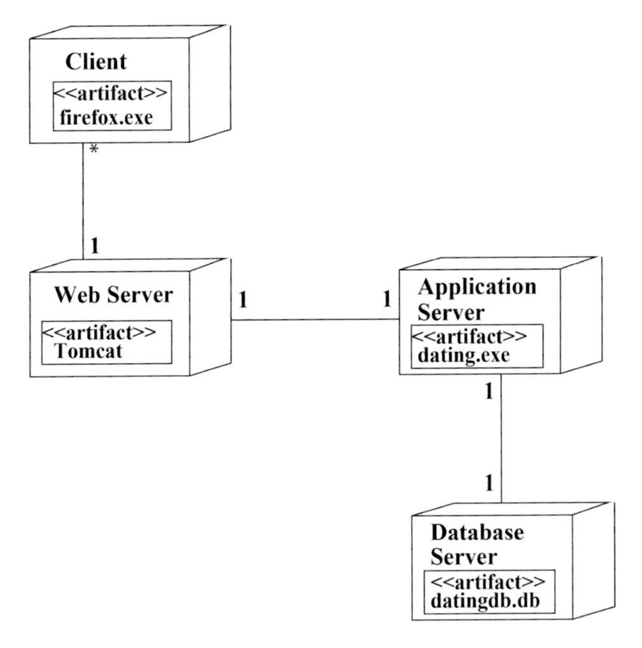

Figure 3.15: Deployment diagram example

The visual representation of a computational device is a 3D box. Connections are shown as lines, and artifacts (software applications and platforms) are written as rectangles and text inside the devices that host them.

A DSL has the following elements:

- An identified application domain where it will be used.
- Representations that are appropriate, visually and conceptually, for modellers in the domain, and for stakeholders who need to review such models.
- A precise abstract grammar, defining the way DSL elements can be combined (e.g., that communication links can connect any two different devices, but not a device to itself). This is known as the *abstract syntax* of the DSL.
- A precise way, usually by means of a model transformation, to map the DSL models to implementations.

In this book, and in UML-RSDS, a DSL is defined by a class diagram that gives the abstract syntax of the DSL as a metamodel. Figure 3.16 shows a possible metamodel for the distributed systems DSL used in Fig. 3.15.

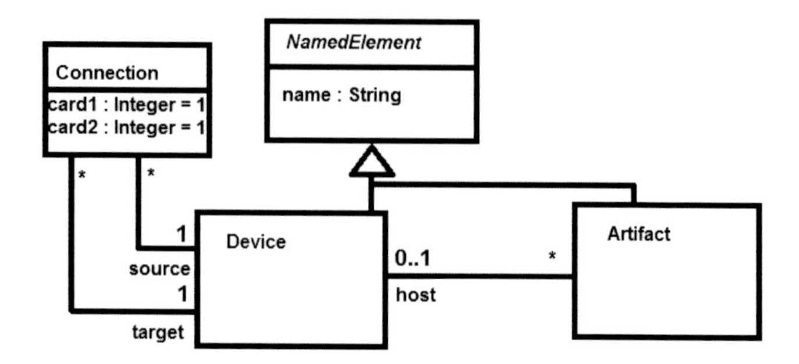

Figure 3.16: Deployment notation DSL

Using such a metamodel, particular DSL models can be specified in a notation such as XML, or in the UML-RSDS textual format for models described in the previous section. For example, the model of Fig. 3.15 could be described by the following textual model in terms of the DSL concepts:

```
pc : Device
pc.name = "Client"
ws : Device
ws.name = "Web Server"
appserv : Device
appserv.name = "Application Server"
dbserv : Device
dbserv.name = "Database Server"
c1 : Connection
c1.source = pc
c1.target = ws
c1.card1 = -1
c1.card2 = 1
c2 : Connection
c2.source = ws
c2.target = appserv
c3 : Connection
c3.source = appserv
c3.target = dbserv
browser : Artifact
browser.name = "firefox.exe"
pc : browser.host
tomcat : Artifact
tomcat.name = "Tomcat"
ws : tomcat.host
app : Artifact
app.name = "dating.exe"
appserv : app.host
db : Artifact
db.name = "datingdb.db"
dbserv : db.host
```

This is an instance model of Fig. 3.16. Notice that default attribute values do not need to be written in the model text file. A transformation (defined in UML-RSDS) can then process such models to generate any necessary configuration files, interfaces for remote invocations, etc., based on the particular network.

A DSL may use the terminology of particular design patterns for a domain. For example, a DSL for enterprise information systems could contain entity types such as *Controller*, *View*, *Model* from the MVC pattern (Chapter 9).

3.5 Generalised application models

For some application domains a DSL may not be necessary, instead a generalised application model may be defined, which defines the common entity types and typical structures of applications in the domain. Such a model can provide systematic and reusable specifications and designs for a number of different developments within the same domain. For example, in the domain of programs for board games (noughts-and-crosses, Sudoku, chess, Scrabble, etc.) there are common concepts (Game, Board, Location, Piece, Move, Score, Player, etc.) and common class diagram structures (a board consists of a fixed aggregation of locations; a game has a number of players, who may make moves using pieces, etc.). This similarity helps a developer to write the general structure and elements of a class diagram, using such concepts, and also helps to make the class diagram comprehensible to stakeholders. Similarly a generalised application model for financial domains could be defined, with concepts such as Option, Market, Share, YieldCurve, etc.

Summary

In this chapter we have described the UML class diagram notations that are used in UML-RSDS, and we have outlined how class diagrams are implemented in code by the UML-RSDS tools. We have also described the concepts of metamodels, domain-specific models and generalised application models.

Chapter 4

Constraints

Logical constraints are used within UML-RSDS to provide more detailed semantics for a class diagram, and to precisely define the functionality of operations and of use cases. Constraints can be translated into formal languages (such as B AMN) to support verification. A subset of the UML Object Constraint Language (OCL) is used as the UML-RSDS constraint language.

The BNF grammar of UML-RSDS OCL expressions, and the abstract syntax of expressions, are given in Appendix A. All expressions have a mathematical semantics ([1]). A significant difference to standard OCL is that we simplify the OCL semantics to use classical 2-valued logic: there are no *undefined* or *null* values in UML-RSDS OCL. In addition, collections are either sequences or sets.

4.1 Constraints in UML-RSDS

OCL and abbreviated OCL constraints can be used in many places in a UML-RSDS specification:

1. Class invariants
2. Global constraints of a class diagram
3. Operation pre and postconditions
4. Use case pre- and post-conditions and invariants
5. State machine state invariants and transition guards.

Constraints are a central facility of UML-RSDS, and enable UML-RSDS specifications to be used both for informal analysis and communication, formal mathematical analysis, and (via code generators) as executable implementations. The key principle of UML-RSDS is:

> *A logical constraint P can be interpreted both as a specification of system behaviour, and as a description of how P will be established in the system implementation.*

Thus an operation post-condition in:

```
static query factorial(i : int) : int
pre: i >= 0
post:
  (i <= 1  =>  result = 1) &
  (i > 1  =>  result = i*factorial(i-1))
```

gives both a declarative and precise description of the factorial function, in a recursive style, and instructs the UML-RSDS code generators to produce recursive implementations based on the postcondition expressions, in Java, C# or C++.

An alternative description could be:

```
static query factorial(i : int) : int
pre: i >= 0
post:
  (i <= 1  =>  result = 1) &
  (i > 1  =>  result = Integer.Prd(2,i,k,k))
```

where Integer.Prd(2,i,k,k) is the product $\Pi_{k=2}^{i} k$ of integers between 2 and i, inclusive. This second description is more efficient, and is of similar clarity to the recursive version.

Figure 4.1 shows the dialog for entering constraints.

A constraint $P \Rightarrow Q$ on entity E is entered by writing E in the first text area, P in the second and Q in the third. We will also write this in text as

$$E ::$$
$$P \Rightarrow Q$$

This has the logical meaning

$$E \rightarrow forAll(P \Rightarrow Q)$$

"for all instances of E, if P holds, then Q holds". Operationally, it means "for all instances of E, if P holds, then make Q hold true, if possible".

The default options on the dialog (system requirement, non-critical, update code, unordered) can be chosen in most cases. E is called the *context* or *scope* of the constraint. Local invariants of class C are specified by naming C in the first field of this dialog. Global constraints forAll-quantified over C also need C to be entered here. Use case constraints based on class C should have the class name entered as context (for the options Add Postcondition, Add Precondition or Add Invariant on the edit use case dialog).

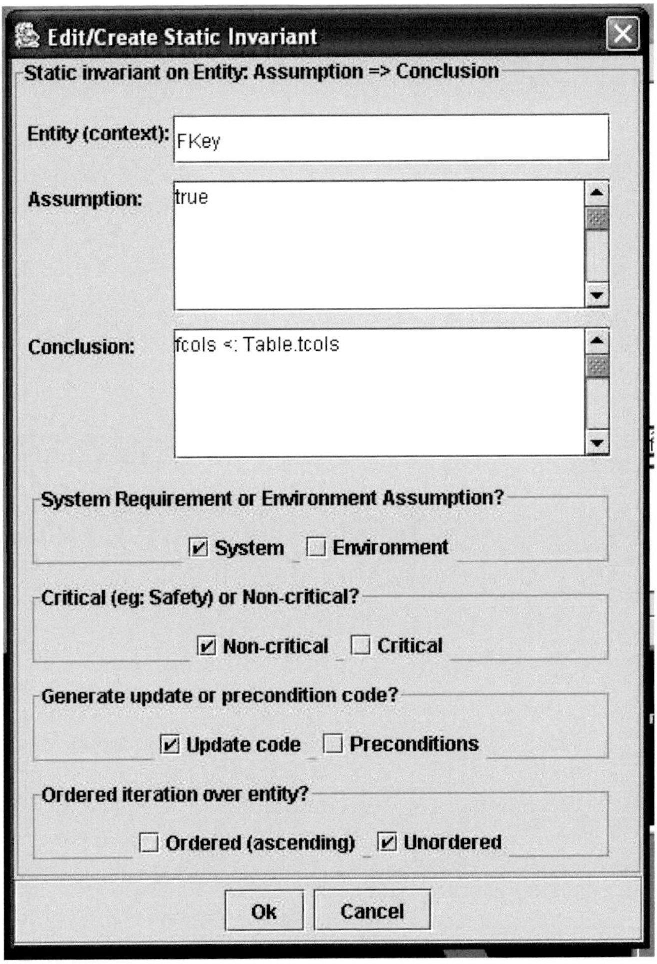

Figure 4.1: Constraint dialog

If the constraint is a local invariant of a class C, or it is a use case constraint based on C, then features of C can be written simply by their names in P or Q. Navigation expressions starting from features of C can also be written. Additional variables x can be introduced by a formula $x : e$ in the constraint antecedent: x must not be the name of a feature of the context class, and it is implicitly universally quantified over the elements of e, which may be an entity type name or other collection-valued expression.

For example:

```
c : courses
```

could be defined as a constraint antecedent P and

```
c.lecturer : staff
```

as a succedent Q. This could be an invariant of class *Department* in our example (Fig. 3.1), expressing that all courses of the department have their lecturer also within the department. c is a quantified variable, whose type is inferred as being $Course$. The same logical constraint can be alternatively expressed without c by writing

```
true
```

as the antecedent P and

```
courses.lecturer <: staff
```

as Q. "The lecturers of all courses are members of staff". Q could also be written in conventional OCL notation as `staff->includesAll(courses.lecturer)`.

Another form of variable that may be introduced in constraint antecedents is the *let* variable, which assigns a constant value of an expression e to the variable to avoid repeated evaluations of that expression:

$$v = e \;\Rightarrow\; P$$

The identifier v may then be used in place of the (possibly complex) expression e in P.

A common simple form of constraint is one that derives the value of an attribute of a class from other features of the class. For example, if we add an integer attribute *numberOfStudents* to *Department* in Fig. 3.1, we can define its value by the invariant constraint

$$numberOfStudents = students.size$$

or

$$numberOfStudents = students{\rightarrow}size()$$

on *Department*.

A constraint has scope or context the entities to which it implicitly applies (features of such entities can be written without object references in the constraint, or with the reference *self*). The most local context is termed the *primary scope* of the expression, and feature references are interpreted as features of this entity by default. Other entities in the scope are termed the *secondary scope*. A local invariant of a class E has primary scope E, as do pre and postconditions of E's instance operations, and state invariants and guards in the entity state machine of E or in its instance operation state machines. A use case constraint with owner entity E also has primary scope E. A predicate P in the body of a select, reject or collect expression without a variable: $e \rightarrow select(P)$, $e \rightarrow reject(P)$, $e \rightarrow collect(P)$, has primary scope the element type of e, if this is an entity type, and secondary scope the scope of the selection expression. Likewise for quantifier expressions without variables: $e \rightarrow forAll(P)$, $e \rightarrow exists(P)$, $e \rightarrow exists1(P)$.

Constraints not only have a declarative meaning at the specification level, defining what conditions should be maintained by a system, but also they may have a procedural meaning, defining code fragments whose execution enforces the conditions. For example, the constraint defining *numberOfStudents* will be used by the UML-RSDS tools to produce code that modifies this attribute in response to any operation *setstudents*, *addstudents* or *removestudents* which affects *students*. For example:

```
public void addstudents(Student studentsxx)
{ students.add(studentsxx);
  numberOfStudents = students.size();
}
```

4.1.1 Basic expressions

Table 4.1 defines the basic expressions which may be used in constraints, and their meanings. *self* corresponds to *this* in programming languages (Java, C# and C++). The notation for attributes, roles and operations is standard. A specific form of basic expression which is used extensively in UML-RSDS is the *object lookup* expression $E[v]$ where E is a class name and v is a string value or a collection of strings. E must have a primary key/identity attribute $att : String$, or an attribute $att : String$ satisfying $E \rightarrow isUnique(att)$. Thus instances of E are uniquely identified by string values: for a string value s there is either 1 or 0 instance x of E at any time that satisfies $x.att = s$. If there is such an instance, it is returned as the value of $E[s]$ (if there is no instance, then $E[s]$ is undefined). This lookup applies for the first (in order of the declarations within the class) attribute that satisfies the primary key/uniqueness property, termed the *principal primary key*. The expression should evaluate in constant time complexity – as with qualified associations, programming language map data structures can be used to ensure this. The returned object can be used in the same way as any other instance of E, in feature application expressions $E[s].f$ or as a parameter value to operations. The expression $E[s]$ could be written as

$$E.allInstances() \rightarrow any(att = s)$$

in standard OCL.

The second variant of the object lookup expression has a collection vs of string values as its argument, and returns the collection of E instances whose att value is in vs:

$$E.allInstances() \rightarrow select(vs \rightarrow includes(att))$$

This is always well-defined. It is a sequence if vs is a sequence, otherwise a set.

Table 4.1: Basic expressions

Expression	*Meaning as query or update*
self	The current object of class E: for constraints with a single scope E
f	Attribute/role f of class E, for constraints with scope E
$op(e1, ..., en)$	Invocation of operation op of E, for constraints with scope E
$f[ind]$	ind element of ordered or qualified role f of current scope E
$E[v]$ v single value	The instance of E with (principal) primary key value v
$E[v]$ v collection	The instances of E with (principal) primary key value in v Sequence-valued if v is
$obj.f$	Data feature f of single object obj
$objs.f$	Collection (set or sequence) of values $o.f$ for $o \in objs$, collection $objs$. This is a set if $objs$ is a set or if f is set-valued
$obj.op(e1, ..., en)$	Apply op to single object obj
$objs.op(e1, ..., en)$	Apply op to each $o \in objs$, collection $objs$ Result value is bag/sequence of individual result values $o.op(e1, ..., en)$ if op has a result
value	Value of enumerated type, numeric value, string value "string", or boolean true, false

Unlike most programming languages, features may be applied directly to collections of objects: $objs.att$ or $objs.f(p)$. These expressions permit concise specification of the collections formed or effects produced by applying the feature on each instance in the collection $objs$.

4.1.2 Logical expressions

Table 4.2 shows the logical operators that can be used in constraints, and their declarative and procedural meanings. The procedural interpretation of a formula P is formally defined as a statement in the activity language (Chapter 6) denoted by $stat(P)$. This is defined in Table 6.2.

Table 4.2: Logical operators

<table>
<tr><th>Expression</th><th>Meaning as query</th><th>Meaning as update</th></tr>
<tr><td>A => B</td><td>If A is true, so is B</td><td>If A holds,
make B true</td></tr>
<tr><td>A & B</td><td>A and B
are both true</td><td>Carry out A,
then B</td></tr>
<tr><td>A or B</td><td>A holds, or B holds</td><td></td></tr>
<tr><td>E->exists(P)
entity E
e->exists(P)
expression e</td><td>An existing instance
of E satisfies P
An existing element
in e satisfies P</td><td></td></tr>
<tr><td>E->forAll(P)
entity E
e->forAll(P)
expression e</td><td>All existing instances
of E satisfy P
All elements
of e satisfy P</td><td></td></tr>
<tr><td>E->exists1(P)
entity E
e->exists1(P)
expression e</td><td>Exactly one existing instance
of E satisfies P
Exactly one element
of e satisfies P</td><td></td></tr>
<tr><td>E->exists(x | P)
concrete entity E
e->exists(x | P)
other expression e</td><td>An existing instance
x of E satisfies P
An existing element
x in e satisfies P</td><td>Create $x : E$
and make P true for x
Select an element of e
and apply P to it</td></tr>
<tr><td>E->forAll(x | P)
entity E
e->forAll(x | P)
expression e</td><td>All existing instances
x of E satisfy P
All elements x
of e satisfy P</td><td>For all existing $x : E$,
make P true for x
Make P true for
all $x : e$</td></tr>
<tr><td>E->exists1(x | P)
concrete entity E

e->exists1(x | P)
other expression e</td><td>Exactly one existing instance
x of E satisfies P

Exactly one element
x of e satisfies P</td><td>If no instance of E
satisfies P, create
$x : E$ and make P true for x
If no $x : e$ satisfies P,
select some $x : e$ and
apply P to it</td></tr>
</table>

There should not be a space between the $-$ and $>$ symbols, or between the $>$ and the operator name in the $\rightarrow op$ operators.

There are some special cases in the operational use of the *exists* and *exists*1 quantifiers:

- In the case of $E \rightarrow exists1(x \mid x.id = v \;\&\; P)$ or $E \rightarrow exists(x \mid x.id = v \;\&\; P)$ where id is the principal primary key of concrete entity E, a lookup for $E[v]$ is performed first, if this object exists then (i) the update effect of P is applied to it (unless P already holds for it). Otherwise, (ii) a new instance of E is created and the formula $x.id = v \;\&\; P$ is applied to it.

- If E is abstract, or of the form F@pre for an entity F, then only the lookup and part (i) of this functionality is performed. Likewise for general expressions e in place of E.

Although checking for existence of an x with $x.id = v$ does affect the efficiency of the implementation, it is necessary to ensure correctness: it would be invalid to create two or more objects with the same id value.

Note that it is invalid to attempt to quantify over infinite sets: $Integer \rightarrow forAll(P)$ is not a valid OCL expression. Only entity type names and other expressions denoting finite collections of objects or values may be used as the first argument of a quantifier expression.

4.1.3 Comparitor, numeric and string expressions

Table 4.3 lists the comparitor operators and their meanings.

Table 4.3: Comparitor operators

Expression	*Meaning as query*	*Meaning as update*
$x : E$ E entity type $x : s$ s collection	x is an existing instance of E x is an element of s	Create x as a new instance of E (for concrete E) Add x to s
`s->includes(x)` s collection	Same as $x : s$	Same as $x : s$
$x \ / : E$ E entity type $x \ / : s$ s collection	x is not an existing instance of E x is not an element of s	 Remove x from s
`s->excludes(x)` s collection	Same as $x \ / : s$	Same as $x \ / : s$
$x = y$	x's value equals y's value	Assign y's value to x
$x < y$ x, y both numbers, or both strings	x's value is less than y's. Likewise for $>$, $<=$, $>=$, $/=$, $!=$ (not equal)	
$s <: t$ s, t collections	All elements of s are also elements of t	Add every element of s to t
`t->includesAll(s)`	Same as $s <: t$	Same as $s <: t$
$s \ / <: t$ collections s, t	s is not a subset of t	Remove all elements of s from t
`t->excludesAll(s)` collections s, t	No element of s is in t	Remove all elements of s from t

As a query expression, $x : E$ for an entity type (classifier) E corresponds to $x.oclIsKindOf(E)$ in OCL [2] and to x *instanceof* E in Java and to x *is* E in C#. The comparitors $<$, $>$, $<=$ and $>=$ can also be used to compare objects of the same class E, provided that E has a public operation

```
query compareTo(obj : E) : int
```

$obj1 < obj2$ is then evaluated as $obj1.compareTo(obj2) < 0$.

Numeric operators for integers and real numbers are shown in Table 4.4. Instead of OCL Integer and Real, we use types `int` (32 bit signed integers) and `double` (signed double precision floating point numbers) which correspond to the types with these names in Java, C++ and C#. `long` may also be used (64 bit signed integers) for code generation. All operators take double values as arguments except as noted. Three operators: ceil, round, floor, take a double value and return an int.

Table 4.4: Numeric operators

Expression	*Meaning as query*
-x	Unary subtraction
x + y	Numeric addition, or string concatenation (if one of x, y is a string)
x - y	Numeric subtraction
x * y	Numeric multiplication
x / y	Integer division (div) if both x, y are integers, otherwise arithmetic division
x mod y	Integer x modulo integer y
x.sqr	Square of x
x.sqrt	Positive square root of x
x.floor	Floor integer of x
x.round	Rounded integer of x
x.ceil	Ceiling integer of x
x.abs	Absolute value of x
x.exp	e to power x
x.log	e-logarithm of x
x.pow(y)	y-th power of x
x.sin	Sine of x (given in radians)
x.cos	Cosine of x (given in radians)
x.tan	Tangent of x (given in radians)
Integer.subrange(st,en)	Sequence of integers starting at st and ending at en, in order

Other math operators in common between Java, C# and C++ are: log10, cbrt, tanh, cosh, sinh, asin, acos, atan. These are double-valued functions of double-valued arguments. Some functions may not be available in old versions of Java (log10, cbrt, tanh, cosh, sinh). The math operators may also be written in the style $e \rightarrow op()$ in cases where e is a bracketed expression or other complex expression. For example: $(x + y) \rightarrow pow(3)$.

Some examples of constraints using numeric operators are:

```
1000->sqrt()->display() & 1728->cbrt()->display() &
0.5->exp()->display() & 3->pow(10)->display()

0->sin()->display() & 0->cos()->display() &
0.5->log()->display() & 1000->log10()->display()

3.45->floor()->display() & 3.45->ceil()->display() &
3.45->round()->display() & -3.45->abs()->display()
```

These produce the following results:

```
31.622776601683793
1.6487212707001282
59049.0

0.0
1.0
-0.6931471805599453

3
4
3
3.45
```

(not including cbrt and log10, which are not available on old Java implementations).

String operators are shown in Table 4.5. Notice that since strings are sequences of characters (single-character substrings), all sequence operations should be expected to be applicable also to strings.

String positions in OCL are numbered starting at 1, as are sequence positions. Some examples of string expressions are given in the following constraints:

```
"faster"->tail()->display() & "faster"->front()->display() &
"faster"->first()->display() & "faster"->last()->display()

"faster"->indexOf("s")->display() & "faster".insertAt(2,"s")->
display() &
"faster"->reverse()->display()
```

These give the results:

```
aster
faste
f
r

3
fsaster
retsaf
```

Table 4.5: String operators

Expression	*Meaning as query*
x + y	String concatenation
`x->size()`	Length of x
`x->first()`	First character of x
`x->front()`	Substring of x omitting last character
`x->last()`	Last character of x
`x->tail()`	Substring of x omitting first character
`x.subrange(i,j)`	Substring of x starting at i-th position, ending at j-th
`x->toLowerCase()`	Copy of x with all characters in lower case
`x->toUpperCase()`	Copy of x with all characters in upper case
`s->indexOf(x)`	Index (starting from 1) of s at which the first subsequence of characters of s equal to string x occurs. Default is 0
`s->hasPrefix(x)`	String x is a prefix of s
`s->hasSuffix(x)`	String x is a suffix of s
`s->characters()`	Sequence of all single character substrings of s in same order as in s
`s.insertAt(i,s1)`	Copy of s with string s1 inserted at position i
`s->count(s1)`	Number of distinct occurrences of s1 as a substring of s
`s->reverse()`	Reversed form of s
`e->display()`	Displays expression e as string on standard output
`s1 - s2`	String s1 with all occurrences of characters in s2 removed
`e->isInteger()`	true if e represents an `int` value
`e->isReal()`	true if e represents a `double` value
`e->toInteger()`	Returns the value if e represents an `int` value
`e->toReal()`	Returns the value if e represents a `double` value

4.1.4 Pre-state expressions

A pre-state expression *f@pre* for feature *f* can be used in operation post-conditions and use case post-condition constraints, but not elsewhere. In an operation post-condition, *f@pre* refers to the value of *f* at the start of the operation, i.e., the same value as denoted by *f* in the operation precondition. Occurrences of *f* without *@pre* in the post-condition refer to the value of *f* at the end of the operation. So, for example:

```
op()
pre: b > a
post: b = b@pre - a@pre
```

decrements b by a. Since a itself is not updated by the operation, there is no need to use *pre* with a, and the operation should be written as:

```
op()
pre: b > a
post: b = b@pre - a
```

In use case postconditions, f@pre denotes the value of f at the start of application of the implementation of the postcondition. Pre-state expressions are used to distinguish between this value and any new value for f which is set by the postcondition. Pre-state expressions should not depend upon additional quantified variables or upon let variables. For example, an expression $E@pre[v]$, where v is a quantified variable, is not valid. The variables of quantifiers or of select/reject/collection expressions should not occur in pre-state expressions. The reason for these restrictions is that pre-forms are evaluated at the start of the application of a constraint, and hence can only involve data which is defined at this point. Internal variables of the constraint are not defined at the initiation of constraint execution.

4.1.5 Collection expressions

There are two kinds of collection in UML-RSDS: sets and sequences. Sets are unordered collections with unique membership (duplicate elements are not permitted), sequences are ordered collections with possibly duplicated elements. If br is an unordered * role of class A, then $ax.br$ is a set for each $ax : A$. If br is an ordered * role of class A, then $ax.br$ is a sequence for each $ax : A$. The further two OCL collection types, ordered sets and bags, can be defined in terms of sequences. Table 4.6 shows the values and operators that apply to sets and sequences.

For sequences s, $s \frown t$ should be used instead of $s \backslash/ t$, which produces the set union of two sets, and $s \rightarrow append(x)$ should be used in preference to $s \rightarrow including(x)$. The implementation of $s \rightarrow including(x)$ for sequence s is to add x to the end of s, and the implementation of $s \rightarrow union(c)$ for sequences s and c is to concatenate c after s. $s \rightarrow at(i)$ for a sequence s is denoted by $s[i]$ in our notation.

Table 4.6: Collection operators

Expression	*Meaning as query*
Set{}	Empty set
Sequence{}	Empty sequence
Set{$x1, x2, ..., xn$}	Set with elements $x1$ to xn
Sequence{$x1, x2, ..., xn$}	Sequence with elements $x1$ to xn
`s->including(x)`	s with element x added
`s->excluding(x)`	s with all occurrences of x removed
`s - t`	s with all occurrences of elements of t removed
`s->prepend(x)`	Sequence s with x prepended as first element
`s->append(x)`	Sequence s with x appended as last element
`s->count(x)`	Number of occurrences of x in sequence s
`s->indexOf(x)`	Position of first occurrence of x in s
$x\backslash/y$	Set union of x and y
$x/\backslash y$	Set intersection of x and y
$x \frown y$	Sequence concatenation of x and y
`x->union(y)`	Same as $x\backslash/y$
`x->intersection(y)`	Same as $x/\backslash y$
`x->unionAll(e)`	Union of y.e for y : x
`x->intersectAll(e)`	Intersection of y.e for y : x
`x->symmetricDifference(y)`	Symmetric difference of sets x, y
`x->any()`	Arbitrary element of non-empty collection x
`x->subcollections()`	Set of subcollections of collection x
`x->reverse()`	Reversed form of sequence x
`x->front()`	Front subsequence of non-empty sequence x
`x->tail()`	Tail subsequence of non-empty sequence x
`x->first()`	First element of non-empty sequence x
`x->last()`	Last element of non-empty sequence x
`x->sort()`	Sorted (ascending) form of sequence x
`x->sortedBy(e)`	x sorted in ascending order of y.e values (numerics or strings) for objects y : x
`x->sum()`	Sum of elements of collection x
`x->prd()`	Product of elements of collection x
`Integer.Sum(a,b,x,e)`	`Integer.subrange(a,b)->collect( x | e )->sum()`
`Integer.Prd(a,b,x,e)`	`Integer.subrange(a,b)->collect( x | e )->prd()`
`x->max()`	Maximum element of non-empty collection x
`x->min()`	Minimum element of non-empty collection x
`x->asSet()`	Set of distinct elements in collection x
`x->asSequence()`	Collection x as a sequence
`s->isUnique(e)`	The collection of values of x.e for x : s has no duplicates
`x->isDeleted()`	Destroys all elements of object collection x Can also be used on individual objects x

Some examples of set and sequence operators are given in the following constraints:

```
s = Set{"a", "h", "bed", "aa"} & t = Set{"bed", "h", "uu", "kl"}  =>
    ( s - t )->display() & s->symmetricDifference(t)->display()

s1 = Set{"a", "h", "bed", "aa"} & t1 = Set{"bed", "h", "uu", "kl"}  =>
    ( s1->intersection(t1) )->display() & ( s1->union(t1) )->display()

p = Set{1,3,5,7,2,9}  =>
    p->any()->display() & p->min()->display() &
    p->max()->display() & p->sum()->display() & p->prd()->display()

p1 = Sequence{1,3,5,7,2,9}  =>
    p1->first()->display() & p1->last()->display() &
    (p1[3])->display() & p1->tail()->display() & p1->front()->display()
```

These produce the following results:

```
["a", "aa"]
["a", "aa", "uu", "kl"]

["h", "bed"]
["a", "h", "bed", "aa", "uu", "kl"]

1
1
9
27
1890

1
9
5
[3,5,7,2,9]
[1,3,5,7,2]
```

$st = sq$ for set st and sequence sq is always false, even if both st and sq are empty. Two sets are equal iff they contain exactly the same elements, two sequences are equal iff they contain exactly the same elements in the same order.

Of particular significance are collection operators which construct new collections consisting of all elements of a given collection which satisfy a certain property, or which derive collections from other collections (Table 4.7).

Any order in the original collection is preserved by the *select* and *reject* operators, i.e., if they are applied to a sequence the result is also a sequence. In the operators *select*, *reject* and *collect* without variables, an element of the first argument, s, can be referred to as *self* (where s contains objects, not primitive values).

Some example constraints using these operators are:

```
Set{1, 5, 2, 3, 7, 0}->sort()->display() &
Set{"a", "g", "ga", "b0", "b"}->sort()->display()

Integer.subrange(1,5)->collect( x | x*x*x )->display()

Set{"a", "aa", "hh", "kle", "o", "kk"}->select( x | x.size >
1 )->display()
```

These produce the output:

```
[0, 1, 2, 3, 5, 7]
[a, b, b0, g, ga]

[1, 8, 27, 64, 125]

[aa, hh, kle, kk]
```

Table 4.7: Selection and collection operators

Expression	*Meaning as query*
`s->select(P)`	Collection of s elements satisfying P
`s->select( x | P )`	Collection of s elements x satisfying P
`s->reject(P)`	Collection of s elements not satisfying P
`s->reject( x | P )`	Collection of s elements x not satisfying P
`s->collect(e)`	Collection of elements x.e for x : s
`s->collect( x | e )`	Collection of elements x.e for x : s
`s->selectMaximals(e)`	Collection of s elements x which have maximal x.e values
`s->selectMinimals(e)`	Collection of s elements x which have minimal x.e values

As with quantifiers, infinite collections cannot be referred to in OCL collection expressions: $String \rightarrow collect(e)$ and other such expressions are not valid.

4.1.6 Navigation expressions

Navigation expressions are applications of features to objects or to object collections, for example $d.courses.lecturer$ in the department data example, where d : *Department*. They can be read backwards, e.g., as "the lecturers of the courses of d". They will denote single objects or values, or collections of values/objects, according to the rules of Table 4.8.

Table 4.8: Navigation expressions

e	feature f	$e.f$
single object	attribute	single value
	1-multiplicity role	single object
	set-valued role	set of objects
	sequence-valued role	sequence of objects
set of objects	attribute	set of values
	1-multiplicity role	set of objects
	set-valued role	set of objects
	sequence-valued role	set of objects
sequence of objects	attribute	sequence of values
	1-multiplicity role	sequence of objects
	set-valued role	set of objects
	sequence-valued role	sequence of objects

Note that long chains of navigation through a model should be avoided in specifications, as these are both difficult to understand, and expensive to compute.

4.1.7 Object deletion

An object or set of objects x of class E can be deleted by using an expression $x \rightarrow isDeleted()$ as an update. This deletion means that the objects are no longer valid elements of E, and leads to several further possible effects:

1. x is removed from any association which has E as an end class.
2. x is removed from any superclass of E.
3. If an association $r : F \rightarrow E$ has multiplicity 1 at the E end, then any F object linked to x by r is also deleted (since each existing F object must have exactly one valid associated E object).
4. If an association $r : E \rightarrow F$ or $r : E \rightarrow Collection(F)$ is an aggregation (composition association) at the E end, then any F objects linked to x by r will also be deleted: this is the 'deletion propagation' semantics of aggregation.

The generated Java carries out this succession of deletion actions via operations *killE*, *killF*, etc., of the *Controller* class. If cascaded deletions in cases 3 and 4 are not intended, then the F objects should be unlinked from x before the deletion of x. If an association end is $\{addOnly\}$, then objects of the class at this end can only be deleted if they do not occur in any role set of this association.

4.1.8 Additional cases of quantification operators

Version 1.4 of UML-RSDS introduced *forAll*, *exists*, *exists*1 quantification over *Integer.subrange*(st, en) sequences, e.g.:

$$result = Integer.subrange(1, 10) \rightarrow exists(x \mid x * x > 50)$$

and $Set(E)$ for entities E, for both query and update uses, e.g.:

$$(s \rightarrow subcollections()) \rightarrow forAll(x \mid x \rightarrow display())$$

exists quantification is supported for query uses.

Note that *select*, *collect* and *reject* cannot be used with collections of numbers or strings, or collections of collections, but only with collections of class instances. Ordered *forAll* quantification over entities can be specified when a constraint is created (the last field in the dialog of Fig. 4.1), and indicates that the executable implementation of the quantification should iterate over the collection elements in ascending order of the specified expression on these elements (this facility is for type 1, 2 and 3 postcondition constraints for general use cases only).

4.2 Type checking

For each expression in a specification, its type and element type (for a collection-valued expression) are determined, according to Tables 4.10, 4.11 and 4.12. The most general numeric type of two numeric values x and y is given by Table 4.9. The least common supertype (LCS) of the types of $x1$, $x2$ is the most specific type which includes both types.

Table 4.9: Numeric type generalisation

x type	*y type*	*General numeric type*
double	double	double
double	int or long	double
int or long	double	double
long	long	long
long	int	long
int	long	long
int	int	int

The OCL standard does not define - or ∩ when the first argument is a sequence, or ∪ for the union of a sequence and a set. For convenience we define these as:

$$\begin{aligned} sq - col &= sq{\rightarrow}reject(x \mid col{\rightarrow}includes(x)) \\ sq \cap col &= sq{\rightarrow}select(x \mid col{\rightarrow}includes(x)) \\ sq \cup st &= sq{\rightarrow}asSet() \cup st \end{aligned}$$

A subtle aspect of the semantics of *collect* is that it always returns a Sequence, even when applied on sets: the result of $st{\rightarrow}collect(e)$ can contain duplicate values for sets st, and the cardinality of the result always equals that of st. This is in accordance with the OCL behaviour of *collectNested*, which should return a Bag of the evaluation results.

Table 4.10: Numeric operator types

Expression	*Type*	*Element type*
x + y	String (if one of x, y is a string), or general numeric type of x, y	–
x - y	General numeric type of x, y	–
x * y	General numeric type of x, y	–
x / y	General numeric type of x, y	–
x mod y	int	–
x.sqr	double	–
x.sqrt	double	–
x.floor	int	–
x.round	int	–
x.ceil	int	–
x.abs	double	–
x.exp	double	–
x.log	double	–
x.pow(y)	double	–
x.sin	double	–
x.cos	double	–
x.tan	double	–
Integer.subrange(st,en)	Sequence	int

4.3 Differences to OCL

UML-RSDS supports most features of the OCL standard library [2]. Omitted are:

- OclAny, OclVoid, OclInvalid: omitted due to absence of suitable semantics. The implementation of OCL in UML-RSDS (and for B, SMV, Z3 and code translations) has a two-valued logic. Association ends of multiplicity 0..1 are treated as sets (or sequences, if ordered) of maximum size 1. There are no explicit null or invalid elements. Null elements cannot be defined, and are not permitted to be members of association ends or other collections.
- OclMessage: replaced by invocation instance expressions on sequence diagrams.
- UnlimitedNatural: not needed for internal use within models.
- Integer, Real: replaced by int, long, double. The reason for this is to ensure that the specification semantics of these types corresponds to their semantics in the generated executable (in Java and C#) and in B.

Table 4.11: Collection operators types

Expression	*Type*	*Element type*
Set{}	Set	–
Sequence{}	Sequence	–
$Set\{x1, x2, ..., xn\}$	Set	LCS of types of $x1$ to xn
$Sequence\{x1, x2, ..., xn\}$	Sequence	LCS of types of $x1$ to xn
`s->including(x)`	Type of s	LCS of type of x and element type of s
`s->excluding(x)`	Type of s	Element type of s
`s - t`	Type of s	Element type of s
`s->prepend(x)`	Sequence	LCS of type of x and element type of s
`s->append(x)`	Sequence	LCS of type of x and element type of s
`s->count(x)`	Integer (int)	–
`s->indexOf(x)`	Integer (int)	–
$x\backslash/y$	Sequence if x, y ordered, otherwise Set	LCS of element types of x, y
$x/\backslash y$	Type of x	Element type of x
$x \frown y$	Sequence	LCS of element types of x and y
`x->union(y)`	Same as $x\backslash/y$	Same as $x\backslash/y$
`x->intersection(y)`	Same as $x/\backslash y$	Same as $x/\backslash y$
`x->symmetricDifference(y)`	Set	LCS of element types of x, y
`x->any()`	Element type of x	–
`x->subcollections()`	Set	Type of x
`x->reverse()`	Sequence	Element type of x
`x->front()`	Sequence	Element type of x
`x->tail()`	Sequence	Element type of x
`x->first()`	Element type of x	–
`x->last()`	Element type of x	–
`x->sort()`	Sequence	Element type of x
`x->sortedBy(e)`	Sequence	Element type of x
`x->sum()`	String if any element of x is a String, otherwise LCS of x element types	–
`x->prd()`	General numeric type of elements of x	–
`Integer.Sum(a,b,x,e)`	Type of e	–
`Integer.Prd(a,b,x,e)`	Type of e	–
`x->max()`	Element type of x	–
`x->min()`	Element type of x	–
`x->asSet()`	Set	Element type of x
`x->asSequence()`	Sequence	Element type of x
`s->isUnique(e)`	Boolean	–
`x->isDeleted()`	Boolean	–

Table 4.12: Selection/collection operators types

Expression	*Type*	*Element type*
`s->select(P)`	Type of s	Element type of s
`s->select( x | P )`	Type of s	Element type of s
`s->reject(P)`	Type of s	Element type of s
`s->reject( x | P )`	Type of s	Element type of s
`s->collect(e)`	Sequence	Type of e
`s->collect( x | e )`	Sequence	Type of e
`s->selectMaximals(e)`	Type of s	Element type of s
`s->selectMinimals(e)`	Type of s	Element type of s
`s->unionAll(e)`	Sequence if s type and e type Sequence, otherwise Set	Element type of e
`s->intersectAll(e)`	Set	Element type of e

- max, min between two values: expressed instead by generalised max, min operations on collections. These apply also to collections of strings.
- oclIsTypeOf(T), flatten(), and equalsIgnoreCase() (to be included in future tool versions).
- The xor logical operator: this can be expressed using the other logical operators. The operators and, implies are written as &, =>.
- Set, sequence and string subtraction uses the - operator, extending OCL in the case of sequences and strings.
- OrderedSet and Bag are omitted: both can be represented by sequences. In the future, sorted sets will be introduced (the {*sorted*} constraint on association ends).
- Tuples and the cartesian product of two collections: these are very rarely used in modelling. The same effect can be achieved by introducing a new class with many-one associations directed to the tuple component entities, and similarly for ternary or higher-arity associations. Sequences can be used to represent tuple values.
- *collectNested* in OCL is the same as *collect* in UML-RSDS, i.e., no flattening takes place.
- The *one* operator of OCL is expressed by $exists1$.

The OCL *substring* and *subsequence* operators are combined into a single *subrange* operation in UML-RSDS.

Sets can be written with the notation $\{x1, ..., xn\}$ instead of $Set\{x1, ..., xn\}$. Intersection and union can use the mathematical operators ∩ and ∪ written as /\ and \/. Likewise, subtraction on collections is written using -.

Additional to standard OCL are the operators $s{\rightarrow}isDeleted()$, $s{\rightarrow}unionAll(e)$, $s{\rightarrow}intersectAll(e)$, $s{\rightarrow}selectMinimals(e)$, $s{\rightarrow}selectMaximals(e)$ and $s{\rightarrow}subcollections()$, $Integer.Sum(a, b, i, e)$, $Integer.Prd(a, b, i, e)$, and the $Integer.subrange(i, j)$ type constructor.

Qualified or composition associations and association classes are not yet fully supported: Java, C#, C++ and B can be defined for these associations, but their values cannot be stored in or retrieved from instance models. Only binary associations are supported.

Summary

This chapter has introduced the OCL notation supported by UML-RSDS, and given examples of specifications using OCL. The rules for expression types and constraint semantics have been defined.

References

[1] K. Lano, *UML-RSDS manual*, http://www.dcs.kcl.ac.uk/staff/kcl/umlrsds.pdf, 2015.

[2] OMG, *Object Constraint Language 2.4 Specification*, www.omg.org/spec/OCL/2.4, 2014.

Chapter 5

Use Cases

Use cases define the services which a system or component provides to its clients or users. In UML-RSDS use cases are precisely specified using OCL, and their effects are defined in terms of the modifications they make to the class diagram data (and possibly calls to external applications). Use cases are shown on the system class diagram together with the classes they relate to.

5.1 Specification of use cases

The use cases of a system represent the externally available services that it provides to clients, either directly to users, or to other parts of a larger system. Two forms of use case can be defined in UML-RSDS:

- EIS use cases – simple operations such as creating an instance of an entity, modifying an instance, etc., intended to be implemented using an EIS platform such as Java Enterprise Edition. These are described in Chapter 20.
- General use cases – units of functionality of arbitrary complexity, specified by constraints (pre and postconditions of the use case). These are particularly used to specify model transformations (Chapter 7). They are visually represented as ovals in the class diagram. They may have their own attributes and operations, since use cases are classifiers in UML. Use case invariants can also be defined for general use cases.

General use cases can be structured and composed using several different mechanisms:

- *extend*: one use case can be declared as an extension of another, to provide additional functionality to the base use case.
- *include*: one use case can be included in one or several other use cases, to define a sub-operation or process of the base use cases. This can be used to factor out common sub-processes.
- *Parameterisation*: a use case can be given typed parameters to make it generic and reusable.
- *Inheritance*: a use case can inherit another, to specialise its effect. Inheritance of use cases is not currently supported by UML-RSDS as we have not found it a useful facility in practice.

The elementary example of a "Hello world" program can be defined as a single postcondition

```
"Hello world!"->display()
```

of a use case on an empty metamodel.

A simple example of a use case for our teaching system (Fig. 3.1) could have the single postcondition (with no entity context):

```
Lecturer->exists( k | k.name = "Mike" )
```

which creates a lecturer object with name "Mike".

The option *Use Case Dependencies* on the *View* menu allows the dependencies between use cases and classes to be seen: a red dashed line is drawn from a use case to each class whose data it may modify, and a green dashed line is drawn from each class read by the use case, to the use case.

5.2 General use cases

A general use case can be introduced into a UML-RSDS system to represent any service provided by that system, or a subpart of such a service. An example is the correlation calculator of Chapter 1.

Figure 5.1 shows the UML-RSDS dialog for creating use cases. For general use cases only the name needs to be entered, along with any parameters.

Figure 5.2 shows the UML-RSDS dialog for modifying use cases.

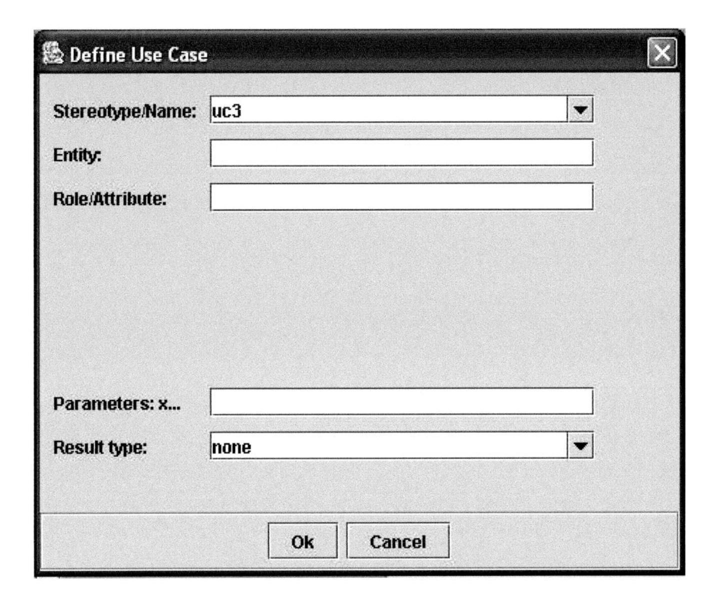

Figure 5.1: Use case dialog

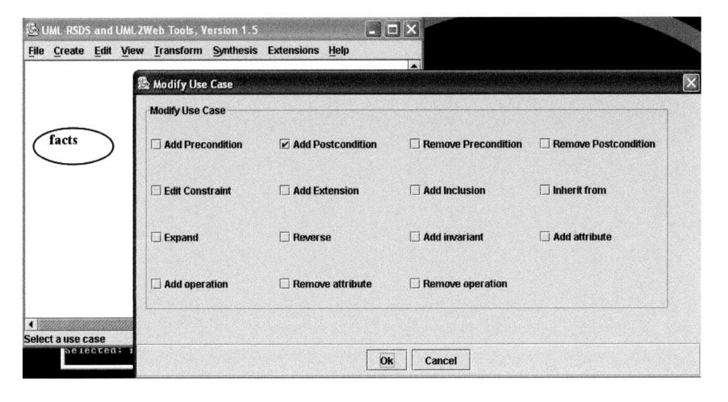

Figure 5.2: Use case edit dialog

A use case has:

- A name: this should be unique among the use cases of the system, and distinct from any class name. It should usually have a lower case initial letter. There is an implicit class associated with any use case: this class has the same name as the use case, but with the first letter capitalised, so this version of the name should also not conflict with any existing class name.
- Parameters, specified by their name and type. These are restricted to numeric, boolean or string types.

- Assumptions (preconditions): constraints Asm defining when the use case may be validly invoked. The default is *true*.
- Postconditions: a sequence of constraints $Post$ defining the effect of the use case. The sequential order of the constraints is used as the order of the corresponding code segments.

Use cases may also have:

- *Invariants*: constraints expressing which properties are preserved during the use case.
- *Activity*: an activity in pseudocode defining the behaviour of the use case, for example, in terms of included use cases. The default, if no explicit activity is defined, is the sequential composition of the included use cases, following the code $stat(Post)$ of the use case itself.
- *Operations and attributes*: these are static features owned by the use case and only available within its postcondition constraints.

5.2.1 Use case post-conditions

Use case post-conditions must be written in particular orders, to ensure that data which is needed by one constraint is available (e.g., because it has been produced by an earlier constraint) at the point where it is used. The correlation calculator example (Section 2.1) illustrates this principle.

For each predicate P we define the *write frame* $wr(P)$ of P, and the *read frame* $rd(P)$ (Table 5.1). These are the sets of entities and features which the procedural interpretation $stat(P)$ of P may update or access, respectively. The *write frame* $wr(P)$ of a predicate is the set of features and entities that it modifies, when interpreted as an activity (an activity $stat(P)$ to establish P). This includes object creation. The *read frame* $rd(P)$ is the set of entities and features read in P. The read and write frames can help to distinguish different implementation strategies for constraints (Chapter 6). $var(P)$ is the set of all features and entity type names used in P.

Table 5.1: Definition of read and write frames

P	$rd(P)$	$wr(P)$
Basic expression e without quantifiers, logical operators or $=, :, E[]$, $\rightarrow includes$, $\rightarrow includesAll$, $\rightarrow excludesAll$, $\rightarrow excludes$, $\rightarrow isDeleted$	Set of features and entity type names used in P: $var(P)$	$\{\}$
$e1 : e2.r$ $e2.r \rightarrow includes(e1)$ r many-valued $e1$, $e2$ single-valued	$var(e1) \cup var(e2)$	$\{r\}$
$e2.r \rightarrow excludes(e1)$ $e1 \ /: \ e2.r$ r many-valued $e1$, $e2$ single-valued	$var(e1) \cup var(e2)$	$\{r\}$
$e1.f = e2$ $e1$ single-valued	$var(e1) \cup var(e2)$	$\{f\}$
$e1.f[i] = e2$ $e1$ single-valued, f sequence-valued	$var(e1) \cup var(e2)$ $\cup \ var(i)$	$\{f\}$
$e1.f \rightarrow last() = e2$ $e1.f \rightarrow first() = e2$ $e1$ single-valued f sequence-valued	$var(e1) \cup var(e2)$	$\{f\}$
$e2.r \rightarrow includesAll(e1)$ $e1 \ <: \ e2.r$ r, $e1$ many-valued $e2$ single-valued	$var(e1) \cup var(e2)$	$\{r\}$
$e2.r \rightarrow excludesAll(e1)$ r, $e1$ many-valued $e2$ single-valued	$var(e1) \cup var(e2)$	$\{r\}$
$E[e1]$	$var(e1) \cup \{E\}$	$\{\}$
$E \rightarrow exists(x \mid Q)$ (E concrete entity type)	$rd(Q)$	$wr(Q) \cup \{E\} \cup$ all superclasses of E
$E \rightarrow exists1(x \mid Q)$ (E concrete entity type)	$rd(Q)$	$wr(Q) \cup \{E\} \cup$ all superclasses of E
$E \rightarrow forAll(x \mid Q)$	$rd(Q) \cup \{E\}$	$wr(Q)$
$x \rightarrow isDeleted()$ x of element type entity type E	$var(x)$	$\{E\} \cup$ all superclasses of $E \cup$ E-valued roles r
$C \ \Rightarrow \ Q$	$var(C) \cup rd(Q)$	$wr(Q)$
$Q \ \& \ R$	$rd(Q) \cup rd(R)$	$wr(Q) \cup wr(R)$

$e2.r \rightarrow excludes(e1)$ for single-valued r is treated as $e2 \rightarrow excludesAll(e2 \rightarrow select(r = e1))$, and $e2.r \rightarrow excludesAll(e1)$ for single-valued r is treated as $e2 \rightarrow excludesAll(e2 \rightarrow select(r : e1))$. $e1.f \rightarrow first() = e2$ has the same read and write frames as $e1.f \rightarrow last() = e2$ (see Section 14.5). In computing $wr(P)$ we also take account of the features and entity types which depend upon the explicitly updated features and entity types of Cn, such as inverse association ends. If there is an invariant constraint φ of the class diagram which implicitly defines a feature g in terms of feature f, i.e.: $f \in rd(\varphi)$ and $g \in wr(\varphi)$, then g depends on f. In particular, if an association end *role*2 has a named opposite end *role*1, then *role*1 depends on *role*2 and vice-versa. Creating an instance x of a concrete entity type E also adds x to each supertype F of E, and so these supertypes are also included in the write frames of $E \rightarrow exists(x \mid Q)$ and $E \rightarrow exists1(x \mid Q)$ in Table 5.1. Deleting an instance x of entity type E by $x \rightarrow isDeleted()$ may affect any supertype of E and any association end owned by E or its supertypes, and any association end incident with E or incident with any supertype of E. Additionally, if entity types E and F are related by an association which is a composition at the E end, or by an association with a mandatory multiplicity at the E end, i.e., a multiplicity with lower bound 1 or more, then deletion of E instances will affect F and its features and supertypes and incident associations, recursively. The read frame of an operation invocation $e.op(pars)$ is the read frame of e and of the *pars* corresponding to the input parameters of *op* together with the read frame of the postcondition $Post_{op}$ of *op*, excluding the formal parameters v of *op*. Its write frame is that of the actual parameters corresponding to the outputs of *op*, and $wr(Post_{op}) - v$. $wr(G)$ of a set G of constraints is the union of the constraint write frames, likewise for $rd(G)$.

A dependency ordering $Cn < Cm$ is defined between constraints by

$$wr(Cn) \cap rd(Cm) \neq \{\}$$

"Cm depends on Cn". A use case with postconditions $C_1, \ldots, C_n$ should satisfy the *syntactic non-interference* conditions:

1. If $C_i < C_j$, with $i \neq j$, then $i < j$.
2. If $i \neq j$ then $wr(C_i) \cap wr(C_j) = \{\}$.

Together, these conditions ensure that the activities $stat(C_j)$ of subsequent constraints C_j cannot invalidate earlier constraints C_i, for $i < j$.

A use case satisfies *semantic non-interference* if for $i < j$:

$$C_i \Rightarrow [stat(C_j)]C_i$$

where $[act]P$ is the weakest-precondition of P with respect to act. Syntactic non-interference implies semantic non-interference, but not conversely. Under either condition we can deduce that the use case achieves the conjunction of its postconditions [1].

The UML-RSDS tools give warnings if the syntactic non-interference conditions fail for two constraints. If both $C_i < C_j$ and $C_j < C_i$ hold, with $i \neq j$, then these constraints are placed in a constraint group $\{C_i, C_j\}$, and implemented as a unit (Chapter 6).

A typical form of postcondition for a refinement or migration transformation is "for all instances of source entity type ST that satisfy condition $Ante$, create an instance of target entity type TT satisfying P": ($R1$)

$$ST :: \quad Ante \;\Rightarrow\; TT{\rightarrow}exists(t \mid P)$$

P typically defines the feature values of the new instance t in terms of the feature values of $self : ST$. The implementation of this constraint is a bounded loop which iterates over all instances $self$ of the context entity ST, checks $Ante$ for $self$, and if this holds, applies the succedent to create a new t corresponding to the source instance. Subsequent constraints in the use case postconditions can then read instances of TT to assist in the construction of other target entity type instances. But later constraints should not write to ST or to its features (in a refinement or migration, the source model is normally read-only in any case).

It would break syntactic non-interference to have a second constraint which also creates and sets TT instances. However this may still be correct with regard to semantic non-interference. For example, if a constraint ($R2$)

$$ST :: \quad not(Ante) \;\Rightarrow\; TT{\rightarrow}exists(t \mid P1)$$

is included together with $R1$, then these do not interfere semantically as only one of $R1$ and $R2$ is executed for each instance of ST, and disjoint collections of TT instances are created for $ST{\rightarrow}select(Ante)$ and for $ST{\rightarrow}select(not(Ante))$.

More complex postconditions may be used for refactorings or other categories of transformation. In such cases the constraints express requirements of the form "for all instances of source entity type ST that satisfy condition $Ante$, update or restructure the model to achieve the condition $Succ$":

$$ST :: \quad Ante \;\Rightarrow\; Succ$$

For a refactoring, $Succ$ may contradict $Ante$, because $Ante$ will express some situation or structure in the model that should be removed by applying $Succ$, as in the example of Section 2.3.

Such constraints will typically be implemented by an unbounded loop

```
while some s : ST satisfies s.Ante
do
  (select such an s;
   apply stat(s.Succ)
  )
```

5.3 Composition and features of use cases

Basic use cases (use cases without activities or included use cases) can be composed together using *extend*. This has the effect of conjoining the postconditions of the extension and extended use cases to form the postconditions of the composed use case, and likewise for preconditions and invariants. This operator acts like a conjunction combinator of use cases (assumed to be mutually consistent). A new use case with the conjoined postconditions is added to the system model. More complex use cases/services can be composed from these basic use cases using *include* and use case activities. If use cases uc1, ..., ucn are included in use case uc, in that order, then the composed effect of uc includes executing uc1, ..., ucn after uc's own effect. An explicit activity for *uc* can alternatively be defined to invoke the included use cases according to an algorithm.

There are the following restrictions on extend/include:

- A use case cannot be an extension of two or more other use cases, nor both an extension and an inclusion.
- Cycles in the include/extend relationships (or cycles involving combinations of include/extend) are not permitted.
- The parameters of an included or extension use case must be a subset of those of the including/extended use case.

An elementary example of a use case is a transformation which calculates the square of some data values (Fig. 5.3). For each instance $a : A$, the use case $a2b$ creates a new $b : B$ instance with $b.y$ equal to $(a.x)^2$. The transformation has postcondition:

$$B \rightarrow exists(b \mid b.y = x * x)$$

on context class A.

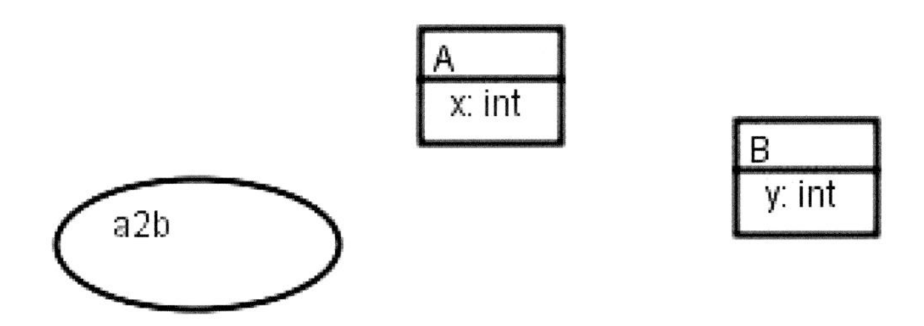

Figure 5.3: A to B Transformation τ_{a2b}

Another use case, $a2c$, could instead compute the cube of the $a.x$:

$$C{\to}exists(c \mid c.z = x * x * x)$$

on context class A, where C is an additional class, with an attribute $z : int$.

5.3.1 Extend-composition

These two use cases could be added as extensions $\ll$ *extend* $\gg$ of a base use case uc which has no postconditions of its own. Effectively, uc has postconditions

$$A :: \ B{\to}exists(b \mid b.y = x * x)$$
$$A :: \ C{\to}exists(c \mid c.z = x * x * x)$$

or (equivalently)

$$A :: \ C{\to}exists(c \mid c.z = x * x * x)$$
$$A :: \ B{\to}exists(b \mid b.y = x * x)$$

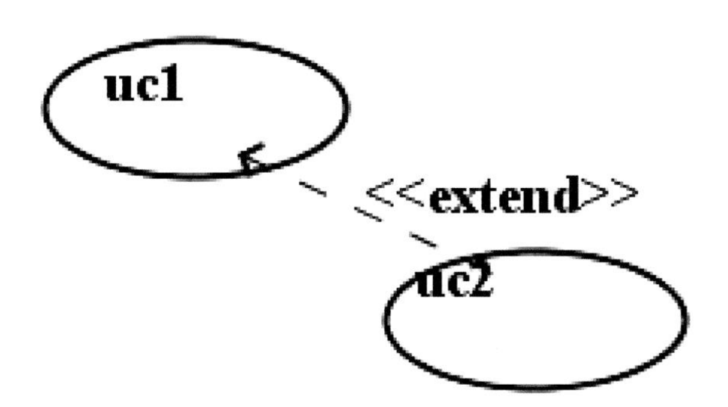

Figure 5.4: Use case extension of $uc1$ by $uc2$

The *extend*-composition of two use cases $uc1$, $uc2$ (Fig. 5.4) is a new use case whose *Post* is a sequence of constraints such that:

- *Post* consists of the constraints of $Post_{uc1}$ and $Post_{uc2}$
- The constraints of $uc1$ appear in the same order in *Post* as in $Post_{uc1}$, likewise for $uc2$.
- If $uc1$ and $uc2$ satisfy syntactic non-interference, and $wr(uc1)$ is disjoint from $wr(uc2)$, and either $wr(uc1)$ and $rd(uc2)$ are disjoint, or $wr(uc2)$ and $rd(uc1)$ are disjoint, then *Post* also satisfies syntactic non-interference.

The assumptions of the composite transformation are the union of the assumptions of its components, likewise for the invariants.

5.3.2 Include-composition

An *extend*-composition is semantically a conjunction or parallel composition of the main use case and its extensions. In contrast, an *include*-composition is a sequential composition of its main use case, followed by the included use cases, in order. For example, a strategy for mapping a class diagram to a relational database schema is to perform the following transformations:

- Replace inheritance by many-one associations
- Add primary key attributes to all classes
- Replace many-many associations by a new class and two many-one associations
- Replace many-one associations by foreign keys.

These could be expressed by separate transformations $uc1$ to $uc4$. The appropriate order for their use is as given above, and they could be defined as $\ll$ *include* $\gg$ use cases of a use case uc in this order, to define an overall transformation that applies the included transformations in this order.

5.3.3 Use case inversion

Given a use case uc whose postconditions are all of type 0 or type 1, an inverse use case $uc^{\sim}$ can be derived. This use case has inverse postcondition constraints $Cn^{\sim}$ for each postcondition constraint Cn of uc, listed in the same order as in uc (Section 7.4). The aim of this use case is to reverse the effect of uc, and permit the recovery of an input model of uc from an output model of its processing. In addition, $Cn^{\sim}$ is a candidate transformation invariant which should be preserved by the transformation steps of $stat(Cn)$. The inverse of a use case can be generated using the option *reverse* on the use case edit dialog.

The inverse of the simple A-to-B example given above is a constraint with the postcondition

$$A \rightarrow exists(a \mid a.x = y \rightarrow sqrt())$$

on context class B.

Constraints with multiple quantifiers, and with additional quantified variables and let variables can be inverted, however all *exists* quantifiers in the succedent should be grouped together. For example:

$$\begin{array}{l} A :: \\ \quad a1 : A \ \& \ a1.x = 2 * x \quad \Rightarrow \\ \qquad C \rightarrow exists(c \mid D \rightarrow exists(d \mid \\ \qquad\qquad c.cId = aId \ \& \ d.dId = a1.aId \ \& \\ \qquad\qquad c.z = x \ \& \ d.w = a1.x \ \& \ d : c.dr)) \end{array}$$

can be inverted to:

$$\begin{array}{l} C :: \\ \quad d : D \ \& \ d : dr \quad \Rightarrow \\ \qquad A \rightarrow exists(ax \mid A \rightarrow exists(a1 \mid \\ \qquad\qquad ax.aId = cId \ \& \ a1.aId = d.dId \ \& \ ax.x = \\ \qquad\qquad z \ \& \ a1.x = d.w \ \& \ a1.x = 2 * ax.x)) \end{array}$$

Transformation and use case inversion is discussed in more detail in Chapter 14.

5.3.4 Use case parameters

Use cases may have parameters of numeric, boolean or string type. The values of the parameters are supplied to the use case when it is executed. The parameters may be used in the assumption and postcondition constraints of the use case. For example, a use case *calculateSqRoot* could have an integer parameter $x : int$, assumption

$$x \geq 0$$

and a postcondition

$$x \rightarrow sqrt() \rightarrow display()$$

5.3.5 Generic use cases

Parameterisation may also be used to substitute part of the specification text of a use case with variant texts. A generic use case can be created from the *Create* menu, and should have a boolean parameter. The name of this parameter can then be used within the constraints of the use case as a subformula. For example, the $a2b$ use case could be made into a generic use case with parameter p. The use case could have postcondition:

$$B \rightarrow exists(b \mid b.y = x * x \ \& \ p)$$

on context class A.

When the use case is instantiated, an actual predicate must be provided as the value for p: this could be some tracing or logging of the use case actions, such as

$$(\text{“}Mapped\text{”} + x + \text{“}to\text{”} + b.y) \rightarrow display()$$

5.3.6 Use case attributes

According to the UML standard, use cases may own attributes, and UML-RSDS also permits this. Such attributes can be used to store the result of a computation performed in one constraint, for use in subsequent constraints. An attribute can be added to a use case using the modify use case dialog. It will be a static attribute. A use case attribute att can be referred to via the notation $Uc.att$ in the post conditions of its use case uc (att actually belongs to the class Uc associated with uc).

5.3.7 Use case operations

Use cases can own operations, to perform some calculation which is used internally in the use case. Such operations can help to avoid duplicated complex expressions in different constraints: an operation can be introduced which evaluates the expression, and this operation is then called in place of the expression within the constraints. The operation will be a static operation, and can be referred to in the postconditions of its use case uc by the notation $Uc.op(params)$.

The following use case is an example of parameterisation and use case operations: display factorial(i) for all i from 1 up to a parameter p. This could be specified as a parameterised UML-RSDS use case $fact(p : int)$ with one postcondition

$$Integer.subrange(1, p) \rightarrow forAll(i \mid Fact.factorial(i) \rightarrow display())$$

where *factorial* is a use case operation

static query $factorial(i \; : \; int) \; : \; int$
pre: $i \; \geq \; 1$
post:
$(i \; \leq \; 2 \quad \Rightarrow \quad result \; = \; i) \; \&$
$(i \; > \; 2 \quad \Rightarrow \quad result \; = \; i * factorial(i - 1))$

and this is defined as a $\ll$ *cached* $\gg$ operation, to avoid repeated evaluation of the factorial for the same argument value.

Summary

In this chapter we have described how system functionality can be specified in UML-RSDS using use cases. The elements of use cases, and the composition operators for use cases, have been explained.

References

[1] K. Lano, S. Kolahdouz-Rahimi and T. Clark, *A framework for model transformation verification*, BCS FACS Journal, 2014.

Chapter 6

Design Synthesis

This chapter examines in detail the design and code synthesis process in UML-RSDS. This process is automated, using a built-in synthesis strategy, but there is scope for user configuration and selection of alternative options in different cases. The process takes as input a logical system specification defined by a class diagram and use cases, and produces a language-independent design in a pseudocode activity language. In turn, this design can be directly mapped into implementations in Java, C#, C++, and potentially many other languages.

6.1 Synthesis of class diagram designs

The structure of the specification class diagram of a UML-RSDS system is directly followed in the design and implementation of the system: for each specification class E there is an implementation class E, with corresponding features: for each specification-level owned attribute $att : T$, the implementation class has an owned attribute att of type T', the implementation type corresponding to T. For each 1-valued role r owned by E, at the F end of a E to F association, there is a F-valued owned attribute of E's implementation. Table 6.1 summarises the specification to implementation mapping for class diagrams. The close structural correspondence between the UML-RSDS specification and the implementation code facilitates debugging of the specification: errors that arise in the code can usually be traced back to an error in the corresponding part of the specification. Of course, such code errors should not be corrected at the code level, but at the specification level.

Table 6.1: Mappings from UML/OCL to programming languages

UML/OCL	Java 4	Java 6	C#	C++
Class E	Class E	Class E	Class E	Class E
Interface I	Interface I + inner class $IOps$	Interface I + inner class $IOps$	Interface I + class $IOps$	Class I
Attribute att : T	instance variable T' att	instance variable T' att	instance variable T' att	instance variable T' att
1-valued role r to class F	instance variable F r	instance variable F r	instance variable F r	instance variable F* r
many-valued role r to class F (unordered)	instance variable List r	instance variable HashSet r	instance variable ArrayList r	instance variable set<F*>* r
many-valued role r to class F (ordered)	instance variable List r	instance variable ArrayList r	instance variable ArrayList r	instance variable vector<F*>* r
instance scope operation op(p : T) : RT of class E	operation public RT' op(T' p) of class E	operation public RT' op(T' p) of class E	operation public RT' op(T' p) of class E	operation public: RT' op(T' p) of class E
Inheritance of class F by class E	class E extends F	class E extends F	class E : F	class E : public F

6.2 Synthesis of constraint code

Constraints (boolean-valued expressions/logical predicates) are used in UML-RSDS both to express a required property which an operation or use case should establish, and to guide the generation of the executable code which *ensures* that the property is established. Thus an equation

$$v = e$$

where v is a variable or other assignable expression, is interpreted as an assignment $v\ :=\ e$ at the design level, and as the corresponding programming language assignment statement at the implementation level. An expression $x : v.r$ is interpreted as a design $v.r\ :=\ v.r{\rightarrow}including(x)$ which adds x to $v.r$, thereby establishing the expression as true. The same principle applies to a wide range of specification constraints. In particular, an existential quantifier $E{\rightarrow}exists(x \mid P)$ for concrete class E corresponds to the creation of an instance x of E.

The basis for the synthesis of operational behaviour from a UML-RSDS specification is the operational interpretation $stat(P)$ of OCL expressions, shown in Table 6.2. This interpretation is an activity which is intended to establish P, i.e., $[stat(P)]P$ holds true, where $[\ ]$ is the weakest-precondition operator.

Updates to association ends may require additional further updates to inverses of the association ends (Table 3.2), updates to entity type extents or to features may require further updates to derived and other data-dependent features, and so forth. These updates are all included in the *stat* activity. In particular, for $x{\rightarrow}isDeleted()$, x is removed from every association end in which it resides, and further cascaded deletions may occur if these ends are mandatory/composition ends.

The clauses for $X{\rightarrow}exists(x \mid x.id = v\ \&\ P1)$ test for existence of an x with $x.id = v$ before creating such an object: this has implications for efficiency but is necessary for correctness: two distinct X elements with the same primary key value should not exist. This design strategy

Table 6.2: Definition of $stat(P)$

P	$stat(P)$	*Condition*
$x = e$	$x := e$	x is assignable, $x \notin var(e)$
$e : x$ $x \rightarrow includes(e)$	$x := x \rightarrow including(e)$	x is assignable, collection-valued, $x \notin var(e)$
$e \; / : x$ $x \rightarrow excludes(e)$	$x := x \rightarrow excluding(e)$	x is assignable, collection-valued, $x \notin var(e)$
$e <: x$ $x \rightarrow includesAll(e)$	$x := x \rightarrow union(e)$	x is assignable, collection-valued, $x \notin var(e)$
$e \; / <: x$ $x \rightarrow excludesAll(e)$	$x := x - e$	x is assignable, collection-valued, $x \notin var(e)$
$x \rightarrow isDeleted()$ (single object x)	; -composition of $E := E \rightarrow excluding(x)$ $y.r := y.r \rightarrow excluding(x)$	Each entity type E containing x each association end $y.r$ containing x
$obj.op(e)$ $objs.op(e)$	$obj.op(e)$ for $x : objs$ do $x.op(e)$	Single object obj Collection $objs$
$P1 \; \& \; P2$	$stat(P1); \; stat(P2)$	$wr(P2) \cap wr(P1) = \{\}$ $wr(P2) \cap rd(P1) = \{\}$
$E \rightarrow exists(x \mid x.id = v$ $\& \; P1)$ $E \rightarrow exists(x \mid P1)$	if $E.id \rightarrow includes(v)$ then $x := E[v]; \; stat(P1)$ else $(x : E;$ $stat(x.id = v \; \& \; P1))$ $(x : E; \; stat(P1))$	E is a concrete entity type with $E \rightarrow isUnique(id)$ E is a concrete entity type, $P1$ not of form $x.id = v \; \& \; P2$ for unique id attribute of E

$e \rightarrow exists(x \mid x.id = v$
$\& \; P1)$
 	
$e \rightarrow exists(x \mid P1)$
	if $e \rightarrow includes(E[v])$ then
	$(x := E[v]; \; stat(P1))$
	else skip
	if $e \rightarrow notEmpty()$ then
	$(x := e \rightarrow any(true);$
	$stat(P1))$ else skip
		Non-writable expression
		e with element entity type E,
		$E \rightarrow isUnique(id)$
		Non-writable expression
		e, $P1$ not of
		above form

$E \rightarrow exists1(x \mid P1)$	if $E \rightarrow exists(x \mid P1)$ then skip else $stat(E \rightarrow exists(x \mid P1))$	E is an entity type or non-writable expression
$E \rightarrow forAll(x \mid P1)$	for $x : E$ do $stat(P1)$	
$P1 \; \Rightarrow \; P2$	if $P1$ then $stat(P2)$ else skip	$P1$ side-effect free $wr(P2) \cap var(P1) = \{\}$

is a case of the well-known principle of 'check before enforce' used in QVT, ETL, ATL and other transformation languages. The write frame of $stat(P)$ is equal to $wr(P)$, the read frame includes $rd(P)$. As an example of these definitions, $stat(R)$ for the postcondition R of the transformation of Fig. 5.3 is:

```
for a : A
do
  (b : B;
   b.y := a.x*a.x;
  )
```

It is invalid to write the same data in both arguments of a conjunction: $x = 5 \; \& \; x = 7$ has no defined procedural form, nor should later conjuncts write data read in earlier conjuncts: $y = x \; \& \; x = 7$ is also invalid. It is acceptable to read data written in earlier conjuncts: $x = 5 \; \& \; y = x + 7$ has procedural interpretation $x := 5; \; y := x + 7$. It is also possible to use pre-forms of data to read their values prior to the effect of the statement being specified: $y = x@pre$ & $x = 7$ is valid and has procedural interpretation $y := x;\ x := 7$.

6.3 Synthesis of use case designs

Each use case is mapped to a global operation of the system, implemented as a public operation of the *Controller* class of the system. The code of the operation for use case *uc* is derived from the use case postconditions. Specific design and implementation strategies are used for use case postcondition constraints, as defined in Table 6.3. The postcondition constraints *Cn* are generally of the form

$$Ante \Rightarrow Succ$$

on some context entity type (classifier) S_i. *Ante* denotes the antecedent (condition) of the constraint, *Succ* the succedent (effect). We include S_i in $rd(Cn)$. The choice of design strategy depends upon the internal data dependencies of the constraint *Cn*: if the constraint modifies a disjoint collection $wr(Cn)$ of model features (entities and their features) to the collection $rd(Cn)$ of model features that it reads, then it can be implemented by a bounded loop (a for-loop in Java, C# or C++). Otherwise a fixed-point iteration is required, which applies the constraint until it is established, using a while-loop.

Table 6.3: Implementation choices for constraints *Cn*

	Data dependencies	*Implementation choice*
Type 0	No contextual classifier	$stat(Cn)$
Type 1	$wr(Cn) \cap rd(Cn) = \{\}$	for loop over S_i
Type 2	Succedent *Succ* has $wr(Succ) \cap rd(Succ) \neq \{\}$ but $wr(Succ) \cap rd(Ante) = \{\}$	while-iteration of for loop over S_i
Type 3	$wr(Succ) \cap rd(Ante) \neq \{\}$	while iteration of search-and-return loop

The higher the constraint type number, the more thorough is the implementation in attempting to establish the constraint – and the more time-consuming and complex is the implementation. Type 2 and 3 constraints are not guaranteed to terminate.

Unless fixed-point iteration is essential for a problem, constraints should be written as type 1 where possible. One technique for enforcing the use of a bounded loop is to use $f@pre$ throughout the constraint, whenever a feature is read. For example, a constraint that increments the $x : Integer$ attribute of every A object could be written (incorrectly) as

$$x = x + 1$$

on context class A. But this reads and writes $A :: x$ and so will have a fixed-point implementation that attempts to make the equality $a.x = a.x + 1$ true for every $a : A$. This would not terminate. The constraint should instead be written as:

$$x = x@pre + 1$$

on context class A. The term $x@pre$ denotes the read-only value of x at the start of execution of the constraint. The write and read frames are now disjoint and a bounded loop implementation is synthesised, which iterates once through the existing instances of A.

If $wr(C_j) \cap rd(C_i)$ and $wr(C_i) \cap wr(C_j)$ are empty, the sequence C_i, C_j of constraints can be implemented by $stat(C_i); \ stat(C_j)$. The structure of the generated design is therefore closely aligned with the specification structure, which helps in relating code generation issues back to the UML specification level.

6.3.1 Type 1 constraints

In the simple case where a constraint Cn of the form $Ante \Rightarrow Succ$ on context class S_i satisfies the type 1 condition:

$$wr(Cn) \cap rd(Cn) \ = \ \{\}$$

the constraint can be implemented by a bounded loop

```
for s : Si do s.opi()
```

over S_i. In S_i we introduce a new instance-scope operation of the form:

```
opi()
post:
  Ante  =>  Succ
```

We refer to this strategy as constraint implementation approach 1. Confluence of this implementation can be shown, if the updates of written data in different executions of the loop body are independent of the order of the executions. The time complexity of the implementation is linear in the size $\#\overline{S_i}$ of the source domain. More precisely the worst case complexity is linear in

$$\#\overline{S_i} * (cost_{eval}(Ante) + cost_{act}(Succ))$$

where $cost_{eval}(e)$ is the time required to evaluate e, and $cost_{act}(e)$ the time required to execute $stat(e)$. If additional source domains D (from quantified variables $v : D$) also need to be iterated over, the cost is also multiplied by their size. This shows that multiple element matching or complex expressions in the constraint should be avoided for efficiency reasons. If ordered iteration over S_i is needed, i.e., an iteration which selects S_i elements in ascending order of the value of an expression e with context S_i, then a sorted version of S_i can be precomputed before the above iteration:

```
sisorted := Si→sortedBy(e);
for s : sisorted do s.opi()
```

The data features occurring in e count as data read in the antecedent of Cn.

6.3.2 Type 2 constraints

A more complex implementation strategy is required for type 2 and type 3 constraints. In the case where

$$wr(Cn) \cap rd(Succ)$$

is non-empty but the other conditions of non-interference still hold (i.e., a type 2 constraint), a fixed-point iteration of the form:

```
running := true;
while (running) do
  (running := false;
   for s : Si do
     if s.Ante then
       [if s.Succ then skip
       else] (s.op(); running := true)
     else skip)
```

can be used, where $op()$ is a new operation of S_i defined as:

```
op()
pre: Ante
post:  Succ
```

In the conditional test within the *for* loop, *Succ* is evaluated in a non side-effecting manner.

We refer to this fixed-point strategy as constraint implementation approach number 2. The conditional test of *Succ* and the code between [and] can be omitted if it is known that $Ante \Rightarrow not(Succ)$. The UML-RSDS tools perform algebraic simplification to check if *Ante* contradicts

Succ. This is the Omit Negative Application Conditions design pattern (Chapter 9). The design synthesiser prompts the user to confirm if this optimisation should be used.

In order to establish termination and confluence, it is necessary to define a measure $Q : \mathbb{N}$ over the source and target models, which is a *variant* for the while loop (2):

$$\forall \nu : \mathbb{N} \cdot Q = \nu \wedge running = true \wedge \nu > 0 \;\Rightarrow\; [body](Q < \nu)$$

where *body* is the body of the while loop.

Q is also necessary to prove correctness: while there remain $s : S_i$ with *Ante* true but *Succ* false, then $Q > 0$ and the iteration will apply *op* to such an s. At termination, $running = false$, which can only occur if there are no $s : S_i$ with *Ante* true but *Succ* false, so the constraint therefore holds true, and $Q = 0$. Confluence also follows if $Q = 0$ is only possible in one unique state of the source and target models which can be reached from the initial state by applying the constraint: this will be the state at termination regardless of the order in which elements were transformed.

The time complexity of the implementation depends on the value of Q on the starting models *smodel*, *tmodel*, and on the size $\#\overline{S_i}$ of the domain. The worst case complexity is of the order

$$Q(smodel, tmodel) * \#\overline{S_i} * (cost_{eval}(Ante) + cost_{eval}(Succ) + cost_{act}(Succ))$$

since the inner loop may be performed Q times. Optimisation by omitting the successor test reduces the complexity by removing the term $cost_{eval}(Succ)$. Again in the case of an ordered iteration, precomputation of a sorted version of the source domain can be used:

```
running := true;
sisorted := Si->sortedBy(e);
while (running) do
  (running := false;
   for s : sisorted do
     if s.Ante then
       [if s.Succ then skip
       else] (s.op(); running := true)
     else skip)
```

The cost of the sorting is then added to the overall execution time.

6.3.3 Type 3 constraints

If the other conditions of non-interference fail (a type 3 constraint), then the application of a constraint to one element may change the elements to which the constraint may subsequently be applied to, so that a fixed *for*-loop iteration over these elements cannot be used. Instead, a schematic iteration of the form:

```
while some source element s satisfies a constraint lhs do
select such an s and apply the constraint
```

can be used. This can be explicitly coded as:

```
running := true;
while running do
  running := search()
```

where *search* is a new static operation of S_i defined as:

```
static search() : Boolean
  (for s : S_i do
     if s.Ante then
       [if s.Succ then skip
       else] (s.op(); return true));
  return false
```

and the new operation *op* of S_i has postcondition *Succ*. We call this approach 3, iteration of a search-and-return loop. As with approach 2, the conditional test of *Succ* and code between [and] can be omitted if it is known that $Ante \Rightarrow not(Succ)$, or if the user confirms this optimisation.

As in approach 2, a Q measure is needed to prove termination and correctness. Termination follows if Q is a variant of the while loop: applying $op()$ to some $s : S_i$ with *Ante* and $not(Succ)$ decreases Q, even if new elements of S_i are generated. Correctness holds since *search* returns false exactly when $Q = 0$, i.e., when no $s : S_i$ falsifying the constraint remain. As for type 2 constraints, confluence can be deduced from uniqueness of the termination state. The worst case execution time complexity is of the order

$$Q(smodel, tmodel) * maxS * (cost_{eval}(Ante) + cost_{eval}(Succ) + cost_{act}(Succ))$$

where *maxS* is the maximum size of $\#\overline{S_i}$ reached during the computation. Again, optimisation can remove the $cost_{eval}(Succ)$ term.

For a sorted iteration, the sorted version of S_i must be recomputed each time S_i or e are modified:

```
static search() : Boolean
  (sisorted := S_i->sortedBy(e);
     for s : sisorted do
     if s.Ante then
       [if s.Succ then skip
       else] (s.op(); return true));
  return false
```

This can result in high time complexities, because the cost of sorting is now included in the cost of each individual iteration. Thus sorted iteration with type 3 constraints should be avoided if at all possible: if neither S_i nor the data of e is written by Cn, then the ordering of the S_i elements will not change, and *sisorted* can be precomputed.

In the case that two or more constraints are mutually data-dependent, i.e.,

$$wr(Ci) \cap rd(Cj) \neq \{\}$$

and

$$wr(Cj) \cap rd(Ci) \neq \{\}$$

then the constraints are combined into a constraint group $\{Ci, Cj\}$, and such groups are amalgamated. Such groups will be implemented by a generalised form of the type 2 or type 3 fixed-point iterations described above. The execution will attempt C_i as many times as possible before attempting C_j, and then repeat this behaviour while either remains applicable to the source model.

6.3.4 Type 0 constraints

Type 0 constraints are those constraints without an entity context. They represent a single action, such as an initialisation step at the beginning of a use case. Even if they do contain expressions of the form

$$E{\rightarrow}forAll(e \mid P)$$

in their conclusion, for an entity type E, this is interpreted as a bounded (for) loop over E, not a fixed-point loop. The interpretation of a type 0 constraint Cn is $stat(Cn)$ as defined in Table 6.2.

A common form of type 0 constraint is an iteration over a finite range of integers, for example:

$$Integer.subrange(1, 10){\rightarrow}forAll(i \mid (i * i){\rightarrow}display())$$

which displays the squares of integers from 1 to 10.

6.3.5 Quantified and let variables

Use case post-condition constraints may have additional quantified variables $v : e$ or let variables $v = e$ in their antecedents, where v is a new identifier not occurring as a feature name of the use case or context class, and distinct from other parameter, let or quantifier variable names of the use case, and e is an expression not involving v.

For example, the constraint (C1) iterating over the *Sudoku* class:

```
Sudoku::
  sq : square & sq.value = 0 & v = sq.possibleValues() & v.size = 1  =>
                              sq.value = v->min() & self.showBoard()
```

from the Sudoku solver of Chapter 2 has quantified variable sq and let variable v. These variables are computed as an additional part of the design and implementation of the constraint. A quantified variable $v : e$ introduces a bounded loop

for v : e
do
  ($stat$)

where $stat$ is the design for the remainder of the constraint following the quantified variable.

A let variable $v = e$ introduces an additional creation and assignment statement:

v : T := e;
$stat$

where T is the type of v deduced from e, and $stat$ is the design for the remainder of the constraint following the let variable.

As discussed previously in this chapter, the use of quantified variables has implications for the efficiency of the constraint implementation, and the use of a pattern such as Restrict Input Ranges (Chapter 9) is recommended to ensure that the smallest possible quantification range e is chosen for an additional quantified variable $v : e$. Let variables $v = e$ will also be recomputed on each application of the constraint, and so the complexity of the expression e defining the variable should also be reduced where possible. Let definitions should be moved before quantified variables unless they depend upon those variables (in $C1$ above, v's definition depends on sq, so this optimisation cannot be made in this case).

6.3.6 *Specification and design guidelines*

The following are useful guidelines to follow when constructing a UML-RSDS specification or design:

Avoid type 2 or 3 constraints if possible: use a pattern such as Replace Fixed-point Iteration by Bounded Iteration (Chapter 9) to avoid writing and reading the same data in a constraint, unless this is necessary for the problem. Type 2 and type 3 constraints are more difficult to analyse than type 0 or type 1 constraints, and cannot generally be inverted.

Avoid collection matching: quantified variables of the form v : $Set(e)$ should be avoided by using a simple quantified variable $x : e$ and defining the required set as some collection based on x: $v = f(x)$. See Sections 2.3 and 13.2 for examples.

Avoid operation calls in succedents of type 2 or 3 constraints: these can lead to incorrect implementations.

Avoid ordered iteration of type 3 constraints: as discussed above, this can lead to highly inefficient implementations.

Summary

This chapter has described how platform-independent designs are synthesised from specifications. The synthesis rules for different cases of constraints have been explained. We have also provided guidelines for improving the efficiency of designs.

Chapter 7

Model Transformations

Model transformations are a central element of model-based development approaches. Transformations can be used to improve the quality of models, to refactor models, to migrate or translate models from one representation to another, and to generate code or other text from models. In UML-RSDS, transformations are defined by use cases and are treated as a special case of software system, to which all the normal development facilities of UML-RSDS can be applied.

7.1 Model transformation concepts

Model transformations operate on one or more models, each model m conforms to a language or metamodel M which defines the structure of the model: what kinds of element it can contain and how these are linked (Fig. 7.1). We write $m : M$ to express that model m conforms to/is an instance of metamodel/language M. Typically a transformation τ maps a single model m which is an instance model of a source language S, to an instance model n of the target language T (a *mapping* or *separate-models* transformation), or updates a model $m : S$ in-place to form a new model $m' : S$ (an *update-in-place* transformation). Transformations operating on more than one language are termed *exogenous*, transformations that only involve a single language are termed *endogenous*. Figure 7.1 shows a *model-to-model* transformation, where source models are mapped to target models. Code generation transformations may instead generate text from models, this is referred to as a *model-to-text* transformation.

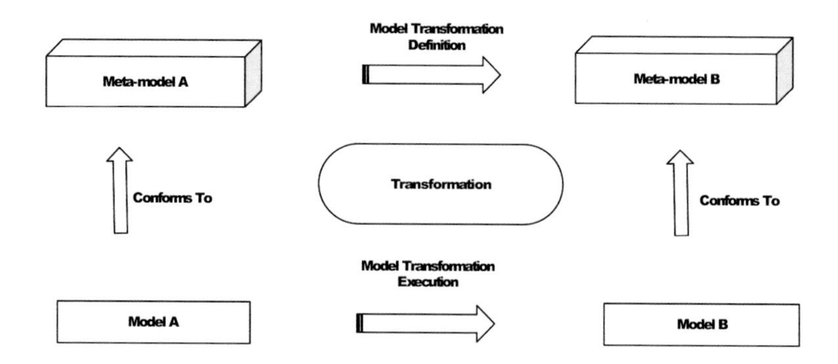

Figure 7.1: Model transformation context

Table 7.1 gives some examples of different kinds of transformations in these categories.

Table 7.1: Examples of transformation applications

	Endogenous	*Exogenous*
Update-in-place	Refactoring Reactive Simulation	–
Separate-models	Enhancement Model copying	Migration, Refinement, Code generation, Reverse-engineering/Abstraction

The categories of transformations include:

- *Code generation/refinement*: these produce a more implementation-specific model or code from an implementation-independent representation. E.g., mapping of UML to Java, or UML to C++, etc.

- *Migration*: these transform one version of a language to another/or to a similar language. E.g., mapping Java to C#, or UML 1.4 to UML 2.3, or from one EHR (Electronic Health Record) format to another, etc.

- *Refactoring/restructuring*: these rewrite a model to improve its structure or quality. E.g., removing redundant inheritance, introducing a design pattern, etc.

- *Enhancement*: these map a model of a system into a model of an enhanced system, e.g., to add access control facilities to a system.

- *Reactive*: transformations which compute a response to input events, e.g., to calculate an optimal move in a game in response to an opponent's move.
- *Reverse-engineering/Abstraction*: producing a more abstract/design or specification-oriented model from an implementation. For example, to produce a class diagram or architecture diagram from source code.

The category of update-in-place exogenous transformations is unusual, one example is the State machine/Petri Net transformation of [4] which uses simultaneous rewrites of the source and target models. Transformations can be *incremental*: applied to incremental changes in a source model to make corresponding changes in a (pre-existing) target model, or *bidirectional*: applicable in either source to target or target to source directions.

A key principle of UML-RSDS is that transformations can be considered to be use cases, and specified and developed using the same techniques as other UML-RSDS applications:

> *Model transformations can be specified as use cases, with use case postconditions expressing the transformation rules.*

The postcondition constraints are both a specification of the postconditions of the transformation, and a definition of the implementation of the transformation rules.

Transformations are naturally expressed as UML use cases: since a Use Case is a Classifier it is also a Namespace for its elements, and may have its own attributes and operations – such as local functions in QVT-R, helper functions and attributes in ATL, etc. Use cases may have invariants, and can be related by inheritance and by relations such as *extend* and *include*, permitting composition of transformations from subtransformations. Most importantly, a Use Case has an associated Behavior, which can represent the synthesised design of a transformation.

7.2 Model transformation languages

A large number of MT languages have been devised – it sometimes seems that each different MT research group or individual researcher has their own particular language – but only a few MT languages have wide usage. MT languages span a wide range of styles, from *declarative languages* expressing transformation rules in a logical high-level manner, most often using a version of OCL, or a graph formalism, to *procedural languages*, defining detailed low-level processing steps. *Hybrid*

MT languages can use both styles. Declarative languages include TGG and QVT-R, while Kermeta and QVT-O are examples of procedural MT languages. Hybrid languages include ATL [1], ETL [3] and GrGen [2]. Our view is that it is not necessary to invent a new specialised language for transformations: UML (with a formalised semantics for class diagrams and use cases) is already sufficient to define all practical cases of transformations. In terms of the classification of MT languages, UML-RSDS is a hybrid language since it contains both declarative logical and procedural imperative facilities: transformations can be defined by the logical postconditions of use cases, together with operations, which may be defined by explicit UML activities. Table 7.2 compares the scope of UML-RSDS with other model transformation languages.

Table 7.2: Comparison of transformation languages

Language	*Mapping*	*Update-in-place*	*Bidirectionality*	*Change-propagation*
UML-RSDS	√√	√√	√	√
ATL	√√	√	×	×
QVT-R	√√	√	√	√
ETL, GrGen	√√	√√	×	×

ATL has a restricted form of update-in-place processing, called *refining mode*. This makes a copy of the source model and then updates this copy based on the (read-only) source model. Thus the effects of updates cannot affect subsequent rule applications. QVT-R also adopts this approach, but repeatedly applies the copy and update process until no further changes occur. In contrast, UML-RSDS and ETL directly apply updates to the source model. Bidirectionality in UML-RSDS is partly supported by the synthesis of inverse transformations (use cases) from mapping transformations. Only transformations consisting of type 0 or type 1 postconditions can be reversed in this way. QVT-R provides the capability to apply a transformation in different directions between the domains (parameters) of the transformation rules. However, as with UML-RSDS, this capability is essentially limited to bijective mapping transformations [6]. QVT-R additionally supports change-propagation, by deleting, creating and modifying target model objects when an incremental change to the source model takes place. UML-RSDS supports change-propagation via the incremental execution mode for use cases (Chapter 14).

Table 7.3 compares ATL and UML-RSDS as transformation languages. Generally, ATL is better suited to separate-models transformations such as migrations, whereas UML-RSDS has full support for update-in-place transformations such as refactorings.

Table 7.3: Comparison of ATL and UML-RSDS

	ATL	*UML-RSDS*
System modularisation	Modules, rules, libraries, helpers	Use cases, operations, constraints, classes
Rule modularisation	Inheritance, called rules	Called operations
Rule scheduling/ priorities	Implicit	Explicit ordering
Multiple input elements in rules	No	Yes
Direct update of input models	No	Yes
Transformation implementation	In ATL interpreter	In Java, C#, C++
Analysis	Runtime rule conflicts detected	Data dependency, confluence, determinacy, rule conflicts

7.3 Model transformation specification in UML-RSDS

Model transformations in UML-RSDS are defined by use cases, the postconditions of these use cases define the transformation effect. Both the source and target metamodels are represented as parts (possibly disjoint and unconnected parts) of the same class diagram. In terms of Fig. 7.1, the source metamodel A and target metamodel B are defined as class diagrams. The transformation definition is given by a use case, consisting of postconditions which define transformation rules. Each use case postcondition constraint Cn usually has context some source metamodel entity E, and defines how elements of E are mapped to target model elements.

Figure 7.2 shows an example where the source metamodel consists of a single entity A with attribute x : int, and the target metamodel consists of entity B with attribute y : int, and there is a use case $a2b$ with a single postcondition constraint:

```
x > 0  =>  B->exists( b | b.y = x*x )
```

on context A. The constraint specifies that a new B object b is created for each $ax : A$ with $ax.x > 0$, and $b.y$ is set to the square of $ax.x$.

Assumptions of a transformation can also be defined, as precondition constraints of its use case. In this example we assume that the target model is initially empty, i.e.:

```
B->size() = 0
```

A valid transformation invariant constraint could be:

```
A->exists( a | a.x > 0 & y = a.x*a.x )
```

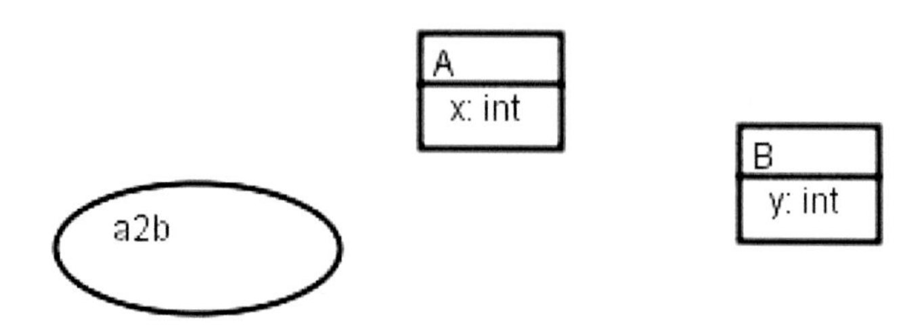

Figure 7.2: A to B transformation

on B. This expresses that every B must have been derived from a corresponding A.

The following steps are taken to define a transformation in UML-RSDS:

- Define the metamodels, as a single class diagram or separate class diagrams.
- Create a general use case, for the transformation.
- Edit the use case, adding pre and postcondition and invariant constraints as required. The order of the postcondition constraints determines the order of execution of the corresponding code segments.
- On the synthesis menu, select Generate Design to analyse the use case and generate a design-level description as operations and activities.
- Select Generate Java (Java 4) to produce an executable version of the transformation, which will be contained in the *output* directory files *Controller.java*, *GUI.java*, *ControllerInterface.java* and *SystemTypes.java*.
- The transformation implementation can be compiled using *javac GUI.java*, and then executed by the command *java GUI*. This opens up a simple dialog to load the input model and execute individual transformations (Fig. 3.13).

  Input models are specified in a file `in.txt`, output models are created in `out.txt`, together with XMI versions in `xsi.txt`.

The instance models Model A and Model B in Fig. 7.1 are represented in the files in.txt and out.txt. The transformation execution is an execution of the GUI class (for Java implementations). As described in Chapter 3, the text format for instance models in `in.txt` and `out.txt` is very simple, consisting of lines of text of three possible forms:

```
e : T
```

declaring that e is an instance of concrete entity T,

```
e.f = v
```

declaring that the attribute or 1-multiplicity role f of e has value v, and

```
obj : e.role
```

declaring that *obj* is a member of the many-valued role *role* of e. For sequence-valued roles, the order in which these membership assertions occur in the file is the order of elements in *role*.

For example, we could have the input model

```
a1 : A
a1.x = 5
a2 : A
a2.x = 3
b1 : B
b1.y = 2
```

for our elementary transformation example (this model does not satisfy the transformation assumption because there is already a B object in the model). Running the transformation on this input model produces the following output model:

```
a1 : A
a1.x = 5
a2 : A
a2.x = 3
b1 : B
b1.y = 2
b2 : B
b2.y = 25
b3 : B
b3.y = 9
```

Qualified associations and association classes cannot be represented in models, however these could be used for internal processing of a transformation, as auxiliary metamodel elements.

7.3.1 Transformation constraints

Four categories of postconditions Cn for transformations are distinguished, corresponding to the constraint categories defined in Chapter 6:

1. Type 0: *Cn* has no context classifier, for example

   ```
   "Hello world!"->display()
   ```

   The constraint dialog field for the context entity is left empty when defining such constraints (the first field in the dialog of Fig. 4.1).

2. Type 1: a context classifier *E* exists, and *wr*(*Cn*), the set of entity types and features written by *Cn*, is disjoint from *rd*(*Cn*), the set that it reads (write and read frames are defined in Table 5.1). Such *Cn* are implemented by a for-loop over *E*. For example, the constraint

   ```
   x > 0  =>  B->exists( b | b.y = x*x )
   ```

   given above is of type 1. *E* (in this case *A*) is specified in the context entity field of the constraint dialog (Fig. 4.1).

3. Type 2: *wr*(*Cn*) has common elements with *rd*(*Cn*), but the context classifier *E* itself is not in *wr*(*Cn*), nor are other entity types/features in the condition/antecedent (LHS) of *Cn*. These are implemented by a simplified fixed-point iteration over *E*.

4. Type 3: as type 2, but *wr*(*Cn*) includes the context classifier or other condition entity types/features. For example, to delete all *A* objects with negative x values, we could write:

   ```
   x < 0  =>  self->isDeleted()
   ```

   on context *A*. For such constraints a general fixed-point implementation is used, however it is often the case that this can be optimised by omitting checks that the conclusion of *Cn* already holds (in the above example, the conclusion obviously never holds for some *self* : *A* before the application of the constraint). A dialog box is presented asking for confirmation of this optimization, during the Generate Design step. Answer y to confirm the optimisation.

The most common form of constraints for migrations and refinements are type 0 or type 1, because such transformations normally only write the target model, and do not update the source model. In constructing the target model normally the data that is read is disjoint from the data that is written, or this separation can be achieved by using a pattern such as Map Objects Before Links (Chapter 9). For refactorings, type 2 or 3 constraints may be necessary because of the update-in-place nature of the transformation processing.

As for other general use cases, distinct postcondition constraints should be ordered so that Cm preceeds Cn if $wr(Cm)$ has common elements with $rd(Cn)$. For the above example this means that the correct order is:

```
x < 0  =>  self->isDeleted()


x > 0  =>  B->exists( b | b.y = x*x )
```

since the first constraint writes A and the second reads A: this order ensures that the invariant identified above actually is maintained by the transformation. Constraint analysis during the Generate Design step identifies any potential problems with the ordering of constraints.

7.3.2 Transformation specification techniques

For migration and refinement transformations, the postcondition constraints are typically of type 0 or type 1, and have informal specifications such as:

> For each instance of source entity Si, that satisfies Cond, instantiate a corresponding instance t of target entity Tj, and set t's features based on the source instance.

Such a mapping requirement would be formalised as a postcondition constraint

```
Si::
  Cond  =>  Tj->exists( t | Post )
```

The a2b example given above fits exactly into this pattern:

```
A::
  x > 0  =>  B->exists( b | b.y = x*x )
```

More realistically, some means of tracing the target elements back to their corresponding source elements is often necessary. To support tracing, we use the principal primary keys of source and target entities (introducing such keys if they do not already exist). If such keys were added to the classes A and B, we could write:

```
A::
  x > 0  =>  B->exists( b | b.bId = aId & b.y = x*x )
```

The A instance corresponding to a B instance b is then $A[b.bId]$.

For comparison, the a2b transformation example could be specified as follows in ATL:

```
rule a2b1 {
  from a : A (a.x > 0)
  to b : B
    ( bId <- a.aId,
      y <- a.x*a.x )
  }
```

ATL is appropriate for the definition of migrations and refinements, although migration between similar languages may involve substantial amounts of copying. In such cases a specialised migration language such as Epsilon Flock may be more appropriate.

Refinements and migrations may involve *entity splitting*, where multiple elements in the target model may be produced from each single source element. An example would be the production of HTML forms, JSP files, beans, and database interfaces from UML classes, as in the UML to web system mappings of Chapter 20. *Entity merging* may also occur, where multiple source elements contribute to the definition of single target elements. The basic form of type 1 constraint given above can also be used for entity splitting and merging, with identity attributes being used to correlate the appropriate sources and targets.

Migrations and refinements are often structured based on the compositional structure of the source language: a source entity $E1$ can be considered to be at a higher compositional level relative to a source entity $E2$ if there is a one-to-many association directed from $E1$ to $E2$. A transformation may then map $E2$ instances to target instances in one constraint, then map $E1$ instances to target instances in a second constraint, which also links the $E1$-targets to their corresponding $E2$-targets. This style of specification is called the Phased Construction pattern (Chapter 9).

For example, the a2b transformation can be elaborated by the addition of another composition level to the languages, and the addition of primary keys to all entities (Fig. 7.3). The specification of ac2bde can be organised as a Phased Construction by relating the entities C and D at one composition level, and then A and B at the next higher level:

$$
\begin{array}{l}
C :: \\
\qquad D{\rightarrow}exists(d \mid d.dId = cId \ \& \ d.w = z + 5) \\
A :: \\
\qquad B{\rightarrow}exists(b \mid b.bId = aId \ \& \ b.y = x * x \ \& \ b.dr = D[cr.cId])
\end{array}
$$

In the second constraint, the D instances corresponding to cr are looked-up by their primary key values: $D[cr.cId]$. The second constraint assumes that all C objects have a corresponding D object – this is ensured by the first constraint.

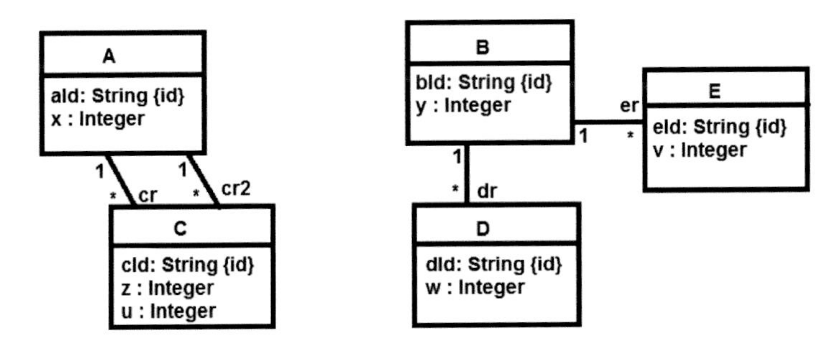

Figure 7.3: Extended a2b example: ac2bde

In ATL, this style of specification is also directly supported, with the linking process being carried out implicitly:

```
rule ac2bde1 {
  from c : C (true)
  to d : D
    ( dId <- c.cId,
      w <- c.z + 5 )
  }

rule ac2bde2 {
  from a : A (true)
  to b : B
    ( bId <- a.aId,
      y <- a.x*a.x,
      dr <- a.cr )
  }
```

In the `to` clause of ac2bde2, the assignment of $a.cr$ to dr actually assigns the D instances corresponding to $a.cr$. This involves an implicit lookup of these instances. We consider migrations in more detail in Chapter 11, and refinements in Chapter 12.

Refactorings and other update-in-place transformations are typically more complex than refinements and migrations. Their postcondition constraints are typically of type 2 or type 3, and have informal specifications such as:

> For each source model situation that satisfies condition Cond, rewrite or replace the situation to achieve condition Post.

Such a refactoring requirement would be formalised as a postcondition constraint

```
Si::
  Cond  =>  Post
```

where *Si* is a source model entity type which is appropriate to use as a *pivot* or main element to refer to the source model structure which is to be rewritten. Usually, *Post* implies *not*(*Cond*), and may involve object deletion using the →*isDeleted*() operator. Such constraints cannot usually be represented in ATL, nor in QVT-R, which does not have explicit object deletion. Unlike refinements and migrations, refactoring transformations may be structured on the basis of the different cases of source model situations/structures which need to be rewritten. In particular, a constraint $C1$ which deals with a specialised case of a situation, should normally precede a constraint $C2$ which deals with a more general case, and constraints which express a higher-priority rewrite for a given situation should precede constraints expressing a lower-priority rewrite. We consider refactoring transformations in detail in Chapter 13.

7.3.3 Pre-state expressions in constraints

Pre-state expressions *e@pre* in use case postcondition constraints refer to the values of entity types or to the values of features of individual objects at the start of execution of individual applications of the constraint. A prestate feature *obj.f@pre* should only be applied to instances *obj* of a prestate entity *E@pre*, unless *E* itself is not written by the constraint. It is bad practice to both read $x.f$@pre and write $x.f$ for the same object x in the same constraint, this can lead to a failure of confluence, and to constraint implementations that do not establish the constraint (or at least, not those parts of the constraint that use f@pre). *f@pre* can be used for features f of the context classifier of the constraint, but cannot be used for features of variables x of *exists*(x | P), etc., *self* of *select*(P), etc., or other internal variables within a constraint. They should not be used in the condition of type 3 constraints (e by itself denotes *e@pre* in such conditions).

An example of the use of prestate entities and features is the computation of the one-step composition of pairs of edges in a graph: if a pair of edges $e1$ and $e2$ are joined head-to-tail at the same node, then create a new edge that directly links $e1$'s source to $e2$'s target (but do not iterate this process any further):

```
e1 : Edge@pre  &  e2 : Edge@pre  &  e2.src@pre = e1.trg@pre  =>
   Edge->exists1( e3 | e3.src = e1.src@pre & e3.trg = e2.trg@pre )
```

The iterations of $e1$ and $e2$ over *Edge@pre* restricts the composition step to operate only upon pre-existing edges, not upon any $e3$ created by the step itself. The use of *src@pre* and *trg@pre* emphasises that the features of pre-existing edges are not changed by the constraint. Taken together, these annotations allow a simple double for-loop implementation of the constraint (because its set of written entities and features are disjoint from its set of read entities and features) rather than a fixed-point implementation. This is an example of the Replace Fixed-point by Bounded Iteration pattern (Chapter 9).

Care is needed in the use of pre-state expressions in constraints, as their use can make constraints harder to understand. Their main use is to serve as annotations to indicate that a type 1 bounded loop implementation should be used, for a constraint which otherwise would be of type 2 or 3 and use fixed-point iteration.

Variables may also be used to express *let* definitions, by placing an equation $v = exp$ in the assumption of a constraint. v can then be used in place of exp in the conclusion. Such v should only be used in read-only contexts after their definition.

7.4 Specifying bidirectional transformations

Support for *bidirectional* transformations is provided by the option to *reverse* a use case (on the edit use case menu). This option generates a new use case whose postconditions are derived as inverses of the postconditions of the original use case: these inverse constraints are often also invariants of the original use case. A type 1 postcondition constraint

$$S_i ::$$
$$SCond(self) \;\Rightarrow\; T_j \rightarrow exists(t \mid TCond(t) \;\&\; P_{i,j}(self, t))$$

is inverted to a constraint of the form

$$T_j ::$$
$$TCond(self) \;\Rightarrow\; S_i \rightarrow exists(s \mid SCond(s) \;\&\; P^{\sim}_{i,j}(s, self))$$

where the predicates $P_{i,j}(s, t)$ define the features of t from those of s, and are invertible: an equivalent form $P^{\sim}_{i,j}(s, t)$ should exist, which expresses the features of s in terms of those of t.

A type 0 postcondition constraint

$$S_i \rightarrow forAll(s \mid SCond(s) \;\Rightarrow\; T_j \rightarrow exists(t \mid TCond(t) \;\&\; P_{i,j}(s, t)))$$

is inverted to

$$T_j \rightarrow forAll(t \mid TCond(t) \;\Rightarrow\; S_i \rightarrow exists(s \mid SCond(s) \;\&\; P^{\sim}_{i,j}(s, t)))$$

Tables 7.4, 7.5 and 7.6 show examples of inverses $P^{\sim}$ of predicates P.

Table 7.4: Inverse of predicates

$P(s,t)$	$P^{\sim}(s,t)$	*Condition*
$t.g = s.f$	$s.f = t.g$	Data features f, g
$t.g = s.f.sqrt$ $t.g = s.f.sqr$	$s.f = t.g.sqr$ $s.f = t.g.sqrt$	f, g non-negative f, g non-negative
$t.g = s.f.exp$ $t.g = s.f.log$	$s.f = t.g.log$ $s.f = t.g.exp$	g positive f positive
$t.g = s.f.pow(x)$ $t.g = s.f.sin$ $t.g = s.f.cos$ $t.g = s.f.tan$	$s.f = t.g.pow(1.0/x)$ $s.f = t.g.asin$ $s.f = t.g.acos$ $s.f = t.g.atan$	x non-zero, independent of $s.f$ $-1 \leq t.g \leq 1$ $-1 \leq t.g \leq 1$
$t.b2 = not(s.b1)$	$s.b1 = not(t.b2)$	Boolean attributes $b1$, $b2$
$t.g = K * s.f + L$ Numeric constants K, L $K \neq 0$	$s.f = (t.g - L)/K$	f, g numeric
$t.g = s.f + SK$	$s.f = t.g.subrange(1, t.g.size - SK.size)$	String constant SK
$t.g = SK + s.f$	$s.f = t.g.subrange(SK.size + 1, t.g.size)$	String constant SK
$t.r = s.f \rightarrow characters()$	$s.f = t.r \rightarrow sum()$	String feature f, sequence r
$R(s,t)$ & $Q(s,t)$	$R^{\sim}(s,t)$ & $Q^{\sim}(s,t)$	

Table 7.5: Inverse of predicates on associations

$P(s,t)$	$P^{\sim}(s,t)$	*Condition*
$t.rr = s.r \rightarrow reverse()$	$s.r = t.rr \rightarrow reverse()$	r, rr ordered association ends
$t.rr = s.r \rightarrow last()$	$s.r \rightarrow last() = t.rr$	rr single-valued, r ordered
$t.rr = s.r \rightarrow first()$	$s.r \rightarrow first() = t.rr$	rr single-valued, r ordered
$t.rr = s.r \rightarrow including(s.p)$ $t.rr = s.r \rightarrow append(s.p)$	$s.r = t.rr \rightarrow front()$ & $s.p = t.rr \rightarrow last()$	rr, r ordered association ends p 1-multiplicity end
$t.rr = s.r \rightarrow prepend(s.p)$	$s.r = t.rr \rightarrow tail()$ & $s.p = t.rr \rightarrow first()$	rr, r ordered association ends p 1-multiplicity end
$t.rr = Sequence\{s.p1, s.p2\}$	$s.p1 = t.rr[1]$ & $s.p2 = t.rr[2]$	rr ordered association end $p1$, $p2$ 1-multiplicity ends
$t.rr = s.r \rightarrow select(P)$	$s.r \rightarrow excludesAll(x \mid P(x)$ & $t.rr \rightarrow excludes(x))$ $\rightarrow includesAll(t.rr \rightarrow select(P))$	r, rr set-valued
$t.rr = s.r \rightarrow reject(P)$	$s.r \rightarrow excludesAll(x \mid not(P(x))$ & $t.rr \rightarrow excludes(x))$ $\rightarrow includesAll(t.rr \rightarrow reject(P))$	r, rr set-valued
$t.rr = s.r \rightarrow sort()$ $t.rr = s.r \rightarrow asSequence()$	$s.r = t.rr \rightarrow asSet()$	r set-valued, rr ordered

Table 7.6: Inverse of predicates (extended)

$P(s,t)$	$P^{\sim}(s,t)$	*Condition*
$t = T_j[s.idS]$	$s = S_i[t.idT]$	idS primary key of S_i idT primary key of T_j
$t = T_j[s.r.idS]$	$s.r = S_i[t.idT]$	idS primary key of S_i idT primary key of T_j
$t.rr = TRef[s.r.idS]$ idS primary key of $SRef$, idT primary key of $TRef$	$s.r = SRef[t.rr.idT]$	rr association end with element type $TRef$, r association end with element type $SRef$
$t.rr = TRef[s.r \rightarrow collect(idS)]$	$s.r =$ $SRef[t.rr \rightarrow collect(idT)]$	As above
$t.g = s.r.idS$ Attribute g	$s.r = SRef[t.g]$	idS primary key of $SRef$, r association end with element type $SRef$
$t.rr = TRef[s.f]$ f is an attribute	$s.f = t.rr.idT$	idT primary key of $TRef$, rr association end with element type $TRef$
$T_j[s.idS].rr = TRef[s.r.idSRef]$ r has element type $SRef$, rr has element type $TRef$	$S_i[t.idT].r =$ $SRef[t.rr.idTRef]$	idS, $idSRef$ primary keys of S_i, $SRef$ idT, $idTRef$ primary keys of T_j, $TRef$

The meaning of assignments of the form $f \rightarrow last() = g$ is dealt with in Chapter 14.

More details of bidirectional transformation specification in UML-RSDS are given in Chapter 14.

7.5 Transformation examples

In this section we describe some simple examples of transformation specification using UML-RSDS.

7.5.1 Refinement example: UML to relational database

The UML to relational database refinement transformation has been used as an archetypical transformation example in many different languages, for example in QVT-R [5]. Figure 7.4 shows the source and target language metamodels of a basic version of this transformation. UML packages are mapped to RDB schemas, classes are mapped to tables, and attributes to columns. In this version the transformation can also be regarded as a structure-preserving migration.

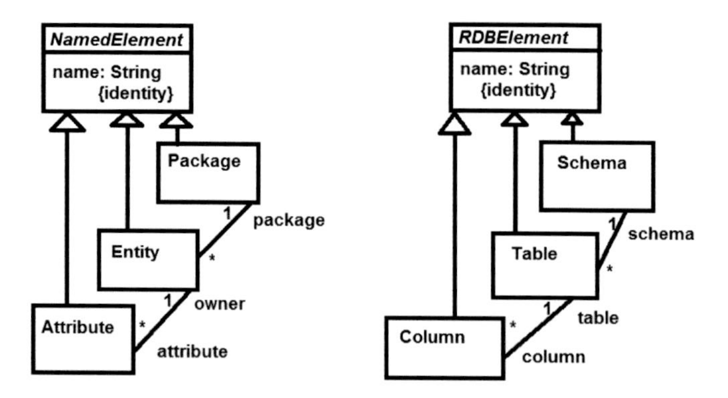

Figure 7.4: UML to relational database metamodels

A specification of the transformation can be written as a use case *uml2rdb* in UML-RSDS as follows. An assumption is that *name* is a primary key for *NamedElement*. This could be explicitly written as:

$$NamedElement \rightarrow isUnique(name)$$

However this is not necessary as *name* is already declared as an *identity* attribute in the source language class diagram.

The property

$$RDBElement \rightarrow isUnique(name)$$

could be asserted as a transformation invariant, that no two target objects with duplicate names should exist/be created. This property is part of the target language theory, and hence is required for the transformation to be syntactically correct (Chapter 18).

The following constraints define the mapping rules performed by *uml2rdb*, and are written in this order in its postconditions:

$$Schema \rightarrow exists(s \mid s.name = name)$$

This constraint has context entity *Package*, and maps packages 1-1 to schemas. For each package p a new schema s is created and its name is set equal to p's name.

$$Table \rightarrow exists(t \mid t.name = name \ \& \ t.schema = Schema[package.name])$$

This constraint has context *Entity*. Each entity instance e is copied to a table t with the same name, and t's schema is set equal to the schema corresponding to e's package: such a schema exists because of the effect of the previous constraint (it guarantees that for every package there is a corresponding schema with the same name).

Finally, the following constraint is iterated over *Attribute* to create columns:

$$Column{\rightarrow}exists(cl \mid cl.name = name \;\&\; cl.table = Table[owner.name])$$

Again, this relies on the preceding constraint to establish that for every class there is a corresponding table, so that $Table[owner.name]$ is well-defined.

This specification structure is an example of the Phased Construction transformation design pattern (Chapter 9), where the source entities at each level of composition are related by a single constraint to target entities, and subsequent constraints can then lookup instances of the constructed target entities to create target instances at a higher (or lower) level of composition. This can be described by a *language mapping*:

$$\begin{array}{l} Package \longmapsto Schema \;\; (via\ constraint\ 1) \\ Entity \longmapsto Table \;\; (via\ constraint\ 2, reads\ Schema) \\ Attribute \longmapsto Column \;\; (via\ constraint\ 3, reads\ Table) \end{array}$$

In general, if the sequence C_1 & ... & C_n of use case postcondition constraints satisfies the syntactic or semantic non-interference conditions, then each constraint may assume that all of its precedessors hold true at its initiation.

7.5.2 Migration example: trees to graphs

This transformation [3] maps tree structures into corresponding node and edge structures, and has the source and target metamodels shown in Fig. 7.5.

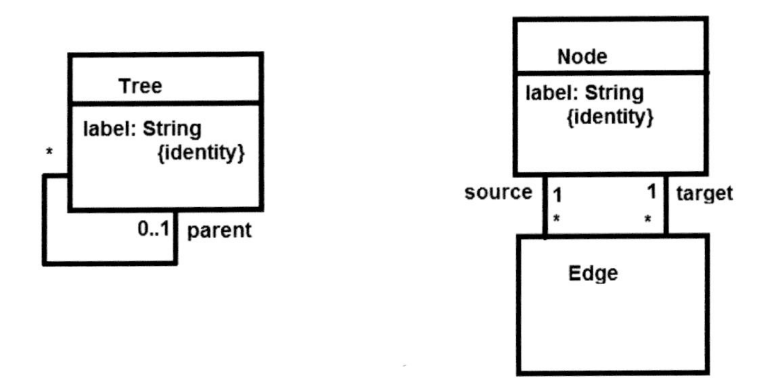

Figure 7.5: Tree to graph metamodels

The transformation is defined as a UML-RSDS use case *tree2graph* with two postconditions. The first constraint has context *Tree*, and creates nodes for each tree:

$$(C1): \quad Node{\rightarrow}exists(n \mid n.label = label)$$

A second constraint then creates edges for each link between parent and child trees. It is a double iteration over *Tree*:

$$\begin{aligned}(C2): \quad & p : parent \;\Rightarrow \\ & \qquad Edge{\rightarrow}exists(e \mid e.source = Node[label] \\ & \qquad\qquad \& \quad e.target = Node[p.label])\end{aligned}$$

For each pair of a tree (*self*) and its parent p, an Edge instance is created corresponding to the link from *self* to p in the source model. This transformation is also an example of a bidirectional transformation, because a reverse mapping *graph2tree* can be directly derived from *tree2graph*. The reverse mapping has postconditions:

$$(I1): \quad Tree{\rightarrow}exists(t \mid t.label = label)$$

on *Node* and

$$\begin{aligned}(I2): \quad & Tree{\rightarrow}exists(t \mid Tree{\rightarrow}exists(p \mid p : t.parent \; \& \\ & \qquad t.label = source.label \; \& \; p.label = \\ & \qquad\qquad target.label))\end{aligned}$$

on *Edge*. These postconditions are derived mechanically from *tree2graph* using Tables 7.4, 7.5 and 7.6.

The specification structure is an example of the Map Objects Before Links transformation design pattern (Chapter 9): trees are mapped to nodes, and then the parent links between trees are mapped by a separate constraint to links (edges) between nodes. Similarly for the inverse transformation.

Summary

We have described how transformations can be defined in UML-RSDS, and we have also compared UML-RSDS with other model transformation languages. Guidelines for defining refinements, migrations and refactorings were given, and simple examples of transformation definitions in UML-RSDS were presented.

References

[1] Eclipsepedia, *ATL User Guide*, http://wiki.eclipse.org/ATL/ User_Guide_-_The_ATL_Language, 2014.

[2] S. Kolahdouz-Rahimi, K. Lano, S. Pillay, J. Troya and P. Van Gorp, *Evaluation of model transformation approaches for model refactoring*, Science of Computer Programming, 2013, http://dx.doi.org/10.1016/j.scico.2013.07.013.

[3] D. Kolovos, R. Paige and F. Polack, *The Epsilon Transformation Language*, in ICMT 2008, LNCS Vol. 5063, pp. 46–60, Springer-Verlag, 2008.

[4] K. Lano, S. Kolahdouz-Rahimi and K. Maroukian, *Solving the Petri-Nets to Statecharts Transformation Case with UML-RSDS*, TTC 2013, EPTCS, 2013.

[5] OMG, *MOF 2.0 Query/View/Transformation Specification v1.1*, 2011.

[6] P. Stevens, *Bidirectional model transformations in QVT*, SoSyM, Vol. 9, No. 1, 2010.

Chapter 8

Case Study: Resource Scheduler

This chapter illustrates the UML-RSDS development process and tools by showing the complete development of a non-trivial example.

8.1 Case study description

This problem was the assessed coursework for the second year undergraduate course Object-oriented Specification and Design at King's College London in 2015/16. The system class diagram and system operations need to be specified in UML-RSDS, and an executable implementation synthesised in Java.

The system is intended to carry out release planning for an agile development process. The development or modification project to be implemented is divided into a number of *Story* objects, which have a *storyId* : *String* unique key. Each story has an ordered list *subtasks* of *Task* objects, which define particular work tasks. Tasks have a unique *taskId* : *String* key, and an Integer *duration*. A task may depend on other tasks (which must be completed before it is started, but could be completed in parallel with each other). A task has a set, *needs*, of *Skill* objects which represent skills needed to carry out the task. In turn, a *Skill* has a unique *skillId* : *String*. An entity type *Staff* represents staff, and has a unique *staffId* : *String*, and an Integer *costDay*. A set, *has*, of skills is associated to each staff object. Finally, the task schedule for an iteration is represented by a class *Schedule*, with an attribute *totalCost* :

Integer, and an ordered list *assignment* of *Assignment* objects, where each *Assignment* has associated staff and task objects.

The required system operations are:

- *allocateStaff*: for each unallocated task t, all of whose *dependsOn* tasks have already been allocated, find an available (unallocated) staff member who has all the skills required by t, and assign the task to the cheapest such staff member, s. Create a new assignment for t and s, and add this to the schedule.

- *calculateCost*: add up the products $s.costDay * t.duration$ for the assignments of the schedule and add this to the *totalCost* of the schedule.

- *displaySchedule*: print the list of assignments in order, with information of the *staffId*, *costDay*, *taskId*, *duration* for each assignment.

The solution is to be evaluated and tested on several test cases of planning problems. The solution is expected to always find a schedule with minimal total cost. Further additional requirements are introduced during the system development: (i) To add a *duration* : *Integer* attribute to *Schedule*, and to compute this as part of the *calculateCost* operation. (ii) To extend the system to calculate schedules for a series of iterations: an iteration schedule is complete when all possible allocations to available staff have been made: that is, when no further allocation of a staff member to a task can be made. Define a use case *nextIteration* to continue a schedule with a further iteration.

8.2 Solution

An initial class diagram, expressing the core data and functionality of the problem, could be defined as in Fig. 8.1.

To display assignments of staff to tasks, a *toString* operation is defined for *Assignment*, overriding the default version of this operation:

```
Assignment::
query toString() : String
pre: true
post:
  result = task.taskId + ", " + task.duration + ", " + staff.staffId
  + ", " + staff.costDay
```

The key use case of the system is *allocateStaff*. We decide therefore to make this the highest-priority development task. The use case can be viewed as a refactoring transformation, which operates to assign as many tasks to staff as possible – this will need a fixed-point iteration,

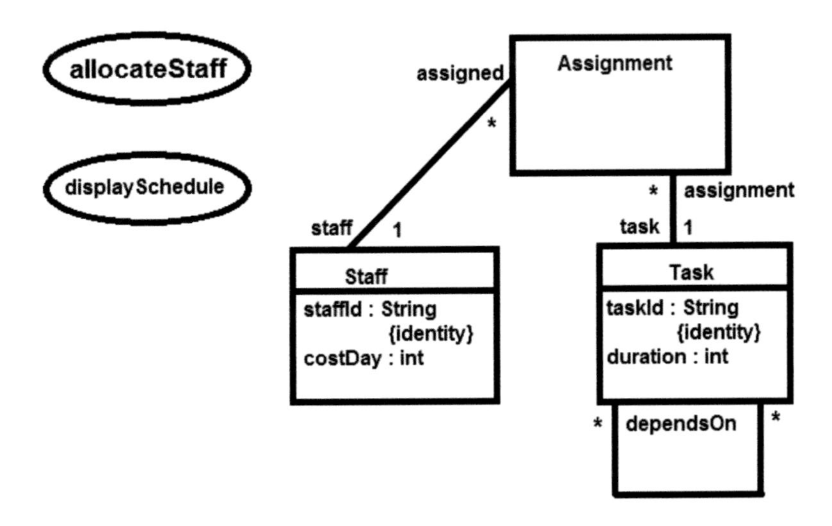

Figure 8.1: Initial schedular class diagram

since allocating one task may then make further tasks (that depend upon the allocated task) allocatable, and will remove the task itself from consideration for allocation. Informally, the scenario handled by this use case could be expressed as:

> If a task *self* is not yet assigned to a staff member, but all of the tasks it depends on have been, and there are some staff who have the skills to work on *self*, then: choose such a staff member of minimal cost and assign *self* to that person, and add the assignment to the schedule.

We specify this scenario using evolutionary prototyping, increasing the functionality of the prototype use case until it achieves all the scenario requirements. In this case the functional requirement is quite clear – if there had been ambiguity or incompleteness in the requirement, then it would be necessary to use requirements elicitation techniques, such as exploratory prototyping and interviews, to identify the true requirements.

Initially, we could write a simple version of the postcondition of *allocateStaff*, on context class *Task*:

```
Task::
  st : Staff & assignment.size = 0 & st.assigned.size = 0   =>
          Assignment->exists( a | a.task = self & a.staff = st )
```

This is a transformation rule that maps *Task* and *Staff* to *Assignment*: for each pair of an unassigned task *self* : *Task*, and an unassigned staff member *st*, the rule creates a new assignment which assigns *st* to work on *self*. It is of type 3 because *assignment* and *assigned* are read in the constraint condition, and implicitly written in the succedent (they are inverse association ends of *task* and *staff*).

This version could map the task and staff data (in *in.txt*):

```
t1 : Task
t1.taskId = "t1"
t1.duration = 10
t2 : Task
t2.taskId = "t2"
t2.duration = 5
t1 : t2.dependsOn
s1 : Staff
s1.staffId = "Mike"
s2 : Staff
s2.staffId = "Amy"
```

to the data (in *out.txt*):

```
a1 : Assignment
a1.task = t1
a1.staff = s1
a2 : Assignment
a2.task = t2
a2.staff = s2
```

If there were 3 tasks and 2 staff, one task would remain unallocated.

This initial version fails to meet all the scenario requirements:

- No account is taken of *dependsOn* or *duration*: a task should only be allocated if all its preceding tasks have already been allocated:

$$dependsOn{\rightarrow}forAll(d \mid d.assignment.size > 0)$$

  or, equivalently:

$$dependsOn.assignment.size = dependsOn.size$$

- We want to assign long-duration tasks first.
- No account is taken of staff skills or cost.

The second prototype addresses the first two points:

```
Task::  (ordered by -duration)
  assignment.size = 0 &
  dependsOn.assignment.size = dependsOn.size &
  st : Staff & st.assigned.size = 0   =>
      Assignment->exists( a | a.task = self & a.staff = st )
```

This considers longer-duration tasks first (because it iterates over *Task* in increasing order of $-duration$, i.e., tasks are considered in *decreasing* order of duration), and gives a task a higher priority than tasks that depend on it. Given the input data

```
t1 : Task
t1.taskId = "t1"
t1.duration = 10
t2 : Task
t2.taskId = "t2"
t2.duration = 15
t1 : t2.dependsOn
t3 : Task
t3.taskId = "t3"
t3.duration = 16
t1 : t3.dependsOn
s1 : Staff
s1.staffId = "Mike"
s2 : Staff
s2.staffId = "Amy"
```

task $t1$ is allocated first, then $t3$, because $t3$ has a longer duration than $t2$.

To base assignments on skills, we need to enrich the simple data model of Fig. 8.1 with representation of skills. Figure 8.2 shows the completed system class diagram after formalisation of the initial and additional requirements.

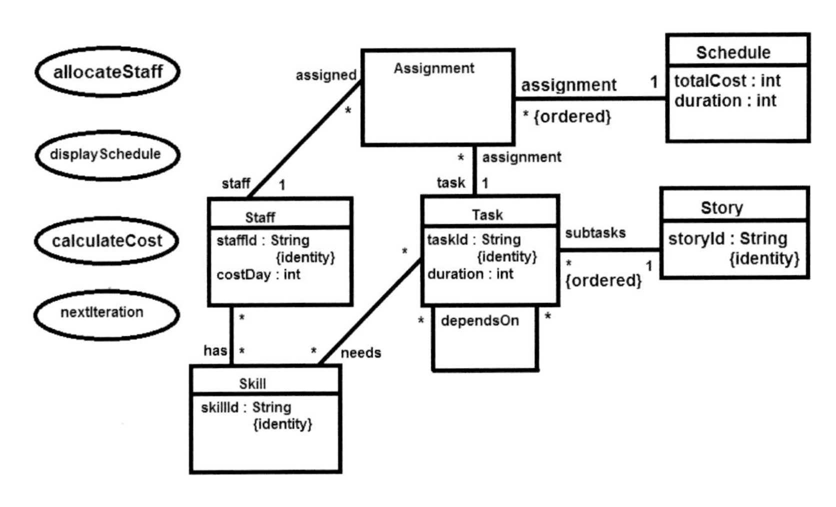

Figure 8.2: Schedular class diagram

Staff skills can now be considered: we should only allocate $st : Staff$ to $t : Task$ if $t.needs \subseteq st.has$, i.e., if st has all the skills that t needs. Instead of $st : Staff \ \& \ st.assigned.size = 0$ in the antecedent, we should therefore use the restricted quantification range

$$st : Staff \rightarrow select(assigned.size = 0 \ \& \ needs \subseteq has)$$

Finally, we want a staff member of minimal cost in this set. One way to do this is to sort the possible staff in increasing order of cost and take the first element of this sorted sequence:

$$availStaff \rightarrow sortedBy(costDay) \rightarrow first()$$

where $availStaff$ is a let variable holding the set $Staff \rightarrow select(assigned.size = 0 \ \& \ needs \subseteq has)$.

The complete scenario specification is as follows. In this version the postcondition refers to the schedule instance (which must exist prior to the execution of *allocateStaff*):

```
Task:: (ordered by -duration)
  s : Schedule & assignment.size = 0 &
  dependsOn.assignment->size() = dependsOn.size &
  availStaff = Staff->select(assigned.size = 0 & needs <: has) &
  availStaff.size > 0  =>
      Assignment->exists( a | a : s.assignment & a.task = self &
                   a.staff = availStaff->sortedBy(costDay)->first() )
```

The constraint only applies to tasks which do not already have allocated staff ($assignment.size = 0$) and whose preceding tasks have all been allocated ($dependsOn.assignment \rightarrow size() = dependsOn.size$). The variable *availStaff* is then calculated as the set of possible staff who could be allocated to the task: staff who have all the needed skills, and who have not yet been allocated. If this set is non-empty, a new assignment a is created (the succedent of the constraint), assigned to the task, added to the schedule, and the staff member chosen for this assignment is someone from *availStaff* with minimum cost. As an optimization, this constraint is iterated in order of $-duration$. This means that the longest tasks are considered first and assigned to cheap developers and scheduled before shorter tasks where possible.

Note that the succedent of this constraint contradicts the antecedent, because the task is assigned (to a) in the succedent, whilst it is required to have no assignment in the antecedent. This means that the constraint will iterate until no task satisfies the antecedent, i.e., until there is no remaining task that can be allocated. At its termination, all possible tasks that can be allocated in this iteration have been allocated, and the constraint is therefore established for all tasks.

The *calculateCost* use case computes the schedule cost (for a single iteration) as the sum of costs for each assignment (the first postcondition), and the schedule duration as the maximum of the assignment durations (the second postcondition):

```
Schedule::
  totalCost = assignment->collect(staff.costDay * task.duration)->sum()
```

```
Schedule::
  assignment.size > 0  =>
      duration = assignment.task.totalDuration()->max()
```

The calculation of schedule durations requires task durations to be calculated, which can be done using the following operation of *Task* :

```
query totalDuration() : int
pre: true
post:
  ( dependsOn.size = 0 => result = duration ) &
  ( dependsOn.size > 0 => result = duration + dependsOn.totalDuration
  ()->max() )
```

The total duration of a task is its own duration, plus the maximum of the durations of the tasks that it directly depends upon. The concept of task and schedule duration was not made clear in the requirements – so some elicitation or research work would be needed to identify the precise functionality required here.

The *displaySchedule* use case displays the assignments of each schedule:

```
Schedule::
  assignment->forAll( a | a->display() )
```

The *nextIteration* use case simply deletes all the tasks of existing assignments (i.e., the tasks which have already been assigned to staff in previous iterations). This also has the effect of deleting all the assignments themselves (because each assignment has a mandatory task and cannot exist without its task), and of making the staff allocated to these assignments available once more:

```
Assignment::
  task->isDeleted()
```

Previously allocated tasks are also removed from the *dependsOn* sets of unallocated tasks. Thus the situation is ready for a further application of *allocateStaff* to compute the next iteration.

8.3 Scheduling examples

The system can be tested with a number of examples. A simple test case involves three tasks, $t1$, $t2$ and $t3$, with $t3$ depending on $t1$, $t2$, and three staff members, with varying skills and costs:

```
s : Schedule
st1 : Story
st1.storyId = "st1"
t1 : Task
t1.taskId = "t1"
t1.duration = 5
t1 : st1.subtasks
t2 : Task
t2.taskId = "t2"
t2.duration = 10
t2 : st1.subtasks
t3 : Task
t3.taskId = "t3"
t3.duration = 7
t3 : st1.subtasks
t1 : t3.dependsOn
t2 : t3.dependsOn
s1 : Staff
s1.staffId = "s1"
s1.costDay = 3
s2 : Staff
s2.staffId = "s2"
s2.costDay = 7
s3 : Staff
s3.staffId = "s3"
s3.costDay = 2
sk1 : Skill
sk1.skillId = "Java"
sk2 : Skill
sk2.skillId = "JSP"
sk3 : Skill
sk3.skillId = "C++"
sk1 : t1.needs
sk2 : t1.needs
sk3 : t2.needs
sk1 : t3.needs
sk1 : s1.has
sk1 : s2.has
sk2 : s2.has
sk3 : s2.has
sk1 : s3.has
sk2 : s3.has
sk3 : s3.has
```

The first allocation assigns $s3$ to $t2$, and then $s2$ to $t1$. Notice that $s2$ could have carried out $t2$, but this would have been a more expensive choice: by allocating the most time-consuming tasks before lower duration tasks, there are a wider choice of developers available to assign to the high duration tasks, and a lower-cost schedule will result.

The output from *displaySchedule* in this case is:

```
t2, 10, s3, 2
t1, 5, s2, 7
```

with a total cost of 55 and a duration of 10 (since the two tasks can proceed in parallel).

After applying nextIteration and performing allocation again, the new iteration

```
t3, 7, s3, 2
```

is displayed, with a total cost of 14 and duration of 7.

A more complex example has four tasks, with a linear dependency of t4 on t3, t3 on t2, and t2 on t1:

```
s : Schedule
st1 : Story
st1.storyId = "st1"
t1 : Task
t1.taskId = "t1"
t1.duration = 5
t1 : st1.subtasks
t2 : Task
t2.taskId = "t2"
t2.duration = 10
t2 : st1.subtasks
t3 : Task
t3.taskId = "t3"
t3.duration = 7
t3 : st1.subtasks
t1 : t2.dependsOn
t2 : t3.dependsOn
t4 : Task
t4.taskId = "t4"
t4 : st1.subtasks
t4.duration = 11
t3 : t4.dependsOn
sk3 : t4.needs
```

The iterations for this example are:

```
t1, 5, s3, 2
---------------------
```

```
t2, 10, s3, 2
t3, 7, s1, 3
---------------------
t4, 11, s3, 2
```

8.4 Optimisation

The solution is functionally correct, however it involves the use of sorted iteration with a type 3 constraint, which is intrinsically quite inefficient: the list of tasks is re-sorted each time the constraint is applied. In this case such re-sorting is unnecessary because the duration of tasks does not change during the *allocateStaff* use case, nor does the extent of *Task*, and so the *allocateStaff* constraint can be optimised by pre-sorting the tasks and using a standard type 3 constraint over this pre-sorted list (this is an application of the Auxiliary Metamodel pattern of Chapter 9). An auxiliary *—* association $sortedtasks : Schedule \rightarrow Task$ is introduced, which is ordered at the *Task* end. This is initialised to $Task \rightarrow sortedBy(-duration)$, and then iterated over by the main allocation constraint:

```
Schedule::
  sortedtasks = Task->sortedBy(-duration)

Schedule::
  t : sortedtasks & t.assignment.size = 0 &
  t.dependsOn.assignment->size() = t.dependsOn.size &
  availStaff = Staff->select(assigned.size = 0 & t.needs <: has) &
  availStaff.size > 0  =>
      Assignment->exists( a | a : assignment & a.task = t &
                     a.staff = availStaff->sortedBy(costDay)->first() )
```

In the same way, the list of staff could be pre-sorted by *costDay*, to avoid sorting the available staff in each constraint application:

```
Schedule::
  sortedtasks = Task->sortedBy(-duration)

Schedule::
  sortedstaff = Staff->sortedBy(costDay)

Schedule::
  t : sortedtasks & t.assignment.size = 0 &
  t.dependsOn.assignment->size() = t.dependsOn.size &
  availStaff = sortedstaff->select(assigned.size = 0 & t.needs <: has)
  & availStaff.size > 0  =>
      Assignment->exists( a | a : assignment & a.task = t &
                                   a.staff = availStaff.first )
```

Table 8.1 shows the effect of these two optimisations, on large scheduling examples.

Table 8.1: Test cases of scheduling

Test case	Unoptimised	Optimised
100 tasks, 100 staff	608ms	515ms
200 tasks, 200 staff	3.8s	2.8s
500 tasks, 500 staff	28.5s	21.4s

Summary

This chapter has illustrated development using UML-RSDS, with development steps shown for a case study of a resource scheduler. Initial exploratory and evolutionary prototyping was used to produce a working solution, then optimisation was carried out to produce an improved production-quality solution.

Chapter 9

Design Patterns and Refactorings

Design patterns are a well-known technique for organising the structure of a software system in order to achieve quality or efficiency goals [3]. Several design patterns are built-in to UML-RSDS and are automatically incorporated into systems synthesised using the tools. Other patterns can be optionally selected by the developer, and there are further patterns which are recommended for UML-RSDS developments, but which need to be manually encoded. Refactorings are system restructurings which aim to improve the quality of the system specification or design, whilst preserving its semantics [2].

9.1 Design patterns

Software design patterns were identified in [3] and in subsequent research work to provide solutions to characteristic problems encountered during software development, and specifically during object-oriented software development. Many of these classical patterns, such as Facade, Singleton, Observer, Model-View-Controller are relevant to UML and to UML-RSDS in order to organise the structure of system class diagrams:

- Facade: defines a 'function-oriented' class which provides a simplified interface to the functionalities of a subsystem by means of its public operations. Client systems can use the Facade class to avoid linking directly to the internal classes of the subsystem.

This pattern is used by the UML-RSDS code generators to define the Controller class of a system as a Facade.

- Singleton: defines a class which has one unique instance, thus providing a globally available resource. Again, the Controller class in UML-RSDS is implemented as a Singleton.
- Observer: provides a structure linking observable objects to view objects which present some view of the observable object data. Updates to observables are automatically propagated to view updates.
- Model-View-Controller (MVC): a generalisation of Observer in which a Controller class co-ordinates the Model-View interaction. This pattern is used in the 'generate web applications' options of UML-RSDS (Chapter 20).

More recently, patterns specific to model transformations have been identified [4, 5]:

1. **Rule modularisation patterns:** Phased Construction; Structure Preservation; Entity Splitting/Structure Elaboration; Entity Merging/Structure Abstraction; Map Objects Before Links; Parallel Composition/Sequential Composition; Auxiliary Metamodel; Construction and Cleanup; Recursive Descent; Replace Explicit Calls by Implicit Calls; Introduce Rule Inheritance.
2. **Optimization patterns:** Unique Instantiation; Object Indexing; Omit Negative Application Conditions; Replace Fixed-point by Bounded Iteration; Decompose Complex Navigations; Restrict Input Ranges; Remove Duplicated Expression Evaluations; Implicit Copy.
3. **Model-to-text patterns:** Model Visitor; Text Templates; Replace Abstract by Concrete Syntax.
4. **Expressiveness patterns:** Simulate Multiple Matching; Simulate Collection Matching; Simulate Universal Quantification; Simulate Explicit Rule Scheduling.
5. **Architectural patterns:** Phased Model Construction; Target Model Splitting; Model Merging; Auxiliary Models; Filter Before Processing.
6. **Bidirectional transformation patterns:** Auxiliary Correspondence Model; Cleanup before Construct; Unique Instantiation; Phased Construction for bx; Entity Merging/Splitting for bx; Map Objects Before Links for bx (Chapter 14).

It is a principle of UML-RSDS development to use such patterns to help to organise and optimize specifications:

> *Use specification patterns where possible to improve the clarity, compositionality and efficiency of UML-RSDS systems.*

The MT design patterns described in [4] and on the MT design pattern repository (http://www.dcs.kcl.ac.uk/staff/kcl/mtdp) form a *pattern language*: the patterns can be combined and composed together to solve problems. For example, Simulate Explicit Rule Ordering and Auxiliary Metamodel are combined in [6]. Figure 9.1 shows the relationships between some of the transformation patterns.

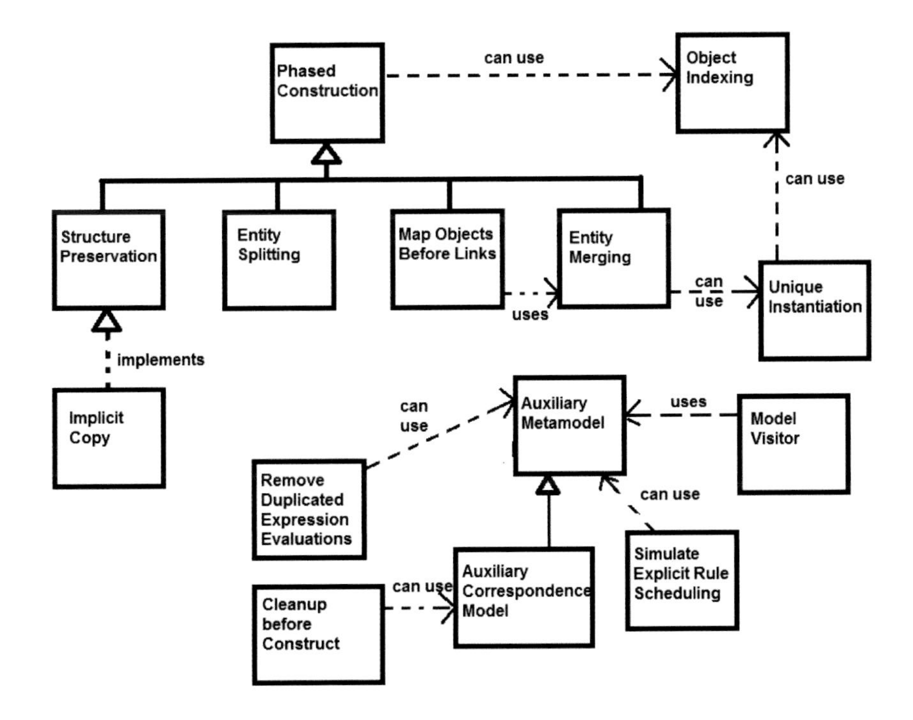

Figure 9.1: Transformation pattern relationships

Phased Construction

This pattern defines a standard organisation of transformation postconditions, based on the composition structure of the source or target languages of the transformation. If there is a composition structure $T1 \longrightarrow_r T2$ of entity types in the target language (that is, an association from $T1$ to $T2$ with a role name r at the $T2$ end), then a transformation creating models of this language could have a rule creating instances of $T2$:

$$S2 :: \quad SCond2 \;\Rightarrow\; T2{\rightarrow}exists(t \mid TCond2 \;\&\; Post2)$$

preceding a rule creating instances of $T1$:

$$S1 : \quad SCond1 \;\Rightarrow\; T1{\rightarrow}exists(t \mid TCond1 \;\&\; Post1 \;\&\; t.r = T2[e.id])$$

The $T2$ instances are created first and then linked to the $T1$ instances by means of a lookup based on the source language features (the object or object collection expression e) and a string-valued primary key id of the element type/type of e. An example of this pattern is the UML to RDB transformation (Section 7.5.1).

Map Objects Before Links

This is a related pattern which applies if there are self-associations $E \longrightarrow_r E$ in the source language which need to be mapped to the target language. In this case the pattern defines an organisation of the transformation into two phases: (1) map E to its target entity T, without considering r; (2) map r to the target model, using lookup of T instances. An example of this pattern is the Tree to Graph example of Section 7.5.2. It is also used in the GMF migration example of Chapter 11.

Entity Merging/Splitting

These patterns are variations of Phased Construction which apply if multiple elements in the source model are to be combined into a single element in the target (for Entity Merging), or if a single element in the source produces multiple elements in the target (for Entity Splitting). An example of Entity Splitting is the generation of multiple web application components from a single logical entity type, as in the synthesis of EIS applications in UML-RSDS (Chapter 20).

Auxiliary Metamodel

This pattern defines entity types and attributes, additional to those in the source or target metamodels, in order to support the transformation processing. The auxiliary elements can be used to record traces or source-target correspondence models, or intermediate transformation results.

An example of this would be an alternative form of the UML to relational database transformation which takes account of inheritance in the source model, and only creates relational tables for root classes in the source model inheritance hierarchy, the data of subclasses is then merged into these tables (Fig. 9.2). This is also an example of Entity Merging.

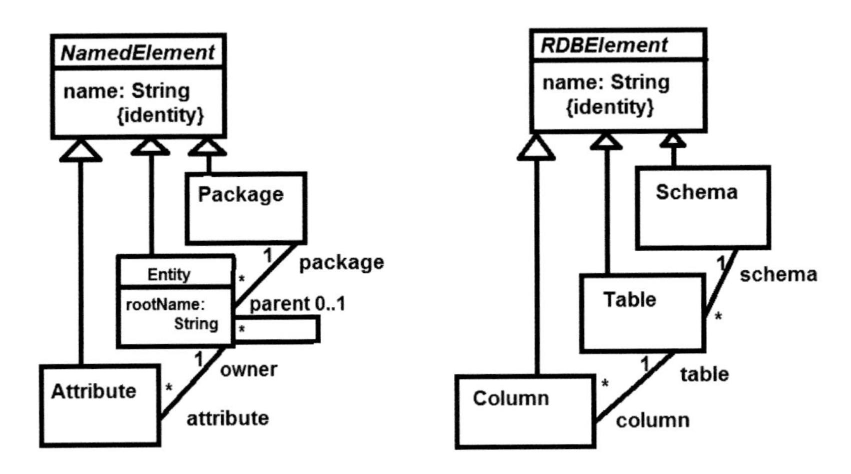

Figure 9.2: UML to relational database metamodels (enhanced)

The attribute *rootName* is a new auxiliary attribute which records the name of the root class of each class. It can be computed using a transformation *setRoots* as follows:

$$Entity ::$$
$$parent = Set\{\} \;\Rightarrow\; rootName = name$$
$$Entity ::$$
$$parent \neq Set\{\} \;\&\; rootName = \text{""} \;\&\; parent.any.rootName \neq \text{""} \;\Rightarrow\; rootName = parent.any.rootName$$

Both postcondition constraints iterate over *Entity*, the second constraint is of type 3, and is iterated until all non-root classes have an assigned non-empty root name.

Sequential Composition

This pattern is used to organise a transformation into a sequential application of modular sub-transformations.

For example, the extended version of the UML to RDB transformation can be split into a sequence of two transformations: (i) *setRoots* to compute the root class name of each class; (ii) *mapUML2RDB* to perform the main mapping, using the auxiliary *rootName* to map subclasses of a root class to the table of the root class.

mapUML2RDB has the postconditions:

$$parent = Set\{\} \;\Rightarrow\; Table{\rightarrow}exists(t \mid t.name = name)$$

on *Entity* creates tables for each root class.

$$a : attribute \Rightarrow \\ \quad Column{\rightarrow}exists(cl \mid \\ \quad\quad cl.name = a.name \;\&\; cl : Table[rootName].column)$$

on *Entity* copies the attributes of each entity to columns of the table corresponding to the entity's root class. Thus the attributes of all subclasses (direct and indirect) of a given root class are all merged together into the single table for the root class.

9.2 Design pattern implementation in UML-RSDS

Several design patterns are provided on the UML-RSDS tools *Transform* menu to assist developers. Some (Singleton and Facade) are applicable to systems in general, whilst others are specialised for transformations:

- *Singleton*: constructs a singleton class, following the classic GoF pattern structure.
- *Facade*: identifies if a Facade class would improve the modularity of a given class diagram.
- *Phased Construction*: converts a transformation postcondition which has a nested *forAll* clause within an *exists* or *exists*1 succedent into a pair of postconditions without the nested clause, and using instead an internal trace association.
- *Implicit Copy*: looks for a total injective mapping from the source language entities and features to the target language and defines a transformation which copies source data to target data according to this mapping.
- *Auxiliary Metamodel*: constructs an auxiliary trace class and associations linking source entity types to the trace class and linking the trace class to target entity types.

Other patterns from [4] are incorporated into the design generation process:

Unique Instantiation: as described following Table 4.2, the update interpretation of $E{\rightarrow}exists(e \mid e.id = val \;\&\; P)$ where *id* is a unique/identity attribute or primary key of E checks if there is already an existing $e : E$ with $e.id = val$ before creating a new object. The design for $E{\rightarrow}exists1(e \mid P)$ checks if there is an existing $e : E$ satisfying P before creating a new object.

Object Indexing: indexing by unique/primary key is built-in: if entity type E has attribute att : $String$ as its first primary key (either inherited or directly declared in E), then lookup by string values is supported: $E[s]$ denotes the unique $e : E$ with $e.att = s$, if such an object exists. Lookup by sets st of string is also supported: $E[st]$ is the set of E objects with att value in st. If there is no identity attribute/primary key defined for E, but some string-valued attribute satisfies $E \rightarrow isUnique(att)$, then lookup by string value can also be used for the first such att.

Omit Negative Application Conditions (NACs): if type 2 or 3 (Chapter 6) constraints $Ante \Rightarrow Succ$ have that $Ante$ and $Succ$ are mutually inconsistent, then the design generator will simplify the generated design by omitting a check for $Succ$ if $Ante$ is true. It will also prompt the user to ask if this simplification should be applied. For example, any constraint with the form

$$Ante \;\Rightarrow\; P \;\&\; self \rightarrow isDeleted()$$

can be optimized in this way, likewise for

$$x : E \;\&\; Ante \;\Rightarrow\; P \;\&\; x \rightarrow isDeleted()$$

Constraints with a *true* antecedent cannot be optimized using this pattern, for example the computation of the transitive closure $ancestor$ of a many-many reflexive association $parent : E \rightarrow_* E$ on an entity E:

```
E::
  parent <: ancestor

E::
  parent.ancestor <: ancestor
```

The first constraint is of type 1, the second is of type 2. The NAC of the second constraint is $not(parent.ancestor \subseteq ancestor)$, and this needs to be checked before attempting to perform $stat(parent.ancestor \subseteq ancestor)$, otherwise the implementation will not terminate.

Note that the use of operation calls within constraints to define update functionality may prevent correct implementation of type 2 or 3 constraints, if NACs are needed. The implementation of a constraint

$$G \;\Rightarrow\; Post \;\&\; op(pars)$$

in such cases tests the NAC G & $not(Post)$ before applying the constraint, meaning that cases where G and $Post$ both hold will not result in application of the constraint, even if the effect of $op(pars)$ has not been established. A partial solution to this problem is to use Auxiliary Metamodel to introduce a new attribute to keep an explicit record of those source elements par for which $op(pars)$ has been completed, and use this to avoid re-application of the constraint to such elements.

Replace Fixed-point Iteration by Bounded Iteration: Again, for type 2 or 3 constraints $Ante \Rightarrow Succ$, which would normally be implemented by a fixed-point iteration, a simpler and more efficient implementation by a bounded loop is possible if the effect of $Succ$ never produces new cases of objects satisfying $Ante$ which did not exist at the start of the constraint implementation. In some cases this can be automatically detected, e.g.:

$$Ante \;\Rightarrow\; self{\rightarrow}isDeleted()$$

on entity E where there is no data-dependency of $Ante$ upon E. The user is prompted to confirm such optimizations.

Developers can enforce the use of bounded iteration by using expressions of the form E@pre, e@pre instead of E, e in read expressions within the constraint. In particular, if there are two mutually data-dependent postcondition constraints $C1$, $C2$ where $C1$ reads entity type E, and $C2$ creates instances of E, we can assert that the instances created by $C2$ are irrelevant to $C1$ by using E@pre in $C1$, denoting the set of E objects created before $C1$ begins execution, thus potentially removing the need for fixed-point iteration of the constraint group $\{C1, C2\}$.

Whenever possible, the use of fixed-point solutions should be avoided. Although it is possible to write a (type 3) constraint such as

$$x < 7 \;\Rightarrow\; x = x + 1 \;\&\; P$$

to mean "increment x until it becomes 7, executing $stat(P)$ for each x value", such a loop is better expressed as a type 1 constraint

$$Integer.subrange(x@pre + 1, 7){\rightarrow}forAll(i \mid P(i)) \;\&\; x = 7$$

which is clearer and easier to understand as well as requiring only a bounded loop implementation.

Factor out Duplicated Expressions: complex duplicated read-only expressions e in constraints can be factored out by defining a let variable v as $v = e$ in the constraint antecedent, and using v in place of the repeated occurrences of e.

9.3 Refactorings

Refactorings are an important technique in agile development [1]. On the *Transform* menu a set of class diagram refactorings are provided to help in improving the structure of a class diagram:

- Remove redundant inheritance: removes generalisations which link classes which already have a generalisation relationship via an intermediate class.
- Introduce superclass: implements the 'Extract superclass' refactoring to identify common parts of two or more classes.
- Pushdown abstract features: used to copy 'virtual' data features of interfaces down to subclasses, in order to implement a form of multiple inheritance.

There are also inbuilt refinement transformations, mainly used for web application construction (transforming a UML class diagram into a relational database schema):

- Express statemachine on class diagram: creates an enumerated type and conditional expressions within the owning entity of a statemachine, in order to express the statemachine semantics.
- Introduce primary key: adds a String-valued identity attribute to the specified classes (Fig. 9.3).
- Remove many-many associations: replaces a many-many association with a new class and two many-one associations.
- Remove inheritance: replaces an inheritance by an association (Fig. 9.4).

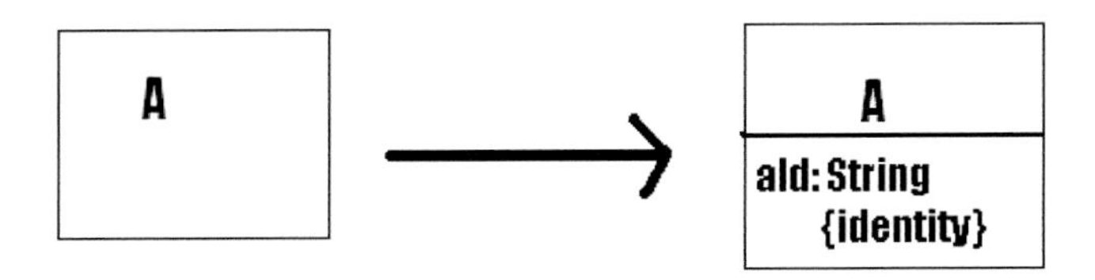

Figure 9.3: Introducing a primary key

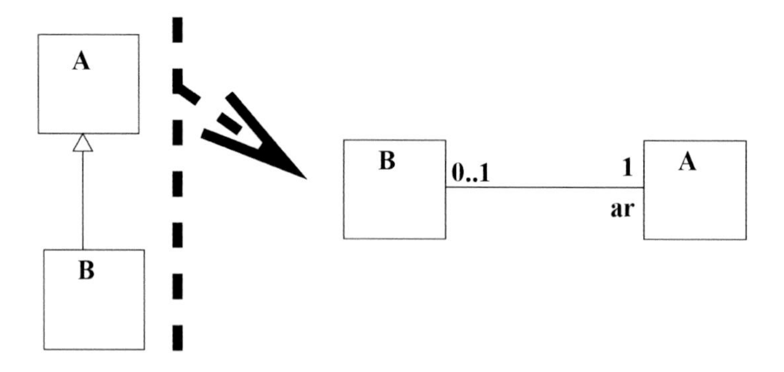

Figure 9.4: Replace inheritance by an association

- Aggregate subclasses: merges all subclasses of a given class into the class, in order to remove inheritance.
- Remove association classes: replaces an association class by two many-one associations.
- Introduce foreign key: define a foreign key to represent a many-one association (Fig. 9.5).
- Matching by backtracking: introduces a search algorithm for constructing a mapping.

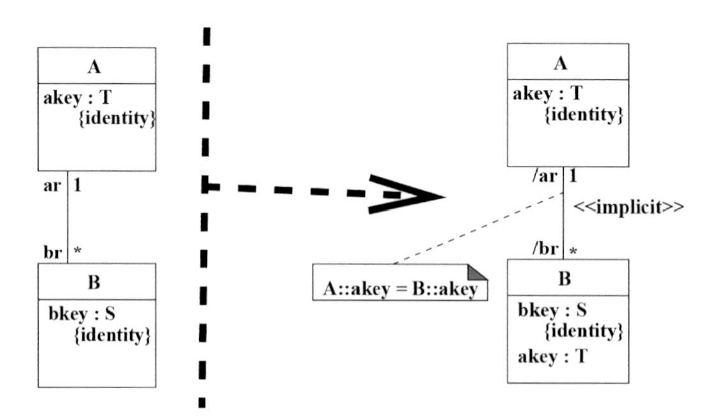

Figure 9.5: Introducing a foreign key

Summary

This chapter has described the design patterns and refactorings that can be used in the UML-RSDS development process. Guidelines for using patterns have been given, and the tool support for applying patterns has been described.

References

[1] K. Beck et al., *Principles behind the Agile Manifesto*, Agile Alliance, 2001. http://agilemanifesto.org/principles.

[2] M. Fowler, K. Beck, J. Brant, W. Opdyke and D. Roberts, *Refactoring: Improving the Design of Existing Code*, Addison-Wesley, 1999.

[3] E. Gamma, R. Helm, R. Johnson and J. Vlissides, *Design Patterns: Elements of Reusable Object-Oriented Software*, Addison-Wesley, 1994.

[4] K. Lano and S. Kolahdouz-Rahimi, *Model-transformation Design Patterns*, IEEE Transactions in Software Engineering, Vol. 40, 2014.

[5] Model-transformation Design Patterns repository, http://www.dcs.kcl.ac.uk/staff/kcl/mtdp, 2015.

[6] W. Smid and A. Rensink, *Class diagram restructuring with GROOVE*, TTC 2013.

Chapter 10

System Composition and Reuse

Medium or large scale systems should ideally be constructed as modular compositions of components, with well-defined interfaces and client-supplier relationships between components. Such components must be at a useful level of granularity, offering a collection of related services to client systems via a convenient interface of operations. To facilitate reuse, the specifications of these operations must be precise and clear, as must the specifications of any data they operate upon. In this chapter we describe ways in which system composition using components can be achieved with UML-RSDS.

10.1 Component-based and service-oriented composition

In component-based design (CBD), a system is formed out of subsystems in a hierarchical manner, with its subcomponents themselves potentially constructed from smaller components. Ideally, as much as possible of the new system should be constructed from reusable components obtained from component libraries. This form of composition and reuse is supported by UML-RSDS, since each UML-RSDS system can be viewed as a component operating on data structured according to the system class diagram, and providing a set of top-level operations characterised by its use cases. These services may be invoked by external client systems in addition to direct users. The external system may be another Java/C#/C++ application which uses the UML-RSDS synthesised system as a subcomponent, or may be an application running on a remote machine and which invokes the use case operations via remote method invocation (RMI) or as web services.

An external system/application M used by another UML-RSDS system is defined in the client system as a class with the stereotype *externalApp*. This signifies that M has been developed as a UML-RSDS application with system name M. The class M acts as a facade or proxy for the supplier system. For Java implementations, the code of system M will be contained in a package with package name M, and should be located in a subdirectory with this name. Calls $M.op(p)$ within the client system are interpreted as $M.Controller.inst().op(p)$ in the complete generated code, and therefore correspond to use cases of the supplier system. The operations of M which are used in the client system should be declared in the M class in the client system class diagram (M may have additional operations – corresponding to use cases of the supplier system M – that are not used in the client, and these do not need to be listed).

Libraries of useful use cases can be developed for reuse as external applications in this manner, for example statistical operations such as the correlation calculator of Chapter 2 could be packaged into a library component for use in other applications. An example of the use of such a library component is the *StatLib* class of the CDO application (Fig. 21.8).

Regarding reuse of components, there is the following principle for UML-RSDS:

> *When a functionality or set of related functionalities have been developed, and which seem to be of potential utility in other systems, construct a reusable component consisting of these functionalities as use cases, supported by the local data. This component can be declared as an externalApp in other UML-RSDS systems.*

Service-based composition, or service-oriented architectures (SOA), achieve reuse and composition by the combination of existing services in workflows (activities) or by other forms of service coordination. UML-RSDS supports this form of composition by enabling use cases to be published as web services (Section 10.4), which can then be externally invoked and combined, including composition with services not developed using UML-RSDS.

10.2 Language extensions and importing libraries

The UML-RSDS language can be extended with new binary and unary $\rightarrow$ operators, and external Java libraries and classes may be declared in the class diagram for use by a UML-RSDS specification. For example, to use Java regular expressions, the `java.util.regex.*` library package would be declared as a new import, and the classes `Pattern` and `Matcher` of this package declared as *external* classes in the class diagram, along with any methods that we wish to use from them, e.g., the static `matches(p : String, s : String) : boolean` operation of `Pattern`. The *external* stereotype on an entity type indicates that the entity type is defined in executable code which is in an external library, or which has been provided by some other source, and that it is not a complete UML-RSDS application. Classes from a UML-RSDS development may be exported as library components via the option "Export system as library" on the Extensions menu. This saves the code of the classes E of the system in separate files *output/E.java*, and these classes can then be used as *external* classes in a different UML-RSDS development.

These extension/import mechanisms are usually language-specific, and there is no support for formal analysis of system elements that depend upon extension/imported components. They are a convenience to support rapid code generation of systems.

10.2.1 External classes and operations

Each external class used by the system should be added to the class diagram, with the stereotype *external*. Any operations of these classes used by the system should also be added to their class. Note that currently, array input and output parameters are not permitted, and only the non-collection types supported by UML-RSDS (boolean, int, long, double, String, enumerations and class types) can be used for parameters of such operations. In expressions, declared operations of external entities may be invoked as with any other operation, e.g., `result = Pattern.matches(".a.", "bar")`. Their form in the generated Java code is the same as in the UML-RSDS expression. Instances of external entity types may be created locally, for example, an activity language statement

```
p : Pattern := Pattern.compile(".a.")
```

defines a local variable p holding a *Pattern* object.

Provided that the external entity type E has a zero-argument constructor, objects of the entity may be created in constraints by using an $E \rightarrow exists(e \mid ...)$ construct. For example:

```
Date->exists( d | d.getTime()->display() )
```

The generated code initialises d with the $Date()$ constructor.

Existing methods of the String class may be called without declaration, e.g., $str.replaceAll(p, s)$ can be used as an expression within a specification to be implemented in Java. Operations of the Java *Math* class (additional to those already provided in UML-RSDS) may be used by defining *Math* as an *external* class, and the necessary operator *op* as a static query operation of *Math*. For example, the $random() : double$ operation for generating random numbers in the interval 0 to 1 could be introduced in this way. The expression $Math.random()$ could then be used within the system and would compile to a call on the Java operator. Figure 10.1 shows a simple example of a system using $Math.random()$ to generate random series, e.g., for a Monte Carlo simulation of share prices.

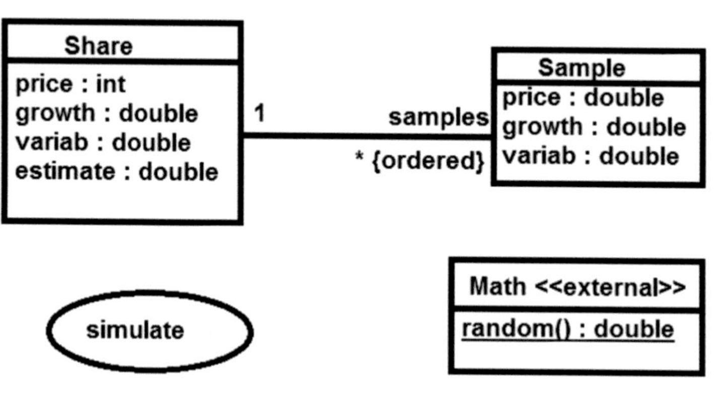

Figure 10.1: System using random numbers

The use case *simulate* has a postcondition

```
Share::
  Integer.subrange(1,10)->forAll( j |
      Sample->exists( s | s.price = price & s.growth = growth &
                          s.variab = variab & s : samples ) )
```

to create 10 samples of the share price trend, and to initialise the share price simulation samples for each share. A second constraint computes the simulations for each sample, using 100 time points, and with random variation in prices simulated by $Math.random()$:

```
Sample::
    Integer.subrange(1,100)->forAll( t |
              price = price@pre + price@pre * growth +
                   price@pre * variab * ((Math.random() * 2) - 1) )
```

This models share price changes as a *stochastic process*, with the price at time t derived from the price at time $t-1$ as the sum of the current price, a growth component, and a variable component subject to random variation (both positive and negative) of degree $varib * price$ about mean 0.

Finally, an overall estimate for the price at time 100 is calculated as the average of the sample prices:

```
Share::
  estimate = (samples.price->sum())/10
```

For example, with a starting price of 100 for a share, growth as 0.1 (10% growth) and variability as 0.05 (5% variability), one run produces a range of price estimates from 105 to 114, and an average of 111. A full Monte Carlo simulation might use 10,000 samples instead of 10.

Another example of an *external* class is *XMLParser* in the XML to code transformation of Fig. 21.2. This class is an external component whose code has been hand-written. No code is generated by UML-RSDS for *external* or *externalApp* classes, because they are assumed to already possess executable implementations.

10.2.2 Adding an import

The Add import option allows the user to add access to a library package to their system, either a standard Java library or a package containing an *external* class or *externalApp* application. The exact text required should be entered, e.g.:

```
import java.util.regex.*;
```

or

```
import mylibrary.*;
```

for a Java import.

10.2.3 Adding a new operator

New unary and binary $\rightarrow$ operators can be added. The dialog asks for the name of the operator, including the arrow, and the result type. For an operator application $\_1 \rightarrow op(\_2)$ the default text in Java and C# is $\_1.op(\_2)$. If an alternative is needed, this should be specified in the following text areas. For example, an operator $str \rightarrow time()$ to compose the current time with a string message could be given the Java template text

```
(_1 + (new Date()).getTime())
```

The operator can then be used as a string expression within constraints.

Operators can be used for query expressions, or as new forms of update expression, in which case the Java/C# form of *stat*() for the operator should be entered. For example:

```
JOptionPane.getInputDialog(_1);
```

for a new unary operator $\rightarrow ask()$.

10.3 Example of library creation and use: mathematical functions

A developer could define an *external* library of useful mathematical functions using UML-RSDS, as follows. They may have found the functions $factorial(n : int) : int$, $combinatorial(n : int, m : int) : int$ "n choose m", $gcd(n : int, m : int) : int$ and $lcm(n : int, m : int) : int$ useful and want to define them in a language-independent manner to use in several different applications. A class should be created with these functions as static operations (Fig. 10.2). Since these functions are to be widely used, particular care should be taken over their efficiency and correctness.

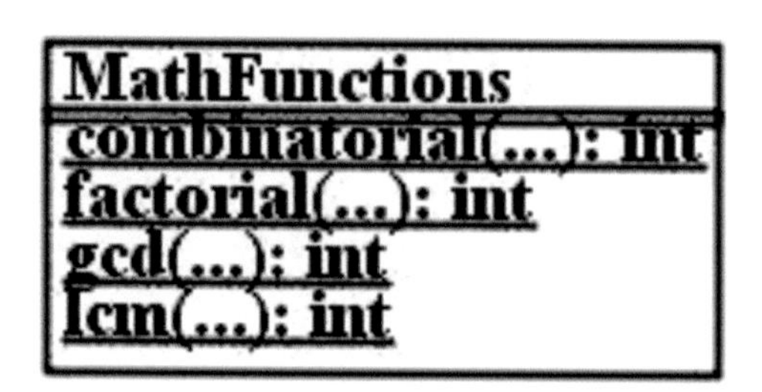

Figure 10.2: Mathematical functions library

For *factorial* and *combinatorial*, the $Integer.Prd(a, b, i, e)$ operator can be used to compute the product $\Pi_{i=a}^{b}\ e$, and to optimize the computation of these operations:

```
static query combinatorial(n: int, m: int): int
pre: n >= m & m >= 0 & n <= 25
post:
  ( n - m < m =>
       result = Integer.Prd(m + 1,n,i,i) / Integer.Prd(1,n - m,j,j) ) &
  ( n - m >= m =>
       result = Integer.Prd(n - m + 1,n,i,i) / Integer.Prd(1,m,j,j) )
```

```
static query factorial(x: int): int
pre: x <= 12
post:
  ( x < 2 => result = 1 ) &
  ( x >= 2 => result = Integer.Prd(2,x,i,i) )
```

The definition of the combinatorial is optimized to avoid duplicate evaluation of products. For example, if $n \geq 2m$, then the second case applies and

$$combinatorial(n, m) \quad = \quad \frac{n*(n-1)*...*(n-m+1)}{m*(m-1)*...*1}$$

An alternative would be to use a recursive computation for *factorial*, and then use this to compute the *combinatorial*, using the equation

$$combinatorial(n, m) \quad = \quad \frac{n!}{m!(n-m)!}$$

But this approach would be less efficient – as recursion is usually more computationally expensive than iteration – and would lead to numeric overflow in the *combinatorial* computation for cases of $n \geq 13$ or $m \geq 13$. The limit on the input value for *factorial* and for the input values of *combinatorial* are expressed in the preconditions: callers of the operations should ensure these preconditions are satisfied. In a more complete library, probably operation versions using *long* integers would also be provided. In addition, *cached* versions of the operations could be provided, to provide additional optimization.

For the gcd, there is a well-known recursive computation:

```
static query gcd(x: int, y: int): int
pre: x >= 0 & y >= 0
post:
  (x = 0  =>  result = y) &
  (y = 0  =>  result = x) &
  (x = y  =>  result = x) &
  (x < y  =>  result = gcd(x, y mod x))  &
  (y < x  =>  result = gcd(x mod y, y))
```

This is considered too inefficient in general for use in the library, and is replaced by an explicit iterative algorithm, defined using the *Create* menu option *Operation Activity*:

```
static query gcd(x: int, y: int): int
pre: x >= 0 & y >= 0
post: true
activity:
```

```
l : int ; k : int ; l := x  ; k := y  ;
while l /= 0 & k /= 0 & l /= k
do
    if l < k then k := k mod l
    else l := l mod k ;
if l = 0 then result := k
else result := l ;
return result
```

This version is derived from the recursive version by the well-known program transformation 'replace recursion by iteration', and the clear relation between the activity and its declarative version increases confidence in its correctness. Formal proof using B refinement could be used if a high degree of assurance was needed. Note that activities should be written without using brackets, and with spaces used around all operators, including the sequence operator ';'.

From the gcd, the lcm can be directly calculated:

```
static query lcm(x: int, y: int): int
pre: x >= 1 & y >= 1
post: result = ( x * y ) / gcd(x,y)
```

To make this system into a library, set the system name to "mathfunctions" and select "Save system as library" after generating the design. This will produce separate Java files MathFunctions.java, SystemTypes.java and Controller.java in the output directory. Move these to a subdirectory called "mathfunctions" and compile them there. The specification file *mm.txt* of a library should also be placed in its directory, for reference by library users.

The library can be imported into another UML-RSDS application by defining an *external* class MathFunctions with the required operations declared (but without activities or detailed postconditions), and then using the *Add import* option to add the library as an imported package:

```
import mathfunctions.*;
```

The library operations can then be used in use cases of the importing system, for example in a postcondition:

```
MathFunctions.combinatorial(10,5)->display() &
MathFunctions.factorial(7)->display() &
MathFunctions.gcd(6,10)->display() &
MathFunctions.lcm(12,20)->display()
```

which tests the operations for different input values.

The output from these tests is:

```
252
5040
2
60
```

which is correct.

10.4 Publishing UML-RSDS applications as web services

Web services are applications which can be accessed by client systems via the internet. They provide operations which, in principle, any client system on any computer connected to the internet can call, independently of the platform and technology used at the client end. Web services are very powerful for connecting applications together and enabling reuse, however they have some limitations:

- Because they are invoked across the internet, there may be delays in web service request and response communications, and hence web services should not be used for time-critical functionalities, nor should frequent fine-grain requests be made. It would be a mistake to try to access a mathematical library such as *MathFunctions* as a web service.
- The data transmitted should be serialisable and not involve platform-specific objects.

UML-RSDS provides two alternative ways to publish an application as a (Java) web service:

REST or *Representational State Transfer* – services are accessed by URLs, and data transfer is by means of the HTTP protocol.

SOAP or *Simple Object Access Protocol* – services are described using a Web Service Description Language (WSDL) specification in an XML file. The services communicate with clients via XML messages in the SOAP format.

In either case, the communication between the client and web service supplier/host involves the client locating the service, constructing a request message to send to the supplier, and then receiving any response message from the supplier.

The share price estimator use case from Section 10.2.1 could be appropriate as a web service. To make it useful to remote callers, all its data needs to be supplied as parameters, so that it is independent of server-side data:

- *currentPrice* : *double*

- *timeDays* : *int* – the length of period considered for the prediction
- *growthRate* : *double*
- *variation* : *double*
- *runs* : *int* – the number of samples to be generated

The result parameter gives the estimated future price as a double. The name of the use case is changed to the more meaningful *estimateFuturePrice*, and the constraints are modified to be:

```
Share->exists( s | s.price = currentPrice & s.growth = growthRate &
                               s.variab = variation )


Share::
  Integer.subrange(1,runs)->forAll( j |
       Sample->exists( s | s.price = price & s.growth = growth &
                              s.variab = variab & s : samples ) )


Sample::
  Integer.subrange(1,timeDays)->forAll( t |
        price = price@pre + price@pre * growth +
                price@pre * variab * ( ( Math.random() * 2 ) - 1 ) )


Share::
  estimate = ( samples.price->sum() ) / runs

  result = Share.estimate->any()
```

The final constraint copies the estimated price of the single *Share* instance to the *result* parameter of the use case. As an alternative to *Math.random*(), the Apache math library class *NormalDistribution* and operation *sample* could preferably be used to obtain samples from the normal distribution N(0,1) [1].

Table 10.1 shows the time complexity of this version, using *initialPrice* = 100, *timeDays* = 50, *growth* = 0.1, *variability* = 0.09, and varying the number of samples. It can be concluded that the service is of adequate efficiency to use in practice.

Table 10.1: Execution time for share price estimator service

Number of samples	*Execution time (Java 4)*
100	40ms
1000	160ms
10,000	5,140ms

10.4.1 REST web service

This is generated by the option *Web Service*, *REST* on the Synthesis menu. This option produces a web interface for the system, consisting of a web page and JSP file for each use case, and a bean class which invokes the system controller. The architecture of Fig. 20.3 is used, with a facade for *Controller* instead of entity beans in the business tier. A web page and JSP for each use case are generated. These should be placed in a server directory *webapps/app/servlets*, where *app* is the system name of the UML-RSDS system.

Figure 10.3 shows the web page for the *estimateFuturePrice* web service. This submits a GET HTTP request to the following JSP:

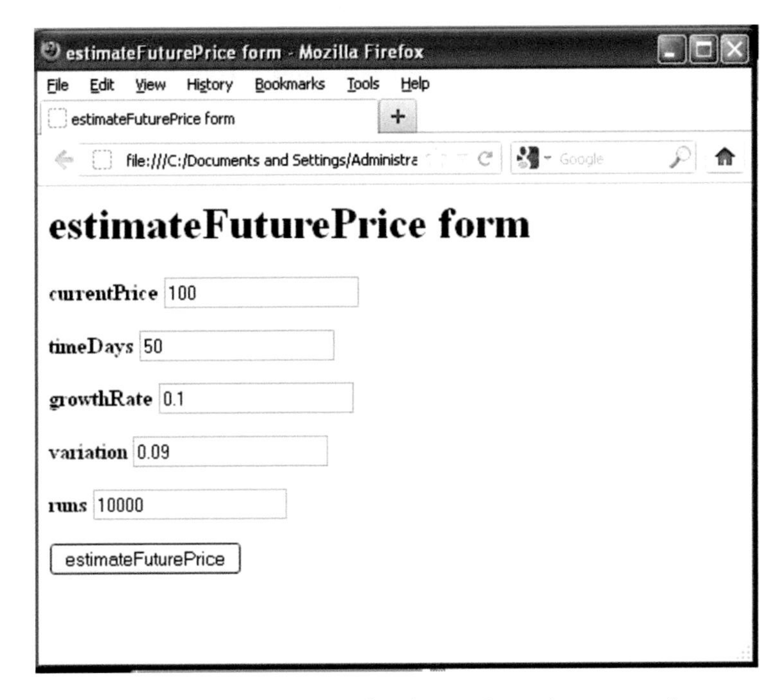

Figure 10.3: Web page for share price estimator service

```
<jsp:useBean id="bean" scope="request"
 class="beans.ControllerBean"/>
<jsp:setProperty name="bean"
property="currentPrice"  param="currentPrice"/>
<jsp:setProperty name="bean"
property="timeDays"  param="timeDays"/>
<jsp:setProperty name="bean"
property="growthRate"  param="growthRate"/>
<jsp:setProperty name="bean"
property="variation"  param="variation"/>
```

```
<jsp:setProperty name="bean"
property="runs"  param="runs"/>

<html>
<head><title>estimateFuturePrice</title></head>

<body>

<h1>estimateFuturePrice</h1>
<% bean.estimateFuturePrice(); %>
<h2>estimateFuturePrice performed</h2>
<strong> Result = </strong> <%= bean.getResult() %>

<hr>

</body>
</html>
```

The request is:

```
http://127.0.0.1:8080/app/servlets/estimateFuturePrice.jsp?
    currentPrice=100&timeDays=50&growthRate=0.1&variation=
     0.09&runs=10000
```

and this request could also be issued programmatically, e.g., using the Java URL class, instead of via a browser.

The JSP calls operations of *ControllerBean* to transfer the service parameters and to invoke the service:

```
package beans;

import java.util.*;

public class ControllerBean
{ Controller cont;

  public ControllerBean() { cont = Controller.inst(); }

  double currentPrice;
  int timeDays;
  double growthRate;
  double variation;
  int runs;
  String result;

  public void setcurrentPrice(String _s)
  {
    try { currentPrice = Double.parseDouble(_s);
     } catch (Exception _e) { return; }
  }
```

```
  public void settimeDays(String _s)
  {
    try { timeDays = Integer.parseInt(_s);
     } catch (Exception _e) { return; }
  }

  public void setgrowthRate(String _s)
  {
    try { growthRate = Double.parseDouble(_s);
     } catch (Exception _e) { return; }
  }

  public void setvariation(String _s)
  {
    try { variation = Double.parseDouble(_s);
     } catch (Exception _e) { return; }
  }

  public void setruns(String _s)
  {
    try { runs = Integer.parseInt(_s);
     } catch (Exception _e) { return; }
  }

  public void estimateFuturePrice()
  { result = "" + cont.estimateFuturePrice(currentPrice,
           timeDays,growthRate,variation,runs); }

  public String getResult() { return result; }
}
```

This class should be located in the *webapps/app/WEB-INF/classes/beans* directory, together with the *Controller* class and other Java files of the system. The *bean* object has *request* scope in the JSP, which means that it only exists for the specific service request, and its data is not retained for subsequent requests.

Figure 10.4 shows the result page returned by the *estimateFuturePrice* web service.

10.4.2 SOAP web service

This is generated by the option *Web Service*, *SOAP* on the Synthesis menu. This produces a SOAP web service specification in Java for the current system, with each use case defined as a web service method. For example, the share price estimator has the following SOAP specification:

```
import java.util.*;
```

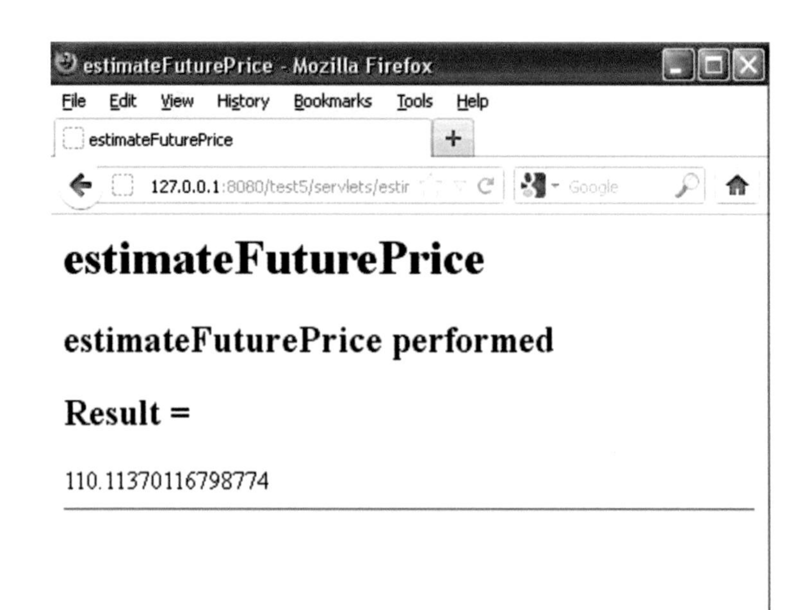

Figure 10.4: Share price estimator result page

```
import javax.jws.WebService;
import javax.jws.WebMethod;
import javax.jws.WebParam;

@WebService( name = "ControllerWebBean",
serviceName = "ControllerWebBeanService" )
public class ControllerWebBean
{ Controller cont;

  public ControllerWebBean() { cont = Controller.inst(); }

  @WebMethod( operationName = "estimateFuturePrice" )
  public String estimateFuturePrice
   (double currentPrice, int timeDays, double growthRate,
                    double variation, int runs)
  { return "" + cont.estimateFuturePrice
   (currentPrice,timeDays,growthRate,variation,runs); }

}
```

This instantiates the system controller and invokes it, in order to carry out requests to the web service. The class annotation @WebService declares ControllerWebBean as a Java web service, and the method annotation @WebMethod declares estimateFuturePrice as a operation which can be called as a web service.

Summary

This chapter has described how UML-RSDS systems can be composed together and how they can use external applications. We have given examples to show how libraries can be created and used, and how UML-RSDS applications can be used as components in other systems, or published as globally-available web services.

References

[1] Apache Commons Math library, http://commons.apache.org/proper/commons-math/apidocs/org/apache/commons/maths3/distribution/NormalDistribution.html.

Chapter 11

Migration Transformations

Migration transformations occur in many situations in MBD: data migration where existing business data stored in legacy repositories needs to be migrated into a new form of storage with an updated data schema; model migration where models structured according to one language/metamodel need to be translated into models structured according to an evolved/updated metamodel. Related operations are data cleansing and model merging, where errors in data need to be detected and removed, or where related data in different models need to be merged and mapped to a new model.

11.1 Characteristics of migration transformations

Migration transformations are typically separate-models transformations, with read-only source models. They should therefore normally consist of type 0 or type 1 constraints only, that is, constraints which write and read separate data items, and have bounded loop implementations. The functional requirements for migration transformations concern how elements of the source language (the entity types and entity type features) should be mapped to elements of the target language, for example "for each instance of source entity type $S1$ there should exist a corresponding instance of target entity type $T1$ such that ...". Such requirements can be expressed at a high level by *language mappings* which define the corresponding target language elements of each source language element in the scope of the transformation. For example, a mapping χ

$$\begin{array}{lcl} S1 & \longmapsto & T1 \\ S2 & \longmapsto & T1 \\ S1 :: f & \longmapsto & T1 :: g \\ S2 :: h & \longmapsto & T1 :: p \end{array}$$

indicates that source entity types $S1$ and $S2$ map to target entity type $T1$ (any instance of the source types should be migrated to an instance of $T1$), and that feature f of $S1$ is represented by feature g of $T1$, and feature h of $S2$ is represented by feature p of $T1$. From such abstract requirements, outline UML-RSDS rules can be derived:

```
S1::
  T1->exists( t1 | t1.g = f )

S2::
  T1->exists( t1 | t1.p = h )
```

Transformation patterns such as Phased Construction, Structure Preservation, Entity Merging, Entity Splitting and Map Objects before Links are particularly relevant for migration transformations. The above example is a case of Entity Merging. Object Indexing and Unique Instantiation are also important, to support tracing of target model elements back to source model elements, and to avoid creating duplicated target elements.

In this chapter we consider two migration examples: (i) migration of Eclipse GMF graphical data models from GMF version 1.0 to GMF version 2.1, and (ii) mapping of ATL transformation specifications to UML-RSDS.

11.2 Case study: GMF model migration

This case study is a re-expression transformation which involves a complex restructuring of the data of a model: indirection between objects is introduced in the target language, so that actual figures in the source model are replaced by references to figures in the target model, and references from a figure to its subfigures are recorded by explicit objects.

Eclipse GMF (Graphical Modeling Framework) is a model-based approach for generating customised graphical editors for modelling languages (www.eclipse.org/modeling/gmp). Between versions 1.0 and 2.1 of GMF there were significant changes in how graphical languages were represented, leading to the need to perform migration from the old to the updated versions of GMF. Figure 11.1 shows the unified metamodels of the source (GMF version 1.0) and target (GMF version 2.1) languages. Since most of the data of a model may remain unchanged by the transformation, we specify the transformation as an update-in-place mapping on this combined metamodel. An alternative would be to copy the unchanged data to the target model, using a pattern such as Structure Preservation. *Figure*1 is the target metamodel version of the *Figure* class, *figures*1 is the target version of the gallery figure list association end.

The migration can be abstractly expressed by the language mapping:

$$\begin{array}{l}
Figure \longmapsto RealFigure \\
Figure :: name \longmapsto RealFigure :: name \\
Figure :: children \longmapsto RealFigure :: children \\
FigureGallery \longmapsto FigureGallery \\
FigureGallery :: figures \longmapsto FigureGallery :: figures1
\end{array}$$

The entity types *DiagramElement*, *Node*, *Compartment*, *Connection*, *DiagramLabel*, *Canvas* and their source language features are mapped identically to the target language. In this transformation there are also refinement aspects, with new entity types (*FigureDescriptor* and *ChildAccess*) being introduced in the target language to hold more detailed information than is present in the source language.

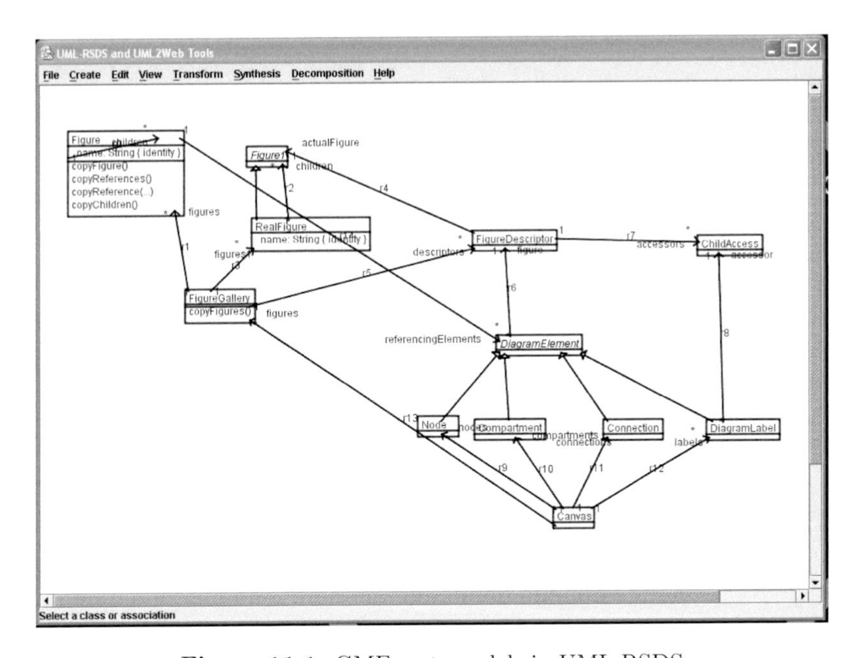

Figure 11.1: GMF metamodels in UML-RSDS

We assume in *Asm* that the input model is a syntactically correct GMF version 1.0 model and that the new entities have no instances:

$$Figure1 = Set\{\}$$
$$FigureDescriptor = Set\{\}$$
$$ChildAccess = Set\{\}$$

We also assume that *name* is an identity attribute (unique identifier) for *Figure* and for *RealFigure*:

$$Figure \rightarrow isUnique(name)$$
$$RealFigure \rightarrow isUnique(name)$$

For simplicity of specification, we decompose the transformation into a first transformation which creates the new data from the old, without deleting any data, and a second transformation which removes the version 1.0 data which is not in version 2.1. This is an example of the Construction and Cleanup design pattern (Chapter 9).

The first transformation is specified by a use case *createTarget* with the following postcondition constraints (C1), (C2), (C3), (C4). (C1) is:

```
Figure::
  RealFigure->exists( rf | rf.name = name &
     FigureDescriptor->exists( fd | fd.actualFigure = rf ) )
```

For each source model figure, there is a unique target model real figure, with an associated figure descriptor.

(C2) is:

```
Figure::
  RealFigure[name].children = RealFigure[children.name]
```

For each source model figure f, the corresponding target model real figure *RealFigure*[f.*name*] has as its children the children (real figures) corresponding to the children of f. This is an example of the Map Objects Before Links pattern: Figures are mapped to RealFigures by (C1), then parent-child links between RealFigures are created from those of Figures by (C2). The pattern is needed because *children* is a reflexive association in the source language. Both (C1) and (C2) are of type 1.

(C3) is:

```
FigureGallery::
  figures1 = RealFigure[figures.name]  &
  descriptors = FigureDescriptor->select(actualFigure : figures1)
```

For each figure gallery, its figures (*figures*1) in the target model are the real figures corresponding to the source model figures of the gallery, its descriptors are the descriptors of these figures. Although in this constraint *figures*1 is both written and read, the update only affects the local data of one *FigureGallery* object *fg*, and no other object is modified, so no other application of the rule is affected. Thus the rule is effectively of type 1: it could be written equivalently (but less efficiently) as:

```
FigureGallery::
  figures1 = RealFigure[figures.name]  &
  descriptors = FigureDescriptor->select(actualFigure :
   RealFigure[figures.name])
```

Another alternative way of expressing this constraint is to use a let variable:

```
FigureGallery::
  fs = RealFigure[figures.name]  =>
                              figures1 = fs &
                              descriptors = FigureDescriptor->
                               select(actualFigure : fs)
```

(C4) is:

```
Figure::
fd : FigureDescriptor & d : referencingElements & fd.actualFigure =
RealFigure[name]   =>
    d.figure = fd &
    (d : DiagramLabel  =>
         ChildAccess->exists( ca | d.accessor = ca & ca :
          fd.accessors) )
```

The figure descriptor *fd* of a diagram element *d* in the target model is that corresponding to the figure which contained the element in the source model. If the diagram element is a label of a nested figure (the condition d : *DiagramLabel*), then an explicit child access object is created to record the access.

The second transformation, *cleanSource*, removes all instances of *Figure* and other source model data which is not needed in the target model:

```
Figure@pre->forAll( f | f.referencingElements = {} )
FigureGallery->forAll( fg | fg.figures = {} )
Figure->isDeleted()
```

11.3 Case study: migration of ATL specifications to UML-RSDS

The UML-RSDS tools provide an option to import ATL modules. This facility enables ATL developers to use UML-RSDS to analyse the ATL transformations, and to generate code for these in Java, C# or C++. It also provides a way to write UML-RSDS specifications (of a restricted kind) using ATL syntax. The translation from ATL to UML-RSDS maps from the abstract syntax of ATL (Fig. 11.2) into the UML-RSDS metamodel (Figs. 11.3, 11.4, and A.2), and can be viewed as a language migration transformation *atl2uml* similar to a programming language migration (e.g., from Visual Basic to Java).

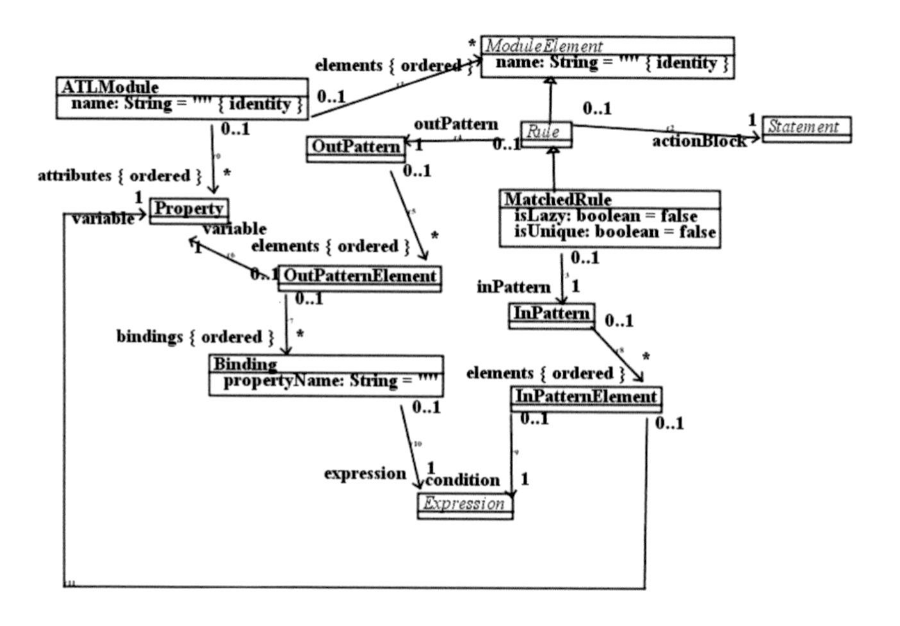

Figure 11.2: ATL metamodel

The informal initial idea of the mapping can be described as follows. Given an ATL module M consisting of a set of rules r1, ..., rn:

```
module M;
create OUT : T from IN : S;
rule r1
{ from s1 : S1, ..., sm : Sm (SCond)
  to t1 : T1 (TCond1), ..., tk : Tk (TCondk)
  do (Stat)
}
...
rule rn { ... }
```

the equivalent UML-RSDS specification is a use case M' with postconditions for each of the ri, and with owned attributes for the local attributes of *M*. A normal MatchedRule r1:

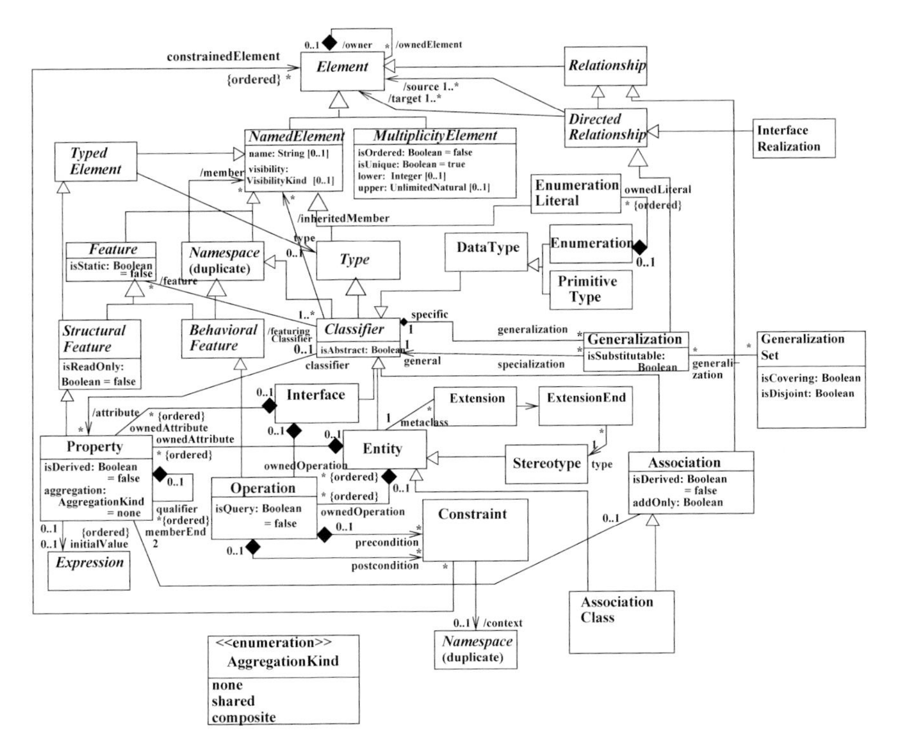

Figure 11.3: UML-RSDS class diagram metamodel

```
rule r1
{ from s1 : S1, ..., sm : Sm (SCond)
  to t1 : T1 (TCond1), ..., tk : Tk (TCondk)
  do (Stat)
}
```

is represented by a postcondition constraint $r1'$ of M':

$$
\begin{aligned}
&S1 :: \\
&\quad s2 : S2 \ \& \ ... \ \& \ sm : Sm \ \& \ SCond' \ \Rightarrow \\
&\qquad T1{\rightarrow}exists(t1 \mid t1.\$id = \$id \ \& \ ... \ \& \\
&\qquad Tk{\rightarrow}exists(tk \mid tk.\$id = \$id \ \& \ TCond1' \ \& \ ... \ \& \\
&\qquad\quad TCondk' \ \& \ s1.opr1(s2, ..., tk))...)
\end{aligned}
$$

where each expression E in ATL is mapped to a corresponding expression E' in UML-RSDS, and $opr1$ is an operation that represents the action block (*do* clause) if this is present. New identity attributes $\$id : String$ are introduced as new principal primary keys for each of the Si and Tj. Lazy and unique lazy rules are translated to operations – such rules are subordinate to matched rules and only execute if invoked from the matched rules.

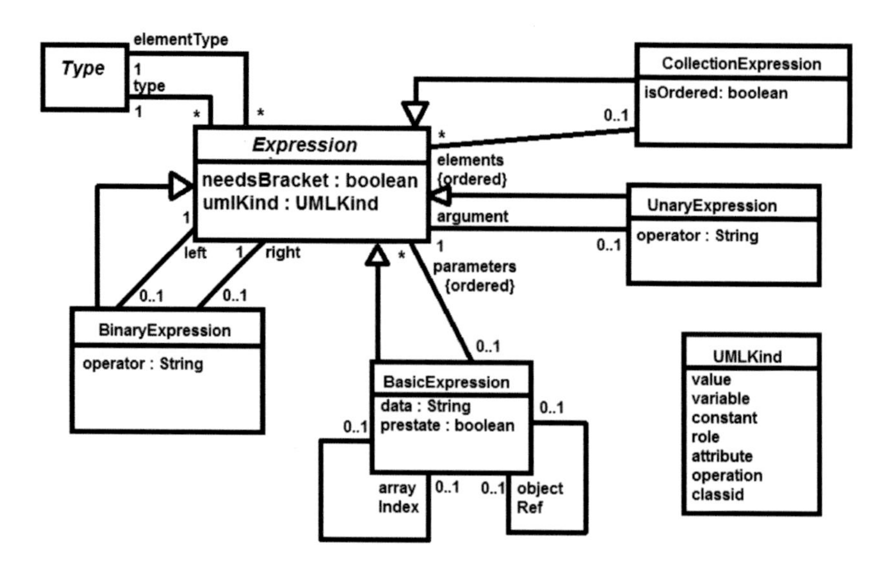

Figure 11.4: UML-RSDS expression metamodel

In order to correctly express the semantics of ATL, each ATL matched rule needs to be translated to two separate constraints in UML-RSDS: the first simply creates the tj objects and sets their $\$id$ values, the second performs the remaining actions of the rule, using object lookup on the $tj.\$id$. This corresponds to the ATL semantics in which a matching (target object creation) phase precedes initialisation (target object linking). This strategy is an example of the Map Objects Before Links pattern, and means that all ATL rules should translate to type 1 constraints in UML-RSDS.

As an example of the migration, the $a2b$ transformation of Section 7.3 expressed in ATL would be written as:

```
module M;
create OUT : T from IN : S;
rule r {
  from a : A (a.x > 0)
  to b : B
    ( y <- a.x*a.x )
  }
```

This would translate into a UML-RSDS use case with two constraints:

$$A ::$$
$$x > 0 \;\Rightarrow$$
$$B \rightarrow exists(b \mid b.\$id = \$id)$$

and

$$
\begin{array}{l}
A :: \\
\quad x > 0 \;\Rightarrow \\
\qquad\qquad B{\rightarrow}exists(b \mid b.\$id = \$id \;\&\; b.y = x * x)
\end{array}
$$

In the second constraint b is looked-up, and not re-created, since the object $B[\$id]$ has been created by the first constraint.

More formally, the migration from ATL to UML-RSDS can be described by a language mapping:

$$
\begin{array}{l}
ATLModule \;\longmapsto\; UseCase \\
MatchedRule \;\longmapsto\; Constraint + Constraint \\
(isLazy = false) \\
MatchedRule \;\longmapsto\; Operation \\
(isLazy = true) \\
InPattern \;\longmapsto\; Expression\ (conjunction\ of\ InPatternElement') \\
InPatternElement \;\longmapsto\; Expression \\
OutPattern \;\longmapsto\; Expression\ (conjunction\ of\ OutPatternElement') \\
OutPatternElement \;\longmapsto\; Expression\ (conjunction\ of\ Binding') \\
Binding \;\longmapsto\; Expression
\end{array}
$$

The output conditions $TCondj$ of an ATL rule are expressed as UML-RSDS expressions $TCondj'$ as follows. A Binding of *propertyName* f of tj to expression e:

$$f \;\leftarrow\; e$$

is interpreted as

$$tj.f = e'$$

where e' is the UML-RSDS interpretation of e. Any implicit type conversion from an expression p in e of a source entity type $SEnt$ to a target entity type $TEnt$ is expressed as

$$TEnt[p.\$id]$$

in e'. This is the target object corresponding to p. Likewise for implicit type conversions from $Set(SEnt)$ to $Set(TEnt)$ and from $Sequence(SEnt)$ to $Sequence(TEnt)$.

If an action block with code $Stat$ is specified, then a new update operation $opri(s2 : S2, ..., sm : Sm, t1 : T1, ..., tk : Tk)$ is introduced to $S1$, and this operation has activity given by the interpretation $Stat'$ of $Stat$ as a UML-RSDS activity. The translation of lazy and called rules r is similar to matched rules, except that the translated constraint r' is used as the postcondition of an operation r of the first input entity type of the rule. Calls $thisModule.r(v1, ..., vm)$ to the rule are interpreted as calls $v1.r(v2, ..., vm)$ of this operation.

This migration transformation *atl2uml* can itself be written in UML-RSDS (this is an example of a *higher-order transformation*: a transformation that produces transformations). The language mapping provides a good guide for the structure of the transformation, which could either be written in a top-down or bottom-up Phased Construction form. We have found that language translations of this kind are more naturally specified in a top-down manner, because there is usually some context information from an enclosing construct which needs to be passed down to the mapping rules for enclosed (subordinate) constructs. E.g., to process a Binding, we need to know the variable of its enclosing OutPatternElement.

At the topmost level, for each ATLModule, a corresponding UseCase must be created:

$$\begin{array}{l} ATLModule :: \\ \quad UseCase{\rightarrow}exists(uc \mid uc.name = name \ \& \\ \quad\quad uc.ownedAttribute = attributes) \end{array}$$

For all non-lazy rules of the ATLModule, a postcondition constraint of the corresponding use case needs to be created:

$$\begin{array}{l} ATLModule :: \\ uc = UseCase[name] \ \& \ r : MatchedRule \ \& \ r : elements \ \& \\ \quad r.isLazy = false \quad \Rightarrow Constraint{\rightarrow}exists(c \mid c.id = \\ \quad\quad r.name \ \& \ c : uc.orderedPostconditions) \end{array}$$

In turn, the InPattern of the matched rule forms the antecedent of the corresponding constraint:

$$\begin{array}{l} MatchedRule :: \\ isLazy = false \ \& \ c = Constraint[name] \ \& \ ipe : inPattern.elements \ \Rightarrow \\ \quad BasicExpression{\rightarrow}exists(varexp \mid varexp.data = ipe.variable.name \ \& \\ \quad\quad varexp.type = ipe.variable.type \ \& \\ \quad\quad varexp.umlKind = variable \ \& \\ \quad\quad BasicExpression{\rightarrow}exists(typeexp \mid typeexp.data = ipe.variable. \\ \quad\quad type.name \ \& \\ \quad\quad\quad typeexp.type = SetType \ \& \ typeexp.umlKind = entity \ \& \\ \quad\quad\quad typeexp.elementType = ipe.variable.type \ \& \\ \quad\quad\quad BinaryExpression{\rightarrow}exists(inexp \mid inexp.operator = \text{“:”} \ \& \\ \quad\quad\quad\quad inexp.left = varexp \ \& \ inexp.right = typeexp \ \& \\ \quad\quad\quad\quad BinaryExpression{\rightarrow}exists(condexp \mid condexp.operator \\ \quad\quad\quad\quad = \text{“\&”} \ \& \\ \quad\quad\quad\quad\quad condexp.left = inexp \ \& \\ \quad\quad\quad\quad\quad condexp.right = ipe.condition \ \& \\ \quad\quad\quad\quad\quad c.conjoinCondition(condexp))))) \end{array}$$

This constraint conjoins the formula $si : Si \; \& \; SCondi$ to the constraint antecedent for the rule. This shows how prolix working in abstract syntax can be: each target language element has to be created and its features set in complete detail (in fact some details are omitted here for readability). Similar (but more complex) mapping rules handle OutPatterns and lazy rules. The completed transformation has been incorporated into UML-RSDS, along with similar transformations for mapping ETL, Flock and QVT-R into UML-RSDS.

Summary

In this chapter we have considered the special aspects of migration transformations, and we have given two examples of how these can be specified in UML-RSDS.

Chapter 12

Refinement and Enhancement Transformations

Refinement transformations are an example of a *vertical* transformation: they map a model at one level of abstraction to a model at a lower level of abstraction. Such transformations are used in MDA and MDD to produce Platform-specific Models (PSMs) from Platform-independent Models (PIMs), and to generate executable code from PSMs. A code generation transformation is a refinement transformation that generates source code text in a programming language. It is an example of a *model-to-text* transformation. A refinement transformation can also be used to enrich a model with more specific and detailed information. For example, a transformation could map syntax-oriented descriptions into semantics-oriented descriptions. In this case it can be considered to be an enhancement transformation and to not change the abstraction level (a *horizontal transformation*).

UML-RSDS makes use of several refinement transformations in its processing: central to the UML-RSDS process is a specification to design transformation which derives a procedurally-oriented version of each use case and operation from their logical specifications (the transformation maps postcondition expressions E to activities $stat(E)$, and determines what algorithm should be used to implement use case postconditions – fixed-point or bounded iterations). There are also code generation transformations which start from the design model and produce Java 4, Java 6, C# or C++ code. Specialised transformations map EIS descriptions to EIS code (Chapter 20).

Refinement transformations are usually separate-models transformations, operating on a read-only source model and producing a target model which is initially empty. Thus their constraints in UML-RSDS should normally be of type 0 or type 1. Each constraint would not normally be re-applied to the same element more than once in a transformation execution (all constraints are implemented by bounded loops). Refinement transformations may be expected to satisfy the property of *conservativeness* (also termed *model coverage*): all information in the target model is derived from the source model, and no new information has been added. To prove such a property, we use transformation invariants to express that all elements of the target model are derived from source model elements. Traceability of target elements back to the source elements they originate from is typically implemented by means of primary key attributes: target element $t : T_j$ corresponds to source element $s : S_i$ if $s.sId = t.tId$ for their respective principal primary keys, where the transformation maps S_i to T_j. In UML-RSDS, the S_i element corresponding to $t : T_j$ is $S_i[t.tId]$, and the T_j element corresponding to $s : S_i$ is $T_j[s.sId]$, when these elements exist.

Enhancement transformations may write new information to the source model – for example to derive all acyclic paths through a graph given by nodes and edges – but leave the initial data unchanged, and so their constraints can also satisfy the type 1 condition. The analysis of refinement and enhancement transformations is therefore technically simpler than for refactoring or bx transformations, however the size and complexity of code-generation transformations in particular requires that these transformations are carefully organised and structured. Transformation patterns such as Phased Construction, Entity Splitting, and Map Objects before Links are relevant for refinements, as are Object Indexing and Unique Instantiation. Sequential Composition, Auxiliary Metamodel and other modularisation patterns are often useful for decomposing a transformation into smaller sub-transformations, and to compose these sub-transformations into a complete transformation. Model-to-text patterns are relevant for code generators.

12.1 Enhancement transformation: computing inheritance depth

This transformation computes the inheritance depth of each class in a UML class diagram, and adds this information as an auxiliary attribute value *depth* to each generalisation. The class diagram metamodel of Fig.

2.5 is used, with *depth* : *int* added to *Generalization*. It is assumed that the generalisation relationship is acyclic.

The first postcondition constraint of this transformation sets the generalisation depth to 1 if the generalisation points to a root class ($C1$):

```
Generalization::
  general.generalisation.size = 0  =>  depth = 1
```

This constraint is of type 1.

If the superclass is not a root class, set the depth to be one more than the maximum depth of the generalisations starting from the superclass of the generalisation ($C2$):

```
Generalization::
  general.generalisation.size > 0  =>
      depth = 1 + general.generalisation.depth->max()
```

This constraint is of type 2 (*Generalization* :: *depth* is both read and written in the succedent), and hence needs to be implemented by a fixed-point iteration. The constraint is applied repeatedly to all generalisations in the model until the implication holds true for all elements. Because termination and confluence are not immediately satisfied by this transformation, a semi-formal proof of these properties is necessary.

It can be argued (i) by induction on the transformation steps that the value of *depth* is always lower than the actual depth:

$$\begin{array}{l} Generalization :: \\ \qquad depth \leq actualDepth \end{array}$$

and that the value of *depth* for a generalisation is always no more than the maximum of the immediately higher generalisations *depth* values, plus 1:

$$\begin{array}{l} Generalization :: \\ \qquad depth \leq 1 + general.generalisation.depth{\rightarrow}max() \end{array}$$

These properties could be proved by invariance proof in B (Chapter 18).

(ii) Each transformation step reduces the difference *actualDepth* − *depth* for at least one generalisation, and does not increase any difference, because some depth is always increased in each step, and none is decreased, and by (i) the depth cannot exceed the actual depth. This means that

$$\Sigma_{g:Generalization}(g.actualDepth - g.depth)$$

is a variant. Since this is an integer quantity bounded below by 0, the transformation terminates.

(iii) Termination coincides with (C2) being true for all non-topmost generalisations, and with the variant being equal to 0. Since there is a unique situation in which this occurs (all depths equal the actual depth), the transformation is confluent.

12.2 Refinement transformation: UML-RSDS to C code generation

UML-RSDS already contains code generators for two versions of Java, and for C# and C++. If a code generator for a new language or language version is required, this can be written as a UML-RSDS transformation from the design (PSM) model of a UML-RSDS application to the abstract and concrete syntax of the new target language. To illustrate this process, we consider the task of generating ANSI C code from UML. Firstly, for each UML-RSDS design language element, the intended C language elements and constructs that should implement this element must be determined (these mapping intents are identified at the informal requirements analysis stage). Various possibilities can be considered, for example the UML-RSDS *boolean* type could be represented by a C `enum`, by a `#define` or by `unsigned char`. Many-multiplicity association ends could be represented by resizable arrays, or by linked list structures. List structures are more flexible, but are less efficient than an array representation. Table 12.1 shows the informal mapping for classes, types and attributes. This shows the schematic concrete grammar for the C elements representing the UML concepts.

Table 12.1: Informal mapping of UML class diagrams to C

UML element e	*C representation* e'
Class E	`struct E { ... };`
Property $p : T$	member `T' p;` of the struct for p's owner, where `T'` represents T

String type
int, long, double types
boolean type
true
false
	`char*`
	same-named C types
	`unsigned char`
	`#define TRUE 1`
	`#define FALSE 0`

Enumeration type	C `enum`
Entity type E	`struct E*` type, and `struct E* newE()` operation
Set(E) type *Sequence*(E) type	`struct E**` (array of E', without duplicates) `struct E**` (array of E', possibly with duplicates)
Operation $op(p : P) : T$ of E	C operation `T' op(E' self, P' p)`

These mappings can then be formalised as UML-RSDS rules, defining the postconditions of a transformation *design2C*. This has input language the UML-RSDS metamodels (Figs. 11.3, A.2, A.1) and output language a simple syntax-directed representation of C programs and data declarations (Fig. 12.1 shows a fragment of this).

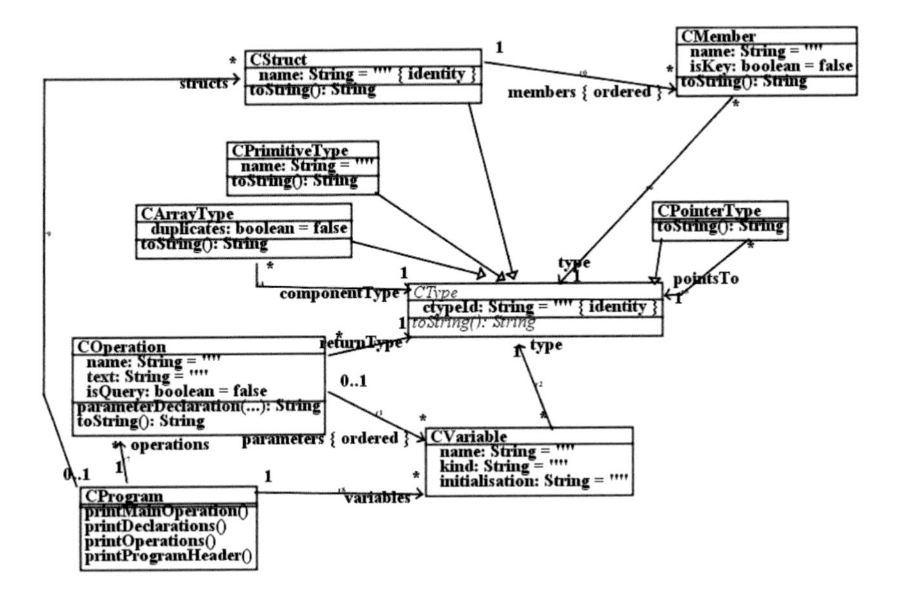

Figure 12.1: C language metamodel

To map classes to struct types, we have the rule:

```
Entity::
  CStruct->exists( c | c.name = name )
```

This rule assumes that class names are unique in the source model.

Then the attributes (and association ends) owned by a class are mapped to members of its corresponding struct:

```
Entity::
  c = CStruct[name] & p : ownedAttribute  =>
        CMember->exists( m | m.name = p.name &
                    m.type = Design2C.type2C(p.type) & m :
                    c.members )
```

where *type2C* is an operation of *design2C* which maps UML types to the text of C types. Note that use case operations and attributes are always static and their names should be preceded by the use case name with an initial capital.

Other rules are also needed to map the C language metamodel elements to text. A model-to-text transformation *genCtext* performs this process:

```
CStruct::
  ("struct " + name)->display() &
  "{"->display() &
  members->forAll( m | ("   " + m.type + "   "
   + m.name + ";")->display() ) &   "};\n"->display()
```

Decomposing the code generator into two sub-transformations improves its modularity, and simplifies the constraints, which would otherwise need to combine language translation and text production. Figure 12.2 shows the resulting transformation architecture.

UML-RSDS
Design
Metamodel
C Language
Metamodel
C Text
design2C
genCtext

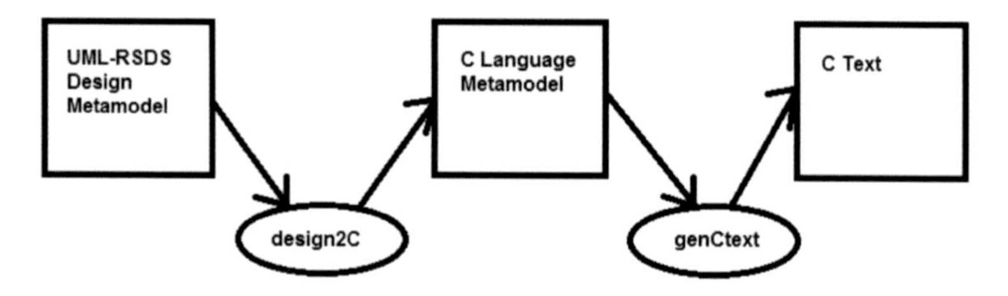

Figure 12.2: C code generator architecture

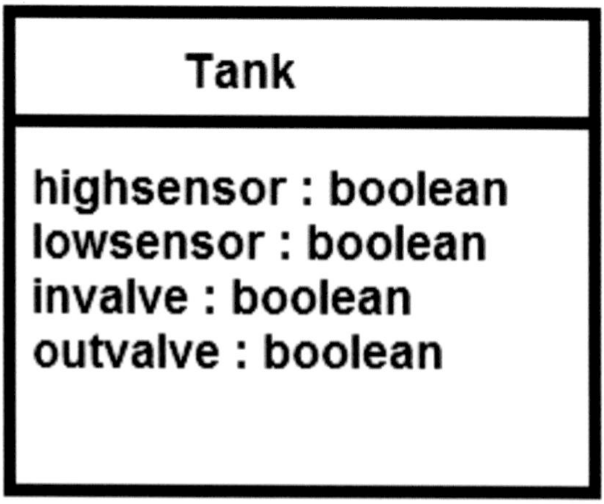

Tank
highsensor : boolean
lowsensor : boolean
invalve : boolean
outvalve : boolean

Figure 12.3: Tank class diagram

For the *Tank* class of Fig. 12.3, the code generator produces the declaration in the following program, which also shows the intended mapping of instance operations and of object creation:

```
#include <stdio.h>
#include <ctype.h>
#include <string.h>
#include <stdlib.h>

#define TRUE 1
#define FALSE 0

struct Tank
{
  unsigned char highsensor;
  unsigned char lowsensor;
  unsigned char invalve;
  unsigned char outvalve;
};

void op(struct Tank* self)
{ if (self->highsensor == TRUE)
  { self->invalve = FALSE;
    self->outvalve = TRUE;
  }
}

int main()
{ struct Tank* t1 = (struct Tank*) malloc(sizeof(struct Tank));
  t1->highsensor = TRUE;
  t1->lowsensor = FALSE;
  op(t1);
  return 0;
}
```

The model data corresponding to Fig. 12.3, and used as input to *design2C*, is the following instance model of the metamodel of Fig. 11.3:

```
Integer : PrimitiveType
Integer.name = "Integer"
Boolean : PrimitiveType
Boolean.name = "Boolean"
Real : PrimitiveType
Real.name = "Real"
String : PrimitiveType
String.name = "String"
SetType : CollectionType
SetType.name = "Set"
SequenceType : CollectionType
```

```
SequenceType.name = "Sequence"
Tank : Entity
Tank.name = "Tank"

highsensor_Tank : Property
highsensor_Tank.name = "highsensor"
highsensor_Tank : Tank.ownedAttribute
highsensor_Tank.type = Boolean
highsensor_Tank.lower = 1
highsensor_Tank.upper = 1
lowsensor_Tank : Property
lowsensor_Tank.name = "lowsensor"
lowsensor_Tank : Tank.ownedAttribute
lowsensor_Tank.type = Boolean
lowsensor_Tank.lower = 1
lowsensor_Tank.upper = 1
invalve_Tank : Property
invalve_Tank.name = "invalve"
invalve_Tank : Tank.ownedAttribute
invalve_Tank.type = Boolean
invalve_Tank.lower = 1
invalve_Tank.upper = 1
outvalve_Tank : Property
outvalve_Tank.name = "outvalve"
outvalve_Tank : Tank.ownedAttribute
outvalve_Tank.type = Boolean
outvalve_Tank.lower = 1
outvalve_Tank.upper = 1
```

The following output (an instance model of Fig. 12.1) is produced from *design2C*:

```
cstructx_0 : CStruct
cstructx_0.name = "Tank"
cmemberx_0 : CMember
cmemberx_0.name = "highsensor"
cmemberx_0.type = "unsigned char"
cmemberx_1 : CMember
cmemberx_1.name = "lowsensor"
cmemberx_1.type = "unsigned char"
cmemberx_2 : CMember
cmemberx_2.name = "invalve"
cmemberx_2.type = "unsigned char"
cmemberx_3 : CMember
cmemberx_3.name = "outvalve"
cmemberx_3.type = "unsigned char"
cmemberx_0 : cstructx_0.members
cmemberx_1 : cstructx_0.members
cmemberx_2 : cstructx_0.members
cmemberx_3 : cstructx_0.members
```

Any code-generator from UML-RSDS designs can be structured into four main parts based on the main UML-RSDS design language divisions:

- Class diagrams (classes, types, attributes, associations, inheritances);
- Expressions;
- Statements (activities);
- Use cases.

Following the usual UML-RSDS code generation approach, UML/OCL expressions should be mapped to C-language strings via two query operations *queryFormC*() : *String* and *updateFormC*() : *String* of *Expression*, which provide the C equivalent of the expression as a query and as an update. Statements may also be mapped to strings, or to a metamodel of C statements, via an operation *updateFormC*() of *Statement*. Use case postconditions will have already been translated to activities via the synthesis process described in Chapter 6.

The UML-RSDS class diagram and use case metamodels are defined in the file `umlrsdscdmm.txt` distributed with the UML-RSDS tools, the expression metamodel is in `umlrsdsexpmm.txt`, and the activity metamodel is in `umlrsdsstatmm.txt`. These metamodels should be loaded into UML-RSDS in this order if the complete language metamodel is needed. Files of specification or design UML-RSDS models (saved using the *Save as* → *Save as model* option on the File menu) conform to this metamodel and can be processed as instance models by transformations such as *design2C*.

In addition to the data features of individual classes, a code generator such as the mapping to C should also handle operations of classes, and generate the *Controller*/Facade code for each application, including:

- Lists of the existing objects of each class;
- Operations to load and save models;
- Indexing maps for classes with identity attributes, to look up their instances by id value;
- Operations for the use cases of the system;
- Global operations on associations (synchronisation of the two ends of bidirectional associations), and on entities (operations to create and delete objects).

As this example shows, developing a code generator is a substantial project, which needs to be organised in a systematic and modular manner based on the source language structure. The three primary considerations for such a transformation are: (i) model-level semantic correctness: does the generated program correctly express the source model semantics? (ii) conformance (syntactic correctness) of the generated text to the target language definition; (iii) efficiency of the resulting program. For critical applications, conservativeness is also required. For new UML-RSDS code generators we recommend following the same translator architecture which has been used for Java/C#/C++ where possible (Fig. 12.4), and to structure the generated code based closely on the source model, so that it is easy for readers of the code to relate code parts to model elements.

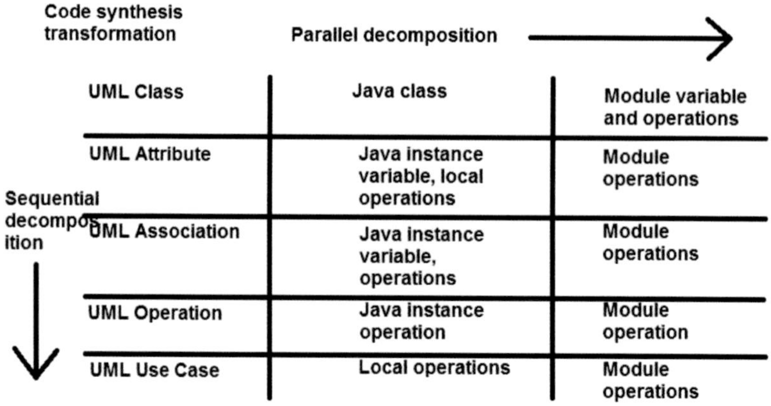

Code synthesis transformation — Parallel decomposition →

Sequential decomposition ↓

UML Class	Java class	Module variable and operations
UML Attribute	Java instance variable, local operations	Module operations
UML Association	Java instance variable, operations	Module operations
UML Operation	Java instance operation	Module operation
UML Use Case	Local operations	Module operations

Figure 12.4: UML-RSDS code generator structure

To establish conservativeness for the UML to C code generator, we define transformation invariants for *design2C*, these invariants express that the C representation is derived entirely from the source UML-RSDS design. They also provide a form of inverse transformation, and can be derived semi-automatically from the forward transformation rules (Chapter 14):

```
CStruct::
  Entity->exists( e | e.name = name )
```

which expresses that every *CStruct* is derived from a corresponding class, and:

```
CStruct::
  e = Entity[name] & m : members  =>
        Property->exists( p | p.name = m.name &
                p.type = C2Design.type2UML(m.type) & p :
                 e.ownedAttribute )
```

which expresses that every member is derived from a corresponding UML Property. *type2UML* is an inverse function for *type2C*.

Summary

We have given examples of enhancement and refinement transformations, and illustrated how UML-RSDS can be extended with new code generators. Guidelines for the organisation of code generation transformations have been provided.

Chapter 13

Refactoring and Update-in-place Transformations

Transformations which update a single model in-place are of wide application, most typically for model refactoring, but also for other uses such as system simulation. Such transformations are usually *horizontal*, because they do not change the level of abstraction of the model they operate upon.

Update-in-place transformations are potentially more complex than separate models transformations, because the source model is both read and written. Conflicts between rules, and failure of confluence are a significant issue for such transformations, and verification is potentially more difficult than for separate-models transformations. Update-in-place transformations may have type 2 and type 3 constraints, with rules that are potentially re-applied multiple times to a single source model element, however in some cases a pattern such as Replace Fixed-point by Bounded Iteration (Chapter 9) can be used to reduce the implementation complexity to a type 1 case. Other optimisation patterns, such as Omit Negative Application Conditions and Restrict Input Ranges, are frequently useful for refactoring transformations. Characteristic of refactoring transformations are constraints whose effect contradicts their condition:

E ::

$$A \Rightarrow B$$

where B logically implies $not(A)$. Such a constraint can only be established by making A false for all applicable instances of E. A good candidate for a variant of the constraint is the number of E instances that satisfy A:

$$E \rightarrow select(A) \rightarrow size()$$

Each application of the constraint should reduce this number.

In this chapter we describe some examples of refactoring transformations, and we give guidelines for reducing the complexity of this form of transformation.

13.1 Case study: repotting geraniums

A simple example of a refactoring is the 'repotting geraniums' example of [4] (Fig. 13.1). This example is however beyond the capabilities of some transformation languages because of its use of nested quantification.

The idea of the transformation is to replace every broken pot p that contains some flowering plant by a new (unbroken) pot p', and to transfer all the flowering plants of p to p'.

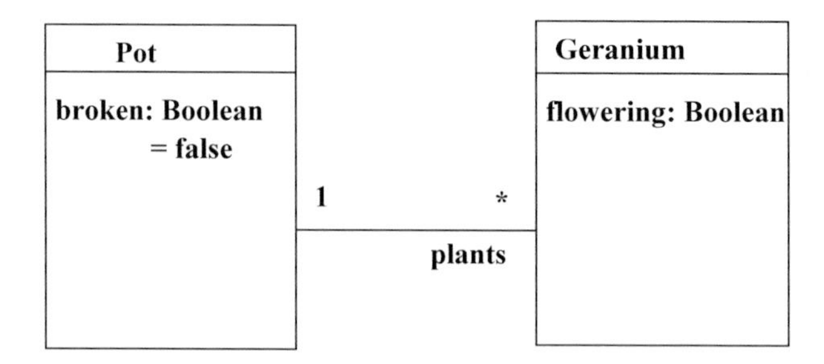

Figure 13.1: Repotting geraniums metamodel

This problem can be specified by a single UML-RSDS rule with context *Pot*:

```
Pot::
  broken = true &
  v = plants->select( flowering = true ) &
  v->size() > 0    =>
         Pot->exists( p1 | p1.plants = v )
```

v is a let-variable which holds the set of flowering plants in the broken pot. It has a constant value throughout the constraint. The introduction of such a variable avoids the need to repeat the complex expression it is defined by, and so helps to make the constraint more readable and more efficient.

Implicit in the effect of the constraint is the removal of the elements of v from *self.plants*: this occurs because of the multiplicity of the *Pot*—*Geranium* association (a plant cannot belong to two pots). Notice that even though new pots are created by applications of the rule, the collection of pots satisfying the application condition (the antecedent of the constraint) is reduced by each application: after the application there is one less broken pot – *self* – containing flowering plants. This means that a bounded loop can be used for the implementation, instead of a fixed-point iteration. Bounded-loop implementation can be enforced by writing the constraint as an iteration over *Pot*@pre, and using pre-forms for the features that are also written in the constraint:

```
Pot@pre::
  broken@pre = true &
  v = plants@pre->select( flowering = true ) &
  v->size() > 0    =>
        Pot->exists( p1 | p1.plants = v )
```

This is an application of the Replace Fixed-point by Bounded Iteration pattern (Chapter 9) and reduces the constraint to type 1.[1] Termination follows, and confluence holds because there is a unique state where the transformation terminates: the state where all originally broken pots containing flowering plants (call this set *brokenflowering*) now have no flowering plants (their other plants are unchanged), no other original pots are modified, and there are new pots for each of the *brokenflowering* pots.

13.2 Case study: state machine refactoring

UML state machines can be restructured to improve their clarity and to reduce their complexity. For example, if several states each have the same outgoing transition behaviour (they each have a transition with the same event label and the same target state), then these states can be placed in one composite state (if they are not already in such a state) and the multiple transitions replaced by a single transition from the composite state (Fig. 13.2).

[1] It is not possible to use @*pre* within a select expression, however.

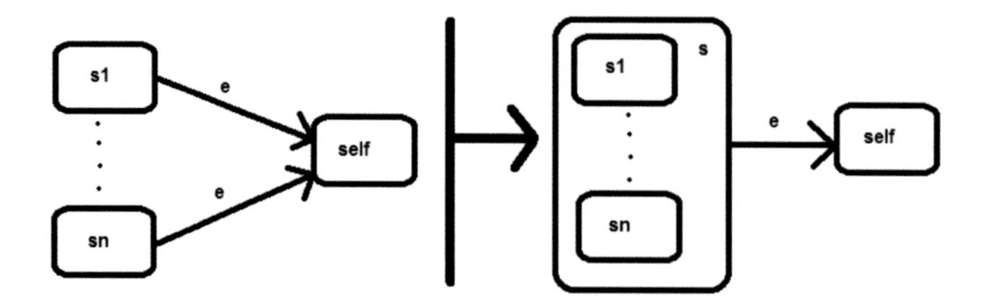

Figure 13.2: Introduce composite state refactoring

The general goal of our state machine refactoring transformation is therefore to introduce composite states to group together states which all have some common outgoing transition. The simplified state machine metamodel of Fig. 13.3 will be used as the source/target language for the transformation.

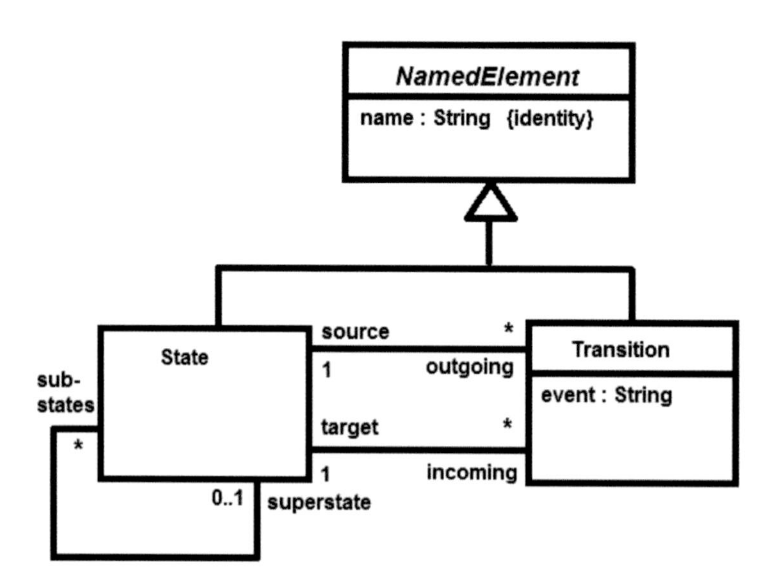

Figure 13.3: Simplified state machine metamodel

Informally, the transformation local functional requirements can be expressed as the following scenario, visually expressed by the concrete grammar sketch of Fig. 13.2:

> If there is a set *tt* of two or more transitions, all of which have the same target state *self*, and the same event label e, and none of the source states of *tt* are contained in a composite state, then: (i) create a new composite state s and make the sources of *tt* the substates of s; (ii) create a new transition t from s to *self*, with event the common event of the *tt*; (iii) delete all the *tt*.

The complexity of the behaviour is evident even in this outline description. One approach to simplify such refactoring rules is to break their action into several steps, implemented as a sequence of separate subrules (e.g., see the class diagram refactoring solution in [5]). However, such an approach requires the introduction of new flag variables or other auxiliary data to enforce a particular flow-of-control, and complicates the implementation and its analysis, so we prefer to implement the informal scenarios by single rules where possible.

There is a non-functional requirement that models of up to 1000 states should be processed within 1 second. Syntactic correctness and termination are essential, but confluence is not required. The informal functional requirements are incomplete (no setting is given for the names of the new state and transition) and vague (should the largest possible set *tt* be taken, or just any set of two or more transitions? Is *self* permitted to be in the set of sources of the *tt*?). After resolving these issues, an initial formalisation can be written as the following prototype rule:

```
State::
  tt : Set(Transition) & tt.size > 1 &
  tt.target = Set{ self } & tt.event->size() = 1 &
  tt.source.superstate->size() = 0  &
  self /: tt.source   =>
        State->exists( s |
            s.name = tt.source.name->sum() &
            tt.source <: s.substates &
            Transition->exists( t |
                t.name = tt.name->sum() &
                t.source = s & t.target = self &
                t.event = tt.event->any() ) ) &
        tt->isDeleted()
```

The intent of the set quantification $tt : Set(Transition)$ is that *tt* should be a maximal set of transitions satisfying the remainder of the antecedent conditions.

Comparing this formalisation to the informal scenario, we can see that all aspects of the intended rule effects are expressed in the formalised version:

- The application condition "a set *tt* of two or more transitions with a common target state *self* and the same event label, and none of the source states of *tt* are contained in a composite state" is expressed by the constraint antecedent conjuncts: $tt.size > 1$ expresses that *tt* has at least two elements; $tt.target = Set\{self\}$ expresses that all the *tt* have *self* as their target; $tt.event{\rightarrow}size() = 1$ expresses that the *tt* all have the same event; $tt.source.superstate{\rightarrow}size() = 0$ expresses that $s1.superstate$ is empty for each $s1 : tt.source$, i.e., that none of these states has a superstate. Note that we have avoided writing quantified formulae for these assumptions, by using navigation expressions and equality instead – in general it is recommended to reduce the number of quantifiers in constraints, in order to improve their comprehensibility.

  In addition to these conditions, we require that *self* is not in the set of source states: $self\ /:\ tt.source$.

- The succedent (effect) of the constraint expresses the three required scenario updates to the model: update (i) is performed by the succedent part:

```
State->exists( s |
        s.name = tt.source.name->sum() &
        tt.source <: s.substates &
```

  The $\rightarrow sum()$ operator applied to a collection of strings concatenates all the strings. It is a convenient means to form new names, but does not guarantee the uniqueness of the composed name, so the syntactic correctness of this constraint (invariance of the state machine language constraints) could not be formally proved.

  Update (ii) is performed by the part:

```
Transition->exists( t |
        t.name = tt.name->sum() &
        t.source = s & t.target = self &
        t.event = tt.event->any() ) )  &
```

  and update (iii) by the part:

```
          tt->isDeleted()
```

We could use this constraint to prototype the transformation and execute it on some simple test cases. While it is correct with respect to the refactoring requirement, this version of the constraint has a number of efficiency problems, and needs to be revised to meet the efficiency requirement.

A quantified variable ranging over an entity type in a constraint antecedent, $q : E$, should be viewed with suspicion, and even more so a variable ranging over collections of E: $qs : Set(E)$. Such quantifications amount to a global search over all the instances of E in the input model, for each instance of the context entity of the constraint, and so can lead to a quadratic or worse time complexity for the constraint implementation. A set quantified antecedent variable (also referred to as *collection matching*) cannot be directly implemented in UML-RSDS for this reason. It can be simulated by using an auxiliary data structure and the Simulate Collection Matching pattern, however this also has severe efficiency limitations. One way to improve the rule is to apply the Restrict Input Ranges pattern (Chapter 9): the quantifier range $tt : Set(Transition)$ can be restricted to the potentially much smaller range $tt : Set(incoming)$, since only transitions incoming to *self* could possibly meet the remaining antecedent conditions. The condition $tt.target = Set\{self\}$ is then redundant and can be omitted. A further improvement is to apply the design principle of *Avoid Collection Matching* (Chapter 6), and replace the search for *tt* by an explicit construction of *tt* based on some incoming transition (the same approach was used in the specification of the class diagram refactoring in Chapter 2). This enables the set quantified variable *tt* to be replaced by an ordinary quantified variable t:

```
State::
  t : incoming &
  tt = incoming->select( event = t.event ) & tt.size > 1 &
  tt.source.superstate->size() = 0  &
  self /: tt.source   =>
        State->exists( s |
            s.name = tt.source.name->sum() &
            tt.source <: s.substates &
            Transition->exists( tr |
                tr.name = tt.name->sum() &
                tr.source = s & tr.target = self &
                tr.event = t.event ) ) &
      tt->isDeleted()
```

tt is the set of incoming transitions which have the same event as the specific transition *t*. Testing of this version shows that it meets the efficiency requirements. For example, a model with 1500 states and 1000 transitions can be processed (with 500 executions of the refactoring rule) in 250 ms. The succedent of the rule contradicts the antecedent (because in the succedent the states of *tt.source* have a superstate, contradicting the antecedent requirement that they do not) so the inbuilt optimisation pattern Omit Negative Application Conditions can be used in the design generation step (the user is prompted if this optimisation should be applied).

Termination of the transformation is clear because each application of the constraint reduces the number of transitions in the model by at least one. Thus $Transition{\rightarrow}size()$ is the basis for a variant. Syntactic correctness would follow if some means of generating unique names for the new states and transitions was used. Confluence fails, as some simple counter-examples show. Model-level semantic correctness could be formulated in terms of a *flattening* semantics: the semantics $Sem(m)$ of a state machine m is taken to be the flattened version $flatten(m)$ where all superstates have been removed and transitions from the superstates are replaced by transitions from each of the immediate substates (this is the reverse process of our refactoring rule). It is clear that the introduce composite state refactoring does not change $Sem(m)$.

Other refactorings such as promote substate transitions (if all substates of a state have transitions with the same event and target, replace these transitions by one from the superstate) could be specified as rules of the transformation in the same manner.

13.3 Case study: simulation of BPMN execution semantics

BPMN (www.bpmn.org) is an OMG standard notation for business process modelling. In order to facilitate analysis and understanding of BPMN models, an executable semantics for BPMN has been defined based on a Petri-net style formalism using tokens within a network. A model transformation can be defined to translate process specifications in BPMN 2.0 notation into their executable semantics, and in addition to simulate the execution of the process using this semantics. This second transformation is an update-in-place transformation, whose computation steps correspond to the process execution (or evolution) steps.

BPMN (Business Process Modeling Notation) [3] is an OMG standard for expressing workflows and business processes, using an elaborated version of UML Activity Diagram notation. The notation can express workflow patterns [1] and can help to systematise business process modelling. In order to avoid ambiguity in the meaning of BPMN diagrams, a formal semantics using Petri-net style token nets has been defined [2]. This gives a precise execution semantics to each valid BPMN diagram in terms of the permitted evolution of token markings of the nets, so that the correctness and behaviour of the process described by the diagram can be analysed.

Figure 13.4 shows the source and target metamodels and use cases of the UML-RSDS specification of the transformation. Note the use of the interface *FlowElementsContainer* and interface inheritance of this by *Process*, in order to permit a limited form of multiple inheritance for *Process*.

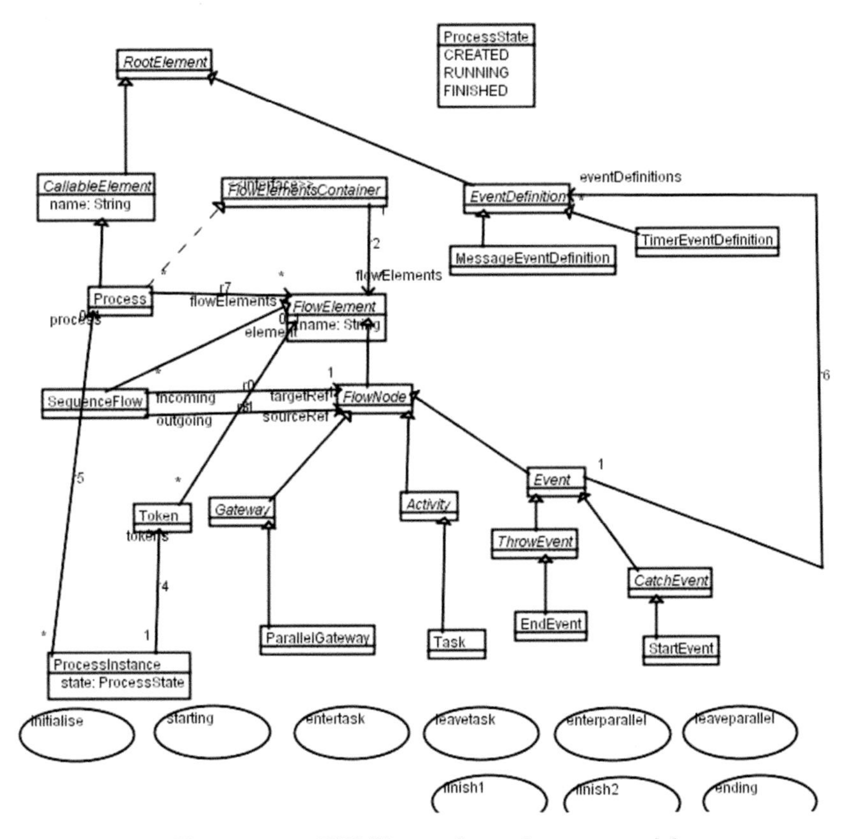

Figure 13.4: BPMN transformation metamodels

The transformation *initialise* maps the BPMN diagram model into its semantic representation in terms of process instances. Separate use cases are also defined for each of the separate situations of token movements, in order to provide interactive control of the execution semantics. The mapping from BPMN to the executable semantics is described by one rule, and the execution behaviour of the semantic representation is defined by several update-in-place rules defining how a process instance may evolve, and how its tokens may move around the process. Process instantiation is formalised by the use case *initialise*. Its postcondition has context *Process*:

```
sn : flowElements & sn : StartEvent &
sn.eventDefinitions->forAll( ed | ed :
 TimerEventDefinition ) =>
  ProcessInstance->exists( pi |
         pi.state = RUNNING &
         self : pi.process &
         Token->exists( t | t : pi.tokens & sn : t.element ) )
```

The effect of this rule can be understood as "If the process has a *StartEvent sn* which has only *TimerEventDefinitions*, create a process instance *pi* for the process, with one token at *sn*". This constraint is of type 1 because its write frame is disjoint from its read frame.

Normal termination of a process is expressed by the postconditions of use cases *finish*1, *finish*2, and these have context *ProcessInstance*:

```
state@pre = RUNNING &
process.flowElements->exists( e | e : EndEvent ) &
tokens@pre->forAll( t | t.element <: EndEvent )  =>
          state = FINISHED & tokens@pre->isDeleted()


state@pre = RUNNING &
process.flowElements->forAll( e | e /: EndEvent ) &
tokens@pre.element->forAll( n | n : FlowNode &
n.outgoing->size() = 0 )  =>
            state = FINISHED & tokens@pre->isDeleted()
```

Either (i) the process has an *EndEvent*, and all its tokens occupy *EndEvent* nodes, or (ii) the process has no *EndEvent*, and all its tokens occupy nodes with no outgoing flow. In either case the process is set to FINISHED and all its tokens deleted. We use `tokens@pre` and `state@pre` in the places where these features are read because we want to enforce a bounded loop implementation: we are only interested in a single process step. Any further steps enabled by this step will be simulated by the user invoking a use case for the step.

A process instance can start (use case *starting*) if it has a token *t* on a start event with at least one outgoing flow:

```
state = RUNNING & t : tokens &
```

```
fe : t.element@pre & fe : StartEvent &
fe.outgoing->size() > 0  =>
          fe.outgoing->exists( sf | t.element = Set{ sf } )
```

The succedent expresses that one of the outgoing flows *sf* of the start event is selected (*sf* is not created, because the argument *fe.outgoing* of the *exists* is not a concrete entity name) and the token *t* is moved to that flow.

If a process instance has a token on a *SequenceFlow* with target node an *EndEvent*, then the token can be moved to the *EndEvent* (use case *ending*):

```
state = RUNNING & t : tokens &
fe : t.element@pre &
fe : SequenceFlow &
fe.targetRef : EndEvent  =>
    t.element = Set{ fe.targetRef }
```

The same step applies if the target is a *Task* (use case *entertask*):

```
state = RUNNING & t : tokens &
fe : t.element@pre &
fe : SequenceFlow &
fe.targetRef : Task  =>
    t.element = Set{ fe.targetRef }
```

A process instance which has a token *t* on a *Task* *fe* can leave *fe* if *fe* has at least one outgoing flow (use case *leavetask*):

```
state = RUNNING & t : tokens@pre &
fe : t.element@pre &
fe : Task & fe.outgoing->size() > 0   =>
    t->isDeleted() &
    fe.outgoing->forAll( sf |
        Token->exists( t1 | sf : t1.element & t1 : tokens ) )
```

t is deleted, and new tokens are created for the process instance on each outgoing flow.

In order to enter a parallel gateway, there must be at most one token for a given process instance on each flow element. The process instance can enter parallel gateway *pg* if it has a token on every incoming flow of *pg*, and there is at least one such flow (use case *enterparallel*):

```
state = RUNNING &
pg : ParallelGateway &
v = tokens->select( t |
```

```
    pg.incoming->exists( sf | sf : t.element ) ) &
v.size > 0 &
v.size = pg.incoming->size()  =>
    Token->exists( t1 | pg : t1.element & t1 : tokens ) &
    v->isDeleted()
```

A single token $t1$ for the process instance on pg is then created, and the set v of the instance tokens on the incoming flows of pg is deleted. In this case the constraint requires fixed-point iteration, as it writes the same data ($Token :: element$) that it reads. The let variable v is used to store the pre-value of the expression it is assigned.

Leaving a parallel gateway is modelled by the following use case *leaveparallel* postcondition on *ProcessInstance*:

```
state = RUNNING & t : tokens@pre &
fe : t.element@pre &
fe : ParallelGateway &
fe.outgoing->size() > 0   =>
  t->isDeleted() &
  fe.outgoing->forAll( sf |
      Token->exists( t1 | sf : t1.element & t1 : tokens ) )
```

"If the process instance is running, and has a token t in a parallel gateway *fe*, with an outgoing flow, then delete t, and create a token for the process instance in each outgoing flow of *fe*."

A simple test case is that described in Fig. 13.5, with four tasks, two parallel gateways and a start and end node.

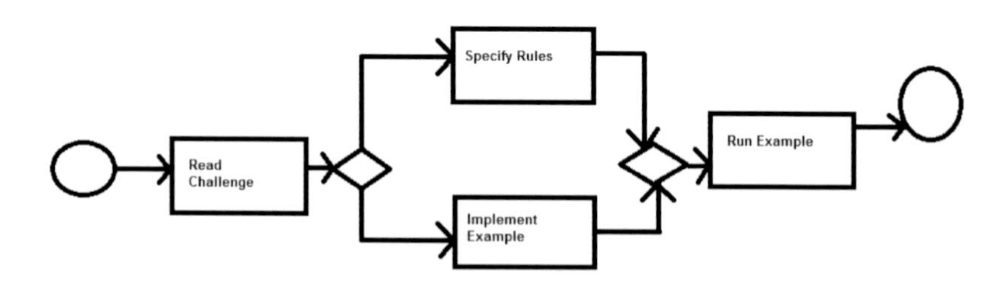

Figure 13.5: BPMN example

The representation of this in the BPMN metamodel is the instance model:

```
p1 : Process
p1.name = "test1"
pg1 : ParallelGateway
pg1.name = "pg1"
pg2 : ParallelGateway
```

```
pg2.name = "pg2"
pg1 : p1.flowElements
pg2 : p1.flowElements
se : StartEvent
se.name = "start event"
se : p1.flowElements
ee : EndEvent
ee.name = "end event"
ee : p1.flowElements
t1 : Task
t1.name = "Read Challenge"
t1 : p1.flowElements
t2 : Task
t2.name = "Specify Rules"
t2 : p1.flowElements
t3 : Task
t3.name = "Implement Example"
t3 : p1.flowElements
t4 : Task
t4.name = "Run Example"
t4 : p1.flowElements
sf1 : SequenceFlow
sf1.name = "startTotask1"
sf1 : p1.flowElements
sf1.sourceRef = se
sf1.targetRef = t1
sf2 : SequenceFlow
sf2.name = "task1Topg1"
sf2 : p1.flowElements
sf2.sourceRef = t1
sf2.targetRef = pg1
sf3 : SequenceFlow
sf3.name = "pg1Totask2"
sf3 : p1.flowElements
sf3.sourceRef = pg1
sf3.targetRef = t2
sf4 : SequenceFlow
sf4.name = "pg1Totask3"
sf4 : p1.flowElements
sf4.sourceRef = pg1
sf4.targetRef = t3
sf5 : SequenceFlow
sf5.name = "task2Topg2"
sf5 : p1.flowElements
sf5.sourceRef = t2
sf5.targetRef = pg2
sf6 : SequenceFlow
sf6.name = "task3Topg2"
sf6 : p1.flowElements
```

```
sf6.sourceRef = t3
sf6.targetRef = pg2
sf7 : SequenceFlow
sf7.name = "pg2Totask4"
sf7 : p1.flowElements
sf7.sourceRef = pg2
sf7.targetRef = t4
sf8 : SequenceFlow
sf8.name = "task4Toend"
sf8 : p1.flowElements
sf8.sourceRef = t4
sf8.targetRef = ee
```

The following shows a trace of the execution of the transformation on this model:

```
Model loaded
Entering startTotask1
Left startTotask1
Entered Read Challenge
Left task Read Challenge
Entered flow task1Topg1
Entered parallel pg1
Left pg1
Entering pg1Totask2
Left pg1
Entering pg1Totask3
Left pg1Totask2
Entered Specify Rules
Left pg1Totask3
Entered Implement Example
Left task Specify Rules
Entered flow task2Topg2
Left task Implement Example
Entered flow task3Topg2
Entered parallel pg2
Left pg2
Entering pg2Totask4
Left pg2Totask4
Entered Run Example
Left task Run Example
Entered flow task4Toend
Leaving task4Toend
Finished process instance
```

Summary

We have described examples of refactoring and update-in-place transformations, and shown how these can be specified in UML-RSDS. We have also given guidelines for the simplification of the specification and analysis of such transformations.

References

[1] W.M.P. van der Aalst, A.H.M. ter Hofstede, B. Kiepuszewski and A.P. Barros, *Workflow patterns*, in: *Distributed and Parallel Databases* 14(1), pp. 5–15, 2003.

[2] R. Dijkman and P.v. Gorp, *BPMN 2.0 Execution semantics formalized as graph rewrite rules*, Eindhoven University of Technology, 2013.

[3] OMG, *Business Process Model and Notation (BPMN) Version 2.0*, www.omg.org/spec/BPMN/2.0/PDF, 2013.

[4] A. Rensink and J.-H. Kuperus, *Repotting the Geraniums: on nested graph transformation rules*, proceedings of GT-VMT 2009, Electronic communications of the EASST vol. 18, 2009, http://dblp.uni-trier.de/db/journals/eceasst/eceasst18.html#RensinkK09.

[5] W. Smid and A. Rensink, *Class diagram restructuring with GROOVE*, TTC 2013.

[6] UML-RSDS toolset and manual, `http://www.dcs.kcl.ac.uk/staff/kcl/uml2web/`, 2014.

[7] YAWL, http://www.yawlfoundation.org, 2014.

Chapter 14

Bidirectional and Incremental Transformations

Bidirectional transformations (bx) are considered important in a number of transformation scenarios:

- Maintaining consistency between two models which may both change, for example, if a UML class diagram and corresponding synthesised Java code both need to be maintained consistently with each other, in order to implement *round-trip engineering* for model-driven development.
- Where a mapping between two languages may need to be operated in either direction for different purposes, for example, to represent behavioural models as either Petri Nets or as state machines [12].
- Where inter-conversion between two different representations is needed, such as two alternative formats of electronic health record [3].

In this chapter we describe specification techniques and patterns for defining bidirectional transformations in UML-RSDS.

14.1 Criteria for bidirectionality

Bidirectional transformations are characterised by a binary relation

$$R : SL \leftrightarrow TL$$

between a source language (metamodel) SL and a target language TL. $R(m, n)$ holds for a pair of models m of SL and n of TL when the models consist of data which corresponds under R.

It should be possible to automatically derive from the definition of R both forward and reverse transformations

$$R^{\rightarrow} : SL \times TL \rightarrow TL$$
$$R^{\leftarrow} : SL \times TL \rightarrow SL$$

which aim to establish R between their first (respectively second) and their result target (respectively source) models, given both existing source and target models.

Stevens [17] has identified two key conditions which bidirectional model transformations should satisfy:

1. *Correctness*: the forward and reverse transformations derived from a relation R do establish R:

   $$R(m, R^{\rightarrow}(m, n))$$
   $$R(R^{\leftarrow}(m, n), n)$$

   for each $m : SL$, $n : TL$.

2. *Hippocraticness*: if source and target models already satisfy R then the forward and reverse transformations do not modify the models:

   $$R(m, n) \Rightarrow R^{\rightarrow}(m, n) = n$$
   $$R(m, n) \Rightarrow R^{\leftarrow}(m, n) = m$$

   for each $m : SL$, $n : TL$.

Hippocraticness is a global property, in practice a stronger local property is desirable: if any part of a target (source) model is already consistent with the corresponding part of the source (target) model, then neither part should be modified. We refer to this as *local Hippocraticness*. In the following, we will consider only *separate-models* transformations, and not *update-in-place* transformations.

14.2 Patterns for bidirectional transformations

Inspection of published examples of bx shows that many rely upon the use of the following patterns:

Auxiliary Correspondence Model: maintain a detailed trace between source model and target model elements to facilitate change-propagation in source to target or target to source directions.

Cleanup before Construct: for $R^{\rightarrow}$, remove superfluous target model elements which are not in the transformation relation R with any source elements, before constructing target elements related to source elements. Similarly for $R^{\leftarrow}$.

Unique Instantiation: Do not recreate elements t in one model which already correspond to an element s in the other model, instead modify data of t to enforce the transformation relation. Use key attributes to identify when elements should be created or updated.

In addition, we have identified the following adaptions of transformation patterns from [15] which can assist in the construction of bx:

Phased Construction for bx: Define the relation between source and target models as a composition of relations between corresponding composition levels in the source and target languages.

Entity Merging/Splitting for bx: Define many-to-one and one-to-many relations between models using links to identify element groups in one model which are related to a single element in the other model.

Map Objects Before Links for bx: Separately relate the elements in the source and target models, and the values of their association ends.

In Section 14.4 we describe these patterns in detail.

14.3 Bidirectional transformation specification in UML-RSDS

As described in Chapter 7, model transformations are specified in UML-RSDS as UML use cases, defined declaratively by three main predicates, expressed in a subset of OCL:

1. Assumptions Asm which define when the transformation is applicable.

2. Postconditions *Post* which define the intended effect of the transformation at its termination. These are an ordered conjunction of OCL constraints (also termed *rules* in the following) and also serve to define a procedural implementation of the transformation.

3. Invariants *Inv* which define expected invariant properties which should hold during the transformation execution. These can be derived from *Post*, or specified explicitly by the developer.

The *Post* constraints are often universally quantified over particular source language entity types, i.e., their context entity. In this chapter we will write these quantifications explicitly (they are not written when writing a transformation specification in the UML-RSDS tools) because this helps to clarify the derivation of inverse constraints and transformations.

For example, an elementary transformation specification τ_{a2b} on the languages S consisting of entity type A and T consisting of entity type B (Fig. 14.1) could be:

- $(Asm): \quad B{\rightarrow}forAll(b \mid b.y \geq 0)$
- $(Post): \quad A{\rightarrow}forAll(a \mid B{\rightarrow}exists(b \mid b.y = a.x{\rightarrow}sqr()))$
- $(Inv): \quad B{\rightarrow}forAll(b \mid A{\rightarrow}exists(a \mid a.x = b.y{\rightarrow}sqrt()))$

The postcondition is written in this case as a type 0 constraint instead of as the equivalent type 1 constraint:

$$A :: \quad B{\rightarrow}exists(b \mid b.y = x{\rightarrow}sqr())$$

The computation steps α of τ_{a2b} are applications of $B{\rightarrow}exists(b \mid b.y = a.x{\rightarrow}sqr())$ to individual $a : A$. These consist of creation of a new $b : B$ instance and setting its y value to $a.x * a.x$. These steps preserve *Inv*: $Inv \Rightarrow [\alpha]Inv$.

In the UML-RSDS tools, both *Post* and *Inv* are entered as constraints using the use case edit dialog. *Post* would be entered as

```
B->exists( b | b.y = x->sqr() )
```

on context entity A, and *Inv* as

```
A->exists( a | a.x = y->sqrt() )
```

on context entity B.

This example shows a typical situation, where the invariant is a dual to the postcondition, and expresses a form of minimality condition on the target model: that the only elements of this model should

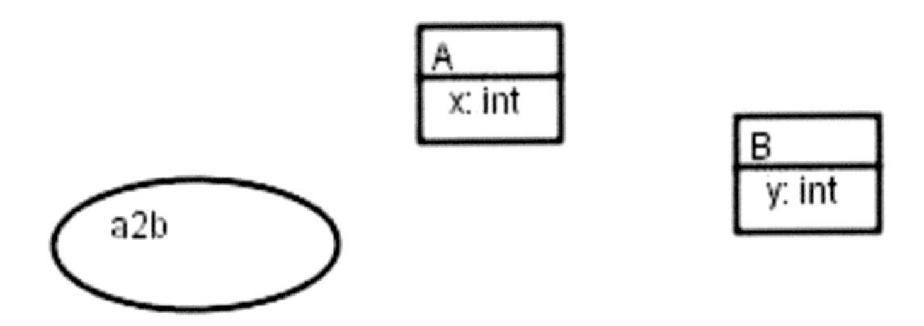

Figure 14.1: A to B transformation τ_{a2b}

be those derived from source elements by the transformation. In terms of the framework of [17], the source-target relation R_τ associated with a UML-RSDS transformation τ is *Post* & *Inv*. As in the above example, R_τ is not necessarily bijective. The forward direction of τ is normally computed as $stat(Post)$: the UML activity derived from *Post* when interpreted procedurally (Table 6.2). However, in order to achieve the correctness and hippocraticness properties, *Inv* must also be considered: before $stat(Post)$ is applied to the source model m, the target model n must be cleared of elements which fail to satisfy *Inv*.

In the $a2b$ example, the transformation $\tau^{\times}_{a2b}$ with postcondition constraints:

$$
\begin{array}{ll}
(CleanTarget1): & B{\rightarrow}forAll(b \mid not(b.y \geq 0) \;\Rightarrow\; b{\rightarrow}isDeleted()) \\
(CleanTarget2): & \\
& B{\rightarrow}forAll(b \mid not(A{\rightarrow}exists(a \mid a.x = b.y{\rightarrow}sqrt())) \;\Rightarrow \\
& \quad b{\rightarrow}isDeleted())
\end{array}
$$

is applied before τ_{a2b}, to remove all B elements which fail to be in R_{a2b} with some $a : A$, or which fail to satisfy Asm.

This is an example of the Cleanup before Construct pattern identified in Section 14.2 above. Additionally, the $E{\rightarrow}exists(e \mid P)$ quantifier in rule succedents should be procedurally interpreted as "create a new $e : E$ and establish P for e, unless there already exists an $e : E$ satisfying P". That is, the Unique Instantiation pattern should be used to implement 'check before enforce' semantics. The forward transformation $\tau^{\rightarrow}$ is then the sequential composition $\tau^{\times}; \; \tau$ of the cleanup transformation and the standard transformation (enhanced by Unique Instantiation).

In the reverse direction, the roles of *Post* and *Inv* are interchanged: elements of the source model which fail to satisfy Asm, or to satisfy *Post* with respect to some element of the target model should be deleted:

$$
\begin{aligned}
&(CleanSource2):\\
&\qquad A{\rightarrow}forAll(a \mid not(B{\rightarrow}exists(b \mid b.y = a.x{\rightarrow}sqr())) \Rightarrow\\
&\qquad\qquad a{\rightarrow}isDeleted())
\end{aligned}
$$

This cleanup transformation is denoted $\tau^{\sim\times}_{a2b}$. It is followed by an application of the normal inverse transformation $\tau^{\sim}$ which has postcondition constraints *Inv* ordered in the corresponding order to *Post*. Again, Unique Instantiation is used for source model element creation. The overall reverse transformation is denoted by $\tau^{\leftarrow}$ and is defined as $\tau^{\sim\times};\ \tau^{\sim}$.

As the above simple example shows, UML-RSDS bx transformations need not be bijective: source models $(\{a1\}, \{a1 \mapsto -3\})$ and $(\{a2\}, \{a2 \mapsto 3\})$ both map to $(\{b1\}, \{b1 \mapsto 9\})$.

In many cases, *Inv* can be derived automatically from *Post* by syntactic transformation, the *CleanTarget* and *CleanSource* constraints can also be derived from *Post*, and from *Asm*. This is an example of a higher-order transformation (HOT) and is implemented in the UML-RSDS tools.

In general, in the following UML-RSDS examples, τ is a separate-models transformation with source language S and target language T, and postcondition *Post* as an ordered conjunction of constraints *Cn* of the form:

$$S_i{\rightarrow}forAll(s \mid SCond(s) \Rightarrow T_j{\rightarrow}exists(t \mid TCond(t)\ \&\ P_{i,j}(s,t)))$$

and *Inv* is a conjunction of dual constraints $Cn^{\sim}$ of the form

$$T_j{\rightarrow}forAll(t \mid TCond(t) \Rightarrow S_i{\rightarrow}exists(s \mid SCond(s)\ \&\ P^{\sim}_{i,j}(s,t)))$$

where the predicates $P_{i,j}(s,t)$ define the features of t from those of s, and are invertible: an equivalent form $P^{\sim}_{i,j}(s,t)$ should exist, which expresses the features of s in terms of those of t, and such that

$$S_i{\rightarrow}forAll(s \mid T_i{\rightarrow}forAll(t \mid P_{i,j}(s,t) = P^{\sim}_{i,j}(s,t)))$$

under the assumptions *Asm*. Tables 7.4, 7.5 and 7.6 show some examples of inverses $P^{\sim}$ of predicates P. The computation of these inverses are implemented in the UML-RSDS tools (the *reverse* option for use cases). More cases are given in [11]. The transformation developer can also specify inverses for particular *Cn* by defining a suitable $Cn^{\sim}$ constraint in *Inv*, for example, to express that a predicate $t.z = s.x + s.y$ should be inverted as $s.x = t.z - s.y$.

Each *CleanTarget* constraint based on *Post* then has the form $Cn^{\times}$:

$$T_j \rightarrow forAll(t \mid TCond(t) \;\&\; not(S_i \rightarrow exists(s \mid SCond(s) \;\&\; P_{i,j}(s,t))) \Rightarrow t \rightarrow isDeleted())$$

Similarly for *CleanSource*.

14.4 Patterns for bx

In this section we give a patterns catalogue for bx, and give pattern examples in UML-RSDS.

14.4.1 Auxiliary correspondence model

This pattern defines auxiliary entity types and associations which link corresponding source and target elements. These are used to record the mappings performed by a bx, and to propagate modifications from source to related target elements or vice-versa, when one model changes.

Figure 14.2 shows a typical schematic structure of the pattern.

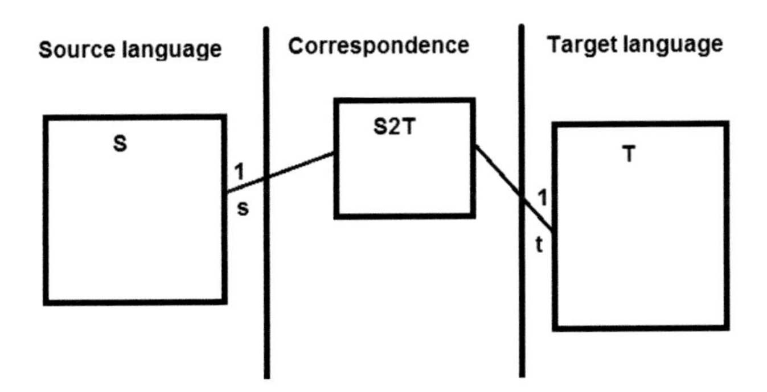

Figure 14.2: Auxiliary Correspondence Model pattern

Benefits:

The pattern is a significant aid in change-propagation between models, and helps to ensure the correctness of a bx. Feature value changes to a source element *s* can be propagated to changes to its corresponding target element, and vice-versa, via the links. Deletion of an element in one model may imply deletion of its corresponding element in the other model.

Disadvantages:

The correspondence metamodel must be maintained (by the transformation engineer) together with the source and target languages, and the necessary actions in creating and accessing correspondence elements adds complexity to the transformation and adds to its execution time and memory requirements.

Related patterns:

This pattern is a specialisation of the *Auxiliary Metamodel* pattern of [15].

Examples:

This mechanism is a key facility of Triple Graph Grammars (TGG) [1, 2], and correspondence traces are maintained explicitly or implicitly by other MT languages such as QVT-R [16]. The pattern could be used to retain intermediate models to facilitate composition of bx transformations [17].

In UML-RSDS the pattern is applied by introducing auxiliary attributes into source and target language entity types. These attributes are primary key/identity attributes for the entity types, and are used to record source-target element correspondences. Target element $t : T_j$ is considered to correspond to source element(s) $s_1 : S_1$, ..., $s_n : S_n$ if they all have the same primary key values: $t.idT_j = s_1.idS_1$, etc. The identity attributes are String-valued.

The existence of identity attributes facilitates element lookup by using the *Object Indexing* pattern (Chapter 9), which defines maps from *String* to each entity type, permitting elements to be retrieved by the value of their identity attribute: $T_j[v]$ denotes the T_j instance t with $t.idT_j = v$ if v is a single String value, or the collection of T_j instances t with $v \rightarrow includes(t.idT_j)$ if v is a collection. This approach is simpler than using a separate auxiliary correspondence model, and is adequate for many cases of bx, including non-bijective bx. Table 7.6 shows inverse predicates based on this approach to correspondence models. In the table S_i elements correspond to T_j elements, i.e., $S_i \rightarrow collect(idS) = T_j \rightarrow collect(idT)$, and likewise *SRef* corresponds to *TRef*.

The pattern can be used to define source-target propagation and incremental application of a transformation τ. For postconditions Cn of the form

$$S_i \rightarrow forAll(s \mid SCond(s) \;\Rightarrow\; T_j \rightarrow exists(t \mid TCond(t) \;\&\; P_{i,j}(s,t)))$$

the following derived constraints Cn^{Δ} are defined for the incremental application of Cn:

$$S_i{\rightarrow}forAll(s \mid s.sId : T_j{\rightarrow}collect(tId) \ \& \ t = T_j[s.sId] \ \& \\ not(SCond(s)) \ \Rightarrow \ t{\rightarrow}isDeleted())$$

This deletes those t which no longer correspond to a suitable s. It is iterated over the $s : S_i$ which have been modified. This constraint is omitted if $SCond$ is absent (i.e., it is the default $true$).

For deleted s, the following constraint is executed:

$$s.sId : T_j{\rightarrow}collect(tId) \ \& \ t = T_j[s.sId] \ \Rightarrow \ t{\rightarrow}isDeleted()$$

A further constraint maintains $P_{i,j}(s,t)$ for corresponding s and t by updating t:

$$S_i{\rightarrow}forAll(s \mid s.sId : T_j{\rightarrow}collect(tId) \ \& \ t = T_j[s.sId] \ \& \\ SCond(s) \ \& \ TCond(t) \ \Rightarrow \ P_{i,j}(s,t))$$

This only needs to be iterated over those $s : S_i$ which have been modified.

$$S_i{\rightarrow}forAll(s \mid T_j{\rightarrow}collect(tId){\rightarrow}excludes(s.sId) \ \& \ SCond(s) \ \Rightarrow \\ T_j{\rightarrow}exists(t \mid TCond(t) \ \& \ P_{i,j}(s,t)))$$

This iterates over modified $s : S_i$ and newly-created $s : S_i$.

The incremental version τ^{Δ} of a transformation τ is defined to have postconditions formed from the above constraints Cn^{Δ} for each postcondition Cn of τ, and ordered according to the order of the Cn in the $Post$ of τ. In a similar way, target-source change propagation can be defined.

14.4.2 Cleanup before construct

This pattern defines a two-phase approach in both forward and reverse transformations associated with a bx with relation R: the forward transformation $R^{\rightarrow}$ first removes all elements from the target model n which fail to satisfy R for any element of the modified source model m', and then modifies or constructs elements of n to satisfy R with respect to m' (Fig. 14.3). The reverse transformation $R^{\leftarrow}$ operates on m in the same manner.

Benefits:

The pattern is an effective way to ensure the correctness of separate-models bx.

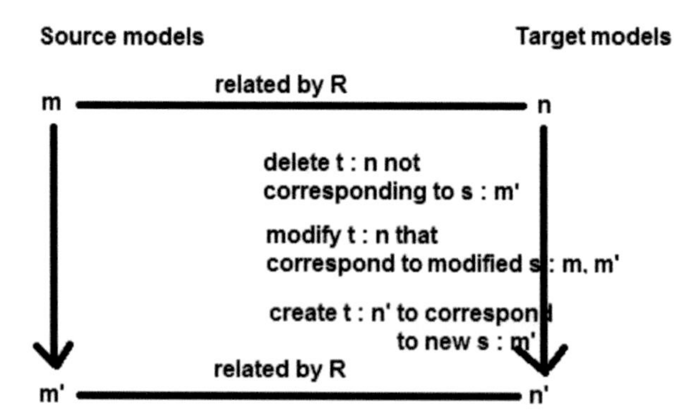

Figure 14.3: Cleanup before Construct pattern

Disadvantages:

There may be efficiency problems because for each target model element, a search through the source model for possibly corresponding source elements may be needed. Elements may be deleted in the Cleanup phase only to be reconstructed in the Construct phase: Auxiliary Correspondence Model may be an alternative strategy to avoid this problem, by enforcing that feature values should change in response to a feature value change in a corresponding element, rather than deletion of elements.

Related patterns:

This pattern is a variant of the *Construction and Cleanup* pattern of [15].

Examples:

An example is the Composers bx [4]. Implicit deletion in QVT operates in a similar manner, but can only modify models (domains) marked as *enforced* [16]. In UML-RSDS, explicit cleanup rules $Cn^{\times}$ can be deduced from the construction rules Cn, for mapping transformations, as described in Section 14.3 above. If identity attributes are used to define the source-target correspondence, then $Cn^{\times}$ can be simplified to:

$$T_j{\rightarrow}forAll(t \mid TCond(t) \ \& \ t.tId \notin S_i{\rightarrow}collect(sId) \ \Rightarrow \\ t{\rightarrow}isDeleted())$$

and

$$T_j{\rightarrow}forAll(t \mid TCond(t) \ \& \ t.tId : S_i{\rightarrow}collect(sId) \ \& \ s = S_i[t.tId] \\ \& \ not(SCond(s)) \ \Rightarrow \ t{\rightarrow}isDeleted())$$

The second constraint is omitted if *SCond* is the default *true* predicate.

In the case that $TCond(t)$ and $SCond(s)$ hold for corresponding s, t, but $P_{i,j}(s,t)$ does not hold, t should not be deleted, but $P_{i,j}(s,t)$ should be established by updating t:

$$S_i \rightarrow forAll(s \mid s.sId : T_j \rightarrow collect(tId) \ \& \ t = T_j[sId] \ \& \\ SCond(s) \ \& \ TCond(t) \ \Rightarrow \ P_{i,j}(s,t))$$

For a transformation τ, the cleanup transformation $\tau^{\times}$ has the above $Cn^{\times}$ constraints as its postconditions, in the same order as the Cn occur in the *Post* of τ. Note that $\tau^{\rightarrow}$ is $\tau^{\times}$; τ, and τ^{Δ} is $\tau^{\times}$; τ incrementally applied.

14.4.3 Unique instantiation

This pattern avoids the creation of unnecessary elements of models and helps to resolve possible choices in reverse mappings. It uses various techniques such as traces and unique keys to identify when elements should be modified and reused instead of being created. In particular, unique keys can be used to simplify checking for existing elements.

Benefits:

The pattern helps to ensure the Hippocraticness property of a bx by avoiding changes to a target model if it is already in the transformation relation with the source model.

Disadvantages:

The need to test for existence of elements adds to the execution cost. This can be ameliorated by the use of the Object Indexing pattern [15] to provide fast lookup of elements by their primary key value.

Examples:

The *key* attributes and check-before-enforce semantics of QVT-R follow this pattern, whereby new elements of source or target models are not created if there are already elements which satisfy the specified relations of the transformation [17]. The $E \rightarrow exists1(e \mid P)$ quantifier in UML-RSDS is used in a similar way. It is procedurally interpreted as "create a new $e : E$ and establish P for e, unless there already exists an $e : E$ satisfying P" [11]. For bx, the quantifier *exists* should also be treated in this way. If a transformation uses identity attributes (to implement Auxiliary Correspondence Model), the quantifier $E \rightarrow exists(e \mid e.eId = v \ \& \ P)$ can be interpreted as: "if $E[v]$ exists, apply $stat(P)$ to this element, otherwise create a new E instance with $eId = v$ and apply $stat(P)$ to it". This ensures local Hippocraticness.

14.4.4 Phased construction for bx

This pattern defines a bx τ by organising R_τ as a composition of relations $R_{Si,Tj}$, which relate instances of entities Si and Tj in corresponding levels of the composition hierarchies of the source and target languages. Figure 14.4 shows the typical schematic structure of the pattern. At each composition level there is a 0..1 to 0..1 relation (or more specialised relation) between the corresponding source and target entity types.

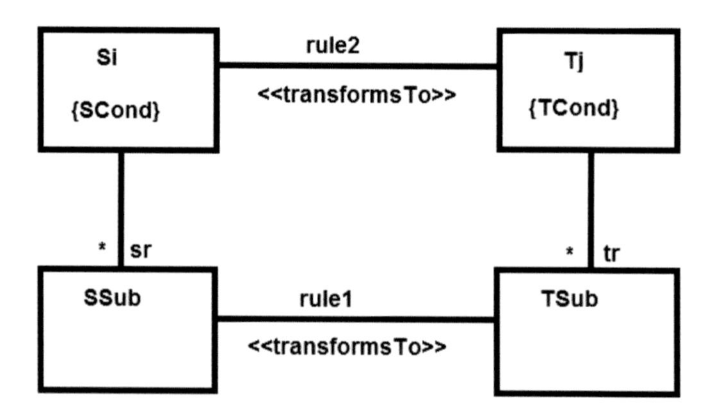

Figure 14.4: Phased Construction pattern

Benefits:

The pattern provides a modular and extensible means to structure a bx.

Disadvantages:

It is sometimes not possible to organise a transformation in this way, if an element in one model corresponds to multiple elements at different composition levels in the other model. In such cases *Entity Merging/Splitting for bx* should be used.

Related patterns:

The pattern is based on the general transformation pattern *Phased Construction* [15].

Examples:

The UML to relational database example of [16] is a typical case, where *Package* and *Schema* correspond at the top of the source/target language hierarchies, as do *Class* and *Table* (in the absence of inheritance), and *Column* and *Attribute* at the lowest level.

In UML-RSDS a transformation defined according to this pattern has its *Post* consisting of constraints *Cn* of the form

$$S_i \rightarrow forAll(s \mid SCond(s) \;\Rightarrow\; T_j \rightarrow exists(t \mid TCond(t) \;\&\; P_{i,j}(s,t)))$$

where S_i and T_j are at corresponding hierarchy levels, and *Inv* consists of constraints $Cn^{\sim}$ of the form

$$T_j \rightarrow forAll(t \mid TCond(t) \;\Rightarrow\; S_i \rightarrow exists(s \mid SCond(s) \;\&\; P^{\sim}_{i,j}(s,t)))$$

No nested quantifiers or deletion expressions $x \rightarrow isDeleted()$ are permitted in *SCond*, *TCond* or $P_{i,j}$, and $P_{i,j}$ is restricted to be formed of invertible expressions.

Each rule creates instances t of some target entity type T_j, and may lookup target elements produced by preceding rules to define the values of association end features of t: $t.tr = TSub[s.sr.idSSub]$ for example, where *TSub* is lower than T_j in the target language composition hierarchy (as in Fig. 14.4) and there are identity attributes in the entity types to implement a source-target correspondence at each level. Both forward and reverse transformations will conform to the pattern if one direction does. The assignment to $t.tr$ has inverse: $s.sr = SSub[t.tr.idTSub]$.

The example of Fig. 14.1 can be elaborated by the addition of another composition level to the languages, and the addition of primary keys to all entity types (Fig. 14.5).

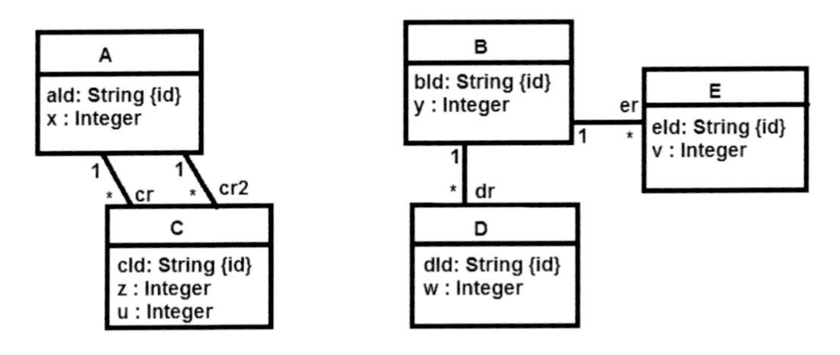

Figure 14.5: Extended a2b example: τ_{ac2bde}

The specification can be organised as a Phased Construction by relating the entity types C and D at one level, and then A and B at the next (higher) level:

$$C\rightarrow forAll(c \mid D\rightarrow exists(d \mid d.dId = c.cId \;\&\; d.w = c.z + 5))$$
$$A\rightarrow forAll(a \mid B\rightarrow exists(b \mid b.bId = a.aId \;\&\; b.y = a.x\rightarrow sqr() \;\&\; b.dr = D[a.cr.cId]))$$

These constraints can be automatically inverted to produce the transformation invariants:

$$D\rightarrow forAll(d \mid C\rightarrow exists(c \mid c.cId = d.dId \;\&\; c.z = d.w - 5))$$
$$B\rightarrow forAll(b \mid A\rightarrow exists(a \mid a.aId = b.bId \;\&\; a.x = b.y\rightarrow sqrt() \;\&\; a.cr = C[b.dr.dId]))$$

These can be used to define the *Post* constraints of the reverse transformation.

Two UML-RSDS bx $\tau : S \rightarrow T$, $\sigma : T \rightarrow U$ using this pattern can be sequentially composed to form another bx between S and U: the language T becomes auxiliary in this new transformation. The forward direction of the composed transformation is $\tau^{\rightarrow}; \sigma^{\rightarrow}$, the reverse direction is $\sigma^{\leftarrow}; \tau^{\leftarrow}$.

14.4.5 Entity merging/splitting for bx

In this variation of Phased Construction, data from multiple source model elements may be combined into single target model elements, or vice-versa, so that there is a many-one relation from one model to the other. The pattern supports the definition of such bx by including correspondence links between the multiple elements in one model which are related to one element in the other.

Benefits:

The additional links enable the transformation to be correctly reversed.

Disadvantages:

Additional auxiliary data needs to be added to record the links. The validity of the links between elements needs to be maintained. There may be potential conflict between different rules which update the same element.

Related patterns:

This pattern uses a variant of Auxiliary Correspondence Model in which there are correspondences between elements in one model in addition to cross-model correspondences. The attributes used to record intra-model correspondences may not necessarily be primary keys.

Examples:

An example of Entity Merging is the Collapse/Expand State Diagrams benchmark of [6]. The UML to RDB transformation is also an example in the case that all subclasses of a given root class are mapped to a single table that represents this class. The Pivot/Unpivot transformation of [3] is an example of Entity Splitting. The forward transformation represents a 2-dimensional table of data as an indexed collection of maps (Fig. 14.6).

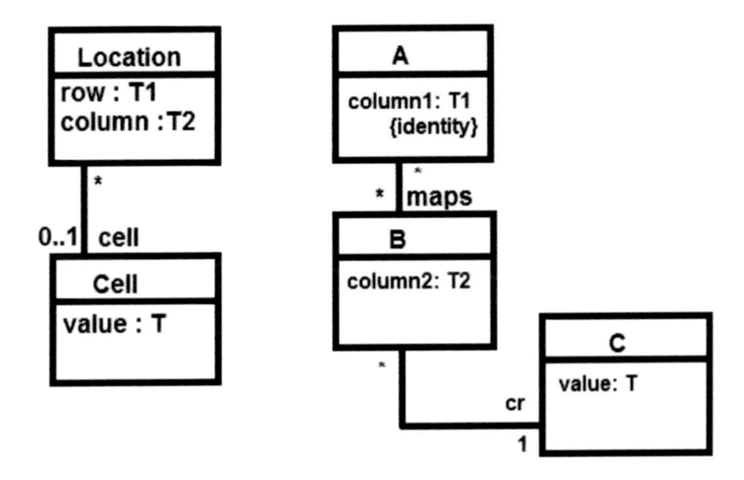

Figure 14.6: Pivoting/unpivoting transformation

The forward transformation has postcondition

$$
\begin{array}{l}
Location{\rightarrow}forAll(l \mid \\
\quad Cell{\rightarrow}forAll(cl \mid l.cell{\rightarrow}includes(cl) \;\Rightarrow \\
\quad\quad A{\rightarrow}exists(a \mid a.column1 = l.row \;\&\; \\
\quad\quad\quad B{\rightarrow}exists(b \mid b : a.maps \;\&\; b.column2 = l.column \;\&\; \\
\quad\quad\quad\quad C{\rightarrow}exists(c \mid b.cr = c \;\&\; c.value = cl.value))))
\end{array}
$$

From this the corresponding invariant can be syntactically derived:

$$
\begin{array}{l}
A{\rightarrow}forAll(a \mid \\
\quad B{\rightarrow}forAll(b \mid b : a.maps \;\Rightarrow \\
\quad\quad C{\rightarrow}forAll(c \mid b.cr = c \;\Rightarrow \\
\quad\quad\quad Location{\rightarrow}exists(l \mid l.column = b.column2 \;\&\; l.row = \\
\quad\quad\quad a.column1 \;\&\; \\
\quad\quad\quad Cell{\rightarrow}exists(cl \mid cl : l.cell \;\&\; cl.value = c.value))))
\end{array}
$$

This illustrates the general case of merging/splitting, where the inverse of Cn:

$$\begin{array}{l} S_{i1}{\rightarrow}forAll(s1 \mid ...\ S_{in}{\rightarrow}forAll(sn \mid SCond(s1,...,sn) \Rightarrow \\ \quad T_{j1}{\rightarrow}exists(t1 \mid ... \\ \qquad T_{jm}{\rightarrow}exists(tm \mid TCond(t1,...,tm)\ \&\ P(s1,...,sn,t1,..., \\ \qquad\quad tm))...))\ ...) \end{array}$$

is $Cn^{\sim}$:

$$\begin{array}{l} T_{j1}{\rightarrow}forAll(t1 \mid ... T_{jm}{\rightarrow}forAll(tm \mid TCond(t1,...,tm) \Rightarrow \\ \quad S_{i1}{\rightarrow}exists(s1 \mid ... \\ \qquad S_{in}{\rightarrow}exists(sn \mid SCond(s1,...,sn)\ \&\ P^{\sim}(s1,...,sn,t1,..., \\ \qquad\quad tm))...))...) \end{array}$$

In UML-RSDS, correspondence links between elements in the same model are maintained using additional attributes. All elements corresponding to a single element will have the same value for the auxiliary attribute (or a value derived by a 1-1 function from that value). In our running example, the entity type C could be split into entity types D and E in the other language:

$$\begin{array}{l} C{\rightarrow}forAll(c \mid D{\rightarrow}exists(d \mid d.dId = c.cId\ \&\ d.w = c.z + 5)) \\ C{\rightarrow}forAll(c \mid E{\rightarrow}exists(e \mid e.eId = c.cId\ \&\ e.v = c.u)) \\ A{\rightarrow}forAll(a \mid B{\rightarrow}exists(b \mid b.bId = a.aId\ \&\ b.y = a.x{\rightarrow}sqr()\ \& \\ \qquad b.dr = D[a.cr.cId]\ \&\ b.er = \\ \qquad\quad E[a.cr2.cId])) \end{array}$$

Again, these constraints can be automatically inverted to produce transformation invariants and an entity merging reverse transformation:

$$\begin{array}{l} D{\rightarrow}forAll(d \mid C{\rightarrow}exists(c \mid c.cId = d.dId\ \&\ c.z = d.w - 5)) \\ E{\rightarrow}forAll(e \mid C{\rightarrow}exists(c \mid c.cId = e.eId\ \&\ c.u = e.v)) \\ B{\rightarrow}forAll(b \mid A{\rightarrow}exists(a \mid a.aId = b.bId\ \&\ a.x = b.y{\rightarrow}sqrt()\ \& \\ \qquad a.cr = C[b.dr.dId]\ \&\ a.cr2 = \\ \qquad\quad C[b.er.eId])) \end{array}$$

14.4.6 Map objects before links for bx

If there are self-associations on source entity types, or other circular dependency structures in the source model, then this variation on Phased Construction for bx can be used. This pattern separates the relation between elements in target and source models from the relation between links in the models.

Benefits:

The specification is made more modular and extensible. For example, if a new association is added to one language, and a corresponding association to the other language, then a new relation relating the values of these features can be added to the transformation without affecting the existing relations.

Disadvantages:

Separate rules (constraints) operate on different features of a single entity type.

Examples:

In UML-RSDS a first phase of such a transformation relates source elements to target elements, then in a second phase source links are related to corresponding target links. The second phase typically has postcondition constraints of the form $S_i \rightarrow forAll(s \mid T_j[s.idS].rr = TRef[s.r.idSRef])$ to define target model association ends rr from source model association ends r, looking-up target model elements $T_j[s.idS]$ and $TRef[s.r.idSRef]$ which have already been created in a first phase. Such constraints can be inverted to define source data from target data as: $T_j \rightarrow forAll(t \mid S_i[t.idT].r = SRef[t.rr.idTRef])$. The reverse transformation also conforms to the Map Objects Before Links pattern.

An example of this pattern is the tree to graph transformation [9], Fig. 14.7.

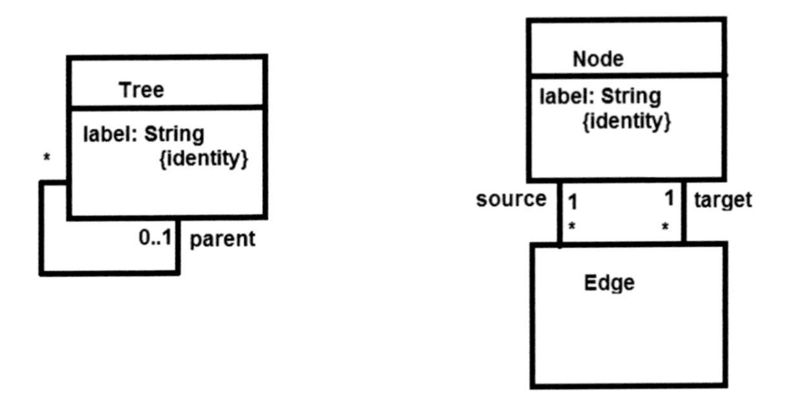

Figure 14.7: Tree to graph metamodels

A first rule creates a node for each tree:

$$(C1): \quad Tree \rightarrow forAll(t \mid Node \rightarrow exists(n \mid n.label = t.label))$$

A second rule then creates edges for each link between parent and child trees:

$$(C2): \quad Tree \rightarrow forAll(t \mid Tree \rightarrow forAll(p \mid p : t.parent \Rightarrow Edge \rightarrow exists(e \mid e.source = Node[t.label] \ \& \ e.target = Node[p.label])))$$

The corresponding invariant predicates, defining the reverse transformation, are:

$$(I1): \quad Node \rightarrow forAll(n \mid Tree \rightarrow exists(t \mid t.label = n.label))$$

and

$$(I2): \quad Edge \rightarrow forAll(e \mid Tree \rightarrow exists(t \mid Tree \rightarrow exists(p \mid p : t.parent \ \& \ t.label = e.source.label \ \& \ p.label = e.target.label)))$$

Inv is derived mechanically from *Post* using Tables 7.4, 7.5 and 7.6, and provides an implementable reverse transformation, since $stat(Inv)$ is defined.

The running example can also be rewritten into this form:

$$C \rightarrow forAll(c \mid D \rightarrow exists(d \mid d.dId = c.cId \ \& \ d.w = c.z + 5))$$
$$C \rightarrow forAll(c \mid E \rightarrow exists(e \mid e.eId = c.cId \ \& \ e.v = c.u))$$
$$A \rightarrow forAll(a \mid B \rightarrow exists(b \mid b.bId = a.aId \ \& \ b.y = a.x \rightarrow sqr()))$$
$$A \rightarrow forAll(a \mid B[a.aId].dr = D[a.cr.cId] \ \& \ B[a.aId].er = E[a.cr2.cId])$$

Again, these constraints can be automatically inverted:

$$D \rightarrow forAll(d \mid C \rightarrow exists(c \mid c.cId = d.dId \ \& \ c.z = d.w - 5))$$
$$E \rightarrow forAll(e \mid C \rightarrow exists(c \mid c.cId = e.eId \ \& \ c.u = e.v))$$
$$B \rightarrow forAll(b \mid A \rightarrow exists(a \mid a.aId = b.bId \ \& \ a.x = b.y \rightarrow sqrt()))$$
$$B \rightarrow forAll(b \mid A[b.bId].cr = C[b.dr.dId] \ \& \ A[b.bId].cr2 = C[b.er.eId])$$

14.5 View updates

If a predicate such as $t.g = s.f \rightarrow last()$ or $t.g = s.f \rightarrow select(P1)$ is inverted, the result is a predicate $s.f \rightarrow last() = t.g$ or $s.f \rightarrow select(P1) =$

$t.g$. These are termed *view updates*, because they specify an update to $s.f$ based on the required value of some view, function or selection of its data. The procedural interpretation $stat(P)$ of such a predicate is a statement which makes P true by making the minimal necessary changes to $s.f$. It can be used to implement target-to-source change propagation for a bx where the target model is constructed from views of the source model.

Table 14.1 shows the view update interpretation for some common view predicates.

In the cases for *tail* and *front*, d is the default element of the element type of f. In the case for *collect*, $e^{\sim}$ is an inverse to e, defined, for example, according to Table 7.4. $s \cap t$ is treated as for $s \rightarrow select(x \mid x : t)$. The operator $f \rightarrow merge(col)$ makes minimal changes to f to add the elements of *col*. It is the same as *union* if f is a set. On sequences it is:

$$sq \rightarrow merge(col) \quad = \quad col \rightarrow asSequence() \frown (sq - col)$$

It has the same ordering and multiplicity of elements as in *col*, for the elements of *col*.

The inverse of an assignment $t.g = D[s.f \rightarrow select(P) \rightarrow collect(bId)]$ is $s.f \rightarrow select(P) = B[t.g.dId]$ if D is the corresponding target entity type for source entity B. The view update definitions can therefore also be used for bx that involve source-target correspondences using identity attributes.

Table 14.1: View update interpretations

P	$stat(P)$
$f \rightarrow last() = x$	if $f.size = 0$ then $f := Sequence\{x\}$ else $f[f.size] := x$
$f \rightarrow front() = g$	if $f.size > 0$ then $f := g \frown Sequence\{f.last\}$ else $f := g \frown Sequence\{d\}$
$f \rightarrow first() = x$	if $f.size = 0$ then $f := Sequence\{x\}$ else $f[1] := x$
$f \rightarrow tail() = g$	if $f.size > 0$ then $f := Sequence\{f.first\} \frown g$ else $f := Sequence\{d\} \frown g$
$f \rightarrow select(P) = g$	$f := f - (f \rightarrow select(P) - g)$; $f := f \rightarrow merge(g \rightarrow select(P))$
$f \rightarrow reject(P) = g$	Same as $f \rightarrow select(not(P)) = g$
$f \rightarrow selectMaximals(e) = g$	$f := f \rightarrow reject(x \mid x.e \geq g.e \rightarrow max() \ \& \ x \notin g)$; $f := f \rightarrow merge(g)$
$f \rightarrow selectMinimals(e) = g$	$f := f \rightarrow reject(x \mid x.e \leq g.e \rightarrow min() \ \& \ x \notin g)$; $f := f \rightarrow merge(g)$
$f \rightarrow collect(e) = g$	$f := f \rightarrow reject(x \mid x.e \ / : g)$; $f := f \rightarrow merge(g \rightarrow collect(e^{\sim}))$
$f \rightarrow any() = x$	$f := f \rightarrow merge(Set\{x\})$
$f.subrange(a, b) = g$	$f := f.subrange(1, a - 1) \frown g \frown f.subrange(b + 1, f.size)$
$f \rightarrow including(x) = g$ set-valued f	$f := f \rightarrow intersection(g)$; $f := f \cup (g - Set\{x\})$
$f \rightarrow including(x) = g$ sequence-valued f	$f := g \rightarrow front()$
$f \rightarrow union(s) = g$ set-valued f	$f := f \rightarrow intersection(g)$; $f := f \cup (g - s)$
$f \rightarrow union(s) = g$ sequence-valued f	$f := g.subrange(1, g.size - s.size)$
$f \rightarrow excluding(x) = g$ set-valued f	$f := f \rightarrow intersection(g \cup Set\{x\})$; $f := f \cup g$
$f \rightarrow excluding(x) = g$ sequence-valued f	$f := f \rightarrow reject(y \mid y \neq x \ \& \ y \notin g)$; $f := f \rightarrow merge(g)$
$f - s = g$ set-valued f	$f := f \rightarrow intersection(s \cup g)$; $f := f \cup g$
$f - s = g$ sequence-valued f	$f := f \rightarrow reject(y \mid y \notin s \ \& \ y \notin g)$; $f := f \rightarrow merge(g)$

14.6 Verification techniques for bidirectional transformations

The properties of correctness and hippocraticness for bidirectional transformations follow by construction for UML-RSDS transformations defined according to the patterns and restrictions given in Sections 14.3, 14.4. Such bx τ should use all of the patterns Auxiliary Correspondence Model, Cleanup before Construct, Unique Instantiation, and at least one of the three patterns Phased Construction for bx, Entity Merging/Splitting for bx, and Map Objects Before Links for bx. The correctness property is ensured by the fact that $\tau^{\rightarrow}$ defined as $\tau^{\times}; \tau$ establishes the *Post* predicate of τ. The reverse transformation $\tau^{\leftarrow}$ defined as $\tau^{\sim\times}; \tau^{\sim}$ likewise establishes *Inv*. Hippocraticness is ensured by the use of Unique Instantiation.

A Phased Construction transformation should satisfy the condition that if a postcondition constraint $R1$ refers to instances of a target entity type $T2$, then any other rule $R2$ which creates $T2$ instances must precede $R1$. This is ensured by the condition of *syntactic non-interference* for a use case: a transformation with rules ordered as $R_1, \ldots, R_n$ should satisfy:

1. If $R_i < R_j$ and $i \neq j$, then $i < j$.

2. If $i \neq j$ then $wr(R_i) \cap wr(R_j) = \{\}$.

Together, these conditions ensure semantic non-interference of the rules: that subsequent rules R_j cannot invalidate earlier rules R_i, for $i < j$. Thus the implementation $stat(Post)$ will establish the conjunction of the postconditions [10]. If a rule r has $rd(r)$ disjoint from $wr(r)$ then it can usually be implemented by a bounded iteration (such as a *for*-loop over a fixed set of elements). Otherwise a fixed-point iteration may be required, in which the rule is applied until no more input elements exist that match its application conditions. If the constraint is *localised* [13], then its implementation is confluent.

For bidirectional transformations, we also require that the reverse transformation based on *Inv* satisfies syntactic non-interference. For Entity Merging transformations, the same target element may be looked-up and updated by different rules of *Post*, so that $wr(C_i) \cap wr(C_j)$ may be non-empty for some $i < j$. However these rules should be semantically non-interfering:

$$C_i \Rightarrow [stat(C_j)]C_i$$

This can be ensured if the rules modify common data in consistent ways, such as both creating instances of a target entity type TE but not deleting instances.

The bx relation *Post and Inv* can encode the semantic equivalence of source and target models, and hence can be used to derive *model-level semantic correctness* [13]. Additional correctness conditions of termination and confluence can be derived by construction, subject to restrictions on the transformation syntax. If each *Post* constraint of τ satisfies the localisation and non-interference conditions of [13], then $stat(Post)$ is terminating and confluent by construction.

Sequential composition is supported for UML-RSDS bx using the Auxiliary Correspondence Model, Cleanup before Construct, Unique Instantiation and one of the other patterns, as shown in Table 14.2.

Table 14.2: Composition rules for UML-RSDS bx

bx Transformation τ	*bx Transformation* σ	*Composed bx* $\tau; \sigma$
Phased Construction	Phased Construction	Phased Construction
Phased Construction	Entity Merging	Entity Merging
Entity Merging	Phased Construction	Entity Merging
Phased Construction	Entity Splitting	Entity Splitting
Entity Splitting	Phased Construction	Entity Splitting
Entity Merging	Entity Merging	Entity Merging
Entity Splitting	Entity Splitting	Entity Splitting

14.7 Incremental transformations

Transformations may need to operate upon data which is presented in a series of increments or model deltas. Transformations of this type are termed *streaming* or *incremental* transformations. A single source model may not exist, because of size limitations, or it may be continually updated (for example, a transformation may need to process tweets in a twitter feed). Bidirectional transformations may need to operate in an incremental manner if their purpose is to maintain consistency between two different models: incremental changes in one model should be propagated to the other without re-executing the transformation on the entire updated model.

In UML-RSDS incremental transformations can be written as use cases defined by postconditions, as for other forms of transformation. The difference in their operation is that models may be loaded for processing at any number of points in their execution, instead of only at the start of execution.

Figure 14.8 shows a typical example of a bidirectional incremental transformation, relating a source entity A to target entity B, where each model holds data which is not represented in the other, but there is related data which must be kept consistent between the models. The postcondition of *bxab* is:

$$A ::$$
$$B \rightarrow exists(b \mid b.z = 2 * x)$$

To implement this transformation as an incremental transformation in UML-RSDS, we need to introduce primary keys $id : String$ into A and B for the Auxiliary Correspondence Model pattern. The postcondition then becomes:

$$A ::$$
$$B \rightarrow exists(b \mid b.id = id \; \& \; b.z = 2 * x)$$

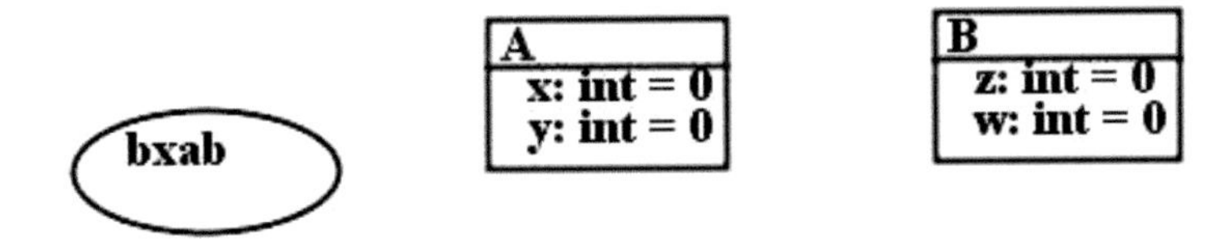

Figure 14.8: Incremental transformation example

The incremental version of *bxab* has the following postconditions (using the Cleanup before Construct pattern from Section 14.4.2) to propagate changes from A to B:

$$B ::$$
$$id \notin A.id \;\Rightarrow\; self{\rightarrow}isDeleted()$$
$$A ::$$
$$id : B.id \;\&\; b = B[id] \;\Rightarrow\; b.z = 2 * x$$
$$A ::$$
$$id \notin B.id \;\Rightarrow\; B{\rightarrow}exists(b \mid b.id = id \;\&\; b.z = 2 * x)$$

This transformation leaves attributes y and w unchanged, and only updates the target model in response to source model changes. The first constraint only needs to be invoked for B instances corresponding to A instances that are deleted in the model increment, the second is only needed for modified A instances, and the third only for new A instances.

A separate-models transformation consisting of type 1 constraints can be given an incremental implementation (for Java 4) by selecting *incremental* as the execution mode option on the use case edit dialog. The *load model* GUI option then loads in.txt incrementally.

The transformation should conform to the Phased Composition or Map Objects before Links patterns, and use identity attributes to implement Auxiliary Correspondence Model. That is, every postcondition constraint should be of the forms

$$Si ::$$
$$Ante \;\Rightarrow\; Tj{\rightarrow}exists(t \mid t.id = id \;\&\; P)$$

or

$$Si ::$$
$$Ante \;\Rightarrow\; Tj[id].tr = TSub[sr.id]$$

The transformation should satisfy syntactic non-interference. Changes to the id values of objects are not possible.

14.8 Related work

The patterns we have described here have been used in a number of different bx examples in different transformation languages. Auxiliary correspondence model is used in TGG (explicitly) and QVT-R (implicitly) by means of correspondence/trace model elements. Unique Instantiation is an important mechanism in QVT-R, and implicit deletion in QVT-R provides a version of Cleanup before Construct. Currently no language provides built-in support for Entity merging/splitting, however correspondence model elements in TGG can be used to implement such one-to-many correspondences.

There are a wide range of approaches to bx [8]. Currently the most advanced approaches [5, 2] use constraint-based programming techniques to interpret relations $P(s, t)$ between source and target elements as specifications in both forward and reverse directions. These techniques would be a potentially useful extension to the syntactic inverses defined in Tables 7.4, 7.5, 7.6 and 14.1, however the efficiency of constraint programming will generally be lower than the statically-computed inverses. The approach also requires the use of additional operators extending standard OCL. Further techniques include the inversion of recursively-defined functions [18], which would also be useful to incorporate into the UML-RSDS approach.

Optimisation patterns such as Restrict Input Ranges [15] are not specific to bx, however they could be used in the design process to make forward and reverse transformations more efficient. Omit Negative Application Conditions applies to the cleanup constraints of the transformations.

Summary

In this chapter we have shown how bidirectional transformations can be implemented in UML-RSDS, based on the derivation of forward and reverse transformations from a specification of dual postcondition and invariant relations between source and target models. We have described transformation patterns which may be used to structure bx, and verification techniques which can be used to show correctness properties for the forward and reverse transformations.

References

[1] A. Anjorin and A. Rensink, *SDF to Sense transformation*, TU Darmstadt, Germany, 2014.

[2] A. Anjorin, G. Varro and A. Schurr, *Complex attribute manipulation in TGGs with constraint-based programming techniques*, BX 2012, Electronic Communications of the EASST vol. 49, 2012.

[3] M. Beine, N. Hames, J. Weber and A. Cleve, *Bidirectional transformations in database evolution: a case study 'at scale'*, EDBT/ICDT 2014, CEUR-WS.org, 2014.

[4] J. Cheney, J. McKinna, P. Stevens and J. Gibbons, *Towards a repository of bx examples*, EDBT/ICDT 2014, 2014, pp. 87–91.

[5] A. Cicchetti, D. Di Ruscio, R. Eramo and A. Pierantonio, *JTL: a bidirectional and change propagating transformation language*, SLE 2010, LNCS vol. 6563, 2011, pp. 183–202.

[6] K. Czarnecki, J. Nathan Foster, Z. Hu, R. Lammel, A. Schurr and J. Terwilliger, *Bidirectional transformations: a cross-discipline perspective*, GRACE workshop, 2008.

[7] E. Gamma, R. Helm, R. Johnson and J. Vlissides, *Design Patterns: Elements of Reusable Object-Oriented Software*, Addison-Wesley, 1994.

[8] Z. Hu, A. Schurr, P. Stevens and J. Terwilliger (eds.), *Report from Dagstuhl Seminar 11031*, January 2011, www.dagstuhl.de/11031.

[9] D.S. Kolovos, R.F. Paige and F. Polack, *The Epsilon Transformation Language*, ICMT, 2008, pp. 46–60.

[10] K. Lano and S. Kolahdouz-Rahimi, *Constraint-based specification of model transformations*, Journal of Systems and Software, vol. 88, no. 2, February 2013, pp. 412–436.

[11] K. Lano, *The UML-RSDS Manual*, www.dcs.kcl.ac.uk/staff/kcl/uml2web/umlrsds.pdf, 2015.

[12] K. Lano, S. Kolahdouz-Rahimi and K. Maroukian, *Solving the Petri-Nets to Statecharts Transformation Case with UML-RSDS*, TTC 2013, EPTCS, 2013.

[13] K. Lano, S. Kolahdouz-Rahimi and T. Clark, *A Framework for Model Transformation Verification*, BCS FACS journal, 2014.

[14] K. Lano and S. Kolahdouz-Rahimi, *Towards more abstract specification of model transformations*, ICTT 2014.

[15] K. Lano and S. Kolahdouz-Rahimi, *Model-transformation Design Patterns*, IEEE Transactions in Software Engineering, Vol. 40, 2014.

[16] OMG, *MOF 2.0 Query/View/Transformation Specification v1.1*, 2011.

[17] P. Stevens, *Bidirectional model transformations in QVT: semantic issues and open questions*, SoSyM, vol. 9, no. 1, January 2010, pp. 7–20.

[18] J. Voigtlander, Z. Hu, K. Matsuda and M. Wang, *Combining syntactic and semantic bidirectionalization*, ICFP '10, ACM Press, 2010.

Chapter 15

Backtracking and Exploratory Transformations

In this chapter we consider examples of backtracking and exploratory transformations, and describe how such transformations can be implemented in UML-RSDS.

15.1 Introduction

In some transformation problems it is difficult to formulate precisely how the transformation (considered as a collection of possibly non-deterministic rewrite rules) should behave. Instead, it is simpler to identify the characteristics of acceptable result models that should be produced by the transformation. This entails that the transformation should not terminate unless a satisfactory model is produced. For example, a program or class diagram refactoring transformation could apply automatically some rewrite rules to the program/model, and it is possible that the applications could lead to a model which is of unacceptably poor quality and the choices which led to this model must therefore be undone, and alternative rewrites applied. Other situations could be multi-objective optimisation problems, or exploratory program design. Figure 15.1 shows the execution behaviour which may occur, with some transformation executions having choice points which may lead to either an acceptable or unacceptable final model being produced, depending on the choice made at the point.

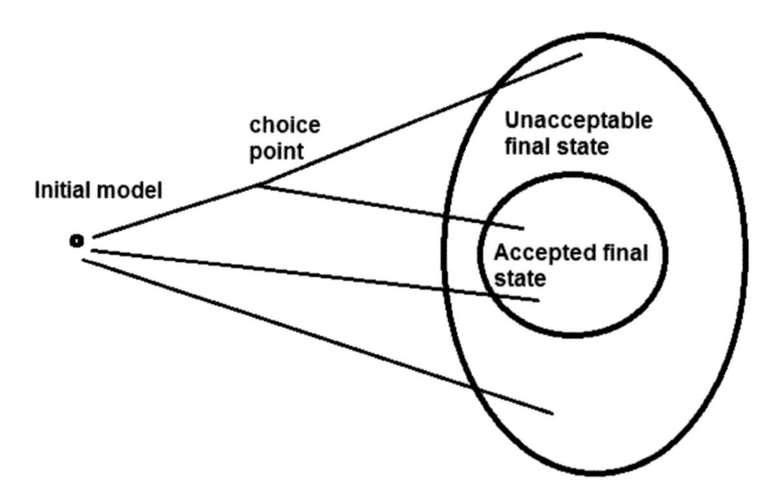

Figure 15.1: Transformation executions with choice points

The ability for a transformation (particularly an update-in-place transformation) to explore alternatives in rule applications, and to undo and redo choices in rule applications is therefore of potential use. In this chapter we consider two simple problems which involve such behaviour: the Sudoku solver, and a maze-solver.

15.2 Implementation of backtracking transformations

In UML-RSDS a backtracking transformation can be specified and implemented using an extension of the development process for standard transformations. Let τ be a transformation for which a backtracking implementation is needed. For simplicity, assume that all of τ's constraints iterate over the same entity type E. The possible executions of τ can be viewed as finite sequences

$$[step_1(ex_1, p1), ..., step_m(ex_m, pm)]$$

of steps, and each step is an application of some postcondition constraint of τ to some instance of E, together with some other parameters – the values of quantified additional variables in the constraint antecedent – the choices of the values for these variables will be backtracked over.

If this sequence of rule applications leads to an unacceptable end state after performing $step_m$, then the rule of this step must be either

(i) *redone* with a different pm value, if more choices for the rule parameter(s) still remain untried at this point in the search, or (ii) *undone* and the search for alternative possibilities attempted at the preceding step $(m - 1)$ instead. If $m = 1$ then the overall search fails. Alternative rules could be attempted at stage (ii) if such are available: the constraint ordering of the postconditions of τ indicates the priority order in which different rules should be attempted.

To organise such an execution model, two auxiliary variables are needed: (i) a variable $\$chosen_r : Set(Value)$ of E for each rule r, which holds for each $ex : E$ the already considered choices for r's other parameter values at this point in the search, and (ii) a variable $\$history : Sequence(Rule * E * Value)$ which records the sequence of completed rule applications. When $r(ex, p)$ completes, p is added to $ex.\$chosen_r$ and (r, ex, p) is added to $\$history$. For each r there will be an expression $ex.possible_r$ which identifies the set of possible parameter values p for which r could be applied with arguments ex, p. Redoing and undoing a step can then be described schematically as:

$redo_r(ex : E,\ p : Value)$
`pre:` $\$history.size > 0\ \&\ (ex.possible_r - ex.\$chosen_r)\rightarrow size() > 0$
`activity:`
  $\$history := \$history\rightarrow front()$;
  undo updates performed by $r(ex, p)$;
  apply r *to* ex *and an element* $p1$ *of* $ex.possible_r - ex.\$chosen_r$

and:

$undo_r(ex : E,\ p : Value)$
`pre:` $\$history.size > 0\ \&\ (ex.possible_r - ex.\$chosen_r)\rightarrow size() = 0$
`activity:`
  $\$history := \$history\rightarrow front()$;
  undo updates performed by $r(ex, p)$;
  clear $ex.\$chosen_r$

An undo action for $v : s$ or $s\rightarrow includes(v)$ for a set s is $s\rightarrow excludes(v)$. Assignments $f = v$ can be undone by assigning the previous (overwritten) value to f, if this is known. Table 15.1 shows some common undo actions for constraint succedent predicates P.

The *redo* and *undo* operations may need to be manually customised for particular constraints. Since the steps

  $\$history := \$history\rightarrow front()$;
  undo updates performed by $r(ex, p)$

are in common to *redo* and *undo*, these can be factored out and placed into *backtrack*. An operation $getPossibleValues()$ of E returns the remaining unchosen possible values:

Table 15.1: Undo actions

P	$undo(P)$	$condition$
$e \rightarrow display()$	$true$	
$e \rightarrow includes(v)$	$e \rightarrow excludes(v)$	Set-valued e, v not originally in e
$e \rightarrow includesAll(v)$	$e \rightarrow excludesAll(v)$	Set-valued e, none of v elements originally in e
$e \rightarrow includes(v)$	$e = e@\mathrm{pre} \rightarrow front()$	Sequence-valued e
$E \rightarrow exists(e \mid e.id = v \,\&\, P1)$	$undo(P1[E[v]/e])$ & $E[v] \rightarrow isDeleted()$	
$v = e$	$v = p$	p is prior value of v, because of antecedent conjunct $v = p$

$$query\ getPossibleValues() \ : \ Set(Value)$$
$$\texttt{post:}\ result \ = \ possible_r \ - \ \$chosen$$

The backtracking operation is then:

```
E::
static backtrack() : Boolean
( while $history.size > 0
  do
  ( sqx : Sequence := $history.last ;
    r : String := sqx.first ;
    ex : E := sqx[2] ;
    p : Value := sqx.last ;
    $history := $history->front() ;
    undo updates performed by r(ex,p) ;

    if ex.getPossibleValues()->size() > 0
    then
    ( ex.redor(p) ;
      return true
    )
    else
    ( ex.undor(p) )
  ) ;
  return false
)
```

Backtracking can fail (the final *return false* statement) if there are no further choices available at any point in the history of the transformation execution. There are two conditions on the state of the search which need to be tested at each step: *Success* indicates that the transformation has reached an acceptable state and the search can terminate, whilst

Backtrack indicates that an unacceptable state has been reached, and that the search must backtrack. Together, all these aspects of a backtracking problem form a DSL (Section 3.4) which could be used as a specification input to the design generation step of UML-RSDS. Figure 15.2 shows the backtracking DSL metamodel.

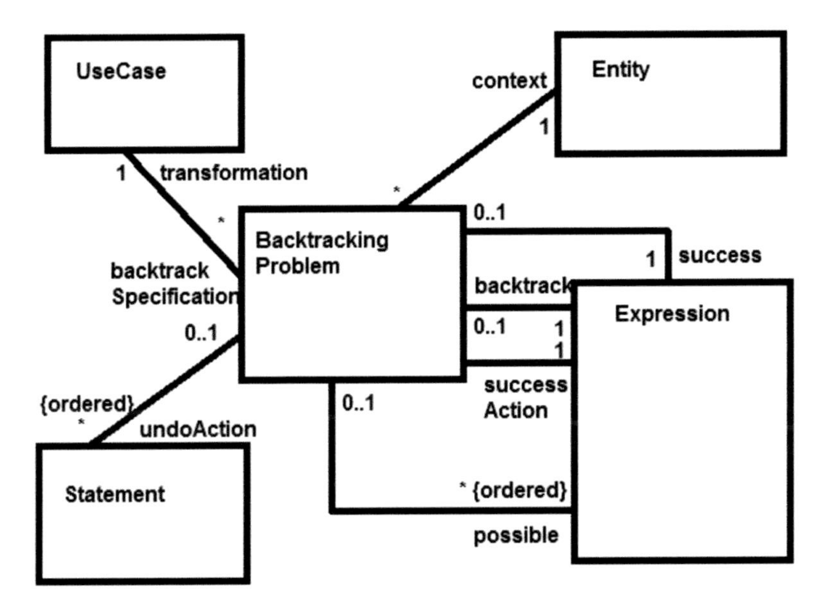

Figure 15.2: Backtracking DSL metamodel

These features are entered using a dialog to specify backtracking execution mode for a use case τ (Fig. 15.3).

If r is the first postcondition of τ, r should have the general form:

$$value : possibler \ \& \ Ante \ \Rightarrow \ Succ$$

with context entity E, where *value* does not occur in *Ante*. Backtracking will be performed over the choices of *value* in *possibler*. The fields of the backtracking mode dialog are filled in based on the use case and r, except for the undo actions and success test, which the developer needs to define. The design generation step augments the transformation with a new static variable: $\$history : Sequence(Sequence(Any))$ of E, and a new instance variable $\$chosen : Set(Any)$ of E. The constraint itself is transformed to:

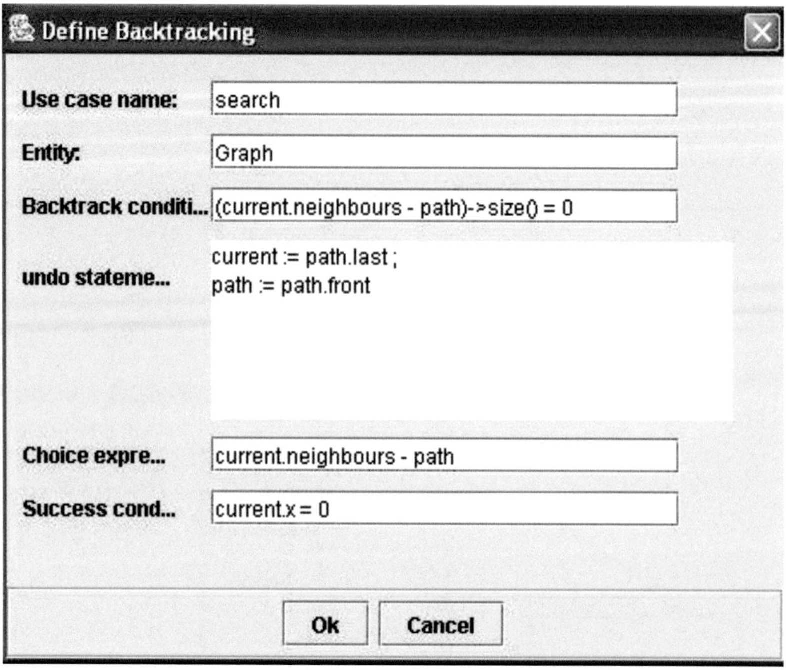

Figure 15.3: Backtracking mode dialog

$$\begin{array}{l} value : getPossibleValues() \ \& \ Ante \ \Rightarrow \\ \qquad\qquad Succ \ \& \ value : \$chosen_r \ \& \ Sequence\{r, self, value\} : \$history \end{array}$$

The generated design for a transformation τ with backtracking behaviour modifies the type 3 search and return loop which iterates through all elements of E:

```
E::
static search() : Boolean
(result : Boolean ;

 for (ex : E)
 do
 ( if ex.rtest() then ex.r()
   else
     ... cases for other rules, in descending priority order ... ;


   if ex.Success then
   ( successAction ; return false ) ;

   if ex.Backtrack then
   ( result = ex.backtrack() ;
```

```
    return result ) ;
  return true
 ) ;
 return false
)
```

This is a static operation of E.

15.3 Case study: maze route finder

This transformation attempts to find a loop-free route (e.g., for a rat) through a maze from a start node to an exit node, moving at each step from the current node to a neighbouring node. Figure 15.4 shows a possible metamodel for the maze problem.

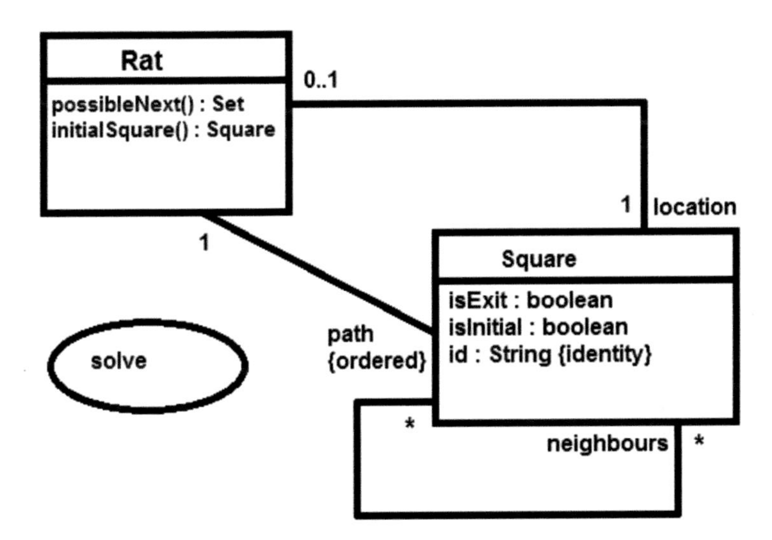

Figure 15.4: Maze solver metamodel

The solution transformation should be broken down into the following subtransformations:

1. Generating a non-trivial maze with alternative paths and dead-ends. Squares are either empty and can be occupied by the solver, or they are filled and cannot be occupied. The allocation of squares as filled or unfilled is fixed at the start of the solution process.

   The maze dimension is fixed at 10 squares in each direction in the test cases, but ideally the transformation should be able to generate problems of any size.

2. An option to display the maze and the location of the solver.

3. A maze solver which explores, step by step, possible paths from the starting square to an exit square. The solver may move one square in any direction to an unoccupied square. The transformation should terminate successfully if it finds a path from start to exit, and should display the path followed.

   If the solver reaches a dead-end with no unvisited squares possible to explore, it should backtrack to the most recent choice point, and attempt an alternative choice.

   If backtracking returns to the initial square, with no further choices available, then the transformation should terminate with a failure message.

The current state of the maze and location of the solver should be displayed upon each change in state/location.

The solution to part 3 of this problem can be implemented using backtracking on the choice of possible next square at each step. An operation *possibleNext* of *Rat* returns the set of possible next squares from a square *sq*: neighbours of *sq* not included in the current path:

```
Rat::
query possibleNext(sq : Square) : Set(Square)
pre: true
post:
  result = sq.neighbours - path
```

The maze solver use case *solve* has the postcondition (*Go*):

```
Rat::
  location.isExit = false & p : possibleNext(location)  =>
                    location = p & p : path & path->display()
```

This defines a single step from the current non-exit location to a neighbour that is not already in the path.

This problem fits the backtracking DSL with the following assignment of DSL features (Table 15.2). The *redoGo*(*p* : *Square*) operation of *Rat* is:

```
redoGo(p : Square)
(Square p1 = (possibleNext(location) - $chosen)->any() ;
 location := p1 ;
 path->includes(p1) ;
 path->display() ;
 $chosen->includes(p1) ;
 $history->includes(Sequence{"Go", self, p1})
)
```

Table 15.2: Maze problem features

transformation	*solve*
context	*Rat*
r	*Go*
$possible_r$	*possibleNext(location)*
Success	*location.isExit = true*
Backtrack	*location.isExit = false & possibleNext(location)→size() =* 0
successAction	("Found exit at " + self)→*display*()
undorAction	*if path.size* > 0 *then location := path.last* ; *path := path.front* *else location := initialSquare*()

An example maze is shown in Fig. 15.5. The data of this maze layout is defined in an instance model file *in.txt*.

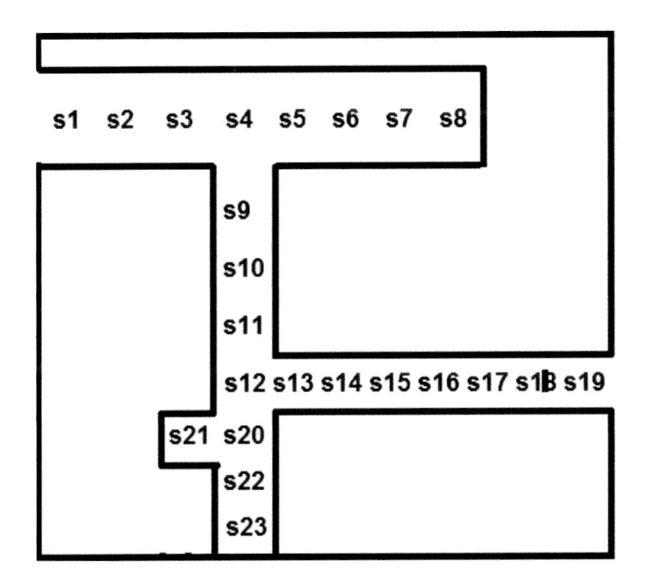

Figure 15.5: Maze example

The maze solver follows the route to s8, then backtracks, starting again from s9, and finally succeeding at s19:

```
....
[(Square) false,0,s1, (Square) false,0,s2, (Square) false,0,s3,
(Square) false,0,s4,
(Square) false,0,s5, (Square) false,0,s6, (Square) false,0,s7,
(Square) false,0,s8]
Backtracking
[(Square) false,0,s1, (Square) false,0,s2, (Square) false,0,s3,
```

```
(Square) false,0,s4,
(Square) false,0,s9]
....
[(Square) false,0,s1, (Square) false,0,s2, (Square) false,0,s3,
(Square) false,0,s4,
(Square) false,0,s9, (Square) false,0,s10, (Square) false,0,s11,
(Square) false,0,s12,
(Square) false,0,s13, (Square) false,0,s14, (Square) false,0,s15,
(Square) false,0,s16,
(Square) false,0,s17, (Square) false,0,s18, (Square) true,0,s19]
Found exit at: (Square) true,0,s19
```

The search hits a dead end at square s8, and backtracks to s4 before finding a route through s9 to the exit.

15.4 Case study: Sudoku solver

The Sudoku solver described in Chapter 2 can be extended to use backtracking to complete 9-by-9 puzzles. For convenience, we add use cases to initialise 4-by-4 and 9-by-9 boards (Fig. 15.6).

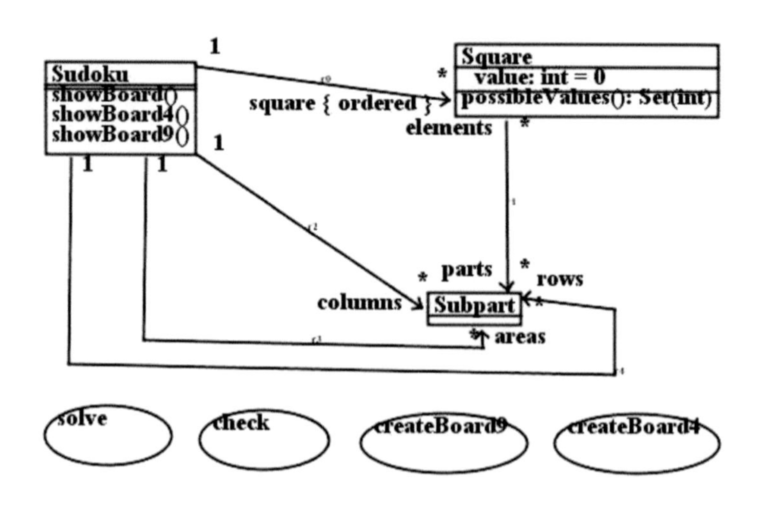

Figure 15.6: Extended Sudoku solver

Initialisation of 4-by-4 boards involves (i) creation of the *Sudoku* game instance; (ii) creating the four rows and the individual squares in them; (iii) creating the four columns and populating them with the appropriate squares; (iv) creating the four 2-by-2 areas and populating them with the appropriate squares. These steps are specified by the following constraints of *createBoard4*:

```
Sudoku->exists( s | true )

Sudoku::
Integer.subrange(1,4)->forAll( i |
    Subpart->exists( p | p : rows &
          Integer.subrange(1,4)->forAll( j |
             Square->exists( sq | sq : p.elements & sq : square ) ) ) )

Sudoku::
Integer.subrange(1,4)->forAll( i |
     Subpart->exists( p | p : columns &
           Integer.subrange(1,4)->forAll( j |
                square[i + ( j - 1 ) * 4] : p.elements ) ) )

Sudoku::
Subpart->exists( ar1 | ar1 : areas &
    square[1] : ar1.elements & square[2] : ar1.elements &
    square[5] : ar1.elements & square[6] : ar1.elements ) &
Subpart->exists( ar2 | ar2 : areas &
    square[3] : ar2.elements & square[4] : ar2.elements &
    square[7] : ar2.elements & square[8] : ar2.elements )

Sudoku::
Subpart->exists( ar3 | ar3 : areas &
    square[9] : ar3.elements & square[10] : ar3.elements &
    square[13] : ar3.elements & square[14] : ar3.elements ) &
Subpart->exists( ar4 | ar4 : areas &
    square[11] : ar4.elements & square[12] : ar4.elements &
    square[15] : ar4.elements & square[16] : ar4.elements )
```

The first constraint is of type 0, the others are of type 1 and iterate over *Sudoku*.

Initialisation of 9-by-9 boards involves (i) creation of the *Sudoku* game instance; (ii) creating the nine rows and the individual squares in them; (iii) creating the nine columns and populating them with the appropriate squares; (iv) creating the nine 3-by-3 areas and populating them with the appropriate squares. These steps are specified by the following constraints of *createBoard9*:

```
Sudoku->exists( s | true )

Sudoku::
Integer.subrange(1,9)->forAll( i |
    Subpart->exists( p | p : rows &
        Integer.subrange(1,9)->forAll( j |
            Square->exists( sq | sq : p.elements & sq : square ) ) ) )

Sudoku::
Integer.subrange(1,9)->forAll( i |
    Subpart->exists( p | p : columns &
```

```
        Integer.subrange(1,9)->forAll( j | square[i + ( j - 1 ) * 9] :
         p.elements ) ) )

Sudoku::
Integer.subrange(0,2)->forAll( j |
    Subpart->exists( ar | ar : areas &
        Integer.subrange(1,3)->forAll( i |
            square[i + j * 3] : ar.elements &
            square[i + 9 + j * 3] : ar.elements &
            square[i + 18 + j * 3] : ar.elements ) ) )

Sudoku::
Integer.subrange(0,2)->forAll( j |
    Subpart->exists( ar | ar : areas &
        Integer.subrange(1,3)->forAll( i |
            square[i + 27 + j * 3] : ar.elements &
            square[i + 36 + j * 3] : ar.elements &
            square[i + 45 + j * 3] : ar.elements ) ) )

Sudoku::
Integer.subrange(0,2)->forAll( j |
    Subpart->exists( ar | ar : areas &
        Integer.subrange(1,3)->forAll( i |
             square[i + 54 + j * 3] : ar.elements &
             square[i + 63 + j * 3] : ar.elements &
             square[i + 72 + j * 3] : ar.elements ) ) )
```

Again, the first constraint is of type 0, the others are of type 1, iterating over *Sudoku*.

The *possibleValues* operation needs to be modified for the case of 9-by-9 games:

```
Square::
query possibleValues() : Set(int)
pre: true
post:
  result = Integer.subrange(1,9) - parts.elements.value
```

The *solve* use case is modified from the version described in Chapter 2 to have a single post-condition r, with context class *Square*:

```
  value = 0 & v : possibleValues()   =>   value = v & Sudoku.showBoard()
```

The constraint is given backtracking behaviour, with the *Backtrack* condition on *Square* defined as:

$$value = 0 \ \& \ possibleValues()\rightarrow size() = 0$$

That is, the square is blank and no possible value can be used to fill it. Table 15.3 shows the features of the Sudoku solver transformation in terms of the backtracking DSL.

The $redor(v : int)$ operation of *Square* is:

Table 15.3: Sudoku problem features

transformation	*solve*
context	*Square*
r	r
$possible_r$	$possibleValues()$
Success	$Square \rightarrow collect(value) \rightarrow excludes(0)$
Backtrack	$value = 0 \ \& \ possibleValues() \rightarrow size() = 0$
successAction	"Game is solved"$\rightarrow display()$
undorAction	$value := 0$

```
redor(v : int)
(int p1 = (possibleValues() - $chosen)->any();
 value := p1;
 Sudoku.showBoard();
 $chosen->includes(p1);
 $history->includes(Sequence{"r", self, p1})
)
```

$undor(v : int)$ is:

```
undor(v : int)
($chosen->clear()
)
```

The display operation needs to be adapted to display boards of either size 4 or size 9:

```
Sudoku::
showBoard()
pre: true
post:
  ( square.size = 16  => showBoard4() ) &
  ( square.size = 81  => showBoard9() )

Sudoku::
showBoard4()
pre: true
post:
  ( square[1].value + " " + square[2].value + " " +
    square[3].value + " " + square[4].value )->display() &
  ( square[5].value + " " + square[6].value + " " +
    square[7].value + " " + square[8].value )->display() &
  ( square[9].value + " " + square[10].value + " " +
    square[11].value + " " + square[12].value )->display() &
  ( square[13].value + " " + square[14].value + " " +
    square[15].value + " " + square[16].value )->display() &
  ""->display()
```

```
Sudoku::
showBoard9()
pre: true
post:
  ( square.subrange(1,9)->collect(value) )->display() &
  ( square.subrange(10,18)->collect(value) )->display() &
  ( square.subrange(19,27)->collect(value) )->display() &
  ( square.subrange(28,36)->collect(value) )->display() &
  ( square.subrange(37,45)->collect(value) )->display() &
  ( square.subrange(46,54)->collect(value) )->display() &
  ( square.subrange(55,63)->collect(value) )->display() &
  ( square.subrange(64,72)->collect(value) )->display() &
  ( square.subrange(73,81)->collect(value) )->display() &
  ""->display()
```

An example execution is as follows, starting from the partially completed board:

```
[8, 1, 0, 0, 0, 0, 0, 6, 0]
[0, 0, 0, 0, 0, 0, 3, 0, 0]
[0, 0, 0, 0, 0, 0, 0, 0, 0]
[0, 0, 1, 0, 0, 0, 0, 0, 0]
[0, 0, 0, 0, 0, 2, 0, 0, 0]
[0, 0, 0, 0, 0, 0, 0, 0, 0]
[0, 0, 0, 0, 0, 0, 0, 0, 0]
[0, 0, 0, 0, 0, 0, 0, 0, 0]
[0, 0, 0, 0, 0, 0, 0, 0, 0]
```

Backtracking behaviour is needed at several points in the solution process, for example:

```
[8, 1, 2, 3, 4, 5, 7, 6, 9]
[4, 5, 6, 1, 2, 0, 3, 8, 0]
[0, 0, 0, 0, 0, 0, 0, 0, 0]
[0, 0, 1, 0, 0, 0, 0, 0, 0]
[0, 0, 0, 0, 0, 2, 0, 0, 0]
[0, 0, 0, 0, 0, 0, 0, 0, 0]
[0, 0, 0, 0, 0, 0, 0, 0, 0]
[0, 0, 0, 0, 0, 0, 0, 0, 0]
[0, 0, 0, 0, 0, 0, 0, 0, 0]

Backtracking, undoing [solve1, (Square) 8, 8]
Backtracking, undoing [solve2, (Square) 2, 2]
[8, 1, 2, 3, 4, 5, 7, 6, 9]
[4, 5, 6, 1, 7, 8, 3, 0, 0]
[0, 0, 0, 0, 0, 0, 0, 0, 0]
```

```
[0, 0, 1, 0, 0, 0, 0, 0, 0]
[0, 0, 0, 0, 0, 2, 0, 0, 0]
[0, 0, 0, 0, 0, 0, 0, 0, 0]
[0, 0, 0, 0, 0, 0, 0, 0, 0]
[0, 0, 0, 0, 0, 0, 0, 0, 0]
[0, 0, 0, 0, 0, 0, 0, 0, 0]

[8, 1, 2, 3, 4, 5, 7, 6, 9]
[4, 5, 6, 1, 7, 8, 3, 2, 0]
[0, 0, 0, 0, 0, 0, 0, 0, 0]
[0, 0, 1, 0, 0, 0, 0, 0, 0]
[0, 0, 0, 0, 0, 2, 0, 0, 0]
[0, 0, 0, 0, 0, 0, 0, 0, 0]
[0, 0, 0, 0, 0, 0, 0, 0, 0]
[0, 0, 0, 0, 0, 0, 0, 0, 0]
[0, 0, 0, 0, 0, 0, 0, 0, 0]

Backtracking, undoing [solve1, (Square) 2, 2]
Backtracking, undoing [solve2, (Square) 8, 8]
[8, 1, 2, 3, 4, 5, 7, 6, 9]
[4, 5, 6, 1, 7, 9, 3, 2, 0]
[0, 0, 0, 0, 0, 0, 0, 0, 0]
[0, 0, 1, 0, 0, 0, 0, 0, 0]
[0, 0, 0, 0, 0, 2, 0, 0, 0]
[0, 0, 0, 0, 0, 0, 0, 0, 0]
[0, 0, 0, 0, 0, 0, 0, 0, 0]
[0, 0, 0, 0, 0, 0, 0, 0, 0]
[0, 0, 0, 0, 0, 0, 0, 0, 0]
```

The first backtracking process is triggered by the last square on the second row becoming unfillable. The *solve*1 step filling the adjacent square with 8 is undone, and the *solve*2 step filling the middle square in the row with 2 is redone, with the value 7 attempted there instead. The second backtracking is also triggered by the last square on row two being unfillable, and the filling of the 6th square on the row with 8 is redone to use 9 instead.

Finally a complete solution is produced:

```
[8, 1, 2, 3, 4, 5, 7, 6, 9]
[4, 5, 6, 1, 7, 9, 3, 2, 8]
[3, 7, 9, 2, 6, 8, 1, 4, 5]
[2, 3, 1, 4, 5, 6, 8, 9, 7]
[5, 4, 7, 8, 9, 2, 6, 1, 3]
[6, 9, 8, 7, 1, 3, 2, 5, 4]
```

```
[1, 2, 3, 5, 8, 4, 9, 7, 6]
[7, 6, 4, 9, 3, 1, 5, 8, 2]
[9, 8, 5, 6, 2, 7, 4, 3, 1]
```

Summary

In this chapter we have described specification and implementation techniques for search-based transformations using UML-RSDS, and we have shown how such transformations can be defined.

Chapter 16

Agile Development and Model-based Development

UML-RSDS may be used with a wide range of development approaches, including traditional plan-based development with strict stages, or with techniques such as pair programming (adapted to become pair modelling). However, we have found in practice that some form of agile development process best suits the approach. In this chapter we describe the concepts of agile development, and identify how agile development can be combined with model-based development to form the concept of agile model-based development (AMBD). Finally, we also describe specific AMBD approaches that can be used with UML-RSDS.

16.1 Agile development

The idea of agile development originated in the recognition that traditional plan-based development, with strict phasing of requirements analysis, specification, design and implementation phases, was unrealistic and counter-productive in situations where requirements may change rapidly and new revisions of a system are required within short periods of time [1].

Even in the early days of software engineering it was recognised that an artificial separation between different software engineering activities was unnecessary, as seen in the following quote from IBM's Watson Research Center in 1969:

> "The basic approach recognises the futility of separating design, evaluation and documentation processes in software design. The design process is structured by an expanding model [...] It is tested and further expanded through a sequence of models that develop an increasing amount of function and detail. Ultimately, the model becomes the system." [8]

Traditional software development processes are plan-based: focussed on prescriptive activities and on a fixed sequence of stages (e.g., Analysis; Design; Coding; Testing in the Waterfall process). Agile processes in contrast attempt to be as lightweight as possible in terms of development process: their primary goal is to deliver a system to the customer that meets their needs, in the shortest possible time, taking account of changes in requirements.

The following principles are key characteristics of agile development [1]:

- Responding to change is more important than following a plan.
- Producing working software is more important than comprehensive documentation.
- Individuals and interactions are emphasised over processes and tools.
- Customer collaboration is emphasised over contract negotiation.

Development cycles are iterative, and build small parts of systems, with continuous testing and integration. Agile development methods include XP [2] and Scrum [13].

Other characteristics of agile development include:

- "Self-selecting teams" – it is argued by agile development proponents that the best architectures, requirements, designs emerge from such teams.
- Agile methodologies are more suitable for smaller organisations and projects.
- eXtreme programming (XP) is more suitable for single projects developed and maintained by a single person or by a small team.

Table 16.1 contrasts agile versus plan-based development. Agile development techniques include:

- *Sprints*: development work which implements specific user requirements, in a short time frame as a step in the production of a new release (Fig. 16.1).

- *Refactoring*: regular restructuring of code to improve its quality, to remove redundancy and other flaws, etc. [3].

Table 16.1: Agile versus plan-based development

Agile development	*Plan-based development*
Small/medium-scale	Large scale (10+ people)
In-house project, co-located team	Distributed/outsourced team
Experienced/self-directed developers	Varied experience levels
Requirements/environment volatile	Requirements fixed in advance
Close interaction with customer	Distant customer/stakeholders
Rapid value, high-responsiveness required	High reliability/correctness required

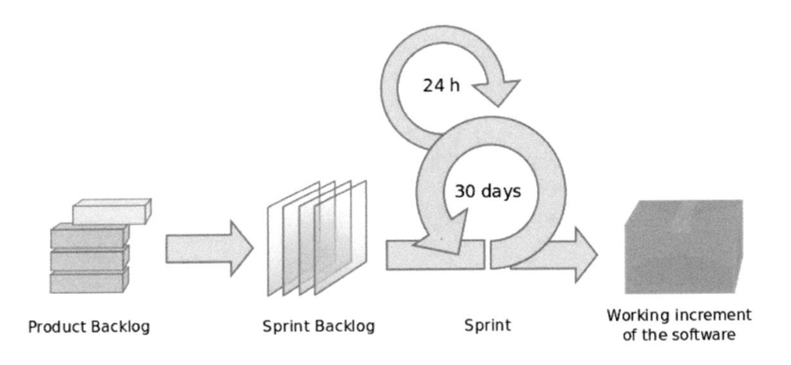

Figure 16.1: Sprints in the Scrum process [13]

16.1.1 Agile development: sprints

Sprints are regular re-occurring development iterations in which a subset of the project backlog work items are completed. This subset is termed the *sprint backlog*. Sprints produce deliverables that contribute to the overall project. Each iteration of a system involves a set of new use cases (termed 'stories' in Scrum) to be implemented, or other required improvements/corrections in the system. The work items to be carried out can be classified by their business value to a customer (high, medium, low), and by the risk/effort required by the developer. High priority and high risk work items should usually be dealt with first. The project *velocity* is the amount of developer-time available per iteration.

16.1.2 Defining a release plan

Taking the factors of work item priority and staff availability into account, the development team can define a *release plan*: a schedule of which work items will be delivered by which iteration and by which developers. A release consists of a set of iterations (e.g., sprints) and produces a deliverable which can be released to the client. The release plan is subject to several constraints:

- If use case $uc1$ depends on $uc2$ via $\ll$ *extend* $\gg$, $\ll$ *include* $\gg$ or an inheritance arrow from $uc1$ to $uc2$, then $uc1$ must be in the same or a later iteration to $uc2$.
- If a functionality $f1$ depends upon a functionality $f2$, then $f1$ must be in the same or a later iteration to $f2$.
- Iterations also depend on the availability of developers with the required skills for development of the use cases/carrying out the work items in an iteration: a work item can only be allocated to an iteration if a developer with the skills required by the work item is available in that iteration.
- Some use cases/work items may be prioritised over others.

Typically the release plan and project backlog is reviewed and reconsidered after completion of each sprint, and may be revised in the light of progress made and any new requirements which have arisen.

Figure 16.2 shows a situation where five new use cases are to be completed in a release, and their development must be scheduled into iterations. There are three developers available, but a high-skilled developer (number 1) is only available for at most 20 days in our project and their time must be blocked in a continuous period. The plan in the lower half of the figure shows a possible schedule meeting all the use case dependency constraints, and the skill and availability constraints of the developers.

16.1.3 Agile development: refactoring

Refactorings are small changes in a system, which aim to rationalise/generalise or improve the structure of the existing system. For example, if we discover that there are common attributes in all subclasses of a class then we can move all the common attributes up to the superclass (Fig. 16.3). Refactoring should not change existing functionality: tests should be rerun to check this. Regular refactoring should be carried out when time is available.

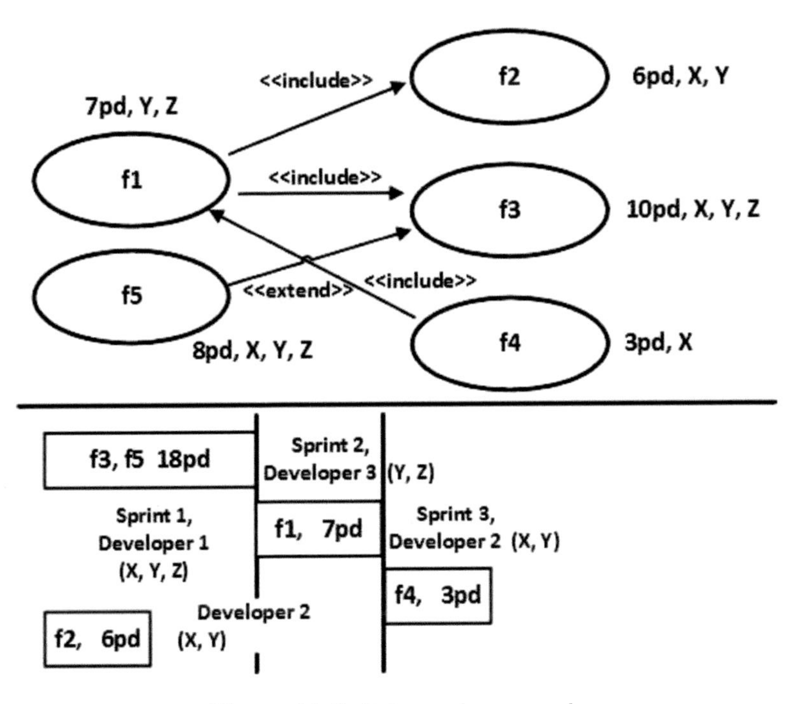

Figure 16.2: Release plan example

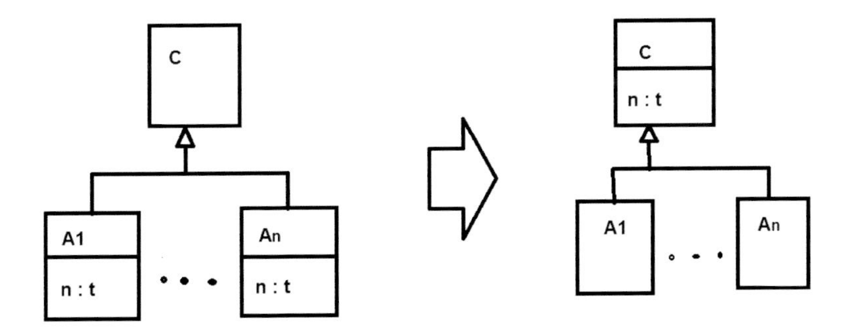

Figure 16.3: Pull-up attribute refactoring

Refactoring improves the maintainability of code by reducing redundancy and rationalising system structure. In the example of Fig. 16.3, changes to the type of attribute n only need to be made in one place after the refactoring, compared with changes to multiple occurrences before the refactoring.

Other agile development techniques include:

- *Test driven development (TDD)*: write unit tests first, then successively extend and refine the code until it passes all tests.
- *Pair programming*: two developers work at the same terminal, with one acting as an observer who has the role to identify problems with and possible improvements to the programmer's code.

TDD is appropriate if there is no clear initial understanding of how to solve the problem. Pair programming can improve quality, and reduce time, whilst increasing cost. Both of these techniques can be used with agile MBD, by using executable models in place of code.

16.1.4 Agile methods: eXtreme Programming (XP)

XP was devised by Beck [2] and other industrial practitioners with the aim of combining best practices in development projects and to focus on agility and code production. It is based on short development cycles and uses the following development practices: pair programming; code reviews; frequent integration and testing; close collaboration with customers; daily review meetings.

XP defines five development phases:

- *Exploration*: determine feasibility, understand the requirements, develop exploratory prototypes.
- *Planning*: agree timing and the use cases for the first release.
- *Iterations*: implement/test use cases in a series of iterations.
- *Productionizing*: prepare documentation, training, etc. and deploy the system.
- *Maintenance*: fix and enhance the deployed system.

XP recommends that iterations are one to two weeks long. The approach is oriented to smaller systems with small development teams, including the case of a single developer. The emphasis is on coding as the key development activity, and as a means of communicating ideas between developers and of demonstrating ideas to customers. Unit testing is used to check the correct implementation of individual features, and acceptance testing is used to check that what has been implemented meets the customer expectations. Integration testing checks that separately developed features operate correctly as part of the complete system.

16.1.5 Agile methods: Scrum

In the Scrum development process [13], a development team maintains a list (the *product backlog*) of product requirements. These consist of new use cases, use case extensions or corrections, or other required work items related to the product. A priority ordering is placed on the product backlog – normally the highest-priority items will be worked on first. Selected work items are removed from the product backlog and placed in the *sprint backlog* for completion in the first sprint (Fig. 16.1). This process is repeated for successive sprints. The product backlog will change from sprint to sprint as work items are moved to sprint backlogs, and also because of new requirements that arise from sprints or externally. A *Scrum board* shows the current list of items in the product backlog, the current sprint backlog and the status of items being worked on or completed in the current sprint (Fig. 16.4).

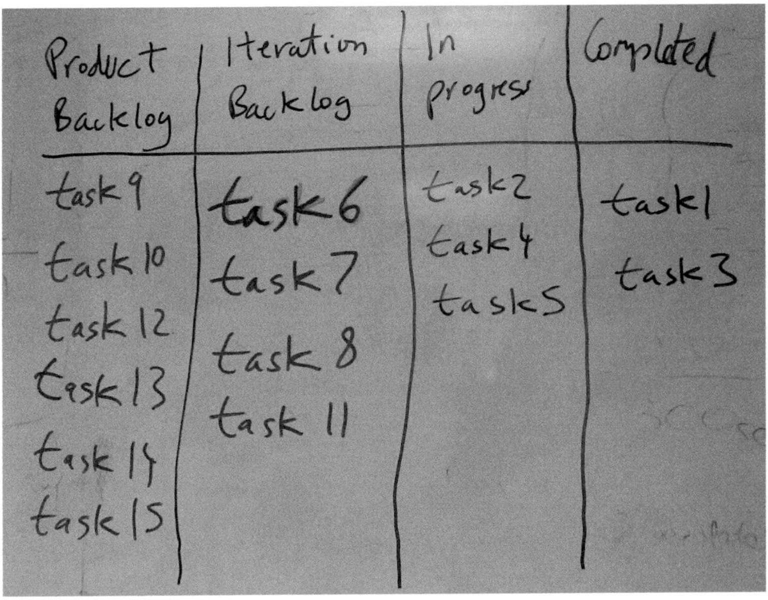

Figure 16.4: Scrum board example

Scrum defines three key roles for members of a software development team using the Scrum method:

- **Product owner**: the customer representative in the team, this role is responsible for liaising between the technical staff and the stakeholders. The product owner identifies required work items, identifies their priority, and adds these to the product backlog.

- **Development team**: the workers who perform the technical work. The team should have all needed skills and be self-organising. Typically there will be between 3 to 9 team members in this role.
- **Scrum master**: the Scrum master facilitates the Scrum process and events, and the self-organisation of the team. This role is not a project manager role and does not have personnel management responsibilities.

Four development phases are defined in Scrum:

- *Planning*: establish the vision, set expectations, develop exploratory prototypes. Identify the product backlog and apply requirements engineering techniques to clarify and refine the requirements.
- *Staging*: prioritise work items and plan the overall release, identify items (the sprint backlog) for the first iteration (sprint), develop exploratory prototypes.
- *Development*: implement the product backlog items in a series of sprints, and refine the release plan. Daily Scrum meetings take place, often at the start of a day, to check progress.
- *Release*: prepare documentation, training, etc. and deploy the system.

Scrum recommends that iterations (sprints) are 1 week to 1 month long.

The main artifacts of the Scrum process are:

- *Story*: a generalised task or work item (e.g., a new or modified use case). It can consist of a set of subtasks.
- *Product backlog*: an ordered (in terms of priority) list of work items required for the product.
- *Sprint backlog*: an ordered list of work items to be completed in a sprint.

The team uses a Scrum board showing the tasks to do, in progress and completed. A *Burndown chart* shows a graph of the remaining work against time.

The key events in Scrum are:

- *Sprint planning*: performed by the Scrum team before the sprint, the team agrees upon the use cases to be worked on in the sprint (the sprint backlog).

- *Daily Scrum*: this organises the activities of the team, reviews sprint progress and deals with any issues. It is time-limited (e.g., 15 mins). The key questions for developers at this meeting are: (i) what did I achieve yesterday? (ii) what will I achieve today? (iii) is there anything blocking me from achieving my work?
- *Sprint review*: at the end of the sprint, this reviews the sprint outcomes, and presents completed work to the stakeholders.
- *Sprint retrospective*: this is performed after the sprint review, and before planning of the next sprint. This is facilitated by the Scrum master and analyses the achievements of the sprint, and ideas for improvement of the process.

It can be seen that Scrum does not necessarily need to be applied at the source code level, and (provided that all required product extensions/changes in the product backlog can be carried out at the specification level) it could instead be applied at the specification model level. For UML-RSDS this means that the Scrum team works on class diagram and use case specifications, instead of on source code. Scrum has become the most widely-used agile method in industry [14].

16.2 Agile model-based development approaches

Various attempts have been made to combine agile development and model-based development [5, 10]. In some ways these development approaches are compatible and complementary:

- Both agile development and MBD aim to reduce the gap between requirements analysis and implementation, and hence to reduce the errors that arise from incorrect interpretation or formulation of requirements. Agile development reduces the gap by using short incremental cycles of development, and by direct involvement of the customer during development, whilst MBD reduces the gap by automating development steps.
- Executable application models (or models from which code can be automatically generated) of MBD serve as a good communication medium between developers and stakeholders, supporting the collaboration which is a key element of agile development.
- Automated code generation accelerates development, in principle, by avoiding the need for much detailed manual low-level coding.
- The need for producing separate documentation is reduced or eliminated, since the executable model is its own documentation.

On the other hand, the culture of agile development is heavily code-centric, and time pressures may result in fixes and corrections being applied directly to generated code, rather than via a reworking of the models, so that models and code become divergent. A possible corrective to this tendency is to view the reworking of the model to align it to the code as a necessary 'refactoring' activity to be performed as soon as time permits. We have followed this approach in some time-critical UML-RSDS applications.

Tables 16.2 and 16.3 summarise the parallels and conflicts between MBD and Agile development.

Table 16.2: Adaptions of Agile development for MBD

Agile practice	*Practice in Agile MBD*
Refactoring for quality improvement	Use application model refactoring, not code refactoring
Test-based validation	(i) Generate tests from application models (ii) Correct-by-construction code generation
Rapid iterations of development	Rapid iterations of modeling + Automated code generation
No documentation separate from code	Application models are both code and documentation

Table 16.3: Conflicts between Agile development and MBD

Conflict	*Resolutions in Agile MBD*
Agile is oriented to source code, not models	(i) Models as code (ii) Round-trip engineering (iii) Manual re-alignment
Agile focus on writing software, not documentation	Application models are both documentation and software
Agile's focus on users involvement in development, versus MBD focus on automation	Active involvement of users in conceptual and system modelling

Agile development has been criticised for its dependence upon highly-skilled developers and its failure to consider non-functional requirements. It also focusses upon the development of a specific product in each process, and does not consider long-term support for the development of product families. An agile MBD approach could help to address these problems as shown in Table 16.4.

Table 16.4: Problems of Agile development

Problem	*Resolutions in Agile MBD*
Cost of customer meetings	Models are more abstract than code, hence easier and faster to review
Lack of design and documentation	Models provide documentation
Requires high-skilled developers	Modelling requires fewer resources than coding
Does not address non-functional requirements	Introduce explicit requirements engineering stage in each iteration
Does not address product-line development	Build libraries and components for reuse within other related products

The key elements of an agile MBD approach are therefore:

- Combines Agile development and MBD.
- The system specification is expressed as an *executable application model* (a PIM or PSM), which can be delivered as a running system, or used to automatically generate a running system, and can be discussed with and demonstrated to customers.
- Incremental development of the system now uses executable models, rather than code. The models serve as documentation, also.
- Alignment of models with code (because of manual modification of generated code) is considered as a high-priority refactoring activity.
- Reuse is considered by creating reusable subcomponents during product developments.

16.2.1 Agile MBD methods

A small number of agile MBD approaches have been formulated and applied:

- Executable UML (xUML) [11].
- UML-RSDS [7].

- Sage [6].
- MDD-SLAP [15].
- Hybrid MDD [4].

Table 16.5 compares these approaches with regard to their support for agility, model-based development, and their domains of use. $\surd$ indicates support, ? indicates partial support, $\times$ no support.

Table 16.5: Comparison of Agile MBD approaches

	xUML	*UML-RSDS*	*Sage*	*MDD-SLAP*	*Hybrid MDD*
Incremental	?	$\surd$	?	$\surd$	$\surd$
Interoperable	?	?	?	$\surd$	$\surd$
Verification	$\surd$	$\surd$	$\times$	$\times$	$\times$
Round-trip engineering	$\times$	$\times$	$\times$	$\times$	$\times$
Model-based testing	$\times$	$\times$	$\times$	?	$\times$
Reuse	$\surd$	$\surd$	?	?	$\times$
Based upon	UML, MDA	UML, MDA	Reactive systems modelling	Scrum + MDD	Parallel agile and MDD
Principal domains of use	High-assurance, embedded	Transformations	Reactive multi- -agent systems	Real-time telecoms	Small medium sized

Both xUML and UML-RSDS use the principle that "The model is the code", and support incremental system changes via changes to the specification. There is a clearly-defined process for incremental revision in UML-RSDS, MDD-SLAP and Hybrid MDD. Interoperability refers to the capability to integrate automatically generated code and subsystems with manually-produced code/systems, and for interoperation of the tools with external tools such as Eclipse. UML-RSDS provides partial support for system integration via the use of proxy/facade classes which can be declared to represent external subsystems, including hand-crafted code modules (Chapter 10). It supports import and export of metamodels and models from and to Eclipse. MDD-SLAP and Hybrid MDD define explicit integration processes for combining synthesised and hand-crafted code.

Explicit verification processes are omitted from Sage, MDD-SLAP and Hybrid MDD. Some support is provided by xUML and UML-RSDS: xUML provides specification simulation and testing facilities, and UML-RSDS supports correctness analysis via a translation to the B formal method. Model checking and counter-example synthesis is also supported via translations to SMV and Z3. By automating code generation, agile MBD approaches should however improve the reliability and correctness of code compared to manual-coding development. None of the approaches support round-trip engineering (the automated interconversion of code and models), which means that synchronisation of divergent code and models is a manual process. MDD-SLAP partially supports model-based test case generation, using sequence diagrams. The Agile MDD approaches provide only limited support for reuse and for product line development, and the issue is not considered in [15] or [4]. xUML supports reuse by means of *domain* (subsystem) definitions, and UML-RSDS supports reuse by definition of reusable components (Chapter 10).

UML-RSDS and xUML are based on modelling using the standard UML model notations, with some variations (action language in the case of xUML, use cases specified by constraints in UML-RSDS), and following a general MDA process (CIM to PIM to PSM to code). Platform modelling is explicitly carried out in xUML but not in UML-RSDS, which uses a single PSM (design) language. Sage uses variants of UML models oriented to reactive system definition using classes and agents. These include environmental, design, behavioural and runtime models. An executable system is produced by integration of these models. MDD-SLAP maps MDD process activities (requirements analysis and high-level design; detailed design and code generation; integration and testing) into three successive sprints used to produce a new model-based increment of a system. Hybrid MDD envisages three separate teams operating in parallel: an agile development team hand-crafting parts of each release; a business analyst team providing system requirements and working with a MDD team to produce domain models. The MDD team also develops synthesised code. MDD-SLAP and Hybrid MDD have the most elaborated development processes. The survey of [5] identifies that Scrum-based approaches such as MDD-SLAP are the most common in practical use of agile MBD (5 of the seven cases examined), with XP also often used (4 of 7 cases).

Most of the development approaches have been applied in a single main domain area. Apart from Sage, there seems no intrinsic reason why the approaches should be restricted in their domains of use, and UML-RSDS and Hybrid MDD are intended to be used widely for any small-medium sized system.

16.2.2 Comparison of xUML and UML-RSDS

These two approaches are both formally-oriented and UML-based MBD approaches. Executable UML (xUML) uses UML class diagrams, state machines and an action language to define explicit platform-independent specifications of systems. Class collaboration diagrams, domains and sequence diagrams are also used. In xUML, classes have state machines and objects interact by sending signals to each other. There is an underlying concurrent execution model for systems: objects synchronise by communication, but otherwise execute independently. xUML targets a wide range of implementation platforms, including C, VHDL, etc. It is a commercial product, marketed by Abstract Solutions Ltd.

In contrast, UML-RSDS uses UML class diagrams, OCL and use cases to define declarative PIM specifications. Individual systems are sequential, although these can interact via remote procedure calls in a distributed implementation. Translations to Java, C#, C++ are provided: it is oriented to this family of languages. It is open source. The system construction process supported by UML-RSDS is shown in Fig. 16.5.

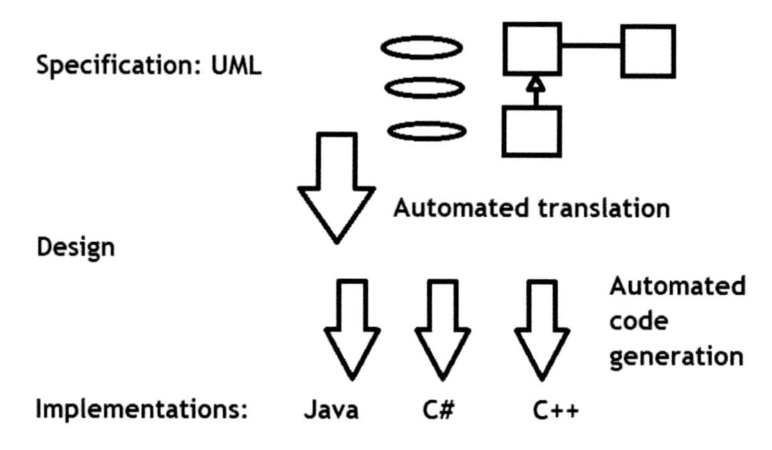

Figure 16.5: UML-RSDS software production process

The architecture of the implemented system in UML-RSDS has a standard structure (Fig. 16.6): a simple user interface component (GUI.java in a Java implementation) invokes operations of a Singleton class (Controller) which has operations corresponding to the use cases of the system, and which acts as a single point of access to the system functionality. Unlike in xUML, it is not possible to vary the design structure.

In the generated UML-RSDS code there are also classes corresponding to each specification class of the class diagram, and their operations are invoked by the controller class, which is the single point of entry (a Facade) for the functional tier of the generated system.

To summarise the differences between the two approaches:

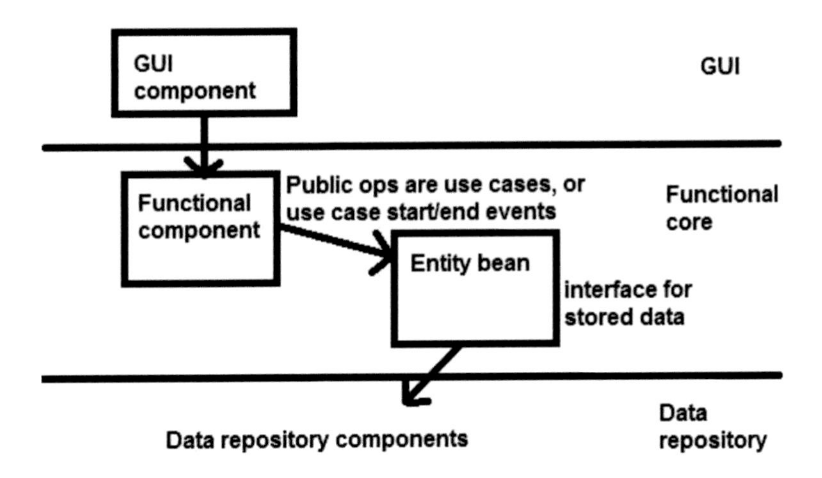

Figure 16.6: UML-RSDS application architecture

- UML-RSDS can be more abstract (tending to the CIM level): use cases are specified purely as logical pre/post predicates, with implicit algorithms. Pseudocode designs are synthesised automatically from these predicates (with some limited user intervention possible).
- UML-RSDS is more restrictive at the PSM level: only the Java-family of languages is supported, and the generated code is aligned class-by-class to the specification. A fixed architectural structure is generated.
- Executable UML requires more work in modelling the execution platform, but permits more variation in the platform and architecture.
- UML-RSDS provides verification capabilities via B AMN, SMV and Z3. xUML provides a specification simulator.

UML-RSDS may therefore be more appropriate for limited and lightweight use of MBD alongside traditional coding, as a first step to adopting MBD. xUML requires a more substantial committment to MBD, and a more substantial revision of development practices.

16.3 Agile MBD processes for UML-RSDS

We can adopt ideas from the Scrum, MDD-SLAP and Hybrid MDD processes to define a general procedure for using UML-RSDS with an agile development approach. The following guidelines for adoption and application of agile MBD are proposed, on the basis of our experiences in case studies:

- **Utilise lead developers**: When introducing MBD to a team inexperienced in its use, employ a small number of team members – especially those who are most positive about the approach and who have appropriate technical backgrounds – to take the lead in acquiring technical understanding and skill in the MBD approach. The lead developers can then train their colleagues.
- **Use paired or small team modelling**: Small teams working together on a task or use case can be very effective, particularly if each team contains a lead developer, who can act as the technical expert. It is suggested in [15] that such teams should also contain a customer representative.
- **Use a clearly defined process and management structure**: The development should be based on a well-defined process, such as XP, Scrum, or the MBD adaptions of these given in this chapter or by MDD-SLAP and Hybrid MDD. A team organiser, such as a Scrum master, who operates as a facilitator and co-ordinator is an important factor, the organiser should not try to dictate work at a fine-grained level but instead enable sub-teams to be effective, self-organised and to work together.
- **Refactor at the specification level**: Refactor application models, not code, to improve system quality and efficiency.
- **Extend the scope of MBD**: Encompassing more of the system into the automated MBD process reduces development costs and time.
- **Look for reuse opportunities**: If a product is part of a programme of development, with previous related systems, there may be existing components that can be reused to assist in the current product development. External libraries (e.g., QuantLib for financial applications) may be available for use. Also look for opportunities to create new reusable components, as discussed in Chapter 10.

A detailed agile MBD process for UML-RSDS can be based upon the MDD-SLAP process. Each development iteration is split into three phases (Fig. 16.7):

- **Requirements and specification:** Identify and refine the iteration requirements from the iteration backlog, and express new/modified functionalities as system use case definitions. Requirements engineering techniques such as exploratory prototyping and scenario analysis can be used, as described in Chapter 17. This phase corresponds to the Application requirements sprint in MDD-SLAP. Its outcome is an iteration backlog with clear and detailed requirements for each work item.

If the use of MBD is novel for the majority of developers in the project team, assign lead developers to take the lead in acquiring technical skills in MBD and UML-RSDS.

- **Development, verification, code generation:** Subteams allocate developers to work items and write unit tests for their assigned use cases. Subteams work on their items in the development phase of an iteration, using techniques such as evolutionary prototyping, in collaboration with stakeholder representatives, to construct detailed use case specifications. Formal verification at the specification level can be used to check critical properties, as described in Chapter 18. Reuse opportunities should be regularly considered, along with specification refactoring.

  Daily Scrum-style meetings can be held within subteams to monitor progress, update plans and address problems. Techniques such as a Scrum board and burndown chart can be used to manage work allocation and progress. The phase terminates with the generation of a complete code version incorporating all the required functionalities from the iteration backlog.

- **Integration and testing:** Do regular full builds, testing and integration in a SIFT phase, including integration with other software and manually-coded parts of the system.

In parallel with these phases, a tooling team may be employed to provide necessary extensions or adaptions to UML-RSDS. For example, to write new data converters or code generators needed by the iteration. After each sprint, a sprint review is held, and planning for the next sprint is carried out, identifying the sprint backlog from the updated product backlog.

The options (ii) (correct-by-construction code generation) and (i) (models as code) from Tables 16.2 and 16.3 are selected for the UML-RSDS agile development process, to increase the automation of development. The explicit requirements phase, active reuse, and the use of models as code addresses the problems identified in Table 16.4.

In the UML-RSDS agile MBD process detailed development activities are organised on the basis of use cases:

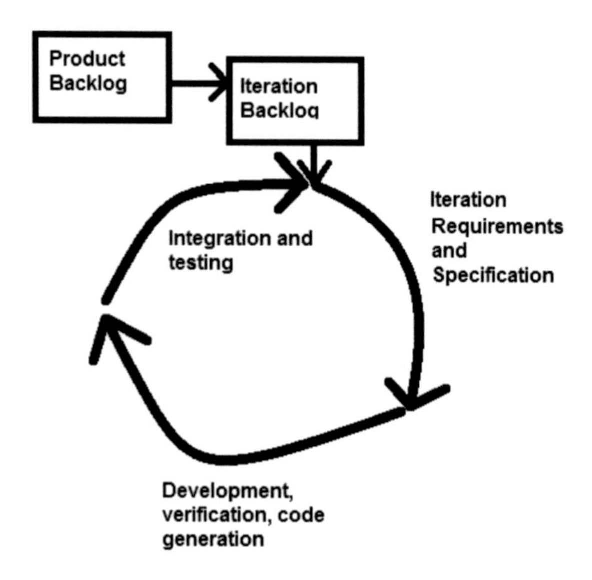

Figure 16.7: UML-RSDS agile MBD process

- Build a baseline class diagram (for a new development) or identify the relevant parts of an existing model (for an enhancement/maintenance development), identify and refine the system use case definitions based on the requirements.

- Developers write unit tests for their assigned use cases. All developers share the same system model (class diagram) but only need access to the use cases they are working on (update access) or that provide functionalities needed by these use cases (read-only access).

- Developers work on their tasks: they should only normally enhance or extend the class diagram, not delete elements or modify existing elements – if any change, including rationalisations, need to be performed, the developer should get the agreement of all the involved team members and team managers.

- The definitions of individual use cases should be stored in separate files where possible, and separate from the main system description.

Good communication within the team is needed. For larger systems the division into separate teams with parallel activities proposed by Hybrid MDD may be necessary. The recognition of reusable components within a product line may result in the establishment of a substantial library of components to support the development of related products. This could also be viewed as a *product line platform* for the domain of systems in the product line.

16.3.1 Examples of UML-RSDS agile MBD

As an example of development, consider that five use cases are to be implemented in a series of sprints for a new release of a system, according to Fig. 16.2. A base class diagram already exists, and is stored in file mm.txt. In each iteration, requirements for the iteration tasks are elicited and formalised in a requirements phase, which leads to their description in terms of textual and semi-formal rules based on the class diagram.

In the development phases, developer 1 takes responsibility for use cases f3 and f5 and writes tests for these. He works on the use cases, and keeps them in files f3.txt and f5.txt. Developer 2 works on the independent use case f2 and keeps its definition in f2.txt. All model and test files are stored in a version control system. At some point in the first sprint, developer 1 identifies necessary enhancements to the class diagram (e.g., that new identity attributes are needed in some classes) and gets the agreement of developer 2 and the team management to make this change.

On completion of the first sprint, f2.txt and f3.txt are passed to developer 3 who needs them (but cannot change them) for development of f1. The definition of f1 is maintained in file f1.txt, and this and f2.txt and f3.txt are passed to developer 2 for the final sprint.

Unit testing in the SIFT phases involves generating executable code for the use case to be tested, and applying the executable to the tests. Integration involves loading mm.txt into UML-RSDS, followed by the files of all use cases to be integrated: if a use case depends on other use cases then all files of (recursively) supplier use cases need to be loaded also. Name clashes in operations will be avoided if the use cases have distinct names, and if any user-defined operations have distinct names from each other and from any use case.

An alternative variant of this approach could assign to particular developers/subteams the responsibility for specific modules, where a module consists of a class diagram and set of use cases operating on that class diagram. In this case:

- Each module has one developer (or a development subteam), who works on its use cases.
- Modules should be maintained as far as possible as independent components, capable of reuse in different systems.

- Developers must agree interfaces of modules, and any shared data.

The developer of one module should never need to see the generated code of another module. They may have read-only access to the class diagram and use case specifications of supplier modules to their module.

An example of such a development could be a financial system development, where there are three modules identified: (i) a main computation component which reads investment datasets and performs calculations on these; (ii) an input-output utility module, supporting data read and write to spreadsheets; (iii) a module containing low-level computation routines, such as combinatorial and statistical functions. Module (i) depends on (ii) and (iii). Alternatively, a module such as (ii) could be supplied by external hand-crafted code that needs to be integrated into the development, using the composition techniques described in Chapter 10.

Summary

This chapter has described agile and model-based software development approaches, and we have identified ways in which UML-RSDS can be used as a tool within such development approaches. The following chapter will consider in detail the requirements engineering and specification of systems in UML-RSDS.

References

[1] K. Beck et al., *Principles behind the Agile Manifesto*, Agile Alliance, 2001. http://agilemanifesto.org/principles.

[2] K. Beck and C. Andres, *Extreme Programming Explained: Embrace change*, 2nd edition, Addison Wesley, 2004.

[3] M. Fowler, K. Beck, J. Brant, W. Opdyke and D. Roberts, *Refactoring: improving the design of existing code*, Addison-Wesley, 1999.

[4] G. Guta, W. Schreiner and D. Draheim, *A lightweight MDSD process applied in small projects*, Proceedings 35th Euromicro conference on Software Engineering and Advanced Applications, IEEE, 2009.

[5] S. Hansson, Y. Zhao and H. Burden, *How MAD are we?: Empirical evidence for model-driven agile development*, 2014.

[6] J. Kirby, *Model-driven Agile Development of Reactive Multi-agent Systems*, COMPSAC '06, 2006.

[7] K. Lano, *The UML-RSDS Manual*, http://www.dcs.kcl.ac.uk/staff/kcl/uml2web, 2015.

[8] G. Larman and V. Basili, *Iterative and Incremental Development: A brief history*, IEEE Computer Society, pp. 47–56, 2003.

[9] I. Lazar, B. Parv, S. Monogna, I.-G. Czibula and C.-L. Lazar, *An agile MDA approach for executable UML structured activities*, Studia Univ. Babes-Bolyai, Informatica, Vol. LII, No. 2, 2007.

[10] R. Matinnejad, *Agile Model Driven Development: an intelligent compromise*, 9th International Conference on Software Engineering Research, Management and Applications, pp. 197–202, 2011.

[11] S. Mellor and M. Balcer, *Executable UML: A foundation for model-driven architectures*, Addison-Wesley, Boston, 2002.

[12] M.B. Nakicenovic, *An Agile Driven Architecture Modernization to a Model-Driven Development Solution*, International Journal on Advances in Software, Vol. 5, Nos. 3, 4, pp. 308–322, 2012.

[13] K. Schwaber and M. Beedble, *Agile software development with Scrum*, Pearson, 2012.

[14] Version One, 9th Annual State of Agile Survey, 2015.

[15] Y. Zhang and S. Patel, *Agile model-driven development in practice*, IEEE Software, Vol. 28, No. 2, pp. 84–91, 2011.

Chapter 17

Requirements Analysis and Specification

Requirements analysis aims to identify the actual requirements for a system, and to express these in a precise and systematic form. This is an essential process for any type of software development: errors or omissions in requirements can be very expensive to correct at later development stages, and they are a significant cause of project failure. Requirements specification defines a precise model which formalises the requirements – in UML-RSDS this model is also the system specification, which is used as the starting point for automated system synthesis.

17.1 Stages in the requirements engineering process

The following four phases have been identified as the main stages in requirements engineering [7]:

- Domain analysis and requirements elicitation: identify stakeholders and gather information on the system domain and the system requirements from users, customers and other stakeholders and sources.
- Evaluation and negotiation: identify conflicts, imprecision, omissions and redundancies in the requirements, and consult and negotiate with stakeholders to reach agreement on resolving these issues.

- Specification and documentation: systematically document the requirements as a system requirements specification, in a formal or precise notation, to serve as an agreement between developers and stakeholders on what will be delivered.
- Validation and verification: check the formalised requirements for consistency, completeness and ability to satisfy stakeholder requirements.

In the following sections we describe the specific techniques that we use for these stages, in UML-RSDS developments of model transformations.

17.2 Requirements engineering for model transformations

Requirements engineering for model transformations involves specialised techniques and approaches, because transformations (i) have highly complex behaviour, involving non-deterministic application of rules and inspection/construction of complex model data; (ii) are often high-integrity and business-critical systems, with strong requirements for reliability and correctness. Transformations do not usually involve much user interaction (they are usually batch-processing systems), but may have security requirements if they process secure data. Correctness requirements which are particularly significant for transformations, due to their characteristic execution as a series of rewrite rule applications, with the order of these applications not algorithmically-determined, are: (i) confluence (that the output models produced by the transformation from a given input model are equivalent, regardless of the rule application orders); (ii) termination (regardless of execution order); (iii) to achieve specified properties of the target model, regardless of execution order: this is referred to as *semantic correctness*.

A transformation is expected to produce models which conform to the target language (this is termed *syntactic correctness*). In addition, correctness of a transformation may include that the semantics of source models is preserved in their target models: this is referred to as *model-level semantic preservation*. For migration, refinement and bidirectional (bx) transformations in particular, *traceability* of the transformation is important: to be able to identify for each target model element which source element(s) it has been derived from. Conservativeness is important for high-integrity code generation and refinement transformations.

The source and target languages of a transformation may be precisely specified by metamodels, or aspects of these may need to be discovered or postulated (e.g., in the case of migration of legacy data where the original data schema has been lost). In addition, the requirements for its processing may initially be quite unclear. For a migration transformation, analysis will be needed to identify how elements of the source language should be mapped to elements of the target: there may not be a clear relationship between parts of these languages, there may be ambiguities and choices in mapping, and there may be necessary assumptions on the input models for a given mapping strategy to be well-defined. There are specialist tools and languages for migrations, such as COPE and Epsilon Flock, which may be selected. The requirements specification should identify how each entity type and feature of the source language should be migrated to the target.

For refactorings, the additional complications arising from update-in-place processing need to be considered: the application of one rule to a model may enable further rule applications which were not originally enabled. Confluence may be difficult to enforce, and may be considered optional (for example, if a refactoring could produce one of a number of different restructurings of a model, of equal quality). The choice of transformation technology will need to consider the level of support for update-in-place processing. Some languages such as ATL and QVT have limited update-in-place support. The requirements specification should identify all the distinct situations which need to be processed by the transformation: the arrangements of model elements and their inter-relationships and significant feature values, and how these situations should be transformed.

Code-generation transformations may be very large, with hundreds of rules for the different cases of modelling language elements to be implemented in code. Effective organisation and modularisation of the transformation, and selection of appropriate processing strategies, are important aspects to consider. Template-based generation of program language text is a useful facility for code generators, and is provided by transformation technologies such as EGL and ATL templates. The requirements specification needs to identify how each source language construct should be translated to code.

The stakeholders of a transformation typically include not only the users of the transformation itself, but also users of the target models or products of the transformation process. For example, a code-generation transformation produces software code for various applications, whilst a refactoring or migration transformation produces refactored/migrated models for use by developers. In the case of a bx transformation both source and target models can be in active use.

Requirements may be functional or non-functional (e.g., concerned with the size of generated models, transformation efficiency or confluence). Another distinction which is useful for transformations is between *local* and *global* requirements:

- Local requirements concern localised parts of one or more models. *Mapping* requirements define when and how a part of one model should be mapped to part of another. *Rewriting/refactoring* requirements concern when and how a part of a model should be refactored/transformed in-place.
- Global requirements concern properties of an entire model. For example, that some global measure of complexity or redundancy is decreased by a refactoring transformation. Invariants, assumptions and postconditions of a transformation usually apply at the entire model level.

17.3 Requirements elicitation

A large number of requirements elicitation techniques have been devised. In the following sections we summarise some of these, and consider their relevance for the requirements analysis of transformations.

Observation

This involves the requirements engineer observing the current manual/semi-automated process used for the transformation.

It is relevant if a currently manual software development or transformation process is to be automated as a transformation. For example, if a procedure for constructing web applications or EIS of a particular architectural structure is to be automated. Observation can capture the operational steps of the manual process currently used by developers, as a basis for the specification of the automated process.

The technique is relevant for all kinds of transformations.

Unstructured interviews

In this technique the requirements engineer asks stakeholders open-ended questions about the domain and current process.

The technique is relevant in identifying the important issues which a transformation should have as goals. For refactorings, these could be what are the important goals for quality improvement of a model/system. For refinements, what are the important properties of the generated code (e.g., efficiency, conformance to a coding standard, readability, etc.). For general transformations, what is the scope of the mapping (which forms of input models are intended to be processed), what semantic/structural properties should be preserved, and what required restrictions there are on the output model structure.

Some possible questions which could be used in this technique are (for source models):

- What size range of input models should the transformation be capable of processing? What are the required data formats/encodings for these models?
- What logical assumptions can be made about the input model(s)? Are there constraints which they should satisfy, in addition to the source language structure constraints? If assumptions can be made, is it the responsibility of the transformation to check them, or will this be done prior to the transformation?
- Should the source model be preserved, or can it be overwritten?
- Are there security restrictions on source model data which should be observed?

Questions regarding the transformation processing could be:

- What category of transformation is needed (refinement, migration, refactoring, bidirectional)?
- Should the transformation be entirely automated, or should there be scope for interaction, e.g., if a choice of different possible refactorings needs to be made?
- How will errors in processing be handled and reported? Should the transformation terminate on the occurrence of an error?
- What are the timing/efficiency requirements for the transformation? What is the maximum permitted time for processing models of specific sizes?
- Should the transformation processing include tracing?
- Is there a requirement for the source and target models to be maintained consistently with each other, i.e., for model synchronisation and change propagation to be supported? What directions of change propagation are required (source to target, target to source or both)?
- Is there a requirement for continuous input data processing, i.e., for a streaming transformation?
- Are there environmental or organisational restrictions on the transformation languages/tools to be used, e.g., that they should operate within Eclipse?
- Is confluence required?

- Should the transformation preserve the source model semantics either in its entirety or partially? (Model-level semantic preservation).

Regarding the target model, possible questions could be:

- What are the required data format(s)/encoding(s) of the target model(s)?
- What logical properties should be ensured for the target models, in addition to conformance to the target metamodel?

This technique is relevant for all kinds of transformations, and for the requirements analysis of the product backlog in an agile development.

Structured interviews

In this technique the requirements engineer asks stakeholders prepared questions about the domain and system.

The requirements engineer needs to define appropriate questions which help to identify issues of scope and product (output model) requirements, as for unstructured interviews. This technique is relevant to all forms of transformation problem.

We have defined a catalogue of MT requirements for refactorings, bidirectional transformations, refinements and migrations, as an aid for structured interviews, and as a checklist to ensure that all forms of requirement appropriate for the transformation are considered (examples from the catalogue are shown in Table 17.1). Model-level semantic preservation and syntactic correctness are relevant global functional requirements for all categories of transformation.

This technique is suitable for all categories of transformations, and for iteration requirements analysis in agile development.

Document mining

This involves deriving information from documentation about an existing system or procedures, and from documentation on the required system.

For transformation problems this is particularly useful to obtain detailed information on the source and target metamodels and their constraints. These metamodels may be defined in standards (such as the OMG standards for UML and BPMN) or more localised documents such as the data schemas of particular enterprise repositories.

Table 17.1: Transformation requirements catalogue

	Refactoring	*Refinement, Migration*	*Bidirectional*
Local Functional	Rewrites/ refactorings	Mappings	Correspondence rules
Local Non-functional	Completeness (all cases considered)	Completeness (all source entities, features mapped)	Completeness (all entities, features considered)
Global Functional	Improvement in quality measure(s); Invariance of language constraints; Assumptions; Postconditions; Model-level semantic preservation; Syntactic correctness	Invariance (tracing target to source); Assumptions; Postconditions; Syntactic correctness; Model-level semantic preservation; Conservativeness	Bidirectionality; Model synchronisation; Assumptions; Postconditions; Hippocraticness; Syntactic correctness; Model-level semantic preservation
Global Non-functional	Termination; Efficiency; Genericity; Confluence; Extensibility; Fault tolerance; Security; Interoperability; Modularity	Termination; Efficiency; Traceability; Confluence; Extensibility Fault tolerance; Security; Interoperability; Modularity	Termination; Efficiency; Traceability; Confluence; Extensibility Fault tolerance; Security; Interoperability; Modularity

Reverse engineering

This involves extracting design and specification information from existing (usually legacy) software. This is useful for identifying data schema/metamodel structures, and as a first step in re-engineering/migration of applications from one programming language/environment to another.

Brainstorming

In this technique the requirements engineer asks a group of stakeholders to generate ideas about the system and problem.

This may be useful for very open-ended/new transformation problems where there is no clear understanding of how to carry out the transformation. For example, complex forms of migration where it is not yet understood how data in the source and target languages should correspond, likewise for complex refinements or merging transformations involving synthesis of information from multiple input models to produce a target model. Complex refactorings such as the introduction of design patterns could also use this approach.

Rapid prototyping

In this technique a stakeholder is asked to comment on a prototype solution.

This technique is relevant for all forms of transformation, where the transformation can be effectively prototyped. Rules could be expressed in a concrete grammar form and reviewed by stakeholders, along with visualisations of input and output models. This approach fits well with an Agile development process for transformations. Some transformation tools and environments are well-suited to rapid prototyping, such as GROOVE. For others, such as ETL or QVT, the complexity of rule semantics may produce misleading results. In UML-RSDS we usually produce simplified versions of the transformation rules, operating on full or simplified versions of the language metamodels. Rules may be considered in isolation before combining them with other rules.

Scenario analysis

In this approach the requirements engineer formulates detailed scenarios/use cases of the system for discussion with the stakeholders.

This is highly relevant for MT requirements elicitation, particularly for local functional requirements. Scenarios can be defined for the different required cases of transformation processing. The scenarios can be used as the basis of requirements formalisation. This technique is proposed for transformations in [2]. We typically use concrete grammar sketches in the notations of the source and target models, or informal rules, to describe scenarios.

A risk with scenario analysis is that this may fail to be complete and may not cover all cases of expected transformation processing. It is more suited to the identification of local rather than global requirements, and to functional rather than non-functional requirements.

Ethnographic methods

These involve systematic observation of actual practice in a workplace.

As for Observation, this may be useful to identify current work practices (such as coding strategies) which can be automated as transformations.

We do not mandate any particular requirements elicitation approach for UML-RSDS, however prototyping and scenario analysis fit in well with the UML-RSDS development approach, and have been used in most UML-RSDS developments. We have also used document mining in several development projects.

17.4 Evaluation and negotiation

Evaluation and negotiation techniques include: exploratory and evolutionary prototyping; viewpoint-based analysis; scenarios; goal-oriented analysis; state machines and other diagrammatic analysis modelling languages; formal modelling languages.

For UML-RSDS, prototyping techniques have been used for evaluating requirements, and for identifying deficiencies and areas where the intended behaviour of the system is not yet well understood. A goal-oriented analysis notation such as KAOS or SysML [1] can be used to document the decomposition of requirements into subgoals. A formal modelling notation such as OCL, temporal logic, or state machines/state charts can be used to expose the implications of requirements, such as inconsistencies, conflicts or incompleteness.

Scenario-based modelling is usually used with prototyping in UML-RSDS. Conflicts between scenarios are identified (cases where more than one scenario could be applicable), and any incompleteness (cases where no scenario applies). Conflicts may be resolved by strengthening application conditions (making the scenario more specialised), or by placing priority orders on scenarios so that one is always applied in preference to another if both are applicable to the same model situation. Incompleteness can be resolved by adding more cases, or by agreement with the customer that the uncovered situations are excluded from the inputs to be processed by the transformation. Scenarios can evolve into rule specifications, with more detail added, and they may be refined into more specialised cases of behaviour.

Global requirements may be refined to local requirements, e.g., by the decomposition of global requirements into a number of cases for different kinds of individual model elements. For update-in-place transformations (reactive or refactoring), syntactic correctness and model-level semantic preservation requirements can be refined to invariance requirements. Initial high-level requirements or goals for a transformation can be decomposed into more specific subgoals, in ways that are characteristic of transformations:

- Model-level semantic preservation for an update-in-place transformation can be refined into an invariance requirement (that the semantics of the model is equal to/equivalent to a constant), and then decomposed into requirements that this invariant is preserved by each rule application.
- Syntactic correctness of update-in-place transformations can be refined to an invariance property that the model satisfies (is conformant with) the metamodel, and then further decomposed into subgoals that each transformation step maintains this invariant.
- A goal to reduce a measure of poor quality in a model, for a refactoring transformation, can be decomposed into cases based on the different possibilities of poor structure in the model, and into subgoals that each transformation rule reduces the measure.
- For reactive systems, a requirement to maintain desired properties of the EUC can be decomposed into specific reactions to input events.
- For migration transformations, a general intended mapping of source to target can be refined into specific source entity-to-target entity mappings based on information elicited from stakeholders of the transformation, and hence decomposed into goals for specific mapping rules.
- Likewise for refinement transformation mapping goals. An example of such a decomposition, for the UML to relational database transformation, is given in [2].

For UML-RSDS we primarily use prototyping at the evaluation stage, with simplified versions of metamodels and rules used to investigate the intended transformation behaviour. This is focussed upon scenarios for local functional requirements, however global requirements could be expressed and evaluated in terms of particular measures (for quality improvements).

17.5 Requirements specification and documentation

Techniques for this stage include: UML and OCL; structured natural language; formal modelling languages.

At the initial stages of requirements elicitation and analysis, the intended effect of a transformation is often expressed by sketches or diagrams using the concrete grammar of the source and target languages concerned (if such grammars exist), or by node and line graphs if there is no concrete grammar. A benefit of concrete grammar rules is that they are directly understandable by stakeholders with knowledge of the source and target language notations. They are also independent of specific MT languages or technologies. Concrete grammar diagrams can be made more precise during requirements formalisation, or refined into abstract grammar rules. An informal mapping/refactoring requirement of the form

> "For each instance e of entity type E, that satisfies condition Cond, establish Pred"

can be formalised as a UML-RSDS use case postcondition

```
E::
  Cond'  =>  Pred'
```

where *Cond′* formalises Cond, and *Pred′* formalises Pred.

An informal mapping/refactoring requirement of the form

> "For each pair of instances e1 of entity type E1, and e2 of entity type E2, that satisfy condition Cond, establish Pred"

can be formalised as a use case postcondition

```
E1::
  e2 : E2 & Cond'  =>  Pred'
```

or as

```
E2::
  e1 : E1 & Cond'  =>  Pred'
```

It is important to avoid explicit assumptions about rule execution orders at this stage, unless such orders are mandated in the requirements. Often it is the case that different rule execution orders are possible (for example, to map attributes to columns before mapping classes to tables, in a UML to relational database transformation; or the reverse order), and the choice between such orders is a design or implementation issue and depends in part upon the MT language chosen for implementation.

Mandated orders of rules can be expressed separately from the rules themselves, either in text, e.g.: "All applications of Class2Table should occur before any application of Attribute2Column" or more formally in temporal logic:

$$Attribute2Column(a) \Rightarrow \blacklozenge \forall c : Class \cdot Class2Table(c)$$

Abstract grammar *transformation cases* are used to formalise MT requirements in [2]. In UML-RSDS the specification of a transformation consists of one or more UML use cases, each consisting of one or more transformation rules, defined by the use case postcondition constraints in OCL. Each externally-required use case from the functional requirements should be expressed as a UML-RSDS use case, but these may be further decomposed into internal use cases which perform sub-steps or phases of the required functionalities. Defining one case for each constraint avoids premature commitment to particular execution orders, enables testing of constraints in isolation, gives the developer greater control over the form of the eventual design and implementation, and can increase the flexibility and reusability of the specification.

Rules should be specified in the simplest manner possible which achieves the required effect, following the principle:

> *Define rules which express the required behaviour in the simplest and clearest form possible.*

Specification patterns such as Phased Construction and Map Objects Before Links should be used to structure and organise rules. Measures of rule complexity are given by the UML-RSDS tools to help guide developers in the simplification of rules. One syntactic measure is the sum of the number of OCL operators plus the number of user identifiers in a rule: a reasonable upper bound on this measure could be 40, and rules exceeding this limit should be decomposed into simpler rules or called operations.

17.6 Requirements validation and verification

Techniques for this stage include: prototyping with testing; formal requirements inspection; requirements checklists; formal modelling and proof/model-checking.

The formalised rules produced by the previous stage can be checked for internal correctness properties such as definedness and determinacy, which should hold for meaningful rules. These checks are performed by the *Generate Design* option of the UML-RSDS tools, and the results are displayed on the console window. A prototype implementation can be generated, and its behaviour on a range of test case input models, covering all of the scenarios considered during requirements elicitation, can be checked.

Global properties of invariance can be checked by proof, using the B formalism (Chapter 18). Proof can also be used to show that model complexity/redundancy measures are decreased by rule applications, for refactoring transformations. Variant expressions can be defined to prove termination of a transformation: if each rule application can be proved to strictly decrease this non-negative integer-valued expression, then the transformation must terminate. Model checkers such as nuSMV [6] can be used to check if required temporal properties are enforced by the implementation.

17.7 Selection of requirements engineering techniques

The following recommendations can be made for the use of the requirements techniques described above:

Initial elicitation stage, with poorly understood/unclear/non-specific requirements: interviews; structured interviews; observation; brainstorming; document mining; reverse engineering.

The stakeholders of the system are identified. Information on the functional and non-functional requirements is elicited from stakeholders and obtained from documentation. Initially expressed general requirements are made more specific and definite, their scope and intent are clarified, and implicit unstated requirements are made explicit. For a migration or refinement transformation from language S to language T, it will be necessary to identify how language elements of S should be represented in terms of T elements. For a refactoring, it will be necessary to identify when and how a general requirement to improve model quality should be put into effect as model rewritings in particular model situations.

Advanced elicitation stage, with decomposed requirements: scenarios; prototyping.

The specific requirements obtained from the first stage are represented as scenarios described by concrete grammar sketches/outline rules. Their functional behaviour can be encoded in prototypes for validation by stakeholders.

Particular migration/refinement/refactoring scenarios can be expressed and validated in this manner.

Evaluation and negotiation: scenarios; prototyping; formal analysis (for high-integrity systems).

Both functional and non-functional requirements are decomposed into more specific goals. The functional scenarios are checked for completeness, consistency and correctness with regard to stakeholder expectations, and enhanced and refined as necessary. Negotiation on the requirements takes place, using the detailed scenarios.

Requirements formalisation (documentation): UML; OCL; formal modelling (for high-integrity systems).

The concrete grammar rules are formalised as abstract grammar rules in OCL. Assumptions, invariants and conformance conditions are also precisely defined as OCL expressions. Non-functional requirements are precisely documented in terms of appropriate measures.

Validation and verification: testing; inspection; formal proof and model-checking (for high-integrity systems).

The formalised specification is tested, inspected and analysed for correctness with respect to stakeholder expectations and for internal consistency. In UML-RSDS this specification will also serve as the basis of the implemented system. The use of optimisation or architectural design patterns (Chapter 9) may be necessary to achieve efficiency/capacity requirements.

17.8 Requirements engineering example: class diagram refactoring

This is an example of an in-place endogenous transformation which refactors class diagrams to improve their quality by removing redundant feature declarations. Figure 17.1 shows the metamodel of the source/target language of this transformation.

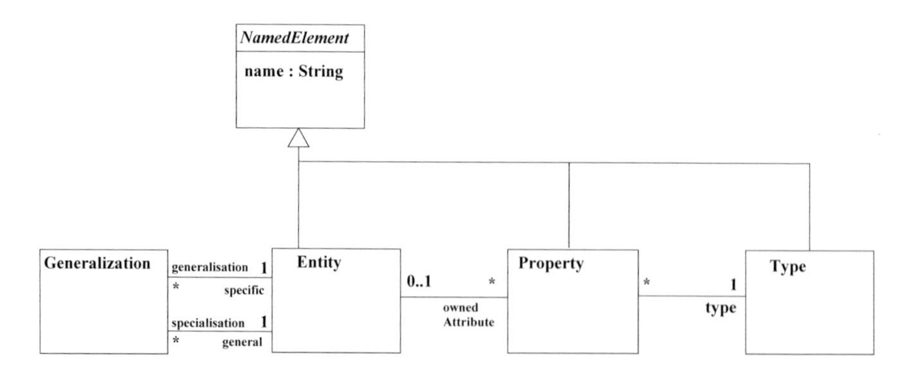

Figure 17.1: Basic class diagram metamodel

17.8.1 Requirements elicitation

The stakeholders of this system are: (i) the company using class diagrams in its development approach, which aims to improve the use of

UML in its development process; (ii) developers using the class diagrams for analysis and modelling; (iii) coders using the class diagrams as input for manual coding.

The initial requirements statement was:

> Refactor a UML class diagram to remove all cases of duplicated attribute declarations in sibling classes (classes which have a common parent).

This statement is concerned purely with the functional behaviour of the transformation. By means of structured interviews with the customer (and with the end users of the refactored diagrams, the development team) we can uncover further, non-functional, requirements:

- *Scope*: all valid input class diagrams with single inheritance and no concrete superclasses should be processed.
- *Efficiency*: the refactoring should be able to process diagrams with 1000 classes and 10,000 attributes in a practical time (less than 5 minutes).
- *Correctness*: the start and end models should have equivalent semantics (model-level semantic preservation/equivalence).
- *Minimality of the target model*: minimise the number of new classes introduced, to avoid introducing superfluous classes into the model.
- *Confluence* would be desirable, but is not mandatory.

The functional requirements can also be clarified and more precisely scoped by the interview process:

- A global functional requirement is the invariance of the class diagram language constraints: that there is no multiple inheritance, and no concrete class with a subclass (syntactic correctness/conformance).
- It is not proposed to refactor associations because of the additional complications this would cause to the developers. Only attributes are to be considered.

By scenario analysis using concrete grammar sketches (in class diagram notation) the main functional requirement is decomposed into three cases: (i) where all (2 or more) direct subclasses of one class have identical attribute declarations (Fig. 17.2); (ii) where 2 or more direct subclasses have identical attribute declarations (Fig. 17.3); (iii) where 2 or more root classes have identical attribute declarations (Fig. 17.4).

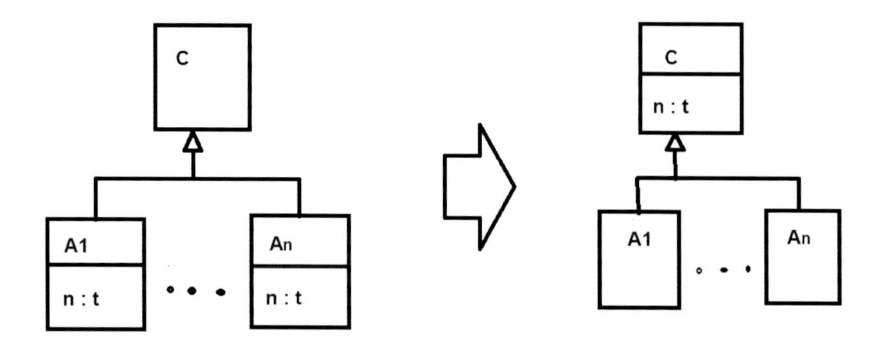

Figure 17.2: Scenario 1

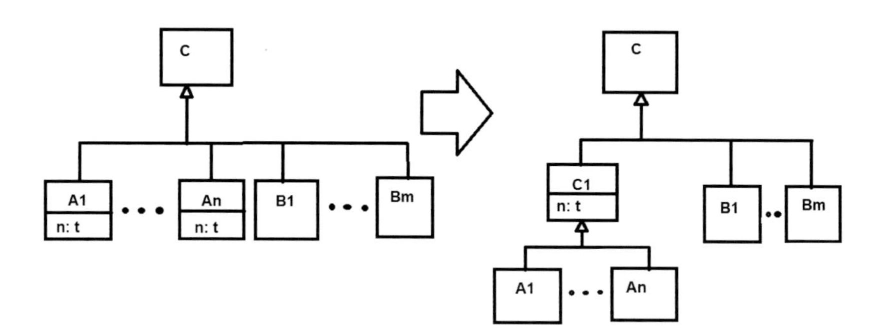

Figure 17.3: Scenario 2

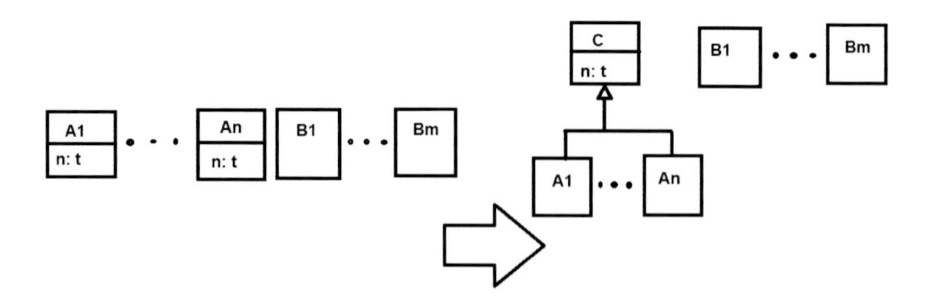

Figure 17.4: Scenario 3

17.8.2 Evaluation and negotiation

At this point we should ask if these scenarios are complete: if they cover all intended cases of the required refactorings. By analysis of the possible structures of class diagrams, and taking into account the invariant of single inheritance, it can be deduced that they are complete. The scenarios are consistent since there is no overlap between their assumptions, however two or more cases may be applicable to the same class diagram, so that a priority order needs to be assigned to them. By exploratory prototyping and testing using particular examples of class diagrams, such as Fig. 2.9, we can identify that the requirement for minimality means that rule 1 "Pull up attributes" should be prioritised over rules 2 "Create subclass" or 3 "Create root class". In addition, the largest sets of duplicated attributes in sibling classes should be removed before smaller sets, for rules 2 and 3.

17.8.3 Requirements formalisation

To formalise the functional requirements, we express the three scenarios as OCL constraints using the abstract grammar of the language of Fig. 17.1.

Rule 1: If the set $g = c.specialisation.specific$ of all direct subclasses of a class c has two or more elements, and all classes in g have an owned attribute with the same name n and type t, add an attribute of this name and type to c, and remove the copies from each element of g (Fig. 17.2).

Rule 2: If a class c has two or more direct subclasses $g = c.specialisation.specific$, and there is a subset $g1$ of g, of size at least 2, all the elements of $g1$ have an owned attribute with the same name n and type t, but there are elements of $g - g1$ without such an attribute, introduce a new class $c1$ as a subclass of c. $c1$ should also be set as a direct superclass of all those classes in g which own a copy of the cloned attribute. Add an attribute of name n and type t to $c1$ and remove the copies from each of its direct subclasses (Fig. 17.3).

Rule 3: If there are two or more root classes all of which have an owned attribute with the same name n and type t, create a new root class c. Make c the direct superclass of all root classes with such an attribute, and add an attribute of name n and type t to c and remove the copies from each of the direct subclasses (Fig. 17.4).

These can then be encoded as three rules in the UML-RSDS constraint language: Each of these operates on instances of *Entity*:

$$
\begin{array}{l}
(C1):\\
a : specialisation.specific.ownedAttribute\ \&\\
specialisation.size > 1\ \&\\
specialisation.specific{\rightarrow}forAll(\\
\quad ownedAttribute{\rightarrow}exists(b \mid b.name = a.name\ \&\ b.type = a.type)) \ \Rightarrow\\
\qquad a : ownedAttribute\ \&\\
\qquad specialisation.specific.ownedAttribute{\rightarrow}select(\\
\qquad\qquad name = a.name){\rightarrow}isDeleted()
\end{array}
$$

$$
\begin{array}{l}
(C2):\\
a : specialisation.specific.ownedAttribute\ \&\\
v = specialisation{\rightarrow}select(\\
\quad specific.ownedAttribute{\rightarrow}exists(b \mid b.name = a.name\ \&\ b.type = a.type))\\
\qquad \&\ v.size > 1 \ \Rightarrow\\
\quad Entity{\rightarrow}exists(e \mid e.name = name + \text{"\_2\_"} + a.name\ \&\\
\qquad a : e.ownedAttribute\ \&\\
\qquad e.specialisation = v\ \&\\
\qquad Generalization{\rightarrow}exists(g \mid g : specialisation\ \&\ g.specific = e))\ \&\\
\qquad v.specific.ownedAttribute{\rightarrow}select(name = a.name){\rightarrow}isDeleted()
\end{array}
$$

$$
\begin{array}{l}
(C3):\\
a : ownedAttribute\ \&\\
generalisation.size = 0\ \&\\
v = Entity{\rightarrow}select(generalisation.size = 0\ \&\\
\quad ownedAttribute{\rightarrow}exists(b \mid b.name = a.name\ \&\ b.type = a.type))\ \&\\
v.size > 1 \ \Rightarrow\\
\quad Entity{\rightarrow}exists(e \mid e.name = name + \text{"\_3\_"} + a.name\ \&\\
\qquad a : e.ownedAttribute\ \&\\
\qquad v.ownedAttribute{\rightarrow}select(name = a.name){\rightarrow}isDeleted()\ \&\\
\qquad v{\rightarrow}forAll(c \mid Generalization{\rightarrow}exists(g \mid\\
\qquad\qquad g : e.specialisation\ \&\ g.specific = c)))
\end{array}
$$

The assumptions/invariants can likewise be formalised as OCL constraints in UML-RSDS:

$$
\begin{array}{l}
Entity{\rightarrow}isUnique(name)\\
Type{\rightarrow}isUnique(name)\\
Entity{\rightarrow}forAll(e \mid e.allAttributes{\rightarrow}isUnique(name))
\end{array}
$$

where *allAttributes* is the collection of declared and inherited attributes of an entity. The condition of single inheritance is expressed by the constraint:

$$generalisation.size \leq 1$$

on *Entity*.

A suitable variant expression is *Property.size*, this is decreased by each rule termination. It is related to the quality measure, which is the number of duplicate attributes in the diagram.

Model-level semantic equivalence can be defined in terms of the semantics *Sem*(*d*) which for a class diagram *d* returns the collection of leaf entities with their complete (owned and inherited) set of attribute descriptors (attribute name and type). This collection should be constant throughout the transformation.

17.8.4 Validation and verification

The functional requirements can be checked by executing the prototype transformation on test cases. In addition, informal reasoning can be used to check that each rule application preserves the invariants. For example, no rule introduces new types, or modifies existing types, so the invariant that type names are unique is clearly preserved by rule applications. Likewise, the model-level semantics is also preserved. Formal proof could be carried out using B.

Termination follows by establishing that each rule application decreases the number of attributes in the diagram, i.e., *Property.size*. The efficiency requirements can be verified by executing the prototype transformation on realistic test cases of increasing size. The Java implementation generated by the UML-RSDS tools shows an above-linear growth of execution time when constructing large sets of elements. Table 17.2 shows the results of the tests on multiple copies of a basic test case with four classes and two attributes with two duplicates for each attribute.

Table 17.2: Execution test results for UML-RSDS

Test case	*Number of classes*	*Number of attributes*	*Execution time*
2*100	400	400	90ms
2*200	800	800	330ms
2*500	2000	2000	2363ms
2*1000	4000	4000	13s
2*5000	20000	20000	156s
2*10000	40000	40000	1137s

It can be observed that confluence fails, by executing the transformation on semantically-equivalent but distinct input models and identifying that the results may not be equivalent. The constraints C2 and C3 do not necessarily select the largest collections of attributes to rationalise at each step, and hence may not satisfy the minimality requirement. This could be corrected by more complex constraint conditions, or by using an Auxiliary Metamodel strategy to explicitly represent alternative families of attributes which are candidates for merging.

Summary

In this chapter we have identified requirements engineering techniques for model transformations. We have also described a requirements engineering process for UML-RSDS, and requirements engineering techniques that can be used in this process.

References

[1] S. Friedenthal, A. Moore and R. Steiner, *A Practical Guide to SysML: The systems modelling language*, Morgan Kaufmann, 2009.

[2] E. Guerra, J. de Lara, D. Kolovos, R. Paige and O. Marchi dos Santos, *Engineering Model Transformations with transML*, SoSyM, Vol. 13, No. 3, July 2013, pp. 555–577.

[3] S. Kolahdouz-Rahimi, K. Lano, S. Pillay, J. Troya and P. Van Gorp, *Evaluation of model transformation approaches for model refactoring*, Science of Computer Programming, 2013, http://dx.doi.org/10.1016/j.scico.2013.07.013.

[4] K. Lano, *The UML-RSDS Manual*, http://www.dcs.kcl.ac.uk/staff/kcl/uml2web, 2015.

[5] N. Maiden and G. Rugg, *ACRE: selecting methods for requirements acquisition*, Software Engineering Journal, May 1996, pp. 183–192.

[6] NuSMV, http://nusmv.fbk.eu, 2015.

[7] I. Sommerville and G. Kotonya, *Requirements Engineering: Processes and Techniques*, J. Wiley, 1998.

Chapter 18

System Verification

Validation and verification of a system are essential activities throughout the development process: validation checks that we are "building the right system": that requirements are correctly expressed in a specification, whilst verification checks that we are "building the system right", that a specification is consistent and internally correct, and that the design meets necessary conditions for termination, confluence and correctness with respect to the specification. In this chapter we describe the validation and verification techniques which are supported by UML-RSDS.

The main emphasis of UML-RSDS is to ensure that code synthesised from models is *correct by construction*: that provided the models satisfy some (relatively easy to check) conditions, the generated implementation will be correct with respect to these models, will terminate, be confluent, etc. This reduces the verification effort compared to post-hoc testing and proof of the implementation. This approach places restrictions on the form of specifications, however (that use cases should only have type 0 or type 1 postconditions, and should satisfy syntactic non-interference, etc.), so verification techniques are still needed for situations (such as refactoring transformations) which fail to meet these restrictions.

18.1 Class diagram correctness

Several checks are made during the construction of a class diagram, to avoid the creation of invalid diagrams:

- Creation of a class with the same name as an existing class: not permitted.
- Creation of an attribute or rolename for a class which already has (or inherits) a data feature of this name: not permitted.
- Creation of cycles of inheritance: not permitted.
- Definition of multiple inheritance: permitted with a warning that the design would be invalid if used for a language other than C++.
- Creation of an inheritance with a concrete class as the superclass: not permitted. Such diagrams should be refactored to use an abstract superclass instead.

18.2 State machine correctness

The following checks are made during the construction of a state machine diagram:

- Creation of a state with the same name as an existing state: not permitted.
- Creation of two transitions triggered by the same event from the same state: permitted with a warning that the developer must ensure the transitions have disjoint (logically conflicting) guards.

The option *Check Structure* can be used to identify potential problems with operation behaviour defined by state machines: cases of unstructured code are identified. States coloured red on the diagram are those states involved in a loop.

18.3 Use case correctness

The correctness of use cases is checked by analysing the definedness, determinacy and data dependencies of the use case postcondition constraints, and by only permitting include/extend relationships which satisfy normal structural restrictions, as described in Chapter 5. Use cases should have distinct names. Their constraints should only refer to classes and features within the same system (or external classes represented via placeholder/proxy classes in the system class diagram). The postconditions should satisfy the data-dependency conditions of syntactic non-interference (Chapter 5), and warnings are issued if these do not hold. Confluence checks are also carried out on type 0 and type 1 constraints, and warnings are given if these fail to be confluent.

Specifiers should also ensure that calls of update operations do not occur in use case invariants, assumptions, postcondition constraint conditions or in other contexts where pure values are expected. It is bad practice to use update operations in postcondition succedents, but this may be necessary in some cases.

18.3.1 Definedness and determinacy of expressions

Usually, constraints in operation or use case postconditions should have well-defined and determinate values. This means that errors such as possible division by zero, attempts to access sequence, set or qualified role values outside the range of their index domain, operation call or feature application to possibly undefined objects, and other internal semantic errors of a specification must be detected. Such errors may also indicate incorrect or incomplete formalisation of requirements, ie., validation errors.

Errors of definedness include:

- Possible division by zero: an expression of the form $e1/e2$ or $e1\ mod\ e2$ where $e2$ may be 0.
- Reference to an out-of-range index:

  $$sq[i]$$

  for a sequence or string sq where $i \leq 0$ or $sq.size < i$ is possible.
- Application of an operator that requires a non-empty collection, to a possibly empty collection:

  $$col{\rightarrow}max()$$

  and likewise for *min*, *any*, *last*, *first*, *front*, *tail*.

Definedness is analysed in UML-RSDS by calculating a *definedness condition* $def(E)$ for each expression E, which gives the conditions under which E is well-defined.

For each postcondition, precondition and invariant constraint Cn of a use case, the definedness condition $def(Cn)$ is a necessary assumption which should hold before the constraint is applied or evaluated, in order that its evaluation is well-defined. For example, if an expression $e{\rightarrow}any()$ occurs in a constraint succedent, there should be an antecedent condition such as $e.size > 0$ to enforce that the collection e is non-empty. Likewise for $e{\rightarrow}min()$, $e{\rightarrow}max()$, etc. Postcondition constraints should normally also satisfy the condition of determinacy. Examples of the clauses for the definedness function $def : Exp(L) \rightarrow Exp(L)$ are given in Table 18.1.

Table 18.1: Definedness conditions for expressions

Constraint expression e	*Definedness condition def(e)*
$e.f$ Data feature application	$def(e)$ & $E.allInstances() \rightarrow includes(e)$ where E is the declared classifier of e
Operation call $e.op(p)$	$def(e)$ & $E.allInstances() \rightarrow includes(e)$ & $def(p)$ & $def(e.Post_{op}(p))$ where E is the declared classifier of e, $Post_{op}$ the postcondition of op
a/b a mod b	$b \neq 0$ & $def(a)$ & $def(b)$
$s[ind]$ sequence, string s	$ind > 0$ & $ind \leq s \rightarrow size()$ & $def(s)$ & $def(ind)$
$E[v]$ entity type E with identity attribute id, v single-valued	$E.id \rightarrow includes(v)$ & $def(v)$
$s \rightarrow last()$ $s \rightarrow first()$ $s \rightarrow max()$ $s \rightarrow min()$	$s \rightarrow size() > 0$ & $def(s)$
$s \rightarrow any(P)$	$s \rightarrow exists(P)$ & $def(s)$ & $def(P)$
$v.sqrt$	$v \geq 0$ & $def(v)$
$v.log$, $v.log10$	$v > 0$ & $def(v)$
$v.asin$, $v.acos$	$(v.abs \leq 1)$ & $def(v)$
$v.pow(x)$	$(v < 0 \Rightarrow x : int)$ & $(v = 0 \Rightarrow x > 0)$ & $def(v)$ & $def(x)$
A & B A *or* B $A \Rightarrow B$	$def(A)$ & $def(B)$ $def(A)$ & $def(B)$ $def(A)$ & $(A \Rightarrow def(B))$
$E \rightarrow exists(x \mid A)$ $E \rightarrow forAll(x \mid A)$	$def(E)$ & $E \rightarrow forAll(x \mid def(A))$

Definedness of an operation call requires proof of termination of the call. Definedness of *or*, & requires definedness of both arguments because different implementation/evaluation strategies could be used in different formalisms or programming languages: it cannot be assumed that short-cut evaluation will be used (in ATL, for example, strict evaluation of logical operators is used [1]). Only in the case of implication is the left hand side used as a 'guard' to ensure the definedness of the succedent. We treat $A \Rightarrow B$ equivalently to OCL *if A then B else true*. Failure of determinacy, where an expression might evaluate to different values, is also usually a sign of an internal error in a specification. As for definedness, an expression $det(E)$ gives conditions under which E is ensured to be determinate. Examples of the clauses for the determinacy condition $det : Exp(L) \rightarrow Exp(L)$ are given in Table 18.2.

Table 18.2: Determinacy conditions for expressions

Constraint expression e	*Determinacy condition det(e)*
$s \rightarrow any(P)$	$s \rightarrow select(P) \rightarrow size() = 1$ & $det(s)$ & $det(P)$
$s \rightarrow asSequence()$	$s \rightarrow size() \leq 1$ & $det(s)$ for set-valued s
Case-conjunction $(E1 \Rightarrow P1)$ & ... & $(En \Rightarrow Pn)$	Conjunction of $not(Ei$ & $Ej)$ for $i \neq j$, and each $(det(Ei)$ & $(Ei \Rightarrow det(Pi)))$
A & B $A \Rightarrow B$	$det(A)$ & $det(B)$ $det(A)$ & $(A \Rightarrow det(B))$
$E \rightarrow exists(x \mid A)$ $E \rightarrow exists1(x \mid A)$	$det(E)$ & $E \rightarrow forAll(x \mid det(A))$
$E \rightarrow forAll(x \mid A)$	$det(E)$ & $E \rightarrow forAll(x \mid det(A))$ Additionally, order-independence of A for $x : E$.

The determinacy and definedness conditions for operation and use case postconditions are displayed in the console window when the *Generate Design* option is selected for a UML-RSDS specification. As discussed above, the antecedent of a constraint should imply the definedness of the succedent:

$$Ante \Rightarrow def(Succ)$$

The overall definedness condition $def(Cn)$ of a postcondition constraint should be incorporated into the preconditions (assumptions) of the use case: if the postconditions are $C_1, ..., C_n$, then the definedness of C_j for $1 \leq j \leq n$ is ensured by the assumption

$$C_1 \text{ \& ... \& } C_{j-1} \Rightarrow def(C_j)$$

because the implementation $stat(C_1, ..., C_{j-1})$ of the preceding constraints establishes their conjunction, and hence, under the above assumption, also the definedness of C_j. An example of this approach to ensure definedness is the correlation calculator problem of Chapter 2.

18.3.2 Confluence checks

The UML-RSDS tools use syntactic checking to check the confluence of type 0 or type 1 constraints (other forms of constraint need confluence proof using variants, as described in Section 18.5). A constraint Cn on class E, of the form

$$Ante \Rightarrow Succ$$

is checked as follows:

1. *self* of type E is added to a set *iterated*, as is each implicitly universally quantified variable $c : C$ of class type occurring in *Ante*. Each distinct combination of bindings to these variables is iterated over exactly once by a bounded-loop implementation of the constraint. (Type 2 constraints may reapply the constraint to particular combinations more than once, and type 3 constraints may also introduce new combinations to be iterated over). Type 0 constraints do not have a context class or implicit *self* variable.

2. A list *created* of the existentially-quantified variables t in $T \rightarrow exists(t \mid pred)$ formulae in *Succ* is maintained. These are the objects which are newly created in each application of the constraint (not looked-up and modified). t is only added to *created* if:

   (a) T is a concrete entity type and not of pre-form,

   (b) either T has no unique/primary key, or its key is assigned by a top-level equation $t.key = c.ckey$ to the primary key of the only element c of *iterated*.

   These conditions ensure that t is genuinely new.

3. Assignments $t.f = e$ for a direct feature f of $t \in created$ are confluent, if $rd(e)$ is disjoint from $wr(Cn)$, as are assignments $self.f = val$ or $c.f = val$ for a value val (with no variables, including *self*) and $c \in iterated$. Assignments $c.f = c.g$ for direct features of $c \in iterated$ are also confluent.

4. Assignments $T_j[sId].g = e$ are confluent if sId is the primary key of the only *iterated* entity type S_j, and if there is a preceding confluent constraint which maps S_j to T_j (so that there is a 1-to-0..1 relation between these entity types based on the primary key values).

5. A formula $t : e.r$ is confluent if $t \in created$, and r is an unordered role. $e : t.r$ and $e <: t.r$ are confluent if $t \in created$ and r is unordered, or if e is an *iterated* variable which has an ordered iteration range (not an entity type). $rd(e)$ must also be disjoint from $wr(Cn)$ in both cases.

6. $e1 \rightarrow includes(e2)$ is treated as for $e2 : e1$, and $e1 \rightarrow includesAll(e2)$ is treated as for $e2 <: e1$.

7. In the case of bidirectional associations, the explicit update to one end, and the corresponding implicit update to the other end must both be confluent.

8. A conjunction $e1$ & $e2$ is confluent if $e1$ and $e2$ are.

9. $e1 \Rightarrow e2$ is confluent if $e2$ is.

10. $T \rightarrow exists(t \mid pred)$ is confluent if *pred* is confluent under the addition of t to *created*, if it is eligible to be added.

11. $T \rightarrow forAll(t \mid pred)$ is confluent if *pred* is confluent under the addition of t to *iterated*.

12. $e \rightarrow display()$ is confluent if there is at most one variable in e, from *iterated*, and with an ordered iteration range.

These rules ensure that data written on previous iterations of Cn cannot affect the execution of the current application of Cn. Set-valued collections, such as the set $E.allInstances$ of objects of a class E, should not be used directly to produce a sequence-valued collection, because the order of the latter will be arbitrary. The sorting operators can instead be used to make the order of the source domain expression determinate.

As an example, the classic UML to RDB transformation is confluent if written as:

```
UMLClass::
  Table->exists( t | t.name = name &
     attributes->forAll( a |
          Column->exists( c | c.name = a.name & c : t.columns &
           c.typename = a.type.name ) ) )
```

If *name* is a primary key for *UMLClass* and *Table*, but not for *Column*, and *attributes* and *columns* are unordered, then t and c are added to *created*, and all updates are confluent.

18.4 Correctness by construction

It is a key principle of UML-RSDS that logical constraints have a dual interpretation both as specifications of required behaviour, and as inputs for code generators which produce programs that are guaranteed to correctly implement this behaviour.

For use cases, this principle means that the use case postconditions both describe the logical contract of the use case: the postconditions can be assumed to be true at termination of the use case, and they also define the code implementation which ensures that this contract is carried out:

If a use case uc has postconditions C_1 to C_n which are all of type 0 or type 1, satisfy syntactic non-interference, confluence and definedness and determinacy conditions, then the implementation code stat(C_1); ...; stat(C_n) of uc is terminating, confluent and correct with respect to its specification [4]. That is, this code establishes the conjunction C_1 & ... & C_n.

The restrictions on the C_i can be relaxed slightly by using semantic non-interference instead of syntactic non-interference: there may be cases where two constraints C_i and C_j both write to the same data, but do so in ways that do not invalidate the other constraint. For example, if both add objects to the same unordered collection.

18.5 Synthesis of B AMN

B is a formal language with powerful analysis capabilities for checking the correctness of specifications [2]. Several commercial or free tools exist for B:

- BToolkit
- Atelier B
- B4Free

The following restrictions on UML-RSDS specifications are necessary before they can be translated to B:

- Attributes of different classes must have different names.
- Update operations cannot have return values or be recursively defined.
- Feature, variable and class names should have more than one letter. Underscore cannot be used in class, variable or feature names.
- Real values cannot be used, only integers, strings and booleans, and enumerated types. Some B implementations only support natural numbers: non-negative integers.

The UML-RSDS type `int` corresponds to `INT` in B, i.e., 32-bit signed integers.

Two options for generating B are provided: (i) separate B machines for each root entity of the class diagram; (ii) one B machine for the entire system. The first can be used if there are not invocations in both directions between two entities (in different inheritance hierarchies), and no bidirectional associations between such entities. In addition, if no entity in one hierarchy contains code creating entity instances in another hierarchy. Otherwise, in the case of such mutual dependencies, option (ii) must be used.

The following verification properties of a UML-RSDS transformation specification τ from source language S to target language T can be checked using B:

1. Syntactic correctness: if a source model m of S satisfies all the assumptions Asm of τ, then the transformation will produce valid target models n of T from m which satisfy the language constraints Γ_T of T.

2. Model-level semantic preservation: if τ maps a source model m to target model n, then these have equivalent semantics under semantics-assigning maps Sem_S and Sem_T:

$$Sem_S(m) \equiv Sem_T(n)$$

3. Semantic preservation: if a predicate φ holds in m, then any target model n produced from m satisfies an interpretation $\chi(\varphi)$ of the formula. χ depends upon τ.

4. Semantic correctness: that a given implementation for τ satisfies its specification.

5. Confluence: that all result models n produced from a given source model m must be isomorphic.

6. Termination: that τ is guaranteed to terminate if applied to valid source models which satisfy Asm.

Proof-based techniques for verifying transformation correctness properties have two main advantages: (i) they can prove the properties for all cases of a transformation, that is, for arbitrary input models and for a range of different implementations; (ii) a record of the proof can be produced, and subjected to further checking, if certification is required. However, proof techniques invariably involve substantial human expertise and resources, due to the interactive nature of the most general forms of proof techniques, and the necessity to work both in the notation of the proof tool and in the transformation notation.

We have selected B AMN as a suitable formalism for proof-based verification, B is a mature formalism, with good tool support, which automates the majority of simple proof obligations. We provide an automated mapping from transformation specifications into B, and this mapping is designed to facilitate the comprehension of the B AMN proof obligations in terms of the transformation being verified. The entity types and features of the languages involved in a transformation τ are mapped into mathematical sets and functions in B. OCL expressions are systematically mapped into set-theory expressions.

A B AMN specification consists of a linked collection of modules, termed *machines*. Each machine encapsulates data and operations on that data. Each transformation τ is represented in a single main B machine M_τ, together with an auxiliary *SystemTypes* machine containing type definitions.

The mapping from UML-RSDS to B performs the following translations:

- Each source and target language (class diagram) L is represented by sets es for each entity type E of the language, with $es \subseteq objects$, and maps $f : es \rightarrow Typ$ for each feature f of E, together with B encodings of the constraints Γ_L of L for unmodified L. In cases where a language entity type or feature g is both read and written by the transformation, a syntactically distinct copy g_pre is used to represent the initial value of g at the start of the transformation. A supertype F of entity type E has B invariant $es \subseteq fs$. Abstract entity types E have a B invariant $es = f1s \cup ... \cup fls$ where the Fi are the direct subtypes of E.

  For each concrete entity type E of a source language, there is an operation $create\_E$ which creates a new instance of E and adds this to es. For each data feature f of an entity type E there is an operation $setf(ex, fx)$ which sets $f(ex)$ to fx.

- The assumptions Asm of the transformation can be included in the machine invariant (using g_pre in place of g for data which is written by the transformation). Asm is also included in the preconditions of the source language operations $create\_E$ and $setf$.

- Each use case postcondition *rule* is encoded as an operation with input parameters the objects which are read by *rule* (including additional quantified variables), and with its effect derived from the rule succedent or from the *behavior* of a design of *rule*. The operation represents transformation computation steps δ_i of the *rule* design.

- Orderings of the steps for particular use case designs can be encoded by preconditions of the operations, expressing that the effect of one or more other *rule'* has been already established for all applicable elements.

- Invariant predicates *Inv* are added as B invariants, using *g_pre* to express pre-state values *g*@pre.

The mapping to B is suitable to support the proof of invariance properties, syntactic correctness, model-level semantic preservation, semantic preservation and semantic correctness by using internal consistency proof in B. A more complex mapping is necessary for the proof of confluence and termination, using refinement proof [3].

The general form of a B machine M_τ representing a separate-models transformation τ with source language S and target language T is:

```
MACHINE Mt SEES SystemTypes
VARIABLES
  /* variables for each entity type and feature of S */
  /* variables for each entity type and feature of T */
INVARIANT
  /* typing definitions for each entity type and feature of S and T */
  GammaS &
  Asm0 & Inv
INITIALISATION
  /* var := {} for each variable */
OPERATIONS
  /* creation operations for entity types of S, restricted by Asm */
  /* update operations for features of S, restricted by Asm */
  /* operations representing transformation steps */
END
```

The machine represents the transformation at any point in its execution. *Asm*0 is the part of *Asm* which refers only to source model data. *SystemTypes* defines the type *Object_OBJ* of all objects, and any other type definitions required, e.g., of enumerated types.

The operations to create and update S elements are used to set up the source model data of the transformation. Subsequently, the operations representing transformation steps are performed. If *Asm*0 consists of universally quantified formulae $\forall s : S_i \cdot \psi$, then the instantiated formulae $\psi[sx/s]$ are used as restrictions on operations creating $sx : S_i$ (or subclasses of S_i). Likewise, operation $setf(sx, fx)$ modifying feature f of S_i has a precondition $\psi[sx/s, fx/s.f]$. All these operations will include the preconditions *Asm*1 from *Asm* which concern only the target model.

As an example, the transformation of Fig. 7.2 can be defined by the following partial machine:

```
MACHINE Mt SEES SystemTypes
VARIABLES objects, as, xx, bs, yy
INVARIANT
  objects <: Object_OBJ &
  as <: objects & bs <: objects &
```

```
  xx : as --> INT & yy : bs --> INT &
  !bb.(bb : bs => #aa.(aa : as & yy(bb) = xx(aa)*xx(aa)))
INITIALISATION
  objects, as, xx, bs, yy := {}, {}, {}, {}, {}
```

The invariant expresses Γ_S and the *Inv* property

$$B{\rightarrow}forAll(b \mid A{\rightarrow}exists(a \mid b.y = a.x * a.x))$$

of the transformation. $\#b.P$ is B syntax for $\exists\, b \cdot P$, $!a.P$ is B syntax for $\forall\, a \cdot P$. & denotes conjunction, <: denotes $\subseteq$ and $--\ >$ is $\rightarrow$ (the total function type constructor). A universal set *objects* of existing objects is maintained, this is a subset of the static type *Object_OBJ* declared in *SystemTypes*. The operations representing computation steps are derived from the rule designs $stat(Cn)$ of the constraints Cn. This modelling approach facilitates verification using weakest precondition calculation, compared to more abstract encodings. If Cn has the form

$$SCond \;\Rightarrow\; Succ$$

on context entity S_i then the operation representing a computation step δ_i of Ci is:

```
delta_i(si) =
  PRE si : sis & SCond & not(si.Succ) &
    C1 & ... & Ci-1 & def(si.Succ)
  THEN
    stat'(si.Succ)
  END
```

where $stat'(P)$ encodes the procedural interpretation $stat(P)$ of P in B program-like statements, AMN generalised substitutions. These have a similar syntax to the programming language described in Appendix A.2, and use the same weakest-precondition semantics. B has an additional statement form $v := e1 \| w := e2$ of *parallel assignment*: the assignments are performed order-independently, with the values of $e1$, $e2$ being simultaneously assigned to v, w. The ANY WHERE THEN statement of B corresponds to a UML-RSDS creation statement, and to let definitions.

If the design of τ defines a non-standard rule design of Ci, this design could be encoded in B in place of the above definition of *delta_i*. If τ's design requires that all constraints $C1$, ..., $Ci - 1$ are established before Ci, this ordering can be encoded by including $C1$, ..., $Ci - 1$ in the preconditions of *delta_i*. $not(Succ)$ can be omitted if negative application conditions are not checked by the design of Ci, as is the case for type 1 constraints. For the mapping to B, $def(Succ)$ includes checks that numeric expressions in *Succ* are within the size bounds of the finite numeric types *NAT* and *INT* of B, and that $objects \neq Object\_OBJ$ prior to any creation of a new object.

The computational model of a transformation τ expressed in M_τ therefore coincides with the definition of transformation computation described in [4]: a computation of τ is a sequence of transformation steps executed in an indeterminate order, constrained only by the need to maintain *Inv*, and, if a specific design *I* is defined, to satisfy the ordering restrictions of *I*'s behaviour.

For the example of Fig. 7.2 the resulting completed B machine M_τ has:

```
OPERATIONS
  create_A(xxxx) =
    PRE xxxx : INT & objects /= Object_OBJ & bs = {}
    THEN
        ANY ax WHERE ax : Object_OBJ - objects
        THEN
            as := as \/ { ax } || objects := objects \/ { ax } ||
            xx(ax) := xxxx
        END
    END;

  setx(ax,xxxx) =
    PRE ax : as & xxxx : INT & bs = {}
    THEN
      xx(ax) := xxxx
    END;

  r1(ax) =
    PRE ax : as & not( #bb.(bb : bs & yy(bb) = xx(ax)*xx(ax)) ) &
      objects /= Object_OBJ & xx(ax)*xx(ax) : INT
    THEN
      ANY bb WHERE bb : Object_OBJ - objects
      THEN
         bs := bs \/ { bb } || objects := objects \/ { bb } ||
         yy(bb) := xx(ax)*xx(ax)
       END
    END
END
```

The assumption *Asm* is $B = Set\{\}$, which is expressed as $bs = \{\}$. $r1$ defines the transformation step of the postcondition constraint. The machine is generated automatically by the UML-RSDS tools from the UML specification of the transformation.[1] UML-RSDS encodes into B

[1] In practice, single-letter feature, variable and entity type names should be avoided in the specification, since these have a special meaning in B AMN.

the semantics of all cases of updates to associations, including situations with mutually inverse association ends.

Using these machines we can verify syntactic correctness and semantic preservation properties of a model transformation, by means of *internal consistency* proof of the B machine representing the transformation and its metamodels. Internal consistency of a B machine consists of the following logical conditions:

- That the state space of the machine is non-empty: $\exists\, v.I$ where v is the tuple of variables of the machine, and I its invariant.
- That the initialisation establishes the invariant: $[Init]I$
- That each operation maintains the invariant:

  $$Pre \wedge I \;\Rightarrow\; [Code]I$$

  where *Pre* is the precondition of the operation, and *Code* its effect.

B machines implicitly satisfy the *frame axiom* for state changes: variables v which are not explicitly updated by an operation are assumed not to be modified by the operation. This corresponds to the assumption made in UML-RSDS that v is unmodified by activity *act* if $v \notin wr(act)$.

Proof of verification properties can be carried out using B, as follows (for separate models transformations):

1. Internal consistency proof of M_τ establishes that *Inv* is an invariant of the transformation.
2. By adding the axioms of Γ_T to the INVARIANT clause, the validity of these during the transformation and in the final state of the transformation can be proved by internal consistency proof, establishing syntactic correctness.
3. Model-level semantic preservation can be verified by encoding the model semantics $Sem(m)$ in M_τ, and proving that this is invariant over transformation steps (for refactoring transformations).
4. By adding φ and $\chi(\varphi)$ to the INVARIANT of M_τ, semantic preservation of φ can be proved by internal consistency proof. Creation and update operations to set up the source model must be suitably restricted by φ.

Termination, confluence and semantic correctness proof needs to use suitable Q variants for each constraint.

Using Atelier B version 4.0, 13 proof obligations for internal consistency of the above machine M_T are generated, of which 10 are automatically proved, and the remainder can be interactively proved using the provided proof assistant tool.

The three unproved obligations are:

```
    "'Local hypotheses'" &
    ax: A_OBJ &
    not(ax: as) &
    bb: bs &
    "'Check that the invariant (!bb.(bb: bs => #aa.(aa: as &
     yy(bb) = xx(aa)*xx(aa)))
     is preserved by the operation - ref 3.4'"
=>
    #aa.(aa: as\/{ax} & yy(bb) = (xx <+ { ax |-> xxx })(aa)*(xx <+
     { ax |-> xxx })(aa))


    "'Local hypotheses'" &
    bb: bs &
    "'Check that the invariant (!bb.(bb: bs => #aa.(aa: as &
     yy(bb) = xx(aa)*xx(aa)))
    is preserved by the operation - ref 3.4'"
=>
    #aa.(aa: as & yy(bb) = (xx <+ { ax |-> xxxx })(aa)*(xx <+
     { ax |-> xxxx })(aa))


    "'Local hypotheses'" &
    bb: B_OBJ &
    not(bb: bs) &
    bb$0: bs\/{bb} &
    "'Check that the invariant (!bb.(bb: bs => #aa.(aa: as &
     yy(bb) = xx(aa)*xx(aa)))
    is preserved by the operation - ref 3.4'"
=>
    #aa.(aa: as & (yy <+ { bb |-> xx(ax)*xx(ax) })(bb$0) =
     xx(aa)*xx(aa))
```

Each of these can be proved by adding additional logical inferences to express why they hold. In the first case, the assumption $bb : bs$ means that

```
#aa.(aa: as & yy(bb) = xx(aa)*xx(aa))
```

holds, from the invariant. From $aa : as$ we can deduce $aa : aa \cup \{ax\}$, and from the assumption $not(ax : as)$ we can deduce $ax \neq aa$ and therefore that

```
(xx <+ { ax |-> xxx })(aa) = xx(aa)
```

and therefore that the required conclusion of the first proof obligation holds.

18.6 Synthesis of SMV

We utilise the SMV/nuSMV language and model checker [5] to analyse temporal properties of systems. We select SMV because it is an established industrial-strength tool which supports linear temporal logic (LTL). LTL defines properties over sequences of states using the operators $\bigcirc$ (in the next state), $\diamond$ (in the current or some future state), $\Box$ (in the current and all future states), $\blacklozenge$ (strictly in the past).

SMV models systems in *modules*, which may contain variables ranging over finite domains such as subranges a..b of natural numbers and enumerated sets. The initial values of variables are set by initialisation statements $init(v) \; := \; e$. A $next(v) \; := \; e$ statement identifies how the value of variable v changes in execution steps of the module. In the UML to SMV encoding objects are modelled by positive integer object identities, attributes and associations are represented as variables, and classes are represented by modules parameterised by the object identity values (so that separate copies of the variables exist for each object of the class). A specific numeric upper bound must be given for the number of possible objects of each class (the bound is specified in the class definition dialog, Fig. 3.2).

A UML-RSDS system is modelled in SMV by modelling system execution steps as module execution steps. The structure of an SMV specification of a class diagram is as follows:

```
MODULE main
VAR
  C : Controller;
  MEntity1 : Entity(C,1);
  .... object instances ....

MODULE Controller
VAR
  Entityid : 1..n;
  event : { createEntity, killEntity, event1, ..., eventm, none };

MODULE Entity(C, id)
VAR
  alive : boolean;
  ... attributes ...
DEFINE
  TcreateEntity := C.event = createEntity & C.Entityid = id;
  TkillEntity := C.event = killEntity & C.Entityid = id;
  Tevent1 := C.event = event1 & C.Entityid = id & alive = TRUE;
```

```
  ...
ASSIGN
  init(alive) := FALSE;

  next(alive) :=
    case
      TcreateEntity : TRUE;
      TkillEntity : FALSE;
      TRUE : alive;
    esac;
```

Each class *Entity* is represented in a separate module, and each instance is listed as a module instance in the main module. System events are listed in the Controller, and the effect of these events on specific objects are defined in the module specific to the class of that object. The value of the *Controller* variable *event* in each execution step identifies which event occurs, and the value of the appropriate *Eid* variable indicates which object of class E the event occurs on.

Associations $r : A \to B$ of 1-multiplicity are represented as attributes

```
r : 1..bcard
```

where *bcard* is the maximum permitted number of objects of B. Events to set and unset this role are included. The lower bound 0 is used for 0..1-multiplicity roles to represent an absent B object. For *-multiplicity unordered roles $r : A \to Set(B)$, an array representation is used instead:

```
r : array 1..bcard of 0..1
```

with the presence/absence of a B element with identity value i being indicated by $r[i] = 1$ or $r[i] = 0$. Operations to add and remove elements are provided. The Controller Aid and Bid variables identify which links are being added/removed. The restrictions of SMV/nuSMV imply that only attributes and expressions of the following kinds can be represented in SMV:

- Booleans and boolean operations.
- Enumerated types.
- Integers and operations on integers within a bounded range.
- Strings represented as elements of an enumerated type. String operations cannot be represented.

SMV keywords must be avoided in specifications to be translated to SMV. These include the temporal operators A, F, O, G, H, X, Y, Z, U, S, V, T, EX, AX, EF, AF, EG, AG, etc. [5]. Classical logic is used in SMV, as in B. The semantics of integer division in nuSMV is now (since version 2.4.0) the standard one used in Java, C#, C++, B and OCL. The temporal logic operators are denoted in SMV by **G** (for $\Box$), **F** (for $\diamond$), **X** (for $\bigcirc$) and **O** (for $\blacklozenge$).

For separate-models transformations, the UML to SMV mapping needs to be modified as follows. Source entities are represented by modules with frozen variables and no assignments: the source model data is fixed at the initial time point. There are no *createE* or *killE* events for any entity. If a type 1 transformation rule r iterates over $s : S_i$ and has the form

$$Ante \;\Rightarrow\; T_j{\rightarrow}exists(t \mid TCond \;\&\; P)$$

then the SMV module for T_j has a parameter of module type S_i, and r is an event of the *Controller*, and $S_i id$ is a Controller variable. The T_j module has a transition defined as

```
Tr := C.event = r & C.Siid = id
```

identifying that the event for r takes place on the S_i object with id equal to the current T_j object, and that the S_i object (statically assigned to parameter S) exists. The updates to features f of t defined in $TCond$, P are then specified by $next(f)$ statements, using Tr as a condition, and $next(alive)$ is set to $TRUE$ under condition Tr. An assumption of the main module expresses that the event r can only take place if $Ante$ is true for the $C.Siid = id$ instance. We assume that source objects are mapped to target objects with the same object identifier number.

The $a2b$ transformation can be encoded as follows, in the case of a model with two A objects and two B objects:

```
MODULE main
VAR
  C : Controller;
  MA1 : AA(C,1);
  MA2 : AA(C,2);
  MB1 : B(C,MA1,1);
  MB2 : B(C,MA2,2);

MODULE Controller
VAR
  Aid : 1..2;
  Bid : 1..2;
  event : { a2b1 };
```

```
MODULE AA(C, id)
FROZENVAR
  x : 0..10;

MODULE B(C, AA, id)
VAR
  alive : boolean;
  y : 0..100;
DEFINE
  Ta2b1 := C.event = a2b1 & C.Bid = id;
ASSIGN
  init(alive) := FALSE;

  next(alive) :=
    case
      Ta2b1 : TRUE;
      TRUE : alive;
    esac;

  init(y) := 0;

  next(y) :=
    case
      Ta2b1 : AA.x*AA.x;
      TRUE : y;
    esac;
```

The possible histories of this SMV system are executions of $a2b1$ on the two B module instances, corresponding to the B instances created from the two A instances. The A instances are initialised with arbitrary integer values x in 0..10.

The invariant property of the transformation can be checked as:

```
LTLSPEC
  G(MB1.alive = TRUE ->  MB1.y = MA1.x*MA1.x) &
  G(MB2.alive = TRUE ->  MB2.y = MA2.x*MA2.x)
```

Counter-examples can also be generated, and are expressed as execution histories, for example:

```
NuSMV > check_ltlspec
-- specification  G (MB1.y < 10 & MB2.y < 10)  is false
-- as demonstrated by the following execution sequence
Trace Description: LTL Counterexample
Trace Type: Counterexample
-> State: 1.1 <-
  MA1.x = 7
  MA2.x = 0
  C.Aid = 1
```

```
  C.Bid = 1
  MB1.alive = FALSE
  MB1.y = 0
  MB2.alive = FALSE
  MB2.y = 0
  C.event = a2b1
  MB1.Ta2b1 = TRUE
  MB2.Ta2b1 = FALSE
-- Loop starts here
-> State: 1.2 <-
  MB1.alive = TRUE
  MB1.y = 49
```

Summary

In this chapter we have described the validation and verification techniques which are supported by UML-RSDS, and which help to provide assurance of system correctness. Syntactic checks of diagram correctness, and of expression definedness and determinacy are provided within the tools. Semantic mathematical analysis via external tools (B, SMV, Z3) is facilitated by automated translations from UML-RSDS to the notations of these tools.

References

[1] F. Jouault, F. Allilaire, J. Bézivin and I. Kurtev, *ATL: A model transformation tool*, Sci. Comput. Program. 72(1-2) (2008) 31–39.

[2] K. Lano, *The B Language and Method*, Springer-Verlag, 1996.

[3] K. Lano, S. Kolahdouz-Rahimi and T. Clark, *Comparing verification techniques for model transformations*, Modevva workshop, MODELS 2012.

[4] K. Lano and S. Kolahdouz-Rahimi, *Constraint-based specification of model transformations*, Journal of Systems and Software, February 2013.

[5] NuSMV, http://nusmv.fbk.eu, 2015.

Chapter 19

Reactive System Development with UML-RSDS

Reactive systems are software systems which have the responsibility to control the state of some equipment under control (EUC): some external devices or other elements which can respond to commands from the reactive system, and whose state can be monitored by the reactive system.

UML-RSDS supports the specification of reactive systems via the use of (i) constraint-based specification of reactive system behaviour; (ii) explicit specification of behaviour via state machines; (iii) specification of temporal properties via interactions; (iv) transformations defining reactions as transformation rules.

19.1 Constraint-based specification of reactive systems

Reactive systems can be modelled abstractly by class diagrams, and their state specified by attributes representing the values of sensors and actuators for the EUC: the sensors provide information about the state of the EUC to the control system, whilst the actuators provide a means for the control system to affect the EUC state. Thus the sensor data is an input to the control system, and the actuator settings are outputs.

The most abstract and declarative means of defining the control system functionality is via constraints

$$Condition \;\Rightarrow\; Response$$

which relate the sensor attribute values (usually given on the LHS) to the actuator attribute values (usually given on the RHS). These constraints are *invariants* of the control system: the system is required to maintain them as true. The invariants of a system not only have a logical interpretation as conditions which should be maintained, but also can be interpreted as procedural instructions for how the invariants should be maintained. For example, if a tank in a chemical processing plant (Fig. 19.1) has a high level sensor *highsensor* and an inlet value *invalve*, an invariant could be:

$$highsensor = true \;\Rightarrow\; invalve = false$$

to express that if the fluid level is high, then the inlet value should be closed.

This could be procedurally interpreted as:

$$sethighsensor(true) \;\&\; highsensor = false \;\Rightarrow\; setinvalve(false)$$

That is, when the event of the high level sensor going on occurs (is detected by the reactive system), the inlet valve is switched off (the reactive system should command the valve to close).

Likewise, a constraint

$$lowsensor = false \;\Rightarrow\; invalve = true \;\&\; outvalve = false$$

has a procedural interpretation to open the inlet valve and close the outlet valve if the fluid level is below the low sensor.

Generally, any event which makes the LHS of a constraint Cn true should also trigger the actions $stat(RHS)$ of the procedural interpretation of the RHS.

Figure 19.1 shows the class diagram of this system. The attributes representing sensor states have the annotation ?, whilst those representing actuators have the annotation !. These are the visual indicators of *sensor* and *actuator* stereotypes for attributes (these stereotypes are entered via the attribute dialog, Fig. 3.3). These stereotypes are used in code generation and in formal analysis (e.g., for translation to SMV). Syntactic completeness and consistency analysis can be performed. Completeness checks that all combinations of sensor values are considered in the constraints:

1. Does each sensor attribute appear in at least one constraint assumption?

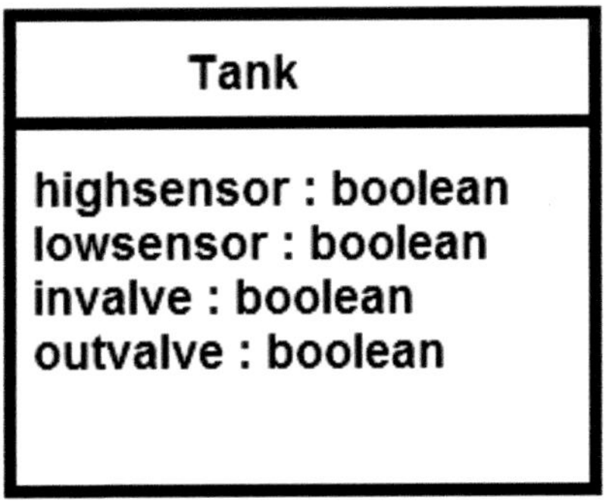

Tank
highsensor : boolean lowsensor : boolean invalve : boolean outvalve : boolean

Figure 19.1: Tank class diagram

2. For each sensor s, and each possible value v of the sensor, does a formula $s = v$ occur in at least one constraint assumption?

For the tank system these checks fail, because no constraint compares either *highsensor* to *false*, or *lowsensor* to *true*. This can be corrected by adding additional constraints.

Consistency checking identifies if two antecedents can both be true at the same time, while both their succedents involve different settings to the same actuator. For the tank there is the inconsistency that if $lowsensor = false$ and $highsensor = true$ then *invalve* has conflicting settings. This can be corrected by making the constraint assumptions more specific, e.g.:

$$highsensor = true \ \& \ lowsensor = true \ \Rightarrow \ invalve = false$$

and

$$lowsensor = false \ \& \ highsensor = false \ \Rightarrow \ invalve = true \ \& \\ outvalve = false$$

Part of the generated Java code for Tank is:

```
class Tank
  implements SystemTypes
{
  private boolean highsensor = false; // sensor
  private boolean lowsensor = false; // sensor
  private boolean invalve = false; // actuator
  private boolean outvalve = false; // actuator

  public void sethighsensor(boolean highsensor_x) { highsensor =
   highsensor_x;
```

```
    if (highsensor == true) { invalve = false; }
  }

  public void setlowsensor(boolean lowsensor_x) { lowsensor =
   lowsensor_x;
    if (lowsensor == false) { invalve = true;
    outvalve = false; }
  }

  ....

}
```

B could be used to prove properties of the specification, as described in Chapter 18. Alternatively, SMV could be used to check if required properties hold. The SMV modules generated from this specification are:

```
MODULE main
VAR
  C : Controller;
  MTank1 : Tank(C,1);
LTLSPEC
  G(((MTank1.highsensor = TRUE -> MTank1.invalve = FALSE) &
     (MTank1.lowsensor = FALSE -> MTank1.invalve = TRUE &
      MTank1.outvalve = FALSE)) ->
      (MTank1.lowsensor = TRUE -> MTank1.outvalve = TRUE));

MODULE Controller
VAR
  Tankid : 1..1;
  event : { createTank, killTank, highsensorFALSE, highsensorTRUE,
            lowsensorFALSE, lowsensorTRUE, none };

MODULE Tank(C, id)
VAR
  alive : boolean;
  highsensor : boolean;
  lowsensor : boolean;
  invalve : boolean;
  outvalve : boolean;
DEFINE
  TcreateTank := C.event = createTank & C.Tankid = id;
  TkillTank := C.event = killTank & C.Tankid = id;
  ThighsensorFALSE := C.event = highsensorFALSE & C.Tankid = id &
   alive = TRUE;
  ThighsensorTRUE := C.event = highsensorTRUE & C.Tankid = id &
   alive = TRUE;
  TlowsensorFALSE := C.event = lowsensorFALSE & C.Tankid = id &
   alive = TRUE;
```

```
  TlowsensorTRUE := C.event = lowsensorTRUE & C.Tankid = id &
   alive = TRUE;
ASSIGN
  init(alive) := FALSE;

  init(highsensor) := FALSE;

  init(lowsensor) := FALSE;

  init(invalve) := FALSE;

  next(invalve) :=
    case
      ThighsensorTRUE : FALSE;
      TlowsensorFALSE : TRUE;
      TRUE : invalve;
    esac;

  init(outvalve) := FALSE;

  next(outvalve) :=
    case
      TlowsensorFALSE : FALSE;
      TRUE : outvalve;
    esac;

  next(alive) :=
    case
      TcreateTank : TRUE;
      TkillTank : FALSE;
      TRUE : alive;
    esac;

  next(highsensor) :=
    case
      ThighsensorFALSE : FALSE;
      ThighsensorTRUE : TRUE;
      TRUE : highsensor;
    esac;

  next(lowsensor) :=
    case
      TlowsensorFALSE : FALSE;
      TlowsensorTRUE : TRUE;
      TRUE : lowsensor;
    esac;
```

Events (changes in attribute values) are only included for the sensor attributes. The invariants of the system are encoded as assumptions in the main module LTLSPEC. To check that other properties are valid, given these assumptions, these can be added as conclusions of the LTLSPEC. In this example, a counterexample to the property $lowsensor = true \Rightarrow outvalve = true$ is generated by SMV to show that this property fails, indicating that it does not follow from the invariants:

```
NuSMV > read_model -i tank.smv
NuSMV > go
WARNING: single-value variable 'C.Tankid' has been stored as a constant
NuSMV > check_ltlspec
-- specification  G (((MTank1.highsensor = TRUE -> MTank1.invalve =
FALSE) & (MTank1.lowsensor = FALSE -> (MTank1.invalve = TRUE &
MTank1.outvalve = FALSE))) ->
(MTank1.lowsensor = TRUE -> MTank1.outvalve = TRUE))  is false
-- as demonstrated by the following execution sequence
Trace Description: LTL Counterexample
Trace Type: Counterexample
-> State: 1.1 <-
  C.event = createTank
  MTank1.alive = FALSE
  MTank1.highsensor = FALSE
  MTank1.lowsensor = FALSE
  MTank1.invalve = FALSE
  MTank1.outvalve = FALSE
  C.Tankid = 1
  MTank1.TlowsensorTRUE = FALSE
  MTank1.TlowsensorFALSE = FALSE
  MTank1.ThighsensorTRUE = FALSE
  MTank1.ThighsensorFALSE = FALSE
  MTank1.TkillTank = FALSE
  MTank1.TcreateTank = TRUE
-> State: 1.2 <-
  C.event = lowsensorTRUE
  MTank1.alive = TRUE
  MTank1.TlowsensorTRUE = TRUE
  MTank1.TcreateTank = FALSE
-> State: 1.3 <-
  C.event = createTank
  MTank1.lowsensor = TRUE
  MTank1.TlowsensorTRUE = FALSE
  MTank1.TcreateTank = TRUE
-- Loop starts here
-> State: 1.4 <-
```

In the final state $lowsensor = TRUE$ but $outvalve = FALSE$, invalidating the conclusion of the LTLSPEC whilst validating its assumption, so the LTLSPEC is false in this state.

19.2 State machines

State machines define the dynamic behaviour of objects and operations. They can be used to give operation definitions (instead of pre and postconditions), and to express the life cycle of objects. For example, a student object could have a linear lifecycle of successive states *Year*1, *Year*2, *Year*3 and *Graduated*.

An editor for state machines is provided, Fig. 19.2 shows the interface for this editor.

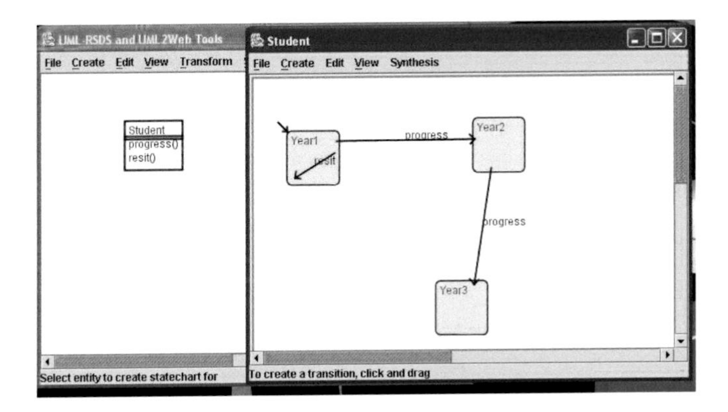

Figure 19.2: State machine editor

There are the following options:

File menu – options to save, load and print the state machine, and options to instantiate a class state machine for a particular object, and to take the product of two state machines. State machine data for model element m is saved and retrieved from the file $output/m.dat$.

Create menu – options to create states and transitions.

Edit menu – options to edit the state machine states and transitions and to resize the diagram.

View menu – options to view the lists of all states, transitions, events and attributes of the model.

Synthesis menu – options to analyse the state machine structure, to generate B and to carry out slicing of the state machine.

On the class diagram editor, the *Transformation* menu option *Express state machine in class diagram* adds variables and expressions to the class diagram to express the meaning of a state machine for an entity.

When an operation state machine is specified, the option *Check structure* on the state machine editor Synthesis menu should be applied before code is generated for the operation (on either the class diagram or state machine editor tool). This option identifies terminal, loop and decision states (displayed as green, red and blue, respectively), and warns the user if the state machine is not in the form of structured code. This analysis is then used to map the state machine into Java, C# or C++ code by the Generate code options on the class diagram Synthesis menu. Currently, only the subset of UML state machine notation with basic states is supported by UML-RSDS, although in principle larger subsets could be encoded [2].

Reactive systems can be specified explicitly by state machines which define the system response to input events, represented as operations. The behaviour of the tank control system of Section 19.1 can be explicitly specified by a state machine which has transitions for the sensor operations *sethighsensor* and *setlowsensor*, and these transitions can invoke actuator operations such as *setinvalve*, *setoutvalve*. Alternatively, each sensor and actuator can be modelled by separate components, each with its own state machine – a more refined specification closer to the actual physical implementation of the system. Figure 19.3 shows the class diagram and sensor state machine for this form of specification of the tank control system. The *TankControl* receives sensor event notifications from the two sensors, maintains its own internal representation of the sensor states, and issues commands to the valves dependent on the sensor states and events.

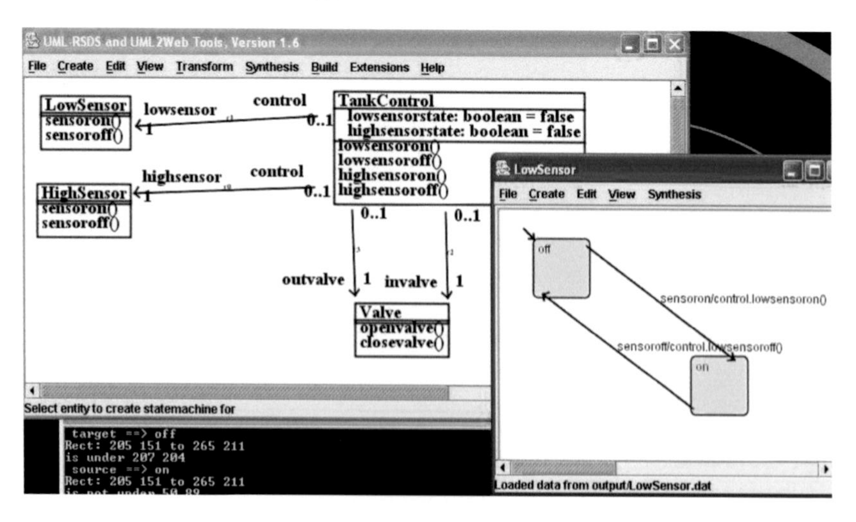

Figure 19.3: Tank control with explicit state machines

The notation *sensoron*/*control*.*lowsensoron*() on the state machine transitions indicates that the sensor event *sensoron* leads to an invocation of the tank control operation *lowsensoron*() when this transition is triggered. In general, a transition can have the annotation $e[G]/action$, indicating that if e occurs when the state machine is in the source state of the transition, and condition G is true, then the *action* is executed. The default for G is *true*. The *action* may be any statement (activity) valid in the context of the class or operation of the state machine.

An example of an operation state machine is the definition of *TankControl* :: *lowsensoroff* (Fig. 19.4).

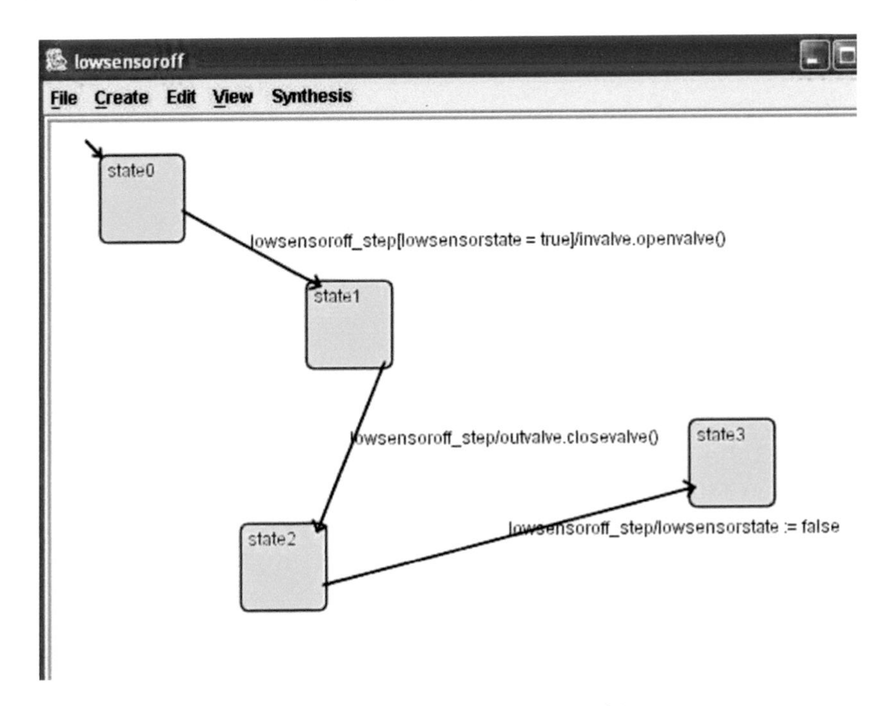

Figure 19.4: Operation state machine

This state machine is then used to generate Java code using the Generate Design and Generate Java 4 options on the Synthesis menu, resulting in the following code in the TankControl class:

```
public void lowsensoroff()
{ TankControl tankcontrolx = this;
  invalve.openvalve();
  outvalve.closevalve();
  lowsensorstate = false;
  return;
}
```

19.3 Interactions

Interactions can be created as UML sequence diagrams, these give examples of system behaviour in terms of object communications. They do not have a formal semantics, and are not used in code generation for systems, but can help to illustrate the processing of specific use cases and to describe expected scenarios of system behaviour, and to agree the details of use cases with customers, during requirements engineering.

The editor for interactions is shown in Fig. 19.5. The vertical lines are object lifelines, showing the time lines of individual objects (time increases from the top to the bottom of the screen). Messages are shown as arrows from sender object to receiver object: the operation invoked is shown on the arrow and must be an operation of the class of the receiver. Operation executions are shown as grey rectangles on the object lifeline, indicating that the object is executing an operation during an interval. A cross at the end of a lifeline indicates destruction of the object.

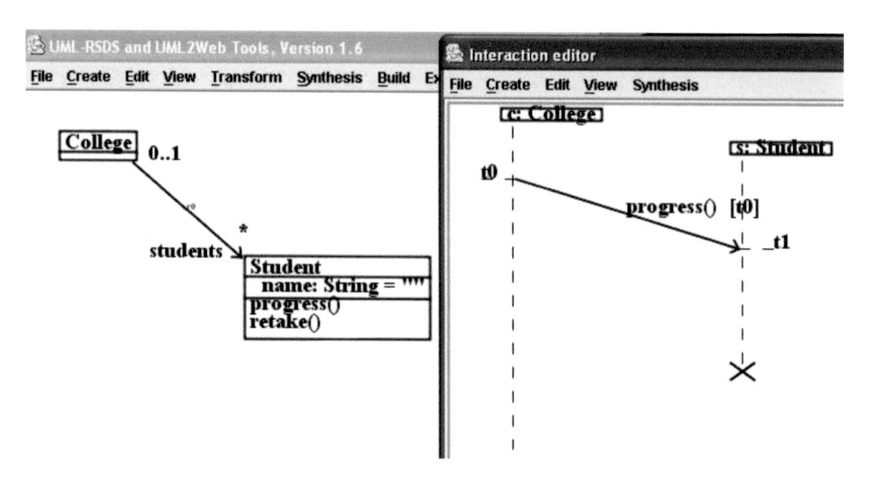

Figure 19.5: Interactions editor

The main interaction editor options are:

File menu – options to save, load and print models.

Create menu – options to create lifelines, messages, states, execution instances, and annotations.

Edit menu – options to edit an interaction.

View menu – options to view lifelines, messages, etc.

Synthesis menu – options to generate the formal real-time logic meaning of the diagram.

19.4 Reactive transformations

Model transformations can be used to define reactive systems: the transformation use case represents the reaction cycle of the reactive controller, and its rules define the responses of the system to given situations in the source model data, which can represent the state of a monitored system (sensor data and an internal representation of the environment). The target model can represent the actuators which the reactive system controls to affect the system under control.

For the tank control system we could write a use case *cycle* to represent the control system reaction cycle: reading all sensor inputs and producing corresponding actuator outputs according to the control invariants. The use case therefore has the control invariants as its postconditions:

$$
\begin{array}{l}
Tank :: \\
\quad highsensor = true \;\Rightarrow\; invalve = false \\
Tank :: \\
\quad lowsensor = false \;\Rightarrow\; invalve = true \;\&\; outvalve = false
\end{array}
$$

As in Section 19.1, the consistency and completeness of these constraints should be checked as part of their validation.

19.4.1 Case study: football player

This case study was the TTC 2014 live case problem [1]: to write a transformation which controls the positions and actions of a football team, and responds to the actions of the opposing team and the position of the ball. The global functional requirement of the system is to score more goals than the opposing team. The transformation communicates with a server via sockets, the server maintains the state of the game, which is effectively the system to be controlled (the EUC). Actions of the teams are sent to the server, and it sends out to each participant (blue team and red team) the updated pitch data with player and ball positions. Data is transmitted as text files in XML format, specifically as EMF XMIResource files. The problem is an example of a reactive transformation: a transformation which is intended to operate repeatedly to monitor an external system (in this case, the football pitch) and to take actions to affect this system. Figure 19.6 shows the class diagram of this system.

The *Update* and *Action* classes describe the responses of the control system (the transformation) to the current state of the *SoccerPitch* (player positions for both teams, and the ball position). Each response may consist of a number of actions for the players of the responding

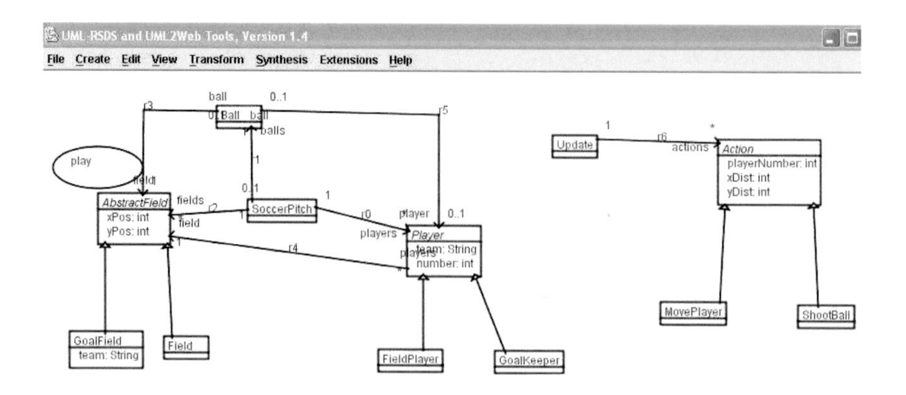

Figure 19.6: Class diagram of football game

team, and can either be an instruction to move a player, or for a player (with the ball) to shoot the ball. Attributes *xDist* and *yDist* are the horizontal and vertical distances for the player to move or for the ball to be kicked.

Figure 19.7 shows the visual interface of the football pitch, which is the equipment/system under control for this reactive system. Numbers denote players, and there are two teams, red and blue, which would normally be controlled by distinct versions of the reactive control system.

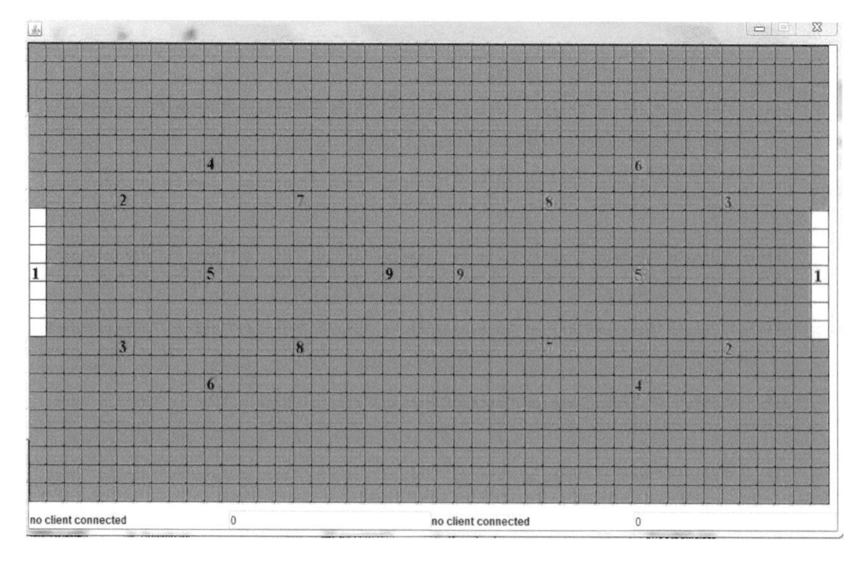

Figure 19.7: Layout of football pitch

There are several restrictions on the allowed moves:

- Each player can only perform at most one action in each turn.
- A player owning the ball can move at most 2 places in x and y directions. A player without the ball can move at most 3 places in x and y directions.
- A goalkeeper can move only on his own team's goal fields.
- Players cannot be moved off the pitch. Nor can the ball be shot off the pitch.
- A shot can move the ball at most 7 places in x and y directions.
- No more than 4 shoot ball actions can take place in each update.

The penalty in each of these cases is a red card – the player involved in the action is removed from the game.

The case study is quite open-ended, in that the global requirement can be refined in many different ways. A wide range of different strategies could be used to play the game, such as defensive or offensive styles of play. There is scope also to build in artificial intelligence techniques to compute suitable moves. One approach is to define specific tactics for each category of player (goalkeeper, defender, midfielder, forward). For example, the goalkeeper specific behaviour is expressed by the operation:

```
Player::
goalkeeperaction(u : Update)
pre: true
post:
  ( ball.size > 0 =>
        ShootBall->exists( a |
              a.playerNumber = number &
              a.xDist = 7 & a.yDist = -3 &
              a : u.actions ) &
        MovePlayer->exists( m |
              m.playerNumber = 5 &
              m.xDist = -3 & m.yDist = -3 &
              m : u.actions )
  )
```

This expresses that if the goalkeeper (of the blue team) has the ball, he should shoot it to the central midfielder number 5, and direct this player to move to the ball. The two actions are added to the update u supplied as a parameter.

The blue team central midfielder behaviour is expressed by:

```
Player::
cmaction(u : Update, theball : Ball)
pre: true
post:
  ( ball.size > 0 & field.xPos < 39 =>
      MovePlayer->exists( m |
          m.playerNumber = number &
          m.xDist = 2 & m.yDist = 0 &
          m : u.actions ) ) &
  ( ball.size > 0 & field.xPos >= 39 =>
      ShootBall->exists( m |
          m.playerNumber = number &
          m.xDist = 44 - field.xPos &
          m.yDist = 0 & m : u.actions ) ) &
  ( theball.blueTeamhasBall() = false =>
                                  moveToBall(u,theball) )
```

If the player has the ball and is not within shooting distance of the red goal, then he should move forward. If he is within shooting distance and has the ball, then he should shoot, otherwise if the blue team does not have the ball, he should move towards the ball.

The reactive behaviour of the blue team controller is expressed by a use case *play*, which has the postcondition:

```
theball = Ball->any() & goalkeeper = GoalKeeper->select(team /= "RED")->any() &
player5 = FieldPlayer->select(team /= "RED" & number = 5)->any() &
player2 = FieldPlayer->select(team /= "RED" & number = 2)->any() &
player3 = FieldPlayer->select(team /= "RED" & number = 3)->any() &
player4 = FieldPlayer->select(team /= "RED" & number = 4)->any() &
player6 = FieldPlayer->select(team /= "RED" & number = 6)->any() &
player7 = FieldPlayer->select(team /= "RED" & number = 7)->any() &
player8 = FieldPlayer->select(team /= "RED" & number = 8)->any() &
player9 = FieldPlayer->select(team /= "RED" & number = 9)->any()  =>
    Update->exists( u | goalkeeper.goalkeeperaction(u) & player5.cmaction(u,theball) &
           player2.defenderaction(u,theball) & player3.defenderaction(u,theball) &
           player4.midfielderaction(u,theball) & player6.midfielderaction(u,theball) &
           player7.midfielderaction(u,theball) & player8.midfielderaction(u,theball) &
           player9.midfielderaction(u,theball) )
```

The antecedent simply defines let-variables to hold the ball and individual players. In the succedent each player adds their own actions to the update. Verification of this use case should check that it generates updates which respect the rules on allowed actions listed above. According to extensive testing this is the case, but no formal proof was carried out.

The implementation of the *play* use case is invoked by a SoccerClient class (manually written), which executes a loop which reads the socket from the soccer server, extracts the SoccerPitch model as an XML-encoded string, and supplies this to a Controller operation *cycle* which constructs the input model from the XML data. The Controller *play* use case is then invoked and the Update data which it produces is then returned to the SoccerClient as an XML file and sent to the soccer server:

```
    int turns = 0;
    while (turns < 400)
    { StringBuffer xmlstring = new StringBuffer();
      s = in.readLine();  // get new pitch model from server

      while (s != null && !(s.equals(END_MARKER)))
      { s = in.readLine();
        xmlstring.append(s);
      }
      String resp = Controller.cycle(xmlstring.toString());

      out.println(START_MARKER);
      out.println(resp);
      out.println(END_MARKER);
      out.flush();  // send updates to server
    }
    in.close();
    out.close();
    client.close();
  } catch (Exception e) { e.printStackTrace(); }
```

The efficiency and response time were satisfactory. The basic playing strategy defined by *play* could be improved by better co-ordination between players and increased use of multi-player moves. Greater use of abstraction in the specification would be beneficial, instead of rules being expressed in terms of specific numeric positions and distances.

Summary

In this chapter we have identified techniques for specifying reactive systems in UML-RSDS, using constraints, state machines, interactions and transformations.

References

[1] T. Horn, *TTC 2014 Live case problem: Transformation tool contest world cup*, TTC 2014.

[2] K. Lano and D. Clark, *Direct semantics of extended state machines*, Journal of Object Technology, Vol. 6, No. 9, 2007.

Chapter 20

Enterprise Systems Development with UML-RSDS

Enterprise information systems hold and manage business-critical data for a company or organisation. They usually implement the core business operations of an enterprise. Examples include accounts data and accounts management operations for a bank. EIS typically involve distributed processing and large-scale secure data storage. The structures and components of an EIS are often of a standard form, independent of the specific application, and EIS platforms such as Java Enterprise Edition (Java EE) and Microsoft .Net provide much of the machinery of data management, data persistence, transaction management and distributed processing which is needed by any EIS. Using UML-RSDS, many of the components of an EIS can be automatically generated from a specification class diagram of the application data, and from identified use cases operating on this data.

20.1 EIS synthesis

An enterprise information system (EIS) implemented with Java technologies typically consists of five tiers of components (Fig. 20.1):

Client tier This contains web pages or other interfaces by which clients use the system. Typical components are HTML files, applets, etc.

Presentation tier Components which construct the GUI of the system and handle requests from the client tier. For example, Java servlets and JSPs.

Business tier Components which represent the business operations (services) offered by the system, and the business entities of the system. These typically include Session Beans and Entity Beans, and Value Objects, which are used to transfer data between the business tier and other tiers.

Integration tier This contains components which serve as a mediating interface between the business and resource tiers: abstracting the resource components. It typically consists of database interfaces and web service interfaces.

Resource tier This tier contains databases and other resources used by the system. Web services provided by external organisations are also included here.

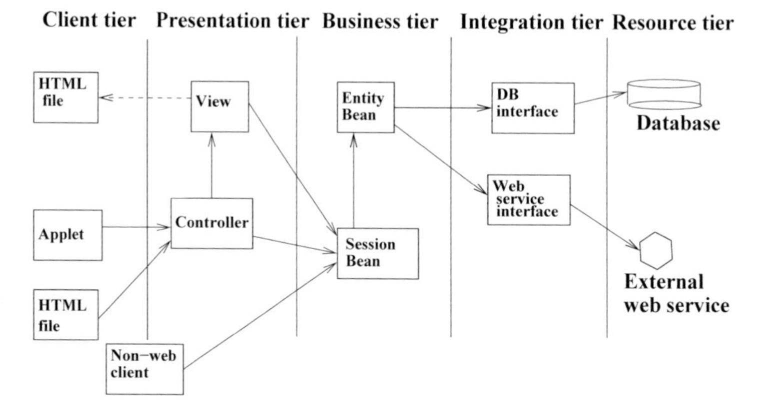

Figure 20.1: Five-tier EIS architecture

Such systems can be defined using many different technologies and techniques. In UML-RSDS we use Java technologies, and three alternative system styles:

- Servlet style: presentation tier is coded using only servlets and auxiliary classes to generate web pages.
- JSP style: presentation tier is coded only using JSPs to handle web requests and to generate web pages.
- J2EE style: Java Entity beans and Session beans are constructed to define the business tier of an application, together with value objects and a database interface class.

In each case, the system specification is given as a class diagram and set of use cases defining the user operations to be provided by the system. These are defined using the first field of the use case dialog (Fig. 5.1) to name the operation, the second field to name the entity, and the third (if needed) to name a specific feature.

EIS use cases can be:

- create E: create an instance of entity E (initializing its attribute values)
- delete E: delete an instance of E
- edit E: edit an instance of entity E (setting its attribute values)
- add E r: add an element to the role set r of an instance of E
- remove E r: remove an element from the role set r of an instance of E
- list E: list all instances of E
- searchBy E a: find instances of E with a given value for attribute a.

These operations are the standard CRUD (create, read, update, delete) actions provided by most data management systems. Any invariant constraints that are defined for a persistent entity are used to generate validation checks on the web pages and in server side functional components. The use cases and class diagram elements map to EIS code components as follows (Table 20.1).

Table 20.1: Mapping of UML-RSDS to web code

Element	*Servlet style*	*JSP style*	*J2EE style*
Class *E*	Database table	Session/Entity bean Value object	Session bean, Entity bean, Value object
Attribute	Table column form field	as Servlet style as Servlet style	as Servlet style as Servlet style
Use case *op*	CommandPage.java (view) opPage.java (form generator) op.html (form) opResultPage.java (view) Dbi.java (data access object)	commands.html op.jsp (view + controller) op.html (form) Dbi.java	as JSP style

In the following section we give an example of the JSP-style EIS architecture.

20.2 Example EIS application: bank accounts system

This example is a simple but typical case of an EIS. The system maintains details of the bank customers, their accounts, and transactions on these accounts (Fig. 20.2).

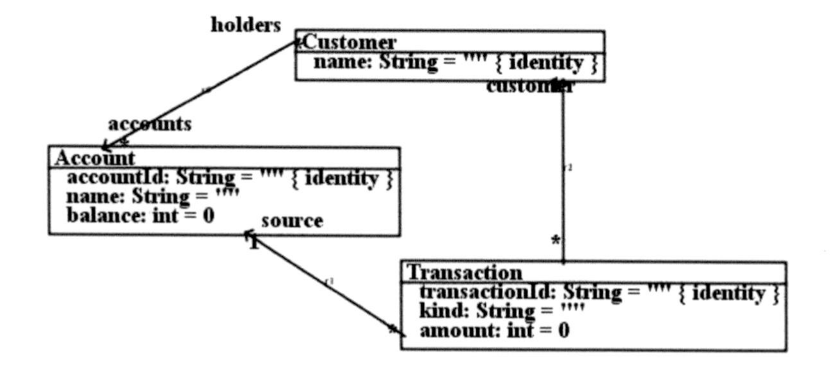

Figure 20.2: Bank account EIS class diagram

The use cases considered are:

- create Account;
- create Customer;
- create Transaction;
- add Customer accounts;
- list Account;
- searchBy Customer name.

An invariant of Account is that $balance \geq 0$.

The *Web system/JSP* option on the *Synthesis* menu then produces a set of files for the client, presentation, business and integration tier of this EIS, according to Table 20.1. The overall architecture of the JSP style of generated web system is shown in Fig. 20.3.

A standard physical organisation of a Java-based web application consists of a subdirectory *app* named after the application, in a directory *webapps* of the web server. Within *app* there are directories (i) *servlets* containing HTML and JSP files (in the case of a JSP-based architecture) and servlets, and (ii) *WEB-INF* containing business-tier Java components in a subdirectory *classes*.

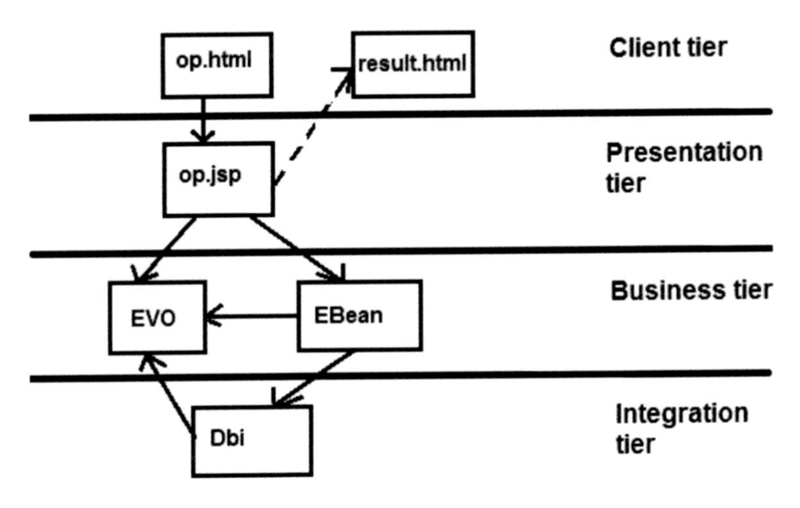

Figure 20.3: JSP style EIS architecture

20.2.1 Client tier

Web pages *op.html* for each use case *op* are synthesised for the client tier. These contain fields for all the input parameters of *op*, and a submit button. They invoke the corresponding *op.jsp* in the presentation tier. The *op.html* web pages should be placed in *webapps/app/servlets*. In our case study we have, for example, *addCustomeraccounts.html* (Fig. 20.4) and *searchBycustomername.html* (Fig. 20.5).

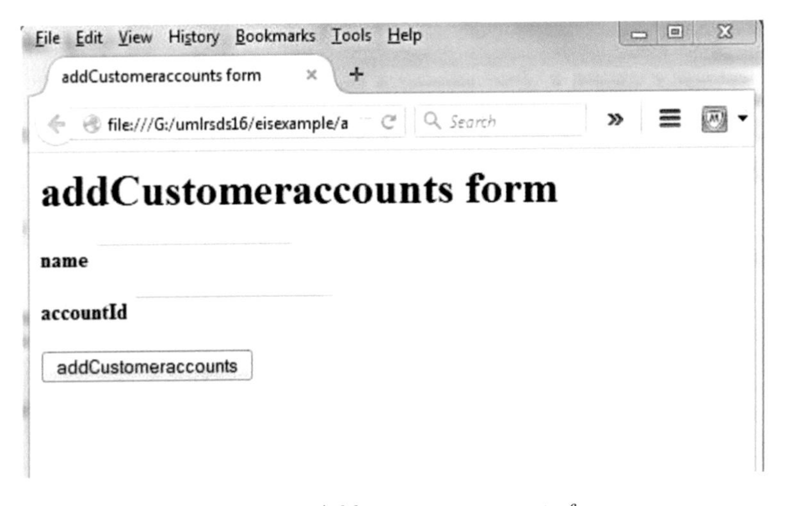

Figure 20.4: Add customer accounts form

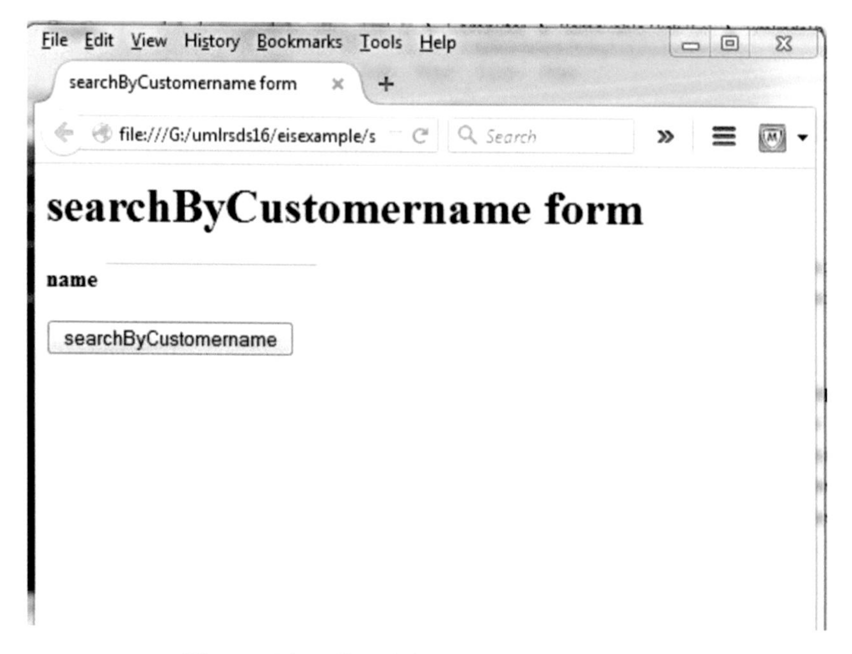

Figure 20.5: Search by customer name form

20.2.2 Presentation tier

In the presentation tier, a JSP *op.jsp* is defined for each use case *op*. This JSP receives the data from *op.html*, calls the session/entity bean responsible for *op*, and then generates a result *opresult.html* page to display the outcome of the operation. An example is the JSP to create an account:

```
<jsp:useBean id="account" scope="session" class="beans.AccountBean"/>
<jsp:setProperty name="account"  property="accountId"  param="accountId"/>
<jsp:setProperty name="account"  property="name"  param="name"/>
<jsp:setProperty name="account"  property="balance"  param="balance"/>

<html>

<head><title>createAccount</title></head>

<body>

<h1>createAccount</h1>

<% if (account.iscreateAccounterror())
{ %> <h2>Error in data: <%= account.errors() %></h2>
<h2>Press Back to re-enter</h2> <% }
else { account.createAccount(); %>
```

```
<h2>createAccount performed</h2>
<% } %>

<hr>

<% include file="commands.html" %>

</body>
</html>
```

The *jsp:useBean* directive links the JSP to the AccountBean instance *account*, the *jsp : setProperty* lines copy the form data to this instance, and the body of the JSP tests if there is an error in this data, and calls *createAccount* on *account* to create the new account (ultimately by adding a new row to the *Account* table in the database) if there is no error. The JSP files should be placed in *webapps/app/servlets*.

20.2.3 Business tier

Entity/session bean classes *EBean* are synthesised for each persistent entity type (class). These beans interact with the Dbi component in the integration tier to store and extract instances of the entity, and to perform data validation checks on instance data based on the attribute types and entity invariants. They have operations for each use case that involves the entity, and error-checking operations for each use case on the entity. For example, *AccountBean* in our case study is:

```
package beans;

import java.util.*;
import java.sql.*;

public class AccountBean
{ Dbi dbi = new Dbi();
  private String accountId = "";
  private String name = "";
  private String balance = "";
  private int ibalance = 0;
  private Vector errors = new Vector();

  public AccountBean() {}

  public void setaccountId(String accountIdx)
  { accountId = accountIdx; }

  public void setname(String namex)
  { name = namex; }

  public void setbalance(String balancex)
```

```
  { balance = balancex; }

  public void resetData()
  { accountId = "";
    name = "";
    balance = "";
  }

  public boolean iscreateAccounterror()
  { errors.clear();
    try { ibalance = Integer.parseInt(balance); }
    catch (Exception e)
    { errors.add(balance + " is not an integer"); }
    if (ibalance >= 0) { }
    else
    { errors.add("Constraint: ibalance >= 0 failed"); }
    return errors.size() > 0; }

  public boolean islistAccounterror()
  { errors.clear();
    return errors.size() > 0; }

  public String errors() { return errors.toString(); }

  public void createAccount()
  { dbi.createAccount(accountId, name, ibalance);
    resetData(); }

  public Iterator listAccount()
  { ResultSet rs = dbi.listAccount();
   List rs_list = new ArrayList();
   try
   { while (rs.next())
     { rs_list.add(new AccountVO(rs.getString("accountId"),
           rs.getString("name"),rs.getInt("balance")));     }
   } catch (Exception e) { }
   resetData();
   return rs_list.iterator();
  }
}
```

Value object classes are generated for each entity, to provide a technology-neutral means of passing data between tiers of the EIS, as in the operation *listAccount* above. In the bank system there are value objects for *Customer*, *Account* and *Transaction*, e.g.:

```
package beans;

public class CustomerVO
```

```
{
 private String name;

  public CustomerVO(String namex)
  {    name = namex;
  }

  public String getname()
  { return name; }

}
```

All of these classes should be placed in the *beans* subdirectory of *webapps/app/WEB-INF/classes* and compiled there.

20.2.4 Integration tier

The integration tier contains the Dbi.java class, which uses Java JDBC to update and read a relational database using SQL commands. This class should also be placed in *webapps/app/WEB-INF/classes/beans*. In the bank system, the Dbi class is as follows:

```
package beans;

import java.sql.*;

public class Dbi
{ private Connection connection;
  private static String defaultDriver = "";
  private static String defaultDb = "";
  private PreparedStatement createAccountStatement;
  private PreparedStatement createCustomerStatement;
  private PreparedStatement createTransactionStatement;
  private PreparedStatement listAccountStatement;
  private PreparedStatement addCustomeraccountsStatement;
  private PreparedStatement searchByCustomernameStatement;
  public Dbi() { this(defaultDriver,defaultDb); }

  public Dbi(String driver, String db)
  { try
    { Class.forName(driver);
      connection = DriverManager.getConnection(db);
      createAccountStatement =
        connection.prepareStatement(
        "INSERT INTO Account (accountId,name,balance) VALUES (?,?,?)");
      createCustomerStatement =
        connection.prepareStatement(
          "INSERT INTO Customer (name) VALUES (?)");
```

```
      createTransactionStatement =
        connection.prepareStatement(
          "INSERT INTO Transaction (transactionId,kind,amount)
           VALUES (?,?,?)");
      listAccountStatement =
        connection.prepareStatement(
          "SELECT accountId,name,balance FROM Account");
      addCustomeraccountsStatement =
        connection.prepareStatement(
          "UPDATE Account SET Account.name = ? WHERE Account.
           accountId = ?");
      searchByCustomernameStatement =
        connection.prepareStatement(
          "SELECT name FROM Customer WHERE name = ?");
    } catch (Exception e) { }
  }

  public synchronized void createAccount(String accountId,
   String name,int balance)
  { try
    { createAccountStatement.setString(1, accountId);
      createAccountStatement.setString(2, name);
      createAccountStatement.setInt(3, balance);
      createAccountStatement.executeUpdate();
      connection.commit();
    } catch (Exception e) { e.printStackTrace(); }
  }

  public synchronized void createCustomer(String name)
  { try
    { createCustomerStatement.setString(1, name);
      createCustomerStatement.executeUpdate();
      connection.commit();
    } catch (Exception e) { e.printStackTrace(); }
  }

  public synchronized void createTransaction(String transactionId,
   String kind,int amount)
  { try
    { createTransactionStatement.setString(1, transactionId);
      createTransactionStatement.setString(2, kind);
      createTransactionStatement.setInt(3, amount);
      createTransactionStatement.executeUpdate();
      connection.commit();
    } catch (Exception e) { e.printStackTrace(); }
  }

  public synchronized ResultSet listAccount()
  { try
```

```
    { return listAccountStatement.executeQuery();
  } catch (Exception e) { e.printStackTrace(); }
  return null; }

  public synchronized void addCustomeraccounts(String name,
   String accountId)
  { try
    {   addCustomeraccountsStatement.setString(1, name);
      addCustomeraccountsStatement.setString(2, accountId);
      addCustomeraccountsStatement.executeUpdate();
    connection.commit();
  } catch (Exception e) { e.printStackTrace(); }
  }

  public synchronized ResultSet searchByCustomername(String name)
  { try
    {   searchByCustomernameStatement.setString(1, name);
      return searchByCustomernameStatement.executeQuery();
  } catch (Exception e) { e.printStackTrace(); }
  return null; }

  public synchronized void logoff()
  { try { connection.close(); }
    catch (Exception e) { e.printStackTrace(); }
  }
}
```

For the final production code, the debugging calls of *printStackTrace* would be removed.

Summary

In this chapter we have described how EIS applications can be synthesised using UML-RSDS, and we have given a detailed example of EIS synthesis.

Chapter 21

Applications of UML-RSDS in Education and Industry

In this chapter we discuss the teaching of UML and model-based development using UML-RSDS. We describe three case studies of UML-RSDS, including two which are suitable for educational use (case studies 1 and 2). An industrial application of a complex financial system is also presented.

21.1 Teaching using UML-RSDS

UML-RSDS has been used for teaching software specification and design using UML, and for practical student projects using model-based and agile development. This has mainly been at the second year level of undergraduate courses. Only a small subset of UML-RSDS features need to be considered for such teaching:

- Introducing UML and the relationship between UML and object-oriented programming languages: the core class diagram notations of classes, attributes, associations, operations and inheritance are needed. More elaborate notations such as association classes, composition, and qualified associations need not be considered. The class diagram editor and code generation facilities of UML-RSDS are used.

- Dynamic modelling: the state machine editor and code generation facilities can be used, together with the interactions editor.
- Model-based development: class diagrams, use cases and constraints are needed, to specify systems and model transformations. The code generators are used to generate executable implementations.

As discussed below, students may find problems using UML-RSDS or other MBD tools, because of the conceptual novelty of writing executable specifications, and because of the tool complexity. We recommend using a clearly defined specification and development procedure which students should follow with the tool, and restricting the notations considered to a sufficient subset. For example, the most often used OCL operators in UML-RSDS specifications are : and $\rightarrow$*includes*, $\rightarrow$*exists*, conjunction and implication, object lookup by identity, numeric operators and comparisons, $\rightarrow$*select*, $\rightarrow$*size*, and $\rightarrow$*forAll*. To solve a particular problem, students can be given a set of operators and language elements to use, which will be *sufficient* for the problem.

21.2 Case study 1: FIXML code generation

This case study was based on the problem described in [18]. Financial transactions can be electronically expressed using formats such as the FIX (Financial Information eXchange) format. New variants/extensions of such message formats can be introduced, which leads to problems in the maintenance of end-user software: the user software, written in various programming languages, which generates and processes financial transaction messages will need to be updated to the latest version of the format each time it changes. In [18] the authors proposed to address this problem by automatically synthesising program code representing the transaction messages from a single XML definition of the message format, so that users would always have the latest code definitions available. For this case study we restricted attention to generating Java, C# and C++ class declarations from messages in FIXML 4.4 format, as defined at http://fixwiki.org/fixwiki/FPL:FIXML_Syntax, and http://www.fixtradingcommunity.org.

The solution transformation should take as input a text file of a message in XML FIXML 4.4 Schema format, and produce as output corresponding Java, C# and C++ text files representing this data.

The problem is divided into the following use cases:

1. Map data represented in an XML text file to an instance model of the XML metamodel (Fig. 21.1).

2. Map a model of the XML metamodel to a model of a suitable metamodel for the programming language/languages under consideration. This has subtasks: 2a. Map XML nodes to classes; 2b. Map XML attributes to attributes; 2c. Map subnodes to object instances.

3. Generate program text from the program model.

In principle these use cases could be developed independently, although the subteams or developers responsible for use cases 2 and 3 need to agree on the programming language metamodel(s) to be used.

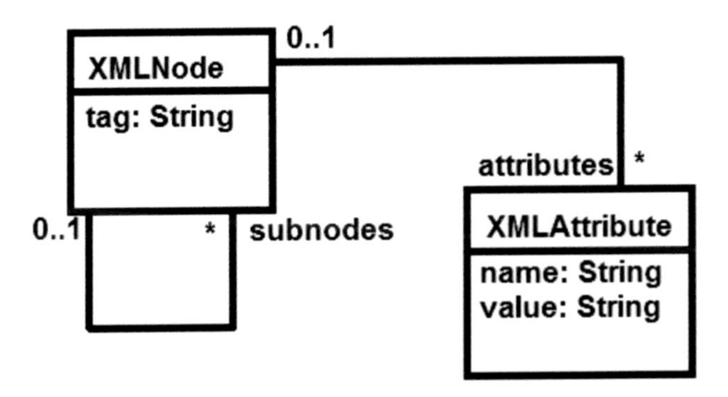

Figure 21.1: XML metamodel

The problem was set as the assessed coursework (counting for 15% of the course marks) for the second year undergraduate course "Object-oriented Specification and Design" (OSD) at King's College in 2013. It was scheduled in the last four weeks at the end of the course. OSD covers UML and MBD and agile development at an introductory level. Students also have experience of team working on the concurrent Software Engineering Group project (SEG). Approximately 120 students were on the course, and these were divided into 12 teams of 10 students each.

The case study involves research into FIXML, XML, UML-RSDS and C# and C++, and definition of use cases in UML-RSDS using OCL. None of these topics had been taught to the students. Scrum, XP, and an outline agile development approach using UML-RSDS had been taught, and the teams were recommended to appoint a team leader. A short (5 page) requirements document was provided, and links to the UML-RSDS tools and manual. Each week there was a one hour timetabled lab session where teams could meet and ask for help from postgraduate students who had some UML-RSDS knowledge.

21.2.1 Solution

The class diagram of a possible solution (specific to Java code output) is shown in Fig. 21.2.

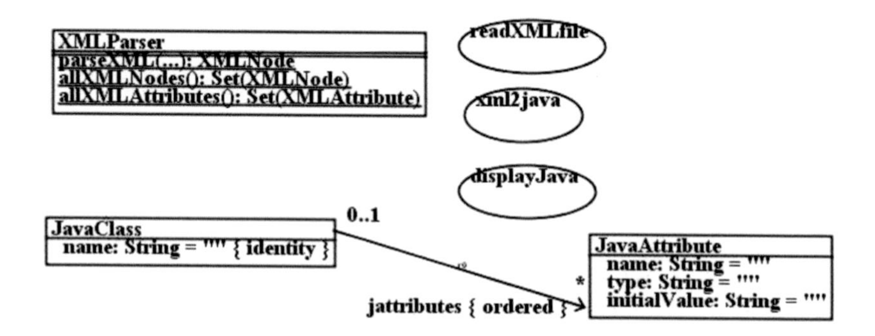

Figure 21.2: FIXML system specification

The class *XMLParser* is an external class (its code is provided by handwritten Java) with the operations

```
query static parseXML(f : String) : XMLNode

query static allXMLNodes() : Set(XMLNode)

query static allXMLAttributes() : Set(XMLAttribute)
```

These parse a given XML document and return the sets of all XML nodes and attributes defined in the document. These operations are used to carry out the first use case, *readXMLfile*, which has the postconditions:

```
XMLParser.parseXML("test1.xml")

XMLParser.allXMLNodes() <: XMLNode &
XMLParser.allXMLAttributes() <: XMLAttribute
```

The second postcondition adds all the parsed XML nodes and attributes from `test1.xml` to the sets of instances of *XMLNode* and *XMLAttribute*, respectively.

The *xml2java* use case has the postcondition constraints:

```
XMLNode::
  JavaClass->exists( jc | jc.name = tag )
```

which creates a Java class for each XMLNode (task 2a). Since *name* is an identity attribute for *JavaClass*, multiple XMLNodes with the same tag will be represented by a *single* JavaClass.

For task 2b there is the postcondition:

```
XMLNode::
  att : attributes & jc = JavaClass[tag] &
  att.name /: jc.jattributes@pre.name@pre  =>
      JavaAttribute->exists( ja |
             ja.name = att.name & ja.type = "String" &
             ja.initialValue = att.value & ja : jc.jattributes )
```

This maps all XML attributes of a given XML node *self* to program attributes of the program class *JavaClass*[*tag*] corresponding to *self*. For each pair of an XML node *self* and attribute *att* : *attributes* a new JavaAttribute is created and added to *JavaClass*[*tag*]. This double iteration is needed because attribute names are not unique: two different XML nodes could both have attributes with a particular name. Thus a pure Phased Construction approach, with attributes mapped first and then looked-up by their key, is not possible. The condition `att.name /: jc.jattributes@pre.name@pre` is needed to check that no program attribute with name *att.name* is already in the program class: an invalid class would result if two attributes with the same name were present. The constraint is of type 1 because pre-forms of *jattributes* and *JavaAttribute*::*name* are used in the antecedent, so that the read frame of the constraint is disjoint from the write frame (a case of the Replace Fixed-point by Bounded Iteration pattern, Chapter 9). It is not confluent, because the unordered association *XMLNode*::*attributes* may be iterated over in an arbitrary order, so that two different attributes with the same name but different initialisations could be processed in either order, resulting in two different *jattribute* collections – because only the first XML attribute to be processed will produce a Java attribute.

For task 2c a similar constraint is used to map subnodes to program attributes:

```
XMLNode::
sn : subnodes  =>
    JavaAttribute->exists( ja |
        ja.name = sn.tag + "_object" + JavaClass[tag].jattributes
         @pre->size() &
        ja.type = sn.tag &
        ja.initialValue = "new " + sn.tag + "()" &
        ja : JavaClass[tag].jattributes )
```

There may be multiple subnodes with the same tag, and each must be separately represented in the code output, so we append the number *JavaClass*[*tag*].*jattributes*@*pre*→*size*() to the Java attribute name to distinguish these. This number increments each time an attribute is added to the class *JavaClass*[*tag*]. Since *subnodes* is unordered, this constraint is not confluent: different orders of iteration through *subnodes* may produce different variable names in the Java class.

The third use case is carried out by the *displayJava* use case:

```
JavaClass::
( "class " + name + " {" )->display() &
jattributes->forAll( ja |
    ( "  " + ja.type + " " + ja.name + " = " + ja.initialValue + ";"
     )->display() ) & "}\n"->display()
```

The same metamodel for Java programs can be used also for C# and C++, and only the *displayX* use case needs to be adapted to print out programs in the syntax of these languages.

21.2.2 Outcome

The outcome of the case study is summarised in Table 21.1.

Table 21.1: Case 1 results

Teams	*Mark range*	*Result*
5, 8, 9, 10	80+	Comprehensive solution and testing, well-organised team
12	80+	Good solution, but used manual coding, not UML-RSDS
4, 7, 11	70–80	Some errors/incompleteness
2, 3, 6	50–60	Failed to complete some tasks
1	Below 40	Failed all tasks, group split into two.

Examples of good practices included:

- Division of a team into sub-teams with sub-team leaders, and separation of team roles into researchers and developers (teams 8, 11).
- Test-driven development (teams 8, 9).
- Metamodel refactoring, to merge different versions of program metamodels for Java, C# and C++ into a single program metamodel.

Because of the difficulty of the problem, teams tended to work together as a unit on each use case, rather than divide into subteams with separate responsibilities. In retrospect the complexity of the task was too high for second year undergraduates, and all the teams struggled both to understand the task and to apply UML-RSDS. With intensive effort, the best teams did manage to master the technical problems and to carry out all the mandated tests on example FIX XML files. The most difficult part of the problem was use case 2, which involved using a particular form of constraint quantification to avoid creating duplicate program features in cases where an XML node has multiple direct subnodes with the same tag name. All teams used UML-RSDS to try to solve the problem, except for team 12, which produced a hand-coded Java solution. The effort expended by this team seemed comparable to that of the successful UML-RSDS teams, but their coding effort was higher whilst their research effort was lower.

Conclusions that can be drawn from this case study are that an excessively complex task is a bad choice as a first project in MBD, and that developers should instead build their expertise using less challenging applications. Only four teams managed to master the development approach, others either reverted to manual coding or produced incomplete solutions. The total effort expended by successful MBD teams was not in excess of that expended by the successful manual coding team, which suggests that the approach can be feasible even in adverse circumstances.

21.3 Case study 2: Electronic health records (EHR) analysis and migration

This case study was the OSD assessed coursework for 2014. It was intended to be somewhat easier than the 2013 coursework. Approximately 140 second year undergraduate students participated, divided into 14 teams of 9 or 10 members.

There were three top level use cases: (1) to analyse a dataset of GP patient data conforming to the language of Fig. 21.3 for cases of missing names, address, etc., feature values; (2) to display information on referrals and consultations in date-sorted order; (3) to integrate the GP patient data with hospital patient data conforming to the EHR language of Fig. 21.4 to produce an integrated dataset conforming to a third model (Fig. 21.5).

Table 21.2 summarises the use cases and their subtasks.

Students were allocated randomly to teams. Teams were advised to select a leader, and to apply an agile development process, although a specific process was not mandated. A short (2 page) requirements document was provided, and links to the UML-RSDS tools and manual. Each week there was a one hour timetabled lab session where teams could meet and ask for help from postgraduate students who had some UML-RSDS knowledge.

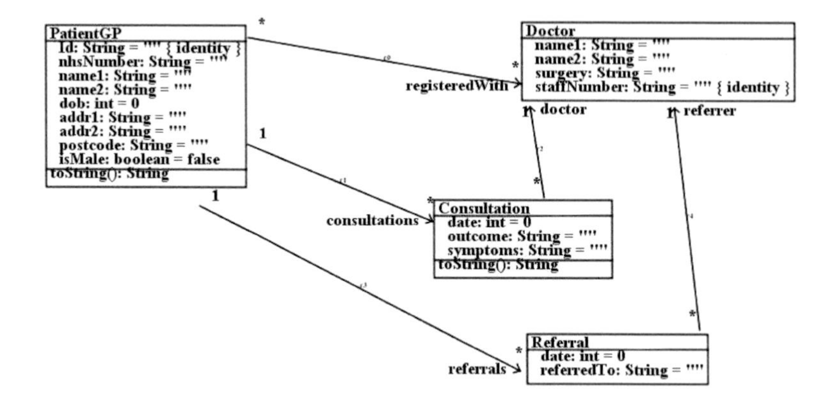

Figure 21.3: GP patient EHR structure gpmm1

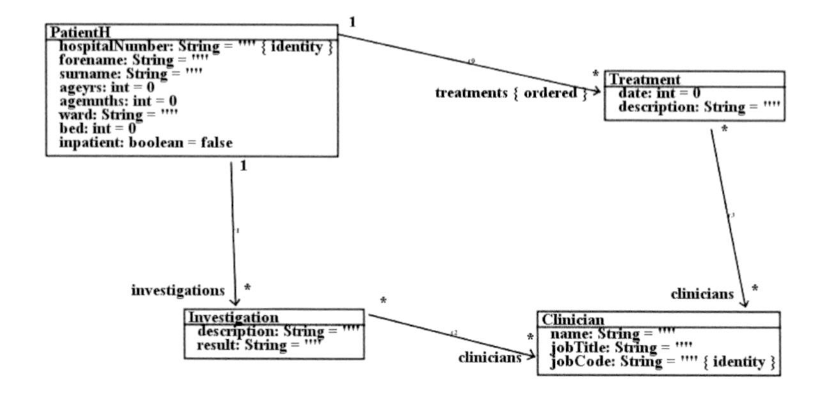

Figure 21.4: Hospital patient EHR structure gpmm2

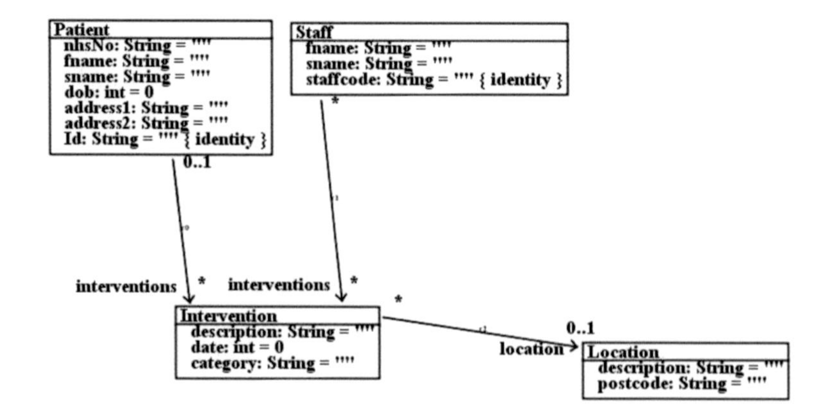

Figure 21.5: Integrated patient EHR structure gpmm3

Table 21.2: Use cases for EHR analysis/migration application

Use case	*Subtasks*	*Models*
1. Analyse data	1a. Detect missing dob, names, addresses in GP dataset	gpmm1
	1b. Detect duplicate patient records	gpmm1
2. View data	2a. Display consultations of each GP patient, in date order	gpmm1
	2b. Display referrals of each GP patient, in date order	gpmm1
3. Integrate data	Combine gpmm1, gpmm2 data into gpmm3	gpmm1, gpmm2, gpmm3

21.3.1 Solution

A possible solution for use case 1 could be the following postconditions of a use case *check*:

```
dob <= 0   =>   ("Patient " + self + " has no valid date of birth")
->display()

name1 = ""   =>   ("Patient " + self + " has no valid first name")
->display()

name2 = ""   =>   ("Patient " + self + " has no valid second name")
->display()

addr1 + addr2 = ""   =>   ("Patient " + self + " has no valid address")
->display()
```

All of these constraints are on the entity context *PatientGP*.

To check for duplicated patients, the following *check* postcondition on entity context *PatientGP* could be used:

```
p : PatientGP & Id < p.Id &
name1 = p.name1 & name2 = p.name2 & dob = p.dob & isMale = p.isMale  =>
          ("Patients " + self + " and " + p + " seem to be duplicates")
           ->display()
```

This performs a double iteration over *PatientGP*, comparing *self* and *p*. The condition $Id < p.Id$ is used, instead of $Id \neq p.Id$, because we only want to consider each distinct pair once.

For the second use case, *viewData*, to display all consultation data for a given GP patient, sorted by date (most recent last), we can use the $\rightarrow$*sortedBy* operator:

```
cons = consultations->sortedBy(date)   =>
                                cons->forAll( c | c->display() )
```

iterated on *PatientGP*, where a suitable *toString*() operation has been added to *Consultation*. Likewise for referrals.

For the third use case to integrate corresponding GP and hospital datasets, we need to convert from integer date of births in the format yearMonthDay (e.g., 19561130) to age in years and months. The age in years is

$$currentYear - (dob/10000)$$

with integer division used on the dob to extract the year. The age in months is

$$currentMonth - ((dob\ mod\ 10000)/100)$$

A solution is then, for December 2014:

```
p : PatientH & name1 = p.forename & name2 = p.surname &
(2014 - (dob/10000)) = p.ageyrs &
(12 - (dob mod 10000)/100) = p.agemnths  =>
    Patient->exists( q | q.Id = Id & q.nhsNo = nhsNumber & q.fname =
     name1 &
          q.sname = name2 & q.dob = dob & q.address1 = addr1 &
           q.address2 = addr2 )
```

on context *PatientGP*.

21.3.2 Outcome

Of the 14 teams, 13 successfully applied the tools and an agile methodology to produce a working solution. Table 21.3 shows the characteristics of the different team solutions. Training time refers to the time needed to learn MBD using UML-RSDS.

Typically the teams divided into subteams, with each subteam given a particular task to develop, so that a degree of parallel development could occur, taking advantage of the independence of the three use cases. Most groups had a defined leader role (this had been advised in the coursework description), and the lack of a leader generally resulted in a poor outcome (as in teams 1, 4, 9, 12, 14).

The key difficulties encountered by most teams were:

- Lack of prior experience in using UML.
- The unfamiliar style of UML-RSDS compared to tools such as Visual Studio, Net Beans and other IDEs.
- Conceptual difficulty with the idea of MBD.

Table 21.3: Outcomes of EHR case study

Team	*Training time*	*Problems,* issues	*Technical outcome*	*Agile process*	*Activities, issues in agile process*
1	> 1 week	Usability problems	8/10	8/10	Disorganised and individual working
2	1 week	None	9/10	8/10	No experience of large teams
3	> 1 week	Lack of tool documentation	8/10	9/10	Used pair modelling, proactive time planning
4	1 week	Inadequate requirements	7/10	8/10	No leader. Parallel working
5	1 week	Usability problems	9/10	8/10	Lead developers
6	1 week	Used model refactoring	8/10	9/10	Used Scrum, sub-team modelling
7	1 week	Started modelling without sufficient tool knowledge	8/10	9/10	Risk analysis, paired modelling
8	1 week	Lead developers trained team	9/10	9/10	Small team modelling
9	> 1 week	Specification difficulties	7/10	7/10	No leader, disorganised
10	1 week	Lead developers trained team	8/10	8/10	Detailed planning, scheduling
11	1 week	Unfamiliar style of tool	9/10	9/10	Used XP
12	> 1 week	Lacked UML knowledge	7/10	5/10	Team split into 2
13	2 weeks	Difficulties using MBD/tools	8/10	8/10	Strong leadership
14	2 weeks	–	0/10	0/10	Failed to work as a team

- Inadequate user documentation for the tools – in particular students struggled to understand how the tools were supposed to be used, and the connection between the specifications written in the tool and the code produced.

- Team management and communication problems due to the size of the teams and variation in skill levels and commitment within a team.

Nonetheless, in 12 of 14 cases the student teams overcame these problems. Two teams (12 and 14) had severe management problems, resulting in failure in the case of team 14.

Particular issues can be seen in the following quotes from the team reports:

> "As with all software there was a learning curve involved in its use, and once we had progressed along this curve and gained some familiarity we found that the software was much easier to use, and every increase in our fluency with the software empowered us to produce higher quality solutions with increasing ease." (Team 8)

> "We found however, that becoming more familiar with UML-RSDS was more of a priority in order to be able to solve further tasks, as it is quite different to anything we had used before." (Team 10)

> "Many group members did not know UML and had to learn it." (Team 12)

The teams were almost unanimous in identifying that they should have committed more time at the start of the project to understand the tools and the MBD approach. This is a case where the agile principle of starting development as soon as possible needs to be tempered by the need for adequate understanding of a new tool and development technique.

Although all students had just attended 8 weeks of an introductory course on UML, some teams had problems with members who had not yet understood UML, which is fundamental to applying MBD with UML-RSDS.

Factors which seemed particularly important in overcoming problems with UML-RSDS and MBD were:

- The use of 'lead developers': a few team members who take the lead in mastering the tool/MBD concepts and who then train their colleagues. This spreads knowledge faster and more effectively than all team individuals trying to learn the material independently. Teams that used this approach had a low training time of 1 week, and achieved an average technical score of 8.66, versus 7.18 for other teams. This difference is statistically significant at the 4% level (removing team 14 from the data).

- Pair-based or small team modelling, with subteams of 2 to 4 people working around one machine. This seems to help to identify errors in modelling which individual developers may make, and additionally, if there is a lead developer in each sub-team, to propagate tool and MBD expertise. Teams using this approach achieved an average technical score of 8.25, compared to 7.2 for other teams. This difference is however not statistically significant if team 14 is excluded.

Teams using both approaches achieved an average technical score of 9, compared to those using just one (8.2) or none (6.9).

Another good practice was the use of model refactoring to improve an initial solution with too complex or too finely-divided use cases into a solution with more appropriate use cases.

The impact of poor team management and the lack of a defined process seems more significant for the outcome of a team, compared to technical problems. The Pearson correlation coefficient of the management/process mark of the project teams with their overall mark is 0.91, suggesting a strong positive relation between team management quality and overall project quality. Groups with a well-defined process and team organisation were able to overcome technical problems more effectively than those with poor management. Groups 3, 5, 7, 11 and 13 are the instances of the first category, and these groups achieved an average of 8.4/10 in the technical score, whilst groups 1, 4, 9, 12 and 14 are the instances of the second category, and these groups achieved an average of 5.8/10 in the technical score. An agile process seems to be helpful in achieving a good technical outcome: the correlation of the agile process and technical outcome scores in Table 21.3 is 0.93.

The outcomes of this case study were better than for the first case study: the average mark was 79% in case study 2, compared to 67.5% for case study 1. This appears to be due to three main factors: (i) a simpler case study involving reduced domain research and technical requirements compared to case study 1; (ii) improvements to the UML-RSDS tools; (iii) stronger advice to follow an agile development approach.

In conclusion, this case study illustrated the problems which may occur when industrial development teams are introduced to MBD and MBD tools for the first time. The positive conclusions which can be drawn are that UML-RSDS appears to be an approach which quite inexperienced developers can use successfully for a range of tasks, even with limited access to tool experts, and that the difficulties involved in learning the tools and development approach are not significantly greater than those that could be encountered with any new SE environment or tools.

21.4 Case study 3: Financial risk evaluation

This case study concerns the risk evaluation of multiple-share financial investments known as *Collateralized Debt Obligations* (CDO), where a portfolio of investments is partitioned into a collection of sectors, and there is the possibility of contagion of defaults between different companies in the same sector [2, 16]. Risk analysis of a CDO contract involves computing the overall probability $P(S = s)$ of a financial loss s based upon the probability of individual company defaults and the probability of default infection within sectors.

The loss estimation function $P(S = s)$ and risk estimation function $P(S \geq s)$ are required. The case study was carried out in conjunction with a financial risk analyst, who was also the customer of the development. Implementations in Java, C# and C++ were required. The required use cases and subtasks are given in Table 21.4. Use case 3 depends upon use case 2.

Table 21.4: Use cases for CDO risk analysis application

Use case	*Subtasks*	*Description*
1. Load data		Read sector data from a .csv spreadsheet
2. Calculate Poisson approximation of loss function	2a Calculate probability of no contagion	
	2b. Calculate probability of contagion	
	2c. Combine 2a, 2b	
3. Calculate risk function		
4. Write data		Write data from 2 or 3 to a .csv spreadsheet

First a phase of research was needed to understand the problem and to clarify the actual computations required. Then tasks 2a, 2b and 2c were carried out in a first development iteration, as these were considered more critical than use cases 1 or 4. Then use case 3 was performed in development iteration 2, and finally use cases 1 and 4 – which both involved use of manual coding – were scheduled to be completed in a third development iteration. A further external requirement was introduced prior to this iteration: to handle the case of cross-sector contagion. This requirement was then scheduled prior to tasks 1 and 4 in a new iteration.

Figure 21.6 shows the system specification of the solution produced at the end of the first development iteration. L is the credit loss per default, in each sector. p is the probability of each default in the sector,

q is the probability of infection in the sector, and n is the number of companies in the sector. *mu* is the Poisson approximation parameter. *test* is the prototype version of use case 2.

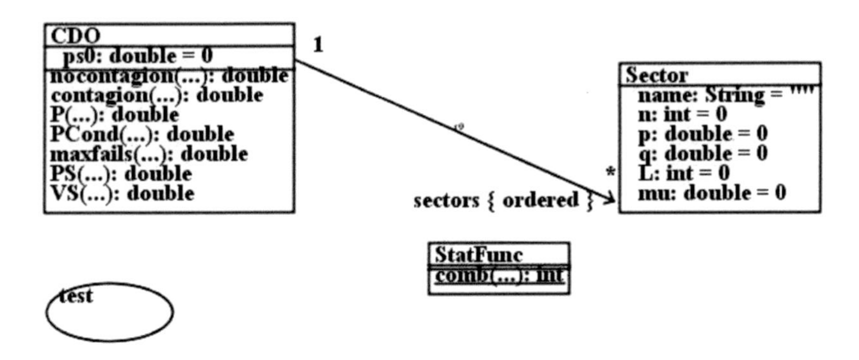

Figure 21.6: CDO version 1 system specification

The following model-based agile development techniques were employed:

- Refactoring: the solutions of 2a and 2b were initially expressed as operations *nocontagion*, *contagion* of the CDO class (Fig. 21.6). It was then realised that they would be simpler and more efficient if defined as Sector operations. The refactoring Move Operation was used. This refactoring did not affect the external interface of the system.
- Customer collaboration in development: the risk analyst gave detailed feedback on the generated code as it was produced, and carried out their own tests using data such as the realistic dataset of [16].
- Replanning and revision of the release plan due to new requirements: the scheduled third iteration was postponed due to a new more urgent requirement being introduced.
- Creation of a library component (*StatFunc*) for potentially reusable general functionalities.

Figure 21.7 shows the refactored system specification.

It was originally intended to use external hand-coded and optimised implementations of critical functions such as the combinatorial function $comb(int\ n, int\ m)$. However this would have resulted in the need for multiple versions of these functions to be coded, one for each target implementation language, and would also increase the time needed

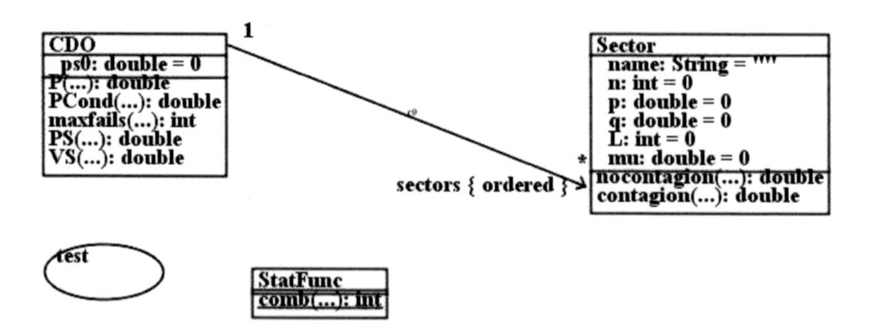

Figure 21.7: CDO version 2 system specification

for system integration. It was found instead that platform-independent specifications could be given in UML-RSDS which were of acceptable efficiency.

The initial efficiency of the loss computation was too low, with calculation of $P(S = s)$ for all values of $s \leq 20$ on the test data of [16] taking over 2 minutes on a standard Windows 7 laptop. To address this problem, the recursive operations and other operations with high usage were given the stereotype ≪ *cached* ≫ to avoid unnecessary recomputation. This stereotype means that operations are implemented using the *memoisation* technique of [17] to store previously-computed results. Table 21.5 shows the improvements in efficiency which this provides, and the results for generated code in other language versions.

Table 21.5: Execution times for CDO versions

Version	*Execution time for first 20 $P(S = s)$ calls*	*Execution time for first 50 $P(S = s)$ calls*
Unoptimised Java	121s	–
Optimised Java	32ms	93ms
C#	10ms	20ms
C++	62ms	100ms

The specification of use case 2 has the following postconditions:

```
CDO::
  s : sectors  =>  s.mu = 1 - ( ( 1 - s.p )->pow(s.n) )

CDO::
  ps0 = -sectors.mu.sum->exp()

CDO::
```

```
  Integer.subrange(0,20)->forAll( s | PS(s)->display() )
```

The first constraint initialises the *mu* attribute value for each sector. The second then initialises *ps*0 using these values. The third constraint calculates and displays $PS(s)$ for integer values s from 0 to 20. The operation $PS(s)$ computes the Poisson approximation of the loss function, and is itself decomposed into computations of losses based on the possible combinations of failures in individual companies. $P(k, m)$ is the probability of m defaults in sector k, $PCond(k, m)$ is the conditional probability of m defaults in sector k, given that there is at least one default:

```
CDO::
query P(k : int, m : int) : double
pre: true
post:
  result = StatFunc.comb(sectors[k].n,m) *
      ( sectors[k].nocontagion(m) + Integer.Sum(1,m - 1,i,sectors[k].
       contagion(i,m)) )
```

```
CDO::
query PCond(k : int, m : int) : double
pre: true
post:
  ( m >= 1 =>
      result = P(k,m) / ( 1 - ( ( 1 - sectors[k].p )->pow
       (sectors[k].n) ) ) ) &
  ( m < 1 => result = 0 )
```

The operation definitions are directly based upon the mathematical specifications of [16]. *Integer.Sum*(a, b, i, e) represents the mathematical sum $\Sigma_{i=a}^{b} e$.

maxfails(k, s) is the maximum number of defaults in sector k which can contribute to a total loss amount s. $PS(s)$ sums over the sectors the loss function $VS(k, s)$, which sums the probability-weighted loss amounts resulting from each of the possible non-zero numbers of defaults in sector k.

```
CDO::
query  maxfails(k : int, s : int) : int
pre: true
post:
  ( sectors[k].n <= ( s / sectors[k].L ) => result = sectors[k].n ) &
  ( sectors[k].n > ( s / sectors[k].L ) => result = s / sectors[k].L )
```

```
CDO::
query cached PS(s : int) : double
pre: true
post:
```

```
( s < 0 => result = 0 ) &
( s = 0 => result = ps0 ) &
( s > 0 => result = Integer.Sum(1,sectors.size,k,VS(k,s)) / s )
```

```
CDO::
query VS(k : int, s : int) : double
pre: true
post:
  result = Integer.Sum(1,maxfails(k,s),mk,
      ( sectors[k].mu * mk * sectors[k].L * PCond(k,mk) * PS(s - mk *
       sectors[k].L) ))
```

PS depends upon *VS*, which in turn depends upon *PS*. This mutual recursion in the definition of the *PS* operation is a strong indicator that optimisation using caching/memoisation is necessary for *PS*.

The following functions of *Sector* implement tasks 2a and 2b:

```
Sector::
query cached nocontagion(m : int) : double
pre: true
post:
  result = ( ( 1 - p )->pow(n - m) ) * ( p->pow(m) ) * ( ( 1 - q )
   ->pow(m * ( n - m )) )
```

```
Sector::
query contagion(i : int, m : int): double
pre : true
post:
  result = ( ( 1 - p )->pow(n - i) ) * ( p->pow(i) ) *
      ( ( 1 - q )->pow(i * ( n - m )) ) * ( ( 1 - ( ( 1 - q )
       ->pow(i) ) )->pow(m - i) ) *
      StatFunc.comb(m,i)
```

Finally, the combinatorial operator $comb(n, m)$ is defined in the utility class *StatFunc*:

```
StatFunc::
query static cached comb(n : int, m : int) : int
pre: n >= m & m >= 0 & n <= 25
post:
  ( n - m < m => result = Integer.Prd(m + 1,n,i,i) / Integer.
   Prd(1,n - m,j,j) ) &
  ( n - m >= m => result = Integer.Prd(n - m + 1,n,i,i) / Integer.
   Prd(1,m,j,j) )
```

This is also cached because it is called very frequently during the computation of *PS*.

The risk calculation function for task 3 is:

```
PLim(v : int) : double
```

```
pre: true
post:
   (v < 0  =>  result = 0) &
   (v >= 0  =>  result = 1 - Integer.Sum(0,v-1,k, PS(k)))
```

The new requirement to handle cross-sector contagion entails a revision to the system class diagram: it is necessary to represent companies (borrowers) which have a presence in several sectors, with a degree of weighting in each sector: the loss amount per sector is a weighted average of the loss due to each borrower (company in the sector):

$$L = sectorborrowers \rightarrow collect(omega * L) \rightarrow sum() \rightarrow round()$$

where *omega* is the weighting factor of the company in the sector. The functionality for tasks 2 and 3 also need to consider this case. Analysis of the requirement identifies that the class diagram should be extended as shown in Fig. 21.8. In addition, the existing versions of computations can be retained, with the loss per sector calculated prior to these computations using the constraint:

$$\begin{array}{l} Sector :: \\ \quad L = sectorborrowers \rightarrow collect(borrower.L * omega * theta) \rightarrow \\ \quad sum() \rightarrow round() \end{array}$$

where *theta* is the fraction of the company participating in this sector.

A preliminary use case, *deriveSectorLoss* computes *Sector*::*L* from the provided *Borrower*::*L* and *BorrowerInSector*::*omega*, *Borrower InSector*::*theta* values, using the above constraint. A further extension could calculate in a similar way the sector probability of default p from the individual probabilities of default of companies in the sector.

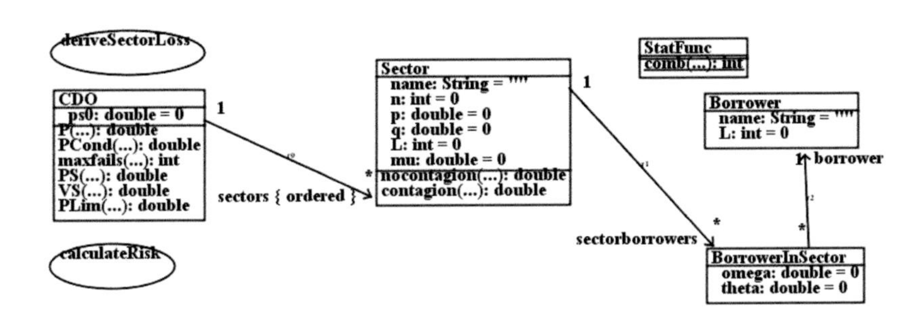

Figure 21.8: CDO version 3 system specification

Our experiences on this case study illustrate the UML-RSDS principles:

- Optimisation and refactoring should be carried out at the specification level in a platform-independent manner where possible, not at the code level.
- The scope of MBD should be extended as far as possible across the system development, reducing the extent of manual coding and integration wherever possible.

In conclusion, this case study showed that a successful outcome is possible for agile MBD in the highly demanding domain of computationally-intensive financial applications. A generic MBD tool, UML-RSDS, was able to produce code of comparable efficiency to existing hand-coded and highly optimised solutions.

21.5 Published case studies

Further examples of the application of UML-RSDS may be found in these publications:

- Migration transformations: UML version 1.4 to version 2.2 migration [9]; GMF migration [11].
- Refinement/enhancement transformations: Large-scale data analysis [15]; graph analysis [10]; UML to relational database [12].
- Refactoring transformations: Class diagram refactoring [7, 14]; State machine refactoring/analysis [8]; Petri-net to statechart mapping [13].

The specifications and executable code of some of these case studies may be found at: `www.dcs.kcl.ac.uk/staff/kcl/uml2web`.

Summary

In this chapter we have described how UML-RSDS can be used to support teaching of UML and MBD. We have also presented several different case studies of UML-RSDS development to illustrate the applications of the approach in education and in industry.

References

[1] K. Beck et al., *Principles behind the Agile Manifesto*, Agile Alliance, 2001. http://agilemanifesto.org/principles

[2] M. Davis and V. Lo, *Infectious Defaults*, Quantitative Finance, Vol. 1, No. 4, pp. 382–387, 2001.

[3] EGL, `www.eclipse.org/epsilon/doc/egl`, 2014.

[4] M. Fowler, K. Beck, J. Brant, W. Opdyke and D. Roberts, *Refactoring: improving the design of existing code*, Addison-Wesley, 1999.

[5] G. Guta, W. Schreiner and D. Draheim, *A lightweight MDSD process applied in small projects*, Proceedings 35th Euromicro conference on Software Engineering and Advanced Applications, IEEE, 2009.

[6] S. Hansson, Y. Zhao and H. Burden, *How MAD are we?: Empirical evidence for model-driven agile development*, 2014.

[7] S. Kolahdouz-Rahimi, K. Lano, S. Pillay, J. Troya and P. Van Gorp, *Evaluation of model transformation approaches for model refactoring*, Science of Computer Programming, 2013, http://dx.doi.org/10.1016/j.scico.2013.07.013.

[8] K. Lano and S. Kolahdouz-Rahimi, *Slicing of UML models using Model Transformations*, MODELS 2010, LNCS vol. 6395, pp. 228–242, 2010.

[9] K. Lano and S. Kolahdouz-Rahimi, *Migration case study using UML-RSDS*, TTC 2010, Malaga, Spain, July 2010. http://is.ieis.tue.nl/staff/pvgorp/events/TTC2010/submissions/final/uml-rsds.pdf.

[10] K. Lano and S. Kolahdouz-Rahimi, *Specification of the "Hello World" case study*, TTC 2011.

[11] K. Lano and S. Kolahdouz-Rahimi, *Solving the TTC 2011 migration case study with UML-RSDS*, TTC 2011, EPTCS vol. 74, pp. 36–41, 2011.

[12] K. Lano and S. Kolahdouz-Rahimi, *Constraint-based specification of Model Transformations*, Journal of Systems and Software, vol. 86, issue 2, February 2013, pp. 412–436.

[13] K. Lano, S. Kolahdouz-Rahimi and K. Maroukian, *Solving the Petri-Nets to Statecharts Transformation Case with UML-RSDS*, TTC 2013, EPTCS, 2013.

[14] K. Lano and S. Kolahdouz-Rahimi, *Case study: Class diagram restructuring*, www.planet-sl.org/community/_/ttc/ttc2013/cases/ClassDiagramRestructuring, 2013.

[15] K. Lano and S. Yassipour-Tehrani, *Solving the Movie Database Case with UML-RSDS*, TTC 2014.

[16] O. Hammarlid, *Aggregating sectors in the infectious defaults model*, Quantitative Finance, vol. 4, no. 1, pp. 64–69, 2004.

[17] D. Michie, *Memo functions and machine learning*, Nature, vol. 218, pp. 19–22, 1968.

[18] M.B. Nakicenovic, *An Agile Driven Architecture Modernization to a Model-Driven Development Solution*, International Journal on Advances in Software, vol. 5, nos. 3, 4, pp. 308–322, 2012.

Appendix A

UML-RSDS Syntax

A.1 OCL expression syntax

UML-RSDS uses both classical set theory expressions and OCL. It only uses sets and sequences, and not bags or ordered sets, unlike OCL. Symmetric binary operators such as $\cup$ and $\cap$ can be written in the classical style, rather than as operators on collections. Likewise for the binary logical operators. There are no null or undefined elements. Table A.1 shows the BNF concrete grammar of UML-RSDS OCL.

A $< unary\_operator >$ is one of: *any*, *size*, *isDeleted*, *display*, *min*, *max*, *sum*, *prd*, *sort*, *asSet*, *asSequence*, *sqrt*, *sqr*, *last*, *first*, *front*, *tail*, *closure*, *characters*, *subcollections*, *reverse*, *isEmpty*, *notEmpty*, *toUpperCase*, *toLowerCase*, *isInteger*, *isReal*, *toInteger*, *toReal*. The mathematical functions *ceil*, *round*, *floor*, *exp*, etc., can also be written as unary expressions, which is convenient if their argument is a function call or a complex bracketed expression.

Other unary and binary operators may be used in a *factor2_expression*, as described in Tables 4.3, 4.4, 4.5, 4.6 and 4.7. Other binary operators are *includes*, *including*, *excludes*, *excluding*, *union*, *intersection*, *selectMaximals*, *selectMinimals*, *includesAll*, *excludesAll*, *append*, *prepend*, *count*, *hasPrefix*, *hasSuffix*, *indexOf*, *sortedBy*. *oclIsKindOf* and *oclAsType* are unusual because the second argument is the name of a UML-RSDS type such as *int*, an entity class name, or *Set* or *Sequence*. Ternary operators are expressed as operation calls, e.g., $s.subrange(i, j)$, $s.insertAt(i, x)$. Likewise for the operators $Integer.Sum(a, b, i, e)$ and $Integer.Prd(a, b, i, e)$.

Table A.1: BNF grammar of UML-RSDS OCL

<table>
<tr><td>< expression ></td><td>::=</td><td>< bracketed_expression > | < equality_expression > |
< logical_expression > | < factor_expression ></td></tr>
<tr><td>< bracketed_expression ></td><td>::=</td><td>“(” < expression > “)”</td></tr>
<tr><td>< logical_expression ></td><td>::=</td><td>< expression > < logical_op > < expression ></td></tr>
<tr><td>< equality_expression ></td><td>::=</td><td>< factor_expression > < equality_op >
< factor_expression ></td></tr>
<tr><td>< factor_expression ></td><td>::=</td><td>< basic_expression > < factor_op >
< factor_expression > |
< factor2_expression ></td></tr>
<tr><td>< factor2_expression ></td><td>::=</td><td>< expression >“->”< unary_operator >“()” |
< expression >“->exists(” < identifier > “|”
< expression > “)” |
< expression >“->exists1(” < identifier > “|”
< expression > “)” |
< expression >“->forAll(” < identifier > “|”
< expression > “)” |
< expression >“->exists(” < expression > “)” |
< expression >“->exists1(” < expression > “)” |
< expression >“->forAll(” < expression > “)” |
< expression >“->select(” < expression > “)” |
< expression >“->select(” < identifier > “|”
< expression > “)” |
< expression >“->reject(” < expression > “)” |
< expression >“->reject(” < identifier > “|”
< expression > “)” |
< expression >“->collect(” < expression > “)” |
< expression >“->collect(” < identifier > “|”
< expression > “)” |
< expression >“->unionAll(” < expression > “)” |
< expression >“->intersectAll(” < expression > “)” |
< expression >“->”< binary_operator >“(”
< expression > “)” |
< basic_expression ></td></tr>
<tr><td>< basic_expression ></td><td>::=</td><td>< set_expression > | < sequence_expression > |
< call_expression > | < array_expression > |
< identifier > | < value ></td></tr>
<tr><td>< set_expression ></td><td>::=</td><td>“Set{” < fe_sequence > “}”</td></tr>
<tr><td>< sequence_expression ></td><td>::=</td><td>“Sequence{” < fe_sequence > “}”</td></tr>
<tr><td>< call_expression ></td><td>::=</td><td>< identifier > “(” < fe_sequence > “)”</td></tr>
<tr><td>< array_expression ></td><td>::=</td><td>< identifier > “[” < factor_expression > “]” |
< identifier > “[” < factor_expression > “].”
< identifier ></td></tr>
</table>

A *logical_op* is one of `=>`, `&`, `or`. An *equality_op* is one of =, / =, >, <, <: (subset-or-equal), <=, >=, :, / : (not-in). A *factor_op* is one of +, /, *, −, \/ (union), $\frown$ (concatenation of sequences), /\ (intersection). An *fe_sequence* is a comma-separated sequence of factor expressions. Identifiers can contain “.” (to denote reference to a feature of an object), but not as the first or last character, and occurrences must be separated by at least one other character. Identifiers can also contain “$”. Spaces, underscores, hyphens, or other whitespace characters should not occur within operators or identifiers.

Valid function symbols are the numeric functions such as *sqrt*, *floor*, *abs*, and OCL operators such as *isReal*, *isInteger*, *toReal*, *toInteger*.

Figure A.1 shows the metamodel for UML-RSDS expressions, which defines the abstract syntax of the OCL expression language used in UML-RSDS. *BinaryExpression*, *BasicExpression*, *UnaryExpression* and

CollectionExpression all inherit from *Expression*. Basic expressions have self-associations *arrayIndex* and *objectRef* to link to their array arguments and object reference (if present), corresponding to the concrete syntax *objectRef.data*[*arrayIndex*] for navigations to sequence elements or navigations of qualified associations.

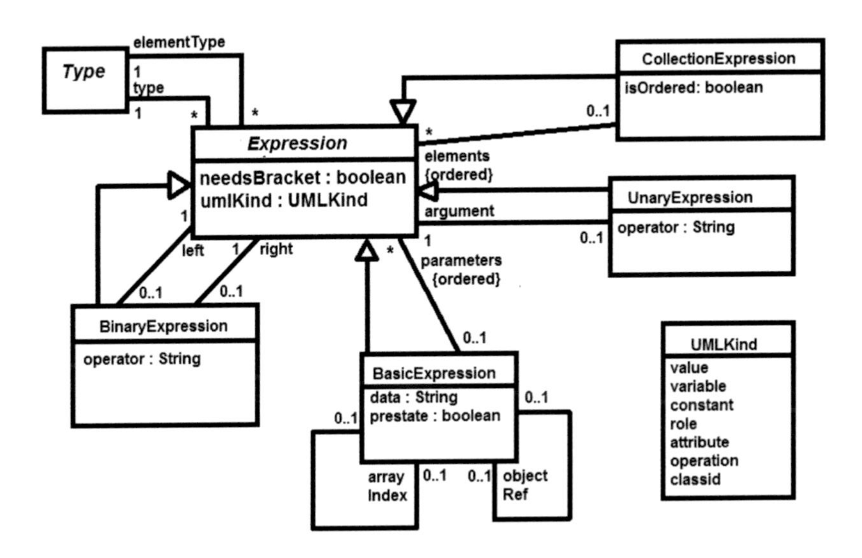

Figure A.1: UML-RSDS constraint language

For example, $x + 5$ is a BinaryExpression with operator +, and left and right BasicExpressions x and 5, of kind variable (or role or attribute) and value, respectively. $sq \rightarrow select(f = 5)$ is a BinaryExpression with operator $\rightarrow select$, whilst $E[x.id]$ is a BasicExpression of kind classid (if E is the name of a class), and with *arrayIndex* the BasicExpression $x.id$, whose *objectRef* is x and data is *id*.

Each expression has both a type and an element type. The element type identifies the type of the elements of a set-valued or sequence-valued expression.

A.2 Activity language syntax

Table A.2 shows the concrete syntax used in UML-RSDS to express UML structured activities.

Table A.2: Activity language syntax

<table>
<tr><td>< statement ></td><td>::=</td><td>< loop_statement > | < creation_statement > |
< conditional_statement > |
< sequence_statement > |
< basic_statement ></td></tr>
<tr><td>< loop_statement ></td><td>::=</td><td>"while" < expression > "do" < statement > |
"for" < expression > "do" < statement ></td></tr>
<tr><td>< conditional_statement ></td><td>::=</td><td>"if" < expression > "then" < statement >
"else" < basic_statement ></td></tr>
<tr><td>< sequence_statement ></td><td>::=</td><td>< statement > ";" < statement ></td></tr>
<tr><td>< creation_statement ></td><td>::=</td><td>< identifier > ":" < identifier ></td></tr>
<tr><td>< basic_statement ></td><td>::=</td><td>< basic_expression > ":=" < expression > |
"skip" |
< identifier > ":" < identifier > ":=" < expression > |
"execute" < expression > |
"return" < expression > | "(" < statement > ")" |
< call_expression ></td></tr>
</table>

Spaces are needed around statement keywords and around symbols such as ';' and ')', '(' when entering activities as text. The abstract syntax of UML-RSDS activities is defined by the metamodel of Fig. A.2. Implicit call statements represent calls of expressions as statements: *execute exp*. These have the effect of $stat(exp)$, which must be defined.

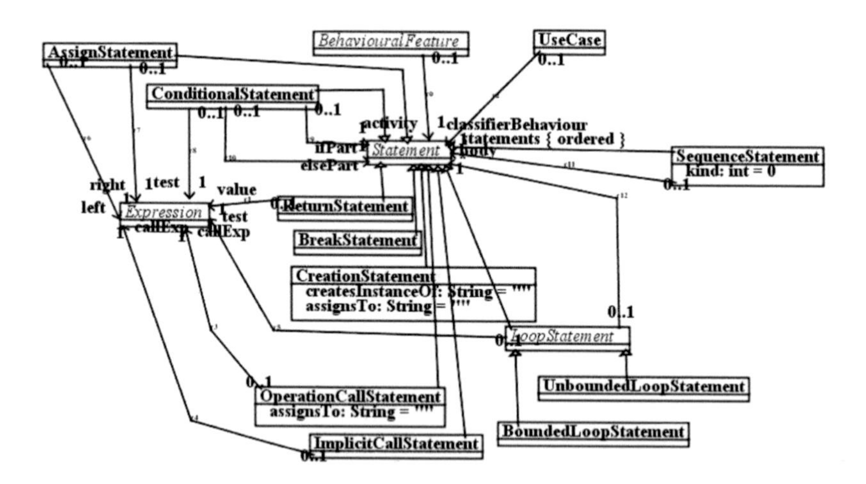

Figure A.2: Statement (activity) metamodel

Appendix B

UML-RSDS tool architecture and components

The UML-RSDS tool architecture is closely based upon the interrelationships between the different UML languages which are supported by the tool: with OCL expressions as a fundamental notation used by all other UML-RSDS languages (class diagrams, activities, state machines, use cases and interactions), and class diagrams as the central notation which the other visual languages and activities depend upon. Figure B.1 shows the tool architecture of UML-RSDS version 1.5. Arrows indicate dependencies between modules.

Users interact primarily with the tools via the class diagram editor, which also manages the creation, deletion and editing of use cases and activities. Auxiliary editors for state machines and interactions are also provided. The *Class diagram* module is likewise the central functional component, dealing with the management, analysis and transformation of class diagrams. A large part of the UML 2 class diagram language is supported. Figure B.2 shows the supported parts of the UML class diagram metamodel.

The *linkedClass* of an association is the class of an association class. *memberEnd* always has exactly 2 elements in UML-RSDS. A use case is a subclass of *Classifier* but is actually considered as an *Entity*, and has *ownedAttribute*, *ownedOperation* and *constraint* links to its attributes, operations and invariants.

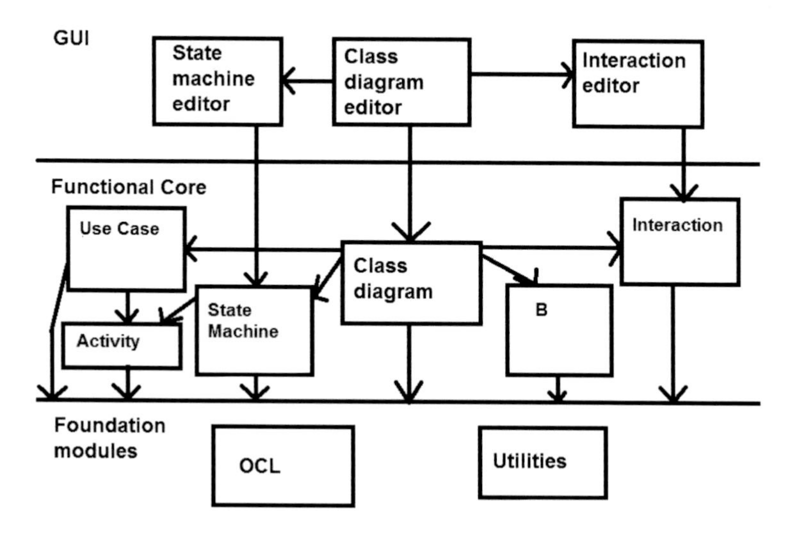

Figure B.1: UML-RSDS tool architecture

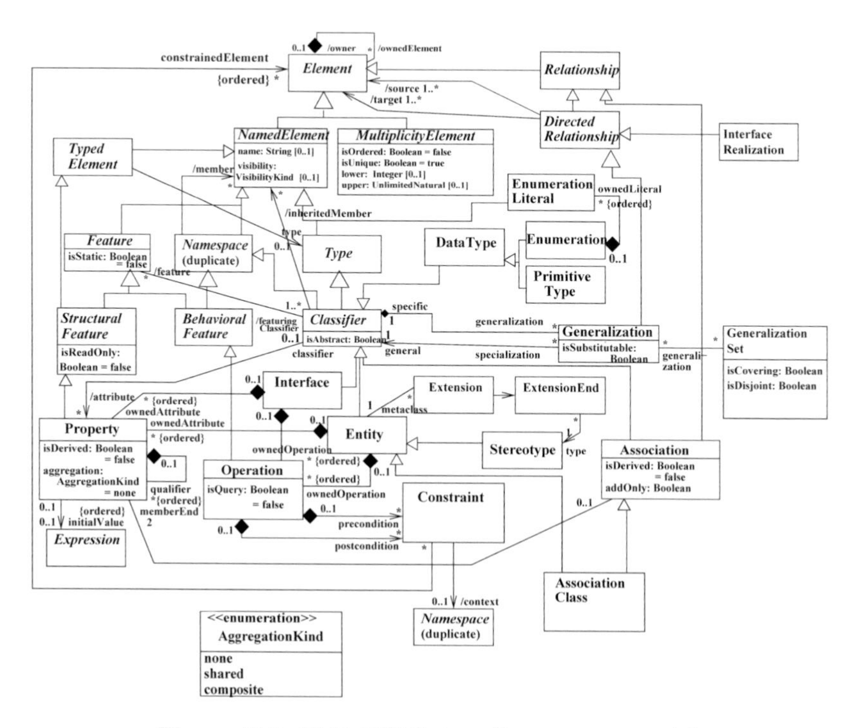

Figure B.2: UML-RSDS class diagram metamodel

Smaller modules deal with state machines, interactions, activities, use cases and the B notation. Finally, the foundational module *OCL* supports the management, analysis and transformation of OCL expressions, and *Utilities* provides general utility classes and operations. The activity language supported by the tools is defined by Fig. A.2. The UML-RSDS tool set is implemented in Java, and the tool classes themselves follow closely the structure of the UML metamodels, in order to facilitate reflexive use of the tools. Some example classes from each module are as follows:

Class diagram editor: UmlTool.java, UCDArea.java, LineData.java

State machine editor: StateWin.java, StatechartArea.java

Interaction editor: InteractionWin.java, InteractionArea.java

Class diagram: ModelElement.java, Entity.java, Association.java, Attribute.java, Generalisation.java, Type.java, BehaviouralFeature.java, Behaviour.java

State Machine: Statemachine.java, State.java, Transition.java, Event.java, StatemachineSlice.java

Use Case: UseCase.java, Extend.java, Include.java

Activity: Statement.java, AssignStatement.java, IfStatement.java

Interaction: InteractionElement.java, LifelineState.java, ExecutionInstance.java, Message.java

B: BComponent.java, BOp.java, BExpression.java

OCL: Expression.java, Constraint.java, BinaryExpression.java, BasicExpression.java, SetExpression.java, UnaryExpression.java

Utilities: Compiler.java, Compiler2.java, VectorUtil.java

Approximately 5000 person hours of development effort have been spent on the tools, about 50% of this on testing and verification.

Appendix C

Key principles of UML-RSDS

We summarise here the central principles of specification and development using UML-RSDS.

> *The specification* is *the system.*

> *A logical constraint P can be interpreted both as a specification of system behaviour, and as a description of how P will be established in the system implementation.*

These mean that a declarative high-level specification can be used for multiple purposes: (i) to express system functionality in a concise form, independent of code; (ii) to support verification at the specification level; (iii) as input for design and code synthesis to produce implementations in multiple target languages, which satisfy the specification by construction.

> *A post-condition constraint P means "Make P true" when interpreted as a specification of system behaviour.*

Post-conditions of operations and of use cases can be interpreted in this way.

In particular, for use cases:

> *If a use case uc has post-conditions C_1 to C_n which are all of type 1, satisfy syntactic non-interference, confluence and*

definedness and determinacy conditions, then the implementation code $stat(C_1)$; ...; $stat(C_n)$ of uc is terminating, confluent and correct with respect to its specification. That is, this code establishes the conjunction C_1 & ... & C_n.

Improve the efficiency of a system at the specification level where possible, whilst keeping a clear and platform-independent specification style.

Because a specification has a dual purpose as a description aimed at human readers, and as an input for automated code generation, it must both be clear and concise, and should avoid computationally inefficient formulations such as the use of duplicated complex expressions.

As for other software systems, patterns can provide systematic solutions for specification and design problems. In UML-RSDS patterns can be applied at the specification level, instead of design or code levels:

Use specification patterns where possible to improve the clarity, compositionality and efficiency of UML-RSDS systems.

Model transformations can be specified as use cases, with use case post-conditions expressing the transformation rules.

A large number of different MT languages exist, but we consider that it is preferable in the long term for transformations to be specified in a single standard language (UML) where possible, making use of the same facilities available for the specification, verification, design and implementation of general software systems. After careful consideration of alternative representations, we came to the conclusion that use cases, with a formalised specification of their behaviour, were the most suitable UML element to serve as descriptions of system functionalities and services, including model transformations.

For transformations, we have the principle:

Define rules which express the required behaviour in the simplest and clearest form possible.

Regarding reuse of components, there is the principle:

When a functionality or set of related functionalities have been developed, and which seem to be of potential utility in other systems, make a reusable component consisting of these functionalities as use cases, supported by the local data. This component can be an externalApp in other UML-RSDS systems.

For agile development, we have the principles:

Models as code: A UML-RSDS specification is both the documentation of a system and a description of its implementation.

UML-RSDS specifications should be used to support communication and collaboration between developers and stakeholders.

Principles which apply to the development process using UML-RSDS are:

Optimisation and refactoring should be carried out at the specification level in a platform-independent manner where possible, not at the code level.

The scope of MBD should be extended as far as possible across the system development, reducing the extent of manual coding and integration wherever possible.

Index